Broadband Communications

IFIP – The International Federation for Information Processing

IFIP was founded in 1960 under the auspices of UNESCO, following the First World Computer Congress held in Paris the previous year. An umbrella organization for societies working in information processing, IFIP's aim is two-fold: to support information processing within its member countries and to encourage technology transfer to developing nations. As its mission statement clearly states,

> IFIP's mission is to be the leading, truly international, apolitical organization which encourages and assists in the development, exploitation and application of information technology for the benefit of all people.

IFIP is a non-profitmaking organization, run almost solely by 2500 volunteers. It operates through a number of technical committees, which organize events and publications. IFIP's events range from an international congress to local seminars, but the most important are:

- the IFIP World Computer Congress, held every second year;
- open conferences;
- working conferences.

The flagship event is the IFIP World Computer Congress, at which both invited and contributed papers are presented. Contributed papers are rigorously refereed and the rejection rate is high.

As with the Congress, participation in the open conferences is open to all and papers may be invited or submitted. Again, submitted papers are stringently refereed.

The working conferences are structured differently. They are usually run by a working group and attendance is small and by invitation only. Their purpose is to create an atmosphere conducive to innovation and development. Refereeing is less rigorous and papers are subjected to extensive group discussion.

Publications arising from IFIP events vary. The papers presented at the IFIP World Computer Congress and at open conferences are published as conference proceedings, while the results of the working conferences are often published as collections of selected and edited papers.

Any national society whose primary activity is in information may apply to become a full member of IFIP, although full membership is restricted to one society per country. Full members are entitled to vote at the annual General Assembly, National societies preferring a less committed involvement may apply for associate or corresponding membership. Associate members enjoy the same benefits as full members, but without voting rights. Corresponding members are not represented in IFIP bodies. Affiliated membership is open to non-national societies, and individual and honorary membership schemes are also offered.

Broadband Communications

Global infrastructure for the information age

Proceedings of the International IFIP-IEEE Conference on Broadband Communications, Canada, 1996

Edited by
Lorne Mason
INRS - Telecommunications
Quebec
Canada

and

Augusto Casaca
IST/INESC
Lisbon
Portugal

Published by Chapman and Hall on behalf of the
International Federation for Information Processing (IFIP)

CHAPMAN & HALL
London · Weinheim · New York · Tokyo · Melbourne · Madras

Published by Chapman & Hall, 2–6 Boundary Row, London SE1 8HN, UK

Chapman & Hall, 2–6 Boundary Row, London SE1 8HN, UK

Blackie Academic & Professional, Wester Cleddens Road, Bishopbriggs, Glasgow G64 2NZ, UK

Chapman & Hall GmbH, Pappelallee 3, 69469 Weinheim, Germany

Chapman & Hall USA, 115 Fifth Avenue, New York, NY 10003, USA

Chapman & Hall Japan, ITP-Japan, Kyowa Building, 3F, 2-2-1 Hirakawacho, Chiyoda-ku, Tokyo 102, Japan

Chapman & Hall Australia, 102 Dodds Street, South Melbourne, Victoria 3205, Australia

Chapman & Hall India, R. Seshadri, 32 Second Main Road, CIT East, Madras 600 035, India

First edition 1996

Printed in Great Britain by Hartnolls Ltd, Bodmin, Cornwall

ISBN 0 412 75970 5

A catalogue record for this book is available from the British Library

Printed on permanent acid-free text paper, manufactured in accordance with ANSI/NISO Z39.48-1984 and ANSI/NISO Z39.48-1984 (Permanence of Paper).

CONTENTS

SESSION SEVEN: RESOURCE ALLOCATION

SESSION EIGHT: PERFORMANCE ANALYSIS

SESSION NINE: TRAFFIC ESTIMATION

Part Two: Broadband Technologies and Network Design

SESSION ONE: ACCESS TECHNOLOGIES

PREFACE

As we approach the new millennium, there is strong evidence that we are about to enter the much heralded "information age", which has been seen by futurists and other commentators, as being of profound consequence to society. The revolutionary change in the workplace and in our personal lives, likely to be ushered in by broadband communications and advanced information technology generally, has been likened in impact to that of the industrial age or even the invention of fire! While such profound change should be of concern to us all, in view of its far reaching social consequences, engineers and scientists have generally left such speculation and hyperbole to others and are concentrating on the more challenging task of designing an effective global network infrastructure to support the coming information age.

"Broadband Communications '96" is an international conference co-sponsored by IFIP Working Group 6.2 and the IEEE Communications Society, which addresses the theme of "Infrastructure for the Information Age". This conference was organized by INRS-Telecommunications and the Canadian Institute for Telecommunications Research (CITR) It is the third in a series, which began with a workshop held in Estoril Portugal in 1992, and followed by a working conference held in Paris, France in 1994, both sponsored by IFIP. This marks the first time the conference has been held in North America and first time that the IEEE Communications Society is a co-sponsor.

One hundred and forty six papers from twenty six countries (five continents) were submitted. The program committee, with the support of some two hundred reviewers, selected fifty two contributed papers for presentation. The selected papers, representing leading edge research results from experts in both industry and academia around the world, are presented in two structured parallel tracks. The theme of track A is "Broadband Traffic Modeling and Control" and is comprised of sessions in Traffic Modeling, Call Admission & Routing, ABR Control Algorithms, Traffic Control, Resource Allocation, Performance Analysis and Traffic Estimation. The theme of track B is "Broadband Technologies and Network Design" and is composed of sessions on Access Technologies, Broadband Switching, Protocols, Traffic Engineering, Network Design, Network Reliability, Testbed Networks and Circuit Emulation.

We wish to express our sincere thanks to the program and organizing committee members and also to all the reviewers who generously carried out the difficult and delicate task of reviewing the contributed papers. Special thanks are due to Otto Spaniol of IFIP TC6, Maeir Blonstein of CITR and IEEE, and Gilles Deslisle, director of INRS-Telecommunications, for their encouragement and support. Finally we wish to thank, the secretarial staff and research assistants at INRS-Telecommunications and in particular Chantal Loiselle, and Dr. Ming Ja who spent many hours effectively carrying out the numerous tasks often under severe time pressure.

Lorne Mason
Augusto Casaca
April, 1996

BROADBAND COMMUNICATIONS '96

REVIEWERS

Aboul-Mayd, O.
Alnuweiri, H.M.
Alonso, Pedro
Alves, Arthur P.
Alvin, C. Francis
Andersen, Allan T.
Andersson, Loa
Antonixder, Neo
Archambault, Sylvain
Awdeh, R.Y
Axell, Jurgen

Barbosa, L.O.
Baumann, Matthias
Bautz, Gregor
Beerends, John G.
Beshai, M.E.
Blaabjerg, S.
Blondia, C.
Boda, Miklos
Bonaventure, O.
Brazio, Jose M.
Breuer, Hans J
Brichet F.
Buhrgard, Magnus

Campbell, Andrew T.
Casaca, Augusto
Cavers, Jim
Chamberland, S.
Chan, Dennis
Chan, Mun Choon
Charzinski, J.
Chow, Keith H.
Christiansen, H.
Cooper, C. A.
Cormier, Doug.
Coudreuse, J.P.
Cox, N.
Cseh, Christian

David, J.
Decreusefond, L.
Delisle, D.
Delisle, Gilles
Deloddere, C.
Desmet, Emmanuel
Despins, Charles
Devetsikiotis, M.
Doverspike, R.
Dupuis, Alain
Dykeman, Doug
Dziong, Z.

Ellinas, Georgios
Eude, Gerard

Falconer, David
Faloutsos, M.
Fan, Y.
Fasbender, Andreas
Friesen, Verna
Funada, Kazushi

Gagnaire, M.
Georganas, N.
Ghani, Nasir
Gilderson, Jim
Girard, Andre
Gonet, P.
Gordon, J.
Gotzer, Martin
Graf, Marcel
Gravey, A.
Gregoire, J-Ch.
Grover, Wayne D.
Groz, Roland
Guillemin, Fabrice
Guo, Duanyang

Harms, Janelle
Hasegawa, Toru
Hashemi, Massoud R.
Hayes, J.F.
Heddes, M.
Hermanns, Oliver
Herrmann, C.
Hoff, Simon
Hopkins, J.
Houdoin
Hubner, M.F.
Hung, Anthony

Indulska, Jadwiga
Inoue, Yuji
Ishkura, Masami
Iversen, Villy

Jain, A. K.
Jakobs, Kai
Jerkins, J.
Jia, Ming
Johansson, Sonny
Johen, Wittas

Kalampoukas, L.
Kant, K.
Kaplan, Michael
Karademir, Sibel
Kaudel, Fred
Kavehrad, M.
Kawai, Shin-Ichiro
Kaye, A.R.
Kesidis, Georges
Khakhar, Dipak
Khorsandi, Siavash
Kimura, H.
Kling, Nils-Gunnar
Knuutila, T.
Kuehn, Paul J.
Kuri, Joy

Lambadaris, I.
Lazar, Aurel A.
Le-Ngoc, Tho
Leh, Robert C.
Leitas, Mario
Letourneau, Eric
Leung, V.C.M.
Levert, Charles
Li, Li
Liao, K.Q.

Lipper, E.
Lubacz, J.

Mac Donald, R.I
Manthorpe, Sam
Mark, J.W.
Mason, Lorne
Mathy, Laurent
Mazumdar, R.
Mcmanus, Jean
Meddeb, Aref
Meempat, G.
Merazo, Luis A.
Mermelstein, P.
Meurisse, Wim
Michiel, H.

Nandikesan, M
Neidhardt, A.
Ngoh, Lek Heng
Niedzwiechi, A.
Nunes, S.

Ohlman, Borje
Olsson, Bengt J.
Ouveysi, Iradj

Palmer, Rob
Petit, G.H.
Pickholtz, R.L.
Pioro, Michal
Predrag, R.

Quernheim, Ulrich

Rasheed, Yasser
Reichl, Peter
Rigault, Claude
Roberts, J.
Rocka, Rui
Rodrigues, J.
Rosenberg, C.
Ruela, Jose
Ryu, Bong

Saltzberg, B.R.
Scotton, Paolo
Semret, Nemo
Senior, D.
Shah, Syed
Shalmon, Michael
Sierens, Chris
Simon, Jean-Louis

Sinha, Rajeev
Spath, Yan
Sridhar, S.

Theimer, Thomas
Thomas, Mathew
Tse, Philip W.

Van As, H.R.
Van Weert, M.J.M.
Van Mieghem, P.
Varma, Anujan
Verma, Sanjeev
Vishnu, M.

Wirkestrand, A.
Wong, J.W.
Woodruff, Gillian

Xiong, Yijun

Yamazaki, Katsuyuki
Yang, O.

Zukerman, Moshe.

PART ONE

Broadband Traffic Modeling and Control

1

Determination of Traffic Descriptors for VPs Carrying Delay-Sensitive Traffic[1]

Michael Ritter
Institute of Computer Science, University of Würzburg
Am Hubland, D-97074 Würzburg, GERMANY
email: ritter@informatik.uni-wuerzburg.de

Jorge García
Departamento de Electronica e Sistemas, Universidade Da Coruña
Campus de Elviña, 15071, La Coruña, SPAIN
email: jorge@sol.des.fi.udc.es

Abstract

In this paper we address the problem of determining source traffic descriptors for virtual path connections formed by a number of delay-sensitive virtual channel connections. In order to obtain a virtual path connection which allows an efficient use of the network resources, traffic shaping should be performed at the origin of the connection to reduce its burstiness. However, a conventional shaping process in conjunction with FIFO multiplexing introduces delays that may be too long, especially for delay-sensitive connections offering a low cell rate, such as phone calls or video telephony.

To solve this trade-off, we propose a shaping algorithm called dual cell spacing, which allows to obtain small shaping delays and an efficient use of the network resources at the same time. Furthermore, we give simple dimensioning rules for the source traffic descriptors of the shaped virtual path. Such a dimensioning is not straightforward if traffic shaping is omitted. Numerical examples are presented to illustrate the feasibility of the proposed approach.

Keywords

Delay Sensitive Traffic, Source Traffic Descriptor, Traffic Shaping, Virtual Path Connection

1 Introduction

In ATM environments, network resources such as link capacity and buffer space are allocated on the basis of a traffic contract, which is negotiated between the user and the network at connection setup. Parameters which form this contract are e.g. Quality of Service (QoS) requirements, traffic descriptors, conformance definitions, as well as the service category.

[1]This work was partially supported by the Spanish ministry of education with grant number TIC95-0982-C02-01.

Several network control and management functions operate on behalf of the parameters agreed on in the traffic contract. The Connection Admission Control (CAC) function, for example, uses the description of the source behavior to allocate network resources and to determine parameter values for the Usage Parameter Control (UPC). The UPC function monitors the cells of a connection to detect violations of the negotiated traffic descriptors in order to maintain the network performance.

In the current state of the standardization process of ATM networks [1, 7], real-time connections are described by a number of source traffic descriptors. These are, in detail, the Peak Cell Rate (PCR), the Sustainable Cell Rate (SCR) and the Burst Tolerance (BT). The specification of the PCR is mandatory, whereas SCR and BT are optional parameters used for VBR connections. An additional specification of a SCR and a BT may allow the network provider to allocate less resources to a connection while still maintaining the requested QoS. Together with the Cell Delay Variation Tolerance (CDVT), which is a network parameter that takes into account the amount of CDV accumulated in the upstream nodes, these parameters are called the connection traffic descriptors. They can be specified for different Cell Loss Priority (CLP) classes. Currently, the priority classes $CLP = 0$ and $CLP = 0 + 1$ are fixed in the standards together with corresponding conformance definitions. In this paper we only focus on the composite cell stream of a connection, i.e. the $CLP = 0 + 1$ class.

During the life-time of a compliant connection, the network provider agrees to guarantee the requested QoS to all conforming cells, where a connection is defined as compliant if the number of non-conforming cells is below a certain threshold set in the traffic contract. To monitor the conformance of cells according to the negotiated connection traffic descriptors, the Generic Cell Rate Algorithm (GCRA) has been proposed [1, 7]. The $GCRA(I, L)$ is a counter scheme based on two parameters, namely the increment value I and the limit value L. For each arriving cell the counter is incremented by I and decreased by one each slot of cell duration. The counter limit is given by $I + L$ and a cell that would cause the counter to overflow is said to be non-conforming. Therefore, the increment parameter I affects the cell rate of a connection and the limit parameter L the burst size. For system modeling, the counter value of the GCRA can be interpreted as the virtual waiting time of a FIFO queue with a deterministic service time I and a virtual waiting time limited by L (cf. [10]).

To monitor a PCR of $1/T$ together with a CDVT of τ, a GCRA with parameters $I = T$ and $L = \tau$ is used. If a SCR of $1/T_s$ is declared additionally in conjunction with a BT of τ_s, a second GCRA with parameters $I = T_s$ and $L = \tau_s + \tau$ is employed to detect violations according to this cell rate (cf. [1, 11]).

In ATM networks, the CAC and UPC functions play an important role for traffic management, especially for congestion avoidance. Therefore, the connection traffic descriptors should be dimensioned carefully, both from the resource allocation and from the conformance testing point of view. In the existing literature, this dimensioning process is well treated for Virtual Channel Connections (VCCs), see e.g. [5, 6, 8, 10, 14]. However, for Virtual Path Connections (VPCs), which are a group of VCCs sharing common paths within a network, only little is published. VPCs are an important component for traffic control and resource management in ATM networks. VPCs can be used, for example, to simplify connection establishment procedures such as routing, to separate traffic types having different QoS demands, or to aggregate user-to-user services such that the UPC can be applied to the aggregate traffic stream.

In this paper we address the problem of the determination of VPC source traffic descriptors which enable an efficient allocation of network resources. Particularly, we focus on VPCs formed by a number of delay-sensitive CBR VCCs having low cell rates. Depending on the way a VPC is formed, the resulting source traffic descriptors enable a resource allocation which is less or more efficient. In section 2, we give general comments on how to form a VPC and on how to find adequate source traffic descriptors for it. The dual spacer algorithm, which was developed in [12] for the shaping of VCCs, is described in section 4 and its application to VPCs is outlined. Simple dimensioning procedures for the source traffic descriptors and numerical examples showing the feasibility of our approach are given in section 5. The paper concludes with a brief summary in section 6. An appendix containing the formulae used for the dimensioning procedure is attached.

2 Forming of virtual path connections

The forming of a VPC out of a number of VCCs and the finding of adequate source traffic descriptors is straightforward if the considered traffic is delay-insensitive or if we have to deal with connections having high cell rates. In case of delay-insensitive traffic, the VPC can be shaped to a cell rate close to the aggregate rate of the VCCs, and the delay introduced by shaping high cell rate connections is rather marginal. However, if the VPC carries traffic from delay-sensitive connections with low cell rates, the determination of traffic descriptors for the aggregate cell stream is more difficult if an efficient allocation of the network resources, which implies a low fare of the VPC, is desired.

The problem of finding suitable traffic descriptors for a VPC is closely related to the way the VPC is formed. In principal, we have the following two possibilities:

(i) Form the VPC by simply multiplexing the traffic from the different VCCs without any additional shaping function. The delay introduced on cells of the individual VCCs is rather marginal in this case, no matter which multiplexing policy is used (FIFO, Round Robin, Weighted Fair Queueing (WFQ), etc.). However, during the busy periods of this multiplexing operation we submit cells into the network at link rate. Thus, either the PCR of this VPC should be equal to the link rate or a large CDVT has to be allocated. Since the network operator has no knowledge about the composition of a VPC, he is not able to estimate a multiplexing gain which might be obtained due to the statistical properties of the VCCs carried in the VPC. The operator has therefore to allocate network resources on the basis of a worst-case behavior of the VPC, which is a major drawback.

(ii) Shape the VPC after multiplexing by enforcing a minimum distance between cells equal to the inverse of the aggregate PCRs of the VCCs. This solution is optimal in the sense that the traffic of the VPC is almost CBR, i.e. the consumption of network resources by this VPC is minimal. However, using this solution, an additional delay due to the shaping process is introduced.

For connections having high cell rates, this delay is quite small due to a cell inter-emission interval which is relatively short. In case of VCCs that are able to tolerate long delays, the additional shaping delay is uncritical anyway. For these two cases, this is the most suitable solution.

If we deal with delay-sensitive connections of low cell rates, the delay introduced by the shaping process can be too long, especially if we look at CBR connections of the same type. As a simple example, let us consider a VPC formed by a small number of CBR connections having a bandwidth demand of a few hundreds of *kbps* or even less. Examples for connections of this type are phone calls or video telephony. If we space the VPC traffic according to the aggregate PCRs, we will enforce a cell inter-emission interval in the order of milli-seconds. This leads to a maximum delay introduced by the shaping function which is in the same order of magnitude.

The solution we propose for delay-sensitive CBR connections with low cell rates is located in between the possibilities *(i)* and *(ii)*. By using the so called *dual spacer algorithm* proposed in

[12], we can shape the VPC according to traffic descriptors which enable an efficient use of the network resources while introducing short delays. Furthermore, the source traffic descriptors PCR, SCR and BT can be dimensioned straightforward for the resulting VPC. Due to the simplicity of this shaping function and the short delays, the algorithm can be implemented at the network access or within network nodes where traffic from different VPCs is merged or splitted.

In [9] and [16] the shaping of VPCs is also addressed. A priority-based solution is proposed when there is a mix of delay-sensitive and loss-sensitive VCCs in the VPC. The trade-off between the shaping delay and the efficiency in use of network resources is studied in [2], however, no solutions to the problem are proposed in this work.

3 A first solution

As mentioned above, a trade-off in forming a VPC occurs, if it carries a number of delay-sensitive CBR connections having low cell rates. On the one hand, delays should be kept small to meet the delay constraints of the VCCs, i.e. traffic shaping with a cell rate close to the aggregate cell rate should be avoided. On the other hand, the determination of suitable source traffic descriptors, which allow an efficient utilization of the network resources, is impossible if traffic shaping is not performed. This trade-off occurs for all multiplexing policies discussed in the literature so far (FIFO, Round Robin, WFQ, etc.), since they only affect the CDV behavior and the fairness among the multiplexed connections but not the major delay behavior itself. Therefore, a way of traffic shaping should be applied to increase the efficiency of the resource allocation by introducing a more deterministic traffic pattern and reducing the burstiness of the cell stream.

A first solution to this problem is simply to enforce a minimum distance between cells of the VPC equal to a value T which is smaller than the inverse of the aggregate cell rate $1/T_{VP}$ of the VCCs gathered in the VPC. In order to minimize the resource consumption of the VPC, which also leads to a minimization of the communication costs, we are interested in finding the maximum value of T that fulfills the delay constraints of the connections carried by the VPC.

The distribution of the delay introduced by the shaping process can be calculated from the distribution of the waiting time of a queue whose input process is the VPC traffic before shaping and whose service time is deterministic with value T. If the VPC is formed by N identical CBR connections with a period of $N \cdot T_{VP}$ and phases of random order, we obtain this distribution by equation (15) in the appendix (cf. [3]). For VCCs with different periods, we do not have an exact formula but we easily derive bounds for the distribution (see equation (16) in the appendix). A worst case dimensioning for uncorrelated CBR connections can be derived by using a Poisson input process (see appendix, equation (17)). Thus, the spacing interval T can be dimensioned easily to meet the delay constraints.

Having declared a value for the PCR larger than $1/T_{VP}$, we are also interested in declaring a SCR and a BT for the VPC. Although tariff structures are unknown until now, users that can specify their traffic in greater detail may incur a lower cost for the connection. Looking at the following scenario, we derive a simple and accurate way to dimension these source traffic descriptors.

Consider a cell stream, e.g. a VPC, with a mean cell inter-arrival time of T_{VP}, which is monitored by a $GCRA(I, L_1)$. Right after the monitoring, the cells pass a shaping device which enforces a minimum cell distance of T. After that, the resulting cell stream is controlled by a second $GCRA(I, L_2)$ (cf. figure 1). We assume that $T_{VP} > I > T$.

Now, let t_1^n and t_2^n denote the time instants just after the arrival of cell number n at the first and second GCRA, respectively. Furthermore, the states of the two GCRAs at time t,

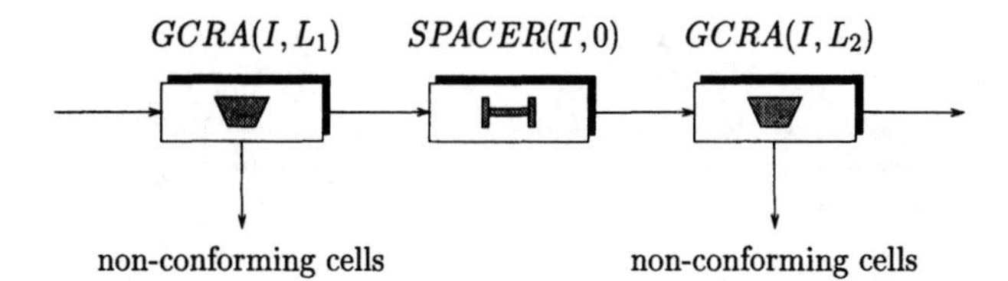

Figure 1: *Scenario for the derivation of the dimensioning procedure for the SCR and BT*

i.e. the virtual waiting time in the corresponding FIFO queues, are described by $S_1(t)$ and $S_2(t)$, respectively. Then, the following property, which is proofed in the first part of the appendix, holds:

$$S_2(t_2^n) = S_1(t_1^n) - (t_2^n - t_1^n) \leq S_1(t_1^n) \quad . \tag{1}$$

Let $L_1 = L_2$. From equation (1) we conclude that the probability of observing a non-conforming cell at the first GCRA ($\Pr\{S_1(t_1^n) > L_1\}$) upper bounds the probability that this cell is declared as non-conforming by the second GCRA ($\Pr\{S_2(t_2^n) > L_2\}$). This bound is tighter for large values of I/T, since it is more likely in this case that $t_2^n = t_1^n$. Table 1 provides simulation results for the dimensioning of L_1 and L_2, where cells are observed as non-conforming with a probability of less than $p = 10^{-3}$. The input traffic consists of the superposition of 6 identical CBR connections with a period of 600 time-slots and random phases. The value of I has been set to 95.

T	I/T	L_1	L_2
5	19.0	330	308
10	9.5	330	298
30	3.2	330	285
50	1.9	330	225

Table 1: *Simulation results showing the tightness of the limit values L_1 and L_2*

The source traffic descriptors T_s and τ_s of a VPC can thus safely be obtained from the distribution of the virtual waiting time in a FIFO queue. The service time has to be deterministic with period T_s and the input process has to be equal to the input traffic of the VPC, before the cell stream passes the shaping device. If T_s is close to T, these values will be too pessimistic, e.g. if $T_s = T$ a BT equal to zero would be sufficient, but it is difficult to compute them exactly for general cases.

To sum up, the procedure for the dimensioning of the source traffic descriptors for this solution is the following:

(1) Chose T as the maximum value so that the delay constraints are fulfilled.

(2) Chose $T_s = \epsilon \cdot T_{VP}$ with ϵ close to 1, e.g. $\epsilon = 0.95$. Using T_s, the BT τ_s can be computed as mentioned above.

Consequently, the way of forming a VPC presented above introduces delays which can be tolerated by the VCCs and allows to determine the source traffic descriptors easily. However, if the VPC is the superposition of independent traffic sources with low cell rates and random phases, we can obtain large values for the BT, i.e. a considerable amount of buffer space must be allocated to this connection. This observation is illustrated by numerical examples in section 5 and leads to a low efficiency in resource allocation.

We can avoid this drawback by the use of the dual spacer algorithm proposed in [12], which allows to declare lower values for the BT while using the same values for PCR and SCR and fulfilling the same delay constraints.

4 The dual spacer algorithm

Conceptually, the dual spacer algorithm is equivalent to two shaping devices operating in series (cf. figure 2). The first one forces the cell stream to be conforming to the SCR $1/T_s$ in conjunction with the BT τ_s, while the second device enforces a minimum inter-cell distance of T which corresponds to the PCR.

DUAL SPACER(T, T_s, τ_s)

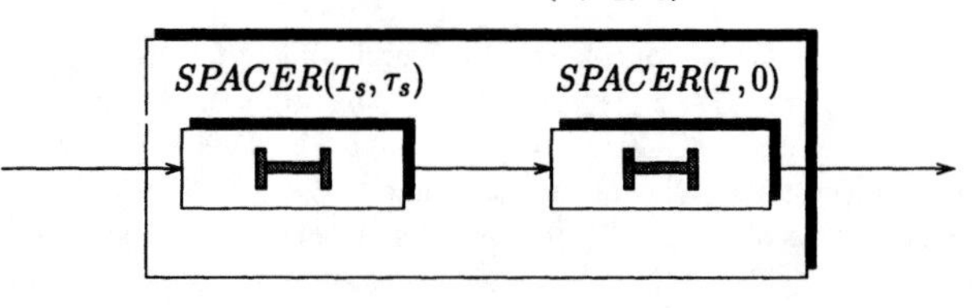

Figure 2: *Conceptual architecture of the dual spacer*

However, from implementation point of view we will have just one shaping device which delays cells in order to obtain a cell stream which is conforming to the three source traffic descriptors T, T_s and τ_s. The complexity of implementation compared to the conventional shaper remains therefore almost the same.

The dimensioning of the traffic descriptors for the dual spacer can be done in the following way. The delay introduced by the dual spacer is equal to the delay introduced by the first shaping device plus the delay introduced by the second device. These delays are not independent of each other and thus, the delay distribution is difficult to obtain for general input traffic. Nevertheless, we can make use of the following approximation. The time D that a cell is delayed by the dual spacer is given by

$$D = \max(D_1, D_2) \quad , \tag{2}$$

where D_1 is the delay required due to the first spacing device and D_2 is the delay required due the second spacing device if the first one is omitted. Assuming independence of the two random variables D_1 and D_2, we arrive at the following equation:

$$\Pr\{D \leq x\} \approx \Pr\{D_1 \leq x\} \cdot \Pr\{D_2 \leq x\} \quad . \tag{3}$$

This approximation is motivated by the fact that we observe periods where either the PCR shaping is decisive (i.e. during bursts), or where the SCR shaping dominates (i.e. between bursts emitted at PCR). A frequent change between these periods is only rarely observed. The distribution of D_1 is given by

$$\Pr\{D_1 \leq x\} = \Pr\{D_1' \leq x + \tau_s\} \quad , \tag{4}$$

where D_1' is the virtual waiting time in a queue with deterministic service time T_s. The distributions of D_1' and D_2 can be derived easily for different traffic scenarios using the formulae presented in the appendix. Numerical results given in the next section will show that the results obtained employing this approximation are in good agreement with the simulation values.

The procedure for the dimensioning of the source traffic descriptors for the dual spacer solution can be summed up as follows:

(1) Chose the same values for T and T_s as in the first solution (cf. section 3).

(2) Dimension τ_s using equation (3) to fulfill the required delay constraints.

As outlined in the next section, the use of the dual spacer algorithm allows to reduce the BT considerably compared to the solution in section 3, while meeting the same delay constraints. Therefore, this algorithm should be preferred for the forming of VPCs, since it provides the more efficient solution from the resource allocation point of view, which may imply a lower cost for the user.

5 Numerical examples

In this section, we present numerical examples to illustrate the dimensioning procedures derived in the sections 3 and 4. With *Solution I* we refer to the procedure developed in section 3 and with *Solution II* we refer to the dual spacer model. We define D as the delay introduced by the shaping device and denote the maximum delay that can be tolerated by Δ. In other words, we want to ensure that $\Pr\{D > \Delta\} < 10^{-x}$ for a given value of x.

With the first dimensioning example we consider an access to a public ATM network at 622 *Mbps* and we want to declare source traffic descriptors for a VPC formed by N CBR connections with a period of $T_{VC} = 6000$ slots of cell duration. That could, for instance, correspond to N 64 *kbps* phone connections. The $1 - 10^{-6}$ quantile of the shaping delay should be limited to 1 *msec*, which corresponds to 1415 cell times in our case. Table 2 shows the corresponding values of T, T_s and the BT τ_s for the *Solutions I* and *II*. In case of *Solution I*, we assumed that the UPC function, which monitors the VPC, declares a cell as non-conforming with a probability of less than 10^{-6}.

N	T	T_{VP}/T	T_s	BT_I	BT_{II}
2	1415	2.12	2850	2850	1415
3	710	2.81	1900	3795	2382
4	491	3.05	1425	4216	2833
5	401	2.99	1140	4370	3079
6	328	2.79	950	4372	2963
8	304	2.47	713	4155	3112
10	268	2.24	570	3872	2823

Table 2: *Comparison of source traffic descriptors for Solution I & II*

From the results for BT_I and BT_{II} it is apparent, that the use of the dual spacer reduces the required value for the BT τ_s considerably. It should be pointed out that for *Solution I* lower values for the BT could be declared while fulfilling the same restrictions for observing non-conforming cells. This holds especially when T_s is close to T. However, these values are difficult to obtain, even approximately, using analytical tools. Note that if we do not perform any shaping, we should declare a PCR equal to the aggregate PCRs of the VCCs (possibly with some safety tolerance) and a CDVT equal to BT_I. For instance, for $N = 5$ sources and a safety tolerance of 0.95 we would declare $PCR = 1/1140$ and $CDVT = 4370$. Thus, 4 back-to-back cells can pass the GCRA transparently, which could result in a waste of network resources if a worst case allocation CAC is used [4].

In the following, we focus on the dual spacer solution. We compare the approximate values computed using equation (3) with simulation results for different traffic conditions. The SCR is set to $T_s = 0.95 \cdot T_{VP}$. Figures 3 to 5 show the $1 - 10^{-3}$ quantiles of the delay distribution as a function of the PCR $1/T$ for different values of τ_s. To obtain figure 3, we used a scenario with 6 homogeneous CBR connections. The inter-cell distance of each VCC is set to $T_{VC} = 6000$.

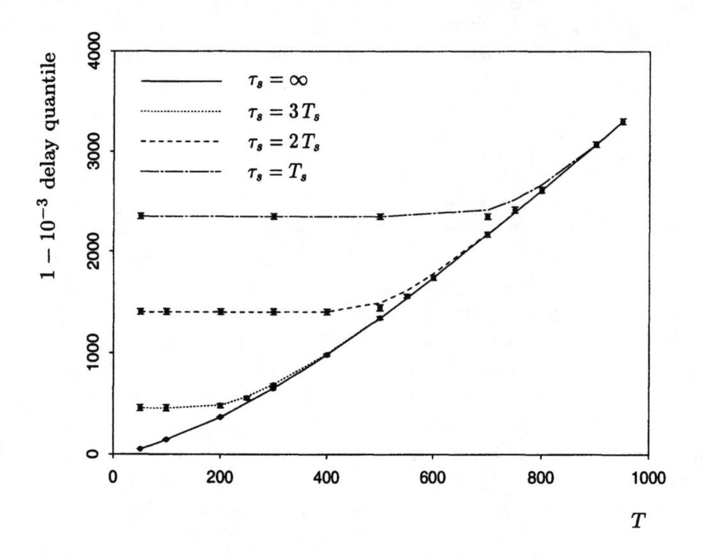

Figure 3: $1 - 10^{-3}$ *quantiles of the delay distribution for homogeneous CBR traffic*

The approximate results are matching well with those obtained by simulation. We observe a slight over-estimation of the simulation results only at the knee of the curves. For small values of T, the quantile of the shaping delay decreases nearly linearly with the increase of τ_s, since the PCR shaping has almost no influence in this case. If T increases, the quantile of the shaping delay remains constant until we approach the curve for $\tau_s = \infty$. At this point, the influence of the SCR shaping is getting lost and thus the curve approaches to that of $\tau_s = \infty$.

Figure 4 shows a similar behavior for the delay introduced by the dual spacer if a heterogeneous traffic scenario consisting of 3 CBR connections with $T_{VC} = 600$ and 30 CBR connections with $T_{VC} = 6000$ is considered. The analytical results, which are upper bounds, correspond to the cases $\tau_s = 285$ and $\tau_s = \infty$. The curve for the infinite BT is already approached for $\tau_s = 2850$, which equals 30 T_s.

A worst case scenario can be investigated by using Poisson traffic. Figure 5 shows results for a Poisson arrival process with a rate of $\lambda = 1/1000$. The analytical results are computed using equation (17) in the appendix. Again, we observe a similar behavior and a good agreement between analytical and simulation results.

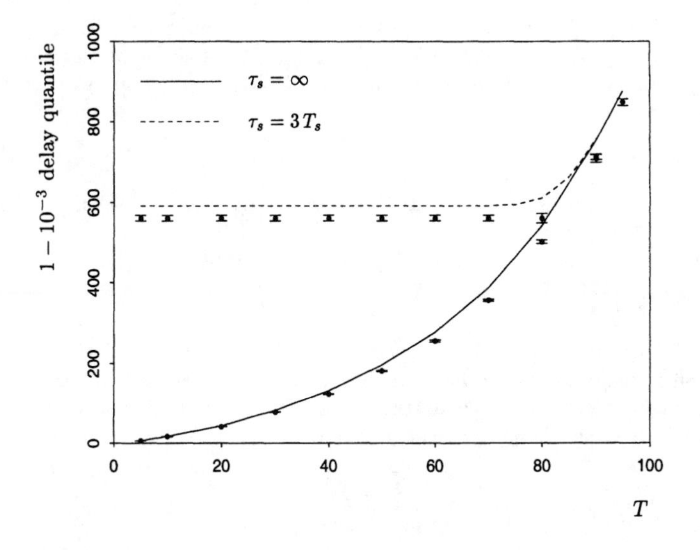

Figure 4: $1 - 10^{-3}$ *quantiles of the delay distribution for heterogeneous CBR traffic*

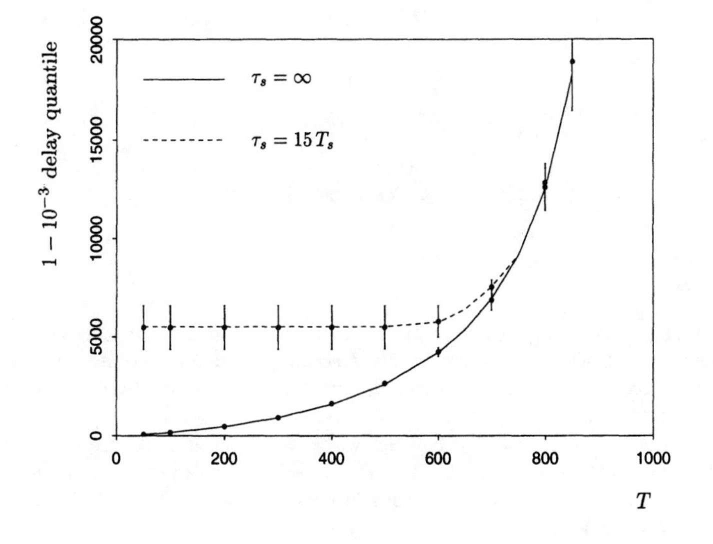

Figure 5: $1 - 10^{-3}$ *quantiles of the delay distribution for Poisson traffic*

If we look more closely at the numerical results for the dual spacer solution (cf. figures 3 to 5), we can think of a rough approximation for the BT if the SCR and the maximum delay Δ regarding the quantiles are given. The smallest value for the BT_{II} which allows to meet the given delay constraints can be approximated by

$$BT_{II}^{approx} = \Delta' - \Delta \quad , \tag{5}$$

where Δ' is the delay experienced for $BT_{II} = 0$. Of course, the delay cannot be shorter than that introduced by PCR shaping.

This approximation is justified by equation (4), since Δ' corresponds to BT_I, i.e.

$$BT_{II}^{approx} = BT_I - \Delta \quad . \tag{6}$$

It is an indication of the reduction of the BT we should expect when using *Solution II.* From table 3 we observe a good agreement between the results obtained with the formulae in the appendix (BT_{II}) and those computed with the simple equation above (BT_{II}^{approx}). The parameter set used to generate the results in table 3 is the same as that described for table 2.

N	BT_I	BT_{II}	BT_{II}^{approx}
2	2850	1415	1435
3	3795	2382	2380
4	4216	2833	2801
5	4370	3079	2955
6	4372	2963	2957
8	4155	3112	2740
10	3872	2823	2457

Table 3: *Accuracy of the rough approximation*

6 Conclusions

We discussed the problem of forming a virtual path connection out of a number of delay-sensitive virtual channel connections, if a high utilization of the network resources is desired. We also focused on the dimensioning of source traffic descriptors, which are required for connection admission control and usage parameter control.

Two possible solutions are identified: The first one, which uses a conventional shaping device to enforce a minimum distance between consecutive cells of the VPC and the second one, which makes use of the so-called dual cell spacer. For both solutions, rules for the dimensioning of the source traffic descriptors are derived.

From numerical examples we conclude, that the solution using the dual spacer leads to a more efficient use of the network resources, especially when delay-sensitive connections with low cell rates are considered. The reason is a burst length reduction compared to the first solution, which results in a smaller value declared for the burst tolerance of the VPC. Another advantage of the dual spacer mechanism is the easy dimensioning of the traffic descriptors for given delay constraints with a simple dimensioning procedure developed in this paper.

Acknowledgement

The authors would like to thank the European Community for the financial support within the framework of the COST 242 project, which enabled us to carry out this study. We also appreciate the support of the Deutsche Telekom AG (FZ Darmstadt).

Appendix

1. Outline of the proof of the property in section 3

We proceed by induction. For the first cell, $t_1^0 = t_2^0$ and a busy period starts simultaneously at both GCRAs. Assume that for cell number $n-1$ of the first busy period of the first GCRA $S_2(t_2^{n-1}) = S_1(t_1^{n-1}) - (t_2^{n-1} - t_1^{n-1})$. The evolution of both GCRAs is given by the following equations:

$$S_1(t_1^n) = I + [S_1(t_1^{n-1}) - (t_1^n - t_1^{n-1})]^+ \quad , \tag{7}$$

$$S_2(t_2^n) = I + [S_2(t_2^{n-1}) - (t_2^n - t_2^{n-1})]^+ \quad . \tag{8}$$

If $T < I$, then we obtain

$$t_2^{n-1} < t_1^{n-1} + S_1(t_1^{n-1}) - I \quad . \tag{9}$$

Note, that the right hand side of the inequation (9) is the departure time of cell $n-1$ from the spacing device if $T = I$. Using this inequation, we can easily see that

$$S_1(t_1^{n-1}) - (t_1^n - t_1^{n-1}) > 0 \tag{10}$$

implies

$$S_1(t_1^{n-1}) - (t_2^n - t_1^{n-1}) > 0 \quad , \tag{11}$$

and consequently

$$S_2(t_2^{n-1}) - (t_2^n - t_2^{n-1}) > 0 \quad . \tag{12}$$

This, on the other hand, implies that

$$S_2(t_2^n) = S_1(t_1^n) - (t_2^n - t_1^n) \quad , \tag{13}$$

i.e. if the first GCRA is in the first busy period then the second GCRA is also in the first busy period.

Let N be the last cell of the first busy period of the first GCRA. The end of this first busy period occurs at the time $t_1^N + S_1(t_1^N)$. At the time $t_2^N < t_1^N + S_1(t_1^N)$ the second GCRA is in the state $S_2(t_2^N) = S_1(t_1^N) - (t_2^N - t_1^N)$ and consequently, the first busy period is finished simultaneously at both GCRAs. The same argument applies for the second busy period, etc.

Thus, we arrive at the conclusion that the busy periods at both GCRAs are coincident and therefore equation (1) is always true.

2. Dimensioning formulae

Let us consider a FIFO queue with deterministic service time T. Let V_i be the amount of work in the system at the end of time slot i and let $A(t-\tau,t)$ denote the amount of work that arrives during $[t-\tau,t]$. In [3], the following expression is derived:

$$\Pr\{V_t > x\} = \sum_{n=\lceil\frac{1+x}{T}\rceil}^{\infty} \Pr\{A(t-(nT-x-1),t) = nT\,,\, V_{t-(nT-x)} = 0\} \quad . \tag{14}$$

If the input process is a superposition of N CBR connections of period D, we derive the delay distribution W_t in the spacing device by

$$\Pr\{W_t > x\} = \sum_{n=\lceil\frac{1+x}{T}\rceil}^{D} \binom{N-1}{n}\left(\frac{nT-x}{D}\right)^{n}\left(1-\frac{nT-x}{D}\right)^{N-1-n}\left(1-\frac{(N-1-n)T}{D-(nT-x)}\right) \quad . \tag{15}$$

In the case of heterogeneous CBR connections, sources of different periods experience different delays. For sources of period D_i we get

$$\Pr\{W_t > x\} < \sum_{n=\lceil\frac{1+x}{T}\rceil}^{\infty} \Pr\{A'(t-(nT-x-1),t) = nT - \left\lfloor\frac{nT-x}{D_i}\right\rfloor\} \quad , \tag{16}$$

where $A'(t-\tau,t)$ is the amount of work that arrives during $[t-\tau,t]$ when we exclude one source of period D_i. Using equation (11) of [13], we obtain an expression for $\Pr\{A'(t-(nT-x-1),t) = nT - \lfloor(nT-x)/(D_i)\rfloor\}$.

If the input process is Poisson with parameter λ, and defining $\rho = \lambda T$, we obtain for $\rho < 1$

$$\Pr\{W_t > x\} = 1-(1-\rho)\sum_{n=0}^{\lceil\frac{x}{T}\rceil-1} e^{-(n\rho-\lambda(x+1))}\frac{(n\rho-\lambda(x+1))^n}{n!} \quad . \tag{17}$$

This expression is susceptible to loss of numerical accuracy when the load is close to 1. In [15], a modified form of the continuous time model is presented, which avoids this numerical problem.

References

[1] ATM Forum. *ATM User-Network Interface Specification 3.1.* September 1994.

[2] Z. Bazanowski, U. Killat. *Trade-Offs in VP-based and VC-based Traffic Shaping.* Proceedings of the 3rd International Conference on Telecommunication Systems, Nashville, 1995, pp. 78-85.

[3] Z. Bazanowski, U. Killat, J. Pitts. *The ND/nD/1-Model and its Application to the Spacing of ATM-Cells.* Proceedings of the International Conference on Informational Networks and Systems '94, Saint-Petersburg, 1994, pp. 109-118.

[4] B. Boyer, F.M. Guillemin, M.J. Servel, J.P. Coudreuse. *Spacing Cells Protects and Enhances Utilization of ATM Network Links.* IEEE Network, September 1992, pp. 38-49.

[5] M. Butto, E. Cavallero, A. Tonietti. *Effectiveness of the Leaky Bucket Policing Mechanism in ATM Networks.* IEEE Journal on Selected Areas in Communications, Vol. 9, No. 3, April 1991, pp. 335-342.

[6] F. Hübner. *Dimensioning of a Peak Cell Rate Monitor Algorithm Using Discrete-Time Analysis.* Proceedings of the 14th ITC, Antibes, 1994, pp. 1415-1424.

[7] ITU-T Study Group 13. *Recommendation I.371: Traffic control and congestion control in B-ISDN.* Geneva, November 1994.

[8] G. Niestegge. *The leaky bucket policing method in the ATM (asynchronous transfer mode) network.* International Journal of Digital and Analog Communication Systems, Vol. 3, 1990, pp. 187-197.

[9] L.K. Reiss, L.F. Merakos. *Shaping of Virtual Path Traffic for ATM B-ISDN.* IEEE INFOCOM '93, San Francisco, 1993, pp. 168-175.

[10] M. Ritter, P. Tran-Gia. *Performance Analysis of Cell Rate Monitoring Mechanisms in ATM Systems.* IFIP 3rd International Conference on Local and Metropolitan Communication Systems, Kyoto, December 1994, pp. 119-139.

[11] M. Ritter, S. Kornprobst, F. Hübner. *Performance Comparison of Design Alternatives for Source Policing Devices in ATM Systems.* 20th IEEE Conference on Local Computer Networks, Minneapolis, October 1995, pp. 132-139.

[12] M. Ritter. *Performance Analysis of the Dual Cell Spacer in ATM Systems.* IFIP 6th International Conference on High Performance Networking, Palma de Mallorca, September 1995, pp. 345-362.

[13] J.W. Roberts, J. Virtamo. *The Superposition of Periodic Cell Arrival Streams in an ATM Multiplexer.* IEEE Transactions on Communications, Vol. 39, No. 1, 1991, pp. 298-303.

[14] C. Rosenberg, G. Hébuterne. *Dimensioning a Traffic Control Device in an ATM Network.* IFIP TC6, Broadband Communications '94, Paris, 1994, paper 12.2.

[15] J. Virtamo. *Numerical evaluation of the distribution of unfinished work in an M/D/1 system.* Electronics Letters, Vol. 31, No. 7, 1995, pp. 531-532.

[16] J. Zeng, L.F. Merakos. *Analysis of a priority Leaky-Bucket regulator for VP Traffic in ATM networks.* GLOBECOM '94, San Francisco 1994, pp. 606-611.

2

Specification of Cell Dispersion in ATM Networks

Fabrice Guillemin*, Catherine Rosenberg†, and Aref Meddeb†

* France Telecom, CNET Lannion A, Route de Trégastel, 22300 Lannion, France.
Tel: + (33) 96-05-13-46, Fax: + (33) 96-05-11-98, Email: guillemi@lannion.cnet.fr.
† Ecole Polytechnique de Montreal, C.P. 6079, succ. A, Montreal, Quebec, Canada H3C 3A7.
Tel: + (514)340-4123, Fax: + (514)340-4562, Email: {cath,aref}@comm.polymtl.ca.

Abstract

So far, cell jitter has been investigated within the standardization bodies for Constant Bit Rate connections only and it has been recognized that cell jitter covers two different topics, namely cell clumping and cell dispersion. Whereas the material for the specification of cell clumping seems stable for the time being, the specification of cell dispersion requires further investigations in particular, because we have shown that the algorithm currently under consideration within the ITU-T for specifying cell dispersion fails to provide useful information for the operation of very simple receiving ATM Adaptation Layer (AAL) mechanisms. Moreover, it exhibits an unstable behavior when cell transfer delays become very large and/or cell loss occurs, and the outcome of the algorithm depends on the specific realization of the observed cell stream. Hence, after a detailed description of the issues related to cell dispersion, we propose an algorithm which computes two quantities of interest for specifying cell dispersion.

Keywords

ATM, Cell Delay Variation, ATM Adaptation Layer, cell dispersion

1 INTRODUCTION

A basic feature of telecommunication networks based upon the Asynchronous Transfer Mode (ATM) is that cells progressing along a connection experience random delays, which may be caused for instance by queuing in multiplexing stages. As a consequence, the initial time structure of any cell stream traversing an ATM network is in general altered by stochastic perturbations.

For example, when observing at some point along a connection an initially periodic cell stream, which has passed through several network elements using asynchronous multiplexing, the cell arrival process is no longer periodic but erratic because of random cell transfer delays. This is the phenomenon of cell jitter, also referred to as Cell Delay Variation (CDV) within the standardization bodies (I.356, 1993) and (I.371, 1995). Roughly speaking, cell jitter on a connection is due to the interaction between this connection and other connections when sharing common resources (transmission capacities, buffers, etc.). So far, cell jitter has been investigated within the ITU-T only for cell streams, which are initially periodic (Constant Bit Rate, CBR, connections). In such a case, it has been recognized that cell jitter is composed of two basic complementary phenomenons, namely cell clumping and cell dispersion (Guillemin and Monin, 1992).

To illustrate the cell clumping effect, consider an initially periodic cell stream with period T passing through an ATM network. When this cell stream is observed at some point along the connection (e.g., the T_B reference point, an inter-Network Node Interface, or the receiving S_B interface), it may happen that the distance between two consecutive cells is less than the initial period T because of random delays. This is precisely what is called cell clumping.

Cell clumping is quantified by using a reference algorithm, namely the Virtual Scheduling Algorithm (VSA). A definition of the VSA is given in the ITU-T Recommendations I.356 (1993) and I.371 (1995) Annex 1. The VSA is also known as Generic Cell Rate Algorithm (GCRA) within the ATM Forum (The ATM Forum, 1994).

Consider a periodic cell stream with period T traversing an ATM network and observed at some point along the connection. It may happen that two consecutive cells experience increasing delays so that their inter-arrival time at the observation point is greater than the initial period T, giving rise to a cell gap in the cell stream. This is the phenomenon of cell dispersion, which is critical for connections supporting a circuit emulation (e.g., ATM Adaptation Layer, AAL, of type 1) and/or when re-sequencing (e.g., two-layer video) has to be performed in the receiving terminal.

In Section 2, we present a detailed formulation of the problems related to cell dispersion in ATM networks. A first specification of cell dispersion has been introduced in Gravey and Boyer (1993) and I.356 (1993). However, the proposed algorithm fails to provide useful information for the operation of very simple receiving ATM Adaptation Layer (AAL) mechanisms (see Guillemin, Rosenberg and Meddeb, 1996). We introduce in Section 3 a new method of characterizing cell dispersion. Similarly to cell clumping, which is specified with respect to a reference algorithm (the VSA), we propose a new algorithm to measure cell dispersion.

2 PROBLEM FORMULATION

2.1 Need for Quantifying Cell Dispersion

Consider a CBR connection with peak emission interval T supporting a circuit emulation and assume that this connection traverses one or several ATM networks. As mentioned above, the cell stream arriving at the receiving terminal may be affected by cell dispersion. This phenomenon is especially critical for AAL 1 mechanism. Specifically, such a mechanism must restore as far as possible the periodic structure of the cell stream. For this purpose, arriving cells are stored in a buffer, usually referred to as *elastic buffer* or *playout buffer*.

Several disciplines may be envisaged to remove cells from an elastic buffer. For instance, the service rate may be such that the elastic buffer may be at any instance half-full. In such a case, the service rate adapts to the instantaneous cell arrival rate, which is time-varying because of cell jitter. Another possibility is to store the first arriving cell in the elastic buffer for a certain time (*fixed playback point*) and then, to read out this buffer with rate $1/T$. This is the most simple AAL mechanism, which was used for instance in the early ATM experiment named PRELUDE (Devault, Cochennec and Servel, 1992). This mechanism will be referred to as *reference AAL mechanism* in the following. In some sense, this mechanism is the worst one since it does not adapt to the instantaneous cell arrival rate and relies mainly on one parameter, namely the buffering delay of the first arriving cell.

With regard to the quality of service (QoS) of the circuit emulation, the key problem is to avoid starvation in the elastic buffer (since overflow can be taken care of by means of large buffers). Indeed, because of cell dispersion, it may happen that the elastic buffer empties (e.g., if cells experience over-large delays), leading to an interruption in the cell disassembling process and then to a degradation of the QoS offered to the application (e.g., clipping in a telephone connection). More precisely, after the initial buffering delay, the elastic buffer is explored either periodically with period T in a reference AAL mechanism or quasi-periodically when the service discipline is more sophisticated. If a cell is present in the buffer at a server exploration time, it is removed. Starvation occurs when there is no cell to remove from the elastic buffer at a server exploration time and hence cell stuffing is required to mask the interruption in the cell disassembling process. The stuffing can be more or less sophisticated depending on the application characteristics and on the acceptable level of complexity at the playout.

In any way, in order to dimension the AAL mechanism to offer the right level of QoS, information is required on the cell dispersion affecting the cell stream (i.e., on the characteristics of the cell stream entering the playout). Means are needed to quantify and measure cell dispersion.

2.2 Differences between Cell Clumping and Cell Dispersion

Cell clumping has a direct impact on the network, for instance on UPC/NPC mechanisms and resource allocation when "pick-up" policing mechanisms are used in the UPC/NPC function (Boyer, Guillemin, Servel and Coudreuse, 1992). According to ITU-T Recommendation I.371, cell clumping quantification is required at each T_B and inter-network NNI interfaces (in particular in order to dimension UPC/NPC mechanisms) and is specified in the traffic contract by means of a CDV tolerance value, usually denoted by τ.

Because of the possible impact of cell clumping on resource allocation, there is a direct incentive for a network operator to limit and even eliminate as far as possible cell clumping at the network access point. This can be achieved by performing cell spacing (Boyer, Guillemin, Servel and Coudreuse, 1992).

In the contrary, cell dispersion on a connection is an end-to-end phenomenon which does not have a direct impact on resource allocation. So, there is no direct incentive for a network to limit or eliminate cell dispersion. Cell dispersion could be eliminated (or at least reduced) by buffering the cells of the connection for a certain time somewhere along the connection to absorb as far as possible the random fluctuations in cell transfer delays. Buffering cells for a certain time would greatly increase the QoS parameter Cell Transfer Delay (CTD, I.371, 1995). This should carefully be examined from an application point of view. If one decides in favor of eliminating part of the cell dispersion effect on a connection (e.g., at the output of each network), this should be done in preference within the ATM layer, since it would be highly desirable not to go beyond the ATM layer in the middle of a connection.

As a consequence, the key difference between cell clumping and cell dispersion is that it may be inappropriate to eliminate cell dispersion somewhere along a connection supporting a real time application while the same operation may be very useful for cell clumping.

2.3 How and Where to Quantify and Measure Cell Dispersion ?

As this point, the necessity to quantify and measure cell dispersion appears clearly. Many questions remain open especially concerning how and where to quantify and measure it. In ways similar to the one followed for cell clumping, there is the need to define an *algorithm* to measure cell dispersion.

An ideal situation would be that no interruptions occur in the cell disassembling process in the receiving terminal. However, since cell transfer delays may be very large, this would require huge buffering delays (this may cause echo phenomenons for telephone connections). A more realistic situation is that starvation occurs with a very low probability and that the AAL mechanism recovers from starvation by producing stuffing cells.

This leads to a statistical characterization of cell dispersion and the necessity to define precisely the stuffing function.

Since cell dispersion is definitely a problem related to the AAL level, we should try to quantify it using some information relevant to AAL mechanisms. However, the existence of many possible AAL mechanisms implies that it will be difficult to find some information common to all AAL mechanisms. Furthermore, even if such information could be found, a more fundamental problem, besides the arbitrary choice of not going beyond the ATM Layer in the middle of a connection would forbid the use of information related to the AAL level to quantify cell dispersion. Namely, the algorithms for measuring cell clumping and cell dispersion should be coordinated in the sense that the cell dispersion algorithm should only consider the conforming cells with respect to cell clumping. Indeed, the UPC/NPC mechanism is entitled to discard any cell, which does not conform with the VSA, whose parameter are the declared peak emission interval and cell clumping tolerance. Thus, the UPC/NPC mechanism by discarding non conforming cells may significantly but legally increase the magnitude of cell dispersion. As a matter of fact, if the algorithm for cell dispersion were to take into account all cells, it would measure a cell dispersion effect smaller than the one it would measure by considering conforming cells only. By discarding non conforming cells, the network should not be held responsible for the increase in cell dispersion.

To illustrate this phenomenon, consider an initially periodic cell stream with period T affected by CDV as depicted in Figure 1. Cell clumping at the observation point is quantified by a CDV tolerance parameter, which allows, say, two back to back cells. But, the cell stream is altered by unexpected excessive cell clumping so that cell clumps are composed of four back to back cells and thus, half of the cells are non conforming. Because of the non conformance of some cells, the stream of conforming cells is affected by cell dispersion, exhibited by the presence of cell gaps. Now, with regard to cell dispersion, if all cells were considered as to be conforming, then it is easily checked that it would be possible, from the jittered cell stream, to restore a perfectly periodic cell stream, for example by smoothing cell clumps in a buffer. In such a case, excessive cell clumping masks cell dispersion. Since a UPC/NPC mechanism may discard non-conforming cells, it may severely increase the magnitude of cell dispersion but this is legal and the network should not be considered as responsible for this increase in cell dispersion. As a consequence, the cell dispersion algorithm must handle only cells that are declared as to be conforming by the cell clumping algorithm and thus, both algorithms must be run simultaneously in a coordinated mode.

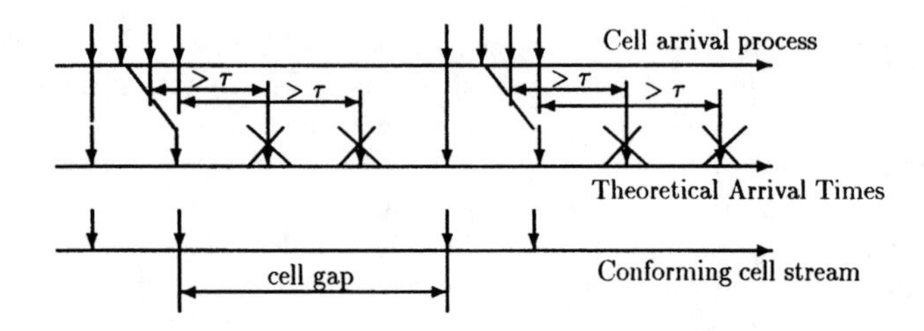

Figure 1 Impact of excessive clumping on cell dispersion.

It follows that cell dispersion should be characterized within the ATM Layer and the algorithms for measuring cell clumping and cell dispersion should be coordinated in the sense defined above.

2.4 Conformance with respect to Cell Dispersion

The problem of cell conformance testing with respect to cell dispersion has not yet been addressed within the standardization bodies but it is believed in this paper that this problem is fundamental, when a connection supports a circuit emulation. Specifically, since the (end-to-end) cell dispersion effect results for a multi-operator connection from the passage through several networks, it will be necessary to create some conformance testing procedure to check that the cell dispersion incurred by the cell stream in each network does not exceed some predefined limits.

As a matter of fact, for a proper operation of the receiving AAL mechanism, the cell dispersion should not exceed some predefined bounds. In the case of a multi-operator connection, a tool, namely a conformance algorithm, should be specified to assess the responsibility in the case of a conflict between a user and the network (at large) with regard to the QoS of a circuit emulation. Roughly speaking, if cells experience over-large transfer delays in a network along the connection, then the QoS of the circuit emulation may be degraded, for instance because of interruptions (or stuffing) due to information starvation in the cell disassembling process in the receiving terminal. In that case, it should be possible, from a *legality* point of view, to determine who is responsible for the QoS degradation, because the user has the right to complain that the different networks involved along the connection have not met their QoS commitments.

Hence, when cells of the connection are taken over from one network to another, the receiving network should be able to check that the delivered cell stream is not too much dispersed. If cell dispersion is too large, then the receiving network should not be considered as responsible for QoS degradation. In addition, each network should behave in such a way that it delivers to the next network a cell stream, which is not too much dispersed.

So even if cell dispersion is an end-to-end problem, the legality question comes down to check that each network behaves correctly with respect to cell dispersion. In fact, cell dispersion conformance testing should be performed at each network access point. Since we have chosen to remain in line with the ATM philosophy regarding the non passage to the AAL level in the middle of a connection, we will have to implement the cell dispersion conformance testing mechanism within the ATM Layer, which precludes the use of AAL information for quantifying the cell dispersion effect.

In summary, the legality problem has two major consequences: the necessity to quantify the cell dispersion effect with information within the ATM Layer and the need for an algorithm able to test conformance of cell dispersion, similar to the VSA used to test cell conformance with respect to cell clumping (I.356, 1993)

2.5 Apportionment of Cell Dispersion

With regard to cell clumping, the user at the T_B interface or the backward network at an inter-network NNI declares a value for the CDV (clumping) tolerance parameter τ that the network can either accept or refuse. An alternative approach would consist in specifying a *reference connection* and in apportioning cell clumping to each component of the connection (the Customer Equipment, each network, etc.). Two reasons why this alternative may be of little interest are that

1. the UPC should be perfectly programmable and thus should allow more flexibility than a reference connection ; in particular, any value for the cell clumping tolerance parameter τ could be declared and may possibly depend on the declared peak cell rate,
2. a network operator can perfectly control cell clumping on a connection via cell spacing (e.g., it can reduce it to that equivalent to one multiplexing stage), and thus, the need for a reference connection to specify cell clumping at each interface is useless ; this last point needs nevertheless further investigations within the standardization bodies since some network operators are still reluctant to perform traffic shaping and in particular cell spacing.

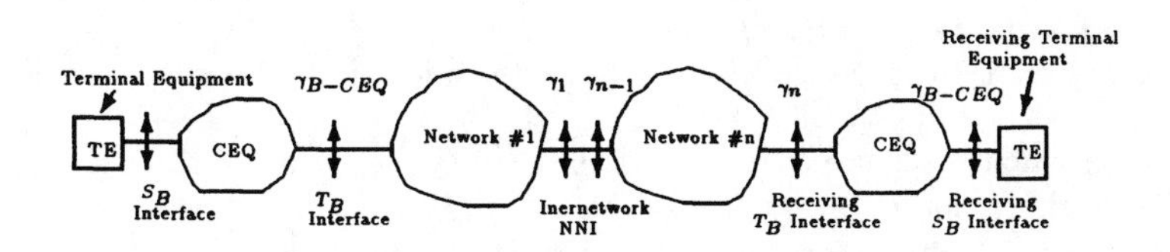

Figure 2 Apportionment of cell dispersion along an ATM connection.

In the contrary, it is of little interest to reduce cell dispersion on a connection. Thus, cell dispersion is mainly a problem to be dealt with at the end-user terminal, hardwired with predetermined AAL characteristics, which cannot be modified on a per connection basis. Since cell dispersion increases as cells progress along a connection and since cell dispersion has an impact on the QoS offered to a circuit emulation, the magnitude of cell dispersion should be kept under appropriate predefined bounds at the end-equipment terminal and thus within each network involved along the connection. A simple method to determine these bounds is to use a reference connection as depicted in Figure 2.

The end-to-end cell dispersion is divided into several components, namely the cell dispersion for the sender Customer Equipment, the cell dispersion for network $\#i$ along the connection, and the cell dispersion for the receiving Customer Equipment, so that the end-to-end dispersion is equal to the sum of all these components.

It is much too early to give an exact definition of a reference connection, given the lack of information on the behavior of switches and AAL mechanisms, in particular on how much cell dispersion is acceptable for guaranteeing a given QoS level. But, under the assumptions that cells are not buffered in the middle of the connection to absorb cell dispersion, we can already assess that the apportionment cannot be linear since cell dispersion is only going to increase along the connection.

2.6 Cell Dispersion Characterization

Besides the points raised above, some questions are still open and will be addressed in Section 3. In particular, the exact definition of the cell dispersion parameters or in other words, the design of a proper algorithm for measuring and testing conformance of the cell dispersion remains an open issue, since we show in Guillemin, Rosenberg and Meddeb (1996) that the algorithm currently under consideration within the ITU-T (I.356, 1993) suffers from many drawbacks.

Before proceeding further, let us summarize the basic points for cell dispersion characterization :

- cell dispersion should not necessarily be absorbed along the connection,
- legality issues imply conformance testing with respect to cell dispersion, which leads to the specification of a conformance algorithm,
- cell dispersion conformance testing requires the quantification of cell dispersion with information within the ATM layer,
- quantification of cell dispersion should be statistical,
- the algorithms for measuring and testing conformance of cell clumping and cell dispersion should be coordinated in the sense that the cell dispersion algorithm should only consider the conforming cells with respect to cell clumping,
- specification of a reference connection and apportionment of cell dispersion are needed.

3 A NEW ALGORITHM FOR THE SPECIFICATION OF CELL DISPERSION

Before addressing the design of an algorithm for the specification of cell dispersion, we should try to better understand which characteristics of cell dispersion are of importance from an application standpoint. As already mentioned, QoS degradation occurs when there is a period of starvation in the playout buffer. Of course, each occurrence of a starvation period may entail a QoS degradation, but the level of the degradation is directly linked to the length of the period.

As a matter of fact, a short starvation period can be masked via cell stuffing, while a long one may well have a disastrous effect on the QoS. In other words, we believe that for a given AAL mechanism, two characteristics of cell dispersion are important from an application standpoint (or that these two characteristics are sufficient to estimate the performances of an AAL mechanism), namely the frequency of starvation periods and the maximum size (or a remote quantile) of a starvation period. If the length of a starvation period is greater than the maximum size, the QoS degradation becomes unacceptable and some drastic actions should be performed by the AAL mechanism to guarantee the QoS of the application.

Roughly speaking, starvation periods of limited size are "natural" in the context of ATM systems and are due to random cell transfer delays through the network. Such starvation periods should be masked by the AAL mechanism (e.g., via cell stuffing) and as long as they do not occur frequently, they have a limited impact on the QoS of the application. This is why the frequency of starvation periods is an important characteristic of cell dispersion. In the contrary, starvation periods of large size may result from error conditions or unexpected traffic configurations within the network, which entail over-large cell transfer delay. Such a starvation period may adversely affect the QoS of the application. For example, the AAL mechanism may be unable to compensate the starvation of cells. For a proper operation of AAL mechanisms, the boundary between starvation periods of small and large size should be specified. This limit is precisely what we have called above maximum size of a starvation period.

If the values of the maximum size of a starvation period and the frequency of starvation periods were known in advance for a particular AAL mechanism, they may be used for dimensioning this AAL mechanism in the receiving terminal equipment. Moreover, the values of these two parameters may depend on the application supported, since they are directly related to the QoS of the application. As a matter of fact, constant bit rate video may have different characteristics and requirements than telephony.

Since a lot of AAL mechanisms are currently studied and implemented, each of them with more or less sophisticated features, we consider in the following the simplest AAL mechanism, namely the reference AAL mechanism introduced in Section 2, which is certainly the most poorly featured mechanism. It is actually expected that the information derived by considering this mechanism is sufficient to make the use of more sophisticated AAL mechanisms possible. Specifically, we are interested in the number of starvation periods N_g as well as in their length in the reference AAL mechanism. Let γ be the maximum allowed size of a starvation period (if the starvation period is greater than γ, the QoS degradation becomes unacceptable).

One question with respect to the algorithm is to decide whether the cell dispersion parameters N_g and γ should be "absolute" (with no closing error) or statistical (i.e., related to some quantile). In the latter case, the magnitude of cell dispersion is specified with some closing error ε. Objective values for ε are typically related to QoS issues. While it is clear that there is no way an application could request to encounter absolutely no starvation period (and thus N_g should be less or equal than a given limit or the ratio $\frac{N_g}{N_c} \leq \varepsilon_g$ where N_c is the number of conforming cells), the size of the starvation period could either be absolutely bounded or statistically bounded by γ. In any case, if a network testing conformance on the dispersion affecting a connection, finds the bound on N_g or the bound on γ exceeded, the network is not responsible for any QoS degradation.

If, the size of the starvation periods is absolutely bounded by γ, i.e., the network cannot create any starvation period of size greater or equal to γ, an algorithm such as the one presently proposed in I. 356 (1993) could check if γ/T (T is the period of the CBR source) is exceeded even if it was not its primary objective and if it has many drawbacks that should preclude its use (Guillemin, Rosenberg and Meddeb, 1996).

We are in favor of statistically bounding the size of the starvation periods, i.e., the proportion of starvation periods of size greater or equal to γ should be less or equal to ε_γ.

Another related issue concerns the ratio of the closing error for cell clumping to that for cell dispersion. For instance, cell clumping could be specified with a 10^{-9} closing error so that a fraction of at most 10^{-9} cells are detected as non-conforming by the UPC/NPC mechanism and cell dispersion could be specified with $\varepsilon_g = 10^{-4}$ and $\varepsilon_\gamma = 10^{-9}$ depending on the admissible QoS, which can be achieved by the receiving AAL mechanism. In this line of investigations on the relationship between peak cell rate control and cell dispersion, another problem, which could be addressed once the algorithm for specifying dispersion has been chosen, is the one concerning cell spacing. Specifically, the general belief that cell spacing increases cell dispersion should be questioned.

So, we expect that, once the reference connection has been defined, that an AAL would have, for this connection, the assurance that the proportion of starvation periods is less than a given ε_g and that for the γ of its choice, the proportion of starvation periods of size greater or equal to γ is less than a given ε_γ.

We propose in this section an algorithm (see Figure 3) for the specification of cell dispersion. This algorithm computes the ratio $\frac{N_g}{N_c}$ where N_c is the number of conforming cells and N_g is the number of starvation periods, and the ratio $\frac{N_\gamma}{N_g}$ where N_γ is the number of starvation periods of size greater than γ.

There are three counters :

- N_c, the number of conforming cells;
- N_g, the number of starvation periods;
- N_γ, the number of starvation periods of size greater than γ.

We have also introduced a variable TST (for Theoretical Service Time) which is the time before which a cell should arrive for the process not to exhibit a starvation period. A cell is conforming and thus taken into account by our algorithm if it is conforming to the VSA(T, τ). For each conforming cell, we compute its TST and compare it with the cell arriving time. If the cell has arrived before its TST (i.e., $TST < t$), there is no starvation period and the algorithm waits for the next conforming cell to arrive. Otherwise there is a starvation period and the algorithm increments the counter N_g, check if the starvation period size is greater than γ (if yes it increments the counter N_γ) and computes in any case the new value of TST. To compute the size of the starvation period and new value of TST, we have introduced the function TST(t) where TST(t) = TST + $\lceil (t - \text{TST})/T \rceil \times T$.

TST(t) is the theoretical service time for a conforming cell which follows a starvation period. The size of the starvation period is computed as the difference between the previously computed TST and TST(t).

At the end of the connection, the proportion $\nu_g = N_g/N_c$ and $\nu_\gamma = N_\gamma/N_g$ can be computed and compared respectively to ε_g and ε_γ.

Note that the big difference between this algorithm and the VSA which is used for characterizing clumping is that this one is inherently discrete in nature.

The value of γ is application dependent and thus the network operators would give for the reference connection introduced earlier the apportioned tolerance for different values of γ.

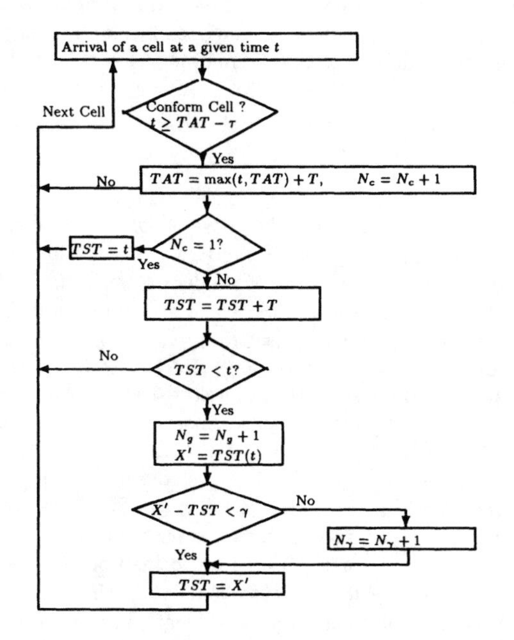

Figure 3 The new algorithm for specifying cell dispersion.

At a given interface, namely the UNI or a NNI, a real time connection is characterized, as far as cell dispersion is concerned, by γ and objective values for ν_g and ν_γ, say, $\varepsilon_g = 10^{-4}$ and $\varepsilon_\gamma = 10^{-9}$, respectively. The connection is said to be conforming if the ratios ν_g and ν_γ computed at the end of the connection for the specified value of γ satisfy $\nu_g \leq \varepsilon_g$ and $\nu_\gamma \leq \varepsilon_\gamma$. In the case of a multiple operator connection, a reference connection should be considered (see Section 2). Each portion of this connection should be assigned an objective for cell dispersion. At each interface, the network operator receiving traffic may check by using the new algorithm whether the connection is conforming with respect to cell dispersion. If the user complains of QoS degradation, the origin of this degradation may be identified.

Note that the algorithm currently considered in ITU-T Recommendation I.356 may be used (see also Guillemin, Rosenberg and Meddeb, 1996), to some extent, as a special case of our algorithm depicted in Figure 3 if $\varepsilon_g = 1$ (i.e, the frequency of cell starvation periods is not considered as a QoS parameter) and $\varepsilon_\gamma = 0$ (i.e., the connection is not conforming as soon as γ is exceeded).

As mentioned in Section 2 and as it can be seen in Figure 3, the cell dispersion algorithm is coordinated with the cell clumping algorithm (only conforming cells are taken into account). Intuitively, cell loss increases the magnitude of cell dispersion. Cell loss may be caused by legitimately discarding cell in the UPC/NPC. In fact, the tolerance τ itself characterizes the magnitude of cell clumping with some closing error ε_c. As a consequence, it may be very unsafe to characterize cell dispersion and cell clumping at an interface with the same accuracy (ε_c and ε_g with the same order of magnitude). It seems more suitable to specify cell dispersion less accurately than cell clumping (say, $\varepsilon_c = 10^{-9}$ and $\varepsilon_g = 10^{-4}$).

4 CONCLUSION

Some issues related to cell dispersion in ATM networks have been discussed in this paper. It turns out that information on cell dispersion is needed for dimensioning AAL mechanisms in receiving terminals. Absorbing cell dispersion in the middle of a connection (e.g., via a circuit emulation) is not suitable and even impossible. Legality problems entail that conformance testing with respect to cell dispersion is necessary. For conformance testing purposes, cell dispersion should be characterized by information within the ATM layer. This characterization should rely on statistical information and should be coupled with that of cell clumping. Namely, only cells, which are conforming with respect to cell clumping, should be considered. In the case of multiple operator connections, cell dispersion should be apportioned by considering a reference connection.

Recognizing that the algorithm currently specified in ITU-T Recommendation I.356 suffers from instability and gives a poor information on cell dispersion, a new algorithm has been proposed in this paper for specifying cell dispersion. Roughly speaking, this algorithm emulates a reference AAL mechanism and counts the occurences of starvation periods and the number of starvation periods whose length is greater than a specified value γ. This two parameters are expected to provide sufficient information for the operation of AAL mechanisms more sophisticated.

This algorithm allows the characterization of cell dispersion. Further investigations are needed to clarify with which accuracy cell clumping and cell dispersion should be specified (respective orders of magnitude of the objectives ε_g and ε_c). Moreover, the general belief that cell spacing adversely increases the magnitude of cell dispersion should be questioned. Namely, the impact of cell spacing on the parameter γ should be clarified. These point will be addressed in forthcoming papers.

REFERENCES

Boyer, P., Guillemin, F., Servel, M., and Coudreuse, J.P. (1992) Spacing cells protects and enhances utilization of ATM network links: *IEEE Communications Magazine*, pages 38–49.

Devault, M., Cochennec, J.Y. and Servel, M. (1988) The PRELUDE experiment : assessments and future prospects: *IEEE Journal on Selected Areas in Communications, Vol 6, n° 9.*

Gravey, A. and Boyer, P. (1993) Cell Delay Variation in ATM networks: *Modeling and Performance Evaluation of ATM Technology*, Martinique.

Guillemin, F. and Monin, M. (1992) Management of Cell Delay Variation in ATM networks: *Globecom'92*, Orlando, FL.

Guillemin, F., Rosenberg, C. and Meddeb, A. (1996) Analysis of cell dispersion in ATM networks. *In preparation.*

I.356. ITU-T Recommendation (1993) B-ISDN ATM layer cell transfer performance, Geneva.

I.371. ITU-T Recommendation (1995) Traffic control and congestion control in B-ISDN, Geneva.

Roberts, J. and Guillemin, F. (1992) Jitter in ATM networks and its impact on peak cell rate control: *Performance Evaluation, Special Issue on Modeling of High Speed Telecommunications Systems.*

The ATM Forum (1994). User-Network Interface Specification, Version 3.1.

3

Long Range Correlation in Multiplexed Pareto Traffic

J. J. Gordon
Bell Communications Research
331 Newman Springs Road, Red Bank, New Jersey 07701, U.S.A.
Phone: +1 908 758 5394, Fax: ext. 4370, email: jgordon@cc.bellcore.com

Abstract

Packet traffic from various sources - ethernet, ISDN, CCSN and VBR video - has been shown to exhibit self-similarity and related properties of long range correlation and slowly decaying variances. The author has previously suggested that self-similar traffic can be modelled by heavy-tailed pareto sources. A traffic stream in which interarrival times are *iid* pareto generates block packet counts which are asymptotically self-similar.

However, single pareto streams do not produce the power law or stretched exponential queue tails observed with real packet traces. This paper shows that multiplexed streams of pareto traffic exhibit long range correlation of block packet counts *and* interarrival times, and that these properties together produce power law queue tails. This result makes pareto sources an attractive framework for modeling self-similar traffic.

Keywords

Packet traffic, self-similar, pareto, ethernet, ISDN, CCS, VBR video.

1 INTRODUCTION

Packet traffic from various sources including ethernet (Leland et. al. 1993, 1994), ISDN (Meier-Hellstern et. al. 1991), CCSN (Duffy et. al. 1994) and VBR video (Beran et. al. 1995) has been shown to be self-similar, and analysis of ATM traffic is currently underway. Self-similar traffic exhibits long range correlation in block packet counts. Let X_k be the number of packets arriving at a buffer in fixed length time interval k. The quantity X_k is referred to as a block packet count (BPC). It is assumed here that the series X_k is covariance stationary with autocorrelation function $\rho_{BPC}(k)$. For self-similar traffic $\rho_{BPC}(k) \sim k^{2H-2}$ with $1/2 < H < 1$ as $k \to \infty$. All 'classical' phase-type models of packet traffic have Hurst parameter $H = 1/2$. Self-similar traffic, with $1/2 < H < 1$, can have significantly different queueing behavior from classical models, and research is being performed to understand the impact of self-similar traffic, and to develop queueing / engineering models which capture the self-similarity and other essential characteristics of real packet traffic.

Perhaps the simplest model of self-similar traffic is fractional brownian motion (FBM) (Norros 1994). In common with other currently proposed packet traffic models, FBM abandons the traditional framework of point arrival processes in favor of modeling the net work process. That is, instead of explicitly modeling packet arrival instants and service times, FBM models the amount of work in excess of the server's capability that arrives in any period of time. In the

FBM model the marginal distribution of the net work arriving per unit time is assumed to be normal. The model has three parameters: the Hurst parameter H, and the mean and variance of the marginal net work distribution (MNWD).

The advantages of FBM are: (i) it is conceptually straightforward, and (ii) in a single server queue it can be shown that FBM produces queue length distributions whose tails are bounded below by stretched exponentials. Specifically, for N large, the probability that the queue length exceeds N is bounded below by an expression of the form $exp(-N^{\kappa})$ where $0 < \kappa < 1$ [6]. Experiments with packet traces suggest that real packet data produces queue length distributions that exhibit stretched exponential, or perhaps even power law, decay. The FBM model is consistent with this observation.

However, the FBM model has several disadvantages. First, methods of generating FBM processes are numerically cumbersome, making FBM difficult to explore through simulation. Second, in real packet traces the marginal net work distribution is often far from normal. It appears to be better modelled by heavy-tailed distributions such as the pareto distribution. Finally, real packet traces having the same Hurst parameter H and similar MNWD means and variances can produce markedly different queueing behavior. This suggests that the three parameters of the FBM model are insufficient to completely characterize real packet traffic. Other essential characteristics of the traffic need to be incorporated into the model through additional or alternative parameters.

This paper argues that multiplexed point arrival processes having pareto distributed interarrival times (IATs) may be better models of self-similar packet traffic than FBM. Multiplexed pareto traffic exhibits long range (power law) IAT correlations. This sort of correlation is difficult to add into a model by hand, and multiplexing of pareto sources therefore provides an attractive way of generating long range correlation. More importantly, long range IAT correlations are not explicitly present in FBM or any other model based on the net work process. To date, models based on the net work process have explicitly included only one type of correlation: correlation in the net work process or, equivalently, in block packet counts.

A model which incorporates both BPC and IAT correlations may be able to explain why packet traces having similar values of H and similar MNWD means and variances can produce markedly different queueing behavior. This paper uses simulation to establish that multiplexed pareto sources exhibit power law BPC and IAT correlations, and investigates the queueing impact of these correlations by comparing equivalent single-source and two-source models. The simulation results show that multiplexed pareto input can produce queue length distributions with stretched exponential or power law tails. These results prove that multiplexed pareto input produces queueing behavior that is qualitatively similar to real packet traces. It remains to be verified through further research whether multiplexed pareto sources can provide quantitatively accurate models of real packet traffic.

2 IAT Correlation in Multiplexed Traffic

The correlation of interarrival times in a multi-source packet stream can in principle be derived from the counting processes for individual sources. For a single source i define

$$g_n^{(i)}(t, \tau) = Pr\begin{pmatrix}\text{at time } t \text{ there have been } n \text{ arrivals} \\ \text{and time since the last arrival is } \tau\end{pmatrix} \quad (1)$$

$$g_n^{(i)}(t) = \int_0^\infty g_n^{(i)}(t, \tau)\, d\tau \quad (2)$$

$$G^{(i)}(z, t) = \sum_{n=0}^{\infty} g_n^{(i)}(t)\, z^n \quad (3)$$

The generating function $G(z, t)$ for the aggregate stream of multiplexed sources is equal to the product of the single source functions

$$G(z, t) = \prod_{i=1}^{N} G^{(i)}(z, t) \quad (4)$$

In the multiplexed stream, let $h_n(t)$ be the probability (density) that the nth arrival occurs at time t, and define $H(z, t) = \Sigma_{n=0}^{\infty} h_n(t)\, z^n$. Then

$$H(z, t) = -G'(z, t) / (1 - z) \quad (5)$$

where differentiation is with respect to t. Finally, note that equation (5) implies the following relation between the Laplace transforms of $G(z, t)$ and $H(z, t)$

$$H_L(z, t) = [1 - s\, G_L(z, s)] / (1 - z) \quad (6)$$

Denote the nth IAT in the multiplexed arrival process by x_n. Further define $\bar{x} = E[x_n]$, $\gamma_{IAT}(k) = cov[x_n, x_{n+k}]$, $\rho_{IAT}(k) = \gamma_{IAT}(k) / \gamma_{IAT}(0)$, $\gamma_{IAT}(z) = \Sigma_{n=1}^{\infty} \gamma_{IAT}(k)\, z^k$ and $\rho_{IAT}(z) =$ $= \Sigma_{n=1}^{\infty} \rho_{IAT}(k)\, z^k$. The mth derivative of $H_L(z, s)$ with respect to s, evaluated at $s = 0$, can be expressed as follows

$$H_L^{(m)}(z, 0) = (-1)^m \sum_{n=0}^{\infty} E[(x_0 + x_1 + \ldots + x_n)^m]\, z^n \quad (7)$$

It follows from equation (7) that

$$\bar{x} = -H_L'(0, 0) \quad (8)$$

$$\gamma_{IAT}(0) = H_L^{(2)}(0, 0) - \bar{x}^2 \quad (9)$$

$$\gamma_{IAT}(z) = \frac{1}{2}(1 - z)^2 H_L^{(2)}(z, 0) - \frac{z\bar{x}^2}{1 - z} - \frac{\gamma_{IAT}(0)}{z} \quad (10)$$

The IAT correlation function $\rho_{IAT}(z)$ can therefore be derived from the aggregate counting function $G_L(z, s)$. Unfortunately, although it may be possible in many cases to calculate the individual source functions $G_L^{(i)}(z, s)$, it is in general difficult to calculate $G_L(z, s)$. One case for which it is possible is the case of two multiplexed hyperexponential sources. Consider two identical hyperexponential sources with parameters p_1, $p_2 = 1 - p_1$, μ_1 and μ_2. The IAT density function for each of these sources is

$$f(t) = p_1\mu_1 e^{-\mu_1 t} + p_2\mu_2 e^{-\mu_2 t} \tag{11}$$

The single source generating function $G_L^{(i)}(z, s)$ is equal to

$$G_L^{(i)}(z, s) = \frac{s + A}{s^2 + Bs + C} \tag{12}$$

where

$$A = p_1\mu_1 + p_2\mu_2 \tag{13}$$

$$B = \mu_1 + \mu_2 - p_1\mu_1 z - p_2\mu_2 z \tag{14}$$

$$C = \mu_1\mu_2(1 - z) \tag{15}$$

Due to the simple form of (12), the two-source function $G_L(z, s)$ can be calculated explicitly

$$G_L(z, s) = \frac{s^2 + sU + V}{s^3 + s^2 W + sX + Y} \tag{16}$$

where

$$A' = B - C/A \qquad W = 3B \tag{17}$$

$$U = A + A' + B \qquad X = 2B^2 + 4C \tag{18}$$

$$V = 2C + 2AA' \qquad Y = 4BC \tag{19}$$

Finally, it can be shown that equations (6), (10) and (16) imply

$$\rho_{IAT}(z) = \frac{\varepsilon z}{1 - \delta z} \tag{20}$$

where

$$\delta = \frac{p_1\mu_1 + p_2\mu_2}{\mu_1 + \mu_2} \tag{21}$$

$$\varepsilon = \left(\frac{p_1 p_2 (\mu_1 - \mu_2)^2}{4\mu_1\mu_2(\mu_1 + \mu_2)^2}\right) \cdot \left(\phi - \frac{(p_1\mu_2 + p_2\mu_1)^2}{4(\mu_1\mu_2)^2}\right)^{-1} \tag{22}$$

$$\phi = \frac{(p_1\mu_2^2 + p_2\mu_1^2)^2 + \mu_1\mu_2(p_1\mu_2 + p_2\mu_1)^2}{2(\mu_1\mu_2)^2(\mu_1 + \mu_2)(p_1\mu_2 + p_2\mu_1)} \tag{23}$$

Equation (20) shows that in a multiplexed arrival stream generated by two identical hyperexponential sources, interarrival times are correlated, and the autocorrelation coefficients decay geometrically with parameter δ. A similar calculation for two pareto sources is difficult, due to the more complex structure of the function $G_L(z, s)$. Section 4 uses simulation to show that IATs generated by two pareto sources are also correlated, and that in the pareto case the autocorrelation coefficients obey a power law decay. The queueing impact of power law IAT correlations is explored in section 4.

3 Description of Simulations

In reference [7] the following form of the pareto IAT density was adopted

$$f(t) = \alpha\beta^{\alpha}/(\beta + t)^{\alpha+1} \tag{24}$$

Finite moments of the density (24) exist only for $k < \alpha$. In particular, for $\alpha > 2$ (24) has finite mean and variance, for $1 < \alpha < 2$ it has finite mean and infinite variance, while for $0 < \alpha < 1$ both mean and variance are infinite. In Gordon (1995) it was shown that for $0 < \alpha < 1$ and $1 < \alpha < 2$ a single pareto stream with IAT density (24) exhibits long range BPC correlation. The BPC autocorrelation function is of the form $\rho_{BPC}(k) \sim k^{2H-2}$ as $k \to \infty$ where

$$H \sim \begin{cases} (1+\alpha)/2 & 0<\alpha<1 \\ (3-\alpha)/2 & 1<\alpha<2 \\ 1/2 & \alpha>2 \end{cases} \tag{25}$$

In the following we consider only cases for which $\alpha > 1$, so that a finite mean IAT exists. In general, BPCs in a multiplexed stream of n *iid* sources possess exactly the same correlation structure as the single-source BPCs. This follows directly from the two-source identity

$$\rho_{BPC}^{(2)}(k) = \frac{cov[X_{n+k} + Y_{n+k}, X_n + Y_n]}{var[X_n + Y_n]} = \frac{2 \cdot cov[X_{n+k}, X_n]}{2 \cdot var[X_n]} = \rho_{BPC}^{(1)}(k) \tag{26}$$

where X_n and Y_n are the single-source BPCs. As shown in section 2, the multiplexed stream also possesses IAT correlations even if the single source streams do not. Since IAT and BPC correlations are independent – one can be varied independently of the other – these types of correlation can be considered to have a separate impact on queueing behavior.

Single hyperexponential (HE) sources produce short range (exponentially decaying) BPC correlations, while multiplexed HE sources also generate short range IAT correlations. In analogy, since single pareto sources produce long range (power law) BPC correlations, it is reasonable to expect that multiplexed pareto sources will also exhibit long range IAT correlations. As noted above, both BPC and IAT correlations can be expected to have an impact on the queueing behavior of pareto sources.

The simulations described below address three questions relating to multiplexed pareto and HE sources: (i) Do IATs in multiplexed pareto streams exhibit long range correlation? (ii) What are the relative impacts of BPC and IAT correlations on the queueing behavior of multiplexed pareto and HE sources? and (iii) Are there conditions under which multiplexed pareto streams can produce power law or stretched exponential tails in queue length distributions? To answer these questions, three HE/M/1 and three P(areto)/M/1 queueing models were simulated for three scenarios, as shown in Table 1. In all cases the mean service time was 1.0 and the mean IAT was 1.25 so that the total utilization was equal to 0.8.

Table 1. Parameter Values for Simulations

model	scenario 1	scenario 2	scenario 3
M1: single pareto	α=1.2, β=0.25	α=1.8, β=1.0	α=2.4, β=1.75
M2: double pareto	α=1.2, β=0.50	α=1.8, β=2.0	α=2.4, β=3.50
M3: single pareto	α=1.4, β=0.50	α=2.6, β=2.0	α=3.8, β=3.50
M4: single 2-stage hyperexponential	p1=0.114216, μ1= 0.182746, μ2=1.417254	p1=0.131374, μ1= 0.210200, μ2=1.389800	p1=0.077423, μ1= 0.123876, μ2=1.476124
M5: double 2-stage hyperexponential	p1=0.114216, μ1= 0.091373, μ2=0.708627	p1=0.131374, μ1= 0.105100, μ2=0.694900	p1=0.077423, μ1= 0.061938, μ2=0.738062
M6: single 3-stage hyperexponential	marginal defined to be identical to M5	marginal defined to be identical to M5	marginal defined to be identical to M5

Consider the three pareto models in scenario 1 of Table 1. Model 1 has Hurst parameter H=0.9 and zero IAT correlation. Model 2 also has H=0.9. However, it is a two-source model so it has non-zero IAT correlation. Model 3 has Hurst parameter H=0.8 and zero IAT correlation. The difference between models 1 and 2 is a change in the IAT correlation structure, while the difference between models 1 and 3 is a change in Hurst parameter from 0.9 to 0.8. Although these changes are accompanied by changes in the IAT and BPC marginal distributions, our basic assumption in the following is that the difference in queueing behavior between the three models largely reflects the impact of the IAT and BPC correlations. Finally, note that models 2 and 3 have identical marginal IAT distributions (this can be proved using (24)) which makes it reasonable to directly compare the queueing behavior of models 2 and 3.

Each set of pareto and HE models in Table 1 follows the above pattern. In each scenario, models 1 and 2 and models 4 and 5 have the same BPC correlation structure but different IAT correlation structure, while models 1 and 3 and models 4 and 6 have the same (identically zero) IAT correlation structure but different BPC correlation structure. In each case, models 2 and 3 and models 5 and 6 have identical marginal IAT distributions. The HE sources in model 4 are equivalent to the corresponding pareto sources in model 1 in the sense of [7]. Specifically, the HE parameters in model 4 are selected so that the IAT distributions in models 1 and 4 have identical means and variances or, in cases where the pareto variance is infinite, identical means and 90th percentiles. The third parameter in the HE model was fixed by requiring the HE sources to have unbalanced means in the ratio 1:1000 (i.e., in the notation of Allen (1990) $(\lambda_1/\mu_1)/(\lambda_2/\mu_2) = 0.001$).

Finally, note that the parameter values in scenarios 1–3 were selected so that the values of α in models 1 and 3 were: (i) both in the range $1 < \alpha < 2$, (ii) in separate ranges $1 < \alpha < 2$ and $\alpha > 2$, or (iii) both in the range $\alpha > 2$. In model 2 these choices correspond to having: (i) $1/2 < H < 1$ and infinite variance IAT marginal, (ii) $1/2 < H < 1$ and finite variance IAT marginal, or (iii) $H = 1/2$ and finite variance IAT marginal. All simulation results reported below are averaged over 10 runs. The HE runs were of duration $1.3\text{x}10^7$ service times (approx. 10^7 arrivals), while the pareto runs were of duration $5.2\text{x}10^7$ service times (approx. $4\text{x}10^7$ arrivals). The very long simulation times were required in order for the pareto statistics to approach equilibrium.

4 Simulation Results

In the following we use the abbreviation SiMj to refer to scenario i model j, as given in Table 1. Figure 1 shows a plot of the IAT correlation coefficients $\rho_{IAT}(k)$ for models 2 and 5. The three solid lines labelled a,b,c are simulation results for model 2 scenarios 1–3 respectively. The three dashed lines labelled d,e,f are theoretical correlation curves for model 5 scenarios 1–3, as given by equations (20)–(23). Superimposed on the dashed curves are several points representing simulation results for model 5. Finally, the circles at the bottom of Figure 1 are the absolute values of the simulated correlation coefficients for model 3. Since model 3 should have zero IAT correlations, the circles at the bottom of Figure 1 represent the accuracy of the simulation results. Results are accurate to about $\pm 10^{-4}$.

Figure 1 is plotted on a log-log scale. The model 2 plots a,b,c appear to be straight lines, suggesting that IAT correlations exhibit power law rather than exponential decay. The model 2 correlation coefficients decay far more slowly than those for model 5, which are known to decay exponentially fast (see equation (20)). The plots for model 5 would be straight lines if the x-axis were on a linear scale. On a log-log scale they exhibit a downward curve characteristic of exponential decay. It is not clear why curve a (scenario 1, model 2) does not tend towards a y-axis intercept of one. The IAT distribution in this case has infinite variance, implying that the autocorrelation coefficients $\rho_{IAT}(k)$ must be defined in terms of limiting rather than equilibrium values. The question arises as to whether the S1M2 curve in Figure 1 is converging to a non-zero limit, or decaying slowly to zero. The simulation results are inconclusive on this issue. Note, however, that even if the limiting value of the S1M2 autocorrelation function is zero, it is still conceivable that power law IAT correlations could have an impact on queueing.

Mean queue lengths for the models in Table 1 are given in Table 2. (Results are time averages. Queue lengths as seen by arrivals are qualitatively the same.) The pareto models all follow the same pattern. In each scenario, the mean queue lengths for models 1–3 are monotonically decreasing. Similarly, within each model mean queue lengths for scenarios 1–3 are monotonically decreasing, as one would expect. For the hyperexponential models it is still true that within each scenario the mean queue lengths for models 1–3 are monotonically decreasing. However, within each model the largest to smallest mean queue lengths are obtained for scenarios 3, 1 and 2 respectively. Since the distinguishing feature of scenarios 1 and 3 is the BPC correlation structure, we conclude that the short range BPC correlations in the HE models are not as important to queueing as long range BPC correlations in the pareto models.

Table 2. Mean Queue Lengths from Simulations

model	scenario 1	scenario 2	scenario 3
M1: single pareto	1.4015E+03	1.3095E+01	7.6230E+00
M2: double pareto	1.1390E+03	1.1819E+01	7.1803E+00
M3: single pareto	5.3021E+01	6.9810E+00	5.4335E+00
M4: single 2-stage HE	8.5635E+00	7.6990E+00	1.1679E+01
M5: double 2-stage HE	8.0377E+00	7.2759E+00	1.0696E+01
M6: single 3-stage HE	5.9395E+00	5.6069E+00	7.0100E+00

Table 2 gives an indication of the relative impact of long range IAT and BPC correlations in the pareto models. Models 1 and 2 differ in their IAT correlation structure, but not in their BPC correlation structure. Conversely, models 1 and 3 differ in their BPC correlation structure, but not in their IAT correlation structure. In general, the difference between mean queue lengths in models 1 and 2 is small (6–22%) compared to the difference between models 1 and 3 (40–2600%). In scenario 1, where the marginal IAT distributions in models 1–3 all have infinite variance, the difference between mean queue lengths for models 1 and 2 and model 3 is extremely large. This difference reflects the impact of long range BPC correlations (i.e., of a Hurst parameter close to one).

The above conclusions regarding the relative impact of IAT and BPC correlations also apply to the HE models 4–6. However, in the case of the HE models both the IAT and BPC correlations are short range, and their impact on queueing is smaller. In the HE models, the difference in mean queue length between models 1 and 2 ranges from 5–9%, while the difference between models 1 and 3 ranges from 37–67%. Queueing behavior is much less extreme than in the pareto models due to the absence of long range correlations.

Queue length distributions for all pareto models are plotted in Figure 2 on a log-linear scale. The correspondence between scenarios, models and plots is: S1M1=a, S1M2=b, S1M3=c, S2M1=d, etc. Similar plots for the HE models are given in Figure 3 with S1M4=a, S1M5=b, S1M6=c, S2M4=d, etc. The fact that all of the plots appear to be straight lines strongly suggests that all of the queue length distributions have exponentially decaying tails. The only plot for which this is not true is the S1M2 plot in Figure 2. In this case the tail of the queue length distribution is extremely long, and in order to see the shape of the distribution it is necessary to expand the y-axis. Figure 4 shows the queue length distributions for S1M1 (a), S1M2 (b) and S1M3 (c) on an expanded y-axis. The S1M1 and S1M3 plots appear to be straight lines, consistent with exponentially decaying tails. However, the S1M2 plot appears definitely not to be linear (i.e., exponential). When re-plotted on a log-log scale in Figure 5, The S1M2 plot appears to start off exponentially and then become a straight line (i.e., power law) after about $N=32$.

This is the shape of a pareto distribution of the form (24). For small t the function (24) will appear to be exponential, while for large t it is power law. The range $[0, t]$ over which (24) appears to be exponential goes to infinity in the limit $\alpha, \beta \rightarrow \infty$ with $\beta / (\alpha - 1)$ held constant. Consider a multiplexed stream of n *iid* pareto sources with parameters $\alpha, n\beta$. The marginal IAT distribution in the multiplexed stream is pareto with parameters $\alpha' = n\alpha - n + 1$ and $\beta' = n\beta$. The mean IAT in the multiplexed stream is $\beta / (\alpha - 1)$ independent of n. It follows that the limit $\alpha, \beta \rightarrow \infty$ with $\beta / (\alpha - 1)$ held constant gives the marginal IAT distribution in a multiplexed stream of n pareto sources, as $n \rightarrow \infty$ with the total utilization held constant.

Figure 5 suggests that the S1M2 queue length distribution is well approximated by a pareto distribution, and that its tail exhibits power law decay. That the S1M2 curve in Figure 5 is qualitatively different from the S1M1 and S1M3 curves is the most interesting result to emerge from the simulations. Since all of the models in Figure 5 have infinite variance IAT marginals, and similar BPC correlation structures, it appears that the pareto-like shape of the S1M2 curve is due to the long range (i.e., power law) IAT correlations. It remains an open question as to why these correlations do not produce the same result in scenarios 2 and 3.

Finally, queue length autocorrelation functions for all pareto and HE models are plotted in Figures 6 and 7 on a log-linear scale, using the same labels as in Figures 2 and 3. All of the plots appear to be straight lines, suggesting that all of the queue length autocorrelation functions decay exponentially. This conclusion applies to S1M2 as well, since expanding the y-axis range in Figure 6 provides no indication that the S1M2 curve is other than linear.

5 Conclusions

The starting point for this paper was the observation that multiplexed streams of point arrival processes exhibit correlation of interarrival times (IATs). The paper addressed three questions relating to multiplexed pareto and hyperexponential (HE) sources: (i) Do interarrival times in multiplexed pareto streams exhibit long range correlation? (ii) What are the relative impacts of block packet count (BPC) and IAT correlations on the queueing behavior of multiplexed pareto and hyperexponential sources? and (iii) Are there conditions under which multiplexed pareto streams can produce power law or stretched exponential tails in queue length distributions?

To answer these questions, six G/M/1 queueing models were simulated across three scenarios, as detailed in Table 1. The simulation results strongly suggest that multiplexed pareto streams do exhibit long range interarrival time correlation across the range of parameter values investigated. This in itself is a potentially useful result, since it provides a straightforward mechanism for generating long range correlation. The simulation results also give some indication of the relative queueing impacts of IAT and BPC correlations. In both pareto and HE models, the impact of BPC correlations is considerably larger than IAT correlations. In pareto models the effect of both types of correlation is magnified because of their long range (i.e., power law) nature. In contrast, in HE models the effect of BPC and IAT correlations is relatively small due to their short range.

The simulation results suggest that for some parameter values, multiplexed pareto sources will produce pareto-like queue length distributions with power law tails, while in other cases they will produce geometrically decaying queue length distributions. In all of the pareto simulations, queue length correlations appeared to decay exponentially fast. All of the simulation results were reviewed to ensure that they represent long run equilibrium results. However, given the variability associated with simulation of power law distributions, the above conclusions need to be confirmed through more extensive simulation or, preferably, through mathematical analysis.

The fact that multiplexed pareto sources can produce pareto-like (i.e., power law) queue length distributions is potentially significant for modeling real packet traffic. As noted in the introduction, real packet traffic produces power law and / or stretched exponential queue length distributions. While fractional brownian motion can reproduce this behavior, it appears to be insufficiently flexible to model real traffic. Specifically, it appears that there are parameters other than the Hurst parameter and the mean and variance of the marginal IAT distribution which need to be incorporated into any realistic traffic model.

Traffic models based on multiplexed pareto sources provide one parameter which is not explicitly present in fractional brownian motion – IAT correlation. Furthermore, it appears that pareto-like marginal distributions are more consistent with IAT distributions seen in real traffic. For this reason we suggest that traffic models based on multiplexed pareto sources may provide more flexible and realistic models of real data traffic than fractional brownian motion. While the above results suggest that pareto models can produce queueing behavior that is qualitatively similar to real packet traces, it remains to be verified through further research whether multiplexed pareto sources can provide quantitatively accurate models of real packet traffic.

6 References

Allen, A.O. (1990) Probability, Statistics and Queueing Theory with Computer Science Applications. *Second Edition, Academic Press, 1990.*

Beran, J., Sherman, R., Taqqu, M.S., Willinger, W. (1995) Variable Bit Rate Video Traffic and Long Range Dependence. *IEEE / ACM Trans. on Communications, 1995.*

Duffy, D.E., McIntosh, A.A., Rosenstein, M., Willinger, W. (1994) Statistical Analysis of CCSN / SS7 Traffic Data from Working CCS Subnetworks. *IEEE J. Select Areas Commun., 1994.*

Gordon, J.J. (1995) Pareto Process as a Model of Self-Similar Packet Traffic. *Proc. IEEE Globecom '95 Conference, Singapore 1995, pp. 2232-2236.*

Leland, W.S., Willinger, W., Taqqu, M.S., Wilson, D.V. (1994) Statistical Analysis and Stochastic Modeling of Self-Similar Datatraffic. *14th International Teletraffic Congres, Juan Les Pins, France 1994, pp. 319-328.*

Leland, W.S., Taqqu, M.S., Willinger, W., Wilson, D.V. (1993) On the Self-Similar Nature of Ethernet Traffic. *Proc. ACM/SIGCOMM'93, San Francisco 1993, pp. 183-193.*

Meier-Hellstern, K., Wirth, P., Yan, Y-L., Hoeflin, D. (1991) Traffic Models for ISDN Data Users: Office Automation Application. *Proc. 13th International Teletraffic Congres, Copenhagen, Denmark 1991.*

Norros, I. (1994) A Storage Model with Self-Similar Input. *Queueing Systems, 1994.*

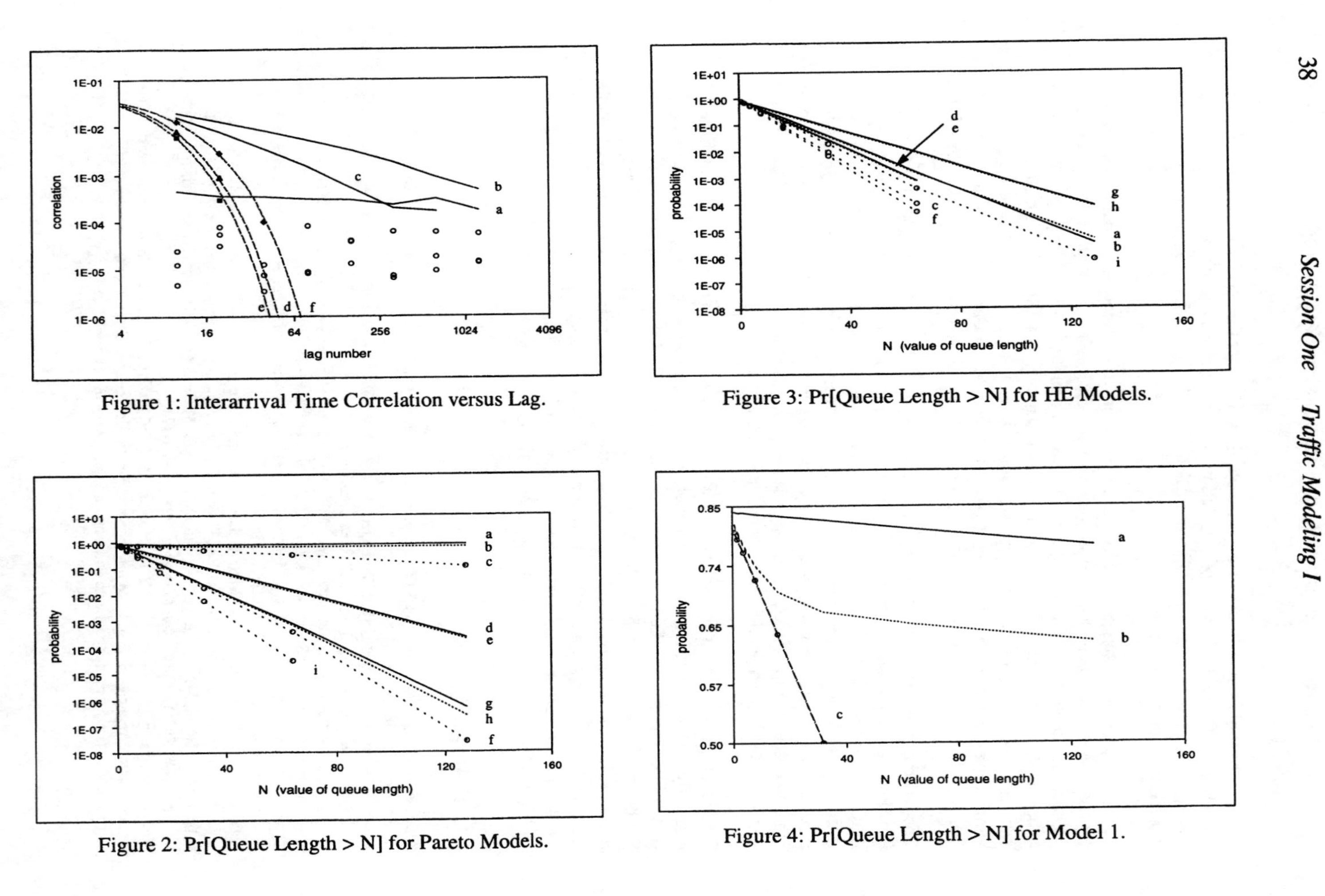

Figure 1: Interarrival Time Correlation versus Lag.

Figure 3: Pr[Queue Length > N] for HE Models.

Figure 2: Pr[Queue Length > N] for Pareto Models.

Figure 4: Pr[Queue Length > N] for Model 1.

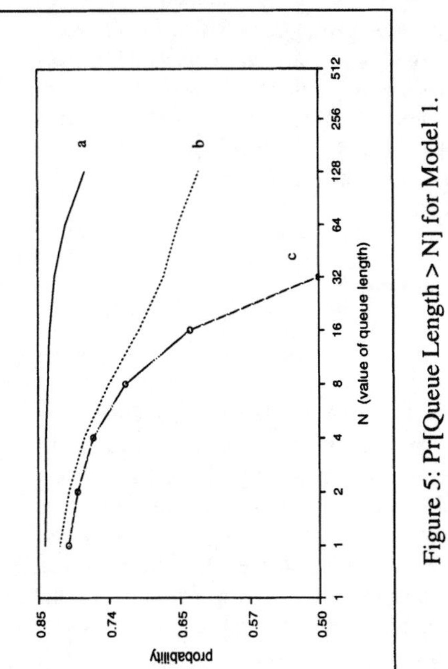

Figure 5: Pr[Queue Length > N] for Model 1.

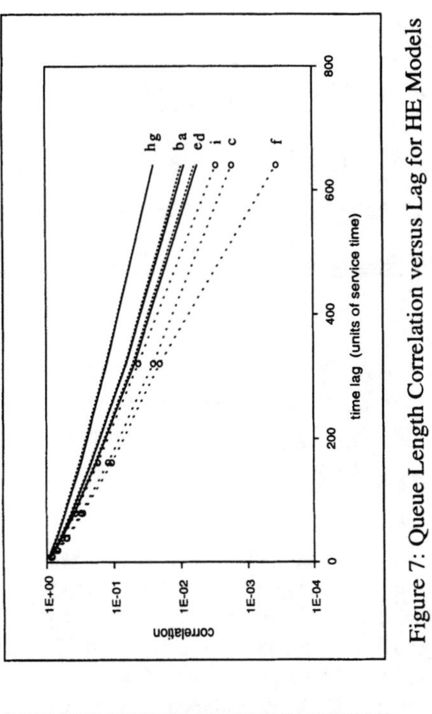

Figure 7: Queue Length Correlation versus Lag for HE Models

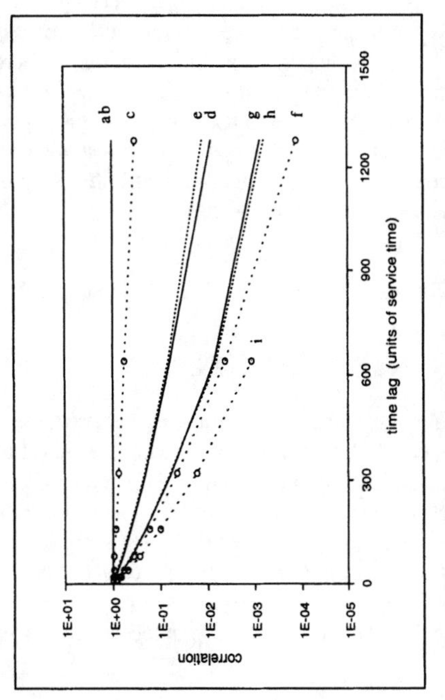

Figure 6: Queue Length Correlation versus Lag for Pareto Models.

4

Correlation properties of MPEG VBR video sources: Relative importance of short and medium term correlation on ATM multiplexer performance

Ragnar Ø Andreassen
Telenor R&D
Mail: P.O. Box 83 Email: ragnar.andreassen@nta.no
N-2007 Kjeller Telephone: +47 63 84 88 51
Norway Fax: +47 63 81 00 76

Abstract

An investigation of correlation properties of MPEG VBR video based on three different video sequences is presented. It is found that the short range properties of these sources resemble those of uncorrelated sources. To investigate the relative importance of short and medium term correlation properties versus the longer term correlation properties of MPEG sources, analytical source models taking into account the periodic properties of the sources are used. Results from these models are compared with trace driven simulations in which long term properties are inherent. One conclusion of this investigation is that for the ATM cell loss probability, interesting regions in terms of buffer capacity and number of sources exist, where long term correlation properties will not influence multiplexer performance. For a number of 12 multiplexed sources, this region extended to buffer lengths of about 10 times the mean video frame size in high loss regions. In low loss regions, no effect of longer term correlation properties was seen at buffer lengths up to 15 times the mean frame lengths, with the same number of sources.

Keywords

ATM multiplexing, MPEG VBR video, correlation properties

1 INTRODUCTION

With the advent of multimedia services like teleconferencing and video distribution, it is expected that video can become a major traffic source in ATM networks. MPEG is the international ISO/ITU standard for coding/decoding of digital video signals, and it is likely that a major part of the video information transmitted will be MPEG signals. In the process of engineering networks for the transmission of such signals, there will be a demand for robust, tractable and precise models for the MPEG VBR video sources. It is by now well established that many video sources display self-similar properties, and it may be seen that the MPEG sources are also autocorrelated over large time scales. The reduction of temporal redundancy as performed by the MPEG algorithm will additionally induce periodic components in the produced information streams, invoking strong periodic correlations at the frame level. In the process of developing models for multiplexer loss behaviour of such sources, it is not evident which properties should be most precisely modelled: short, medium or long term correlations. This question may not be given a simple answer, as the parameters of buffer length and number of sources, and the considered loss probability region, will determine which properties of the source will dominate.

The aim of this paper is to shed some light on the questions outlined above. To this end we perform an investigation of correlation properties of MPEG sources using three different approaches: autocorrelation functions, time-variance plots and leaky bucket loss contour diagrams. To determine which correlation properties dominate loss behaviour of multiplexed MPEG sources, we compare trace driven simulations, inherently containing long term correlation properties, with Markov models of MPEG sources. By using Markov models in the investigation, it will be possible to utilize certain source properties in a controlled manner. To aid our purpose, a Markov model is presented which makes possible the modelling of the periodic properties of multiplexed MPEG sources.

Inspirations for the current paper are the results published in (Garret, 1993), (Garret, 1994) and (Beran, 1995) demonstrating the self-similar properties of video sequences, and the work documented in (Reininger, 1994) and (Huang, 1995) developing simulator models for MPEG sources taking into account source correlation properties. The Markov model presented in a later section is related to the ones found in (Hübner, 1994), and (Landry, 1994). In the treatment of the periodic property of MPEG sources, a combinatorial approach is presented which may be employed both in the context of simulations and analytical models. Such an approach is, to the knowledge of the author, not documented by others.

2 GENERATION OF VBR MPEG VIDEO TRAFFIC

Three frametypes are defined in the MPEG algorithm: intra (I) frames, predicted (P) frames and interpolated (B) frames. A group of pictures (GOP) is composed of a regular pattern of frame types headed with an I frame succeeded by B and P frames. In the coding process, I and P frames may be used as references: I frames are coded with no reference to other frames, P frames are coded with reference to the previous reference frame and B-frames may be coded with references to both the previous and the succeeding reference frame. Hence the B-frames usually contain the least information, P frames somewhat more than B frames, and I frames more than both B and P frames. Typically the sizes of I versus P/B frames will differ by about a factor of 3 to 10, but the relative sizes of the frametypes may vary considerably between sequences. A source may keep its GOP phase during a connection, or it may change phase at scene boundaries in an effort to reduce information transmitted. Thus the phase constellations at a node in the network will vary in an arbitrary manner. The period and structure of the GOP cycle are coding parameters that may be chosen.

The quantization parameter (q), controls the lossy part of the coding algorithm, in effect regulating the coarseness of the decoded picture. Active use of the q-parameter during encoding may be employed as a flow regulation mechanism. In the sequences used here, fixed quality video is investigated, and the q-parameter is set at a constant value for each frame type.

We shall use statistics obtained from three film sequences, each consisting of 40 000 frames, corresponding to about half an hour of realtime video. One is an action film (James Bond), referred to as Bond, another is a 'psycho thriller' (The silence of the Lambs), referred to as Lambs, and the last contains episodes from a cartoon series (Simpsons), referred to as Simpsons. All sequences are coded with a GOP size of 12 frames, following the pattern of **IBBPBBPBBPBB**. Further documentation concerning production of the traces may be found in (Rose, 1995). In table 1 we display maximum (a_{max}) and expected number of cells (E(a)) in frames, where indices I, P and B refer to the frame types

Table 1 Trace statistics

Trace	a_{max}	E(a)	E(a_I)	E(a_P)	E(a_B)
Bond	694	69.1	236.6	117.7	29.9
Simpsons	682	52.8	210.4	61.2	29.9
Lambs	381	20.8	108.0	21.1	9.7

Here and in all subsequent numeric computations, a cell payload of 44 bytes is assumed.

3 CORRELATION PROPERTIES OF VBR MPEG VIDEO

Temporal redundancies in frame contents and temporal correlation of frame sizes in a sequence of frames are related quantities: If 100% removal of temporal redundancy was achieved, only 'new' information should be transmitted in each frame. The number of bits required to code this 'new' information should be unpredictable, otherwise the information would not be 'new'. So, a perfect (in this respect) coding scheme will result in an uncorrelated sequence of frames. On the other hand, a completely redundant stream, such as repeating the same frame, will obviously result in perfect correlation. It then seems reasonable to expect that in reducing temporal redundancy, the temporal correlation properties should also be weakened in a manner reflecting the properties of the redundancy removing mechanism. We shall in the following perform a qualitative investigation of temporal properties of MPEG sequences by three different methods, to see if some compliance with the above informal argument may be observed.

Firstly, we will consider the autocorrelation functions. In Figure 1, autocorrelation at the frame and group levels are displayed for the different traces, restricting the lag range to 10 000 frame periods as to cut off noise effects which may be observed at longer ranges. Lag scale for group level curves is adjusted so as to match the frame level curves. The periodic property of the sources is reflected in the autocorrelation functions, and the correlation amplitudes at different lags are related to the relative mean sizes of frames in the GOP structure. This can be observed by comparing the graphs with the numbers in Table 1. In a deterministic periodic sequence with the above group structure, the amplitude at lags corresponding to distances between B-frames and reference frames (lag values $i+3 \cdot j$, $i=1,2$, $j=0,1,2,...$) would be negative. A strong positive relative correlation between neighbouring frames of different types will tend to raise the above correlation values, and may explain the lack of negative amplitude as can be observed in two of the sequences. Group autocorrelations fall off very slowly with increasing lags, indicating the contents of GOPs to be strongly correlated over large time scales. As redundancy reduction is confined to groups, this behaviour is as expected considering the results in (Garret, 1994) and (Beran, 1995).

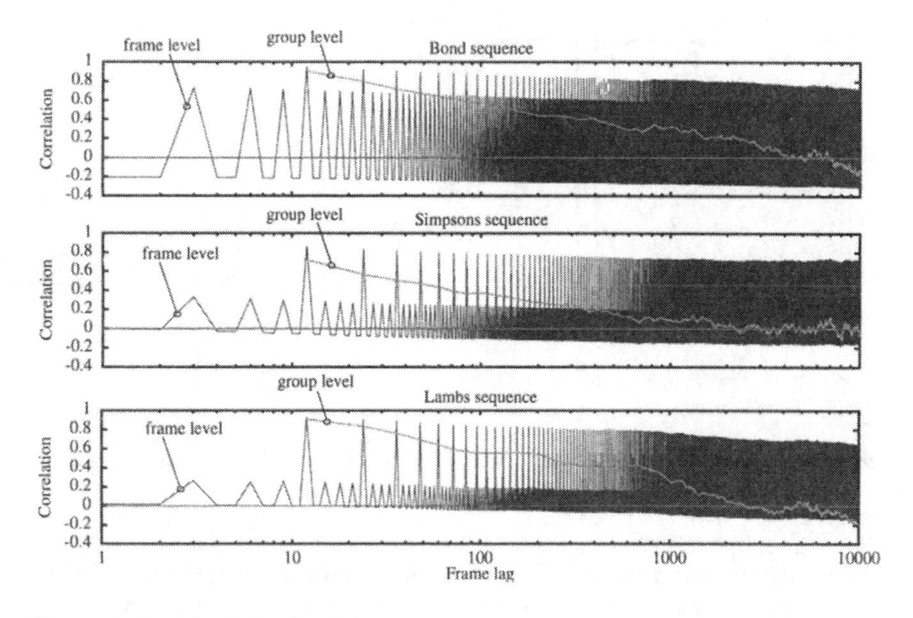

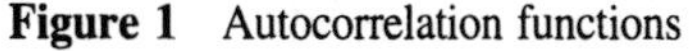

Figure 1 Autocorrelation functions

As another way of investigating temporal properties of MPEG sources, we shall employ the method of time-variance plots. Following (Garret, 1993), sequences of lengths S may be partitioned into a number of blocks $T = \lfloor S/M \rfloor$ for some block size M. Defining the block mean stochastic variable $Y_k = \frac{1}{M}\sum_{i=(k-1)M+1}^{kM} a_i$, where a_i denotes frame sizes, one may form the general expression $s_Y^2 = M^{-\beta} s_a^2$, where s_Y^2, s_a^2 denotes the sample variances. This may be expressed as $ln(s_Y^2)/ln(s_a^2) = 1 - \beta ln(M)/ln(s_a^2)$. Plotting the empirical left hand side of this equation against $ln(M)/ln(s_a^2)$ should result in a linear curve with unity negative slope in the case of an independent time series. i.e. $\beta = 1$. For an exactly self-similar sequence, β should remain constant in the range of [0,1> over all time scales, whereas for an asymptotically self-similar process, β will converge towards a constant value in the same range as the sample sizes increase. In Figure 2 are plotted the time-variance graphs for the sequences (denoted 'Original'). In the same diagrams are plotted the $\beta = 1$ line (denoted 'Independent'), and the time-variance curve for an artificial sequence (denoted 'Group') obtained by subsequently drawing independently the size of the 12 frames of the GOP structure from their respective marginal distributions. The following properties of these diagrams may be noted: The first few points up to $M \approx 10$ follow the independent-curve more or less closely. In this region, the periodic property will also influence the time-variance graph, which can be noted by comparing the 'Group' curve and the 'Original' curve. This effect is strongest in the Bond sequence. As block levels increase, the two curves depart due to the lack of correlations above frame level in the 'Group' sequences. In many sequences

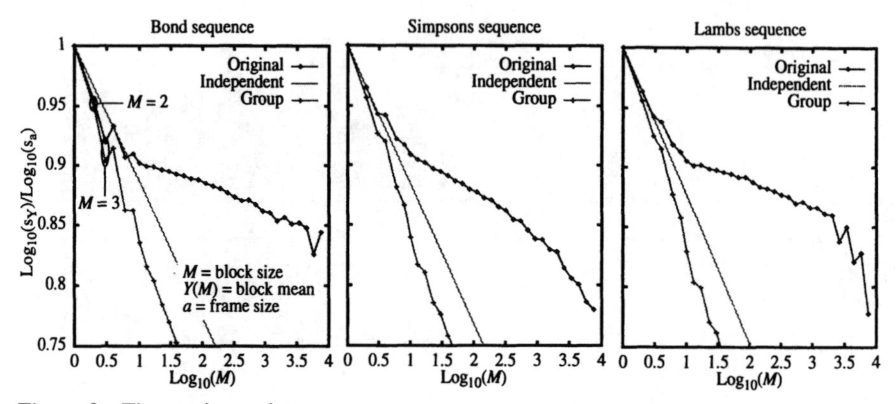

Figure 2 Time variance plots

we have investigated, the rest of the graph will be partitioned in two fairly distinct regions, a middle region with very low β-value, and a region comprising the larger blocks showing a somewhat higher β-value. This behaviour is most clearly seen in the Simpsons sequence.

The last investigation in this section will be devoted to leaky bucket loss contour diagrams. Such diagrams have been proposed in (Lucantoni, 1994) and (Garret, 1993) as an alternative way of characterizing the statistical properties of data sources, and will give an operational picture of source properties. The points of the curves in the diagram were obtained by fixing leaky bucket sizes, then adjusting the load of a trace driven simulator until a specified loss probability was achieved. In each diagram in Figure 2, curves are shown for the real source (denoted 'Original') and an artificial sequence generated by independently drawing frame sizes from the marginal distributions (denoted 'Independent'). In all diagrams can be observed a region in which the two traces show a similar behaviour. As buffer lengths increase, low and medium loss iso-lines of real sequences depart abruptly from the loss iso-lines of the independent sequences, indicating the effect of correlations at the group level.

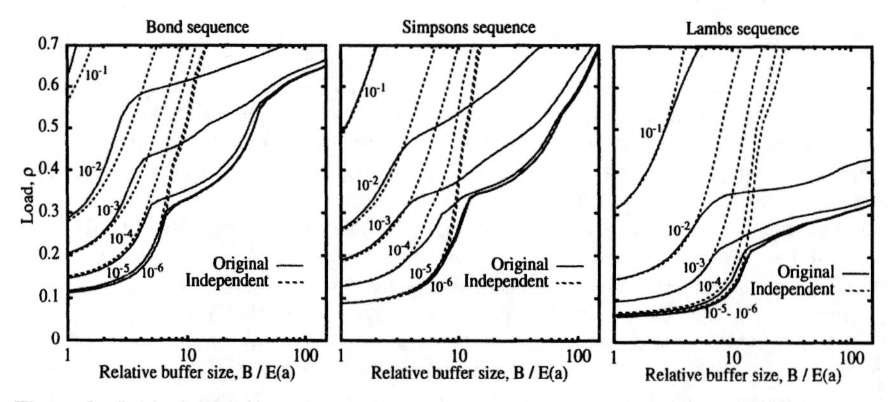

Figure 3 Leaky bucket loss contour diagrams

We draw the following conclusions from these investigations: At time scales smaller than the GOP period, the single MPEG sources investigated display behaviour resembling that of uncorrelated sources. This agrees very well with the operational principles of the MPEG algorithm: removing temporal redundancy within each GOP, but not across GOPs.

4 SOURCE MODELS

4.1 Single source

This system consists of one source generating frames at a fixed rate into a buffer served with constant output. Frames are segmented into cells when fed into the buffer. We shall consider the case where cells are spaced evenly through the inter frame period ('frame stretching'). The number of cells in subsequent frames follows the marginal distribution of cells per frame from a real source. The following definitions will be needed:

- n_k r.v. for the number of cells in the buffer immediately before arrival of the first cell of the *k*th frame at time t_k
- a_k r.v. for the number of cells in the *k*th frame.
- d The number of cells that may be serviced in the inter frame period. When cell period is set at unity, d denotes the inter frame period.
- b Buffer limit

It is assumed that the time intervals t_k - t_{k-1} are integral cell periods, which may be justified by observing that the frame period is much longer than the cell period. As frames are stretched, and assuming service before arrival, the system may reach the state of b-1 at the end of the frame period. This situation will happen if the system overflows, or the cell flow into the buffer exactly fills it up during the frame period. At instants t_k, the following recursion relation will then apply:

$$n_{k+1} = Max(Min(n_k + a_k - d, b - 1), 0) \tag{1}$$

It may be noted that this way of modelling the arrival process will correctly handle short term autocorrelation in the cell interarrival times caused by the frame stretching mechanism. An interesting feature of the above relation is a certain scaling property. If, by performing the necessary multiplications with the cell payload, the problem is recast using information content rather than cell numbers, we get the relation:

$$\eta_{k+1} = Max(Min(\eta_k + \alpha_k - \delta, \beta), 0) \tag{2}$$

Where the obvious identifications of symbols for information content rather than cell numbers may be performed. Going to the fluid limit by letting payload size P diminish towards zero, keeping $\eta \equiv n \cdot P$ constant, it may be seen that the resulting problem is completely scalable insofar that for the information loss ratio function Λ, we have that $\Lambda(\delta, \beta) = \Lambda'(\delta', \beta')$ if $\delta' = s\delta$, $\beta' = s\beta$, $\alpha'_k = s\alpha_k$. An important implication of this observation is the invariance of the solution to the relation $\beta / E(\alpha)$, i.e. the ratio between buffer size and expected frame size. The discrete cell case is not accurately scalable as a scaling of cell size will change the granularity of the involved stochastic variables. The above result will nevertheless give the natural scaling when results from different sources are compared, and as no assumptions about the nature of the sources other than the frame stretching mechanism is made, will have a broad range of application.

We now embed a Markov chain at instants t_k letting n_k denote the states of the Markov model. To satisfy the Markov condition, arrivals a_k must then be assumed to be independent and identically distributed. Omitting sequence indexes, the stationary state probabilities π_j may be expressed as:

$$\pi_j = \begin{cases} 1-(F*\pi)_{(b-1)+d-1} & , \quad j=b-1 \\ (A*\pi)_{j+d} & , \quad 0<j<b-1 \\ (F*\pi)_d & , \quad j=0 \end{cases} \tag{3}$$

Where

$$A_j = Pr(a=j) \qquad F_j = Pr(a \le j) \qquad \pi_j = Pr(n=j) \qquad \sum_j \pi_j = 1$$

And the $*$ operator denotes the discrete convolution operation consistent with the ranges of the operands. If the steady state equation (3) is written out in matrix form as $\pi = \pi \boldsymbol{P}$, the transition probability matrix will be of order b, the buffer length.

Generating function approaches exist both for infinite buffer length (Lin, 1990) and finite buffer lengths (Østerbø, 1995). We have however chosen a numeric solution method iterating on the probability transition matrix. This method turned out to be very fast.

Cell loss probability R can be determined from the ratio between overflowed cells and arrived cells:

$$R(\rho, b) = \frac{1}{E(a)} \sum_{j=1}^{a_{max}-d} j \cdot (A*\pi)_{(b-1)+d+j} \tag{4}$$

The load ρ is given by $\rho = E(a)/d$.

A generalisation of this model to a Markov model consisting of a group of N frames with independent but different framesize distributions is possible. Embedding the Markov chain at the group level rather than at the frame level, the resulting equilibrium equation may be expressed as

$$\pi^{(k+N)} = \pi^{(k)} \boldsymbol{P}_0 \boldsymbol{P}_1 \ldots \boldsymbol{P}_{N-1} \tag{5}$$

Where the $\boldsymbol{P}_k$ matrixes are obtained in a similar manner as previously, but with the use of the structural arrival probabilities. The total loss probability is then given by

$$R(\rho, b) = \frac{\sum_{i=0}^{N-1} R_i E(a_i)}{\sum_{i=0}^{N-1} E(a_i)} \tag{6}$$

where the individual loss ratios may be found by equation (4).

4.2 Multiplexed sources

Two logical multiplexing models will be considered, one in which the periodic GOP property of sources is taken into account, and a simpler one neglecting this property. In the simpler model (denoted 'Frame' in figures), all sources are aligned at frame boundaries and send uncorrelated sequences of frames following the marginal frame size distribution of the original source. The resulting queuing problem is as for the single source, replacing the frame size distribution with the K-fold convolution of the same. The assumption of frame aligned sources greatly simplifies the resulting queuing problem, as the general case implies the treatment of a telescoping recursion relation. In a previous paper (Andreassen, 1995) the consequences of the frame alignment assumption are investigated. For the number of sources and buffer lengths considered here, we have experienced that a model based on frame alignment will upper bound loss probabilities of free running sources by approximately an order of magnitude in the low loss regions ($\sim 10^{-10}$ - $\sim 10^{-6}$), less in high loss regions.

In the other logical multiplexing model that is considered (denoted 'Periodic' in figures) each of the K sources occupies one of N distinct phase values relative to the GOP cycle. Phase constellations change on a single source basis, and the duration of a constellation of phases is assumed to be long enough such that steady state behaviour will dominate over phase change transients. Thus, the treatment of different constellations may be decoupled. A symmetrical phase change model is assumed, such that the probability for a single source to be in a specific state is uniformly distributed over the phases. Each source sends frames according to a periodic pattern of marginal distributions. This model will take into account the periodic correlation property of the sources, but neglecting other correlation properties.

We now consider the distribution of sources in the N discrete phases of the GOP cycle. A specific constellation of sources in the phase space may be identified by a stochastic constellation vector $\boldsymbol{K}$, where the elements K_i denote the number of sources in state i, $i = 0, 1, ..., N\text{-}1$. As there are N possible phase values, and $K = \sum_{i=0}^{N-1} K_i$ independent sources, the probability of a specific phase constellation is given by the multinomial distribution:

$$f(\boldsymbol{k}) \equiv Pr(\boldsymbol{K}=\boldsymbol{k}) = \frac{K!}{\prod_{i=0}^{N-1} k_i!} \prod_{i=0}^{N-1} p_i^{k_i} \tag{7}$$

Where $p_i = Pr($A single source being in phase state $i)$

With the uniformity assumptions made above, we have that $p_i = 1/N$. As arrival rates do not change between constellations, a mean cell loss probability may in principle be found by taking the expectation of the loss function $R(\rho, b, \boldsymbol{k})$ under the multivariate density $f(\boldsymbol{k})$. In practise, such computations will not be possible, and further simplifications are necessary. To this end, the concept of nearly equivalent phase constellations will be defined.

This is a simplification based on the intuitive idea that the number of in-phase sources in a constellation is more important than the actual phases of the constituent sources in the constellation. Thus, we approximate $R(\rho, b, \boldsymbol{k}) \approx R(\rho, b, perm(\boldsymbol{k}))$, where $perm(\boldsymbol{k})$ denotes an arbitrary permutation of the vector $\boldsymbol{k}$, and note that $f(\boldsymbol{k}) = f(perm(\boldsymbol{k}))$. A set of all vectors that can be permuted into each other form a permutation group A, formally defined as

$A = \{k,j | k,j \in A \Leftrightarrow k = perm(j)\}$. All different permutation groups given the GOP length N and the number of sources K make up the set E, which is the set to be traversed using the above simplification. To clarify the above concepts, consider the simple example of $K = 3$ sources with $N = 2$ different phases. Possible constellation vectors are (3,0), (2,1), (1,2) and (0,3), so $A_1 = \{(3,0), (0,3)\}$, $A_2 = \{(2,1), (1,2)\}$, $E = \{A_1, A_2\}$.

To determine the number of members in each permutation group, let the constraint on the vectors be that $\sum_{i=0}^{N-1} k_i = K$, which may be written as $\sum_{i=0}^{K} i \cdot m_i = K$, where an order summation is performed instead of the element summation. The order corresponds to a number of in-phase sources, whereas the factors m_i denote the multiplicity of the ith order in the summation, i.e. how many times a number of in-phase sources are repeated in the constellation vector. An ordered set of m_i-values will uniquely identify a permutation group and vice versa. The m_i factors may take values in the range 0,1,..., K, with the restriction that $\sum_{i=0}^{K} m_i = N$. It may then be seen that the sought number can be expressed by the multinomial weight in the m_i's. To compare with throwing dice, K would be the number of faces of the die, m_i would be the number of dice showing face value i, and N would be the number of dice. The complete expression for the loss probability will be:

$$\hat{R}(\rho, b, K, N) = \sum_{A \in E} \left(\frac{N!}{\prod_{i=0}^{K} m_i!} f(k_A) R(\rho, b, k_A) \right) \tag{8}$$

The composite source used in the above computation may generally be expressed by the N different marginal densities in the MPEG cycle. As we do not want to use any information from the constellation other than whether sources are in phase or not, a natural approach will be to use only two different densities for frame types, one for I-frames and another for P/B-frames. For a given constellation this will give:

$$\hat{A}_i = A_I^{(k_i)} * A_{PB}^{(K-k_i)} \qquad i = 0, 1, \ldots, N-1 \tag{9}$$

Where $A^{(k)}$ denotes the k-fold convolution of the A distribution. The above distributions may then be used in the relations (5) and (6) giving solutions for the periodic group. For each permutation group upper and lower bound on loss probabilities may be computed, assuming that the two cases will occur for the respective maximum concentration and maximum dispersion of the I-frames of the constellations in the permutation group over the GOP period.

To complete the analysis, it will be necessary to find the members of the set E. The problem can be described as the number of ways to form the sum K out of N natural numbers without using more than one member of each permutation group. This problem lends itself to solution by a recursive algorithm, noting that forming subconstellations in a constellation may be done in an invariant manner.

4.3 Simulation model

To compare model results with those from real sources, a trace driven simulator is employed. As the frame alignment assumption is used to get comparable results, the recursion relation

(1) applies in the calculation of buffer occupancy and cell losses. The simulator uses the same algorithm as outlined in the previous section. This will allow estimates of low loss probabilities as compared to a more direct approach. To determine confidence, a loss estimator based on equation (8) is used: Each estimate is based on a weighted mean of simulation results from all permutation groups, where constellations inside a permutation group are chosen at random. To avoid correlation between sources, each source is uniformly offset over the sequence while preserving group constellation.

5 RESULTS AND DISCUSSION

In the following, some results from analytical and simulation models will be compared. All figures display the behaviour of 12 sources, a number which is chosen so as to give a nontrivial situation, tractable computations and simulations, and not to strain the frame alignment assumption too hard. For each source, three scenarios are displayed, each comprising buffer lengths of 5, 10 and 15 times the expected frame lengths of the sources. For the simulation curves, errorbars indicate the 95% confidence interval obtained by performing 100 independent simulations. For the periodic model, an errorbar indicates the upper and lower bound on losses, the line being the mean value of these.

To evaluate results, the following properties of the different models should be kept in mind: The 'Frame' source model is memoryless at the frame level, the 'Periodic' model takes into account the periodic property of the sources, but is otherwise memoryless. The 'Simulation' model contains the naturally occuring correlation properties of the sources, and accounts also for the way different phase constellations are formed.

Results from the Bond, Simpsons and Lambs sources are displayed in Figure 4. In the low loss region, the simulation and model curves fit reasonably well for all buffer lengths displayed. It seems that loss in this region is not related to correlation properties above the frame level of the MPEG sources. Rather the losses are due to the rare situations in which several sources send large frames simultaneously, the scenario which is explicitly modelled both by the constellation simulation model and the Markov models.

The Bond source features a negative-correlation property, easily discernible in the autocorrelation and time variance graphs of Figures 1 and 2. In the latter figure, blocks of sizes two and three frames have a low variance in their sample means as the frames constituting the blocks are negatively correlated. In the loss contour diagram, this property will manifest itself by showing smaller losses for the original source than for the independent source in the upper left regions of that diagram. A similar effect is seen in the case of multiplexed sources. At buffer lengths of 347 and 695 (i.e. five and ten frame lengths), loss probabilities of the real source are less or equals those of the models. Only in the case of buffer length corresponding to 15 frame lengths (b=1043), will the effect of correlations beyond the frame level dominate behaviour in the high loss region, giving higher loss than the Markov models.

The Lambs source is the one most precisely modelled by Markov models in terms of relative buffer lengths. This property is not easily deducible from the autocorrelation and time-variance plots, as these diagrams seem very similar for the Simpsons and the Lambs sources at the short ranges. Considering however the loss contour diagrams, a closer match between an independent and the original source for the Lambs sequence can be observed.

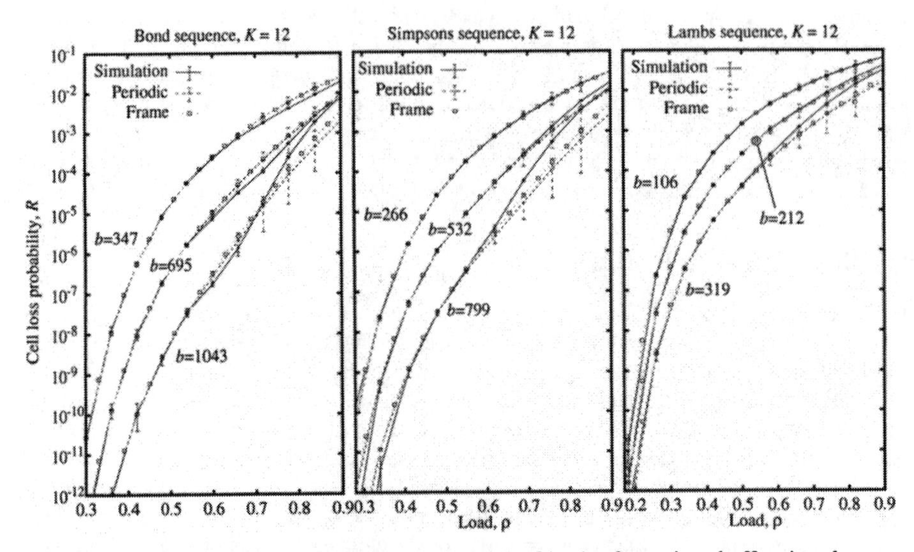

Figure 4 Multiplexer loss probabilities R as function of load ρ for various buffer sizes b

For a number of 12 sources and buffer lengths considered, the periodic property of the sources does not dominate multiplexer behaviour. For all sequences, the simple 'Frame' model gives results well within the uncertainty bounds of the 'Periodic' model. It is our experience that difference in results between these models may be large when only a few sources are multiplexed. This effect seems to diminish as the number of sources increases.

6 CONCLUSIONS AND FURTHER WORK

We have in this paper investigated correlation properties of MPEG VBR video sources, and what impact these properties will have on cell loss probabilities of multiplexed sources. Such sources exhibit properties at short time scales which in some respects resemble those of uncorrelated sources, a phenomenon which is found consistent with the nature of the MPEG coding algorithm. At larger time scales, the sources were however seen to exhibit strong autocorrelation properties. By employing Markov models utilizing only certain periodic properties of the sources, an investigation of which properties would dominate with respect to loss probability of multiplexed frame aligned sources was made possible. It was found that for a number of 12 sources, the short term properties were dominant for buffer lengths up to about 10 times the expected frame lengths of the video sources considered. In low loss regions, no effect of longer term correlation properties was seen at buffer lengths up to 15 times the mean frame lengths, with the same number of sources. Concluding that the longer range properties of the sources are not important in the above considered regions, may however be a hasty conclusion. The cell loss probability measure, even though widely applied, is not the best performance measure of the multiplexer. In order to determine the effect on services, the time distribution of cell losses will be significant, and correlation properties may certainly influence these distributions. An investigation of these questions will be an interesting topic for further research.

Acknowledgements

The author would like to thank Oliver Rose at the University of Würzburg for making the video traces available. Several useful discussions with professors Peder J. Emstad and Tore Riksaasen at the Norwegian University of Science and Technology is also gratefully acknowledged. The traces are available at: ftp-info3.informatik.uni-wuerzburg.de:/pub/MPEG

7 REFERENCES

Andreassen, R., Emstad, P., Riksaasen, T. (1995) Cell losses of multiplexed VBR MPEG sources in an ATM multiplexer, in *Proc. 12th Nordic Teletraffic Seminar*, 83-95

Beran, J., Sherman, R., Taqqy, M. S., Willinger, W. (1995) Long Range Dependence in Variable-Bit-Rate Video Traffic, *IEEE Transactions on Networking*, **43**, (2/3/4), 1566-1579

Garret, M.W. (1993) *Contributions towards real-time services on packet switched networks*, Ph.D. Thesis Columbia University.

Garret, M.W., Willinger, W. (1994) Analysis, Modelling and Generation of Self-Similar VBR Video Traffic, in *Proc. ACM Sigcomm 94*, ACM press, 269-280.

Huang, C., Devetsikiotis, M., Lambadaris, I., Kaye, A. R. (1995) Modeling and Simulation of Self-Similar Variable Bit Rate Compressed Video: A Unified Approach, in *Proc. ACM Sigcomm 95*, 114-125

Hübner, F. (1994) Dimensioning of a Peak cell Rate Monitor Algorithm Using Discrete-Time Analysis, in *Proc. ITC 14*, 1415-1424

Landry, R., Stavrakakis, I. (1994) Non-Deterministic Periodic Packet Streams and Their Impact on Finite-Capacity Multiplexer, in *Proc. IEEE Infocom 94*, 224-231

Lin, A. Y. M., Silvester, J. A. (1990) Queueing analysis of an ATM Switch with Multichannel Transmission Groups, *Performance Evaluation Review*, **18**, (1), 96-105

Lucantoni, D.M., Neuts, M. F., Reibman, A. R. (1994) Methods for Performance evaluation of VBR Video Traffic Models, *IEEE/ACM Transactions on Networking*, **2**, (2), 176-180

Reininger, D., Melamed, B., Raychaudhuri, D. (1994) Variable Bit Rate MPEG Video: Characteristics, Modeling and Multiplexing, in *Proc. ITC 14*, 295-306

Rose, O. (1995) *Statistical Properties of MPEG video traffic and their impact on traffic modelling in ATM systems*, (University of Würzburg Institute of Computer Science Research Report Series), Report No. 101

Østerbø, O. (1995) *Some Important Queueing Models for ATM*, (TF Rapport 22/95), Kjeller, Norwegian Telecom Research

5

On the Relevance of Long Term Correlation in MPEG-1 Video Traffic

Marco Conti, Enrico Gregori, Andreas Larsson
CNUCE, Institute of National Research Council
Via S. Maria 36 - 56126 Pisa - Italy, Phone: +39-50-593111
Fax: +39-50-589354, e-mail: man@cnuce.cnr.it

Abstract

Variable Bit Rate, *VBR*, video is expected to become increasingly important with the large scale deployment of Broadband-Integrated Services Networks, *B-ISDN*s, over the next few years. Although the modeling of VBR video sources has recently received significant attention, there is currently no widely accepted model which lends itself to mathematical analysis. Furthermore, new video compression standards, such as the *MPEG* family, are emerging. On the basis of results of a detailed statistical analysis of a long sample of a movie encoded with the MPEG-1 algorithm, an analytically tractable model is developed and analyzed in detail. The model is able to capture both the distributional and temporal characteristics of this kind of traffic. The model was validated using a two-hour long sequence generated by the MPEG coding of the movie "Star Wars". We show that our model is a flexible tool to study network issues such as bandwidth allocation and statistical multiplexing.

Keywords

VBR video, MPEG, Markov chain, simulation, ATM, statistical multiplexing

1 INTRODUCTION

Recent technological advances in fiber optics and switching systems have provided the technological basis for the development of high capacity Broadband-Integrated Services Digital Networks (B-ISDNs), which are capable of supporting transmission speeds of several hundred Mbps [1]. This enormous potential for fast and massive information transport should be able to support not only the traditional data and voice services, but also a variety of new applications, including the transport of

---

Work carried out in the framework of CNR coordinated projects. Andreas Larsson's research was performed while visiting the CNUCE institute in the framework of the ERASMUS project. He is now with Ericsson Hewlett-Packard Telecommunications AB.

images, teleconferencing, moving video, and large volumes of interactive computer data. Asynchronous Transfer Mode (ATM) is the transfer technique for the implementation of such B-ISDNs, due to its efficiency and flexibility [1].

Variable Bit Rate (*VBR*) video is by far the most interesting and challenging application. VBR video traffic is highly variable and dependent on the coding scheme adopted and the activity of the movie. A variable bit rate encoder attempts to keep the quality of video output constant at the price of changing the bit rate. A better utilization of network resources is also obtained, since only the real amount of information has to be transferred.

While the modeling of VBR video sources has recently received significant attention [1, 2, 3, 9, 10, 11, 12, 13], there is currently no widely accepted model which lends itself to mathematical analysis. Furthermore, while most of the previous studies focus on teleconferencing traffic, the *MPEG* standards for VBR encoding of moving images are currently under development [6, 7], and there is no proof that previous models can also be used for characterizing *MPEG* traffic (e.g. movies coded with the *MPEG* algorithm). This work focuses on the analysis and characterization of the traffic generated by an MPEG-1 encoder. Specifically, Section 2 presents the characteristics of an MPEG-1 source relevant for our investigation. An MPEG-1 analytical model is presented and validated in Section 3. The relevance of the long term correlation in MPEG-1 modeling is pointed out in Section 4.

2 MPEG-1 VIDEO SOURCE

MPEG-1 is a specification for coding moving pictures, developed by the ISO Joint Motion Pictures Experts group. MPEG-1 is an interframe coder. Coders in this class exploit, in addition to intraframe coding, the temporal redundancy between adjacent frames by predicting the next frame from the current one. A key feature that distinguishes MPEG-1 from previous coding algorithms is *bidirectional temporal prediction*. For this type of prediction, some of the frames are encoded using two reference frames, one in the past and one in the future. This results in higher compression gains.

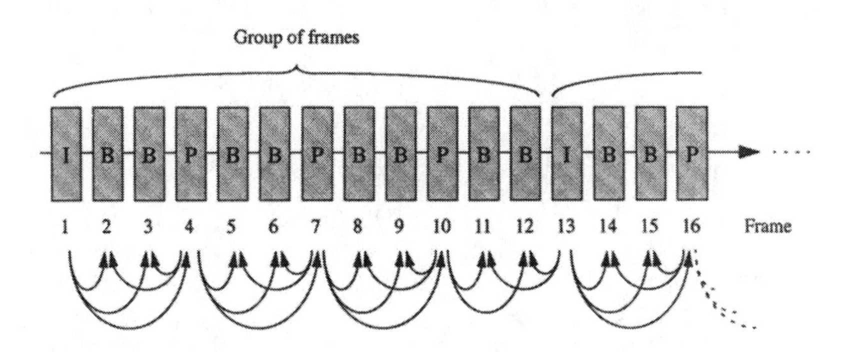

Figure 1 A sequence of MPEG-1 frames and their relationship.

As indicated above, when applying MPEG-1 to video, one of three different coding modes can be used for each frame. The terminology used for the resulting frame is related to the mode used and is as follows:

- *I-frame*: intra frame coded.

- *P-frame*: predictive coded with reference to a past picture.
- *B-frame*: bidirectional predictive coded.

I-frames provide access points for random access but only with moderate compression. Predictive coded frames are also generally used as a reference for future P-frames. Type B frames provide the highest amount of compression but require both a past and future reference prediction. In addition B-frames are never used as reference frames.

In the encoded sequence, the frames are arranged into *groups*, as shown in Figure 1. In this case a group consists of 12 frames - one I-frame, three P-frames and eight B-frames. Figure 1 also shows the relationship between the frames. We can see that I-frames are independent, P-frames are predicted, and B-frames bidirectionally predicted.Statistics of MPEG-1 coded movie

This section presents a statistical analysis of an MPEG-1 encoded movie. The source for the analysis is a bit per frame trace, released by M. Garret at Bellcore, obtained from an MPEG-1 encoder fed with approximately two hours of the movie "Star Wars". Specifically, frames are coded in groups of twelve frames as defined in Figure 1 (i.e. the frame pattern is IBBPBBPBBPBB).

Several basic statistics of the MPEG-1 coder trace are shown in Table 1. The table contains statis-

Table 1 Basic statistics for the MPEG-1 "Star Wars" movie

Measure	Original Sequence	Aggregate Sequence
Mean bandwidth, μ	15598 bits/frame	187185 bits/group
Standard deviation, σ	18165 bits/frame	72468 bits/group
Coefficient of variation, μ/σ	1.16	0.39
Peak bandwidth	185267 bits/frame	932710 bits/group
Minimum bandwidth	476 bits/frame	77754 bits/group
Peak/mean bandwidth	11.88	4.98

tics related to the following sequences

- *original sequence*: the sequence of the number of bits per frame;
- *aggregate sequence:* the sequence of the number of bits per group i.e. IBBPBBPBB-PBB.

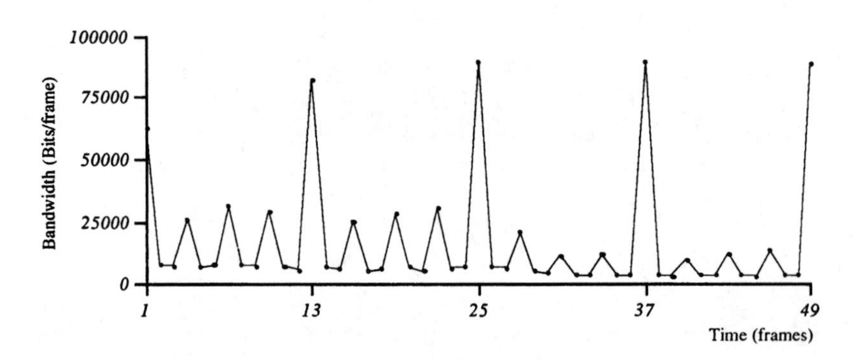

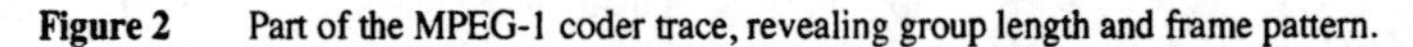

Figure 2 Part of the MPEG-1 coder trace, revealing group length and frame pattern.

As shown in Figure 2 the bandwidth varies greatly due to the different frame-types, I, P and B. An MPEG encoder can therefore be seen as a source generating three different kinds of traffic, each according to the individual characteristics of a frame-type. In order to understand the level of interdependence between these three subsequences, we measure the correlation $Corr[x(k), y(k)]$, where $x(k)$ and $y(k)$ are the kth values in each sequence. Specifically, to use the correlation for our purposes we let k correspond to the kth group in the original sequence. Within each group, we sum the number of bits generated for each kind of frame. The correlation is then computed among the three different sequences, i.e., $I(k)$, $\sum P(k)$ and $\sum B(k)$. The resulting correlation values of the

Table 2 Correlation between frame-types.

$Corr[I(k), \sum P(k)]$	0.35
$Corr[I(k), \sum B(k)]$	-0.13
$Corr[\sum P(k), \sum B(k)]$	0.71

MPEG-1 coder trace are summarized in Table 2. They tell us that dependencies exist between the three sequences, and therefore they cannot be represented with three independent processes.

3 MPEG-1 MODELING

As indicated by the statistical analysis presented in the previous section, the output of an MPEG-1 encoder should be described by three partially correlated submodels where each submodel describes the output process corresponding to one frame-type. Obviously this leads to a model with a very large state space.

The model space complexity is reduced by avoiding a separate representation for the different frame-types. Specifically, this is obtained by considering a different time scale in which the time unit is the group (i.e. a sequence IBBPBBPBBPBB) and the bit rate per time unit is the sum of the amount of bandwidth generated by all the frames in a group. In this case one group is equal to 12 frames and each frame is generated every 1/24th second. The resulting sequence is hereafter named *aggregate sequence*.

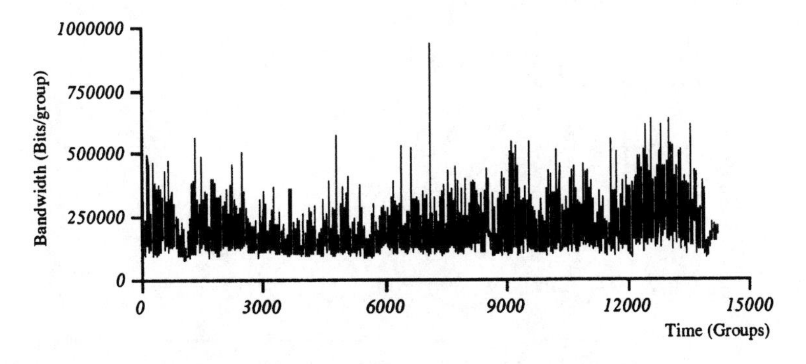

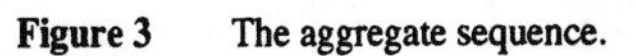

Figure 3 The aggregate sequence.

Figure 3 shows a plot of the *aggregate sequence* generated by an MPEG-1 coder with the movie Star Wars as a source. A time unit, on the x-axis, is equal to 0.5 seconds, i.e. a group interarrival.

The bit rate of consecutive frames shows that the bandwidth changes in a rapid but bursty way. However, there is also a slowly changing underlying structure. This *low frequency* underlying struc-

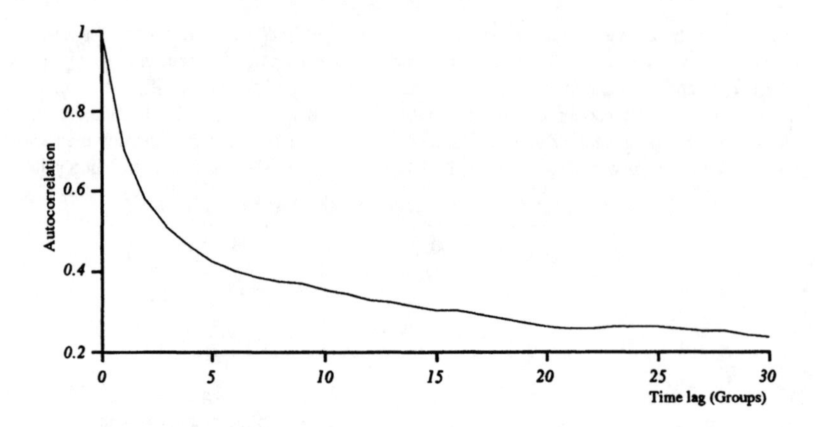

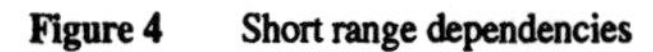

Figure 4 Short range dependencies

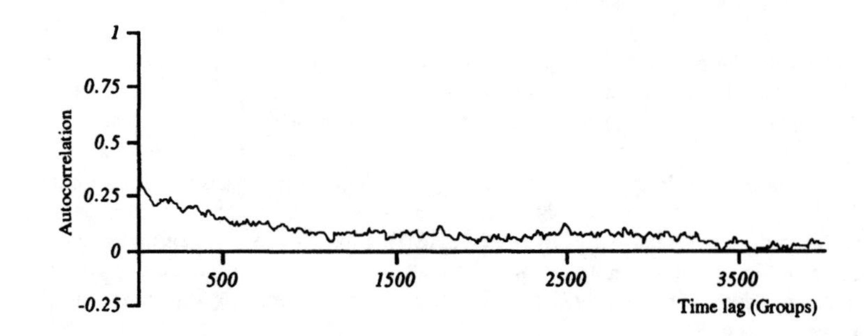

Figure 5 Long range dependencies

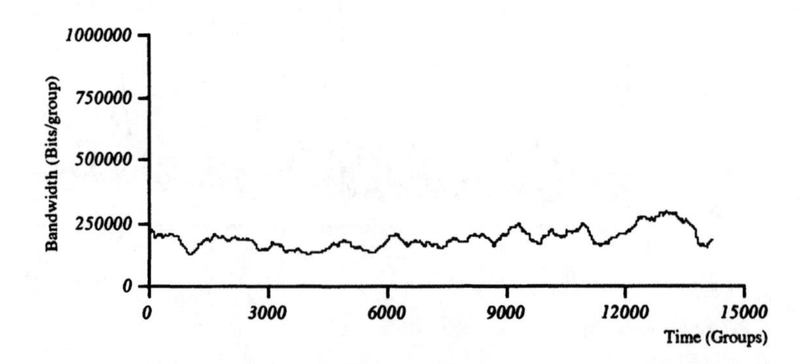

Figure 6 Low frequency component of the aggregate sequence.

ture of the sequence can be better highlighted by passing the aggregate sequence through a moving average filter of length *W*. The result with a *window size W*=300 groups (i.e. two and a half minutes), is shown in Figure 4

The autocorrelation function for the aggregate sequence is plotted in Figures 5 and 6, showing the short $(0 \le n \le 30)$ and long range $(n > 30)$ dependencies, respectively. Figure 5 shows the existence of a strong short-range dependence for time lags below approximately 30 groups, which corresponds to 15 seconds. In this range the autocorrelation function drops quickly (the autocorrelation with $n = 30$ is about 0.2). However, after this sharp initial decrease, as shown in Figure 6, it takes a very long time before the autocorrelation function drops to zero. Specifically, Figure 6 highlights the existence of a significant long-range dependence which lasts for time lags of up to 3500 groups, i.e. about 29 minutes. The tail of the autocorrelation function decreases slowly.

3.1 Description of the Model

Figures 5 and 6 show that in the aggregate sequence there are both short-range dependencies which last for around 20-30 groups (some seconds), and long-range dependencies which last for thousands of groups (some minutes). In order to capture both types of dependencies a bidimensional Markov chain $\{L_k, H_k, k \ge 0\}$ is used, in which $\{L_k | k \ge 0\}$ is used to represent the long term correlation, while $\{H_k | k \ge 0\}$ represents the short term correlation. Specifically, in our model the process $\{H_k | k \ge 0\}$ describes the bit rate per group of an MPEG-1 encoder. To avoid unnecessary complexity (in the state space $\{H_k | k \ge 0\}$) we quantize the bit rate information into a number of levels. The number of quantization levels for the process $\{H_k | k \ge 0\}$ will hereafter be denoted by *N* (i.e. $H_k \in \{0, 1, 2, \ldots, N-1\}$). The question of which quantization method should be used is not discussed here. For us it seemed natural to use uniform quantization. For this reason, let *max* and *min* denote the maximum and minimum bit rates observed in the aggregate sequence. The possible bit rates between *max* and *min* are quantized with a constant step size $\Delta = (max - min)/N$, resulting in the actual bit rate of the source equal to $j \cdot \Delta + min$ where *j* is the quantization level holding the property $0 \le j \le N-1$.

To represent the low-frequency component of an MPEG source, a modulating process $\{L_k | k \ge 0\}$ is included in the model as well $(L_k \in \{0, 1, 2, \ldots, M-1\})$.

How is L_k controlled? To this end let the random variables X_i denote the time in which the low-frequency process has stayed in level *i*. If the process has stayed in a level *i* for *t*-1 slots, there is the probability of leaving *i* at slot *t* equal to

$$P(X_i = t | X_i > t-1) = 1 - q_i(t) \tag{3.1}$$

and a probability of staying in *i* more than *t* equal to $q_i(t)$. Hence to control the transitions in the $\{L_k\}$ process, the variable *t* is added to the Markov chain state, and the transitions between the low frequency levels are managed according to (3.1).

It is still impossible to tell which distribution the "real" X_i has to be. The general approach would entail estimating the mass function from the real sequence. Since we only have a single trace, the estimated statistics may be biased. To minimize the biasing effect, in this work we assume a Geometric distribution for the duration of a low frequency state which only entails estimating a single parameter, i.e., the average time the $\{L_k\}$ spends in each state. The model is thus named *Geometric Model*.

The process we now want to model takes the form $\{L_k, H_k, k \ge 0\}$, where $L_k \in \{0, \ldots, M-1\}$ is the status of the low frequency process corresponding to the *k*th group, and $H_k \in \{0, \ldots, N-1\}$ is the corresponding state in the high frequency process.

The source is modelled with a Markov chain whose transition probabilities are

$$p_{ij,\,lm} = p^{(1)}{}_{ij,\,lm} = P(L_k = l, H_k = m | L_{k-1} = i, H_{k-1} = j). \tag{3.2}$$

The procedure for fitting it to a real source (i.e., to construct the transition probabilities, $p_{ij,\,lm}$, of the Markov chain starting from a real source) is shown in [4].

3.2 Model Analysis

In this section we analyse the characteristics of the model to see whether it can imitate the behavior of the real source well.

To compute the statistics of our model (e.g., average, variance, peak/average ratio, autocorrelation function) we first need to compute the steady-state probabilities $\pi_{i,j} = P(L_k = i, H_k = j)$ of our Markov chain.

Several basic statistics which can be immediately calculated from the steady-state probabilities are shown in Table 3. The results were obtained with $M = 8$ and $N = 8$.[1] We see that the values coincide well.

Table 3 Comparison of some basic statistics for the real source and the model.

Measure	*Real source*	*Model*
Mean bandwidth level, μ	0.526	0.527
Standard deviation, σ	0.712	0.712
Coefficient of variation, σ/μ	1.354	1.351
Peak bandwidth level	7	7
Minimum bandwidth level	0	0
Peak/mean bandwidth level	13.307	13.283

We will now examine whether the model is successful in capturing the time correlation structure of our source. The autocorrelation function $r(n)$ of the model is thus needed:

$$r(n) = \frac{E[H_k H_{k+n}] - \mu^2}{\sigma^2} . \tag{3.3}$$

Figure 7 shows a plot of $r(n)$ for the real source and four different models constructed with $M = 8, N = 8$ and various moving average window lengths, W. Time lags used for the calculation range from 0 to 30 groups. The plot thus compares the short-range dependence of the real source, and different parametrizations of the model.

The model constructed with W=20, has a stronger short-range dependence than the real source. It has, however, a faster decay. Even though $r(n)$ of the model is still above the real source one for a time lag equal to 30 groups, the difference is smaller than for $n = 10$.

As the value of W is increased, the autocorrelation function of the model tends to fall off at the beginning but it decreases slower. The model with $W = 40$ is a good example to emphasize this behavior. For n less than 7, the short-range dependence of the model takes on values lower than the real source. For time lags beyond this point, the plot shows that $r(n)$ of the model decay slower than for the real source. The autocorrelation functions for models constructed with $W = 60$ and $W = 80$ follow the same pattern.

We know that the long-range dependence of MPEG-1 coded VBR video is very strong. Figure 8 compares the autocorrelation function of the model and the real source for time lags of 0 up to 2000 groups. Several values of W have been used. A model constructed with $W = 100$ has a long-range dependence which is not as strong as that of the real source for n greater than 100. It also reaches zero at a time lag approximately equal to 1100, which is much earlier than in the real source.

---

1. The choice of the value 8 is shown in [4] to be a good compromise between accuracy of the model and complexity of the state space in the Markov chain.

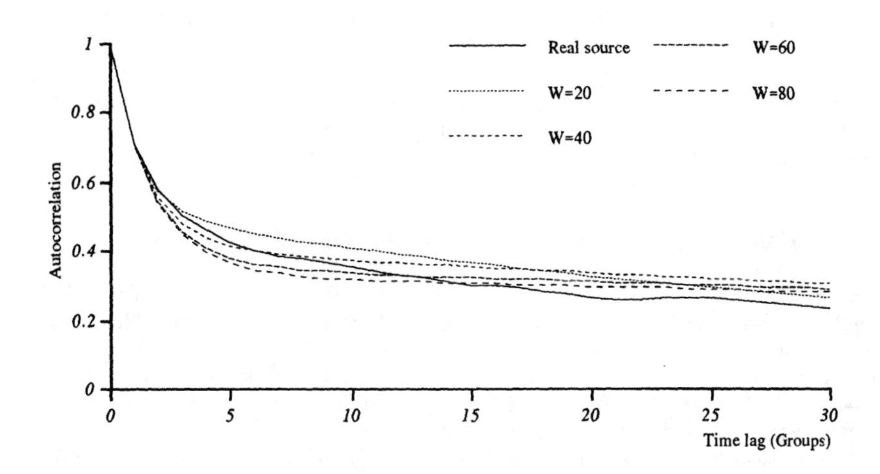

Figure 7 Comparison of the real source's and the model's short-range dependencies.

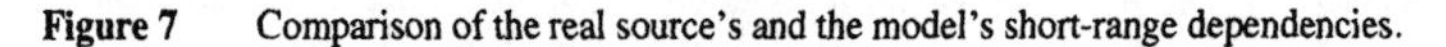

$r(n)$ for the other models plotted in Figure 8 tells us that the long-range dependence of the model tends to get stronger, and thus approaches the real source, as W is increased. For example, a model constructed with $W = 300$ matches the long-range dependence of the real source better than if it is created with $W = 200$. At the same time we know, from the previous subsection, that a higher value of W implies a weaker short-range dependence.

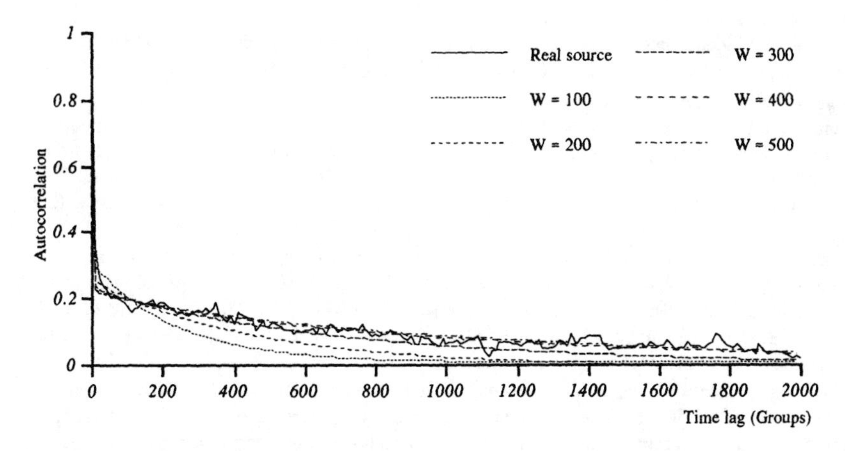

Figure 8 Comparison of the real source's and the model's long-range dependencies.

We can make the following conclusions from the model analysis:

- Distributional properties of a source, i.e. steady-state probabilities and basic statistics, are captured well by the model.
- The parameter W can be used to construct source models with different time correlation structures. The rules to follow are: stronger short- and weaker long-range depend-

encies for lower W-values. On the other hand, weaker short-range and stronger long-range dependencies are obtained for higher W-values.

Validation of the model is currently underway by exploiting the new traces which have been recently released [14]. Preliminary results of this validation process [5] show that the model can be tuned to fit different kinds of MPEG sources (e.g. movies, sports events, talk shows).

4 IMPORTANCE OF THE LONG-TERM DEPENDENCIES

At the moment ATM networks assign peak rate bandwidth to real time applications, that is, by avoiding multiplexing and thus utilizing the residual bandwidth for non real time traffic. This implies that high priority must be assigned to real time applications. Data services, on the other hand, tolerate delays and can compensate for loss by retransmission. Low priority can thus be assigned to these applications. In Section 4.1 the effect of an MPEG-1 source on the quality of service experienced by low priority traffic is investigated.

However, since the ratio peak/average for VBR video traffic may be greater than four, the peak rate approach for bandwidth allocation significantly reduces the number of VBR applications that can concurrently use the network. For this reason the peak rate allocation will be abandoned as soon as the problem of characterizing and multiplexing the traffic generated by VBR video is addressed successfully. In Section 4.2 the potential gains which can be achieved by multiplexing several MPEG video sources are investigated.

4.1 Peak rate allocation

We assume a peak rate allocation for VBR video and we consider a network with two categories of traffic: i) VBR video which has high priority and peak rate allocation; ii) data traffic which is transmitted utilizing the bandwidth reserved but not used by VBR video traffic (low priority traffic).

We consider the system which consists of a VBR video source, a data source, a priority queueing system and a transmission channel, or *server*. The server has the capacity C, to transfer a certain amount of information per time unit. Two queues, each associated with one class of traffic, are used to buffer arriving information from the sources when the server's capacity is exceeded. Items from the buffers are transmitted according to their priority. Inside each queue, transmission is first in first out (*FIFO*).

The VBR video source is described by the Geometric model. For each time slot, equal to the length of a group, the VBR video source generates traffic corresponding to one of the eight levels (between 0 and 7) of the aggregate bit rates of the MPEG-1 coded movie Star Wars. Thus for each time slot the VBR video requires between 1 and 8 *units* of bandwidth. The channel has a static allocation which implies that C is equal to 8 units/group.

Data is generated according to a Poisson process with rate λ. On average data can use a percentage p of the units not used by video. This means that the average number of data units in a group is equal to $\lambda = p(7-\mu)$ units/group, where μ is the average bandwidth level for the video source. Hereafter, p is set to 0.9, and therefore $\lambda = 0.9 \cdot (7-0.53) = 5.823$ units/group ($\mu = 0.53$ is obtained from Table 1).

Three simulations with different sources representing the VBR video were run. Simulation I was a trace-driven simulation with the quantized aggregate sequence of the "Star Wars" movie. Using a real trace in a simulation has some drawbacks. First of all, using only one realisation does not make it possible to draw any conclusions on the accuracy of the simulation. The behavior of the resulting video traffic is not general either, and thus not very representative. Handling a long trace is, furthermore, impractical. We do, however, include the trace-driven simulation here to have some sort of comparison with the other results.

In simulations II and III we used the Geometric instance model as a source, first constructed with W=20 and then with W=300. Simulative estimates are obtained with a confidence level of 90%.

We will now look at the simulation results. Specifically, we investigate the mean queue length μ_{ql} and mean queueing time μ_{qt}, in relation to the size of the low priority queue, and the time that data has to wait there, respectively. For the case where the real trace is used, i.e. simulation I, we can obtain meaningful estimates only for the first moments of queue length and queueing time. This is due to the fact that we only have one realisation of such a real trace. Table 4 compares these values obtained from simulations I, II and III. The results show that only estimates obtained for the long-range model are close to those obtained from real data.

Table 4 Data queue-length and queueing time

Measure	*Real trace*	*Model, W=20*	*Model, W=300*
μ_{ql}	21.77	11.56	23.01
μ_{qt}	29.96	15.88	31.63

In simulations II and III, on the other hand, we used a number of realisations which made accurate estimations on the queue length and queueing time. Figure 9 highlights that the long-range dependence impact on queueing time is significant. For example, as shown in Figure 9, when W=300 is used for high-priority traffic model the 99th percentile of the queueing time is about three times higher than that obtained with W=20.

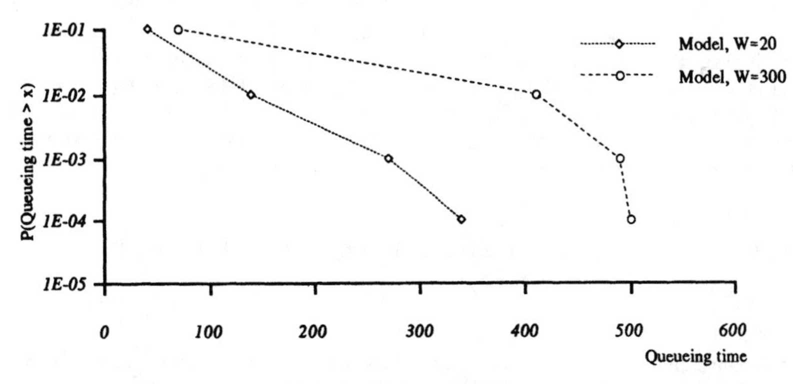

Figure 9 Data queue-length tail distribution for the two models.

The only difference between the two models studied here is their temporal behavior. Furthermore, as shown in Table 4, the model with long-range dependence was shown to be more similar to the real trace. Consequently, as a conclusion from the simulation results, one can say that by neglecting the long-range correlation one would obtain a rather optimistic estimate on queue length and queueing time.

4.2 Multiplexing of i.i.d sources

Mutliplexing of VBR video sources is complex, as these applications have low tolerance towards network congestion. Although sufficient buffer capacity may be available, excessive buffering may not be possible, due to the resulting unacceptable delays. In this section we therefore investigate the queueing time distribution experienced by VBR video traffic as a function of the bandwidth reserved for each source. As shown before (see Table 1), the peak rate for our MPEG source corresponds to a bandwidth level equal to *c*=7, while the average is about bandwidth level equal to 0.53. Below we

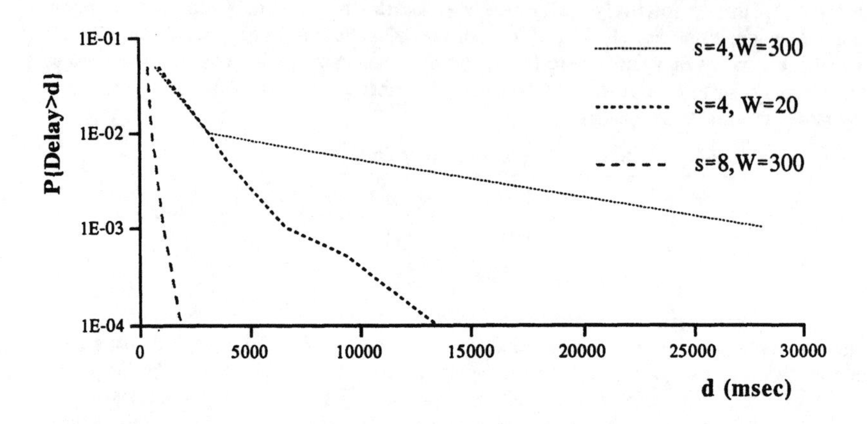

Figure 10 Tail of delay distribution, c=1.0

investigate the delay experienced by VBR video traffic by assuming that the bandwidth allocated for each source is about twice the average, i.e., *c*=1. The results reported in Figure 10 were obtained by studying via simulation the queueing delay distribution in a single server queueing system with a deterministic service time, FIFO, and input traffic generated by *s* independent and identically distributed MPEG-1 sources. Figure 10 shows that a 75% network utilization and acceptable delays can be achieved if at least eight sources are multiplexed. In addition, the figure clearly indicates that the tail estimated with *W*=20 is extremely underestimated in the region (1E-04,1E-02).

Acknowledgements

The authors wish to express their gratitude to the Mark Garret (Bellcore) for releasing the Star Wars trace.

References

[1] R. O. Onvural, *Asynchronous Transfer Mode Networks: Performance Issues*, Artech House, Inc., Norwood, MA, 1994.

[2] N. Ohta, *Paket Video: Modeling and signal processing*, Artech House, Inc., Norwood, MA, 1994.

[3] D.Heyman, A.Tabatabai and T. LAKshaman, "Statistical Analysis and Simulation Study of Video Teleconference Traffic in ATM Networks", *IEEE Transactions on Circuits and Systems for Video Technology*, Vol. 2, No. 1, March. 1992, pp. 49-59.

[4] A. Larsson, "Traffic Engeneering of Broadband Networks: Modelling of MPEG 1 coded VBR Video", Master's Thesis, Linkoping University 1995, performed at CNUCE, Pisa, Italy, in the framework of the ERASMUS project.

[5] M. Conti, E. Gregori, M. Nava, "Analysis and Modeling of an MPEG-1 Video Source", Third IFIP Workshop on Performance Modelling and Evaluation of ATM Networks, 2nd-6th July, 1995, Ilkley, West Yorkshire, U.K.

[6] D. Le Gall, MPEG: A Video Compression Standard for Multimedia Applications, *Communications of the ACM*, Vol. 34, No. 4, April 1991, pp. 46-58.

[7] L. Chiariglione,"The development of an integrated audiovisual coding standard: MPEG", *IEEE Proceeding*,Vol. 83, No. 2, February 1995, pp. 151-157.

[8] M. Nomura, T. Fujii and N. Ohta, "Basic Characteristics of Variable Rate Video Coding in ATM Environment", *IEEE Journal on Selected Areas in Communications*, Vol. 7, No. 5, June 1989, pp. 752-760.

[9] G.Ramamurthy, B. Sengupta, "Modeling and Analysis of a Variable Bit Rate Video Multiplexer", Proceedings of IEEE INFOCOM, Florence 1992, pp.817-827.

[10] P. Skelly, M. Schwartz and S. Dixit, "A Histogram-Based Model for Video Traffic Behavior in an ATM Multiplexer", *IEEE/ACM Transactions on Networking*, Vol. 1, No. 4, Aug. 1993, pp. 446-459.

[11] M. R. Frater, P. Tan and J. F. Arnold, "Variable Bit Rate Video Traffic on the Broadband ISDN: Modeling and Verification", *ITC 14*, 1994, pp. 1351-1360.

[12] M. W. Garret, *Contributions Toward Real-Time Services on Packet Switched Networks*, Ph.D. Dissertation CU/CTR/TR 340-93-20, Columbia University, New York, N.Y., May 1993.

[13] M. W. Garret,W. Willinger., "Modeling and Generation of Self-Similar VBR Video Traffic", *SIGCOMM'94*, London, Sept. 1994, pp.269-280.

[14] O.Rose, " Statistical Properties of MPEG Video Traffic and Their Impact on Traffic Modeling in ATM Systems", University of Wurzurg, Research Report No.101, February 1995.

6

Multiple Time Scales and Subexponentiality in MPEG Video Streams

Predrag R. Jelenković, Aurel A. Lazar and Nemo Semret
Center for Telecommunications Research
Columbia University, New York, NY 10027, USA.
Telephone: (212)-854-2399. Fax: (212) 316-9068.
email: {predrag, aurel, nemo}@ctr.columbia.edu

Abstract

We develop a practical, multiple time scale model for MPEG video traffic whose accuracy and relatively low computational complexity make it well suited for real-time traffic generation experiments on broadband networks. The major feature of our approach is the decomposition of the frame size sequence into simple slow and fast time scale components. This accurately captures aspects of queueing behavior that are difficult to model otherwise. The model also exploits the existence of deterministic patterns that are due to the MPEG coding scheme.

We also present a novel modeling approach based on spatial renewal processes (SRP). This model gives *exact* matches to any desired marginal distribution and any convex non-increasing autocorrelation function. In particular, it can match subexponentially decaying autocorrelations (i.e., can capture long range dependence), something no other model of comparable complexity can do. A SRP is suited for on-line model construction, since it involves no search in parameter spaces, and matches aggregated streams as easily as single streams. The SRP approach yields an analytically tractable queueing behavior, and thus provides a basis for admission control policies that take the dependence structure of video streams into account.

The models are validated by queueing simulations.

Keywords

ATM, VBR, traffic modeling, time scales, subexponentiality, dependence structure, real-time traffic generation, admission control.

1 INTRODUCTION

The objective of this work is to gain understanding of the statistical properties of MPEG video traffic, and in particular to identify and accurately characterize those properties that have an impact on queueing behavior when one or many video sources generate traffic to be transmitted over a network. Understanding these properties plays a crucial part in, for example, admission control policies that ensure efficient utilization of network resources while providing quality of services guarantees.

In section 2, we examine our data set, a set of MPEG-I video traces. We compute the essential statistics, such as marginal distributions and autocorrelation, for full sequences as well as component subsequences.

Then in section 3, we consider the problem of modeling MPEG video on multiple time

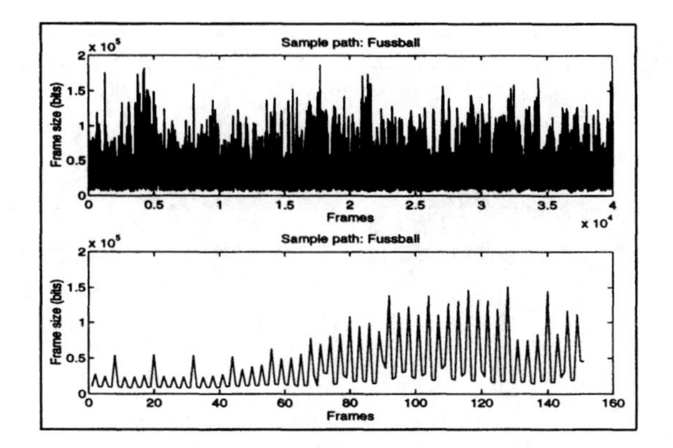

Figure 1 Sample path of Fussball sequence

scales. We develop a model which accurately matches the real data, and is well suited for real-time traffic generation. Our model consists of three deterministically interleaved independent slow time scale processes (each having positive autocorrelation and corresponding to the three frame types in MPEG), plus a single fast time scale correlated "noise" process. Through simulations we demonstrate how this separation helps account for different (temporal as well as distributional) aspects of the queue length behavior.

In section 4, we review the basic concept of subexponentiality, and introduce the Spatial Renewal Process approach to modeling. This new approach allows for matching the statistical characteristics of a broad class of processes exactly (up to second order statistics), including processes with subexponential characteristics, and can be used as a building block for more complex models. It also has the advantage of being analytical tractable, and may be very useful as a basis for implementing admission control policies.

2 DESCRIPTION OF THE MPEG TRACE DATA SET

The data set consists of sequences of MPEG-I frame sizes created at the Institute of Computer Science at the University of Würzburg (Rose (1995)). In all, 17 sequences (sportscasts, movies, music videos, newscasts, talk shows, cartoons and set top) of 40,000 frames each are available.

From the raw video frames, the MPEG coder produces three types of frames at its output: **I** frames, where only spatial redundancies are exploited; **P** frames, where motion compensation with respect to the previous I frame is used to achieve further compression; and **B** frames, where both the previous and the next I or P frames are used to minimize temporal as well as spatial redundancies. The frame types occur in a fixed periodic pattern. In this data set, the period is 12 frames, and the pattern is: IBBPBBPBBPBB. Such a segment is called a group of pictures (GOP).

Figure 1 shows the trace of frame sizes for one of the 17 sequences in our data set, a soccer game. The close-up view of a portion of the sequence clearly shows the deterministic pattern induced by the GOP pattern IBBP...: a large peak, followed by two small values, then one medium peak, etc. It also reveals that the sequence seems modulated by slower variations: a small envelope for the first 60 frames, followed by a larger one.

In Figure 2 the marginal distributions of the different types of frames show the expected differences between the three types of frames. On average, P frames are approximately twice the size of B frames, and one third the size of I frames. The autocorrelation function (ACF) of the whole sequence exhibits a complex pseudo-periodic structure. However,

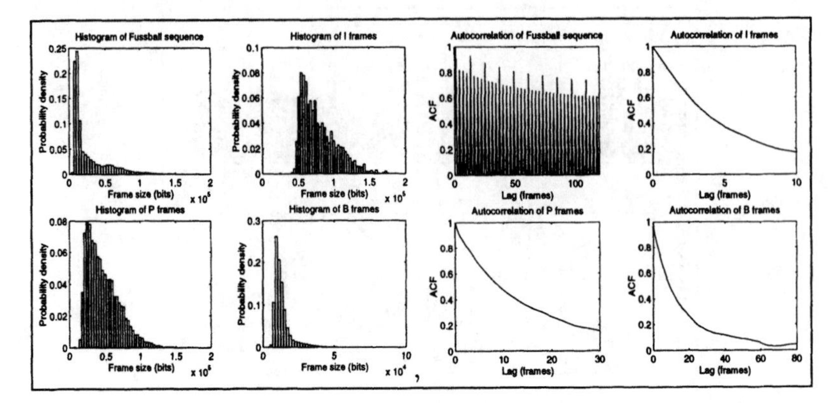

Figure 2 Statistics of Fussball MPEG sequence frame sizes

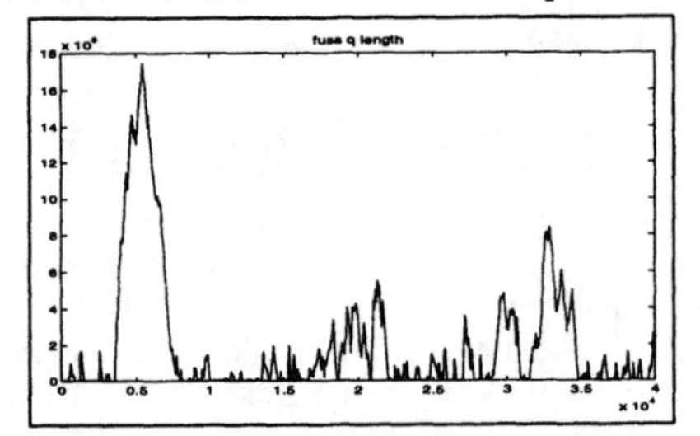

Figure 3 Queue length as a function of time: FCFS queue, utilization = 0.9

focusing on each type of frame separately, we see that the dependence there is simple: a monotonically decaying autocorrelation. Moreover, as can be seen from Figure 2, where the ACFs are plotted on the same time scale (indeed in 120 frames, there are 10 I frames, 30 P frames and 80 B frames), the time range of dependence is similar for the three types of frames. Thus we see that the complex and very heavy dependence seen in the ACF of the whole sequence (autocorrelation of 0.6 for a lag of 120 frames, i.e., 5 seconds) is to a large extent deterministic. It is due to the GOP pattern of the MPEG coder which forces every 12th frame to be of roughly the same size when compared to the overall average. The stochastic dependence that is inherent in the source (characterized by the autocorrelation function of the I, P and B frames) is much simpler: monotonically decaying in time.

Figure 3 shows the queue length (buffer size) as a function of time when our MPEG stream is fed to a FCFS single server queue with constant service rate. The utilization is 0.9. Figure 3 will be used as a basis of comparison for validating our models.

3 MULTIPLE TIME SCALE MODELING OF MPEG VIDEO

The two important features of our modeling approach are the separation of multiple time scales, and the recognition of the deterministic MPEG frame pattern.

In the literature, the existence of multiple time scale statistics in multimedia traffic has been observed from different perspectives. Li and Hwang (1993) argue, from the frequency domain point of view, that the low frequency band of the autocorrelation's Fourier transform (long term correlation) has the most significant impact on queueing. The notion of long range dependence, which has been the subject of much recent work

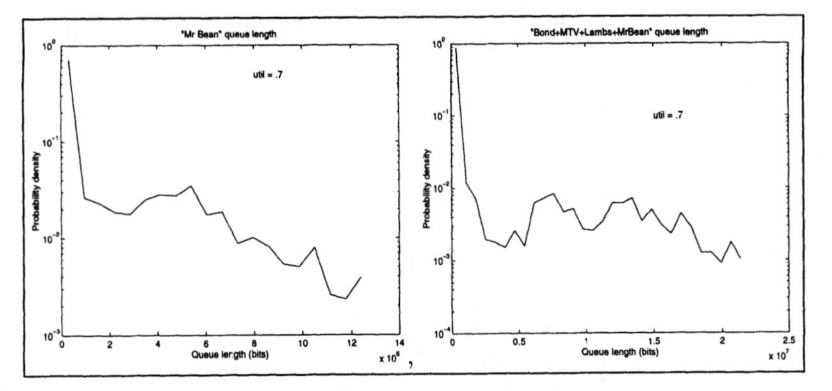

Figure 4 Queue length distribution of MPEG streams

in video traffic (see, e.g., Beran et al. (1995)), is equivalent to the existence of large time scale components. Lazar et al. (1994) develop video models for the slice and frame time scales, and show that the time scales relevant to scheduling depend on the quality of service requirements. Frost and Melamed (1994) survey a wide range of approaches to traffic modeling, several of which take multiple time scales into account, for example, self-similar or fractal models. These essentially attempt to capture an infinite number of time scales, and for that reason, they generally suffer from high run-time complexity, which makes them impractical for real-time generation. Fractal models are also relatively difficult to treat analytically in terms of queueing behavior. More recently, Landry and Stavrakakis (1995) have presented a modeling approach based on multiple time scales that is appropriate for cell and slice levels of video traffic. The types of dependency investigated (i.e., the shape of the autocorrelation), however, is restricted by the constraining *periodic* Markov chain underlying their model. Jelenkovic and Lazar (1995a) have analytically shown that when a stream with multiple time scales passes through a queue, the queue length distribution has multiple decay rates. In the current work, we find that precisely this behavior is found in the actual MPEG streams (see Figure 4). A model without multiple time scales can only capture one of the decay rates of the queue length distribution (see Figure 6 and the attendant explanation).

Other approaches that explicitly model the deterministic structure of MPEG video have been proposed recently, but they either do not capture any second order or time dependence properties at all (see Krunz et al. (1995)), or if they do, do not accurately model the cross-correlation between different frame types (see Ismail et al. (1995)).

3.1 Constructing the model

If we decompose the video stream into two time scales, the GOP pattern is naturally associated with the slow time scale, via three mutually independent (but correlated) slow time scale processes corresponding to the I, P and B substreams. Indeed, I frames are coded independently of all other frames, and therefore, their size depends only on the "complexity" of the individual images. On the other hand, B frames exploit temporal redundancy as much as possible, and therefore, their size depends more on the "action" in the sequence. Thus, it can be argued that these aspects should be modeled separately, because in general the level of action may be independent of the degree of complexity*.

*Imagine filming a painting hanging on a wall with a fixed camera: the images are complex and difficult to compress, so I frames will be large. But there is no motion at all, so B frames will be very small. At

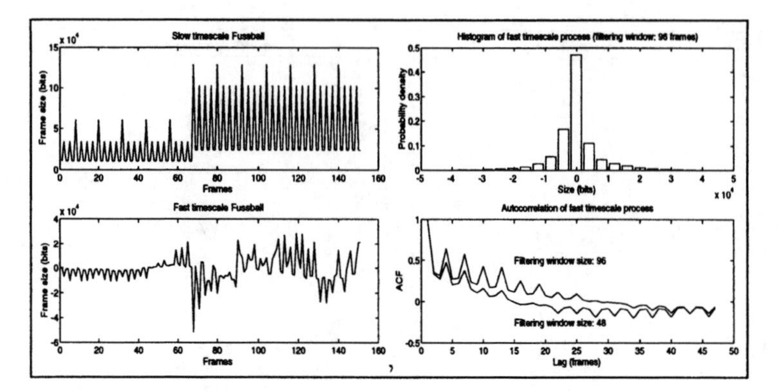

Figure 5 Decomposition into slow and fast time scale processes – averaging window of 96 frames

Also it is natural to adjust parameters that are directly tied to a visible feature of the video stream. On the fast time scale, since we would like to capture the very short term correlation (for lags as small as 1 frame) and the deterministic GOP pattern has been captured in the slow time scale, having three independent I, P and B fast time scale processes is not desirable. Instead, a single correlated fast time scale process is chosen.

We begin the development of our model by analyzing the multiple time scale statistics of an arbitrary video stream. To filter out the fast time scale process, we average the frame size sequence over non-overlapping windows. Note that we average the three types of frames separately. The averaged process gives us the slow time scale component. Figure 5 shows, for a window size of 96 frames (8 GOPs), the resulting decomposition into slow and fast time scale processes, the latter being obtained by subtracting the former from the original sequence.

To model the slow time scale process, we proceed as follows. At the beginning of a window of, say, 96 frames, we generate one frame of each type. The three frames are then repeated following the GOP pattern. The frame size sequences are generated by three independent TES (transform-expand-sample) processes, each generating one sample every 96 frames. A complete description of the TES method can be found in Melamed et al. (1991, 1992), and a detailed description of how we used it in Semret (1995). Briefly, TES methods can match the marginal distribution exactly, and the autocorrelation is approximated by tuning some model parameters. We generate samples via a fast algorithm using integer-only arithmetic operations. For the slow time scale I frames, we created a TES process, with histogram and ACF matching those of the subsequence formed by every 96th frame of the averaged process, starting from the first I frame. The TES processes for B and P frames are created similarly.

The statistics of the fast time scale process are shown in Figure 5. From the autocorrelation, it is apparent that the dependence due to the GOP pattern is still present. The fast time scale shows the variations from the long term average, which is why the histogram is symmetric about zero. Note that I frames are not only bigger than the P and B frames, they also have a larger variance. This explains the residue of the pseudo-periodic structure in the ACF. As shown in the figure, this can be reduced by averaging over windows of smaller size. In fact, how to choose the window size needs further investigation. More

---

the other extreme, a music video with a rapid fire sequence of simple but unrelated pictures would cause relatively small I frames, but large B frames.

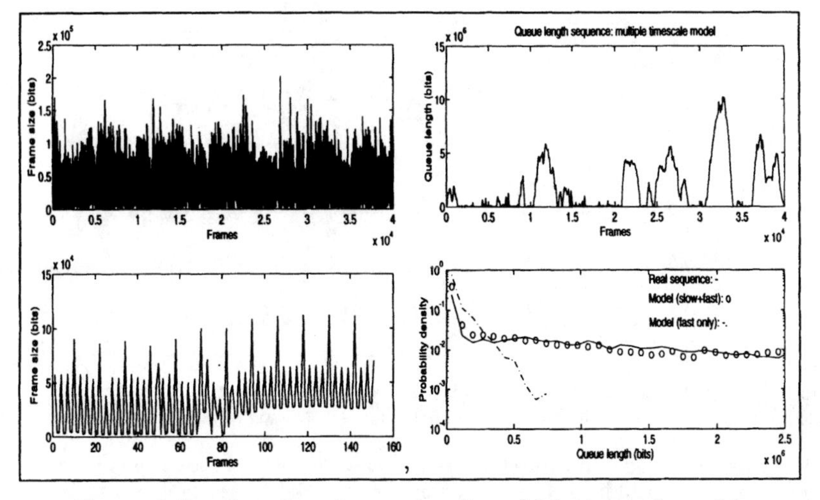

Figure 6 Sample path and queue length: multiple time scale model

analysis would have to be done to determine the optimal separation between "slow" and "fast" time scales.

In any case, compared to the slow process, the fast time scale process has a rapidly decaying ACF, so the small pseudo-periodic bumps may be ignored. Thus we choose to model the fast time scale as a simple TES process, with smoothly decaying ACF. The fast time scale process can be viewed as a small perturbation or noise with short term correlation which is added on to all frames (no distinction is made between I, P and B).
Extending the model to multiple sources: This can be done without increasing the run-time complexity of the model. Due to space limitations the reader is referred to Jelenkovic et al. (1995).
Experiments: Figure 6 (left hand side) shows a sample path generated by our model. The two time scales as well as the GOP pattern are clearly discernible. To validate our model, we conducted some queueing simulations (right hand side of Figure 6). The model **very accurately matches the real sequence in terms of the queue length distribution**. Also, the queue length sample path (top right of Figure 6) is realistic both in height and width of bursts (compare with Figure 3).

3.2 The advantage of separating time scales

In addition to its accuracy, the advantage of our approach is that the explicit separation of time scales helps account for different aspects of queueing behavior. By plotting the queue length distribution for the fast time scale alone[†], we get a distribution with the same slope at small queue lengths (bottom right of Figure 6), but not at the tail. We conclude that the good match of the tail of the queue length density is due to the slow time scale component of the model.

Now, focusing on a time period where the queue is small, we see that the real video stream causes small queue buildups (top of Figure 7, around frame numbers 50 and 250),

[†]By fast time scale alone, we mean the fast time scale TES process plus a constant equal to the overall average of the real sequence. Thus the overall average of the frame sizes and the fast dynamics remain the same, but the slow variations are removed.

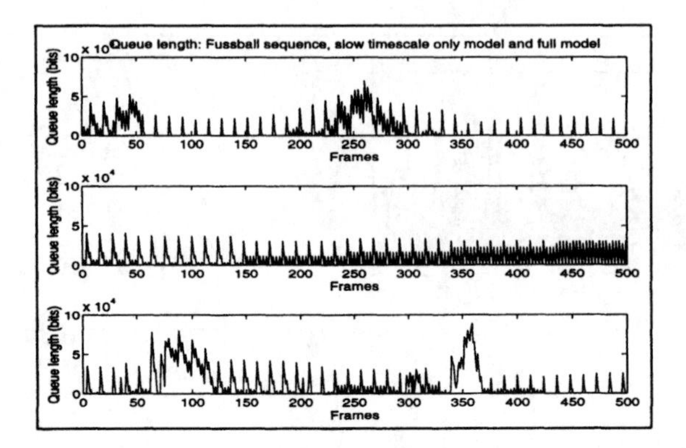

Figure 7 Queue length: Fussball MPEG sequences (top), slow time scale model only (middle), multiple time scale model (bottom). Slow time scale alone is not good enough!

which cannot be reproduced by feeding the model with slow time scale components only through the queue (middle of Figure 7). But the full model does capture this behavior (bottom), thanks to the short term correlation of the fast time scale component.

Thus either the slow or the fast time scale components (or both) may be necessary for capturing the range of queueing behavior one is interested in. That depends on the quality of service requirements. Consider a video source with mean frame size of 26 Kbits, and a rate of 24 frames per second; the utilization is 0.9 (i.e., a server with a capacity of $24 \times 26000/0.9 = 700$Kbits/sec). These were the values used in the simulations shown in Figures 6 and 7. If the maximum acceptable delay is 70ms, then the maximum allowed queue length is about 50 Kbits. In this buffer range the dominant effect on the queue is that of the fast time scale, assuming that the large bursts (which are due to the slow time scale) occur rarely. Thus, as can be seen in Figure 7, the fast time scale model is the relevant one. On the other hand if the maximum delay is ten times larger, then we are only interested in the situations where the queue length approaches 0.5 Mbits, i.e., in the big bursts. Then the fast time scale process becomes negligible, and the slow time scale should be modeled accurately.

Thus, our multiple time scale approach provides a framework for making the correct trade-offs first in what kind of data to gather, and second, what model parameters to carefully tune.

4 SUBEXPONENTIALITY OF THE VIDEO TRAFFIC

The class of subexponential distributions was first introduced by Chistakov (1964). The basic property of this class is that if $X_1, X_2, \ldots, X_n$ are independent, identically and subexponentially distributed random variables,

$$P(X_1 + X_2 + \ldots + X_n > x) \sim nP(X_1 > x), \tag{1}$$

as $x \to \infty$ (see Jelenkovic and Lazar (1995b)). In words, the big peaks tend to be isolated. Some examples of subexponential distribution functions are the Pareto family, the lognormal distribution and the Weibull distribution.

In the case of video traffic, subexponentiality can manifest itself in the marginal distributions or its time-dependent structure. In Jelenkovic and Lazar (1995b), it is shown that when either the marginal or the ACF of the input processes are subexponential, the tail

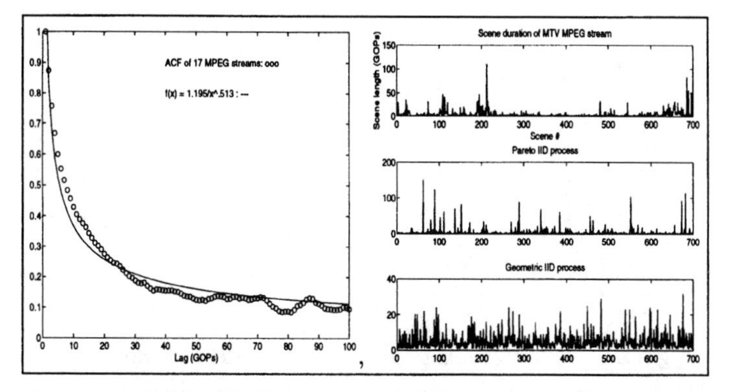

Figure 8 ACF and scene duration of MPEG streams is subexponential (matches Pareto function)

of the queue length distribution is proportional to, respectively, the tail of the marginal distribution, or the ACF. Thus, if real time video traffic exhibits subexponentiality in either form, it is crucial to capture it in modeling. As will be made clear by (4) below, the ACF is directly related to the distribution of the duration of "scenes" or "plateaus" in the video stream.

Figure 8, on the left, shows that the ACF of an MPEG video source (17 streams multiplexed) matches the (subexponential) Pareto function $f(t) = \frac{1.195}{t^{0.513}}$. As mentioned above, another way of showing the same intrinsic subexponential character is by looking at "scene durations". The right hand side of Figure 8 shows a sequence of scene durations[‡] for the MTV MPEG stream (top), and for rough comparison, the sample paths generated by i.i.d. processes with Pareto (middle) and geometric distributions (bottom). Clearly, the scene durations have a subexponential character, as does the Pareto process, where the large peaks tend to be isolated in time, as suggested by (1). Note that this is unlike the geometrically distributed process depicted in Figure 8.

In the sequel, we introduce a model called the *spatial renewal process* (SRP) with arbitrary marginal distribution. We show that its autocorrelation function is equal to the integrated tail of the renewal time distribution. This result allows constructing an SRP with a given marginal distribution function and *any* convex decreasing autocorrelation function. (In particular, the model can very simply match subexponential ACFs.) The model is validated by comparing sample paths with MPEG streams, and by comparing the results of simple queueing simulations. We begin by introducing the spatial renewal process.

4.1 Constructing a Spatial Renewal Process

Consider a point process $T = \{T_0 \leq 0, T_n, n \geq 1\}$ such that $T_n - T_{n-1}, n \geq 1$, are i.i.d. with distribution function F, and mean $m = \int_0^\infty uF(u)du = \int_0^\infty [1 - F(u)]\, du$. Further, let $X = \{X_n, n \geq 0\}$, be a sequence of i.i.d. random variables with a distribution $\mathbb{P}[X_n \leq x] = G(x)$. X and T are jointly independent processes.

Now we define the process $A = \{A_t\}$ by:

$$A_t = X_n \quad \text{for} \quad T_n \leq t < T_{n+1}. \tag{2}$$

[‡]We define a scene change to occur when there is a jump of more than 50 Kbits in the GOP size.

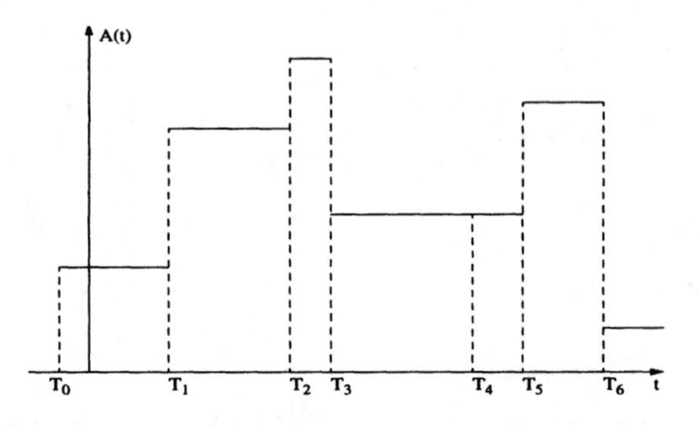

Figure 9 A possible realization of a Spatial Renewal Process.

This is called a Spatial Renewal Process (SRP). A typical sample path of such a process is given in Figure 9. In order to make the process stationary (see Cinlar (1975), section 9.3) we choose the residual time at zero until the first jump to be distributed as an integrated tail of F, i.e.,

$$F_1(t) = \mathbb{P}[\mathrm{T}_1 \leq \mathrm{t}] = \mathrm{m}^{-1} \int_0^{\mathrm{t}} [1 - \mathrm{F(u)}]\,\mathrm{du}. \tag{3}$$

Now it is easy to prove (see Jelenkovic et al. (1995)) that the autocorrelation function $R(t) \stackrel{def}{=} (\mathbb{E}\mathrm{A}_0\mathrm{A}_t - (\mathbb{E}\mathrm{A}_0))^2/\sigma(\mathrm{A}_0)^2$ of the SRP process satisfies the relation

$$R(t) \;=\; 1 - F_1(t). \tag{4}$$

4.2 SRP Video Traffic Model

Using the model of section 4.1, we can generate a process A_t that matches *exactly* a pair (G, R) of an arbitrary marginal distribution and an ACF which is convex, non-increasing and tends to zero for infinite lags. This can be achieved by simply generating two i.i.d. sequences τ_n and X_n, with distributions F and G respectively, corresponding to the widths and heights of the rectangles in Figure 9.

The distribution F is obtained from R. For the purpose of generation, we assume R is given as a sequence of discrete time values $R(0), R(1), \ldots,$ and we seek the values $F(1), F(2), \ldots,$ which will constitute the histogram of the distribution of τ_n. From (3) and (4), we have,

$$1 - R(t) = m^{-1} \sum_{i=0}^{t-1} [1 - F(i)]\,.$$

Setting $t = 1$ and requiring that $F(0) = 0$, we get

$$m = \frac{1}{1 - R(1)}, \tag{5}$$

and $[1 - R(t+1)] - [1 - R(t)] = m^{-1}\,[1 - F(t)]$ yields

$$F(t) = 1 - m\,[R(t) - R(t+1)]\,. \tag{6}$$

Conditions on R: For F to be a probability distribution, it must be non-decreasing,

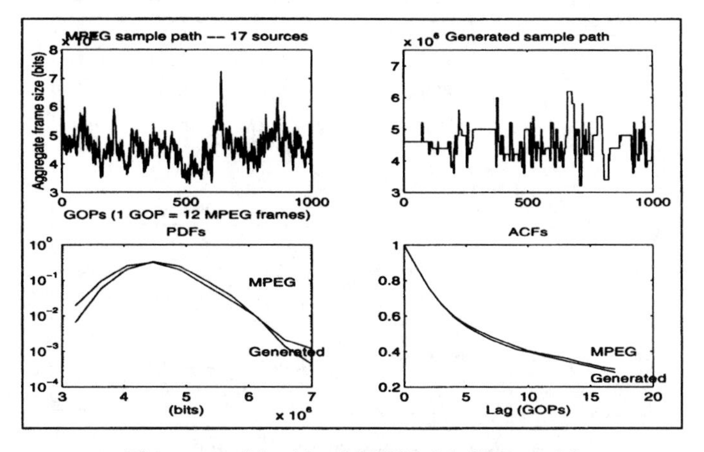

Figure 10 Matching MPEG with SRP model

which implies that R must be **convex**, and $F \leq 1$ implies that R must be **non-increasing**. Finally, consistency calls for $m = \sum_{t=0}^{\infty}[1 - F(t)] = \sum_{t=0}^{\infty} m\,[R(t) - R(t+1)]$. Since $R(0) = 1$, this implies that we must have $\lim_{t \to \infty} R(t) = 0$.

Modeling aggregate sources: This can be done by simply matching the aggregate (G, R); the aggregate marginal distribution function is given as a convolution of the individual marginal distribution functions, and the aggregate ACF is (due to independence) the weighted average of the individual ACFs (see Jelenkovic et al. (1995)).

Strengths of the SRP modeling approach: The main strengths of the SRP modeling approach are the following. First, it provides an exact match of the first and second order statistics, a property which is not achieved by any other known approach. In addition, a very broad and realistic class of autocorrelation functions can be matched: for example, Markovian models such as the Markov Modulated Poisson Process of Skelly et al. (1992), or the TES-based models, can only match exponentially decreasing ACFs. An SRP model can match *subexponential* ACFs, and as such it is a uniquely efficient and simple model capturing long range dependence. The run-time complexity is exceedingly low; the main operations involved are evaluations of F^{-1} and G^{-1}, which can be essentially done by a sequence of comparisons. The number of comparisons to generate one sample is $O(N_G + N_R)$, where N_G is the number of bins on the input histogram for the marginal G, and N_R the number of points on the graph of the ACF R. Second, matching an SRP model involves no tuning, i.e., no heuristic or algorithmic search in the parameter space. Thus, computationally, the model is very inexpensive to construct. Finally, the process has a structure which is analytically tractable, and thus constitutes a useful traffic characterization for queueing analysis and admission control. This is further explored in section 5.

4.3 SRP Traffic Model: Experiments

In this section, we use the SRP model to generate traffic matching that of 17 MPEG sources multiplexed (summed) into one stream. For clarity, we consider the stream at the GOP level, where the measured ACF exhibits the desired (convex, decreasing) properties. Of course, the SRP can also serve as a building block for composite models, capturing more complex dependence structures, exactly as the TES method was used for multiple time scale models in section 3.

Figure 10 shows portions of sample paths for the real and generated traffic, and how

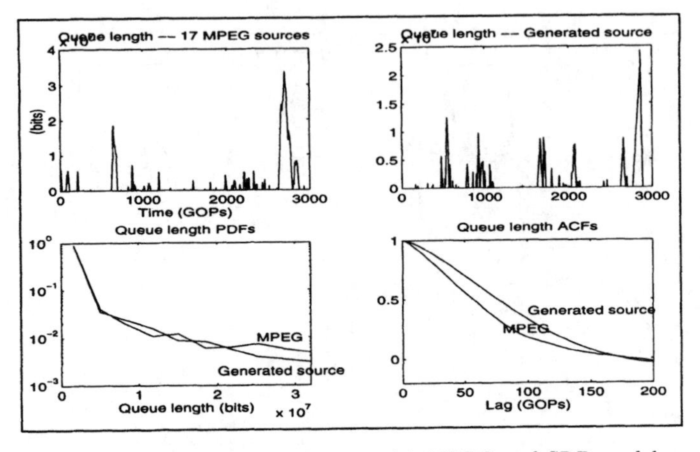

Figure 11 Queueing Simulations with MPEG and SRP model

well the statistics match. More important validation is obtained when both streams (real and generated) are run through queueing simulations, as shown in Figure 11.

The cases shown here are for a utilization of 0.9. As can be seen from the figures, the model also provides an excellent match of queueing behavior: the queue length process in time looks similar, and the queue length distribution is accurately matched.

5 SINGLE NODE ADMISSION CONTROL USING THE SRP CHARACTERIZATION

We assume that the quality of service (QOS) parameters of interest, and hence the admission control decision, can all be directly derived from the queue tail distribution $P(Q > x)$[§]. Note that since $P(Q > x)$ is for the aggregate stream, we can only guarantee QOS in the aggregate and not on the individual streams. Whether an individual stream may find its QOS violated depends on several other factors, including the scheduling policy.

Admission control policy: *(i) given a set of input streams, evaluate the aggregate (G, R), (ii) construct the corresponding SRP model, (iii) compute the queue length distribution, and (iv) determine if QOS is violated.*

Thanks to the simple form of the SRP, several theoretical tools can be used to calculate the queue length distribution. The queue length process at the renewal times T_n follows Lindley's recursion for the GI/GI/1 queue. When the queue increment process, $Y_n \stackrel{def}{=} (T_{n+1} - T_n)(X_n - C)$, is bounded we can numerically obtain the queue length distribution by well known z-transform techniques. However, for sources with heavy tailed autocorrelations (i.e., long range dependence) or marginals, Y_n can be large, and these techniques break down. In such cases, specifically when the distribution of Y_n is **subexponential** asymptotic results from Jelenkovic and Lazar (1995b) can be used for approximating the queue length distribution. (Again, due to space limitation, the reader is referred to Jelenkovic et al. 1995.)

[§]Indeed, since the service rate is constant, cell delay and cell loss probabilities can be directly derived from $P(Q > x)$.

6 CONCLUSION

We have presented two approaches to traffic modeling. The first emphasizes the multiple time scales inherent to video streams, and is geared toward real-time traffic generation. The second, the SRP model, has the advantage of being able to exactly match first and second order statistics, and of being analytically tractable and thus useful for admission control. It should be noted that the SRP model is equally efficient for real-time generation, and can replace TES in the multiple time scale model.

The admission control based on SRP as we have presented it here is complete for the single node case. As such, it is directly applicable to admission control for, e.g., a video on demand (VOD) server. Currently, we are conducting further research into the use of SRP models for admission control, in particular into extensions to the network case.

REFERENCES

J. Beran, R. Sherman, M. S. Taqqu, and W. Willinger (1995). Long-range dependence in variable bit-rate video traffic. *IEEE Trans. Commun.*, 43:1566–1579.

V. P. Chistakov (1964). A theorem on sums of independent positive random variables and its application to branching random processes. *Theor. Probab. Appl.*, 9:640–648.

E. Cinlar (1975). *Introduction to Stochastic Processes.* Prentice-Hall.

V. S. Frost and B. Melamed (1994). Traffic modeling for telecommunication networks. *IEEE Communications Magazine*, pages 70–81, March 1994.

M. R. Ismail, I. Lambdaris, M. Devetsikiotis, and A. R. Kaye (1995). Modeling prioritized MPEG video using tes and a frame spreading strategy for transmission in ATM networks. In *Proc. IEEE Infocomm.*, pages 762–769, April 1995.

P. Jelenkovic, A.A. Lazar, and N. Semret (1995). Multiple time scales and subexponentiality in MPEG video streams. Technical Report CU/CTR/TR 430-95-36, Columbia University. `http://www.ctr.columbia.edu/comet/publications`.

P. R. Jelenković and A. A. Lazar (1995a). On the dependence of the queue tail distribution on multiple time scales of ATM multiplexers. In *Conference on Information Sciences and Systems*, pages 435–440, Baltimore, MD, March 1995. (`http: //www.ctr.columbia.edu/comet/publications`).

P. R. Jelenković and A. A. Lazar (1995b). Subexponential asymptotics of a markov-modulated G/G/1 queue. *Submitted to Journal of Appl. Prob.*.

M. Krunz, R. Sass, and H. Hughes 1995. Statistical characteristics and multiplexing of MPEG streams. In *Proc. IEEE Infocomm.*, pages 455–462, April 1995.

R. Landry and I. Stavrakakis (1995). Multiplexing ATM traffic streams with time-scale-dependent arrival processes. Preprint.

A. A. Lazar, G. Pacifici, and D. E. Pendarakis (1994). Modeling video sources for real-time scheduling. *Multimedia Systems*, 1(6):253–266.

S. Qi Li and C.-L. Hwang (1993). Queue response to input correlation functions: Discrete spectral analysis. *IEEE/ACM Trans. Networking*, 1(5):317–329.

B. Melamed (1991). TES: A class of methods for generating autocorrelated uniform variates. *ORSA Journal on Computing*, 3(4):317–329.

B. Melamed, D. Raychaudhuri, B. Sengupta, and J. Zdepski (1992). TES-based traffic modeling for performance evaluation of integrated networks. In *Proc. IEEE Infocomm.*, May 1992.

O. Rose (1995). Statistical properties of MPEG video traffic and their impact on traffic modeling in ATM systems. Technical Report 101, Institute of Computer Science, University of Würzburg.

N. Semret (1995). Characterization and modeling of MPEG video traffic on multiple timescales. `http://www.ctr.columbia.edu/~nemo/mmn.ps`.

P. Skelly, S. Dixit, and M. Schwartz (1992). A histogram based model for video traffic behaviour in an ATM network node with an application to congestion control. In *Proc. IEEE Infocomm.*, May 1992.

Biographies of the authors can be found at `http://www.ctr.columbia.edu/~{predrag, aurel, nemo}`.

7

A Combined Fast-Routing and Bandwidth-Reservation Algorithm for ATM Networks

Raphael Rom *   *Yuval Shavitt*
Technion — Israel Institute of Technology
Department of Electrical Engineering, Technion city, Haifa 32000, Israel.

Abstract
Traditionally, the task of establishing a connection in connection-oriented networks is divided into two subtasks. First, a route is selected based on a topology database that is maintained in every node according to some optimization rule, such as, shortest path, minimal delay, etc. After the route is selected, the source tries to allocate resources (such as bandwidth, buffers, etc.) along the selected route. A failure to reserve resources along the selected route causes the source to give-up on the connection or to retry, i.e., look for another route and then try to reserve resources along it. The delay imposed by such a failure, especially for long haul ATM networks, might be too long for many bursty applications, e.g., queries to distributed databases.

In this work we suggest a more efficient scheme where the search for the optimal route and the resource allocation along it are combined. We present (few members of) a family of algorithms that enables the application planner to select the suitable trade-off between optimal route selection and the speed of establishing the route. Asymptotic complexity analysis is given for the presented algorithms.

Keywords
distributed shortest path, routing, bandwidth reservation, ATM, mutual exclusion, resource allocation.

1 INTRODUCTION

ATM networks are aimed to span the entire globe and to serve as a transport to virtually every electronic communications format from e-mail to phone calls to movies. Communication in ATM is connection oriented, i.e., before data can be transferred a connection should be established. To ensure the appropriate quality of service to a connection, network resources have to be reserved (De Prycker, 1993).

Connections are defined in two levels: *virtual path* (VP) and *virtual channel* (VC). A virtual path is,

*Currently on leave at SUN Microsystem Labs, Mountain View, CA 94043-1100

similar to a virtual circuit in traditional networks, a concatenation of physical links that serves as a transmission pipe for many connections. A virtual channel is a *unidirectional* concatenation of one or more virtual paths, and serve as the basic type of connection between two switching points. Changes in the VP topology are rare. VCs, on the other hand, are established on a per-need basis, which makes the VC topology dynamic.

Traditionally, the task of establishing a connection in connection-oriented networks is divided into two subtasks. First, a route is selected based on a topology database that is maintained in every node (Humblet and Soloway, 1988/9). The route is selected according to some optimization rule, such as, shortest path, minimal delay, etc. After the route is selected, the source tries to allocate resources (such as bandwidth, buffers, etc.) along the selected route (Boyer and Tranchier 1992; Cidon, Gopal, and Segall 1990; Hui 1988). In case of a failure, the source can try another route (that might be precomputed) usually one that is the most disjoint from the first one, or the source can stall and try again at a later time.

Much work was devoted to solve the distributed shortest path problem (e.g., Chapter 5 in (Bertsekas and Gallager, 1992)), which is closely related to this work. Assuming that topology changes are sufficiently infrequent (the quasi-static model) distributed algorithms are proposed to calculate the shortest path (Abram and Rhoads, 1978), or to maintain topology databases (Humblet and Soloway, 1988/9; Awerbuch, Cidon, and Kutten, 1990; Vishkin, 1983). These algorithms can only follow the slow changes in the VP topology and perhaps be updated about the link cost, but in the dynamic ATM environment these are incapable of giving an accurate picture about the availability of resources along the selected route. In networks with low reliability where changes are frequent, shortest path looses its meaning. In this case, circuit style store and forward routing algorithms have been suggested to reliably carry data packets between two endpoints while maintaining the packets in order and avoiding multiplications (Awerbuch, Kutten, and Peleg, 1991; Baratz, Gopal, and Segall, 1994), this approach is certainly inappropriate for the ATM environment. Given that a (shortest) path is selected, we are left with the problem of distributed resource allocation since many connections compete for networks resources. This problem did receive some attention, however, in the model that is generally used (Styer and Peterson, 1988; Chandy and Misra, 1984) there is no selection among several optional resources as in our case, and the solutions are limited to distributed mutual exclusion.

Many applications, e.g., queries to distributed databases or the WWW, are sensitive to the connection set-up time while others are more sensitive to the optimality of the route itself. For these applications, we suggest a family of fast connection-establishment algorithms that search for the route while reserving resources along it. The ability to reserve resources while the search is in progress eliminates retries caused by the failure to allocate resources along the optimal route. The family of algorithms offers a trade-off between the path optimality and the set-up speed.

The algorithms are based on a flooding algorithm that tries to reserve bandwidth along several possible routes. Generally, searching for a route between two nodes in the entire network is inefficient in terms of communication cost and set-up time. Thus, we assume that a topology-update algorithm works in the network which informs the nodes about the (slow) changes in the VP topology and about the cost of the VPs. When a node wishes to establish a connection (a VC), it searches for the optimal route in a subgraph of the VP graph that contains links that are "on the way" to the destination and that have a "reasonable" cost. We call this subgraph a *diroute*. The selection of the diroute can be made by the source node or, in a distributed manner by the nodes on the graph. To avoid reservation of resources in the entire diroute until the optimal route is chosen, the algorithms release resources from segments of the diroute as soon as it is learned that these segments are inferior to another segment where reservation was made. The implementation of this *early release* of bandwidth is possible since a node in the diroute that receives two or more reservation messages from different links, can locally select the optimal one, and can locally decide to release the bandwidth from the suboptimal incoming paths.

The suggested algorithms share the same operation principle but differ in the implementation. Three algorithms are presented: a fast algorithm which might be suboptimal in terms of route selection and which has an exponential message cost but works fast; a slow algorithm that selects the optimal route with the optimal message cost; and a logarithmic algorithm that improves the message complexity of the fast algorithm at the expense of finding a path with inferior cost compared to that found by the fast algorithms.

In the special case, where the diroute is a collection of disjoint paths, the routing part of the presented algorithms resembles Maxemchuck's dispersity routing (Maxemchuck, 1993) (the (k,1) strategy). But, the presented algorithms can efficiently handle routes that share either nodes or links in any combination. In fact, a diroute which is a collection of disjoint paths is the worst selection since it makes a poor use of the early release mechanism. Cidon, Rom, and Shavitt (1995) suggested deflection routing and reservation in the physical links of a VP. Their idea can be adapted to a higher level of deflection routing atop the VP structure, but it limits the structure of the diroute to contain only local alternatives to a preferred route. From the resource allocation timing point of view, the algorithm of (Cidon, Rom, and Shavitt, 1995) performs better since it first selects the preferred local route segment and only then allocates resources along the selected segment, while in this work first resources are allocated along all the candidate (local) routes and only after the selection of the preferred one the unnecessary resources are released. The ability of (Cidon, Rom, and Shavitt, 1995) to do the selection first stems from the restrictions that they imposed on the diroute structure. Bahk and El-Zarki (1994) have recently studied the problem of route selection for VCs with consideration of resource availability in VPs. Their solution requires the source node to know of the load in the VPs of the entire network; their algorithm has a separated path-selection phase and resource-allocation phase, and fails to include timing consideration.

In the following two sections we describe the network model and present the essence of the algorithms. The following section compares the asymptotic complexity of these algorithms. Resilience to failures is shortly discussed in the next section, and then some simulation results are presented.

2 NETWORK MODEL AND PROBLEM DEFINITION

We consider an ATM network where unidirectional VPs are pre-established. Changes in the VP topology are assumed rare due to the low probability of link failures. Unidirectional VCs are established on a per-need basis as a concatenation of one or more VPs. Thus, the network is modeled as a directed graph (the *VP-digraph*) where the VC switches are the nodes and the VPs the (directed) links. The establishment of a VC is done by VC switches (nodes) that exchange signaling messages along dedicated signaling VCs (control VCs). Note, that although VPs are unidirectional for data cells two signaling VCs, one in each direction, connect the two VP endpoints. For a VC to be operational it must be allocated network resources, such as bandwidth and buffer space.

A routing algorithm in an ATM network is one that finds a route in the VP-digraph between the source and destination nodes. The algorithms we present are executed only on a subgraph of VP-digraph, called a *diroute*. A diroute contains all the links (VPs) that might be used in the route and their associated nodes.

The diroute can be computed locally by the source node based on topology information. In this case, the algorithm messages can carry the diroute structure. Another possibility is to periodically compute an incoming directed acyclic subgraph for every destination in the network, and to store it in dirouting tables in the nodes. The use of these dirouting tables is similar to the use of shortest path routing tables (Bertsekas and Gallager, 1992, Chapter 5) only here the table for a destination can point to more than one link. A third approach is to calculate the diroute in a distributed manner based on the local topology database that the nodes maintain.

Each link in the diroute is associated with a cost, available resources, and processing/transmission delays. While the cost of all the links might be known before the algorithm starts its operation, the rapidly changing availability of resources and the processing delay are unknown in advance. In addition, several connections might compete for resources and even slight timing differences among the reservation messages will dictate which connections will reserve resources successfully and which will fail. Under the above constraints, we wish to find (and reserve) a path in the diroute between the source and the destination nodes that has a favorable cost and sufficient resources.

3 THE ALGORITHM DESCRIPTION

The suggested algorithms share the same operation principle but differ in the implementation. All algorithms guarantee to find a feasible path within the diroute if one exists. Three algorithms are presented:

A fast algorithm where the reservation message travels to the destination as fast as possible, but the selected route might be suboptimal. This algorithm provides a fast negative response if no feasible path exists. The message complexity of this algorithm is exponential.

A slow algorithm where the reservation message travels to the destination at the speed of the slowest route, but the selected route is optimal and the message complexity is linear in the number of diroute links.

A logarithmic algorithm that improves the message complexity of the fast algorithm while maintaining its speed of operation. The selected route, however, might be suboptimal.

A pseudo-code formal description of the algorithms can be found in (Rom and Shavitt, 1996).

3.1 The Fast Algorithm

The main thrust of the algorithm is to reach the destination with a feasible path as fast as possible (using a flooding approach), updating and correcting the path if better alternatives are found in time, and releasing superfluous reserved bandwidth as soon as it is identified.

We next describe the algorithm's operation. Since the algorithm is executed on the diroute, the term "all outgoing links" refers only to links of the diroute. A link is called *eligible* if it has enough available resources to potentially be part of the requested route.

The source node starts the algorithm by reserving the needed resources in all eligible outgoing links and then sending a *Request* message along these links. The *Request* message carries a specification of the resources needed for the connection and the total cost of the path from the source to the current node.

A node that receives the first *Request* message on some incoming link l reserves the needed resources in all the eligible outgoing links and forwards the *Request* message on all these links. The node marks link l as the link to the father; this marking will be used to forward the positive acknowledgment message, *Accept*, or the negative acknowledgment message, *Reject*, back to the source. This establishes the fast flooding towards the destination node. A node that receives another *Request* message with a better path-cost forwards the message on all the eligible outgoing links, changes the father marking, and sends a *Reject* message to the old father. This action replaces the old route with a better one and causes the *early release* of reserved bandwidth on the previously marked link which is now known to be suboptimal. Receiving a *Request* message with a worse path-cost then the current one is responded to with *Reject*. Thus, with the early release mechanism a nodes always keeps, at most, a single incoming path with resource reservation.

A node sends a *Reject* messages if it has no outgoing link with reserved resources. This situation can happen if the node receives a *Request* message and finds no eligible link, or if the node receives *Reject* messages for all the *Request* messages it sent. If the source receives *Reject* messages from all its outgoing links the reservation algorithm failed and the connection can not be established.

A node that receives a *Request* message with a better path cost forwards this message on *all* its eligible outgoing links, even those for on which a *Reject* message was previously received. The reason is that the better path that was just discovered my make previously rejected paths eligible again. This means that new *Request* message and old *Rejects* my be mixed and cause confusion since the node cannot know if the *Reject* message is a reply to the first or the second *Request*. To avoid this ambiguity, a *Reject* message carries the value of the optimal path cost that will be answered with a *Reject* by the sending node. The number that is received by the last *Reject* on every link is stored and is used to save the transmission of *Request* messages with path-cost values that are known to be rejected.

The destination node answers with *Accept* to the first *Request* it receives, and with *Reject* to all the other subsequent *Request* messages. This ensures the fastest establishment of a connection with the information known thus far. The *Accept* message is immediately forwarded by every node to its father until it reaches the source, at which time it can start transmitting data. A node that forwarded an *Accept* message will respond with *Reject*(0) to every subsequent *Request* it receives to signal the termination of the route establishment process.

When the source receives the *Accept* message, it knows that a reserved path exists to the destination. However, the source does not know when all the nodes in the diroute terminate their part of the algorithm (which is irrelevant to the task of path establishment) and more important, when the algorithm (message) terminates, i.e., when no more messages belonging to the algorithm exist in the network, a node that is not part of the selected path can never detect this situation. The problem with inability to identify termination is that nodes cannot release the variables allocated to the algorithm (note, that this has nothing to do with the resources required from the network which are guaranteed to be released). Thus, we introduce a termination mechanism that is started by the source node when it receives *Accept*. The source starts a flooding algorithm (such as PIF (Segall, 1983)) with the *Terminate* message, i.e., it sends *Terminate* on all its outgoing links. A node that receives *Terminate* for the first time marks this link as the preferred link and sends *Terminate* on all its outgoing and incoming links but the preferred link. In addition, if the node has some network resources reserved on any of its outgoing links, it releases these resources. This release would have been done by the algorithm without the *Terminate* message, but at a later time. After the node receives the *Terminate* message, it disregards any *Reject* or *Request* messages it receives. When the node receives *Terminate* from all its outgoing and incoming links, it sends one on the preferred link and releases all the variables of the algorithms. When the source node receives *Terminate* from all its outgoing links, it knows that the algorithm terminated and that all the nodes know this fact.

A slow link in the diroute halts the early bandwidth release process on all the paths that lead to it. This disadvantage of the algorithm can be counterbalanced by the selection of the links that form the diroute.

3.2 The Slow Algorithm

The fast algorithm attempts to establish a connection fairly fast at the cost of path optimality, i.e., the established path may not be optimal. For those cases where path optimality is important the fast algorithm

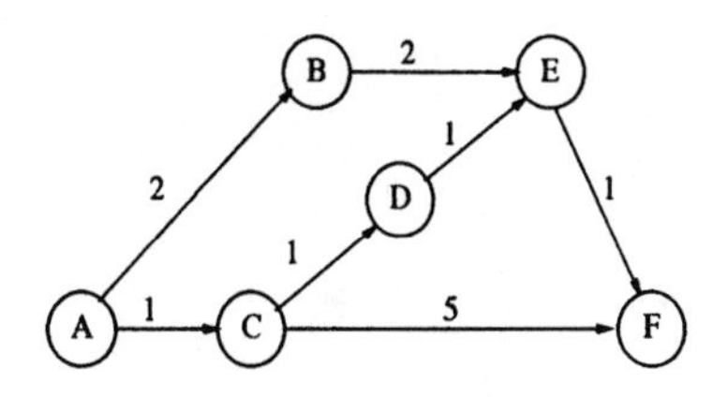

Figure 1 A diroute example

will therefore not do and we offer here a slower algorithm that guarantees path optimality at the cost of connection establishment latency.

The main difference between the two algorithms is the decision when to send the *Request* message. In the fast algorithm, a node sends a *Request* immediately after it receives the first *Request*, and sends additional *Requests* when it learns about better paths. In the slow algorithm, a node sends a *Request* message only after it knows the optimal path from the source to itself, i.e., only after it receives *Request* messages from all the incoming links. The early release mechanism is used here, in the same way as with the other two algorithms. When a node receives an additional *Request* message it sends a *Reject* message along the inferior incoming path.

It is therefore clear that the algorithm progresses according to the slowest path, which is typical to this type of flooding algorithm (Segall, 1983). Furthermore, depending on implementation, it might be necessary to send messages even on non-eligible links to assure proper progress of the algorithm.

The algorithm as described above may cause either deadlocks or livelocks unless the diroute is acyclic. Heuristics are needed to transform the diroute, which is a general directed graph, to a directed acyclic graph without deleting any link that belongs to the minimal feasible path.

A node that receives messages from all its outgoing links can release all the algorithms variables as it will not receive any more messages. As opposed to the fast algorithm, no *Terminate* message is needed here.

3.3 The Logarithmic Algorithm

The fast and slow algorithms present two options for path establishment that trade off path optimality and establishment delay. The two algorithms also differ in the number of messages generated; we shall show in the next subsection that the fast algorithm message complexity is exponential in the number of nodes in the diroute, even for acyclic diroutes. We wish to devise an algorithm that compares to the fast algorithm, enables us to slow-down if better paths are needed, and has an acceptable message complexity.

We base the new algorithm on the fast algorithm, but do not forward every *Request* message that improves the path. Instead, the node maintains a counter that is set to 1 when the first *Request* message arrives, and is increased by one for every arrival of a *Request* message with a better path cost (actually, we can count all the *Request* messages and still get an acceptable complexity). *Request* messages are sent only if the arriving *Request* advances the counter to a value that is a natural power of two, i.e., only the first, the second, the forth, the eighth, etc. arriving *Request* messages are forwarded. Essentially, the number of messages sent by a node is logarithmic compared to the number of messages that would be sent by the fast algorithm in the same network. Note that the early release of bandwidth is performed for all the arriving *Request* messages. For the message complexity analysis, see (section 4).

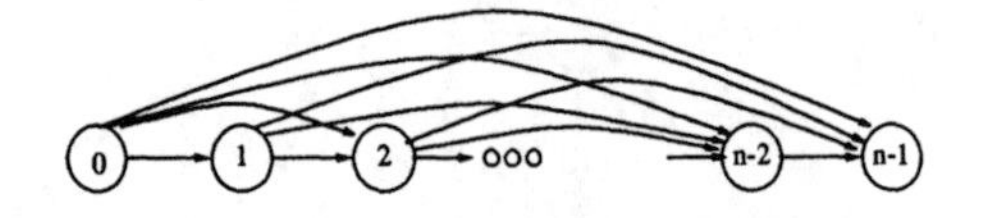

Figure 2 The worst case acyclic diroute

4 ASYMPTOTIC ANALYSIS

In this section we analyze the message complexity of the presented algorithms. We start with the trivial analysis of the slow algorithm, move to the more complex analysis of the fast algorithm and finally present bounds on the complexity of the logarithmic algorithm. In all algorithms, the number of *Reject* and *Accept* messages is at most equal to the number of *Request* messages. Thus, throughout the analysis we count only *Request* messages. $|E|$ and $|V|$ denote the number of links and the number of nodes in the diroute, respectively.

In the **slow algorithm**, exactly one *Request* is sent on every link. This translates to a message complexity of $\Theta(|E|)$ which is optimal for this problem.

Intuitively, the message complexity of the **fast algorithm** is exponential since, in the worst case, a message can be sent for every path in the diroute. We'll show now that this is true even for the case where the diroute is acyclic.

Consider the acyclic diroute of figure 2 where node i has a link to node j if $i < j$. Node 1, receives a single *Request* message. Node i receives all the messages received by node $i-1$ directly and, in addition, the messages sent by node $i-1$ to node i. In the worst case, node $i-1$ forwards all the messages it receives to node i. Thus the number of messages received by node i is twice the number of messages received by node $i-1$, and the total amount of messages that are sent by the fast algorithm is

$$\sum_{i=1}^{|V|-1} 2^{i-1} = O(2^{|V|}) \tag{1}$$

To analyze the **logarithmic algorithm**, we first observe that if node i receives χ_i messages it transmits $\lfloor \log_2 \chi_i \rfloor + 1$ messages on every outgoing link. Let j be the node that transmit the maximum number of messages during an execution of the algorithm. Using the above observation, we can write (denoting $n = |V|$)

$$\chi_j \leq \lfloor \log_2 \sum_{i \neq j} \chi_i \rfloor + 1 \leq \lfloor \log_2 (n-1)\chi_j \rfloor + 1 \leq \log_2 (n-1)\chi_j + 1 \tag{2}$$

Now, assume that the number of messages that are sent by node j is proportional to n^α for some $\alpha > 0$ (i.e., $\chi_j = O(n^\alpha)$). Substituting in (2) we get

$$n^\alpha \leq \log_2(n-1) \cdot n^\alpha + 1 \leq (2+\alpha)\log n \tag{3}$$

But, n^α is asymptotically greater than $(2+\alpha)\log n$ for every positive α; which implies that the number of messages that are transmitted by node j on every link is clearly bounded by $o(n^\alpha)$. The message complexity of the algorithm is, thus, in $o(|V|^{2+\alpha})$ for every positive α. Since our algorithm sends at least one message on every link $\Omega(|E|) = \Omega(|V|^2)$ is a lower bound for the message complexity. This leaves us with a gap that remains an open problem.

For the special case where the in-degree of all nodes in bounded by the constant D, (2) takes the form

$$\chi_j \leq \lfloor \log_2 D\chi_j \rfloor + 1 \leq \log_2 D\chi_j + 1 \tag{4}$$

Now, assume that the number of messages that are sent by node j is $\log^{(m)} n$ for some natural m ($\log^{(m)} n$ is the logarithm function applied m times in succession, starting with argument n). Substituting in (4) we get

$$\log^{(m)} n \leq \log_2 D \log^{(m)} n + 1 = \log^{(m+1)} n + \log_2 2D \tag{5}$$

But, $\log^{(m)} n$ is asymptotically greater than $\log^{(m+1)} n + \log_2 2D$ for every natural m, which implies that the number of messages that are transmitted by node j on every link is $o(\log^{(m)} n)$. Thus, for the case where the in-degree is bounded by a constant, the message complexity of the algorithm is in $o(|V| \log^{(m)} |V|)$ for every natural m. Since the number of links, for this case, is linear in $|V|$ the lower bound for the number of messages is $\Omega(|E|) = \Omega(|V|)$. Thus, a gap remains for this case too, although it is narrower than the gap for the general case.

For acyclic diroutes, we can derive a tighter upper bound by counting the number of messages that are sent for the worst case diroute shown in figure 2. Let χ_i be the number of messages received by node i. Since node i receives all the messages received by node $i-1$ plus the ones transmitted by node $i-1$ we can write

$$\chi_i = \chi_{i-1} + \log_2 \chi_i + 1 \tag{6}$$

It can be shown by induction that $\chi_n = O(n \log n)$.

5 HANDLING VP (LINK) FAILURES

A failure of a link is a situation where no messages can go through the link for a period of time, and both ends of the link can detect this situations (Baratz and Segall, 1988). In (Rom and Shavitt, 1996), we show how the presented algorithms can be modified to handle link failures. Due to space limitations, we shall only give a concise discription of the types of solutions that can be found in (Rom and Shavitt, 1996).

The high reliability of ATM networks may tempt us to handle the rare VP failures with a simple reset algorithm that terminates the original algorithm operation. Such algorithms where studied in the past (see (Afek, Awerbuch, and Gafni, 1987) and the references therein) and their application to all the algorithms presented in (section 3) is discussed in (Rom and Shavitt, 1996). However, since diroutes contain more than a single path between source and destination, it is desired that a failure in one (or more) of the links in the diroute does not automatically abort a progressing connection-establishment process. In (Rom and Shavitt, 1996) we describe mechanisms that are added to the three algorithms to facilitate link-failure handling.

The approach taken in (Rom and Shavitt, 1996) for link-failure handling is to conserve the operation principles and the message complexity of the algorithms. It makes no sense to try and 'save at any cost' the reservation process when a failure occurs; adding complicated code, not only makes the algorithm harder to implement but might increase dramatically the algorithm's complexity (note that for some failures, e.g., those that dissect the diroute, nothing can be done to save the reservation process).

It is important to note that once the connection is established a link failure cannot be viewed as part of the reservation algorithm, but should be treated by a general connection take-down algorithm (Rom and Shavitt, 1995).

6 SIMULATION RESULTS

In this section, we report results from simulations that compare the fast algorithm with two other algorithms. The first algorithm tries to reserve bandwidth along a preselected shortest path, i.e., first the source node selects a shortest route, then it tries to reserve bandwidth along this route, and upon a success the burst is transmitted. The second algorithm also routes along the shortest route, but if a link is blocked and another link that belongs to a route with the same length is available the reservation is deflected to this route. Figure 3 depicts the network topology that consists of two sites that are connected by a long link (its delay is five times more than the delay of the other links). The numbers below the links in the figure are the link capacities. Due to symmetry, we only simulated traffic from sources in the left site to destinations in the right site. No in-site traffic was considered to shorten the simulation run time.

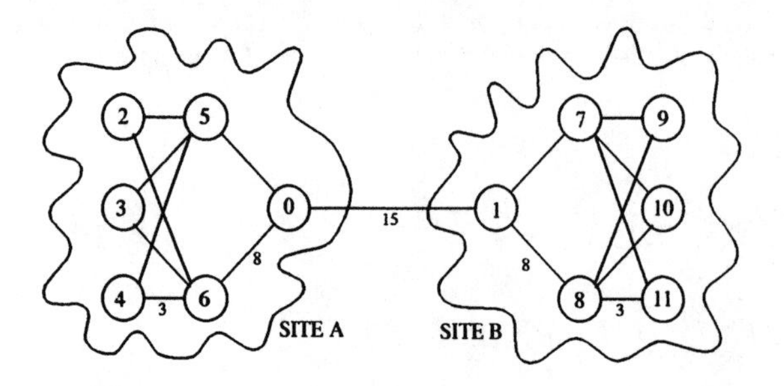

Figure 3 The network topology

Bursts are generated according to a Poisson process. For every burst, the source is selected uniformly among nodes 2, 3, and 4 and the destination among nodes 9, 10, and 11. Burst duration is exponentially distributed with mean 100. The link cost for the fast algorithm is set to 1. If the reservation process fails (for either algorithm) a new retry time is selected according to an exponential random variable with mean 20. If a burst cannot be transmitted after 100 time units it is discarded.

Two parameters were checked: the percentage of bursts that are successfully transmitted (figure 4), and this percentage when the threshold time for discarding a burst is changed (figure 5 depicts the probability for a burst to be delayed more than the X-axis value). Figure 4 shows that the probability of a burst to be transmitted are higher when the fast algorithm is used than this probability when the other two algorithms are used. Figure 5 shows that this gap in the performance of the algorithms remains if the threshold value when a burst is discarded is decreased. In other words, for every time limit on the reservation duration process the fast algorithm succeeds to reserve bandwidth for a higher portion of the generated bursts.

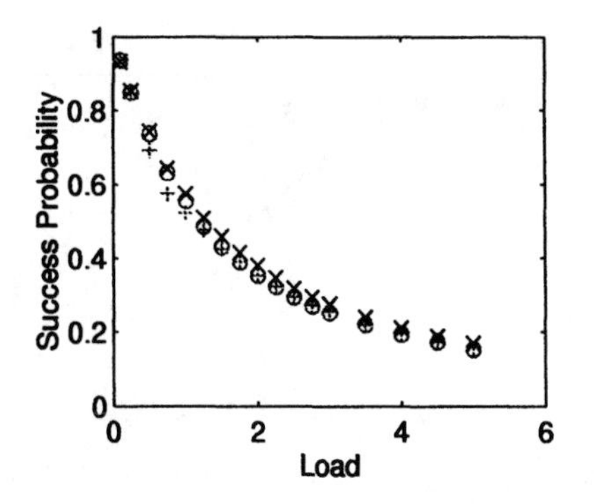

Figure 4 A comparison of the reservation success probability of the fast algorithm (Xs) with algorithms that tries to reserve only along a single route (Os and +s).

7 CONCLUDING REMARKS

We presented a family of algorithms that search for a shortest path and reserve resources along the path while the search is in progress. In this paper we examined the algorithms by the pace of their progress and their message complexity. Observing additional aspects of the problem is currently under work and will lead to development of more algorithms, so the future implementor will be presented with a menagerie of algorithms to chose from.

In the fast and the logarithmic algorithms we may delay the selection of the best route by the destination node. This will allow for the possible generation of better routes (the source node can even indicate in the message the amount of such a delay). The message can also collect estimates on the delays it experiences along its journey which can be used to calculate a time-out that is based on the round-trip delay and the delay threshold limit the source application can tolerate. Another variation is to send a cost threshold value in the *Request* message. The destination node waits until the first *Request* with a path cost arrives and than sends an *Accept*. Yet another twist, is to wait for a *Request* from all the incoming links and then to send the first *Accept*. Note that in contrast to the slow algorithm, here an optimal selection is not guaranteed.

Given the topology database, the selection of the diroute has a substantial influence on the performance of the algorithms. The different algorithms have different and sometimes contradictory requirements from the diroute structure. For example, the addition of a link with low cost and large delay can slow down the slow algorithm since it advances at the speed of the slowest path. However, adding this link to the diroute when the fast algorithm is employed can result is a selection of a path with a better cost; in case this added link is not part of the chosen path it can not delay the algorithm. Developing algorithms or heuristics for selecting links to a diroute remains open.

We presented our algorithms to be used for the establishment of VCs in an ATM network with a given VP topology. However, these algorithms can be used for fast path selection and establishment in any connection-oriented network where reservations of resources are needed.

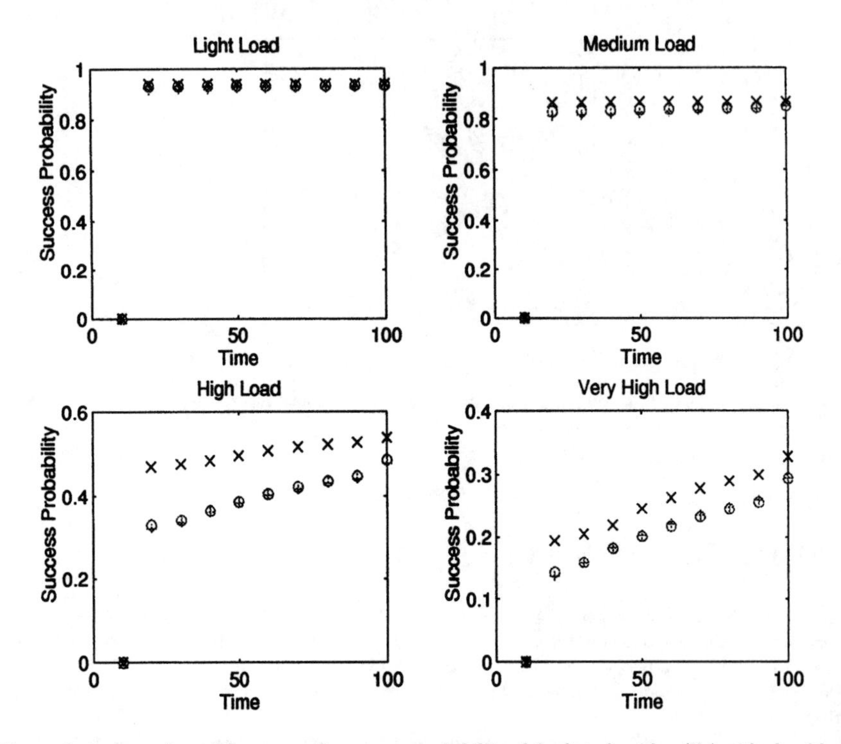

Figure 5 A comparison of the reservation success probability of the fast algorithm (Xs) with algorithms that tries to reserve only along a single route (Os and +s) as a function of the threshold time for selected loads.

REFERENCES

Afek Y., Awerbuch B., and Gafni E. (1987) Applying static network protocols to dynamic networks. *FOCS'87*, pp. 358–370.

Awerbuch B., Cidon I., and Kutten S. (1990) Communication-optimal maintenance of replicated information. *FOCS'90*, pp. 492–502.

Awerbuch B., Kutten K., and Peleg D. (1991) Efficient deadlock-free routing. *ACM PODC'91*, pp. 177–188.

Abram J.M. and Rhoads I.B. (1978) A decentralized shortest path algorithm. *16th Annual Allerton Conference on communication, control, and computing*, pp. 271–7.

Bahk S. and El-Zarki M. (1994) Congestion control based dynamic routing in ATM networks. *Computer Communication*, **17**, 826–835.

Bertsekas D. and Gallager R. (1992) *Data Networks*. Prentice Hall, second edition.

Baratz A., Gopal I., and Segall A. (1994) Fault tolerant queries in computer networks. *IEEE Transactions on Communications*, **COM-42**, 100–9.

Baratz A.E. and Segall A. (1988) Reliable link initialization procedures. *IEEE Transactions on Communications*, **COM-36**, 144–152.

Boyer P.E. and Tranchier D.P. (1992) A reservation principle with applications to the ATM traffic control. *Computer Networks and ISDN Systems*, **24**, 321–334.

Cidon I., Gopal I., and Segall A. (1990) Fast connection establishment in high speed networks. *ACM SIGCOM'90*, pp. 287–296.

Chandy K.M. and Misra J. (1984) The drinking philosophers problem. *ACM Transactions on Programming Languages and Systems*,**6**, 632–646.

Cidon I., Rom R., and Shavitt Y. (1995) Fast bypass algorithms for high-speed networks. *INFOCOM'95*, pp. 1214–21.

Humblet P.A. and Soloway S.R. (1988/9) Topology broadcast algorithms. *Computer Networks and ISDN Systems*, **16**, 179–186.

Hui J.Y. (1988) Resource allocation for broadband networks. *IEEE Journal on Selected Areas in Communications*, **6**, 1598–1608.

Maxemchuk N.F. (1993) Dispersity routing on ATM network. *INFOCOM'93*, pp. 347–357.

De Prycker M. (1993) *Asynchronous Transfer Mode solutions for Broadband ISDN*. Ellis Horwood Limited, second edition.

Rom R. and Shavitt Y. (1995) Efficient bandwidth release after failures in ATM networks. Technical Report CC PUB #116, Technion — Israel Institute of Technology.

Rom R. and Shavitt Y. (1996) A combined fast-routing and bandwidth-reservation algorithm for ATM networks. Technical Report CC PUB #135, Technion — Israel Institute of Technology.

Segall A. (1983) Distributed network protocols. *IEEE Transactions on Information Theory*, **IT-29**, 23–35.

Styer E. and Peterson G.L. (1988) Improved algorithms for distributed resource allocation. *ACM PODC'88*, pp. 105–116.

Vishkin U. (1983) A distributed orientation algorithm. *IEEE Transactions on Information Theory*, **IT-29**, 624–9.

8

A Fast Routing and Bandwidth Management Method in ATM Networks

Mitsuaki Kakemizu, Masahiro Taka
Ultra-high Speed Network and Computer Technology Laboratories
2-6 Toranomon 5-Chome, Minato-ku Tokyo 105, Japan
Telephone: +81-3-3578-9361 Fax:+81-3-3578-8184
E-mail:kake@magical.egg.or.jp

Abstract

Current supercomputers and workstations operate at 100 GFLOPS-TFLOPS and at 10-100 MFLOPS, respectively, and these processor speeds will increase in the future. This has led to an increasing demand for the real-time transfer of scientific high-dimensional image data, and ATM networks would therefore have to provide user-network interfaces with bandwidths in the gigabit range from now on. Because computer communication usually produces unpredictable burst traffic, it is difficult for the bandwidth reservation method based on statistical multiplexing to use bandwidths efficiently. If the connectionless server approach were adopted, these servers would cause throughput bottlenecks in gigabit applications. As a result, this paper proposes a fast routing and bandwidth management method that can effectively transport gigabit traffic. It also undertakes performance evaluation by comparing the proposed method with the P-NNI method that the ATM Forum proposes.

Keywords

Routing, bandwidth management, virutual channel connection, hierarchy, ATM

1 BACKGROUND

The Asynchronous Transfer Mode (ITU-T, 1990) is expected to be the basic technology providing multimedia services in a B-ISDN. Its features include bit-rate flexibility and statistical multiplexing capability, and ATM networks are also expected to be used in transferring scientific image data generated by supercomputers. Such an application is characterized by high-speed burst traffic, however, and when the amount of burst traffic exceeds the traffic-handling capability of a network using statistical multiplexing, degradation in service quality such as cell loss occurs. To prevent network congestion, statistical multiplexing requires that each traffic source notifies all QOS parameters such as peak cell rate, average cell rate, and cell loss ratio in the call set-up phase. Only a few computer communication applications, however, are able to accurately specify average cell rate in the call set-up phase. This is because the average cell rate changes dynamically and cannot be predicted. As a result, it may not be appropriate to introduce statistical multiplexing for handling such high-speed burst traffic in ATM networks.

ITU Recommendation I.364(ITU-T, 1992) proposes that ATM networks make use of connectionless servers (CLSs) to support connectionless services in the megabit range. These servers would be located at specific ATM nodes, and the source data terminal equipment (DTE) accommodated in one of the nodes would send cells to a CLS by default routing. The CLS would then route the cells to the other CLS at the destination ATM node. Consequently, the CLS would cause a bottleneck in gigabit applications because the ATM connection must always be terminated at each CLS and a Connectionless Network Access Protocol or Connectionless Network Interface Protocol must be processed for the first cell belonging to the protocol data unit of the ATM adaptation layer (AAL)-3/4 in each CLS.

2 BASIC CONCEPT

One of the most effective methods is FRP (Fast Reservation Protocol) (Boyer, 1992), which selects a route in the call set-up phase and reserves bandwidths along the route on a burst-by-burst basis in the call. This method, however, would increase burst blocking probability especially in gigabit applications and would result in long connection delay for back-off timer delay because network states during each burst reservation phase are different from states in the call set-up phase due to the dynamic behavior of available bandwidth resources in networks.

Another effective method is the P-NNI(Private NNI) Draft Specification(ATM Forum, 1995), which calculates the optimal route on a burst-by-burst basis and reserves bandwidth on the optimal route on an in-series and link-by-link basis. However, this method would also result in long connection delay because it requires the optimal route to be calculated on a burst-by-burst basis.

We therefore used an approach that allows the optimal route to be rapidly selected and reserves bandwidth along the route on a burst-by-burst basis. We propose a routing and bandwidth management method (Kakemizu, 1995) that efficiently supports gigabit computer communications applications such as the real-time transfer of scientific high-dimensional image data and medical image data. The main features of this method are as follows:

l Separation between the VCC establishment procedure and the management of routing tables.
As this method periodically calculates the optimal route between necessary nodes without being synchronized with the VCC establishment procedure, route selection can be performed rapidly when establishing a VCC because the only procedure required is to refer to the predetermined route.

l Division of ATM networks into several subnetworks and the concept of hierarchical subnetworks.
This method not only allows ATM networks to be divided into several subnetworks, but also allows larger subnetworks to be divided into smaller subnetworks if necessary. VCC establishment processing is performed based on the scope of each subnetwork independent of other subnetworks, and the management of routing tables is also performed independent of all lower subnetworks and other subnetworks in different upper subnetworks. The P-NNI method also adopts the division and hierarchy of networks. However, the extent of routing table in each subnetwork using this method is smaller than that in each ATM node in each subnetwork called the peer group in the P-NNI method. This is because the P-NNI method requires each ATM node in networks to manage not only the routing table in the lowest peer group that it belongs to but also the routing tables in all upper peer groups that it belongs to. This is even though all the peer groups and links in all upper peer groups that it belongs to are logically abstracted. This

distributed processing in the proposed method would therefore make it more suitable for large-scale networks than the P-NNI method.

Parallel execution of route selection and bandwidth reservation procedures in each lower subnetwork in compliance with request from the common upper subnetwork.

As shown in figure 1, although each VCC establishment is performed in-series from upper subnetworks toward lower subnetworks between the source and destination DTE, route selection and bandwidth reservation procedures for VCC establishment in all subnetworks in a common upper subnetwork are executed in parallel in compliance with requests from the common upper subnetwork. As a result, this method allows the establishment period of all VCs in each subnetwork to be almost constant independent of the sum of subnetworks/nodes in the subnetwork while conventional methods (including P-NNI method and FRP method) result in a VCC establishment time that is proportional to the sum of the nodes between the source DTE and destination DTE. This is because these conventional methods execute processing on an in-series and link-by-link basis.

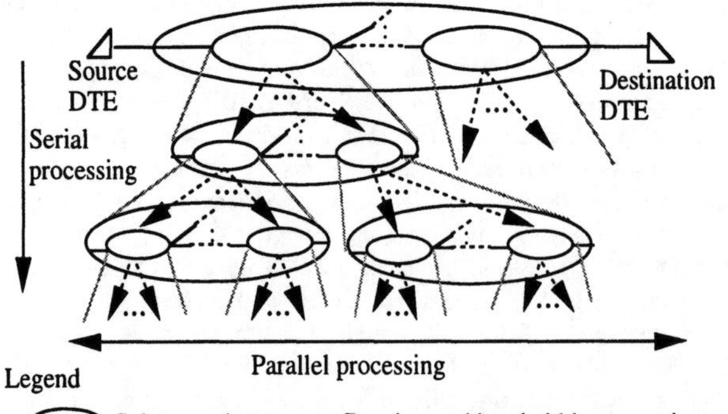

Figure 1 Conceptual procedure for VCC establishment

Simplification of signalling procedures by sending RM-cells through dedicated PVCs.

Permanent Virtual Channels (PVCs) have been established in advance between all DTEs that may establish VCCs based on the proposed method and the subscriber node accommodating these DTEs, and between all adjacent nodes. Because these PVCs are only used to transfer resource management cells (RM-cells) for VCCs establishment based on this method, the proposed method can prevent the protocol overheads of AAL-5 and signalling protocols such as Q.2931(ITU-T, 1993), UNI Signalling 4.0(ATM Forum, 1995) and Q.2764(ITU-T, 1995). This method can also simplify signalling procedures.

3 RAPID METHOD FOR ESTABLISHING VIRTUAL CHANNEL CONNECTIONS

3.1 Network model

Figure 2 shows the network model on which our proposed method is based. Networks can be divided into several virtual subnetworks, and each subnetwork can be divided into several smaller virtual subnetworks. Thus, there are hierarchical relationships between subnetworks, and the network model is recursive.

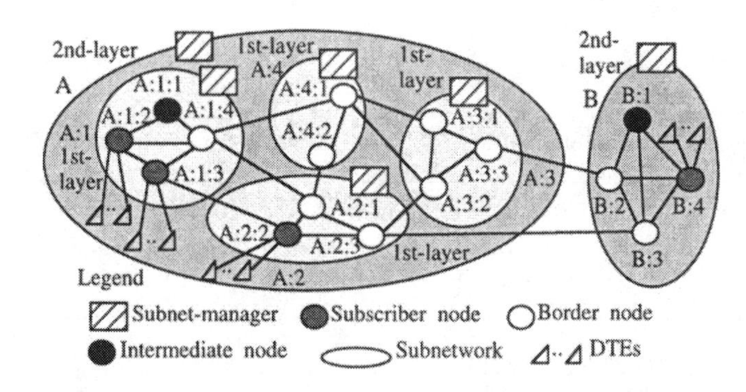

Figure 2 Network model.

One subnet-manager, which is responsible for routing and requesting VCC establishment, is assigned in each subnetwork. Each subnet-manager may be implemented in any physical node in the subnetwork that the subnet-manager belongs to or may be implemented in a dedicated node independent of other physical nodes in the subnetwork. PVCs for carrying routing information and signalling messages are established in advance between each node and all subnet-managers receive routing information from this node, between upper and lower subnet-managers, and between each subscriber node and all subnet-managers accommodating it. PVCs for exchanging routing information are also established in advance between subnet-managers that belong to subnetworks in the common upper subnetwork. VCC establishment processing is performed by each subnet-manager independent of the subnet-managers in other subnetworks, and the management of routing tables is also performed by each subnet-manager independent of the subnet-managers in other subnetworks in the different upper subnetwork. As noted in Section 2, such distributed processing makes this method suitable for large-scale networks.

3.2 Management of routing tables

Routing tables consist of an inter-subnetwork routing table and an intra-subnetwork routing table. The former is managed by all subnet-managers unless there is only one subnetwork in the network, and the latter is managed only by subnet-managers that belong to the lowest subnetworks. The inter-subnetwork routing table shown in Table 1-(a) is used to select an optimal route between any two subnetworks extracted from all subscriber subnetworks and all border subnetworks in the common upper subnetwork. The subscriber subnetworks shown as 'A', 'A:1', 'A:2' or 'B' in Figure 2 are subnetworks accommodating at least one subscriber node that itself accommodates at least one DTE. The border subnetwork shown as 'A:3' in Figure 2 is a subnetwork connected to other subnetworks contained in a different upper subnetwork and is not a subscriber subnetwork. The intra-subnetwork routing table shown in Table 1-(b) is used to select an optimal route between any two nodes extracted from all subscriber nodes and all border nodes in the common lowest subnetwork. The border node shown in Figure 2 stands for a node connected to other nodes contained in a different lowest subnetwork and is not a subscriber node. The optimal physical link in Table 1 is a link with the maximum available bandwidth of all physical links between subnetworks. Each bottleneck bandwidth in Table 1 is the minimum available bandwidth for all links on the route between the source subnetwork and the destination subnetwork or between the source node and the destination node. The optimal route shown in this paper is defined as the route with the least bottleneck bandwidth consisting of the fewest hops.

Table 1 Examples of routing tables

(a) Inter-routing table for each subnetwork A:1-4 in subnetwork A in Figure 2.

Source/Destination subnetwork id	Number of hops on route	Bottleneck bandwidth (Gbps)	Subnetwork id on route	Optimal phisical link
1-2	1	2.5	1-2	1:3-2:2
	2	2	1-4-2	1:4-4:1, 4:2-2:1
	3	3	1-4-3-2	1:4-4:1, 4:1-3:2, 3:2-2:3
1-3	...	...	...	...
2-3	...	...	...	...

(b) Intra-routing table for subnetwork A:1 in Figure 2.

Source/Destination node id	Number of hops on route	Bottleneck bandwidth (Gbps)	Node id on route
2-3	1	1.5	
	2	2	2-4-3
	3	2.3	2-1-4-3
2-4	...	...	...
3-4	...	...	...

Each subnet-manager is periodically notified of the available bandwidth for all physical links between the subnetwork that it belongs to and other subnetworks directly connected to it from all nodes that manage their links. Each subnet-manager also periodically exchanges this information with other subnet-managers in the common upper subnetwork. After this, each subnet-manager calculates the optimal route by arranging all routes between both source/destination subnetworks as mentioned above, in ordering fewer hops and a broader bandwidth to create less bottlenecking. Similarly, after each individual subnet-manager belonging to the lowest subnetwork has received information on the available bandwidth for each physical link between all pairs of nodes in the subnetwork from all nodes that manage their links, it calculates the optimal route by arranging all routes between both source/destination nodes as mentioned above, in ordering fewer hops and a broader bandwidth to create less bottlenecking. Because these arrangements are executed periodically, without being synchronized with the VCC establishment procedure, route selection when establishing a VCC can be performed rapidly as the only procedure required is a reference to the most desirable predetermined route of the routes arranged last. There is also another option: that each node does not send the available bandwidth periodically but only when the difference between it and the last available bandwidth sent exceeds a certain threshold.

3.3 VCC establishment procedures

As shown in Figure 3, a VCC is established as follows:

I. The source node:

a) receives VCC_req and reserves the link bandwidth between the source node and source DTE.

b) sends VCC_req_ack to the source DTE.

c) analyzes both the network prefix of the source DTE address and the network prefix of the destination DTE address, and sends VCC_req to one of the *n*th-layer SMs called the *n*th-layer responsible SM. Address format in the proposed method is identical to the one in the ATM Forum (ATM Forum, 1995), and includes a network prefix that represents the configuration for each DTE or node or subnetwork in a hierarchical structure for networks, and this network prefix assists the subnet-manager to select the optimal route on a burst-by-burst basis. This SM is assigned by the subnetwork accommodating the source node in the lowermost common layer where both the subnetwork accommodating source node and the subnetwork accommodating destination node coexist.

2. The nth-layer responsible SM:
   a) receives VCC_req and selects the optimal route between the subnetwork that the nth-layer responsible SM belongs to and the other subnetwork that the message transmitting SM shown in Figure 3 belongs to by referring to the inter-routing table.
   b) sends VCC_req to all nodes as shown in Figure 3 and the destination node through the message transmitting SM on the optimal route. Then, each of these nodes in-parallel independently reserves bandwidth on the link designated by the VCC_req.
   c) sends Conn_Ack to all these nodes except the destination node and sends VCC_req to some (n-1)th-layer SMs, termed (n-1)th-layer responsible SMs, after receiving Conn from all nodes. Each of these (n-1)th-layer responsible SMs is separately assigned by the (n-1)th-layer subnetwork in each nth-layer subnetwork on the optimal route. Then, each of these nodes except the destination node in-parallel independently establishes the virtual channel.
   d) sends Conn_Ack to all the (n-1)th-layer responsible SMs after receiving Conn from them.
3. Until each lowest-layer SM receives VCC_req from the same layer SM, each (n-1)th-layer responsible SM in-parallel and recursively:
   a) receives VCC_req and selects the optimal route between the subnetworks designated by the VCC_req by referring to the inter-routing table.
   b) sends VCC_req to all the nodes as shown in Figure 3 on the optimal route. Then, each of these nodes in-parallel independently reserves bandwidth on the link designated by the VCC_req.
   c) sends Conn_Ack to all these nodes and sends VCC_req to (n-2)th layer SMs that are (n-2)th-layer responsible SMs, after receiving Conn from all nodes. Each of these (n-2)th-layer responsible SMs is separately assigned by the (n-2)th-layer subnetwork in each (n-1)th-layer subnetwork on the optimal route. Then, each of these nodes in-parallel independently establishes the virtual channel. If there is no (n-2)th-layer SM, the (n-1)th-layer SM sends VCC_req to all same-layer SMs on the optimal route.
   d) sends Conn_Ack to all the (n-2)th-layer responsible SMs after receiving Conn from them.
   e) sends Conn to the nth-layer responsible SM.
4. After each lowest-layer SM receives VCC_req from the same-layer SM, each lowest-layer SM, in-parallel:
   a) selects the optimal route between the nodes designated by the VCC_req by referring to the intra-routing table.
   b) executes the same procedure as shown in 3.b).
   c) sends Conn_Ack to all these nodes after receiving Conn from these nodes, and then each of these nodes in-parallel independently establishes the virtual channel.
   d) sends Conn to the SM having received VCC_req.
5. The nth-layer responsible SM sends Conn to the source DTE through the source node and sends Conn_Ack to the destination node through the message transmitting SM.
6. The source node receives Conn_Ack from the source DTE and establishes the virtual channel.
7. The destination node establishes the virtual channel and sends Conn_Ack to the destination DTE.

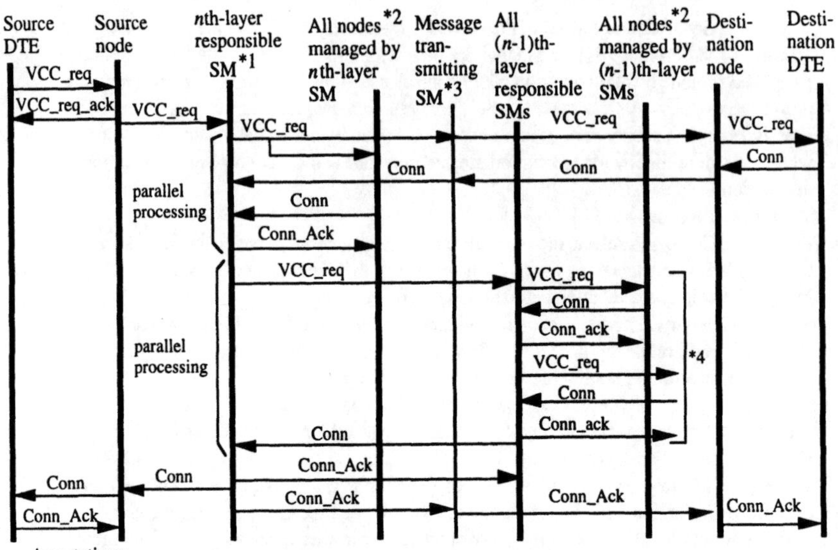

Annotations

*1: 'SM' is short for 'subnet-manager'.

*2: These are source nodes that manage the physical link between nth-layer subnetworks or (n-1)th-layer subnetworks on the optimal route.

*3: This is the nth-layer SM accommodating destination node, and has only message transmitting capability in this sequence, although it can execute route selection and bandwidth reservation request procedures in other sequences.

*4: These procedures are recursively executed until each lowest-layer SM selects the optimal route by referring to the intra-routing table and interacts with all nodes on that route.

Figure 3 Sequence chart for VCC establishment.

As mentioned above, the proposed method allows VCCs to be established more rapidly than conventional methods because route selection and bandwidth reservation procedures for VCC establishment are executed in-parallel in all subnetworks and all nodes between source DTE and destination DTE. In contrast, conventional methods have procedures on a link-by-link basis and in-series.

3.4 Comparison with other methods

Because the FRP method must always reserve bandwidth on the only route selected in the call set-up phase on a burst-by-burst basis, this method tends to increase burst blocking probability especially in gigabit applications. Moreover, this method reserves bandwidth on a link-by-link basis and in-series. As a result, this method would result in long connection delays. On the other hand, the P-NNI method reserves bandwidths on the optimal route on a burst-by-burst basis and decreases burst-blocking probability like the proposed method. However, the P-NNI method takes more time in establishing VCCs than the proposed method because it needs to calculate the optimal route just after VCC establishment request from the source DTE or the source peer group. It also needs to reserve bandwidth on an in-series and link-by-link basis while the VCC the establishment procedure for the proposed method refers to the optimal route calculated just before VCC establishment request from the source node or upper SM, in-parallel to reserve bandwidth on the route.

4 PERFORMANCE EVALUATION

4.1 Simulation model

Figure 4 shows the traffic model that was used to evaluate the proposed method by simulation. Burst duration denotes the duration for any terminal to send burst data consisting of many ATM cells to the subscriber node accommodating the terminal, and idle duration denotes the duration for any terminal not sending any ATM cells to the subscriber node. The necessary bandwidth denotes the bandwidth to send burst data for the burst duration, and is assumed to be exponentially distributed on a burst-by-burst basis and to be constant for each burst duration. Burst duration and idle duration are also assumed to be exponentially distributed with average length of 100[ms] and 900[ms] on a burst-by-burst basis, respectively. Inter-burst arrival and burst duration distributions for each terminal are independent. The number of reattempts for each blocked burst is equal to zero. The aggregate arrival process is assumed to be a Poisson distribution.

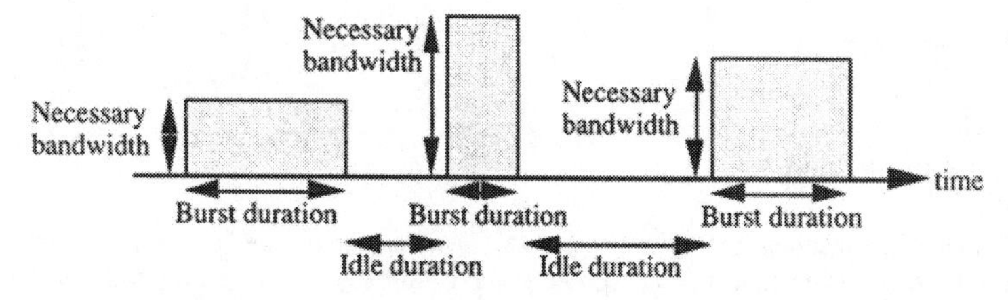

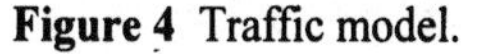
Figure 4 Traffic model.

Figure 5 shows the network topology used to evaluate the proposed method by simulation. Each Source 1-5 and each Destination 1-5 consists of 16 terminals independent of one another, respectively. Each terminal in each Source 1-5 establishes and releases VCC on a burst-by-burst basis to send each terminal in each Destination 1-5 burst data based on the burst generation style in Figure 4 through the network in Figure 5. Propagation delay between the subscriber nodes Switch 1 or Switch 8 and each terminal in each Source 1-5 or Destination 1-5 is assumed to be zero, and each propagation delay between nodes or between the subnet-manager and each node is assumed to be a value in proportion to a distance of 2[km]. Each link bandwidth between the subnet-manager and all ATM nodes is 1[Mbps] and each link bandwidth between ATM nodes and each link bandwidth between the subscriber nodes and 16 terminals in each Source 1-5 or Destination 1-5 are respectively shown in Figure 5.

The proposed method has only been compared with the P-NNI method because of the increased connection time as a result of high burst blocking probability with FRP. Performance of the suggested method is evaluated in the next section through use of a simulator that includes not only the traffic model and network topology in Figures 4-5 but also the functions of both the proposed and P-NNI methods.

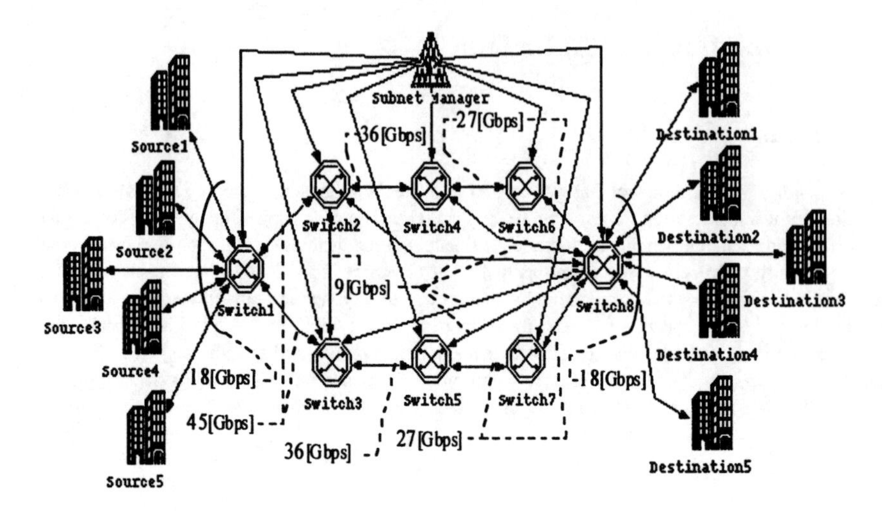

Figure 5 Network topology.

4.2 Consideration

Figure 6 shows the execution results for the simulator mentioned in the previous section and compares VCC establishment time between the proposed and P-NNI methods. The transverse axis shows the average necessary bandwidth per burst requested to be sent by each terminal in Source 1-5. That is, the average necessary bandwidth B_{av} can be defined by the following

$$B_{av} = \frac{1}{n}\sum_{i=1}^{n}\left(\frac{1}{k}\sum_{j=1}^{k} B_{ij}\right)$$

where n $(=5*16)$ is the number of source terminals, k is the sum of burst data that each source terminal requests the network to send to the corresponding destination terminal, and B_{ij} is the necessary bandwidth of the jth-burst data requested (including failure to send) by the ith-source terminal. The vertical axis shows the average VCC establishment time for all burst data, that is, the average VCC establishment time T_{av} can be defined by the following

$$T_{av} = \frac{1}{n}\sum_{i=1}^{n}\left(\frac{1}{m}\sum_{j=1}^{m} T_{ij}\right)$$

where m is the total number of burst data intervals successfully sent by each source terminal to the corresponding destination terminal, and T_{ij} is the VCC establishment time of the jth-burst data successfully sent by the ith-source terminal. Figure 6 shows that reference to the predetermined route, simplification of signalling procedures, and parallel execution of route selection and bandwidth reservation procedures shorten the average VCC establishment time in the proposed method in comparison to the P-NNI method. In particular, the average VCC establishment time for the P-NNI method increases with increasing average necessary bandwidth, while it is almost constant when using the proposed method. This therefore proves the effectiveness of parallel execution of route selection and bandwidth reservation procedures because routes with more hops are often selected in establishing VCCs due to the lack of bandwidths on routes with fewer hops.

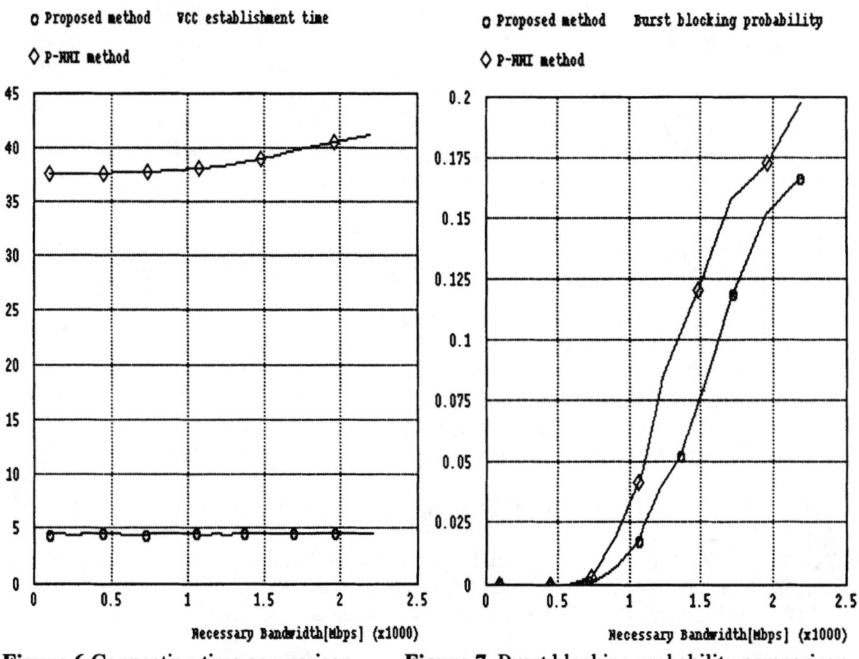

Figure 6 Connection time comparison. **Figure 7** Burst blocking probability comparison.

Figure 7 shows the results of execution for the simulator mentioned in the previous section and compares burst blocking probability between the proposed method and the P-NNI method. The transverse axis is identical to the one in Figure 6. The vertical axis shows average burst blocking probability for all burst data. That is, average burst blocking probability P_{av} can be defined by the following

$$P_{av} = \sum_{i=1}^{n} F_i \Big/ \sum_{i=1}^{n} R_i$$

where R_i is the sum of burst data requested to be sent by the ith-source terminal, and F_i is the sum of burst data that the ith-source terminal fails to send to the corresponding destination terminal. Figure 7 shows that the proposed method can more efficiently use the bandwidth resource in the network shown in Figure 5 than the P-NNI method. This is not only because VCC establishment time is shortened, but also because VCC release time is shortened by the parallel execution of bandwidth release procedures and the simplification of signalling procedures. As a result, the proposed method can rapidly utilize bandwidth resources.

5 CONCLUSIONS

This paper has proposed a fast routing and bandwidth management method for ATM networks. The proposed method has four features: separation of the VCC establishment procedure from the management of routing tables, division of ATM networks into several hierarchical subnetworks, parallel execution of route selection and bandwidth reservation procedures in each subnetwork, and simplification of signalling procedures by sending RM-cells through dedicated PVCs. Performance evaluation showed that the proposed method enables each VCC to be established faster than with the P-NNI method. It can be deduced that the proposed method also enables each VCC to be established faster than by the FRP method. This is because the FRP method causes higher burst blocking probability especially in gigabit applications than the proposed method and reserves bandwidth on an in-series and link-by-link basis. Performance evaluation also showed that the proposed method is able to make more efficient use of the bandwidth resource than the P-NNI method.

6 REFERENCES

ATM Forum (1995) UNI Signalling 4.0, ATM Forum/94-1018R7.

ATM Forum (1995) P-NNI Draft Specification, ATM Forum/94-0471R13.

Boyer P.E. et al. (1992) A reservation principle with applications to the ATM traffic control, Computer Networks and ISDN Systems, Vol. 24.

ITU-T (1990) Broadband Aspects of ISDN, Recommendation I.121 (Draft).

ITU-T (1992) Support of Broadband Connectionless Data Service on B-ISDN, Recommendation I.364 (Draft).

ITU-T (1993) B-ISDN DSS2 Signalling for Basic Call/Connection Control, Recommendation Q.2931 (Draft).

ITU-T (1995) B-ISDN, B-ISDN User Part - Basic Call Procedures, Recommendation Q.2764 (Draft).

Kakemizu M. et al. (1995) A Fast Method for Establishing Virtual Channel Connections in Gigabit ATM Networks, Asia Pacific Conference on Communications '95, pp. 780-784.

9

Admission Control and Dynamic Routing Schemes for Wide-Area Broadband Networks: Their Interaction and Network Performance*

D. Medhi & I. Sukiman
Department of Computer Networking
University of Missouri–Kansas City
Kansas City, MO 64110 USA
Email: {dmedhi, isukiman}@cstp.umkc.edu

Abstract

We consider wide-area broadband networks where multiple services with differing bandwidth requirements are offered (with peak rate allocation for the duration of a connection in a loss network environment). We present several dynamic routing schemes in such a multi-service setting along with an admission control policy. The routing schemes cover issues such as crankback, periodicity of computation of routing, and what type of information is used in routing decisions. We present results on their performance under normal as well as overloaded network conditions. We observe that in an integrated environment, for low-bandwidth service (such as voice), the various routing schemes provide similar performance; this is not so for high-bandwidth service (such as video) — we observe noticeable differences in the network performance for high-bandwidth service depending on whether the routing scheme has crankback, and how the routing decision is made. We also observe that the proper setting of the admission control scheme can reduce the discrepancies in blocking for different services.

Keywords

B-ISDN; multi-rate loss networks; dynamic routing; admission control; network overload; network performance.

1 INTRODUCTION

Future broadband networks will be required to provide a variety of services with different bandwidth requirements for each service while at the same time each service may have different grade-of-service[1] (GoS) requirements. In this paper, we limit ourselves to considering services that require peak rate allocation of bandwidth for the duration of a call (Class 1 service in B-ISDN) while the peak rate may be different for different services; thus, the performance issue here is GoS, i.e., call blocking for each service type. We consider such services in wide-area networks with dynamic routing capabilities.

For peak rate allocation in single-service (voice) networks, dynamic routing has been studied extensively in the last two decades (see, for example, [1, 2, 4, 7, 16]); several routing schemes,

---

* Research supported by the University of Missouri Research Board under grant K-3-40605 and by US National Science Foundation under grants CDA-9422092 and NCR-9506652.

[1] The term GoS here refers to acceptable call blocking probability.

such as AT&T's Dynamic Non-Hierarchical Routing (DNHR) [2] and Real-Time Network Routing (RTNR) [4], Northern Telecom's Dynamically Controlled Routing (DCR), British Telecom's Dynamic Alternate Routing (DAR), have been implemented and deployed. See Girard [7] and references therein for more details.

Several issues are important in the context of single-service dynamic routing schemes: (a) trunk reservation (also known as state protection), (b) maximum number of links allowed to connect a call, (c) crankback feature, (d) the manner in which the routing is computed and a route is chosen, (e) periodicity of routing table update or (near) real-time computation of routing and impact on call set-up time.

Of these issues, trunk reservation, apparently first presented by Weber [18], and setting the maximum number of links to two, are the common features found in almost all single-service dynamic routing schemes. The essence of trunk reservation is to protect direct routed traffic if the number of idle channels falls below a certain tolerance on the direct link which is needed primarily in overloaded conditions and for network stability. The restriction on the maximum number of links to two has to do primarily with the trade-off between the complexity of multi-link calls as introduced in switching software and in signalling requirements, and the marginal improvement in network throughput from two-link to more than two-link call routing [2]. (Recently, Krishnan and Cardwell [12] have observed that restriction to two-link routes in virtual-path based ATM networks imposes no penalty on network cost for large networks.) Crankback refers to the ability to offer a call blocked at the intermediate node in a two-link call to another route. For example, while crankback is implemented in DNHR and RTNR, it is neither a feature in DCR nor DAR. Item (d) above has been addressed differently for different routing schemes: off-line computation based on an optimization model (e.g., DNHR), least-loaded path, either based on available capacity (e.g., RTNR) or based on offered load and available capacity (e.g., Bellcore's state-dependent routing [7]), probabilistic choice of alternate path based on availability of bandwidth (e.g., DCR), and random choice of alternate route (e.g., DAR). Finally, the issue of routing decision when a call arrives: ideally, it is preferable to compute it based on the network status at the time of arrival of the call; however, actual real-time computation may increase the call setup time – this may not be acceptable from a network provider's viewpoint. Thus, near real-time computation is much more acceptable than actual real-time computation to reduce the impact on call setup time; this has led to either off-line computation of routes (and some correction through network management update [1], e.g. DNHR), frequent/periodic update (less than a minute) of the routing table (e.g. DCR), or using the routing information available at the time of the previous call (e.g. RTNR).

In the case of multi-service wide-area broadband networks, different services with different peak rates for the duration of a call are going to be offered. It is expected that the issues discussed above for single-service networks will remain in the multi-service networks too. It has previously been observed for mixed bandwidth traffic in a *single-link* system (no routing) that if there is no admission control, then the traffic type with lower bandwidth requirement per call has less blocking than the traffic type with higher bandwidth requirement per call [11]. There has been, in fact, considerable work on admission control and performance of a single-link system where multiple services (multi-rate) are offered (see, for example, [10, 11, 14]). Thus, if a certain GoS is to be provided to a service type, then some form of admission control may be required to address the GoS issue in a network setting as well. Further, this may need to be triggered under *normal* network operating conditions. (We will discuss this in section 4.) We will consider the case of equitable GoS for all services, e.g., 1% blocking GoS for all services for clarity and simplicity (obviously, different GoSs may be required for different services). Thus, two important issues to be considered for multi-service networks (in addition to the issues discussed earlier) are: admission control[2] and service protection under normal network operating conditions. Thus, in a multi-service broadband network setting, we propose two phases associated with any newly arrived call: admission control phase – to determine if the call is to be admitted to the network (not connected yet), and routing phase – to determine if there is capacity (subject to trunk reservation) in the network to connect the call once it passes the admission control phase. Thus, a call may be rejected in either phase.

Multi-service dynamic routing has been addressed in recent years by several researchers (see, for example, [3, 4, 5, 6, 8, 9, 13, 15]); of which, only a few of them have considered admission control with routing. The purpose of our present paper is to understand the implication and interaction of admission control with various dynamic routing schemes in a multi-service setting with different bandwidth requirement for each service type (beyond the work presented in [15]). Towards this end, we have proposed various dynamic routing schemes in a multi-service setting addressing the issues stated earlier, and a simple admission control policy (adapted from a similar one for a single-link system [14]), and have studied their interaction and network performance through network simulation.

Finally, we discuss the notion of a basic bandwidth unit (BBU) before we start the main body of this paper. We use the term basic bandwidth unit [11] to refer to a base rate, and assume that all services require some multiple of this base rate. For example, if we consider 64Kbps to be the basic bandwidth unit, then a 64Kbps peak rate voice call requires one BBU, while a 384Kbps peak rate video call requires six BBUs. Also, we will use service unit (SU) to refer to the unit for a particular service per connection which is given in terms of BBUs. Note our work is not limited to using BBUs to be 64Kbps. Some idea of BBU may be desirable for bandwidth quantization in broadband networks both for network planning and service deployment. Quantization has been recently addressed by Lea and Alyatama [13] where they have reported that a proper quantization may not result in significant loss in network throughput.

2 DYNAMIC ROUTING METHODS

In the following discussion, we assume that we have a set of services $\mathcal{S}$; each service $s \in \mathcal{S}$ requires a peak rate allocation of w_s BBUs for the duration of the call. Two nodes i and j in the network may be connected by a direct link (i, j). The traffic pair between nodes i and j is denoted by $[i, j]$. A call for traffic pair $[i, j]$ is connected either on the direct link (i, j) or is on an alternate route via at most one more node; this limits any call to use at most two links to connect. The network is assumed to be nearly fully interconnected. We present below six multi-service dynamic routing schemes.

2.1 Routing Rule: MACRPC

For each switching node pair, an arriving call for each service type that passes the admission control (described in section 3) first tries the direct traffic link. If there is capacity on the direct link to serve the bandwidth requirement for this call, the call is connected. If there is no free capacity on the direct link or the direct link does not exist, then the call first tries the first alternate via node given in the routing table computed periodically; if it cannot find any available trunks, especially in the second leg of this two-link alternate route, (subject to trunk

[2] It can be argued that trunk reservation is an admission control feature; however, in our discussion, we consider trunk reservation to be a part of routing feature, and we classify admission control to be the issue of whether to admit a call to the network *before* it goes to the routing phase.

reservation [18], also known as state protection [7]), then the call is cranked back ([2]) and is tried using the next alternate via node as given in the routing table. If the call cannot find any available capacity after trying all the alternate routes given in the routing table, then the call is blocked. Like dynamic non-hierarchical routing (DNHR) [2] and trunk status map routing [1] for single-service networks, this routing has the crankback feature; while DNHR used an off-line computed routing (with some real-time network management added routes in case of overload [1]) and TSMR used DNHR with some added routes computed regularly, the routing used here updates the entire routing table at regular intervals, somewhat similar to DCR. Note that the routing we use attempts various alternate routes in the order given in the routing table using crankback, if needed, while DCR uses probabilistic values to pick the alternate route from the routing table; additionally, DCR does not have crankback. We call this routing scheme – Maximum Available Capacity Routing with Periodic update and Crankback (MACRPC). Note that MACRPC is aimed for multiple services, and thus, the routing table computation takes into account w_s (=the peak rate bandwidth required per connection by calls of a particular service type). This is described next.

Consider the traffic link (i,j) for the node pair $[i,j]$ with end nodes i and j. Let

$t_{(i,j)}$:= Total number of BBUs on link (i,j)

$o_{(i,j)}$:= Number of BBUs on link (i,j) that are presently allocated to active calls of all types

$r_{(i,j)}$:= Number of BBUs on link (i,j) reserved (trunk reservation) for its own direct traffic

Then, the available capacity for pair $[i,j]$ via node $v(\neq i,j)$ at the instant of computation of the routing table is given by:

$$z^v_{[i,j]} = \min \{ t_{(i,v)} - o_{(i,v)} - r_{(i,v)}, t_{(v,j)} - o_{(v,j)} - r_{(v,j)} \}. \tag{1}$$

Now, for each service type s, we consider the candidate list

$$\mathcal{V}^{(s,\beta_s)}_{[i,j]} = \left\{ v \;\middle|\; z^v_{[i,j]} \geq \max\{1, \beta_s w_s\} \right\}, \tag{2}$$

where β_s satisfies $0 \leq \beta_s \leq 1$. If $\beta_s = 0$ for all service types s, then the candidate routing list is the same for each service type (chosen with at least one unit of BBU available). If $\beta_s = 1$ for all s, then the candidate list for each service type contains the alternate paths which have at least w_s units of BBU available (at the time of computation) to connect a call for that service type. If β_s is a value closer to 1, then a path is considered in the candidate list even if it does not have enough bandwidth to connect a call for that service type at the time of routing table update, but is close enough such that it merits consideration in case any existing calls are completed. The candidate list is sorted in descending order (of available capacity as given by $z^v_{[i,j]}$ in (1)) to determine the routing table for service type s for traffic pair $[i,j]$ subject to the condition given in (2). Note that although the routing *rule* is the same for each service type for a specific traffic pair, the routing *table* can be different for different service types depending on the value of β_s (and w_s). As noted earlier, this updating is done periodically/frequently. This computation can be done either in a centralized manner where each switch sends its information regularly to a central processor, or in a distributed manner where the busy capacity information is exchanged between switches using the signalling network.

2.2 Routing Rule: OCORC

This routing scheme is inspired by RTNR [4]. In this routing, a traffic pair has an alternate via node available ("stored" via node) at any time. Once a call passes admission control, it first tries the direct link. If there is capacity available, the call is connected. If there is no capacity on the direct link or the direct link does not exist, then the stored via node is tried for alternate routing; at the same time a process is spawned to compute a new via node. If the call is completed using the stored via node, then the newly computed via node becomes the stored via node for the next call that arrives for the same traffic pair and service type. If the call cannot be completed via the originally stored via node due to the unavailability of enough capacity, then the call is cranked back and waits for the completion of the spawned process which computes the new via node and then tries to route it using this newly computed via node. The idea behind this concept of "one call old" routing is to minimize the call set-up time for most calls [4]. The computation of the new via node has similarities to the computation of the periodic routing table in MACRPC. In this case, we use the rule given in (1) along with (2) (with $\beta_s = 1$ for all s) in a distributed manner to determine the available via nodes and picks the via node with the maximum available capacity to be the newly computed via node on a per call basis. Thus, a switching node pair requests all its possible via nodes to send availability of capacity for computing the best-via-path that is used for crankback if needed and for the next call that arrives for this same traffic pair and service type. We will refer to this routing scheme as One Call Old Routing with Crankback (OCORC). Note that OCORC is conceptually similar to RTNR [4]; however, instead of computing sets of via nodes based on network status bit map as in RTNR, OCORC uses the information on available capacity on each link for computation of alternate routes.

2.3 Routing Rule: FOCORC

This scheme is similar to OCORC except that the determination of the via node to attempt at the instant of a call arrival for *some* services is different than OCORC. Usually, when a new service is introduced in the network, the amount of traffic for such services is significantly lower than existing services. As such, the requests for connection for such new services can arrive quite infrequently. This leads to the situation in which the network state may have changed significantly since the alternate route was computed in the last call for this service type as done in OCORC; i.e., the information may be outdated for a newly arrived call for the same service [15]. Thus, in such a situation it may be preferable to force computation of the via node on each call basis at the *expense* of possible increase in call set-up time for new services. The computation of the via node availability is done as discussed for OCORC. Thus, in this scheme, some services use "one call old" routing for alternate routing while others use freshly computed via node for alternate routing. We refer to this routing scheme as *Forced* One Call Old Routing with Crankback (FOCORC).

2.4 Routing Rule: PACRP

In this scheme, we will consider two variations. For both cases, the routing update is done periodically as in MACRPC; however, this scheme does not have crankback. Route choices are done here in a probabilistic manner which is similar to DCR; however, the procedure is modified from DCR to work for multi-service cases with differing bandwidth requirements. As with previous schemes, an arriving call first goes through the admission control check before it is handed to routing phase check; once it is in routing phase check, the direct link is tried first.

As in MACRPC, periodically, $z^{v}_{[i,j]}$ is computed using the expression given in (1). Further, for each service, the set $\mathcal{V}^{(s,1)}_{[i,j]}$ is determined by considering β_s to be 1, i.e., due to no crankback, we prefer to have a set which shows the paths with sufficient bandwidth availability. Then the probability of choosing a via node v for service type s for the pair $[i,j]$ is computed as given below:

$$q_{[i,j]}^{v,s} = z_{[i,j]}^{v} \Big/ \sum_{m \in \mathcal{V}_{[i,j]}^{(s,1)}} z_{[i,j]}^{m}, \qquad v \in \mathcal{V}_{[i,j]}^{(s,1)}. \tag{3}$$

Thus, a routing table is periodically prepared based on availability which contains the routing via nodes and the probability value for each alternate path as computed using (3). We refer to this scheme as Probabilistic Available Capacity Routing with Periodic update (PACRP). We present here two variations of PACRP for actual alternate route selection based on the above probabilistic "goodness" of a path. In the first approach, to be referred to as PACRPa, when a call for service type s arrives for pair $[i,j]$ and enters the routing selection phase, the call is first attempted on the direct path (if available); if the direct path is not available, an alternate path via node v is chosen at random with the probabilities given by the above expression. If the call cannot be completed on this path, the call is lost (blocked and cleared).

In the second variation, PACRPb, the first alternate route selection is done as in PACRPa. However, this time, if the call cannot find available capacity on the first leg of the two-link path, then a second path is tried at random with a modified probability computed for the rest of the routes in the routing table. Suppose, the first via node selected is x; then the probability values for the rest of the routes in the routing table is computed as follows:

$$\tilde{q}_{[i,j]}^{v,s} = z_{[i,j]}^{v} \Big/ \sum_{m \in \mathcal{V}_{[i,j]}^{(s,1)} \setminus \{x\}} z_{[i,j]}^{m}, \qquad v \in \mathcal{V}_{[i,j]}^{(s,1)} \setminus \{x\}. \tag{4}$$

This process is continued for all alternate paths that have positive probabilities listed in the routing table (obtained from last periodic update) and a path is eliminated from probabilistic choice of routing if it has been tried once for this call; note that this is done as long as the call cannot find available capacity only on the first leg of two-link route. This modified probability is computed for the present call only (i.e., paths are not eliminated from the routing table for any future calls). If, however, for any route chosen, there is capacity available in the first leg, but none available in the second leg, then the call is lost (blocked and cleared) and there is no crankback.

2.5 Routing Rule: POCOR

This routing scheme draws certain features from OCORC and PACRPa. In this routing scheme, the routing decision is probabilistic based on the information available for the last call for the same traffic pair and the same service type. Thus, in this routing scheme, when a call arrives in the routing phase, it first tries the direct link. If there is no bandwidth available on the direct link, then the decision for routing this call is made based on the information about the routes available from the last call. Unlike OCORC, in this routing the available bandwidth for possible alternate routes are mapped to probability values similar to PACRPa. Instead of periodic computation of the probabilities as in PACRPa, the computation for the available bandwidth information is done in the last call; no crankback is performed as described for PACRPa. Thus, like OCORC, when a call arrives a process is spawned to obtain the availability of capacity on candidate two-link paths; this information is used for the next call that arrives for this traffic pair and service type. We refer to this routing scheme as Probabilistic One Call Old Routing (POCOR).

2.6 Routing Rule: FPOCOR

Finally, this scheme combines some features from FOCORC and PACRPa. For some services, the routing decision is made as in POCOR. However, for certain new emerging services, instead of using the information for the last call as in POCOR, the routing decision is based on querying every switch about available bandwidth for the arrived call as is done for FOCORC. Here, for the available bandwidth information obtained, a success probability is computed for each alternate routes using the expression given by (3), and then the decision for choosing a path is made at random using probabilities computed for each path, similar to PACRPa; only one alternate path is chosen and as such no crankback is done. We will refer to this routing scheme as *Forced* Probabilistic One Call Old Routing (FPOCOR). Thus, with FPOCOR, services for which the forced computation is done have an impact on call setup time due to time required for querying and decision making, as was the case with FOCORC.

3 ADMISSION CONTROL

Our admission control policy is probabilistic in the sense that depending on the amount of free capacity on the direct link available (within a specified range) at the instant a call for a service type arrives, it is accepted (not connected yet) to the network with a certain probability. The admission control is an extension of an acceptance policy described in the context of a single-link system for heterogeneous traffic [14]; the admission control in the present form is presented in [15] and is discussed here for completeness. If the call is not accepted, then it is blocked and cleared. If the call is accepted in the admission control phase, then it goes to the routing phase. The admission control can be given for service $s \in \mathcal{S}$ for traffic pair $[i,j]$ by the following acceptance function:

$$\alpha^s_{[i,j]} = p^s_{[i,j]}, \text{ if } L_{(i,j)} \leq t_{(i,j)} - o_{(i,j)} < U_{(i,j)}; \quad 0, \text{ otherwise}, \tag{5}$$

where $L_{(i,j)}$ and $U_{(i,j)}$ are the lower and upper bounds in BBUs, respectively, on free capacity for probabilistic acceptance, and $0 \leq p^s_{[i,j]} \leq 1$. If $p^s_{[i,j]} = 1, s \in \mathcal{S}$ for all traffic pairs, then there is no admission control in the network, which is sometimes known as complete sharing [10]. It should be noted that the admission control decision is local; it is not based on the complete network information, instead, the decision is based on the occupancy of the direct link at the time of the arrival of a call. Specific instances of the admission control policy can be considered using (5) by specifying different values for the parameters, p^s, L and U.

4 RESULTS

To study network performance, we have developed a network simulator where the above routing schemes and the admission control policy are implemented. In our simulator, two service types are considered: the first service type $s = 1$ requires $w_{(s=1)} = 1$ BBU for a call, while the second service $s = 2$ requires $w_{(s=2)} = 6$ BBUs for a call. Often, the first service will be called the voice (narrowband) service and the second will be called the video (wideband) service. We have considered a ten switching node sample network for computational work. The data for this network is based on a real public switched network spanning the continental US [15]. The offered load for voice traffic is extracted from this network. Due to nonavailability, the offered load for the second traffic type is generated using a uniform random number generator by picking a value between 0% and 5% of the offered load for the voice traffic, separately for each traffic pair in the network. Further, in our study we use two different load hours (with asymmetric traffic data for each hour) to reflect variation of traffic during the day; these two load hours will be referred to as hr-A and hr-B. In our case, total offered loads (in erlangs) for the network for load hours hr-A and hr-B for voice traffic are 2826.16 erl and 3224.08 erl, respectively; offered loads for video traffic for load hours hr-A and hr-B are 72.96 erl and 84.34 erl, respectively. Note that effective offered (weighted) load for video service is six times more than the loads listed here for video since an SU for video requires six BBUs for the duration of a call. The voice service is considered as a "mature" service while the video service is considered as a "newly deployed emerging" service due to the loads used. As such for *forced* routing schemes, FOCORC and FPOCOR, we set video service to do forced routing computation while computation for voice is done as in OCORC and POCOR, respectively. The bandwidth required on each link in the network so as to provide 1% blocking GoS to each service is determined by a multi-hour dimensioning procedure presented in [15]. This dimensioning procedure generated 4,815 total BBUs of capacity for the network. We consider three different scenarios for traffic loads: "normal load" where each traffic type has offered load as stated above, i.e., normal network operating conditions; "video-only overload" where video traffic has 5% overload while voice traffic is at normal load; and "both overload" where both voice and video traffic has 5% more traffic (uniformly) than their respective normal load; in the figures and the rest of the discussion they are indicated by Lf-1.0, Lf-Vid, and Lf-1.05, respectively.

In our simulation, the arrival of a call is assumed to be Poisson and the mean call holding time to be exponentially distributed. The mean call holding time for voice is assumed to be three minutes. The mean call holding time for the video service is set to 7.5 minutes, 2.5 times more than that of voice service. The frequency of routing table update for MACRPC and PACRP are set to 15 seconds. For each case we considered, we have run ten independent replications, and all the results reported are based on computing the average of the ten runs. We have also computed 95% confidence intervals, and have found that, typically, for voice traffic the range of confidence interval is less than a tenth of a percentage of blocking, and for video traffic, the range of the confidence interval is often less than half a percent of blocking; the average value for each point is on the graphs (in all but one) so as not to clutter with vertical lines for confidence intervals. In each run, we have thrown away the first six hours of simulation data to take into account the simulation warm up period, and then the simulation was run for another twenty hours to collect the network statistics. Trunk reservation, $r_{(i,j)}$, discussed in (1) is dynamically set based on pairwise blocking; this is similar to [4] and is done in the same manner for all routing schemes. It is worth mentioning that in RTNR [4], trunk reservation is done separately for each service; however, in our work we have used a shared trunk reservation approach for all services for a particular traffic pair while letting admission control address service specific GoS.

We consider a simple instance of the admission control given by (5) for our study:

AC-1 :: for video, $p^2_{[i,j]} = 1$ in all cases; for voice, admit with probability $p^{(s=1)}_{[i,j]} = p^1$ when $L_{(i,j)} := 0$; $U_{(i,j)} := 1$ video SU $= w_{(s=2)} = 6$ BBUs

This means that when the number of free BBUs on a direct link falls below six, then the voice call is admitted for this traffic pair with a certain probability (p^1) while all video calls are admitted so as to give some preference to video service (see [15] for another instance).

In our study, we considered the following scenarios: 1) the impact due to different load hours, 2) the impact due to three load scaling scenarios (Lf-1.0, Lf-Vid, Lf-1.05). These two scenarios are considered for the six routing schemes presented earlier. From our preliminary study, we observed that the various parameter values for MACRPC have little difference: for the results reported here, we use the value $\beta_s = 0.5$ when the routing table is updated. Out of two variations of PACRP, we have chosen PACRPb for comparison with other schemes since from simulation we observed that PACRPb provides lower blocking than PACRPa for video service.

In Fig. 1, we plot the six routing schemes for normal load (Lf-1.0) for load hour, hr-B, for three values of p^1 for the admission control policy, AC-1: $p^1 = 0.8, 0.9, 1.0$. Recall that $p^1 = 1.0$ refers to *no* admission control. We observe that under no admission control, voice blocking is virtually zero while video blocking varies from about 1.5% to over 6%. This shows the disparity in GoS for different bandwidth services under no admission control for each of the routing schemes; as the value of p^1 is decreased from 1.0 to 0.8 to activate admission control, we see that for each routing scheme, there is an intersection point where the voice blocking line crosses with video blocking line giving the point of equal blocking for each traffic stream. Observe that the equal blocking point corresponds to about 1% blocking for routing schemes MACRPC, OCORC, FOCORC and FPOCOR; however, this cross point is found to be about 2% and 4% blocking for routing schemes PACRPb and POCOR, respectively. This suggests that for probabilistic routing the routing information from the last call may be too old for a newly emerging service, even when the information is updated periodically but with no crankback feature, as in PACRPb, it results in higher blocking. Note however that if forced computation of routing information is used for the probabilistic choice with new information (FPOCOR), the blocking improves significantly as to obtain a performance along the routing schemes with crankback. Further, for low-bandwidth calls (i.e., voice), the blocking is essentially indistinguishable among all the routing schemes, however for higher-bandwidth calls (i.e., video), there is a noticeable difference. Note also that there is no difference in network performance due to one call old routing with crankback and forced computation for video (i.e., OCORC and FOCORC), except for the possible impact on an increase in the call setup time due to FOCORC. For the same case with hr-A, we observe similar behavior as with hr-B with one exception. The video blocking with FPOCOR is slightly lower than with OCORC in hr-A while it is slightly higher in hr-B; the reverse effect is true for voice blocking — this is only noticeable primarily for the 'no admission control' ($p^1 = 1.0$) case. This suggests that depending on the network traffic, forced probabilistic routing as in FPOCOR sometimes does a somewhat better job of spreading high-bandwidth calls than the choice of taking the maximum available capacity path as in OCORC.

Now we discuss the results for a 5% overload of video traffic (Lf-Vid). In Fig. 2, the performance of various routing schemes are plotted for load hour, hr-A with 5% overload for video, and the same is done for load hour, hr-B, in Fig. 3. We observe a similar pattern as in the case of normal load except that the video blocking gap between PACRPb and the rest of them (except POCOR) has widened for hr-B. The main observation to notice here is that 5% overload in video traffic has virtually no impact in increasing the blocking for voice traffic, but for video traffic the blocking increase is noticeable. This minimal impact on voice traffic may partly be due to the fact that even with video overload, video traffic has not increased significantly. Two interesting questions arise when the overload for one traffic stream occurs in a network: a) should the network be geared to provide equal blocking to each traffic type even when one traffic class is overloaded, or b) should the blocking for unaffected traffic type remain the same as under normal network operating condition, i.e., 1% blocking. If the answer to the first question is yes, then the admission control should be used with p^1 set to be around 0.95 to provide equal blocking of about 1.5% to both services (for hr-A and routing MACRPC, OCORC, FOCORC, FPOCOR); if the answer to the second question is yes, then using a higher value of p^1 (around 0.97) is acceptable, which corresponds to providing 1% blocking for voice and over 2% blocking for video, thus letting the overloaded traffic class suffer higher blocking. Obviously, the choice of one over the other depends on the network's objective. Nevertheless, our simple admission control can effectively interact with the routing schemes to respond to either scenario.

In Fig. 4, the network blocking is plotted for 5% network-wide overload (Lf-1.05) over normal network traffic corresponding to hr-A; the corresponding graph for hr-B is given in Fig. 5. The pattern very much is the same as before; however, we discuss here the salient points. Observe that under no admission control, while the blocking for voice service increases by about 1% blocking for all routing schemes, the blocking increases significantly for video traffic; for example, for OCORC video blocking increases from about 2% blocking under normal load to over 8% blocking with 5% network-wide traffic overload; for POCOR, the change for video is from 6% blocking to over 12% blocking. However, with the activation of admission control, this difference is less pronounced. This suggests that our simple admission control can reduce discrepancies in network performance when combined with any of the dynamic routing schemes. To illustrate this aspect, we have plotted all three load situations (Lf-1.0, Lf-Vid, Lf-1.05) for routing MACRPC for hr-B in Fig. 6 (short vertical lines indicate 95% confidence intervals). Notice the big change in video blocking under no admission control which drops to almost zero as $p^1 \to 0.8$, while the gap for voice blocking continually remains at about 1% blocking. It is also interesting to observe from this figure that p^1 which provides the point of equal blocking decreases as the network load increases. This indicates that in a network with changing network load and for overloaded situation, an adaptive scheme to update p^1 may be desirable.

5 SUMMARY

In this paper, our effort has been to understand the interaction of admission control and dynamic routing, and its impact on network performance. We have presented here six routing schemes. Based on our study of a ten-node switched network with two heterogeneous service types, we can summarize the following observations:

- Various routing schemes are indistinguishable in terms of network performance for low bandwidth service (i.e. voice).
- For higher bandwidth service (video), from the performance of routing schemes MACRPC and PACRPb, we observe that the crankback feature seems to provide lower blocking compared to routing that has no crankback feature, if the routing update is done frequently.
- There is virtually no difference in performance between OCORC and FOCORC, i.e, one call old information with crankback is as good as forced computation of route at the arrival of a call for high bandwidth call; we believe this is due to the fact that OCORC does crankback if the old information is not good thus using new information. The only difference is that on average FOCORC will have to impose a higher call setup time for the high-bandwidth service than that for OCORC on more calls.
- Probabilistic choice of a route without crankback based on information available in the last call seems to provide the highest blocking for the newly emerging service. On the other hand, if a forced computation is performed for the newly emerging service even with probabilistic choice, the performance improves significantly and is at par with routing with the crankback feature, although this has the impact of increase in the call set-up time.
- A simple admission control scheme with local information as presented here can reduce the difference in network blocking between voice and video service. Further, with the proper value of the admission control parameter, an equitable grade-of-service can be provided to both services. If equitable grade-of-service is the network provisioning objective, then admission control is required *even* in normal network operating conditions.
- The impact of admission control parameter value on network performance behavior under overload situation suggests that an adaptive approach to updating the admission control parameter can provide desirable effect on the performance for various services.

Nevertheless, we have made several observations about how services are affected due to different routing schemes and bandwidth requirements which have not been observed to our knowledge. We have also pointed out above the possible increase in call set-up time if the latest network information is used. While this may not be acceptable for some services, it may possibly be acceptable (within a reasonable limit) for some other services; e.g., with video conferencing services, some delay in set-up time may be tolerable since the actual conference may not start until all the participants arrive (this is obviously a subjective issue). Another issue is the adaptive nature of the admission control scheme. We have tried to understand this by considering two different load hours and network overloading. Obviously, a better understanding of admission control in the presence of nonstationary traffic is still desirable (see [17] for admission control under nonstationary traffic on a single-link system). The performance impact on various services under severe network failure or mass calling, however, remains to be seen and understood.

REFERENCES

[1] Ash, G. R (1995) Dynamic Network Evolution, with Example from AT&T's Evolving Dynamic Network. *IEEE Communications Magazine*, **33(7)**, 26-39.

[2] Ash, G. R., Cardwell, R. H. and Murray, R. P. (1981) Design and Optimization of Networks with Dynamic Routing. *Bell Sys. Tech. Journal*, **60**, 1787-1820.

[3] Ash, G. R. and Chang, F. (1993) Management, Control and Design of Integrated Networks with Real-Time Dynamic Routing. *Jrnl. Network & Systems Mngmt.*, **1**, 237-54.

[4] Ash, G. R., Chen, J.-S., Frey, A. E. and Huang, B.-D. (1991) Real-time Network Routing in a Dynamic Class-of-Service Network. *Proc. 13th Intl. Teletraffic Congress*.

[5] Dziong, Z. and Mason, L. (1994) An Analysis of Near Optimal Call Admission and Routing Model for Multi-service Loss Networks. *IEEE Trans. Comm.*, **42**, 2011-22.

[6] Gersht, A., *et al* (1994) Dynamic Bandwidth Allocation, Routing and Access Control in ATM Networks. in *Network Management and Control*, **2**, Plenum Press, pp. 131-149.

[7] Girard, A. (1990) *Routing and Dimensioning in Circuit-Switched Networks*, Addison-Wesley, Readings, Mass.

[8] Gupta, S., Ross, K. and El Zarki, M. (1995) On Routing in ATM Networks. in *Routing in Communications Networks*, M. Streenstrup (ed.), pp. 49-74.

[9] Hwang, R., Kurose, J. and Towsley, D. (1992) State-Dependent Routing for Multirate Loss Networks. *Proceedings of IEEE GLOBECOM'92*, pp. 565-70.

[10] Kaufman, J. S. (1981) Blocking in a Shared Resource Environment. *IEEE Trans. on Comm.*, **COM-29**, 1474-81.

[11] Kraimeche, B. and Schwartz, M. (1985) Analysis of Traffic Access Control Strategies in Integrated Service Networks. *IEEE Trans. on Comm.*, **COM-33**, 1085-93.

[12] Krishnan, K. R. and Cardwell, R. H. (1994) Routing and Virtual-Path Design in ATM Networks. *Proceedings of IEEE GLOBECOM'94*, pp. 765-9.

[13] Lea, C.-T. and Alyatama, A. (1995) Bandwidth Quantization and State Reduction in the Broadband ISDN. *IEEE/ACM Trans. on Networking*, **3**, 352-60.

[14] Medhi, D., van de Liefvoort, A. and Reece, C. S. (1995) Performance Analysis of a Digital Link with Heterogeneous Multislot Traffic. *IEEE Trans. on Comm.*, **43**, 968-76.

[15] Medhi, D. and Guptan, S. (1994/95) Network Dimensioning and Performance of Multi-service, Multi-Rate Loss Networks with Dynamic Routing. submitted to *IEEE/ACM Trans. Networking*, December 1994, revised December 1995. (available by anonymous ftp; `ftp://ftp.cstp.umkc.edu/papers/dmedhi/mg_tr_6_r1_95.ps`)

[16] Nakagome, Y. and Mori, H. (1973) Flexible Routing in the Global Communication Network. *Proc. 7th Intl. Teletraffic Congress*, pp. 426.1-426.8.

[17] Qian, Y., Tipper, D. and Medhi, D. (1996) A Nonstationary Analysis of Bandwidth Access Control Schemes for Heterogeneous Traffic in B-ISDN. *Proc. of IEEE INFOCOM'96*, pp. 6c.1.1-8.

[18] Weber, J. H. (1964) A Simulation Study of Routing and Control in Communications Networks. *Bell Sys Tech Journal*, **43**, 2639-76.

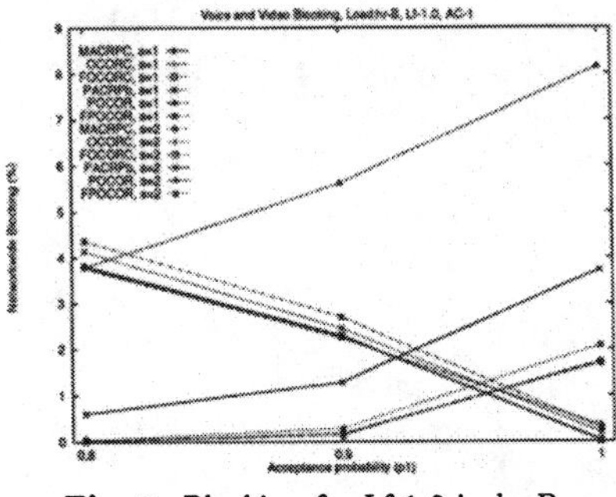

Fig. 1: Blocking for Lf-1.0 in hr-B

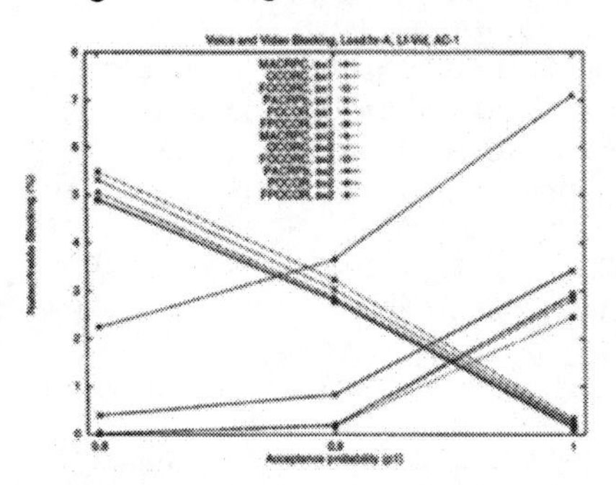

Fig. 2: Blocking for Lf-Vid in hr-A

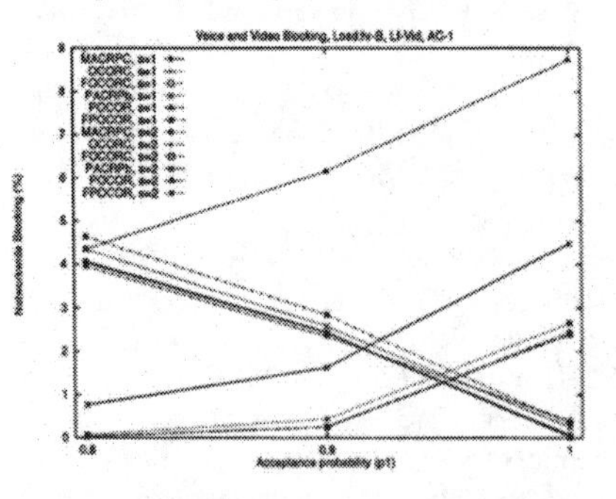

Fig. 3: Blocking for Lf-Vid in hr-B

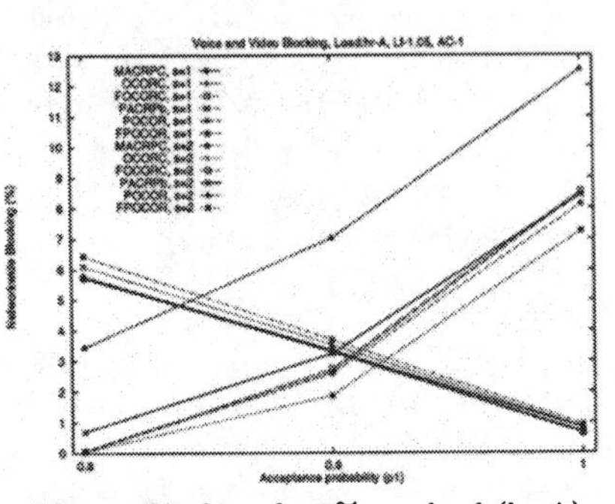

Fig. 4: Blocking for 5% overload (hr-A)

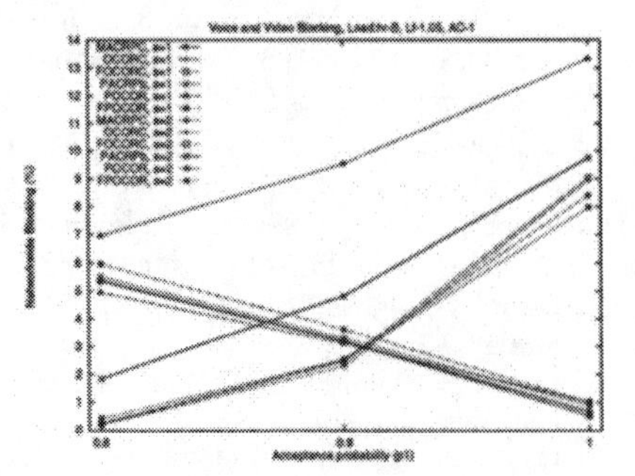

Fig. 5: Blocking for 5% overload (hr-B)

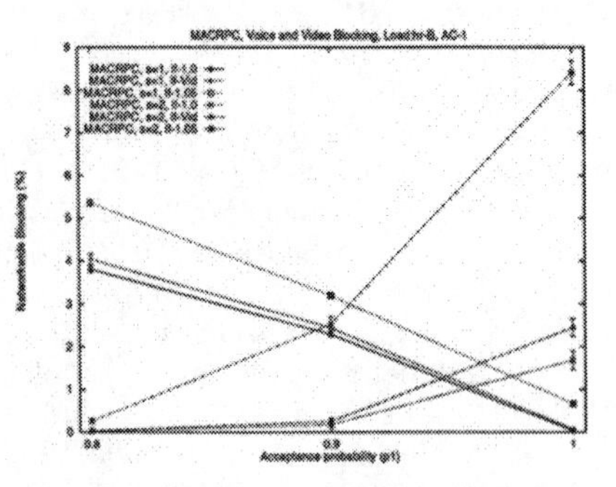

Fig. 6: Blocking for MACRPC (hr-B)

10

Dynamics of an Explicit Rate Allocation Algorithm for ATM Networks

*L. Kalampoukas, A. Varma**
Computer Engineering Department
University of California, Santa Cruz, CA 95064, USA
E-mail: {lampros,varma}@cse.ucsc.edu

K. K. Ramakrishnan
AT&T Bell Laboratories, Murray Hill, NJ 07974, USA
E-mail: kkrama@research.att.com

Abstract

In this paper we study the performance of an explicit rate allocation algorithm for ATM networks using the available bit-rate (ABR) class of service. We examine the behavior of ABR traffic with simple cell sources, and demonstrate that the allocation algorithm is fair and maintains network efficiency in a variety of network configurations. We also study the behavior of TCP/IP sources using ABR service in a network of switches employing the rate allocation algorithm; the results show substantial improvements in fairness and efficiency in comparison with the performance of TCP in an underlying datagram-based network. We study the performance of ABR traffic in the presence of higher-priority variable bit-rate (VBR) video traffic and show that the overall system achieves high utilization with modest queue sizes in the switches, and the ABR flows adapt to the available rate in a fairly short interval. We also demonstrate the scalability of the rate allocation algorithm with respect to the number of connections.

Keywords

Explicit rate allocation, congestion control, TCP over ATM.

1 INTRODUCTION

Asynchronous Transfer Mode (ATM) networks are being developed with the intent of providing a single common technology to carry voice, video and data traffic. Networks based on ATM combine the flexibility of packet-switched networks with the service guarantees and predictability offered by circuit-switched networks.

Several service classes have been defined in the context of ATM networks. The *Available-Bit-Rate* (ABR) service class [Giroux, 1995], defined to support delay tolerant best-effort applications, uses rate-based feedback mechanisms to allow them to adjust their transmission rates to make full utilization of the available bandwidth [Bonomi, 1995]. Compliant connections are also assured of a low loss rate, and if needed, a minimum bandwidth allocation. The ATM Forum Traffic Management Committee is currently defining a rate-based

*Supported by the Advanced Research Projects Agency (ARPA) under Contract No. F19628-93-C-0175 and by the NSF Young Investigator Award No. MIP-9257103.

congestion control framework to meet this objective. This framework allows a number of options for the switches to signal their congestion state to the source. With the *explicit-rate marking* option that is the focus of our work here, the source of each connection periodically transmits a special *resource management* (RM) cell. The source specifies the bandwidth demand and the current transmission rate of the connection in each transmitted RM cell. With the explicit rate scheme, switches communicate in the RM cell, the amount of instantaneous bandwidth it can allocate to each connection to the source of the connection. The goal of the allocation is to arrive at an efficient allocation that is also *max-min fair* [Bertsekas, 1992].

Several rate allocation algorithms using the explicit-rate option have been proposed [Charny, 1994, Kalampoukas, 1995a, Jain, 1995]. In this work we study the dynamics and evaluate the performance of the rate allocation algorithm proposed in [Kalampoukas, 1995a We consider the behavior of the rate allocation algorithm in both ATM-layer-generated ABR traffic and TCP-controlled ABR traffic. In the first case we show that, in the network configurations being analyzed, the algorithm converges to a steady state, allocates the available bandwidth fairly among competing connections, and has modest buffer requirements. We demonstrate its scaling capabilities by increasing the number of active connections by a factor of more than 10.

We also study the behavior of TCP-controlled ABR traffic. We demonstrate that the use of an explicit rate allocation scheme enhances the fairness achieved for TCP/IP traffic compared to its performance in traditional datagram networks.

The paper is organized as follows: Section 2 briefly reviews the rate-based congestion control framework and describes the proposed rate allocation algorithm. Section 3 provides a description of the simulation models used in this work. Section 4 discusses the dynamics of the described algorithm in a network configuration consisting of connections with widely-different round-trip times. We study the behavior of the network first with only ATM-layer generated traffic, and subsequently with ABR traffic that is flow-controlled by TCP. Section 4.3 studies the performance of ABR connections when mixed with VBR traffic. Section 4.4 investigates the performance of TCP in a dynamic environment, where the total bandwidth available to TCP connections is varied dynamically. Finally, Section 5 summarizes the results and proposes directions for future work.

2 SOURCE AND SWITCH BEHAVIOR

We describe in this section the source and destination algorithms used in our study. Due to lack of space we provide just an outline of the source and switch behavior. Detailed information on the ATM Forum's source policy as well as a complete description of the rate allocation algorithm under investigation can be found in [Bonomi, 1995, Kalampoukas, 1995b].

According to the ATM Forum's framework, the source of a connection (VC) transmits cells at a rate allowed by the network, termed the *allowed cell rate* (ACR), until it receives new information in an RM cell that it had transmitted previously. The source sends an RM cell every $Nrm - 1$ data cells transmitted. This proportional transmission of RM cells is to ensure that the amount of overhead for RM cells is a constant, independent of the number of VCs in the network or their rates.

The switches use the *explicit rate* option. Whenever an RM cell from connection is received at a given switch, the switch determines the allocation for the VC based on the bandwidth being requested in the explicit rate (ER) field. If the maximum bandwidth that can be allocated is less than the value in the ER field, then the ER field is updated to reflect the new maximum possible allocation on the path of connection so far.The explicit rate option assumes the existence of an algorithm within the switch that allocates the available bandwidth on each outgoing link among the connections sharing it. In this paper we consider the rate allocation algorithm described in [Kalampoukas, 1995a].

When the RM cell returns back to the source, the transmission rate (allowed cell rate, ACR) is updated based on the value indicated by the ER field in the returned RM cell. The value in the ER field reflects the bandwidth that can be allocated at the bottleneck link in the path of the connection. If the current transmission rate is above the value in the ER field, the source immediately reduces its rate to this value. However, if the current rate is less than the returned ER value, the transmission rate ACR is increased gradually by a constant amount ($Nrm \cdot AIR$) to the current ACR. In addition, the source is never allowed to exceed the rate specified by the ER field of the returned cell. Thus, if $r(t)$ is the transmission rate of the source the instant just before the arrival of an RM cell, and $r(t+)$ the rate after the update, then

$$r(t+) = \min(r(t) + Nrm \cdot AIR, ER).$$

3 SIMULATION ENVIRONMENT

In this section, we provide an overview of the simulation models used in the paper. More detailed description of a specific topology used in a simulation will be given in the corresponding sections describing the simulation results. The simulations were performed using the OPNET tool.

The links in the network are full-duplex with a capacity of 155 Mbits/sec each, unless otherwise specified. The switches are nonblocking, output-buffered crossbars. There is one queue per output port for ABR traffic and its scheduling policy is FIFO, with each output queue being shared by all the virtual circuits (VCs) that are sharing the outgoing link. We assume that all the switches support the explicit rate allocation algorithm of Section 2 and the sources follow the source policy described. The parameters for the source-end systems are set as follows:

- Nrm = 32 cells.
- ICR = PCR/50 ≈7300 cells/sec. This is the initial cell rate of the source
- AIR = 180 cells/sec per cell transmitted (about 5760 cells/sec maximum rate increase every Nrm cells).
- PCR = 365,566 cells/sec. This is the peak rate for all VCs (link rate).

An important observation we make is that the same set of parameters are used for both WAN and LAN configurations which have widely different characteristics.

When simulating TCP over ATM, we use the ATM Adaptation Layer Type 5 (AAL 5) [I363]. AAL 5 performs segmentation and re-assembly between IP packets and ATM cells. Each IP packet is extended by eight bytes to accommodate the AAL header. Thus, the number of ATM cells produced by the original IP datagram is given by $\left\lceil \frac{\text{IP packet size} + 8}{48} \right\rceil$.

The model for TCP used in the simulations is based on the TCP-Reno version. It supports the congestion control mechanism described by Jacobson [Jacobson, 1988], ex-

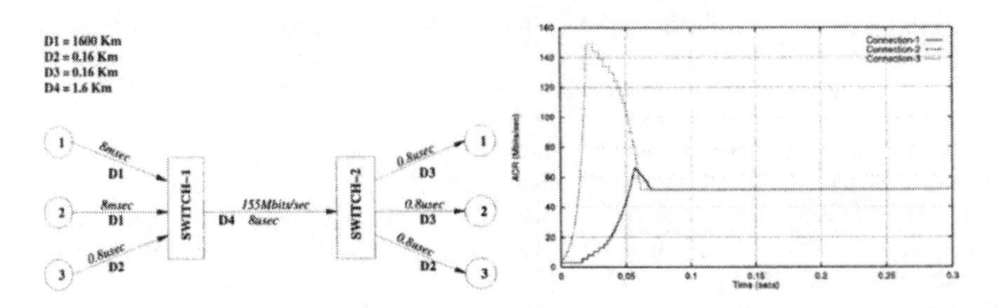

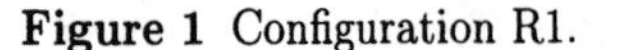
Figure 1 Configuration R1.

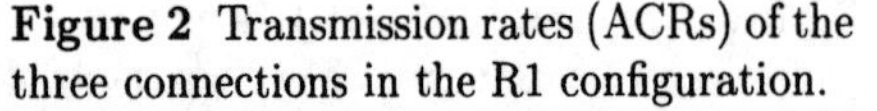
Figure 2 Transmission rates (ACRs) of the three connections in the R1 configuration.

ponential back-off, enhanced round-trip (RTT) estimation based on both the mean and the variance of the measured RTT, and the *fast retransmit and fast recovery* mechanisms.

We focus on the following performance measures: 1) The ACR, or transmission rate, at the source. In all the graphs we show the ACR value on every change, without any filtering of the collected information. 2) The utilization of the links. We present the utilization averaged over 5 millisecond intervals. 3) The queue length at the switch for individual links. We present in the plots the queue length, which is the maximum value observed during a 5 millisecond interval. 4) In the case of TCP traffic, we also show the TCP sequence number growth for each individual TCP connection and the corresponding window size in bytes, where appropriate.

4 SIMULATION RESULTS

In this section we consider the dynamics of the rate allocation algorithm in a network configuration where connections with widely different feedback delays interact. We first study the behavior of the algorithm with ABR traffic from cell sources, and subsequently characterize its behavior with TCP-generated ABR traffic.

4.1 Performance with ABR Traffic

We begin the evaluation of our rate-allocation algorithm with a simple configuration, referred to from now on as the *R1 configuration* , shown in Figure 1. It consists of three connections which open simultaneously and request peak bandwidth. Data flows from the end-systems on the left to the ones on the right. Connections are set up between corresponding end-systems, identified by the same index within circles. The link propagation delays and capacities are as shown in the figure. The reason we find the configuration R1 interesting is because of the large difference (three orders of magnitude) between the round-trip times of the different connections. The round-trip delay of connection 3 (to be referred from now on as the *short connection*) is 11.2 μseconds while that of connections 1 and 2 (from now on to be referred as *long connections*) is 16.076 milliseconds. We expect D4 to be the bottleneck link in the configuration. All the sources follow the source policy outlined in Section 2. The sources are assumed to be greedy, that is, they always set the ER field of every transmitted RM cell to the peak link capacity of 155 Mbits/sec.

Because of the large difference in the feedback delay between the long and short connections, and the small initial rates of the flows (ICR), we expect that the short connection (connection 3) will quickly ramp up to acquire a larger than fair share of the bottleneck

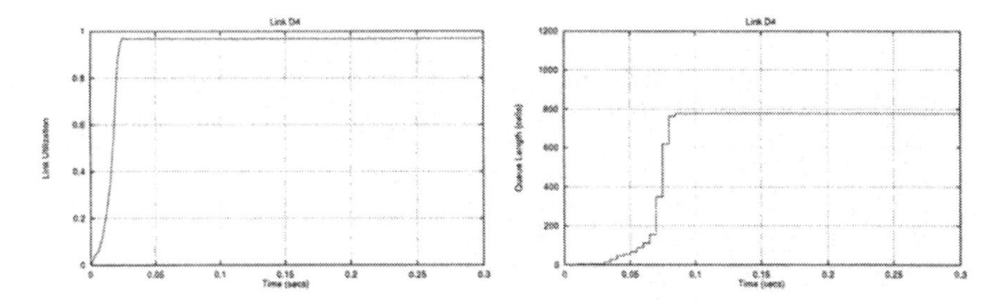

Figure 3 Total utilization of link D4 in the R1 configuration with three connections.

Figure 4 Queue length at the bottleneck link in switch-1.

link bandwidth. This initial start-up transient is clearly seen in Figure 2 which shows the exact evolution of the transmission rate at the source (ACR) for the three VCs. However, as time passes, the returned RM cells for the long connections allow them to acquire their fair share, and the short connection releases the bandwidth it acquired during the slow start-up of the long connections. Eventually, all the rates converge to their final allocation of one-third of the link bandwidth (about 51.5 Mbits/sec per connection).

The behavior of the rate decrease for the short connection during the start-up phase is gradual, rather than in a small number of discrete steps. This is due to the fact that the transmission rates for the long connections, which are being carried by the CCR field, increase gradually, thus resulting in a gradual decrease to the available rate to the short connection until the steady state is reached.

It is important to note the small overshoot in the transmission rates of the long connections before convergence is finally reached. This overshoot is a direct result of the allocation based on CCR values of the connections, and can be explained as follows: Assume that each of the two long connections transmits at time t_1 a forward RM cell with its CCR field containing transmission rates $r_1(t_1)$ and $r_2(t_1)$, respectively. These RM cells arrive at Switch 1 at time $t_2 = t_1 + 8$ msecs. Let $A_3(t_2)$ be the current allocation for the short connection in Switch 1 at that time. A computation for rate allocation is performed at time t_2 for each of the long connections. Assume that the RM cell from Connection 1 is the first seen by the switch. The updated ER value in its RM cell will now be $B - (A_3(t_2) + r_2(t_1))$, where B is the link capacity. It is easy to observe that, if $(A_3(t_2) + r_2(t_1))$ is less than $2B_{eq}$, the ER value signaled to Connection 1 can be larger than B_{eq}. The same ER value is also signaled to Connection 2 when its RM cell is processed. When these RM cells reach the sources of the long connections, the sources attempt to gradually increase their rates to the new ER values signaled, resulting in the rates exceeding the fair value B_{eq} temporarily. This overshoot is soon corrected when the increased CCRs of the long connections reach the switch, which clamps their allocations to B_{eq}. However, because of the long feedback delay, convergence to B_{eq} occurs slowly at the sources of the long connections.

Notice here that the transient bandwidth over-allocation may be reduced if we update the ER field of the RM cells going in the backward direction also. In that case, if the value in the ER field carried by an RM cell is larger than the most recent value of A_{max}, we update the ER field with the new A_{max}. This might improve the convergence of the rate allocation process in the general case; however, the modification would have little effect in this specific example because the congested switch is very close to the destination.

In Figure 2, the transmission rates of the sources converge to their final values within

70 msecs. Considering the 16 msecs round-trip delay of the long connections, this is quite reasonable, After the rate allocation process is completed, the transmission rates remain constant and the overall behavior is stable as long as the network state remains unchanged.

Although the feedback delay affects responsiveness of long connections to network changes, the utilization of the congested link is less affected. This is because of the short connection is able to utilize the excess bandwidth of the link while the long connections are gradually increasing their rates. Figure 3 shows the utilization of the link D4. Note that the utilization reaches its maximum value within approximately 25 msecs and remains constant thereafter. The maximum link utilization reached is about 97%, the theoretical maximum achievable after accounting for the overhead due to RM cells.

The transient bandwidth over-allocation causes a queue build-up in switch-1, as illustrated by Figure 4. The length of the queue is a function of the amount of the bandwidth over-allocation and the duration of the transient phase. As shown in Figure 4, however, even in a network with a round-trip delay of the order of 16 msecs, the built-up queue was relatively small, approximately 780 cells (approx. 40 Kbytes). Once built up, the queue size remains steady until a change in network state occurs, because our target link utilization is set at 100 %.

To examine how the allocation algorithm scales with the number of connections, especially with respect to its convergence time, the required amount of buffering and the bottleneck link utilization, we slightly extended configuration R1. The new configuration (the figure is omitted due to space constraints) contains 5 nodes on each side. Every source node on the left now originates eight connections, thus increasing the total number of VCs to 40. The round-trip delay for the two new sources was chosen identical to that of the long connections in Figure 1. Thus, the configuration consists of 32 connections with 16 msecs round-trip delay (long connections), and 8 connections with very small (about 11.2 μsecs) round-trip delay (short connections).

The transmission rates (ACR) for the connections in this modified configuration are shown in Figure 5. For simplicity, we have plotted only the transmission rate for a single connection in the set of connections originating at each source node. The behavior of the transmission rates is almost identical to that of the original R-1 configuration with three connections. However, convergence of the transmission rate to the final values is faster than before, taking only about 45 msecs (compared to 70 msecs with 3 connections). Therefore, the convergence time scales well with increasing number of connections.

As before, setting the target link utilization at 100% can lead to queue buildup at the bottleneck switch during the transient phase. However, the behavior of the queue size at the bottleneck link in switch 1, shown in Figure 6, indicates that the queue at the bottleneck link does not grow rapidly with the number of active connections. An increase by a factor of about 13 for the number of connections results in increasing the queue size by only a factor of 3. When we increased the number of short connections rather than the number of long connections in the R1 configuration, the increase in queue size was even smaller, about 30%.

The utilization of the bottleneck link achieves its maximum value somewhat quicker when the number of connections is increased. This is done in about 15 msecs, compared to 25 msecs for the same configuration with three connections. This is due to a combination of the relatively large aggregate initial rate for all the long connections and the fast ramp-up by the short connections due to their small feedback delay.

From our observations of the limited increase in the queue size, fast convergence and

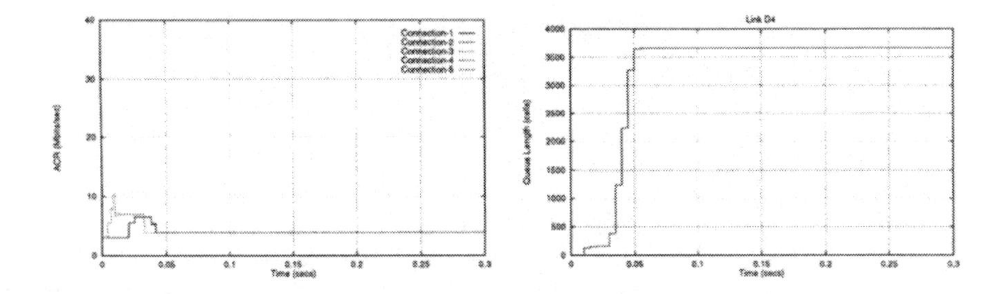

Figure 5 Transmission rates for the five sets of connections in the modified R1 configuration with 40 connections.

Figure 6 Queue length for the bottleneck link in switch-1 in the modified R1 configuration with 40 connections .

maintenance of high utilization, we believe that the allocation algorithm scales well with the number of connections.

4.2 Dynamics of TCP Traffic over ABR Service in a Network Configuration with Unequal Feedback Delays

The ABR traffic will not, in general, consist of ATM-layer-generated data only. Many applications use a transport protocol to provide reliable end-to-end transmission of data. Since TCP is currently the most widely used reliable transport protocol, ATM will likely be used widely as the datalink layer for the TCP/IP Internet as a means of evolving from the current infrastructure. In this subsection we study how the rate allocation algorithm at the ATM layer influences the behavior of TCP.

Of particular interest is to study how the TCP congestion control mechanisms affect the behavior of the rate allocation algorithm. The TCP congestion control algorithm is based on end-to-end windows and consists of several components. Key components are the *slow-start* algorithm, a congestion-avoidance mechanism, and an algorithm to estimate round-trip delays [Jacobson, 1988]. The TCP Reno Version, introduced in 1990, added the *fast retransmit and fast recovery* algorithm to avoid performing slow-start when the level of congestion in the network is not severe to warrant congestion recovery by slow-start.

For this study, we use the same R1 configuration considered in the previous subsection with two long connections and one short connection. The only difference is that the traffic of each ABR connection is now flow-controlled by TCP.

In addition to studying the initial start-up phase and the steady-state behavior of the connections, we also examine the dynamics of the connections when a packet loss occurs. This is achieved by dropping a cell from a TCP segment from connection 1 at time $t = 0.5$ seconds. In this case the AAL5 layer at the receiving end will detect a corrupted packet and discard all the remaining cells from that packet. The segment loss is later detected by the TCP source, which then retransmits the segment.

Figure 7 shows the ACR values at the sources of the three connections. The source rate behavior during the start-up phase is similar to that with cell sources, except for the more abrupt increase and decrease steps. This change in behavior is due to the TCP slow-start algorithm which increases the window size by doubling it every round-trip time. This produces intervals of time during which sources have no data to transmit. Since the source rate is allowed to increase only on the receipt of an RM cell, the idle intervals produce

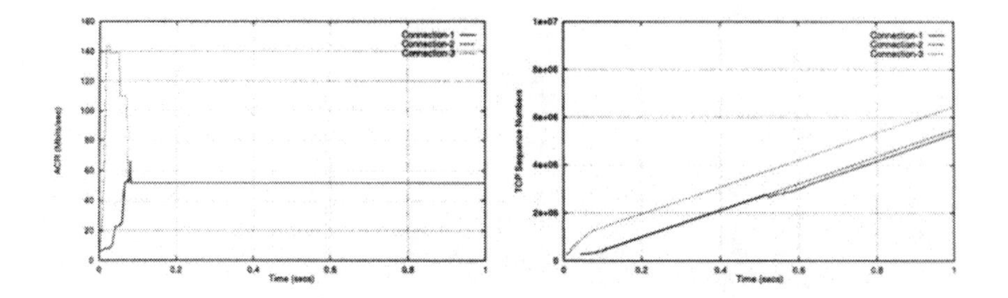

Figure 7 ACRs for the three connections in R1 configurations with TCP-generated traffic (in Mbits/second).

Figure 8 TCP sequence numbers for the three connections.

breaks in the rate increase process. However, the ACRs eventually converge to the fair values and remain steady. Note also that once the rate allocation algorithm converges, the behavior is similar to each connection operating over a dedicated link with no interference from other connection.

Figure 8 shows the increase in the sequence numbers of TCP segments transmitted by the three connections as a function of time. The plot for the short connection has a substantially higher slope during the slow-start phase, owing to its much smaller round-trip delay. However, in steady state, the rate of increase for all the connections is identical, demonstrating the effectiveness of the rate allocation scheme in providing fairness among connections with widely different round-trip delays.

The simulation results regarding the congestion windows (the graphs are omitted due to lack of space) show that the congestion windows for all three connections open to their maximum size of 150 Kbytes within 100 ms and remain at that state until 0.5 secs when a packet loss is simulated by discarding a single ATM cell from connection 1. The source TCP of connection 1 soon detects the segment loss and enters the fast-retransmit and fast-recovery phase, while the other two connections remain completely unaffected owing to the isolation provided by the rate allocation algorithm.

The simulation results for the overall utilization of the bottleneck link D4 (again the graphs are omitted due to lack of space) show that the maximum utilization of 97% is reached within 200 ms after start-up. The long period required for the link utilization to get maximized is caused by the TCP slow-start process. At time 0.5 second when the simulated cell loss happen, the utilization is transiently reduced because the ATM source policy we have implemented does not incorporate any provisions for recovering bandwidth from an idle source and therefore, the unused bandwidth of connection 1 during its recovery phase is not made available to other connections.

In summary, the simulation results in this subsection show that substantial improvements in fairness and efficiency in the operation of TCP can be obtained by the use of ABR service in conjunction with our rate allocation algorithm.

4.3 Performance of ABR Traffic Mixed with VBR Traffic

Up to now, we have focused on the rate allocation process when considering only ABR traffic. It is important to examine the ability of the scheme to adapt to changes in the available bandwidth, when there is a mix of high-priority traffic such as video and voice.

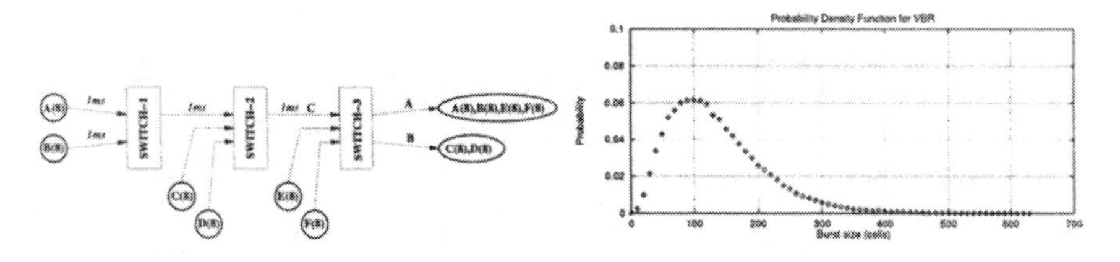

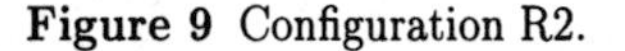

Figure 9 Configuration R2.

Figure 10 Probability density function for the burstiness of VBR traffic.

In this section, we study the effects of variable-rate real-time traffic (VBR traffic) on ABR traffic that is rate-controlled using our rate allocation algorithm.

In this set of simulation we use configuration R2 shown in Figure 9. In this configuration each source consists of four VBR and four ABR connections. Each of the sources originates eight connections, providing a total of 48 connections. All the ABR connections request peak bandwidth. The VBR connections have an allocated bandwidth, which is based on the average rate for the video data generated by the application.

The VBR traffic is based on the model described in [Heyman, 1992]. One frame of video data is generated approximately every 1/25 seconds (the model assumes a PAL system, not an NTSC system which transmits 30 frames/sec) and the size of the frame expressed in number of cells follows the probability density function given in Figure 10. The resulting process has a distribution that generates data with an average rate of about 1.5 Mbits/secs. We reserve, in all the links on the path from the source to the destination, a bandwidth of 1.9 Mbits/sec for each VBR connection. Thus, the average utilization of the reserved bandwidth for a VBR connection is expected to be about 75%.

The VBR traffic generated with the model described earlier may exhibit very bursty behavior. In order to limit the burstiness of each VBR connection, we shape its traffic using a token bucket. The bucket size is set to 50 cells and the rate of token arrival was set to be equal to the bandwidth reserved to each connection, that is 4,500 cells/sec=1.9 Mbits/sec. In order to avoid any synchronization between the video streams as much as possible, the corresponding VBR connections open at random times that are uniformly distributed in the interval (0,50 msecs).

We assume that the VBR and ABR classes of traffic are buffered in the switches in separate queues. Scheduling between the two classes is based on static priorities, with the VBR traffic always taking higher priority. Thus, the switch transmits an ABR cell only when the VBR queue is empty. Since we are primarily interested in the effect of VBR service on the ABR class, we use a single FIFO queue for all VBR traffic. As always, we assume that the ABR traffic share a single common FIFO queue at each outgoing link of a switch.

Figure 12 shows the ACR for a representative ABR connection from each source node (other connections from the same source exhibit similar behavior). The rate allocation process converges quickly (within 50 msecs) to the final allocation. After convergence, the rate allocated to each connection is a fair share of the available bandwidth: this is the total bandwidth minus the bandwidth reserved for VBR traffic. Here, the bandwidth available to ABR traffic on links A and C is 124.6 Mbits/sec and therefore each ABR connection has an available capacity of about 7.8 Mbits/sec.

The expected link utilization is close to 100% for link A and 50% for link B. However, the instantaneous link utilization (averaged over 5 msec intervals) exhibits spikes as shown

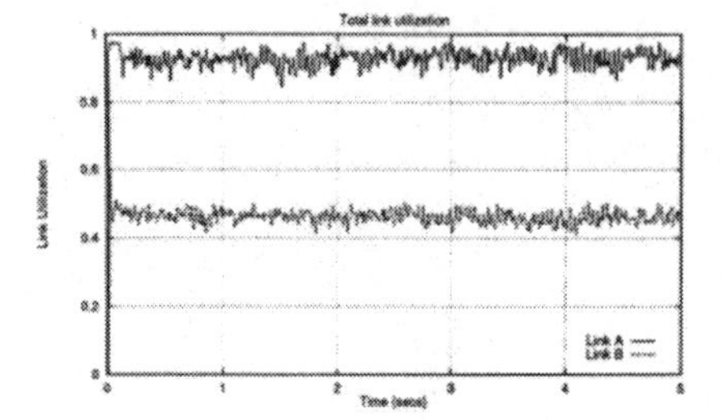

Figure 11 Utilization of links A and B.

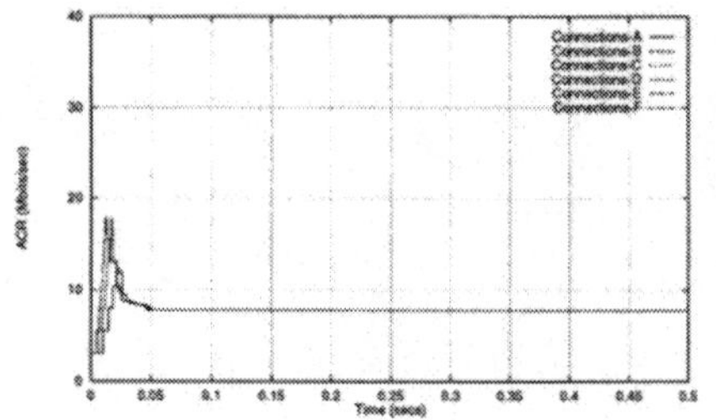

Figure 12 ACR for each connection set.

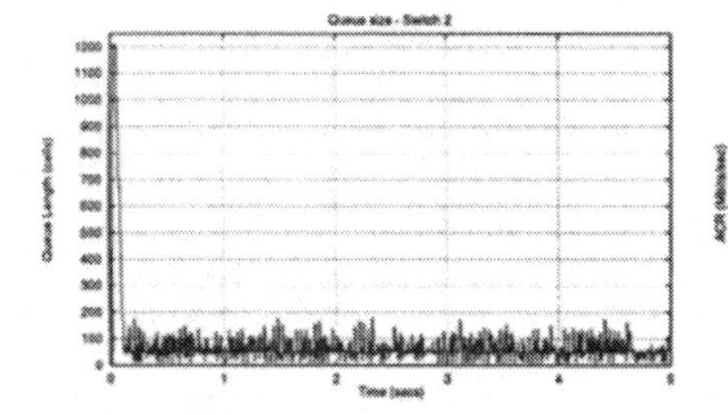

Figure 13 Queue size for ABR traffic – Switch 2.

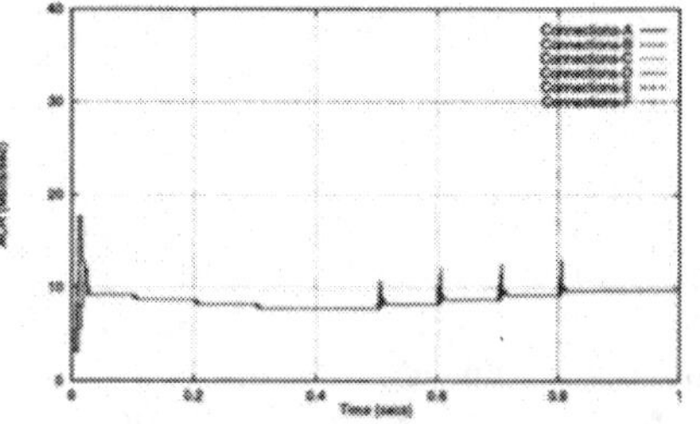

Figure 14 ACR for each connection set with staggered opening and closing of VBR connections at 100 millisecond intervals.

in Figure 11. Accounting for the overhead of 3% of the available bandwidth for RM cells, the utilization of the links is close to the maximum attainable for both the links A and B. The difference is simply because of the over-allocation of bandwidth that we did for the VBR connections: on the average, the VBR traffic should utilize only 75% of its allocated bandwidth. However, this conservative over-allocation for the VBR connections has the desirable side effect of maintaining the queue sizes small. Although the utilization is kept high, the queue sizes for ABR traffic remain small even in the presence of VBR traffic. The queue length for ABR traffic in switch 2 is shown in Figure 13. The queue behavior for the other switches is similar. The queue sizes have an average of about 50 cells (which is also the size of the token bucket of a VBR connection) and a maximum of about 200 cells in steady state.

Figure 14 shows the behavior of the ACRs of the ABR connections when the VBR connections open in a staggered fashion. Six VBR connections (one from each source) open every 100 msecs starting at time $t = 0$ seconds. Then, at time $t = 0.5$ seconds, these VBR connections start closing in a similar manner. As shown in the figure, the convergence to the final allocation after each change in available bandwidth is rapid. In the worst case, the rate allocation process is completed within 20 msecs.

Our results suggest that the rate allocation algorithm will perform well even in the presence of different service classes. For the specific configuration considered, its performance was efficient and scaled well with the number of connections. However, further work is needed to understand in greater detail the dynamics of the algorithm in more general network configurations.

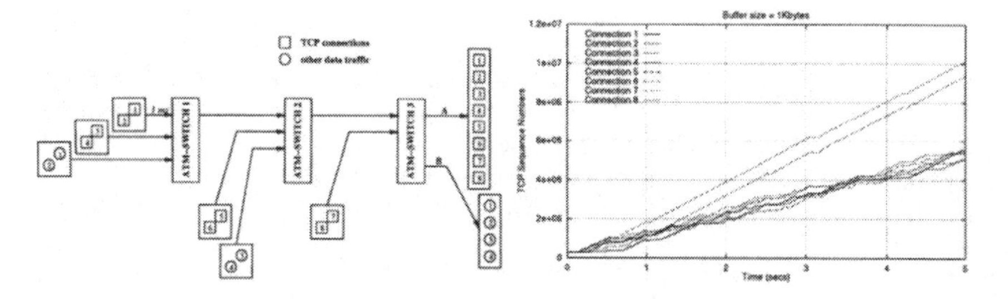

Figure 15 The R3 network configuration. **Figure 16** TCP sequence numbers (1 Kbytes buffer size).

4.4 TCP in a Dynamic Environment

In this section, we study the performance of TCP in a network configuration in which the bandwidth allocated to TCP connections is not constant.

The configuration we use is shown in Figure 15 and will be referred to as *R3*. Two types of connections are simulated: TCP-controlled connections and ABR cell sources. The TCP connections are kept open throughput the simulation. To simulate a dynamic environment, the ABR cell sources open and close frequently according to an exponential ON/OFF model. The mean duration for both the ON and OFF periods is set to 100 ms.

The random opening and closing of the ATM-layer connections trigger frequent re-computations of the allocations at the switches. To observe the effect of possible transient over-allocations during the convergence of the algorithm, we chose a very small buffer size of 1 Kbyte for the switches, making a packet loss very likely if a transient over-allocation occurs during convergence of the algorithm after a change in the connection states.

Figure 16 shows the progress of the sequence numbers for the eight TCP connections. All of the connections make steady progress. Connections 7 and 8, being closest to the destination, are able to use more than their fair share during transient periods. Note that this behavior does not represent any inherent unfairness in the rate allocation algorithm, but is due to the delay of the control loop. The aggregate throughput sustained by the TCP connections even with only 1 Kbyte of buffering is about 60% of the maximum attainable.

5 CONCLUSION

In this paper, we evaluated the performance of the explicit rate allocation algorithm presented in [Kalampoukas, 1995a] in a variety of network configurations and with a diverse range of workloads. Traditionally, feedback-based congestion control mechanisms have shown a bias towards flows with shorter round-trip times when they co-exist with long round-trip time flows. Because of the rate allocation mechanism we use here, this bias is eliminated, and we see a fair allocation even under the extreme condition where the round-trip times differ by three orders of magnitude.

We also show that the proposed allocation algorithm retains several desirable characteristics as we scale in the number of connections. Increasing the number of the connections by a factor of 13, the queueing requirements go up only by a factor of 3 with the increase from 2 to 32 of the long-round-trip time connections (with 16 milliseconds round-trip time).

An important requirement for the ABR service is that it mesh well with traditional higher-layer protocols such as TCP/IP. It is well known that TCP exhibits unfairness when multiple TCP connections sharing a bottleneck link have widely different round-trip times. We show that TCP running over ABR avoids this unfairness, there is a dramatic reduction in the queueing requirements at the bottleneck link, and no packet losses occur due to congestion.

Another important need is for ABR flows to operate well when there are higher-priority VBR flows co-existing with the ABR traffic. In the *parking lot* configuration with 48 connections (equally divided as 24 ABR and 24 VBR flows), we achieved fairness, and full utilization of the bottleneck links. In spite of the burstiness of the VBR traffic and the greediness of the ABR flows, the queue sizes were of the order of only 100 cells, which is quite reasonable.

In the future we plan to study the interaction of the proposed rate allocation algorithm with non-cooperative sources. Also, we would like to examine the system performance with not all flows use their stated bandwidth, as specified in the CCR field.

REFERENCES

[Giroux, 1995] N. Giroux and D. Chiswell, "ATM-layer traffic management functions and procedures," in *Proceedings of INTEROP '95 Engineer Conference*, March 1995.

[Bonomi, 1995] F. Bonomi and K. W. Fendick, "The Rate-Based Flow Control Framework for the Available Bit Rate ATM Service," *IEEE Network*, vol. 9, no. 2, pp. 25–39, March/April 1995.

[Bertsekas, 1992] D. Bertsekas and R. Gallager, *Data Networks.* Prentice Hall, 2nd ed., 1992.

[Charny, 1994] A. Charny, "An Algorithm for Rate Allocation in a Packet-Switching Network with Feedback," Master's thesis, Massachusets Institute of Technology, May 1994.

[Kalampoukas, 1995a] L. Kalampoukas, A. Varma, and K. K. Ramakrishnan, "An efficient rate allocation algorithm for ATM networks providing max-min fairness," in *Proceedings of 6th IFIP International Conference on High Performance Networking, HPN'95*, September 1995.

[Jain, 1995] R. Jain, "Congestion Control and Traffic Management in ATM Networks: Recent Advances and A Survey." submitted to Computer Networks and ISDN Systems.

[Kalampoukas, 1995b] L. Kalampoukas, A. Varma and K. Ramakrishnan, "Dynamics of an Explicit Rate Allocation Algorithm for Available Bit-Rate (ABR) Service in ATM Networks," Tech. Rep. UCSC-CRL-95-54, University of California, Santa Cruz, December 1995.

[I363] CCITT, *Draft Recommendation I.363.* CCITT Study Group XVIII, Geneva, January 1993.

[Jacobson, 1988] V. Jacobson, "Congestion avoidance and control," in *Proceedings of ACM SIGCOMM'88*, pp. 314–329, 1988.

[Heyman, 1992] D. P. Heyman, A. Tabatabai, and T. V. Lakshman, "Statistical analysis and simulation study of video teleconference traffic in ATM networks," *IEEE Transactions on Circuits and Systems for Video Technology*, vol. 2, no. 1, pp. 49–59, Mar. 1992.

11

Comparison of ER and EFCI Flow Control Schemes for ABR Service in Wide Area Networks

Aleksandar Kolarov and G. Ramamurthy
C&C Research Laboratories, NEC USA Inc.
4 Independence Way, Princeton, NJ 08540, USA.
email: kolarov,mgr@ccrl.nj.nec.com

Abstract

An adaptive end-to-end rate based congestion control scheme to support a class of best effort service known as Available Bit Rate Service (ABR) is being proposed by the ATM Forum. In this paper we investigate two variants of this control scheme. In the first scheme known as the Explicit Forward Congestion Indication (EFCI) scheme, a single bit is used to convey to the source the state of the network. In the second scheme, known as the Explicit Rate (ER) scheme, the network informs the sources from time to time, the maximum rate at which they can transmit. We investigate the steady state and transient performance of the controls in a wide area multihop network, including the presence of high priority variable bit rate traffic. To improve the performance of the EFCI scheme in multihop networks, we propose a priority based EFCI scheme. The EFCI scheme exhibits a robust behavior, and ensures fair share of the bandwidth for all VC's in the long run, regardless of the number of hops they traverse. However, the EFCI mechanism is fundamentally oscillatory in nature and can lead to large cell loss. The ER scheme is very stable, even under extreme loading conditions, and ensures fair sharing of resources.

Keywords

Flow Control, ABR ATM Service, Broadband WAN

1 INTRODUCTION

Asynchronous Transport Mode (ATM) can simultaneously switch traffic with real time constraints such as voice or video along with data traffic, and thus provides an ideal platform for supporting multi-media based applications. There is tremendous interest in the development of ATM-based broadband networks that will provide high bandwidth connectivity between work-stations and servers, ATM Hubs and WAN's that will interconnect geographically separated LAN's.

ATM-based switches have been generally characterized by small internal buffers, ranging from a few hundred cells to a few thousand cells. Further, to provide a fast and lean transport, all link level controls have been eliminated, and one has to resort to end-to-end based controls [Prycker 1993] . If there are traffic hot spots (i.e., traffic from several bursty sources are directed to a single output port), statistical multiplexing of ATM cells with no mechanism to control the source rate will lead to severe cell loss. This loss will be aggravated if the input port speeds are comparable to the output port speeds and the burst sizes are large. Such cell loss will lead to even larger frame loss compared to other technologies that have a media access control. Since ATM interleaves cells from different connections, when cell loss occurs, it can lead to loss of all the frames that are currently being transmitted. Frequent retransmissions due to cell loss increases the effective load on the system, resulting in an end-to-end throughput that is several times less than that of a shared medium network [Kolarov 1994a].

The ATM Forum has defined a new service class for data applications called the Available Bit Rate Service (ABR) [ATM Forum 1995]. Users of this service dynamically share the available bandwidth. While this service does not provide any strict guarantees, it attempts to minimize the cell loss at the expense of delay. The dynamic sharing of bandwidth between competing users has to be achieved via appropriate set of distributed controls. Two end-to-end rate based feedback congestion control mechanisms have been proposed. In the first scheme known as the Explicit Forward Congestion Indication (EFCI), a single bit is used to indicate if a switch in a virtual channel's (VC) path is congested or not. The switches use the ECI bit in the header of ATM data cells to notify the destination if they are congested. The destination filters this information and signals the source through a special control cell. This binary scheme is similar to the DEC bit scheme for packet-switched networks [Ramakrishnan 1990]. In the second scheme known as the Explicit Rate (ER) scheme, instead of a single bit feedback, the switches explicitly specify the maximum rate each VC is allowed to use. The switches will compute the rate for each VC based on the state of its queues, the available capacity for ABR service and the number of active sources.

While the thrust for ABR service comes from LAN providers, such services should also be capable of being operated across wide area networks. The effectiveness of any feedback control scheme is limited by the latency of the feedback loop. Hence, such end-to-end controls are likely to be less effective as the propagation delay and the bandwidth of the network increase. In particular, when there are virtual channels that traverse several hops, extreme unfairness can result. Virtual channels whose feedback delays are smaller and thus have more up to date information can have an unfair advantage over virtual channels that have larger feedback delays. In a wide network environment, the latency of the feedback loop coupled with the fact that the amount of buffers at each node can be less than the bandwidth delay product, can lead to significant degradation in the network performance. In fact, the network throughput can collapse at high offered loads if one take into account the retransmission traffic. In this paper we compare the performances of the EFCI and ER based end-to-end congestion control schemes in a WAN environment. We investigate the performance under both steady state and transient conditions, including the presence of high priority Variable Bit Rate (VBR) cross traffic.

2 REFERENCE MODELS

In order to compare the performance of different competing control schemes, we define a reference network model and a reference traffic model.

2.1 Reference Network Model

Figure 1 shows one of the reference network models, adopted by the ATM Forum [Kolarov 1994b, Ramamurthy 1994] that consists of a multi node network, with three local switches (Switch 1, Switch 3 and Switch 4), and one or more transit switches (Switch 2). Terminals generating traffic are attached to the input ports of all four switches, while terminals receiving traffic are attached to the output ports of switches 3 and 4 only. The terminals are connected to the switches through links running at 155 Mbits/sec. The switches are interconnected via links running at 155 Mbits/sec. Switches receive traffic from terminals that are attached to their input ports. Switches 1, 3 and 4 also receive traffic from upstream switches. The routing of traffic is such that traffic originating at switch 1 traverses 3 hops, traffic originating at switch 2 traverses 2 hops and traffic originating at switches 3 and 4 traverse 1 hop. Further, all $n(n = 1,2,3)$ hop traffic compete with $m(m = 1,2,3,\ m \neq n)$ hop traffic. Each transmit terminal has one virtual channel that is terminated on a unique receive terminal. Virtual channels constituting the $n(n = 1,2,3)$ hop stream are grouped into two groups of 8 and 4 virtual channels respectively. The distance between the terminals and their respective access switch is $D_1 = D_4 = 25$ km. The distance D_2 between switch 1 (representing a LAN) and switch 2 (representing a transit switch), and the distance D_3 between switch 2 and the terminating switches 3 and 4, is 1000 km. When there are two or more transit switches, the distance between them is also 1000 km.

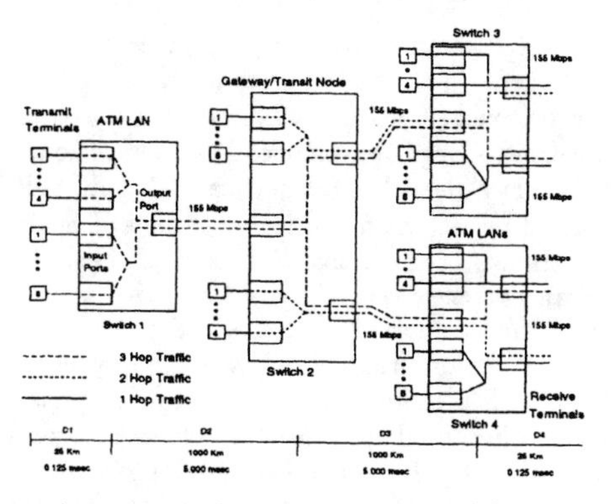

Figure 1 Reference network model

This reference model captures:

- The interference between traffic traveling 1, 2 and 3 hops.
- The effect of large propagation delay on the effectiveness of the feedback control.
- Fairness between 1, 2 and 3 hop traffic.

2.2 Switch Architecture

To evaluate the steady state performance, each switch is modeled as a generic input buffered switch with 1000 cell buffers at each input port. Each output port has 128 cell buffers. Internally, the switches use a random in random out (RIRO) scheduling to move cells from the input buffers to the appropriate output buffers. This RIRO scheduling eliminates the head of line (HOL) blocking effect common to most input buffered switches.

Further, selective flow control between the output and input buffers eliminates cell loss at the output buffers.

2.3 Source Model

Each terminal generates traffic based on 3 -state model (Figure 2), also adopted by the ATM Forum [Kolarov 1994b, Ramamurthy 1994]. A source can be either in an ACTIVE

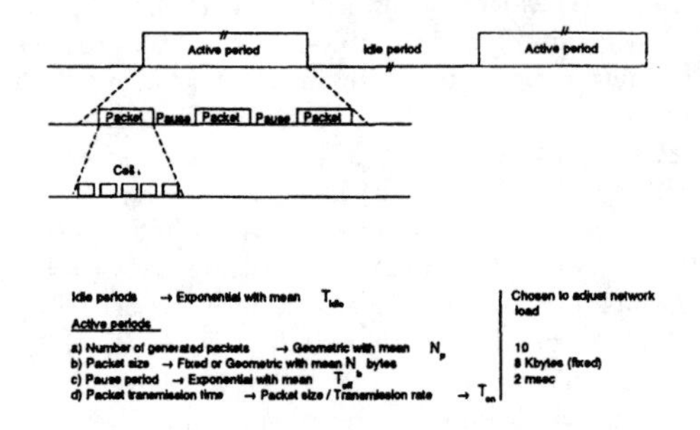

Figure 2 Source traffic model

or an IDLE state. In the IDLE state no traffic is generated. While in an ACTIVE state, the source generates a series of packets or bursts which are interspersed by short pauses. Each packet can either be fixed (8 or 64 Kbytes) or can be a truncated exponential. Each pause period is drawn from a negative exponential distribution. The number of packets generated during an ACTIVE period is geometrically distributed. The IDLE periods can have any general distribution. In the numerical examples of Section 4 where we examine the steady-state performance of the congestion control mechanisms, the pause periods between packets have an average value of 2 msec and the packets have a fixed length of 8 Kbytes. The length of the idle period is determined by the offered load on the link.

We use a simple retransmission protocol in case of cell losses, which occur during periods of congestion. A packet is presumed to be lost even if a single cell is lost. Packets that are received by the receive terminal with missing cells are retransmitted by the transmit terminal until successful delivery. Packet retransmission is scheduled with a back of delay equal to twice the current (estimated) round trip delay. The useful throughput represents the actual throughput of packets (in Mbits/sec) that are eventually delivered to the destination without cell loss (after retransmission if necessary).

To evaluate the transient response of the control, we use persistent sources with infinite backlog.

2.4 Basic Source and Destination Operations [ATM Forum 1995, Bonomi 1995]

The rate at which an ABR source is allowed to schedule cells for transmission is denoted by Allowed Cell Rate (ACR). The ACR is initially set to the Initial Cell Rate (ICR) and is always bounded between the Minimum Cell Rate (MCR) and the Peak Cell Rate (PCR). Transmission is initiated by the sending of a Resource Management (RM) cell followed by data cells. The source will continue to send RM cells after every $(N_{RM} - 1)$ data cells are transmitted. The source rate is controlled by returning RM cells, which contain information about the state of the network. The source places the rate at which it is currently transmitting (the ACR) in the CCR field of the RM cell, and the rate at which it wishes to transmit cells (usually the PCR) in the ER field. The RM cell travels forward through the network, providing the switches in its path with information about the state of the source. The switches use this information to determine the allocation of bandwidth among ABR connections. The switches can also decide to reduce the value of the Explicit Rate (ER) indicated by the ER field, or set the Congestion Indication (CI) bit to 1. Switches only supporting the EFCI mechanism will ignore the content of the RM cell. When the RM cell arrives at the destination, the destination changes the direction bit in the RM cell and returns the RM cell to the source. If the destination is congested and cannot support the rate in the ER field, the destination can reduce the value of ER to whatever rate it can support. When returning an RM cell, if the destination had observed that the last data cell received had its EFCI bit set, then it should set the RM cell's CI bit to indicate congestion. As the RM cell travels back to the source through the network, each switch may examine the cell and determine if it can support the rate ER for this VC. If ER is too high, the switch can reduce it to a rate that it can support. No switch can increase the value of ER in the RM cell since prior switch congestion information would be lost.

When the RM cell arrives back at the source, the source should modify its rate, ACR, based on the information carried back by the RM cell. If the congestion indication bit is not set (i.e., CI = 0), then the source is allowed to increase its rate ACR by an amount $N_{RM} * AIR$, where AIR is additive increase in rate. ACR can be increased up to the ER value contained in the last received RM cell, but never exceeding PCR. If the congestion indication bit is set (CI = 1), then the source should decrease its rate by at least $ACR * N_{RM}/RDF$ where RDF is Rate Decrease Factor (an exponential decrease). ACR is further decreased to the returned ER value, although never below the MCR.

When a source starts transmitting after being idle, if the elapsed time since the last RM cell was sent is greater than $T_{rm} = 100$ msec, and during this interval if the source did not send more than $M_{rm} = 2$ cells, then the next cell to be sent out is an RM cell. Before sending an RM cell, if the time T that has elapsed since the last RM cell was sent is greater than $T_{of} * N_{RM}$ (Time Out Factor $T_{of} = 2$) cell intervals (of 1/ACR), and if ACR is greater than ICR, then the ACR should be reduced to at most the rate obtained as follows: $\frac{1}{ACR_{new}} = \frac{1}{ACR_{old}} + \frac{T}{RDF}$. If the new rate is smaller than ICR, then it should be set to ICR. The last constraint aims at protecting the network from the impact of sources that, having gone idle at a high ACR, do not claim large bandwidth as soon as they become active, and lead to possible congestion.

3 SWITCH BEHAVIOR

We compare the performance of two different types of switches. The first type of switches are EFCI-based switches which set the EFCI bit in data cells to indicate congestion. The second type of switches are Explicit Rate-based switches which modify the ER field of the RM cell to indicate the rate at which a VC may transmit. In this paper we assume that the ER setting is performed on backward RM cells. We now describe in detail a control algorithm for each type of switch.

3.1 Switch Control with EFCI Marking

The EFCI signal is generated at a switch on a per VC basis by setting the EFCI bit in the cell header of data cells when the output port is congested. The output port is marked congested if the total number of ABR cells waiting at all the input ports and are destined to be routed to this output port exceeds a high water mark H_2 (500 cells). We will refer to these waiting ABR cells as the global queue fill with respect to a given output port. The output port remains in the congested state until the global queue fill falls below the low water mark L_2 (300) cells). This hysteresis ensures that the oscillations are minimized. When a cell arrives at an input port, the EFCI bit is set if the output port to which the cell must be routed is congested. The cell is then routed towards its destination.

In [Kolarov 1994a, Kolarov 1995] we show that with end-to-end based controls, VC's traversing two or more hops can see a throughput collapse at high loads. To overcome the unfair advantage that VC's traversing small number of hops have over VC's traversing larger number of hops, we proposed an access priority mechanism where transit traffic from upstream switches are given priority of service over new traffic entering the network at each switch.

3.2 Switch Control with ER Marking

The primary consideration in the design of a congestion controller for ER based switches is to ensure good dynamic characteristics, high utilization and fairness in resource allocations amongst competing VC's [Benmohamed 1993]. We call our explicit rate controller a Predictive Explicit Rate Controller (PERC), since we use a predictive control law for the operation of the congestion controller. The main idea is to compute a rate which will bring the queue fill of a given output port to a desired threshold level x^0 in a fixed number of update intervals denoted by D.

The level of congestion at the output port is estimated by monitoring the difference between the queue fill x and a queue set point x^0. Based on this difference, the congestion controller associated with each output port periodically calculates an explicit rate that is common to all VC's using the given output port. The control algorithm updates the explicit rate once every T msec. In the examples presented in this paper, T is equal to the time required to transmit 400 cells at rate 155 Mbits/sec, which is approximately 1 msec. In the numerical example, the queue set point x^0 is set equal to 90 cells. We choose D to be 10 which corresponds to an average round trip time of 10 msec. We assume that time is slotted, where each slot is T msec long. The rate controller equation is given by:

$$R(n) = min\left\{\hat{C}, \left[\hat{C} - \frac{(x(n) - x^0)}{D}\right]^{\dagger}\right\}\frac{1}{\gamma\hat{N}}, \qquad (1)$$

where $R(n)$ is the explicit rate for each VC, computed at the n-th time slot ($n = 0, 1, ...$) $\hat{C}$ is the estimated link capacity available for ABR service, $x(n)$ is the queue occupancy at the n-th time slot, x^0 is the desired level of queue occupancy, D is a constant described above, γ is the link utilization with respect to ABR service ($0 < \gamma \leq 1$), and $\hat{N}$ is the estimated number of active ABR VC's using the given port. The available capacity for ABR service can be computed as the difference between the total port capacity and the bandwidth reserved for other services, such as CBR and VBR. The estimate $\hat{N}$ can be obtained by monitoring the arrivals of RM cells in the forward direction, on a time interval that is longer than the basic update interval T (e.g. $4T$)). At each switch the ER field in the backward RM cell is modified as $ER = min\{R(n), ER^*\}$ where ER^* is the value of ER in the received RM cell at the switch.

4 NUMERICAL RESULTS

4.1 Steady State Performance

To investigate the performance over larger distances as in WAN's, we set the inter switch distances D_2 and D_3 to 1000 km (5 msec propagation delay). First, we compare the EFCI schemes with and without priority for transit traffic. The source parameters are chosen as follows: $PCR = 365$ cells/msec (155 Mbits/sec), $ICR = 25$ cells/msec (10 Mbits/sec), $MCR = 4$ cells/msec (1.5 Mbits/sec), $N_{rm} = 32$, $RDF = 1024$, and $AIR = 0.15$ cells/msec (0.06 Mbits/sec). This choice is in line with the ATM Forum recommendation. Figure 3 shows the plots of 1, 2 , and 3 -hop source useful throughput (at the packet level) in the case of EFCI scheme with and without priority for transit traffic, as a function of end-to-end delay. The numbers in parenthesis indicate the overload factor on each link. For example an overload factor of 1.4 means that the average offered load on a link is 1.4 times its maximum capacity. As the offered load increases, the packet level throughput decreases and the end-to-end delay increases.

From Figure 3 we see that without priority for transit traffic the 2 and 3 -hop VC's experience a throughput collapse under overload. However, the 1 -hop source throughput is high with low delay even under overload. Thus, VC's traversing two or more hops have significant performance degradation. This performance degradation results from the fact that a VC traversing a larger number of hops is more likely to loose a cell due to blocking at some intermediate switch, compared to VC's traversing fewer hops. In particular, the 3 -hop source throughput collapses when the loading factor of the link is 0.8 . Note that the end-to-end delay for 3 -hop VC's at a low link load is close to the propagation delay of 10 msec. At a high load of 1.4 times the link capacity, the end-to-end delay increases to 100 msec (a 10 fold increase) while the useful throughput reduces by 33 % from its peak throughput. The 1 -hop VC's incur the least degradation. While the 1 -hop throughput monotonically increases, the end-to-end delay even under overload only shows a small increase. On the other hand, with priority for transit traffic, the 1, 2 and 3 -hop traffic throughputs are nearly equal at all offered loads. For example, at an offered load of 0.8 times the link capacity, the throughput of all VC's are around 8 Mbits/sec. Note that the end-to-end delays are different since the propagation delays are different. Even with an overload factor of 1.4, the 2 and 3 hop VC's do not show any throughput collapse, and the end-to-end delay shows only a marginal increase. In the rest of the text, we will only consider the EFCI scheme with priority for transit traffic, and compare it with the PERC scheme.

In order to allow a more aggressive source behavior in networks with ER based switches, we change the two source parameters: $RDF = 512$ and $AIR = 0.70$ cells/msec (0.3 Mbits/sec). Figure 4 shows the plots of 1, 2 , and 3 -hop source useful throughput (at the

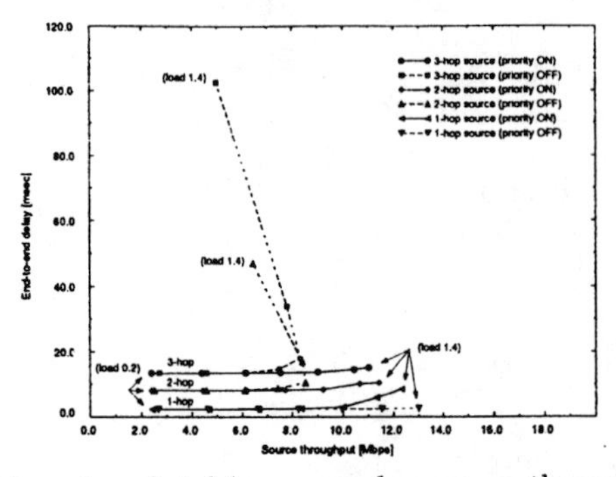

Figure 3 End-to-end packet delay vs. per hop source throughput under EFCI

packet level) in case of the EFCI scheme with priority for transit traffic and the PERC scheme, as a function of end-to-end delay. Note that in the PERC scheme, transit traffic do not have priority over traffic entering the network. We observe that at higher loads, all

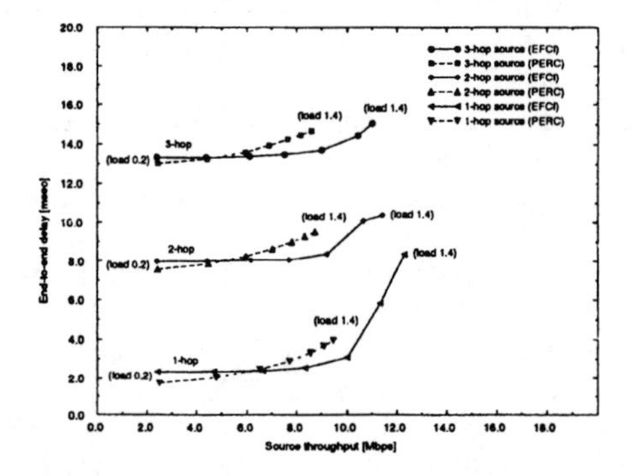

Figure 4 End-to-end packet delay vs. per hop source throughput under EFCI and PERC

VC's under the EFCI scheme have a higher throughput than under the PERC scheme. On the other hand, the end-to-end delay characteristics are worse in case of the EFCI scheme. It is important to emphasize the PERC scheme shows more fairness in bandwidth allocation amongst the competing VC's than the EFCI scheme. In case of PERC, the 1, 2 and 3 -hop traffic throughput are nearly identical at all offered loads. The reason why VC's have a lower throughput under the PERC scheme, is due to the fact that the Allowed Cell Rate (ACR) is obtained as a minimum function of explicit rates obtained from all switches on the VC route. Since at any given time some switch is likely to be in congestion, the source rate is always constrained by the bottleneck rate on its path. The PERC scheme is closer to a pure rate control and does not fully utilize the store and forward feature of packet switched networks. In attempt to minimize cell loss by keeping the queues small, at the expense of overall utilization. The EFCI scheme on the other hand uses the buffers better and can achieve higher utilization at the expense of cell loss. One must note that we are comparing the EFCI scheme with priority with the PERC scheme without priority.

The examples considered so far assume that all the link capacity was available for ABR service. In reality, there will be CBR and VBR traffic that have higher priority than ABR traffic. To reflect this condition, at each output port of switches 3 and 4 we replace one 1 -hop ABR source by a VBR source. We model the VBR traffic as a first order autoregressive process [Ramamurthy 1993]. We assume that the VBR traffic arrival rate changes every 33 msec, and the rate $R_{vbr}(n)$ in the n -th frame is related to the rate $R_{vbr}(n+1)$ in the $(n+1)$-th frame through:

$$R_{vbr}(n+1) = \alpha R_{vbr}(n) + y(n), \tag{2}$$

where $y(n)$ is normally distributed. In our example, we choose the VBR mean arrival rate R_{vbr}^{peak} = 60 cells/msec (25 Mbits/sec), the peak rate R_{vbr}^{peak} = 365 cells/msec 155 Mbits/sec), the correlation coefficient $\alpha = 0.9$, and the squared coefficient of variation in the arrival rate $C^2 = 1$. With these parameters the VBR source peak rate can be as high as the line rate. The VBR source has an average link utilization of 16%. Comparing Figures 4 and 5, we see that the ABR VC performance (in Figure 5) degrades for both schemes, especially with regard to the end-to-end delay characteristics. However, the performance degradation is much less in case of the PERC scheme which better uses the knowledge about the available bandwidth for ABR service.

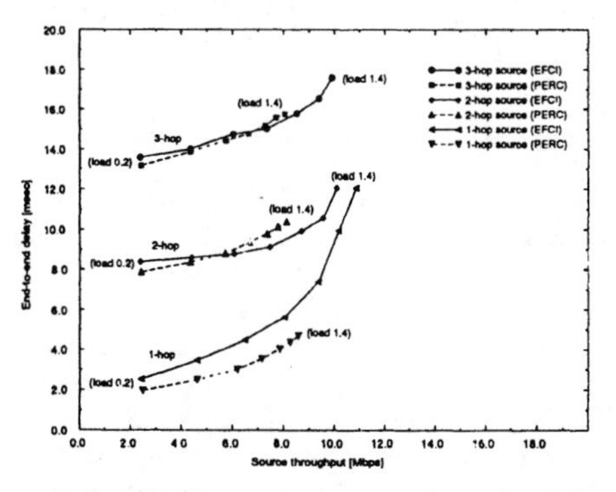

Figure 5 End-to-end packet delay vs. per hop source throughput (VBR case)

4.2 Transient Responses

In this section we study the transient performance of the two controls. There are two key performance for consideration.

- The response of the control when the load on the network changes suddenly. Of particular interest is the time it takes for the system to reach a stable state (if at all it does), and the loss that results under transient overload.
- Fair sharing of the bandwidth between competing users, independent of the number of hops each connection traverses.

Figure 6 shows a peer to peer configuration, where two switches SW1 and SW2 are interconnected by a 1000 km link. There are 3 groups of persistent sources (with infinite backlog). Group A has one source, while group B and C have two sources each. The source in group A starts transmission at time $t = 0$ msec with an initial rate of 10 Mbits/sec. Sources in group B start at time $t = 100$ msec while group C sources start at time $t = 500$ msec. Figure 7 and 8 show the source rate for each source, with the EFCI scheme and the PERC scheme respectively. From Figure 7 we observe that as soon as group B sources

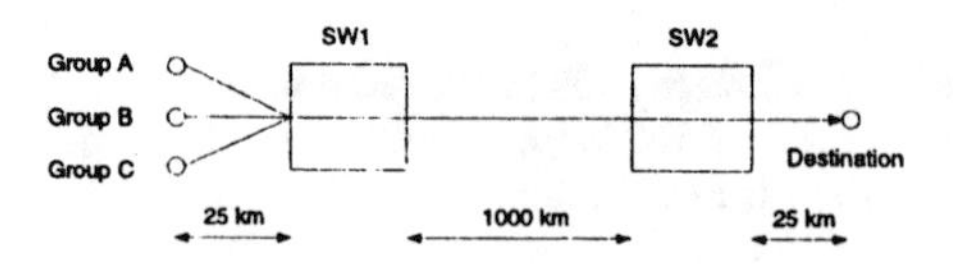

Figure 6 Peer-to-peer network configuration

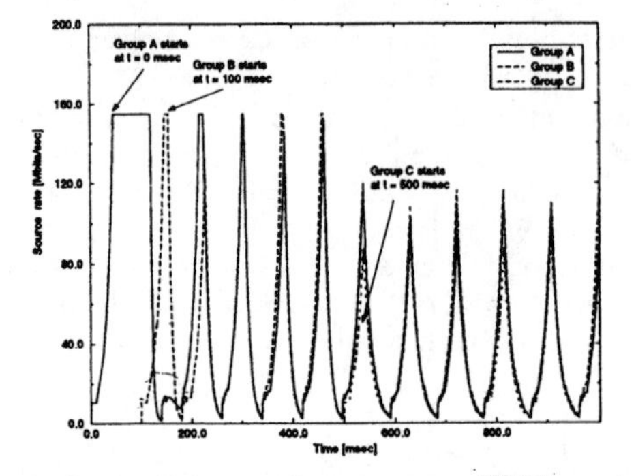

Figure 7 Source rate transient response (EFCI control)

start transmitting, the source rates start oscillating. Further, the source rate behavior of all sources are synchronized. The source rates oscillate between a minimum rate equal to the minimum cell rate $MCR = 1.5$ Mbits/sec, and a peak value that progressively decreases as the number of active sources increase. We also notice that after sufficient time has elapsed since the last source becomes inactive, all sources achieve nearly the same maximum rate, even though they start at different times. In contrast, from Figure 8 we observe that with the PERC scheme, each time the load on the network changes (with new sources becoming active), the control achieves stability quickly with very little oscillations. The EFCI scheme on the other hand continues to oscillate without damping long after a load change has occurred. From Figure 8 we observe that each time a new source becomes active, the link capacity is shared equally by all the active sources. Figure 9 and 10 show the transient performance of the EFCI and PERC schemes respectively, when a high priority VBR source (Eq. 2) is added to the peer to peer configuration. The starting times of the five persistent sources are the same as in the previous example. Comparing Figure 7 and 9 we observe that in the case of the EFCI scheme, when the

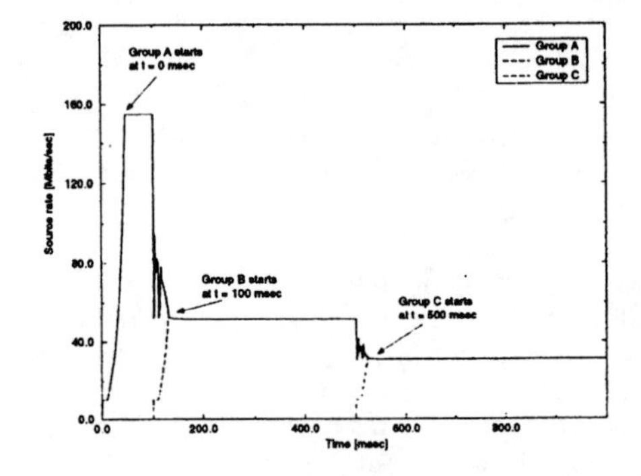

Figure 8 Source rate transient response (PERC control)

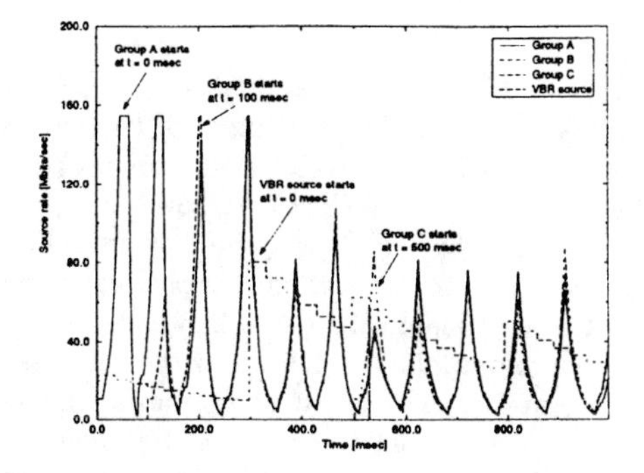

Figure 9 Source rate transient response in the presence of VBR source (EFCI)

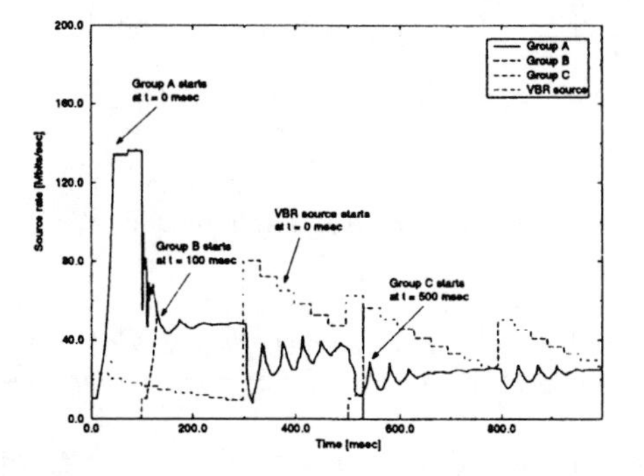

Figure 10 Source rate transient response in the presence of VBR source (PERC)

VBR source is added, the individual source rates continue to oscillate as before with diminishing amplitude as more sources become active. The peak source rate also reduces as the VBR rate increases. Further, we observe that, while the source rate behavior of all sources are synchronized, all sources do not reach the same peak value. Comparing Figure 8 and 10 we observe that, in the case of the PERC scheme, the source rates of the persistence sources track the available capacity well, with all sources receiving their fair share.

The EFCI based scheme is a reactive control that allows congestion to set in first and then reacts to it by reducing the load. Each congestion period is then followed by an underload period which causes the control to increase the load again. The large propagation delays directly contribute to the amplitude of oscillations in source rate and queue fill. The PERC scheme on the other hand is conservative and always acts in a preventive mode. While the conservative approach leads to a small reduction in utilization, it is very stable in its operation, with very low loss.

5 CONCLUSION

We investigate the performance of two end-to-end rate based flow control mechanisms to support ABR service that is being proposed by the ATM Forum. We use a multihop network model with VC grouping to evaluate the performance of this control. In the case of the EFCI scheme, we observe that VC's traversing two or more hops can see a throughput collapse at high loads. The unfair advantage that VC's traversing small number of hops have over VC's traversing larger number of hops can be overcome by a priority mechanism where transit traffic from upstream switches are given priority of service over local traffic at each switch. The EFCI scheme however is oscillatory in nature and can lead to large cell loss if the bandwidth delay product is large compared to the number of buffers. The explicit rate control scheme on the other hand requires no priority mechanism. It has a good transient response with low latency and quick settling time. The PERC mechanism for ER also achieves min-max fairness without excessive computation. However its steady state performance can lead to slightly lower throughput compared to the more aggressive EFCI scheme. From these studies one may conclude that the simpler EFCI based method with access priority will be more than adequate in a LAN environment where the propagation delays are small. The more sophisticated explicit rate based switches can be deployed in WAN's to obtain stable performance.

REFERENCES

Prycker, M.D. (1993) *Asynchronous Transfer Mode: Solutions for BISDN.* [Ellis Horwood, New York].

Kolarov, A. and Ramamurthy, G. (1994a) Comparison of Congestion Control Schemes for ABR Service in ATM Local Area Networks. *Proceedings of GLOBECOM'94*, 913–918.

ATM Forum (1995) *Baseline Text for Traffic Management Group: Specification of the ABR Service.*

Ramakrishnan, R.R. and Jain, R. (1990) A Binary Feedback Scheme for Congestion Avoidance in Computer Networks. *ACM Trans. on Computer Systems*, **8**, 158–181.

Kolarov, A. and Ramamurthy, G. (1994b) Evaluation of Congestion Control Schemes for ABR Service. *ATM Forum TMWG Meeting, Lake Tahoe.*

Ramamurthy, G., Kolarov, A., Ikeda, C. and Suzuki, H. (1994) Evaluation of Rate Based Congestion Control Schemes for ABR Service in Wide Area ATM Networks. *ATM Forum TMWG Meeting, Munich.*

Bonomi, F. and Fendick, K. (1995) The Rate-Based Flow Control Framework for the Available Bit Rate ATM Service. *IEEE Network*, **9**, **no.2**, 25–39.

Kolarov, A. and Ramamurthy, G. (1995) End-to-end Adaptive Rate Based Congestion Control Scheme for ABR Service in Wide Area ATM Networks. *Proceedings of ICC'95*, 138–143.

Benmohamed, L. and Meerkov, S.M. (1993) Feedback Control of Congestion in Packet Switching Networks: The Case of a Single Congested Node. *IEEE/ACM Trans. on Networking*, **1**, 693–707.

Ramamurthy, G. and Sengupta, B. (1993) A Predictive Hop-by-Hop Congestion Control Policy for High-Speed. *Proceedings of INFOCOM'93*, 1033–1040.

12

Effect of Bursty Source Traffic on Rate-Based ABR Congestion Control Schemes

Madhavi Hegde *mhegde@stratacom.com*

W. Melody Moh *melody@cs.sjsu.edu*

Dept. of Mathematics and Computer Science
San Jose State University, San Jose, CA 95192, U.S.A.

Abstract

Congestion control is an important function of traffic management in Asynchronous Transfer Mode (ATM) networks. Available Bit Rate (ABR) is a category of service for ATM networks wherein the source adapts its generation rate to network conditions based on feedback. In this paper we study closed-loop ABR congestion control using special cells called Resource Management (RM) cells to relay feedback to the source. In particular, we investigate both the effectiveness of rate-based ABR congestion control in the presence of bursty source traffic and the relationship between the burst time scale and the ABR control time scale. Two ABR congestion control schemes, the ABR Explicit Forward Congestion Indication (EFCI) and ABR Congestion Indication (CI) schemes, are compared with Unspecified Bit Rate (UBR) transport which makes no effort to control congestion. Traffic sources of various burst lengths of 100, 1000, 10000, and an equal mix of 100 and 10000 cells are used in simulations. It is found that ABR congestion control effectively controls low frequency, medium to long term traffic load transients, but does not control high frequency, short term load transients. In the latter case, ABR control is not necessary since short term transients do not require large amount of buffering. Of the two ABR schemes considered, the more sophisticated ABR CI scheme performs significantly better than the ABR EFCI scheme in terms of delay and buffer occupancy.

1. Introduction:

The effectiveness of congestion control is governed by the capacity of the network, traffic density, and the propagation delay in informing the traffic source to change its traffic generation rate. Since the link speeds are very high and propagation delays are long in ATM networks, traditional window flow control techniques may not be suitable.

There are five service categories for ATM networks: Constant Bit Rate (CBR), real-time and non-real-time Variable Bit Rate (VBR), Unspecified Bit Rate (UBR), and Available Bit Rate (ABR). UBR and ABR are the two categories that most closely match the needs of data traffic, which is both bursty and unpredictable (ATM Forum, 1995).

The UBR category does not involve any feedback from the network since it offers very minimal traffic-related service guarantees. All connection requests are accepted and sources generate cells at rates that are not dependent on network conditions i.e., there is no feedback from the network to inform the sources of congestion. The ABR category, however, expects a low cell loss rate. Sources of this category have to adapt their rates in response to changes in network conditions.

There are two kinds of congestion control - open-loop and closed-loop. Open-loop control involves assigning bandwidth to a connection based on its declared traffic parameters, such as the Peak Cell Rate (PCR), cell delay variation, etc. (Woodruff, 1990), (IEEECM, 1991). Open-loop control is not suitable for data transfer since it is difficult for a connection to take advantage of bandwidth assigned to another connection, even when it is not being utilized (Ohsaki, 1995).

Closed-loop or reactive congestion control limits the number of cells in the network by constant monitoring and sending feedback to sources. The sources, upon receiving feedback from the network, adapt their cell generation rates to the prevalent conditions of the network. There are two kinds of closed-loop control schemes, credit-based and rate-based. The credit-based scheme involves a large data buffer and a credit counter per connection.

This scheme is based on a link-by-link window flow control (Kung, 1994). Each link has an independent flow control. A connection has to reserve resources for its cells at all the links in its route before it is allowed to transmit. This is the credit it accumulates at each node. A connection is not allowed to transmit if it fails to gain credit from the next node. The credit counter accumulates credits at a rate initiated from the network and is depleted every time a cell is sent. The number of credits cannot be reduced below a threshold level. Transient congestion is controlled very well by this scheme. There is almost no cell loss since cells are not transmitted if there are no credits. However, a very large buffer is needed and complex buffer management has to be performed at the switch.

The rate-based scheme involves controlling the source's rate according to feedback from the network, using special cells called Resource Management (RM) cells. The number of cells in the network is contained by having the sources themselves reduce their rate of transmission, thus simplifying control. There is no need for large buffers, which reduces cost. There are many such rate-based schemes which vary in complexity and performance. A detailed description of existing rate-based control schemes is presented in the next section.

Two important types of traffic sources that have been studied are: (1) Persistent - this type of source always has cells to transmit, and (2) Bursty - this type of source generates bursts of traffic. The burst length may assume a geometric or a uniform distribution, or a deterministic distribution. The period between two bursts usually follows an exponential distribution. Many ABR simulation and analysis studies have focussed on the simulation of persistent sources because these allow a good understanding of the behavior without interference of statistical fluctuations caused by random sources (Ohsaki, 1995).

This paper investigates the effectiveness of ABR congestion control in the presence of *non-persistent* traffic sources. We present results of simulating three control schemes: UBR, ABR Explicit Forward Congestion Indication (EFCI), and ABR Congestion Indication (CI) with per-VC accounting in the presence of bursty sources. To see the effect of these schemes on sources of bursty traffic, we vary the burst length of the sources. Short, medium, long, and an equal mix of short and long burst lengths are used. We study the performance of the congestion control schemes in terms of Available Cell Rate (ACR) of the sources and the buffer queue length at the switch. We also measure mean queuing delay, cell loss, and link utilization under the different scenarios.

Section 2 of this paper describes ABR traffic control in detail and presents various algorithms that switches could implement to achieve it. Section 3 presents the network configuration and the parameters used for the simulation. Section 4 presents the results obtained and discussions of these simulation results. The conclusions drawn from these results appear in Section 5. The Appendix contains expansions of the acronyms used in this paper.

2. ABR Traffic Control

The ATM Forum has adopted rate-based control for ABR traffic and defined reference ABR source and destination behaviors (ATM Forum, 1995). ABR flow control occurs between two endsystems (refer to Figure 1). A connection between such a pair is bi-directional in that both endsystems are each a source and a destination. An ABR endsystem always has to implement both the source and the destination behaviors. For simplicity, we will consider information flow only in one direction, from the source endsystem to the destination endsystem. The forward direction is that from the source endsystem to the destination endsystem and the backward direction is that from the destination endsystem to the source endsystem.

ABR flow control involves closed-loop control implemented by RM cells which originate at the source endsystem, reach the destination endsystem, and are turned back to the source endsystem. RM cells are distinguished by the value six in the Payload Type Indicator (PTI) field of their ATM header. They also have a bit field DIR, to indicate their direction of travel - forward or backward with reference to their point of origin. The RM cells are generated by the source either every Nrm (number of cells between two RM cells) or every Trm (time between two RM cells). They carry information about the network conditions to the source endsystem after being turned around by the destination endsystem.

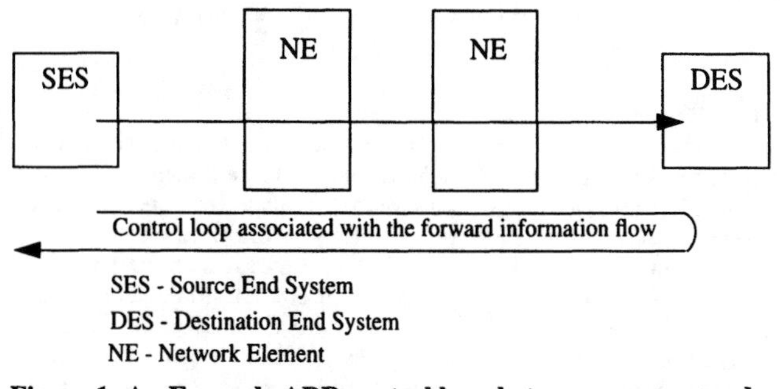

SES - Source End System
DES - Destination End System
NE - Network Element

Figure 1: An Example ABR control loop between a source and a destination

The Message Type field of RM cells contains the fields the Congestion Indication (CI), the No Increase (NI), and the Explicit Rate (ER) that are used by various rate-based congestion control schemes to alter source rates. The rate-based ABR congestion control schemes use the Explicit Forward Congestion Indication bit in the data cells too.

The network feedback information is written into RM cells by network elements. The network elements may do one of three things - (i) directly insert information into RM cells when they pass or, (ii) indirectly indicate congestion conditions to the source endsystem by setting EFCI bits in data cells, in which case the destination endsystem inserts information into RM cells before turning them around or, (iii) generate RM cells themselves.

The control loop may be segmented into multiple control loops by using a network element which implements virtual source/virtual destination behavior. This segmentation would isolate control segments and increase link utilization (Bonomi, 1995).

The sources have to declare the maximum required bandwidth and the minimum bandwidth necessary for the connection at the time of connection establishment. These connection parameters are called the Peak Cell Rate (PCR) and the Minimum Cell Rate (MCR). The actual rate at which the source can generate traffic is the Allowed Cell Rate (ACR).

The ABR traffic control maintains an ACR for each source, which will be equal to the guaranteed MCR or higher, but never higher than PCR, throughout the lifetime of the connection. It "provides rapid access to unused network bandwidth at up to PCR, whenever the network bandwidth is available" (ATMF, 1995). Low cell loss is expected to result from ABR traffic control if the source and destination endsystems follow reference behavior.

The switch algorithm is not defined by the ATM Forum and hence many algorithms are possible. Using a standardized format for the RM cell and control algorithms, different switch architectures will be able to coexist and interwork with each other. The switches can control the rate of the sources by one of several ways:

Relative Rate marking

The source sends RM cells with their CI bits reset either a certain number of cells (Nrm) apart or a certain amount of time (Trm) apart. The destination turns around the RM cells it receives from the source and sends it back to the source, using the same path. The switches on the path of the connection set the CI bit in these backward RM cells when there is congestion in the forward direction. When the source receives a backward RM cell with its CI bit set, it reduces its ACR by ACR * RDF. The resulting ACR, if lower than the MCR, is replaced by MCR. If the source receives backward RM cells with reset CI bits, it increases its ACR by (ACR * RIF) (ATMF, 1995).

Explicit Forward Congestion Indication (EFCI) marking

The source sends its data cells with the EFCI bit reset at a rate of one cell every 1/ACR time. It also sends RM cells. When a switch in the path of the connection detects congestion, it sets the EFCI bit in the data cells. The destination, on receiving data cells with their EFCI bits set, will set the CI bit in the very next forward RM cell that it receives. It then turns the RM cell around. The source will react to RM cells with their CI bits set by reducing its ACR. The source will increase its ACR if it receives RM cells with their CI bits reset (ATMF, 1995), (Hluchyj, 1994).

Explicit Rate (ER) marking

The source writes the rate it desires in the RM cells it sends between data cells at regular intervals. The switch computes the fair share of bandwidth that should be allocated to each VC, and traffic load by monitoring the queue length. It updates the ER field of backward RM cells with the calculated fair share. The source will change its ACR to the value in the ER field of returning RM cells (ATMF, 1995), (Barnhart, 1995).

There are many variations of these schemes. A separate queue can be used at the switch for RM cells, to speed up response. This queue could be given priority over the data queue in order that RM cells are not blocked for too long due to congestion. Deciding when to set the EFCI bit or CI bit or by how much to reduce ER is implementation specific. There are two common methods of congestion detection: queue length-based and link utilization-based (Bonomi, 1995). The queue length-based congestion detection is implemented by monitoring the queue length at the switch. If the length is greater than a certain threshold, congestion is declared. The link utilization-based congestion detection is based on the change in queue length, monitored at some fixed intervals.

Several works have appeared in literature on rate-based congestion control. Lee et al. formally represent ABR source and destination behavior using an extended finite state machine (Lee, 1996). Bonomi and Fendick compare the EFCI with the ER scheme in terms of fairness (Bonomi, 1995). Ohsaki et al. quantitatively evaluate the performance of all these three algorithms in terms of maximum queue length for persistent traffic. They vary propagation delay and the number of VCs to show the effectiveness of rate-based congestion control (Ohsaki, 1995). Chang et al. show that EFCI and ER switches can interoperate provided that the switch implementations conform to reference behavior in terms of congestion notification and usage of RM cells (Chang, 1995).

The source rate, in all of the above schemes, is controlled by an ABR rate scheduler. There are many parameters used to implement the congestion control which the ABR scheduler keeps track of, such as the ACR, PCR, etc. The scheduler uses the value of ACR to schedule a cell every 1/ACR time.

3. Simulation Configuration and Parameters:

We present results of simulating ABR EFCI, ABR CI and UBR congestion control. ABR EFCI congestion control scheme is simulated with FIFO queue length congestion triggering. This is the simplest of all those mentioned above. It is chosen to compare with ABR control against UBR control since switches implementing this scheme are currently available. We would expect more advanced methods to out-perform the EFCI based results presented here. It is also the least expensive scheme to implement.

The ABR EFCI scheme simulated involves a separate queue at the switches for RM cells, to avoid their being blocked during congestion conditions. The more sophisticated ABR CI scheme involves setting CI bits in backward RM cells by switches if congestion is detected in the forward direction. The switches perform per-VC accounting. i.e., they keep count of cells belonging to each VC and set the CI bits in the RM cells of those VCs that have exceeded the per-VC threshold.

To gain a better understanding of the issues being studied, this paper investigates the simple configuration which is described below.

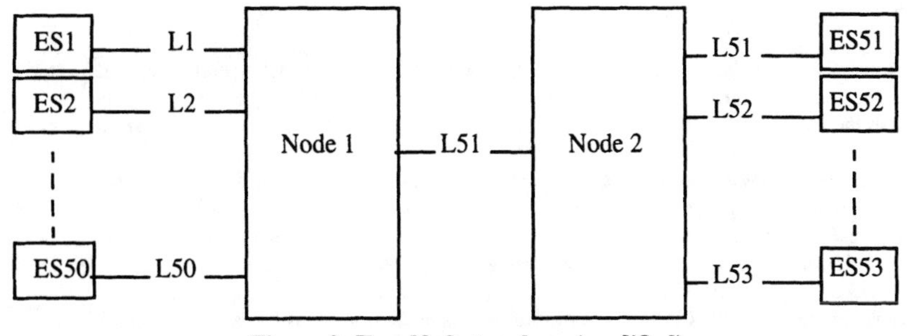

Figure 2: Two-Node topology (modified)

The backbone link between the two nodes and the expected point of congestion is an OC3 (150 Mbps) link. There are 50 endsystems connected to each of the nodes. The endsystems are connected to the node by DS3 (40 Mbps) links. The propagation delay between the two nodes is 10 ms (unless specifically noted). The propagation delay between the endsystem and the node is 0.01ms. There are 50 virtual connections between these endsystems, all passing through the backbone link.

The simulation is repeated for four source traffic scenarios.

1. All sources with a short mean burst length (100 cells)
2. All sources with a medium mean burst length (1000 cells)
3. All sources with a long mean burst length (10000 cells)
4. Half sources with a short mean burst length and half with a long burst length (100 cells and 10000 cells)

The burst length of the sources follows a geometric distribution around the mean value. The period between two bursts follows an exponential distribution.The activity fraction was 1/16 for each bursty source, resulting in a long-term traffic load of almost 85% (50 VCs * 96 cells/ms * (1/16) / 353 cells/ms, since there are 50 VCs initiated by sources with PCR 96 cells/ms and the capacity of the backbone link is 353 cells/ms). The observed load for each simulation is a statistical sample only and may be higher or lower than this value, as indicated by the results.

ABR Scheduler Parameters:

PCR (Peak Cell Rate) is 96 cells/ms or 40 Mbps.
ICR (Initial Cell Rate) is the PCR.
MCR (Minimum Cell Rate) is 0.
Nrm (Number of data cells between two RM cells) is 32.
Trm (Time between rate updates) is 20ms.
RIF (Rate Increase Factor) is 1/256.
RDF (Rate Decrease Factor) is 1/16.
The congestion threshold was set at 5000 cells for ABR EFCI and 1250 cells for ABR CI.

4. Performance:

Results of simulating UBR, ABR EFCI and ABR CI control schemes with 50 bursty sources for 10 seconds are collated here. The random sample does not accurately reflect long-term behavior and so the numerical values here may be better or worse than typical steady state performance of ABR control. For example, the observed load has a wide range of values, even though the calculated long term load is 85%. Hence, the simulation results cannot be used to make a quantitative assessment. However, the transient characteristics can illustrate important aspects of ABR behavior.

4.1. Short length bursty sources

Figures 3 and 4 show the buffer queue length in UBR, ABR EFCI and ABR CI control simulations with 50 bursty sources, all with a mean burst length of 100 cells. All these cases used identical random traffic. Table 1 summarizes the maximum buffer length, mean queuing delay, cell loss, and link utilization characteristics of these simulations.

The mean queuing delay values shown here resulting from ABR control are much higher than that from the UBR control. However, it does not incorporate the delay caused by retransmission by the upper layers.

From Figures 3 and 4 we can see that ABR control does not improve the maximum queue length resulting from UBR control for short length bursts. The transient congestion causes ACR of the sources to be adjusted down from their PCR. However, the ACR is not kept at that level for too long and is increased to restore its value to PCR almost immediately. ABR CI control neither reduces the ACR to levels as low nor as often as ABR EFCI does and the ACR recovers faster, resulting in a shorter delay. This is largely because the ABR CI control scheme is acting on RM cells travelling in the backward direction and so has a faster response time.

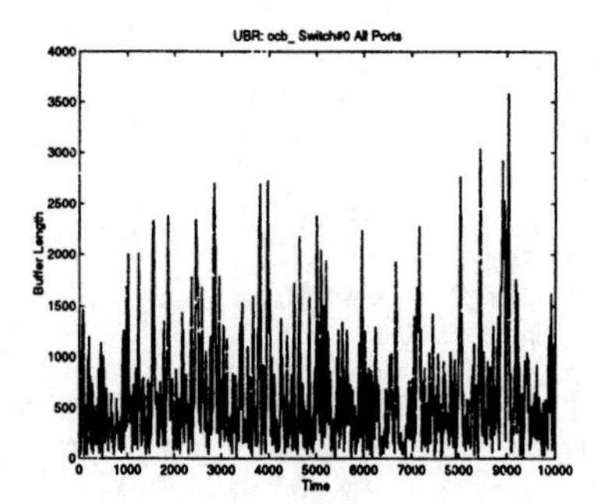

Figure 3: UBR for 50 bursty sources with mean burst length 100 cells - Buffer queue length vs. time

Table 1: Maximum buffer length, mean queuing delay and link utilization for 50 bursty sources with mean burst length of 100 cells

Congestion control scheme	Maximum buffer length in cells	Mean queuing delay in ms	Number of cells dropped	Link Utilization (%)
UBR	3600	1.1722	0	85.24
ABR EFCI	3600	2.7295	0	85.23
ABR CI	3700	1.8392	0	85.24

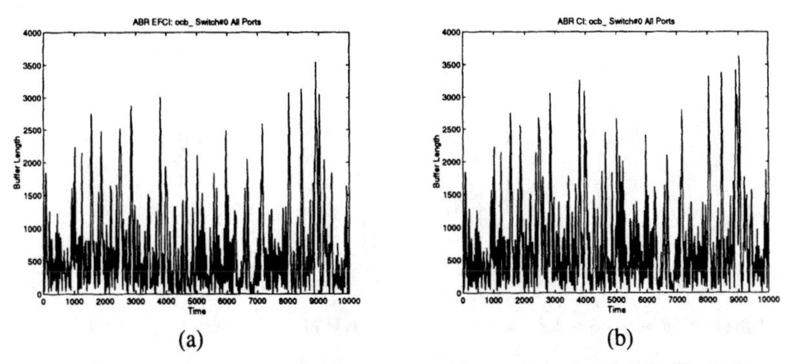

Figure 4: ABR EFCI and CI for 50 bursty sources with mean burst length 100 cells - Buffer queue length vs. time

4.2. Medium length bursty sources

Figures 5 and 6 show the buffer queue length in the UBR, ABR EFCI, and ABR CI control simulations with 50 bursty sources, all with a mean burst length of 1000 cells. All these cases used identical random traffic.

It can be seen from comparing Figures 5 and 6 and table 2 that for the medium length bursts, the maximum buffer length with ABR EFCI is significantly less than that with UBR control. The ACR of the sources is reduced sometimes to very low values to contain the number of cells in the network. ABR CI results in using the same maximum buffer length as ABR EFCI, though the average buffer occupancy is much lower than that of ABR EFCI, as can be seen from Figures 5 (a) and 5 (b). It results in a significantly shorter delay for comparable link utilization, since it does not lower the ACR to very low levels as often as ABR EFCI does.

Table 2: Maximum buffer length, mean queuing delay and link utilization for 50 bursty sources with mean burst length of 1000 cells

Congestion control scheme	Maximum buffer length in cells	Mean queuing delay in ms	Number of cells dropped	Link Utilization (%)
UBR	17000	7.1380	0	83.61
ABR EFCI	11000	53.7136	0	82.74
ABR CI	10700	19.9260	0	83.37

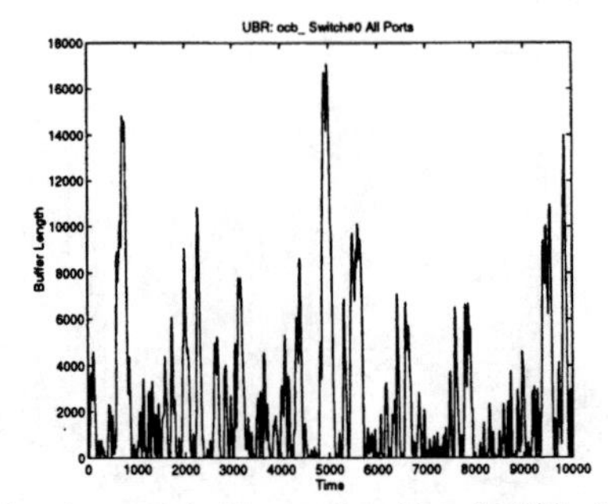

Figure 5: UBR with 50 bursty sources with mean burst length 1000 cells - Buffer queue length vs. time

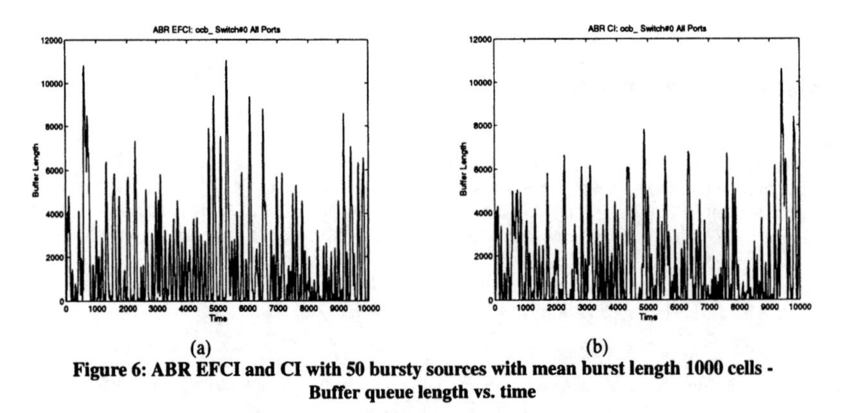

Figure 6: ABR EFCI and CI with 50 bursty sources with mean burst length 1000 cells - Buffer queue length vs. time

4.3. Long length bursty sources

Figures 7 and 8 show the buffer queue length in the UBR, ABR EFCI and ABR CI control simulations with 50 bursty sources, all with a mean burst length of 10000 cells.

Figure 7 shows that for long length bursts, even a maximum buffer length of 50000, which is a reasonable number for OC3 links, is not adequate for UBR control, resulting in significant cell loss. On the other hand, ABR EFCI and ABR CI perform extremely well in inhibiting buffer occupancy and result in zero cell loss, the most important QoS (Quality of Service) criterion for ABR traffic. Again, ABR CI performs better than ABR EFCI in terms of delay and buffer occupancy, as can be seen from table 3.

We have determined that the additional queuing delay for ABR control is a consequence of using binary rate control with a moderately large propagation delay. With a 10 ms propagation delay between the switches, it is not possible to choose AIR and RDF to result in full utilization, even in steady state. A mathematical derivation for this is given in (Ikeda, 1995). When the burst lengths are big and the steady state throughput is less than 100%, the traffic will bank up in the endsystem creating a pseudo-persistent source, as reflected by the regular patterns in Figure 8. To demonstrate that this is the issue, we have presented the results of ABR control simulations with a 0.01 ms propagation delay between the switches in the last two rows of table 3. The queuing delay in these cases are much less than for the cases with a 10 ms propagation delay between the switches. Although not verified here, we would expect much better delay performance from an explicit rate scheme.

Table 3: Maximum buffer length, mean queuing delay and link utilization for 50 bursty sources with mean burst length of 10000 cells

Congestion control scheme	Maximum buffer length in cells	Mean queuing delay in ms	Number of cells dropped	Link Utilization (%)
UBR	50000	49.6040	61694	85.49
ABR EFCI (10 ms)	6900	765.8332	0	70.14
ABR CI (10 ms)	1450	525.1764	0	75.46
ABR EFCI (0.01 ms)	6000	105.3194	0	85.1
ABR CI (0.01 ms)	950	99.2834	0	85.25

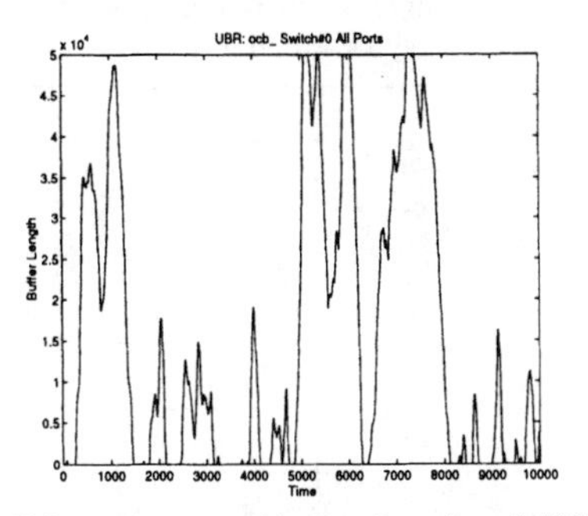

Figure 7: UBR with 50 bursty sources with mean burst length 10000 cells - Buffer queue length vs. time

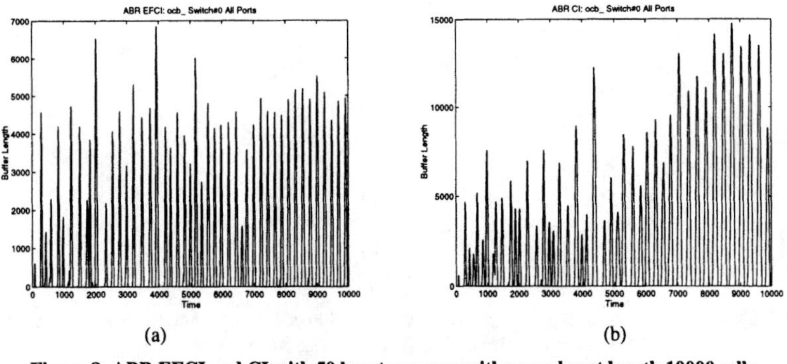

Figure 8: ABR EFCI and CI with 50 bursty sources with mean burst length 10000 cells - Buffer queue length vs. time

4.4. Mixture of short and long length bursty sources

Figures 9 and 10 show the buffer queue length in UBR, ABR EFCI and ABR CI control simulations with 25 bursty sources with a mean burst length of 100 cells and 25 bursty sources with a mean burst length of 10000 cells. All these cases used identical random traffic. Table 4 summarizes the maximum buffer length, queuing delay, cell loss and link utilization characteristics of these simulations.

It can be seen from Figure 9 and table 4 that UBR control results in significant cell loss since the maximum buffer length of 50000 cells is not sufficient. Again, both ABR EFCI and ABR CI work extremely well in curbing switch buffer occupancy, as observed from Figures 10 (a) and 10 (b).

However, it is most important to note from table 4 that the ACR of the sources with short burst length are not much altered, as can be deduced from the reasonable delay they face. The sources with long bursts, on the other hand, face long delays, indicating that ABR control has been instituted.

As before, we present the results of simulating ABR control with a 0.01 ms propagation delay between the switches in the last two rows of table 4. It can be seen that the queuing delay in these cases are much lower than in the cases with a propagation delay of 10 ms between the switches, thus showing the limitation of binary rate control schemes.

Table 4: Maximum buffer length, mean queuing delay and link utilization for 25 bursty sources with mean burst length of 100 cells and 25 bursty sources with mean burst length of 10000 cells

Congestion control scheme	Maximum buffer length in cells	Mean queuing delay (100 cells) in ms	Mean queuing delay (10000 cells) in ms	Number of cells dropped	Link Utilization (%)
UBR	50000	33.1059	41.1097	30588	86.83
ABR EFCI (10 ms)	6500	4.3085	823.6975	0	79.34
ABR CI (10 ms)	11900	5.2705	439.3031	0	83.37
ABR EFCI (0.01 ms)	8600	6.3523	129.9396	0	86.54
ABR CI (0.01 ms)	17000	11.7227	105.9821	0	86.68

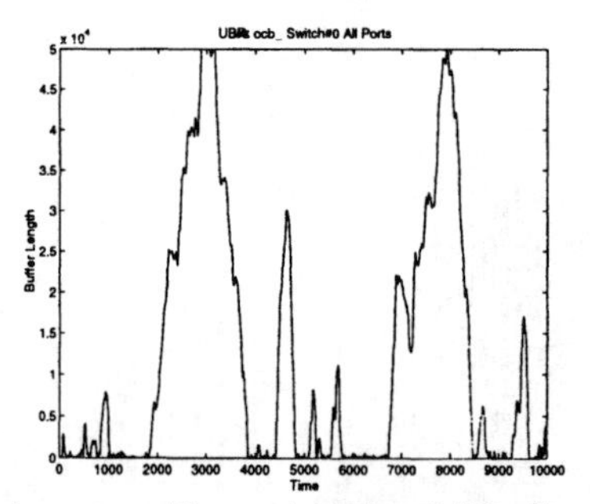

Figure 9: UBR with 25 bursty sources with mean burst length 100 cells and 25 bursty sources with mean burst length 10000 cells - Buffer queue length vs. time

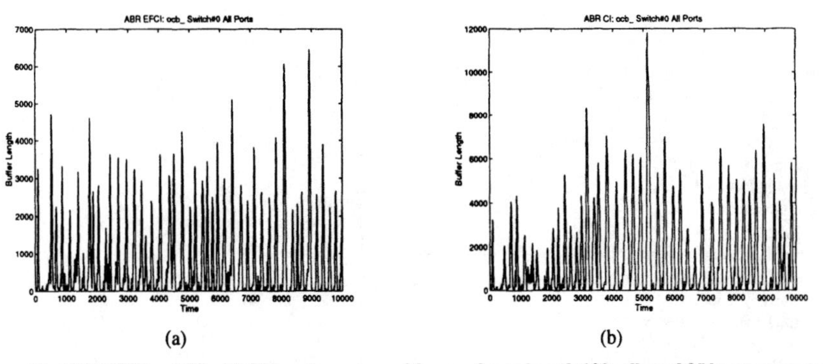

(a) (b)

Figure 10: ABR EFCI and CI with 25 bursty sources with mean burst length 100 cells and 25 bursty sources with mean burst length 10000 cells - Buffer queue length vs. time

5. Conclusion:

Rate-based congestion control is an architecturally flexible approach for handling ABR traffic in ATM networks. Since all the different schemes proposed for the control use standardized RM cells, switches implementing them can interoperate. Closed-loop control is necessary for ABR traffic since its bursty feature can be utilized to service more connections by allocating unused bandwidth of one connection to other connections that can use it.

We study the performance of various ABR congestion control algorithms in the presence of bursty traffic sources since they are very likely to occur in reality. Results of simulating ABR EFCI and ABR CI control schemes show that ABR control is extremely effective in controlling low frequency, medium to long length bursts. This is very desirable since it is the low frequency behavior which results in extended periods of overload and corresponding cell loss. By controlling the low frequency behavior, ABR control reduces trunk queue lengths and minimizes overflow and cell loss. ABR control does not control high frequency, short length traffic bursts. However, since short bursts do not require a large amount of buffering, ABR control is not required in these cases. Another limitation of these two binary rate control schemes is that they result in large queuing delay when the propagation delay is large. We believe that explicit rate (ER) ABR control schemes will perform much better in such cases.

ABR EFCI control is the simplest and most cost-effective ABR control to implement. Our results show that it is effective in minimizing buffer occupancy. ABR CI performs significantly better than ABR EFCI in terms of delay and maximum buffer length, since network feedback information reach the traffic sources faster.

Acknowledgment:

Special thanks are due to Dr. David Hughes for many discussions on this work. Melody Moh is supported in part by CSU Research Grant, SJSU Foundation Research Grant, and SJSU Affirmative Action Faculty Award.

Appendix:

Acronyms used in the paper:

ABR - Available Bit Rate
ACR - Allowed Cell Rate
AIR - Additive Increase Rate
AIRF - Additive Increase Rate Factor

ATM - Asynchronous Transfer Mode
CCR - Current Cell Rate
CI - Congestion Indication (bit in RM cell)
DES - Destination End System
DIR - Direction (bit in RM cell)
EFCI - Explicit Forward Congestion Indication
ER - Explicit Rate
FIFO - First In First Out
MCR - Minimum Cell Rate
NI - No Increase
Nrm - Maximum number of data cells between RM cell generation
PCR - Peak Cell Rate
PTI - Payload Type Indicator
QL - Queue Length (not used by ABR Forum)
RDF - Rate Decrease Factor
RIF - Rate Increase Factor
RM - Resource Management cell
SES - Source End System
SN - Sequence Number (not used by ABR Forum)
Trm - Time for RM generation
UBR - Unspecified Bit Rate
VC - Virtual Connection
VCI - Virtual Connection Identifier
VP - Virtual Path
VPI - Virtual Path Identifier

References:

ATM Forum Traffic Management Specification (1995) Version 4.0, Release 9.

Barnhart, A. W. (1995) Example Switch Algorithm for Section 5.4 of TM Specification, *ATM Forum Contribution 95-0195.*

Bonomi, F. and Fendick, K. W. (1995) The Rate-Based Flow Control Framework for the Available Bit Rate ATM Service, *IEEE Network Magazine*, vol. 9, 25-39.

Chang, Y., Golmie, N. and Su, D. (1995) Study of Interoperability between EFCI and ER Switch Mechanisms for ABR Traffic in an ATM Network, *IC3N 95, 1995 International Conference on Computer Communication Networks*, 310-315.

Hluchyj, M. and Yin, N. (1994) On Closed-loop rate control for ATM networks, *Proceedings of INFOCOM '94*, 99-108.

Ikeda, C., Suzuki, H., Osaki and Murata M. (1995) Recommendation Parameter Set for Binary Switch, *ATM Forum Contribution 95-1482.*

Kung, H. T. and Chapman, A. (1994) Credit-based flow control for ATM networks: Credit update protocol, adaptive credit allocation, and statistical multiplexing, *The Proceedings of SIGCOMM '94*, vol. 24, 101-114.

Lee, D., Moh, W. M., Ramakrishnan, K. K. and Shankar, A. U. (1996) An Extended Finite State Machine Representation of the Source/Destination Behavior, *ATM Forum Contribution 96-0231.*

Ohsaki, H., Murata, M., Suzuki, H., Ikeda, C., and Miyahara, H. (1995) Rate-Based Congestion Control for ATM Networks, *ACM SIGCOMM Computer Communication Review*, 60-72.

Special issue on 'Congestion control in high speed networks', (1991) *IEEE Communications Magazine*, vol. 29.

Woodruff, G. and Ksitpaiboon, R. (1990) Multimedia traffic management principles for guaranteed ATM network performance, *IEEE Journal on Selected Areas in Communications*, vol. 8, 437-446.

13

Refinements to Rate-Based Flow Control with Extensions to Multidrop Applications

S. Pejhan, M. Schwartz and D. Anastassiou
Department of Electrical Engineering, Columbia University
New York, NY 10027, USA, {sassan,schwartz,anastas}@ctr.columbia.edu

Abstract
We present a rate-based, explicit feedback flow control mechanism. Congestion control is distributed among all the nodes on the path from the sender to the receiver. This leads to a more accurate control mechanism and a scalable design. Our mechanism is shown to be effective in scenarios where the bottleneck can vary among the intermediate nodes without leading to buffer overflows during transitional periods. Overhead is kept low by using the same control packet to relay control information and measure internode delays. We analyze both an end-to-end and a hop-by-hop approach, as well as multidrop scenarios.

1 INTRODUCTION

Rate-based congestion control schemes gained popularity with the introduction of fiber optics, and the ensuing development of "light-weight" transport protocols [5]. Distributed multimedia applications and the need for integrated services also gave new impetus to rate-based schemes*, though credit-based schemes are still under serious consideration [9] (mostly for LANs [13]).

Rate-based control uses feedback from the network to adjust the source transmission rate so as to prevent congestion. Earlier schemes proposed having one or more "congestion thresholds" and adjusting the source rate to a "high" or "low" level accordingly (e.g. [16]). The choice of appropriate thresholds and high/low transmission rates in such scenarios often boiled down to intelligent guesses. More sophisticated schemes are described in [2, 17] where a linear increase/exponential decrease algorithm is used to adjust the source rate, again according to threshold based feedback information. Rate-based schemes proposed for ATM networks use Explicit Forward Congestion Indication (EFCI): when a packet encounters congestion at an intermediate node—defined as the queue size at the node buffer exceeding some threshold—it will set the congestion bit to 1. The receiver checks this field and if congestion was experienced notifies the sender [10]. Again, there is the problem of how to define the threshold, but more importantly, by having intermediate nodes relay congestion information through the receiver the system will be slow to respond. Furthermore, the sender has no information regarding the magnitude of the congestion.

A number of publications propose using explicit feedback to capture the magnitude of congestion [8, 7, 1, 6]. The solutions in [8] and [7] follow an end-to-end solution, while a hop-by-hop

*The ATM Forum has selected a rate-based flow control mechanism to support ABR service [4].

approach has been proposed in [11]. Closed-loop rate control mechanisms may also be used in conjunction with open-loop mechanisms [17].

In Section 2, we propose an end to end flow control scheme, stating the controller objectives and deriving the control mechanism. Simulation results compare the proposed scheme with others proposed in the literature. Section 3 describes the hop-by-hop version of the proposed schemes. Extensions of the control mechanism to a multidrop scenario are then described (Section 4), followed by concluding remarks (Section 5).

2 END TO END APPROACH

2.1 Network Model and Controller Objectives

Our model consists of the sender, the receiver and the network cloud. The bottleneck is defined as the node with the lowest processing rate and could be any of the intermediate nodes between the sender and the receiver, or even the receiver itself. The bottleneck node could vary with time. The parameters of interest are as follows:

- $\lambda_s(t)$: The source transmission rate. We assume that the sender is able to match the transmission rate to the rate computed by the control mechanism.[†].
- $\lambda_R(t)$: The rate at which packets enter the bottleneck. Note that this is *not* the same as $\lambda_s(t)$ above, although it is related to it. We shall elaborate on this later on.
- $x(t)$: The buffer occupancy at the bottleneck. This is the variable that we want to control through the sender's transmission rate. It is bounded by 0 and some maximum buffer size, B. We are assuming that each flow is allocated its own separate buffer.
- $\mu_b(t)$: The processing rate allocated to the connection of interest at the bottleneck. This is a parameter over which we assume no control[‡], but varies within bounds (§2.4).

The design strategy that is to aim for a target queue size, x^*, at the bottleneck. Congestion control strategies aim at increasing throughout, or reducing packet delay and packet loss. As pointed out in [8], the choice of the set-point reflects a trade-off between these three parameters: the lower the set-point, the lower the packet delay and packet loss, but the more underutilized the bandwidth. We shall not concern ourselves with the choice of the set-point[§]. It is usually set to $B/2$, half the maximum buffer size.

The buffer occupancy at the bottleneck is governed by equation (1) below, which states that the buffer occupancy at time $t + t_0$ is equal to that at time t plus what came in and minus what left during the time interval $[t, t + t_0]$:

$$x(t+t_0) = x(t) + \int_t^{t+t_0} (\lambda_R(\tau) - \mu_b(\tau))d\tau \qquad \text{subject to } 0 \le x(t+t_0) \le B. \qquad (1)$$

[†]In practice, it is difficult to accurately achieve this for images and video, as the produced bit-rate is very much source dependent. An elaborate discussion is beyond the scope of this paper.

[‡]These are required to be controllable in a hop-by-hop design as explained in Section 3.

[§]A more elaborate, quantitative discussion appears in [8].

The objective of the controller is to adjust the transmission rate of the sender, $\lambda_s(t)$, in such a way so that the buffer occupancy at the bottleneck, x, remains close to x^*—the two are related via $\lambda_R(t)$ and equation (1).

2.2 Controlling site

The controller needs to frequently obtain information regarding the buffer occupancy $x(t)$, the incoming packet rate $\lambda_R(t)$ and the processing rate $\mu_b(t)$ of the bottleneck. The simplest, and most accurate, form of obtaining this information is from the bottleneck itself. The controller also has to have a model for predicting the future values of these bottleneck parameters due to the non-zero propagation delays (it takes time for information to propagate from the bottleneck to the sender; it will also take time for any change in the transmission rate to affect the buffer occupancy at the bottleneck). These models are discussed in more detail in later sections.

In [7] the bottleneck measures the buffer occupancy and processing rate, and relays this information to the sender. In [8] a packet-pair is sent by the sender, and by measuring the time differential between the acknowledgements, the sender can estimate the processing rate at the bottleneck. This method is somewhat inaccurate, however, as pointed out by the author. It also adds unnecessary overhead, since four packets are generated for each estimate. In both cases, however, the incoming packet rate at the bottleneck is assumed to be identical to the source transmission rate. Meanwhile, [6] and [1] assume a fixed processing rate for the bottleneck, while assuming the incoming packet rate to be equal to a time-shifted version of the source transmission rate.

In all cases except [1], however, the sender acts as controller. One of the main differences in our model is that the bottleneck is the controlling site. This way the controller can use up-to-date information regarding the three parameters of interest. In [7] and [6], the information is already inaccurate by the time it reaches the source. Unlike these parameters, the source transmission rate is fixed for the duration of a control interval. So this information is still valid when it reaches the bottleneck/controller in our design.

Each intermediate node in our model will assume that *it* is the bottleneck. It will compute its desired source transmission rate according to the control mechanism described in §2.4. When the node becomes the bottleneck, its model for predicting its future processing rate would be valid. The same cannot be said for source-based controllers since when bottlenecks change, the source's model for estimating these parameters will not be valid (initially). This is another argument for locating the controller at the bottleneck.

2.3 Control Packet

The receiver will send a high-priority control packet periodically—at intervals of t_0—to the source, indicating its own processing rate (PROC_RT), as well as the rate at which it wants the source to transmit (DES_RT), as computed by its control mechanism (§2.4). Along the path, a switch will modify those parameters under two conditions only. If its own processing rate is lower than PROC_RT in the control packet, this means that the current switch is slower than all those downstream. Hence it will replace both parameters (PROC_RT and DES_RT) with its own processing and desired source transmission rates, as in [7]. If its processing rate is higher than PROC_RT but lower than DES_RT, it will replace the *latter* only with the value of its current processing rate.

This second condition is one of the key refinements that we have made: there is no point for a downstream 'bottleneck' node to request a source transmission rate that is higher than what can be delivered by any of the intermediate upstream nodes¶. With the schemes described in [7] and [8], there can be temporary buffer overflow at upstream nodes. Our mechanism will prevent these. This will be illustrated in our simulations (§2.5).

The sender does not need to have explicit knowledge about which node is the bottleneck: by the time the control packet arrives at the source, PROC_RT will be set to the processing rate of the slowest node. At the sender, the packet is acknowledged immediately, with the new transmission rate (which is set to DES_RT) indicated in the acknowledgement. Upon receipt of the sender's acknowledgement, intermediate nodes will note the new transmission rate, which they will need to compute the desired transmission rate for the next update. They will pass on the control packet to the neighboring downstream node without modifications except if their processing rate is less than the source transmission rate. This will happen only at the bottleneck, in which case the source transmission rate indicated in the control packet is replaced by the node's processing rate.

2.4 Control Mechanism

Control Law

For each intermediate node an 'interval' begins when it receives a control packet going downstream. The intervals at the different nodes are thus not synchronized. At the beginning of the kth interval, the controller records its buffer occupancy, $x^{(k)}$. Using equation (1), it can then estimate its buffer occupancy, $x^{(k+1)}$, at the end of the current interval. This, however, requires knowledge of the incoming packet rate and node processing rates during the current (kth) interval. The incoming rate is provided by the control packet at the very beginning of the interval.

If the buffers at upstream nodes are not empty, the incoming packet rate for a particular node will depend on the processing rate of its upstream nodes. Each node will assume that its incoming packet rate is equal to the transmission rate indicated in the control packet (the propagation delay is accounted for by the fact that the control interval of the intermediate node is delayed by that much with reference to the control interval of the source). Let us denote the 'indicated' transmission rate by $\lambda_{S'}$ to distinguish it from the actual source transmission rate. The two will be equal for the bottleneck and all preceding nodes, but $\lambda_{S'}$ will equal the bottleneck processing rate for all nodes downstream of the bottleneck. When the buffer occupancies at all nodes are 0 except at the bottleneck, this assumption is accurate: for nodes preceding the bottleneck, the incoming rate is indeed a delayed version of the transmission rate, while for nodes located after the bottleneck the incoming rate is a delayed version of the bottleneck's processing rate.

During the 'transition' periods (i.e. at session start-up and just after changes in the location of the bottleneck), this assumption will underestimate the actual incoming packet rate (for those nodes located downstream of the old bottleneck). As long as the buffers at the different nodes are equal in size, however, this will not pose a problem (i.e. buffer overflows will not occur): the new bottleneck can accommodate all the packets held by the previous bottleneck, after which the system will reach a 'steady state'. Again, this will be illustrated by the simulations.

Using equation (1), and the foregoing discussion, we can estimate the buffer occupancy at the beginning of the next $(k+1)$ interval using a discrete-time model (since the source transmission rate is piece-wise constant):

$$\hat{x}^{(k+1)} = x^{(k)} + t_0 \left[\lambda_{S'}^{(k)} - \hat{\mu}_b^{(k)}\right] \qquad \text{subject to } 0 \leq \hat{x}^{(k+1)} \leq B \tag{2}$$

¶This situation will occur if the bottleneck node's buffer occupancy is significantly below its target.

$\hat{\mu}_b^{(k)}$ is estimated as described in section 2.4. The next task is to set $x^{(k+2)}$ equal to x^* and solve for the desired source transmission rate for the interval $k+1$, $\lambda_S^{(k+1)}$.

$$x^* = \hat{x}^{(k+1)} + t_0 \left[\lambda_{S'}^{(k+1)} - \hat{\mu}_b^{(k+1)}\right] \tag{3}$$

We estimate $\hat{\mu}_b^{(k+1)}$ using the method of section 2.4. We then derive the desired value of $\lambda_{S'}^{(k+1)}$ using equation (3),

$$\lambda_{S'}^{(k+1)} = \frac{x^* - \hat{x}^{(k+1)}}{t_0} + \hat{\mu}_b^{(k+1)} \tag{4}$$

The control mechanism—equation (4)—will require the source to transmit at a higher rate than the current processing rate if the buffer occupancy is below the target, and vice versa. The control mechanism as it stands now can lead to instability (see [8] for proof). To rectify this problem, we need to add a gain factor, δ, to equation (4), as shown below:

$$\lambda_{S'}^{(k+1)} = \delta \left(\frac{x^* - \hat{x}^{(k+1)}}{t_0} + \hat{\mu}_b^{(k+1)}\right) \tag{5}$$

where $0 < \delta < 1$. If δ is too small, the system will be too slow to respond. A value of 0.9 seems to be a good choice for δ [8].

Estimating the Processing Rate at the Bottleneck

The processing rates at intermediate nodes vary because new connections are established and old ones broken continuously. [8] and [7] both use basically the same model to estimate the future processing rate. The processing rate during the kth interval, $\mu^{(k)}$, is estimated by equation (6):

$$\hat{\mu}^{(k)} = (1-\theta)\tilde{\mu}^{(k-1)} + \theta\hat{\mu}^{(k-1)} \tag{6}$$

where $\tilde{\mu}^{(k-1)}$ is the last observation of the processing rate[||]. Both cited references concur that θ should be a variable. A small value of θ would attach too much importance to the last observation made, which could be corrupted by either observation noise, or uncharacteristic spikes in the processing rate. A large value of θ will make the system slow to respond. In both schemes θ is defined so as to track abrupt changes in the service rate, but not be affected by small changes or noise. While [8] uses a fuzzy predictor to this end[**], a simpler scheme is proposed by [7]:

$$\begin{aligned} \theta &= 0.25E^2/\sigma^{(k+1)} \\ E &= \mu^{(k)} - \hat{\mu}^{(k)} \\ \sigma^{(k+1)} &= 0.25E^2 + (1-0.25)\sigma^{(k)} \end{aligned} \tag{7}$$

where E and σ are the estimation error and the estimate for the squared estimation error, respectively. The number 0.25 was derived empirically. Both references indicated that this model performed satisfactorily. Hence we shall use the same model as in [7].

[||] In [7] θ is defined the other way around.

[**] A random walk model is also proposed for the processing rate, but is discarded in favor of the fuzzy predictor.

In our model, since each intermediate node carries out its own estimation for its future processing rate, θ will have a different value for each of them. With the source-based controller in [7] and [8], if the bottleneck changes, their model for predicting the future processing rate of the bottleneck will be incorrect during some transition period since the information regarding past values of $\mu^{(k)}$, E and $\sigma^{(k)}$ is inaccurate and pertaining to a different node.

Control frequency

Our analysis assumes that the receiver sends control packets once per round-trip time. Increasing the frequency of control makes the system respond faster. On the other hand, this implies that the current incoming packet rate has not yet been affected by the few most recent changes in the source transmission rate. Conversely, the node cannot change the next few changes in the transmission rate, and has to aim for affecting it several steps down the road. That, in turn, requires estimating λ_S, μ and x for several steps down the road. Those estimates will have to rely on other estimates, rather than actual values, which carries the danger of propagating errors.

2.5 Simulation results

The simulation model, consisting of a sender, a receiver and three intermediate nodes is shown in Figure 1. The propagation delay for each link is also shown.

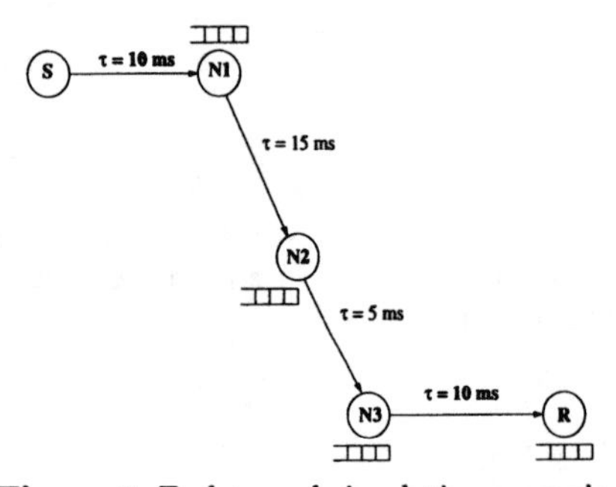

Figure 1 End to end simulation scenario

The simulation parameters are summarized in Table 1. The inter-packet delay at the source was assumed to be 0 (i.e. the source was assumed to always have packets to send), and packets were assumed to have fixed packet sizes (on the order of ATM cell sizes). Congestion was induced by introducing abrupt changes in the processing rates of the intermediate nodes and the receiver at the times shown in Table 2 (the bottleneck node for each time interval is highlighted). The numerical values chosen (in the order of 1000–2000 packets/s or 0.5–1.0 Mbits/s assuming ATM cells) could represent MPEG-1 streams for example.

Figures 2 and 3 show the buffer occupancies at nodes N1 through R for our scheme and that of [7], respectively. The source transmission rate for the two schemes is compared in Figure 4. There are no buffer overflows during transition periods with the proposed scheme—hence no packets are lost. With the scheme proposed in [7], buffer overflows resulted in 23 packets being dropped at node N1 and 159 packets at node N3. This was despite the fact that a relatively low gain factor (0.5) was used. Higher gain factors led to larger oscillations in the transmission rate and higher packet losses.

During the first period (0 to 4.0 seconds) the buffer at the bottleneck, R, rises very quickly

Simulation Period	Control Frequency	Buffer Size	x^*	δ	gain factor for [7]
30 seconds	once every 80 ms	100	50	0.9	0.5

Table 1 Parameters for end to end simulation

time (s)	0.0	4.0	7.6	12.0	16.0
N1	1000	1000	1000	750	750
N2	1200	1200	1200	**720**	950
N3	800	**800**	800	800	**650**
R	**700**	900	**780**	780	780

Table 2 Processing rates (packets/s) for end to end simulation

to its target level (Figure 2). The incoming rate to our bottleneck is limited to 800 packets/s by node R, hence the net rate of increase in buffer occupancy is 100 packets/s at most. During the third period (4.0 seconds to 7.6 seconds), when R again becomes the bottleneck, it rises to its target level at a much slower pace. This is because its processing rate is just under that of $N3$ (780 vs 800 packets/s), hence its buffer is increasing at the net rate of only 20 packets/s (thus the almost 2.5 second rise time to its target level). Now this is true in both schemes. The difference is that in our proposed scheme, the transmitter does not unnecessarily transmit at a higher rate than 800 packets/s whereas it shoots to over 1000 packets/s in the scheme proposed in [7]. This is why we get buffer overflow at $N3$ with the [7] scheme at $t = 7.6$ seconds despite the fact that $N3$ is *not* the bottleneck during this period. The same phenomenon can be observed at $t = 12$ seconds, when $N2$ becomes the bottleneck and we have buffer overflow at $N1$.

Note also the behavior of the source transmission rate at $t = 4.0$ seconds, when $N3$ becomes

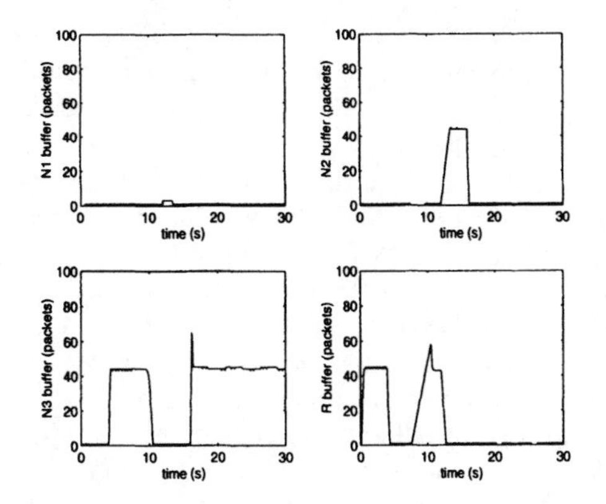

Figure 2 Buffer occupancies with proposed end to end scheme

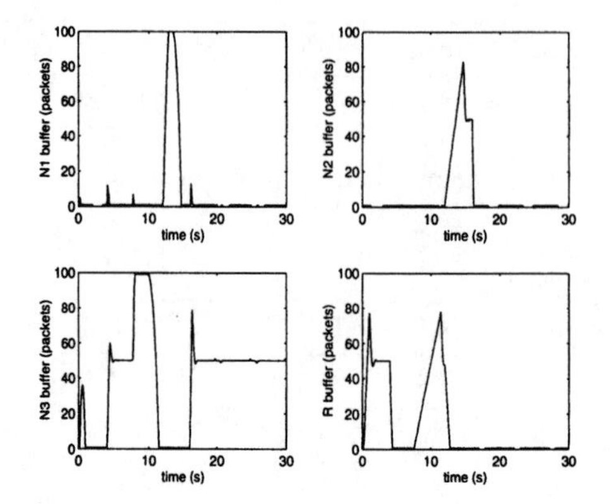

Figure 3 Buffer occupancies with scheme proposed by [7]

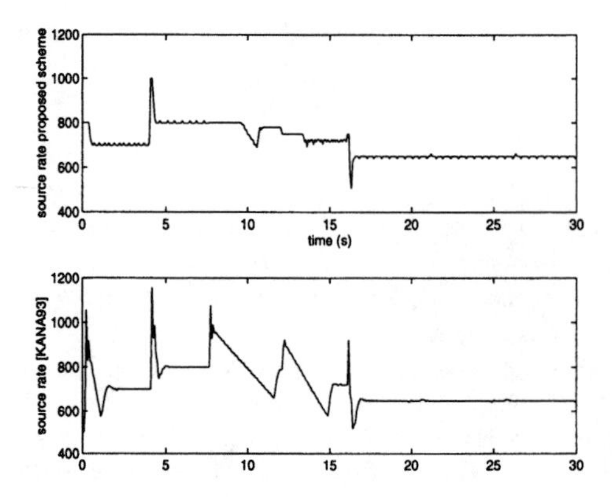

Figure 4 Source rate variations for the proposed scheme and that of [7]

the bottleneck. Its buffer is empty so it asks for a high transmission rate (close to 1200 packets/s as shown in the curve corresponding to the scheme in [7]). In our scheme, however, the source transmission rate is limited to 1000 packets/s by node $N1$.

At $t = 10$ seconds, the occupancy of the bottleneck, R has reached its target level, but it is still increasing at a net level of 20 packets/s because $N2$, which had previously been the bottleneck, is still transmitting the packets left in its buffer at 800 packets/s. At $t = 10$, $N2$ is finally done, but $N3$ has overshot its target, so the source transmission rate is lowered and then adjusted back to 780 packets as the target is reached (upper graph in Figure 4).

A second set of simulations was run using the same configuration as before, but with the processing rates increased by an order of magnitude (representing MPEG-2 rates, for example). The only other difference was that the buffer sizes were also increased by an order of magnitude. In general, the buffer sizes should be on the order of the number of packets generated during one control interval. Smaller buffer sizes lead to heavy packet losses for both schemes. The results were very similar to the first set. Again, packet loss due to buffer overflow was 0 for the proposed scheme while 139 packets were lost at node N3 for the scheme in [7] (see [12] for details).

3 HOP BY HOP APPROACH

One of the main drawbacks of end-to-end schemes using a feedback mechanism is the problem of feedback implosion at the source, under a large multicast scenario. This has led many researchers to reject closed-loop schemes altogether (e.g. [15]). Others have used a combination of probabilistic querying (receivers send feedback information with some probability), randomly delayed responses (receivers randomly delay their feedback control messages) and expanding scoped search (transmitter gradually increases the scope of its control messages) to deal with this problem ([3]). We will develop a hop-by-hop approach (based on the end-to-end scheme devised earlier) because it can be applied to large scale multicast applications and can respond to changes faster, as the propagation delay between adjacent nodes is smaller than that between the two end-points of a connection [11, 9].

In a hop-by-hop model, the sender and the receiver are connected via a number of intermediate nodes. The major difference is that the processing rates at the intermediate nodes (μ_i) are controllable. The maximum processing rate, μ_i^{max} say, is allocated depending on the number of connections going through the node. This maximum value corresponds to μ_i in the end-to-end approach. In the hop-by-hop approach, the actual processing rate can be less than or equal to this value depending on the rates requested by the downstream node (this is elaborated below).

3.1 New control mechanism

The sender and every intermediate node adjusts its transmission rate to the requests made by the immediately downstream node. Separate control packets will be exchanged between adjacent nodes. For each pair of nodes, one control packet is exchanged per round-trip time.

The control law itself is exactly the same as before. All nodes will aim for a target buffer occupancy x_i^*. An 'interval' begins upon receipt of a control packet from an upstream neighbor. At the beginning of each interval a node will estimate its buffer occupancy at the end of the interval using equation (2), and its desired incoming rate for the next interval using equation (5). It will send this desired rate to the upstream neighbor via the control packet.

The upstream neighbor will adjust its transmission rate according to the following rule: μ_{i-1} will be set to the desired rate if the latter is less than μ_{i-1}^{max}; otherwise it will be set to μ_{i-1}^{max}. Regardless of what it sets its rate to, the ability of the upstream node to deliver packets at that rate depends on its own incoming rate and the number of packets in its buffer. The maximum rate at which the upstream rate can actually deliver is given by $\lambda_{i-1} + x_{i-1}/t_0$ (where λ_{i-1} is the incoming rate of the upstream neighbor). If this quanitity is less than the rate at which it wants to send, this will result in the downstream node overestimating its incoming rate. As the simulations will show, however, this scenario is not likely if the target buffer occupancies are chosen carefully (i.e. targets are not too close to 0, so that there is always a pool of packets that can be used if the downstream node requests an increased rate). Note that only those nodes located upstream of the bottleneck will actually be able to reach their targets (x_i^*).

3.2 Simulations

For ease of comparison, the same network topology as before was used to simulate the hop-by-hop control mechanism, but with the rates and chain of events shown in Table 3. The buffer occupancies and processing rates are shown in Figures 5 and 6. As can be seen, all buffers upstream of the bottleneck maintain their target occupancies. This allows for a faster convergence when the bottleneck parameters change. As before, no packets were dropped.

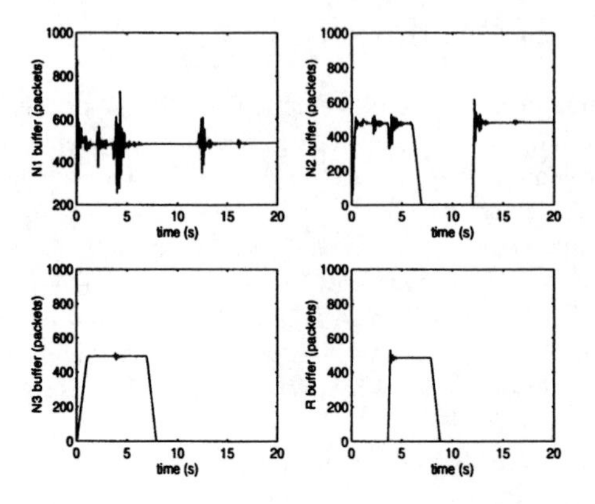

Figure 5 Buffer occupancies for hop-by-hop scheme

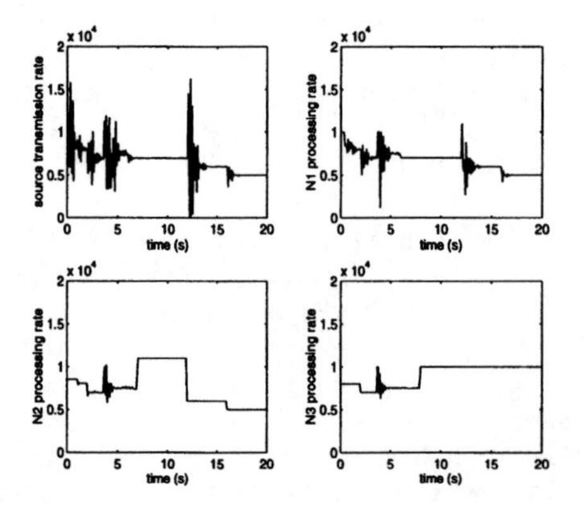

Figure 6 Processing rates for hop-by-hop scheme

Looking at Figures 5 and 6, we see that nodes closer to the source exhibit greater oscillations than those closer to the receiver. Node N3 adjusts its rate to that of the receiver. The latter changes infrequently (once in our simulation). Node 2, however, has to adjust itself to the rate of Node 3, which is slightly more volatile (since that itself is being adjusted to the rate of the receiver), and so on. Each upstream node will be adjusting its rate to a more volatile target. Since delta is less than 1 (0.9), it will not lead to instability though. An interesting instance of this phenomenon occurs at $t = 3.6$ seconds. At this point, R becomes the bottleneck. $N3$ adjusts its rate after a few slight oscillations (in both its processing rate and buffer occupancy). $N2$ has to go through greater oscillations (in both parameters), and $N1$ through even bigger oscillations. This is why there are wide fluctuations in $N1$'s buffer occupancy (and processing rate) at $t = 3.6$ seconds.

Nevertheless, to reduce the oscillations, we gradually reduced delta from 0.9 (receiver) to 0.6 (N1). This resulted in lower oscillations but slightly slower convergence [12].

4 MULTIDROP SCENARIO

time	0.0 s	2.0	3.6	6.0	12.0	16.0
N1	10000	10000	10000	**7000**	11000	5500
N2	8500	11000	11000	11000	**6000**	**5000**
N3	**8000**	**7000**	10000	10000	10000	10000
R	10000	10000	**7500**	7500	7500	7500

Table 3 Processing rates (packets/s) for hop-by-hop simulation

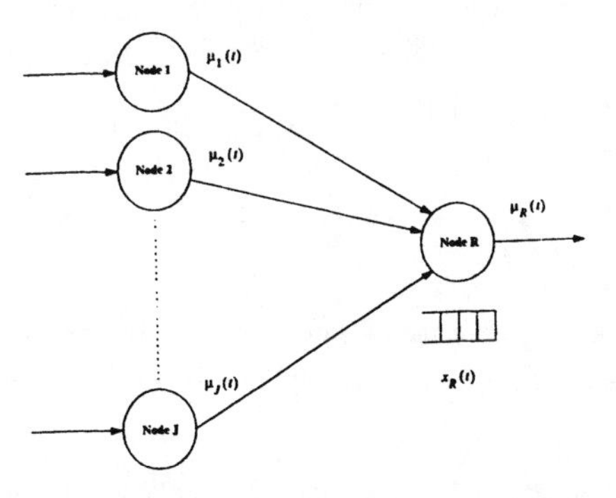

Figure 7 Model for Multidrop Scenario

Figure 7 depicts the case where an intermediate node has more than one upstream neighbor (in the hop-by-hop case) or more than one source transmitting to it (in the end-to-end case). There are J streams flowing into the same buffer at node R. To calculate the buffer occupancy at node R, we need to modify equation (2) as shown:

$$\begin{aligned} x_R(t+t_0) &= x_R(t) + \int_t^{t+t_0} \Big(\sum_{j=1}^{J} \mu_j(\tau - d_j) - \mu_R(\tau)\Big)d\tau \\ \hat{x}_R^{(k+1)} &= x_R^{(k)} + \sum_{j=1}^{J} d_j \mu_j^{(k-1)} + \sum_{j=1}^{J}(t_0 - d_j)\hat{\mu}_j^{(k)} - t_0\hat{\mu}_R^{(k)} \qquad , 0 \le x_R^{(k+1)} \le B_R \end{aligned} \tag{8}$$

d_j is the propagation delay between nodes R and j. We are assuming that the control period is longer than (or equal to) all d_j. Note the explicit inclusion of the delays in equation (8). If the delays between all sources and the bottleneck were assumed to be equal (as in [14]) then indeed we could use the same logic as before and equation (8) would reduce to equation (2). But given the unequal delays, the length of the interval is now set to that of the farthest away source. The bottleneck's interval will thus be synchronized with that of the farthest away source, and overlap two control periods for the other sources.

The control packet can be used to measure the propagation delay between intermediate nodes and the sender. In [7], it is only stated that the delay (measured in multiples of the control frequency interval) is computed using time-stamps included in the control packets, while [8] and [6] assume that the delay is given.

In our mechanism, each intermediate node computes the delay between itself and the source. Upon receipt of the control packet, the node registers the time it passes the packet upstream. As the sender's acknowledgement is then received by each node, it can compute the round-trip (propagation) delay between itself and the source. This is another advantage of having the controller at the bottleneck. It would not be practical to have the source measure the round-trip delays, as it would a) not be a scalable mechanism, and b) introduce too much overhead since the source would have to send a control packet separately to each intermediate node, receive its corresponding acknowledgment, and compute the delay using time-stamp information.

We proceed as before and set $x_R^{(k+2)}$ equal to x_R^*.

$$x_R^* = x_R^{(k+1)} + \sum_{j=1}^{J} d_j \mu_j^{(k)} + \sum_{j=1}^{J} (t_0 - d_j)\widehat{\mu}_j^{(k+1)} - t_0 \widehat{\mu}_R^{(k+1)} \tag{9}$$

which leads to a single equation with J unknowns:

$$\sum_{j=1}^{J} (t_0 - d_j)\widehat{\mu}_j^{(k+1)} = x_R^* - x_R^{(k+1)} - \sum_{j=1}^{J} d_j \mu_j^{(k)} + t_0 \widehat{\mu}_R^{(k+1)} \tag{10}$$

To solve equation (10), we can apply a fairness criterion and have all J nodes transmit at the same rate, say $\mu_f^{(k+1)}$, and then solve for this single unknown, which, after replacing $x_R^{(k+1)}$ using equation (8), yields:

$$\mu_f^{(k+1)} = \frac{x_R^* - x_R^{(k)} - \sum_{j=1}^{J} d_j \mu_j^{(k-1)} + t_0[\widehat{\mu}_R^{(k+1)} + \widehat{\mu}_R^{(k)} - \sum_{j=1}^{J} \mu_j^{(k)}]}{\sum_{j=1}^{J}(t_0 - d_j)} \tag{11}$$

The problem with the fairness criterion is that it assigns equal transmission rates to all J nodes regardless of their capabilities or requirements. A more sophisticated scheme is to have each of the J nodes designate a desired target transmission rate, say λ_j^*, selected according to some criteria[††]. The target values could be communicated to the downstream controller in the control packet. The controller would then attempt to solve equation (10) with the aim of minimizing the difference between the allocated transmission rate and the target transmission rate over all J nodes. We can express this mathematically as:

$$Min \left(\sum_{j=1}^{J} (\lambda_j^* - \mu_j^{(k+1)})^2 \right) \qquad \text{subject to satisfying equation (10)}$$

The above term is minimized (see [12] for proof) for:

$$\mu_j^{(k+1)} = \lambda_j^* - \frac{(t_0 - d_j)K}{\sum_{j=1}^{J}(t_0 - d_j)} \tag{12}$$

where K is equal to the right hand side of equation (10).

Equation (12) is biased towards more remote transmitters. The denominator of the second

[††] In the hop-by-hop case, the targets would simply be μ_i^{max}.

term on the right is the same for all transmitters, as is the term K, but $t_0 - d_j$ is smaller for more remote transmitters, hence their allocated rates are closer to their desired target rates. This is a positive feature, since more remote receivers are slower to adapt to network changes at node R anyway. So it is beneficial to have them deviate less from their desired targets, and have the closer, faster responding, transmitters do most of the work to achieve the desired buffer occupancy.

5 CONCLUSIONS

We presented a rate-based congestion control scheme, with both an end-to-end approach and a hop-by-hop approach. The scheme differed from traditional techniques in that control was distributed among all the hosts on the path from the source to the receiver (as opposed to a source-based approach). This led to a more accurate and scalable design. Compared with similar schemes, our design was better able to deal with varying bottlenecks, preventing buffer overflow during the transition periods. We also showed how the same control packet could be used to relay control information and serve as a means of measuring the inter-node delay for all intermediate nodes. Finally, we showed how the scheme could be extended to multipoint scenarios.

REFERENCES

[1] L. Benmohamed and S. M. Meerkov. Feedback Control of Congestion in Packet Switching Networks: The Case of a Single Congested Node. *IEEE/ACM Transactions on Networking*, 1(6):693–708, December 1993.

[2] J. Bolot and T. Turletti. A Rate Control Mechanism for Packet Video in the Internet. In *Proceedings of the IEEE INFOCOM*, pages 1216–1223, Toronto, Canada, June 1994.

[3] J. Bolot, T. Turletti, and I. Wakeman. Scalable Feedback Control for Multicast Video Distribution in the Internet. In *Proceedings of the ACM SIGCOMM*, pages 58–67, London, UK, August 1994.

[4] F. Bonomi and K. Fendick. The Rate-Based Flow Control Framework for the Available Bit Rate ATM Service. *IEEE Network Magazine*, 9(2):25–39, March/April 1995.

[5] W. A. Doeringer, D. Dykeman, M. Kaiserswerth, B. W. Meister, H. Rudin, and R. Williamson. A Survey of Light-Weight Transport Protocols for High-Speed Networks. *IEEE Transactions on Communications*, 38(11):2025–2039, November 1990.

[6] Kerry W. Fendick, Manoel A. Rodrigues, and Alan Weiss. Analysis of a Rate-Based Control Strategy with Delayed Feedback. In *Proceedings of the ACM SIGCOMM '92*, pages 136–148, Baltimore, MD, August 1992.

[7] H. Kanakia, P.P. Mishra, and A. Reibman. An Adaptive Congestion Control Scheme for Real-Time Packet Video Transport. In *Proceedings of the ACM SIGCOMM '93*, pages 21–31, San Fransisco, CA, September 1993.

[8] Srinivasan Keshav. A Control-Theoretic Approach to Flow Control. In *Proceedings of the ACM SIGCOMM '91*, pages 3–15, Zürich, Switzerland, September 1991.

[9] H. T. Kung and K. Chang. Receiver-Oriented Adaptive Buffer Allocation in Credit-Based Flow Control for ATM Networks. In *Proceedings of the IEEE INFOCOM*, pages 239–252, Boston, MA, April 1995.

[10] D. E. McDysan and D. L. Spohn. *ATM: Theory and Applications.* McGraw Hill, 1995.

[11] P.P. Mishra and H. Kanakia. A Hop by Hop Rate-based Congestion Control Scheme. In *Proceedings of the ACM SIGCOMM '92*, pages 112–123, Baltimore, MD, August 1992.

[12] S. Pejhan. *Protocols for Multipoint, Multimedia Communications.* PhD thesis, Columbia University, 1995.

[13] K. K. Ramakrisnan and P. Newman. Integration of Rate and Credit Schemes for ATM Flow Control. *IEEE Network Magazine*, 9(2):49–56, March/April 1995.

[14] C. Roche and N. T. Plotkin. The Converging Flows Problem: an Analytical Study. In *Proceedings of the IEEE INFOCOM*, pages 32–39, Boston, MA, April 1995.

[15] R. Yavatkar and L. Manoj. Optimistic Strategies for Large-Scale Dissemination of Multimedia Information. In *ACM Multimedia '93*, pages 12–20, Anaheim, CA, August 1993.

[16] N. Yin and M.G. Hluchyj. A Dynamic Rate Control Mechanism for Source Coded Traffic in a Fast Packet Network. *IEEE Journal on Selected Areas in Communications*, 9(7):158–181, September 1991.

[17] N. Yin and M.G. Hluchyj. On Closed Loop Rate-Control for ATM Cell Relay Networks. In *Proceedings of the IEEE INFOCOM*, pages 99–108, Toronto, Canada, June 1994.

14
A Minimal-Buffer Loss-Free Flow Control Protocol for ATM Networks

Siavash Khorsandi and Alberto Leon-Garcia
University of Toronto
Department of Electrical and Computer Engineering,
University of Toronto, Toronto, Ont., Canada M4Y 1R5,
Email: khorsand@comm.utoronto.ca

Abstract

Flow control is essential in ATM networks to prevent service quality degradation during periods of network congestion. In particular, best effort services such as data file transfer have stringent cell loss requirements. Use of backpressure in conjunction with buffer reservation is a promising approach. Buffer reservation on per-VC basis require large buffers. Besides, fairness and high network utilization cannot be supported at low buffer sizes. In this paper, we propose a new scheme based on buffer reservation for groups of VCs rather than for individual VCs. A general framework for credit-based link-by-link flow control is developed by disassociating buffer reservation from credit allocation. Optimality conditions for a credit allocation mechanism are found and an adaptive credit allocation algorithm is designed. The buffer requirement of this scheme is close to one round trip time worth of cells per group. Simulation results indicate that fairness is maintained in a robust manner.

Keywords

Flow control, Credit-based, ATM

1 INTRODUCTION

In ATM networks due to statistical multiplexing, service quality degradation will occur during periods of network congestion [Chiabaut 1994]. Best-effort services such as data file transfer have stringent cell loss requirements. In gigabit networks, the end-to-end feedback based congestion control mechanisms cannot react quickly enough to short-term congestion due to delay between the sender and receiver [Maxemchuck 1990]. Use of selective backpressure based on a distributed flow control protocol is an alternative approach [Mishra 1992, Kung 1993]. In order to avoid possible deadlock and unfairness problems it is necessary to reserve buffers for each active connection or flow [Tanenbaum 1989]. In the credit-based flow control, by combining selective backpressure per flow and buffer reservation mechanisms, it is possible to guarantee that cells are never dropped due to congestion. Furthermore, quick reaction to release of resources results in a better network utilization.

A possible approach for buffer reservation is to strictly partition the available buffer space at each node among active flows. This scheme requires a memory equal to one round-trip delay worth of cells for each connection. To reduce the memory size, it is possible to periodically evaluate the requirements of individual connections and dynamically reallocate buffers to meet those requirements [Ozveren 1994, Kung 1994]. These schemes still require a memory equal to several round-trip delays worth of cells for every group of connections.

In this paper, we propose a credit-based flow control algorithm which differs from the previous approaches mainly in buffer reservation process. We perform the buffer reservation for groups of VCs rather than for individual connections. As a result, the buffer requirement of the scheme is close to the minimum of one round trip delay worth of cells per group. Two levels of flow control is applied. One to control the aggragate flow of a group and the other to control the flow of individual VCs inside a group. Hence, credit allocation of individual VCs can be optimized independent form their actual buffer usage to achieve good transient response and high network throughput.

The rest of this paper is organized as follows. In Section 2, we develop the framework of the proposed flow control mechanism and the concept of group-level buffer reservation is established. In Section 3, the adaptive credit allocation mechanism for flow controlled connections is discussed and a rate-based algorithm is presented. In Section 4, we study the properties of the proposed scheme. Finally, Section 5 contains the simulation results in which transient and steady state behavior of the protocol is studied.

2 FLOW CONTROL MECHANISM AND GROUP-BASED BUFFER RESERVATION

2.1 Basic operation of credit-based flow control

The operation of the credit-based flow control protocol is shown in Figure 1 for a single hop of an ATM virtual circuit (VC). The protocol is applied on a hop-by-hop basis. At the upstream node, U, every VC is allocated a number of credits, C_i , $\forall i$. Each time a cell is forwarded on this VC by U, it increases a counter that keeps track of the number of outstanding cells, $O_i(t)$. At any time, the number of available credits of a VC is equal to $C_i - O_i(t)$. As long as the number of available credits remains positive, the upstream node can forward cells on that VC. At the downstream node, D, the data for each VC is queued separately. Every τ seconds, the number of cells forwarded from each VC's buffer is acknowledged to the upstream node where the number of outstanding cells is reduced by the same amount.

Let 'd' be the link propagation delay between U and D and R to be the round-trip time (RTT) defined as $2 * d + \tau$ both in seconds. We also denote the raw link rate (between U and D) in cells/second by B. Then, the maximum long-term average transmission rate of a VC is equal to $\lambda_i^s = \min(B, \frac{C_i}{R})$. This is referred to as *maximum sustainable rate* of the connection. We also define $G(t)$ and $H(t)$ as the cumulative cell departure process at the upstream and downstream nodes respectively (Figure 1). During the transmission of a cell, $G(t)$ and $H(t)$ increase at a constant rate of B cells/second.

In conjunction with the credit-based mechanism, a per VC fair queuing scheme is necessary to provide each VC with a fair share of the bandwidth. The primary goal in fair queuing is to serve sessions in proportion to some prespecified service shares, independent of the queuing load presented by the sessions [Parekh 1993].

2.2 Adaptive credit allocation

In static credit allocation, C_i remains constant during the life time of a connection. In order to prevent buffer overflow at D, each VC must then have a reserved buffer space for C_i cells

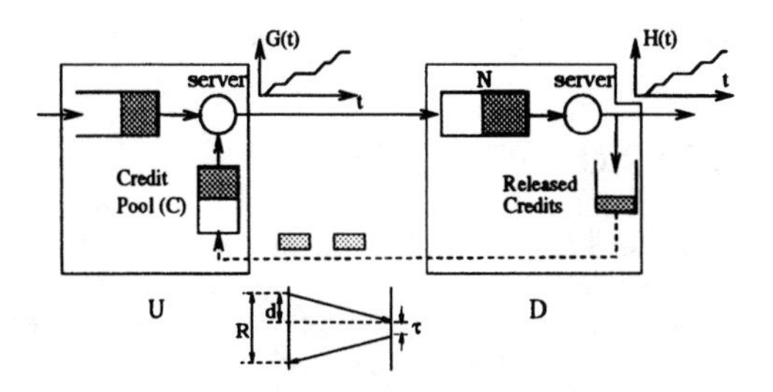

Figure 1 The operation of a credit-based flow-controlled connection.

which can be in the order a RTT worth of cells which is wasteful and expensive. Since the bandwidth requirement of the connections is not constant, in this work we periodically evaluate the requirements of individual connections and adjust their credit allocation to meet those requirements. This could be done only among those connections that can dynamically share the buffer at D. We define a set of such VCs as a *group* and it is denoted by $\mathcal{G}$. If the cells are buffered at a shared memory or at the input of the switch at D, a group includes all the VCs in the same input link. Otherwise, the VCs going from U to D may form several groups depending on their output port at D.

The time between two credit adjustments, τ_c, is called a *control interval.* The credit allocation of a VC during kth control interval is denoted by $C_i(k)$. Otherwise, all the previous formulations still apply. The time at the beginning of kth control interval at U is denoted by t_k. Also, for brevity we denote $t_k + d$ by t_k^d which is the start of kth control interval at D (Figure 2-b).

2.3 Group-level buffer reservation

In per VC buffer reservation, the available buffer space for a group, N, is strictly partitioned among VCs in that group and the credit allocation of a connection corresponds to its buffer allocation. Hence, we have $\sum_{i\in\mathcal{G}} C_i(k) \leq N$. As we will demonstrate later, this approach suffers from slow-start phenomenon and requires a buffer size of at least two RTT worth of cells.

Our approach is based on buffer reservation at group level. The buffer allocated for a group at D is not pre-partitioned among the VCs by the upstream node. The credit allocation of VCs does not necessarily correspond to their buffer allocation and hence the restriction on $\sum_{i\in\mathcal{G}} C_i(k)$ is removed, that is, $\sum_{i\in\mathcal{G}} C_i(k) \lg N$. This allows the credit allocation of VCs to be optimized independent from their actual buffer utilization. Besides, full buffer sharing is provided by allowing the actual buffer utilization of connections to be determined through contention among them. However, it may also result in buffer overflow at D since the availability of a credit may no longer imply the availability of a buffer space. To prevent this, a group-level flow control is applied. Figure 2-a demonstrates the operation of the proposed flow control mechanism.

The group-level flow control regulates the aggregate flow of a group. Therefore, lossless transmission is provided regardless of individual VCs credit allocations. The group-level flow control works similar to VC-level flow control. The credit allocation of a group is denoted by C_g which is equal to its buffer size, N. The number of outstanding cells of a group at time t is also denoted by $O_g(t)$. The number of available credits of the group is equal to $C_g - O_g(t)$. The upstream node can forward a cell on a VC belonging to $\mathcal{G}$ only if the number of available credits for that group is positive. As the acknowledgements for individual VCs in the group arrive, O_g is also reduced. Therefore, the group-level flow control does not require any extra transfer of information between nodes. In the next section, we will address the adaptive credit allocation for VCs.

3 CREDIT ALLOCATION PROCESS

Although the credit allocation of a VC does not necessarily correspond to its buffer allocation, it sets an upper bound on its buffer utilization. The credit allocation of VCs must be optimized to achieve objectives such as preventing a possible deadlock, maintaining fairness, maximizing network throughput and achieving good transient response.

If a VC is targeted to transmit at the maximum sustainable rate of λ_i^s, we must have $C_i(k) \geq \lambda_i^s R$. The credit allocation of a VC is equal to the maximum number of cells that it can transmit over a RTT. To proceed, we define the following average rates:

$$\lambda_i^R(k) = \frac{G(tk+R) - G(t_k)}{R}, \quad \mu_i^R(k) = \frac{H(t_k^d + R) - H(t_k^d)}{R}, \quad \gamma_i^R(k) = \frac{H(t_k^{-d} + R) - H(t_k^{-d})}{R} \tag{1}$$

As depicted in Figure 2-b, $\lambda_i^R(k)$ and $\mu_i^R(k)$ denote the average transmission rate over a period of R seconds at U and D respectively starting with the kth control interval. Also, $\gamma_i^R(k)$ is the average rate of credit arrivals at U during the same period. We also define $\hat{\mu}_i^R(k)$ as the maximum value of $\mu_i^R(k)$ given that the connection is able to transmit at its maximum sustainable rate, λ_i^s, at U. For simplicity of formulation, we assume that a RTT is an integer multiple of the control interval, that is, $R = m \cdot \tau_c$ where m is a positive integer. Nevertheless, the algorithm itself is robust and does not depend on this assumption. It can be seen that $\gamma_i^R(k) = \mu_i^R(k-m)$. We also define $n_i(t)$ to be the number of cells of VC #i in the buffer at D at a given time t.

Lemma 1: Under credit-based flow control with adaptive credit allocation, the following conditions are necessary to maximize network throughput:

a) $$C_i(k) \geq \hat{\mu}_i^R(k) \cdot R\ , \ \forall i$$

b) $$\sum_{i \in \mathcal{G}} [C_i(k) - \hat{\mu}_i^R(k) \cdot R]^+ \leq N - B \cdot R \tag{2}$$

where $[x]^+$ is equal to x if $x \geq 0$ and is 0 if $x < 0$.

Proof: a) The proof is straightforward. $\hat{\mu}_i^R(k) \cdot R$ is the number of cells that can be forwarded at D on this connection over the next R seconds. In order for the VC to utilize the available bandwidth, its credit allocation must be equal or greater than $\hat{\mu}_i^R(k) \cdot R$. On the other hand suppose $C_i(k) < \hat{\mu}_i^R(k) \cdot R$. The throughput of an active connection will be throttled by the flow control mechanism while available bandwidth at D is underutilized.

b) An active connection utilizing more than $\hat{\mu}_i^R(k) \cdot R$ cells over a RTT will eventually build up a queue of size $C_i(k) - \hat{\mu}_i^R(k) \cdot R$ at D. Taken over all the connections, a queue size of $\sum_{i \in \mathcal{G}} [C_i(k) - \hat{\mu}_i^R(k) \cdot R]^+$ will be built at D. If this is greater than $N - B \cdot R$, the total number of available group credits at U over a RTT is less than $B \cdot R$. Therefore, full link capacity cannot be utilized even if there are active connection with available bandwidth at D.□.

Combining conditions (2-a) and (2-b), the credit allocation of a connection is determined as follows:

$$C_i(k) = \hat{\mu}_i^R(k) RTT + N_i^+(k) \tag{3}$$

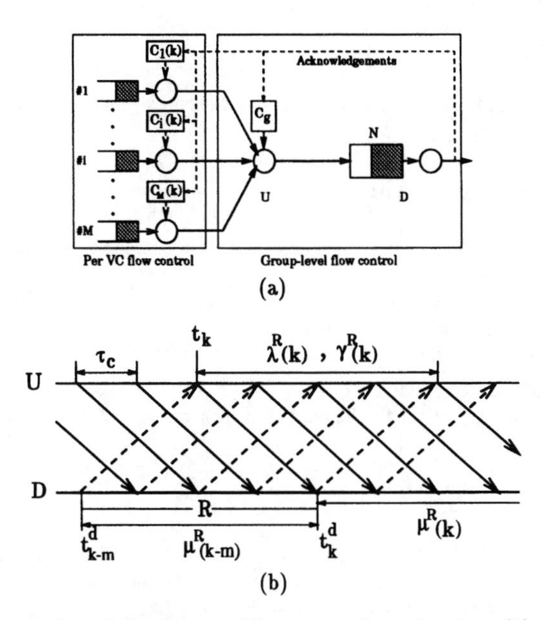

Figure 2 (a) Operation of the proposed flow control mechanism. (b) Time diagram of flow control procedure. We have omitted the subscript i for simplicity.

where $N_i^+(k)$ is called the credit *overallocation* of connection i. From (2-b), we have $\sum_i N_i^+(k) \leq N - B \cdot R$. Equation 3 is our credit allocation *law*. At the beginning of a control interval k, $\hat{\mu}_i^R(k)$ needs to be predicted based on the current and past observations of the cell departure rate at D.

The number of available credits at t_k is $C_i(k) - O_i(t_k)$ and the number of credits received over a RTT is $\gamma_i^R(k)R$. Hence, if the credit allocation remains equal to $C_i(k)$, the maximum number of cells forwarded by U over this period is $C_i(k) - O_i(t_k) + \gamma_i^R(k)R$. By replacing $C_i(k)$ from (3), we have:

$$\lambda_i^R(k) \leq \frac{1}{R}[\hat{\mu}_i^R(k)R + N_i^+(k) - O_i(t_k) + \gamma_i^R(k)R] \quad (4)$$

To complete the credit allocation process in (3), we need to calculate $N_i^+(k)$ and $\hat{\mu}_i^R(k)$. We will address these two problems later in this section.

An alternative formulation

It is instructive to present an alternative formulation of credit allocation process as a rate control mechanism. We show that the control law in (3) is equivalent to a rate-based flow control which aims to keep the buffer occupancy at a level less than or equal to $N_i^+(k)$. We have

$$n_i(t_k^d + R) = n_i(t_k^d) + \lambda_i^R(k)R - \mu_i^R(k)R. \quad (5)$$

From Figure 2-b, it is easy to show that

$$n_i(t_k^d) = O_i(t_k) - \gamma_i^R(k)R \tag{6}$$

Substituting (6) in (5) and setting $n_i(t_k^d + R)$ at less than or equal to $N_i^+(k)$, equation (4) is obtained.

3.1 Credit overallocation

The overallocation is necessary for a connection to keep a backlog in the buffer at D. If the departure rate from the buffer increases, the backlog is drained and the rate increase will be sensed at U by an increase in the rate of credit (acknowledgements) arrival. The credit allocation of the connection is then increased accordingly. The increase in the rate of credit arrivals over a RTT depends on the size of credit overallocation. For a fixed $N_i^+(k)$ over a RTT, maximum increase in the rate of the connection is equal to $\frac{N_i^+(k)}{R}$. We use the following formula to update the credit overallocation of a VC in the beginning of each control interval:

$$N_i^+(k) = \eta \cdot \gamma_i^{T_c}(k-1) \cdot R + \frac{N^+}{\#active\ VCs} \tag{7}$$

where $N^+ = [N - (1+\eta)B \cdot R]^+$ is the buffer space equally divided among active VCs, and $\gamma_i^{T_c}(k-1)$ is the average rate of credit arrival at U over the past control interval. A VC i is considered active if it has a backlog at U or if $\lambda_i^R(k-m) > 0$. The first term in (7) enables a VC to increase its rate to $(1+\eta) * 100\%$ of its current rate in one RTT. The rate increase is significant if the current rate is large. Hence, the first term favors high rate VCs. The second term, on the other hand, lets a VC to increase its rate by a fixed portion of link rate every RTT which is especially significant for low-rate VCs. Also, the second term ensures that every VC has a minimum buffer reservation. This protects the network from a possible deadlock.

3.2 Rate prediction

To predict the departure rate $\hat{\mu}_i^R(k)$, we use exponential averaging which is a first order autoregressive filter:

$$\hat{\mu}_i^R(k) = (1-\alpha)\hat{\mu}_i^R(k-1) + \alpha \cdot \gamma_i^{T_c}(k-1) \tag{8}$$

where $\gamma_i^{T_c}(k-1) = \mu_i^R(k-m-1) - \mu_i^R(k-m)$ is the latest observation of the average departure rate over a control interval. The predictor is controlled by parameter α which is the weight given to the latest observation relative to the past history. Although more complex predictions can be used, this predictor has proved effective and is simple enough to be implemented in ATM switches.

Parameter α

Parameter α can be dynamically determined using Kalman filtering if the perturbation in the departure rate is modeled as a white Gaussian noise, $\mu_i^R(k) = \gamma_i^{T_c}(k-1) + w_k$, and the variance of noise is known [Keshav 1991]. However, this is not usually the case. An alternative approach is to use an adaptive scheme to adjust α based on the prediction error. Intuitively, the value of α must be small when the transmission rate is steady to give more weight to the past history and filter out transient changes in $gamma_i^{T_c}(k)$. However, when the transmission rate is quickly changing, α must be large to give the emphasis to the recent observations since past observations are too old.

Speed of Convergence

Starting with an initial state $\hat{\mu}_i(0) = 0$, a VC can increase its transmission rate only by $\frac{N_i^+(k)}{R}$ every RTT. To increase the speed of rate increase, the following actions are taken:

a- Instead of starting with $\hat{\mu}_i(0) = 0$, the departure rate of a VC can be predicted using the transmission rate of other active VCs in the same virtual path. An alternative is to start from a fixed nonzero initial state or develop a statistical method to predict the available transmission rate through the network.

b- If $\gamma_i^{Tc}(k-1) > \hat{\mu}_i^R(k-1)$ and $\lambda_i^R(k-1) > \hat{\mu}_i^R(k-1)$, this indicates the VC is in ramp up mode, that is, it requires higher transmission capacity which is available both at U and D. Hence, the prediction of departure rate is scaled up by a factor of β:

$$\hat{\mu}_i^R(k) = (1-\alpha)\hat{\mu}_i^R(k-1) + (\alpha+\beta)\gamma_i^{Tc}(k-1) \tag{9}$$

4 PROPERTIES OF THE PROPOSED SCHEME

4.1 Credit allocation

Due to separation of credit allocation from buffer reservation, we have:

1- It is possible to have $C_i(k) < O_i(t_k)$ which means that in-use credits of a VC can be revoked and assigned to other VCs. Consequently, as acknowledgements arrive for these credits, they can be immediately used by other VCs.

2- Credit overbooking is possible, that is, $\sum_{i \in \mathcal{G}} C_i(k) > N$. Hence a buffer space can be simultaneously assigned to several VCs. While cell loss is prevented by group-level flow control, the actual buffer usage is determined by contention among VCs. In the absence of contention, a VC can use a larger share of the buffer resulting in a better transient response and better resource utilization.

4.2 Buffer requirement

The minimum buffer requirement of a group is obtained by substituting (3) in (2-b) and taking the equality, $N = B_l \cdot RTT + \sum_{i \in G} N_i^+(k)$. Combining this with (8), we have $N = B_l \cdot RTT + \eta RTT \sum_{i \in G} \gamma_i(k-1) + N^+$. In the worst case, we have $\sum_i \gamma_i(k-1) = B_l$, then

$$N = (1+\eta) \cdot B \cdot R + N^+ \tag{10}$$

A typical value for γ is 0.2 which allows a VC to immediately increase its rate by 20%. The optimal value of N^+ depends on the network condition. However, if we let $N^+ = 0.3B \cdot R$. Then the total buffer requirement of a group is equal to $1.5B \cdot R$, that is, one and half round trip delay worth of cells. This scheme can work with buffer sizes very close to the minimum value of $B \cdot R$. A minimum buffer size of $B \cdot R$ is required to allow a group to fully utilize the link capacity with no cell loss occurring due to buffer overflow.

4.3 Stability and fairness

It can be shown that the flow control mechanism is stable for $\alpha < 1$ and that an steady state is achieved in a finite time. Besides, since the group-level flow control does not interfere with the scheduling mechanism, it can be proved that the bandwidth is fairly divided among VCs. Due to space limitation, we omit the proof of these statements.

4.4 Comparison with a per-VC buffer reservation scheme

An adaptive credit allocation scheme based on per VC buffer reservation is proposed by ATM-Forum in [Chiabaut 1994]. Using our notations, the credit allocation is done as follows:

$$C_i(k) = \hat{\mu}_i^R(k)R + \frac{N^+}{\#active\ VCs}$$

However, this credit allocation has to be modified based on the following two constraints: $C_i(k) \geq O_i(k)$ and $\sum_{i \in \mathcal{G}} C_i(k) \leq N$. The first constraint does not allow $C_i(k)$ to be arbitrarily reduced and the second one does not allow it to be arbitrarily increased. Due to these constraints, the credit allocation algorithm has to be invoked every time an acknowledgement message arrives. This results in an increased computational complexity compared to our scheme where credit adjustment is performed once in a control interval and the size of control interval can be arbitrarily set. Besides, this scheme requires a memory of at least $2B \cdot R$.

5 SIMULATION RESULTS

Extensive simulations have been carried out to evaluate the transient and long-term performance of the proposed scheme. To demonstrate the importance of buffer reservation at group level, we have also simulated the per VC buffer reservation (PVBR) scheme described in Section 4.4. using the same prediction algorithm and similar set of parameters.

The following choice of parameters are made: τ is set to $R/10$ to limit the bandwidth usage of acknowledgement messages to 2% of the link rate, τ_c is set to be equal to τ in order to make our scheme compatible with the PVBR, and η and β are both set to 0.2. In our experiments the distance between every two ATM switch is 330km and the transmission links operate at 1.2Gbps. The simulations are done for various buffer sizes. However, for PVBR scheme, the total buffer allocation of each group is always set at $2B \cdot R$.

5.1 Transient response

A three-hop network scenario shown in Figure 3 is used to evaluate the transient behavior of the protocol. The performance of four source-destination pairs, SD1-SD4, is monitored. Cross traffic is generated by high-speed and low-speed sources in bunches of 10. Low speed sources are added to test the sensitivity of the protocol to presence of large number of active VCs. Peak rate of high-speed and low-speed sources are B and $0.01B$ respectively. Destination node D1 receives 1,2 and 3 hop traffic. For SD1 and SD2, C-D1 link is the bottleneck while for SD3 and SD4 the bottleneck is B-C link.

Connections SD1 to SD4 are activated according to the pattern of Figure 4. Therefore, SD1 and SD2 will have enough time to fill up the group buffer before SD3 and SD4 become active. This is a particularly difficult scenario to handle since the buffer is occupied by the connections which face congestion two hops farther down the line. As the last source to become active, SD4 is expected to receive the worst performance.

Credit allocation

The process of credit allocation of connection SD4 at node A is shown in Figure 5. The credit allocation starts from an initial state and gradually adapts to the network dynamics. When the buffer size is reduced from $2B \cdot R$ to $1.5B \cdot R$, the allocation process goes through an oscillatory period but still stays around the steady state range of allocation. On the other hand, the credit allocation in PVBR scheme has to start from zero and remains oscillatory. In the PVBR, a VC cannot increase its allocation before other VCs' allocations are decreased. Meanwhile, these VCs continue to use a larger share of bandwidth. Larger delay in adapting to the network dynamics results in oscillation.

Network throughput

Figures 6 and 7 compare the throughput of SD4 and the overall throughput for various cases. The reference point is the proposed scheme with a buffer size of $2B \cdot R$ in both cases. By decreasing the buffer size, the throughput also decreases in a transient period. On the other hand, too large a buffer is also problematic. Using a buffer size of $10B \cdot R$, the overall throughput is reduced up to 30%. The reason lies in the fact that due to a large buffer size, connections SD1 and SD2 continue to consume a large portion of the bandwidth while they face congestion at node C. However, for smaller buffer sizes, SD1 and SD2 are stopped faster and the bandwidth is used by SD3 and SD4 which does not face congestion and hence the network throughput increases. This effect is more explicitly shown in Figure 8 which depicts the overall cumulative throughput after 10 and 15 RTTs as a function of buffer size. The peak throughput is achieved for a buffer size of around $2B \cdot R$. In general, optimal buffer size depends on the number of connections and the network scenario.

Fairness issue

Two similar connections, SD3 and SD4, should receive the same amount of bandwidth when both are backlogged. Figure 9 shows the difference in the throughput of SD3 and SD4 after SD4 becomes active. The proposed scheme demonstrates a robust form of fairness. The throughput SD3 and SD4 remain very close to each other regardless of the buffer size. However, under PVBR the difference between their throughput is up to 80%.

5.2 Long-term performance

The average network throughput and burst delay over a long term have been measured in a statistically symmetric multihop network scenario as a function of the buffer size and average network load.

Source model

We have used On-Off source model with exponentially distributed on and off periods and the average burst size is 100 Kbytes. Each link is shared by 40 VCs and we change the utilization factor of the sources in order to adjust the network load.

We have constructed a statistically symmetric multihop network (Figure 10) where every two switches are 330km apart, each ATM switch is connected to ten local terminals, and all the virtual connections traverse four network hops. Therefore, each link is shared with 1,2,3 and 4 hop traffic.

The average burst queuing delay is shown in Figure 11. The long-term average network throughput, ρ, in Figure 11-a is 0.6 and in Figure 11-b, it is 0.85. As it is seen the network throughput of $\rho = 0.85$ is supported with a group buffer size of as low as $1.2B \cdot R$. In terms of average burst delay, it is seen that the proposed scheme with a buffer size of $1.3B \cdot R$ works better than PVBR scheme with a buffer size of $2B \cdot R$. The minimum burst delay is achieved with a buffer size of $3B \cdot R$. In fact, in the heavy load condition, the delay tends to rise for buffer sizes beyond $3B \cdot R$ although the increase is not considerable.

6 CONCLUSION

In this paper, we developed a distributed credit-based flow control mechanism in which the buffer allocation is handled at group level while adaptive credit allocation is performed at VC level. The group level flow control mechanism provides strictly lossless transmission regardless of individual VCs credit allocation. As a result, credit allocation can be optimized independently. We developed a rate based credit allocation scheme and we proved that it is necessary to maximize the network throughput.

The technique developed in this paper can work with buffer sizes very close to one round trip delay worth of cells while maintaining the network throughput at a relatively high level. We investigated the transient and long-term performance of the technique. According to our results, in a multihop network scenario, peak network utilization is achieved with a buffer size of around $1.2B \cdot R$ per input link. We have also compared the proposed technique to an adaptive credit allocation technique based on per VC buffer reservation and the proposed scheme was clearly superior.

REFERENCES

J. Chiabaut (1994) Flow control for ATM networks, Bell-Northern Research, Technical document, Ref. CRL-94134

S. Keshav (1991) A control-theoretic approach to flow control, *SIGCOMM'91 Conference, Computer Communication Review*, Vol.21, No. 4, Sept. 1991

H.T. Kung, R. Morris, T. Charuhans and D. Lin (1993) Use of link-by-link flow control in maximizing ATM networks performance: simulation results, *IEEE Hot Interconnect Symposium'93*, Palo Alto, CA., August 1993

H.T. Kung, T. Blackwell and A. Chapman (1994) Credit-based flow control for ATM networks: credit update protocol, adaptive credit allocation and statistical multiplexing, *Computer Comm. Review*, Vol. 24, Oct. 1994, pp. 101-114

N. Maxemchuck and M. El Zarki (1990) Routing and flow control in high speed, wide area networks, *Proceedings of the IEEE*, Vol. 78, No. 1, Jan. 1990

P.P. Mishra and M. Kanakia (1992) A hop-by-hop rate-based congestion control scheme, *SIGCOMM'92*, pp. 112-123

C. Ozveren, R. Simcoe and G. Varghese (1994) Reliable and efficient hop-by-hop flow control, *Computer Comm. Review*, Vol. 24, Oct. 1994, pp. 89-100

A.K. Parekh and R.G. Gallager (1993) A generalized processor sharing approach to flow control in integrated services networks: the single-node case, *IEEE/ACM Transaction on Networking*, Vol. 1, No. 3, 1993, pp. 344-357

A. Tanenbaum (1989) *Computer Networks*, Prentice Hall, 2d edition

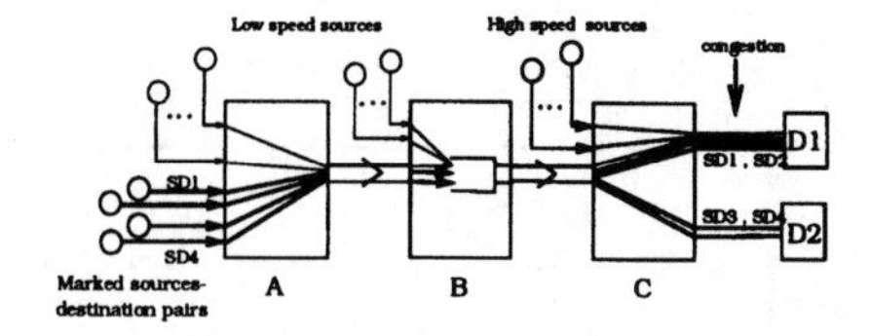

Figure 3 A network scenario used in evaluating the transient responses.

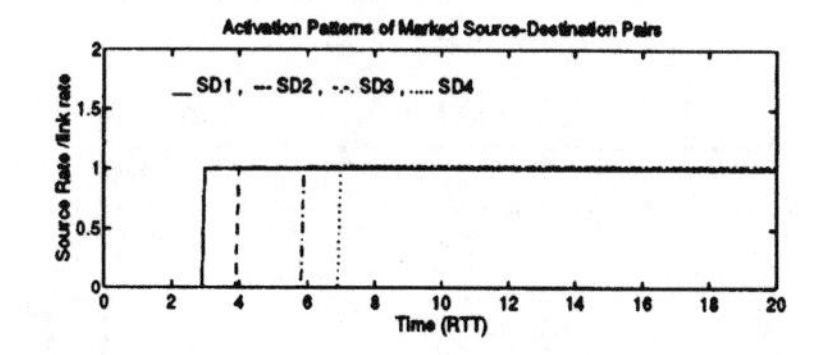

Figure 4 The activation patterns of marked source-destination pairs.

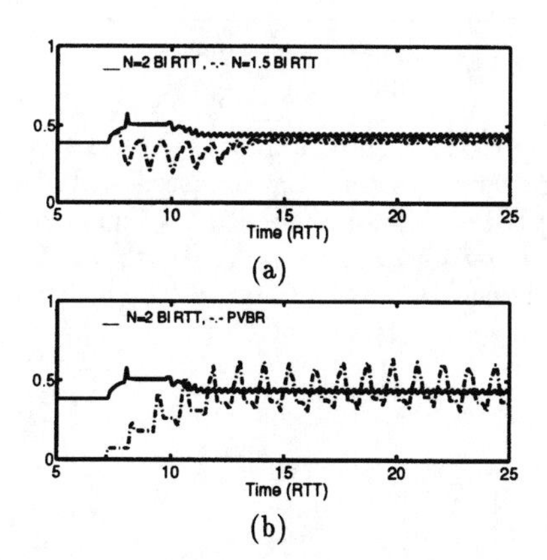

Figure 5 Credit allocation process for SD4 : a) for two buffer sizes $1.5B \cdot R$ and $2B \cdot R$ b) for the proposed and the PVBR schemes with buffer size $2B \cdot R$.

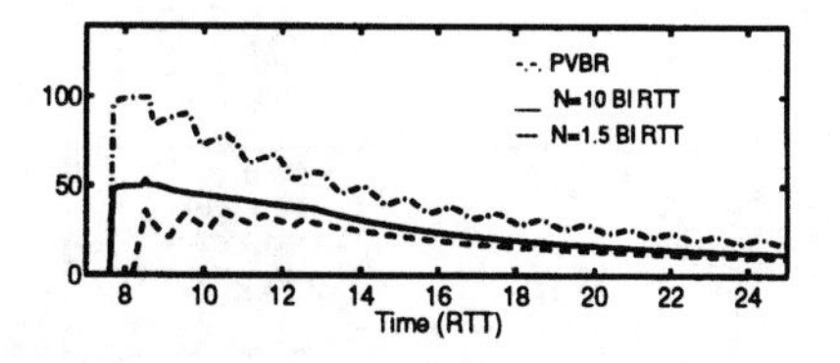

Figure 6 Percentage of loss in the cumulative throughput of connection SD4 relative to the proposed scheme with group buffer size of $2B \cdot R$.

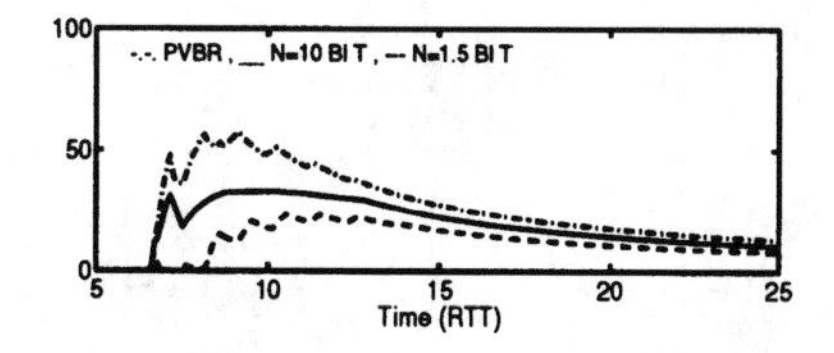

Figure 7 Percentage of loss in the overall cumulative throughput of marked connections relative to the proposed scheme with group buffer size of $2B \cdot R$.

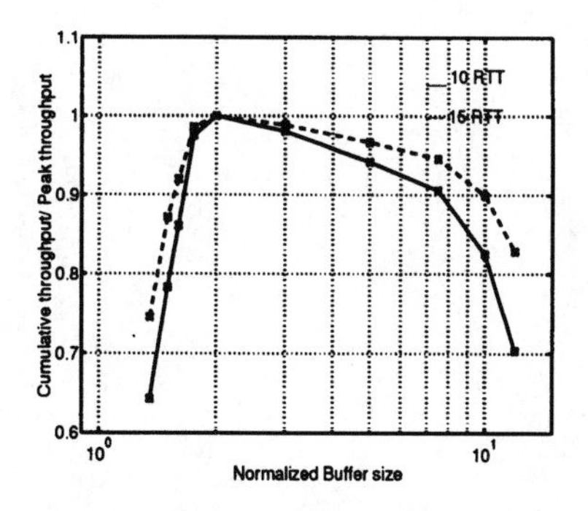

Figure 8 The overall cumulative throughput of the marked connections 10 and 15 RTTs after activation of SD1.

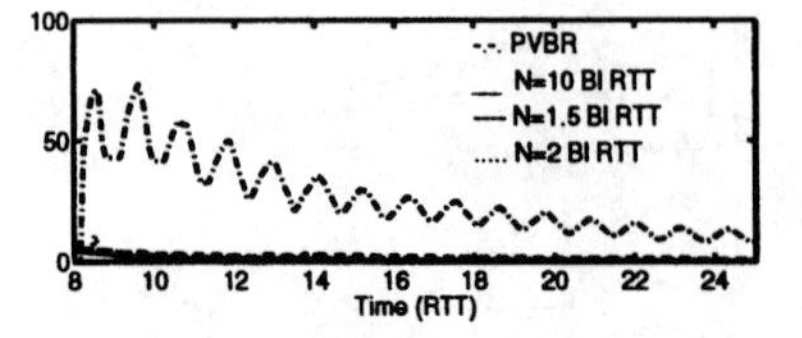

Figure 9 Percentage of difference between the cumulative throughput of connection SD3 and SD4 when both were active.

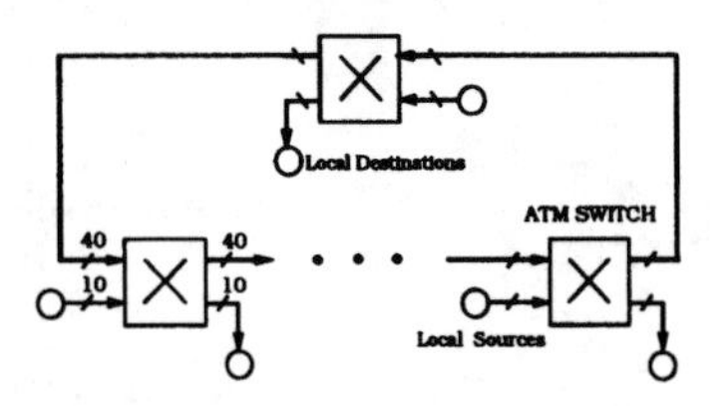

Figure 10 A symmetric multihop network scenario used in evaluating the long-term performance of the protocol.

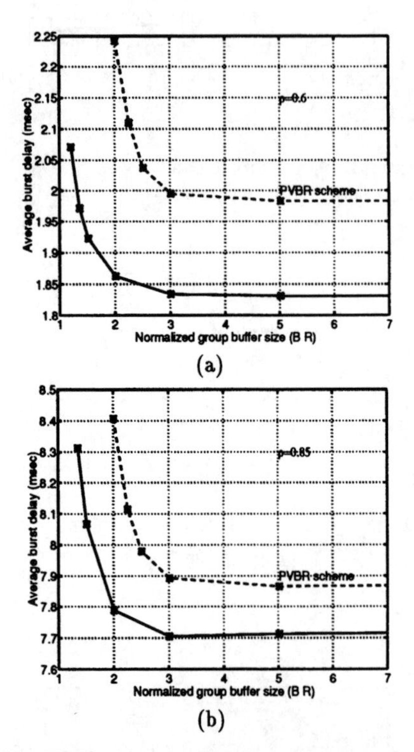

Figure 11 Comparison between the average burst delay of the proposed and the PVBR schemes as a function of group buffer size in a 4-hop symmetric network: (a) for a moderate activity factor of 0.6 (b) for a high activity factor of 0.85.

15

Flow Control Mechanisms for SAAL Links

K. Kant
Bell Communications Research
331 Newman Springs Road, Red Bank, NJ 07701, USA
Tel: (908)758-5384, Email: kant@perf.bellcore.com

Abstract

Signalling protocols in telecommunication networks provide for link-level flow-control procedures to ensure that the signalling traffic brought in by a link does not overwhelm the processing capacity on the receive end. This paper examines the flow-control procedures available in the signalling ATM adaptation layer (SAAL). The SAAL receiver effects flow control by granting an appropriate amount of *credit* to the SAAL transmitter; however, the details of how much credit is granted and when, are not standardized and left to implementation. This paper proposes two simple credit allocation schemes and provides guidelines for their parameterization. Since the receive buffer size depends on the credit mechanism, the paper also proposes buffer sizing guidelines.

The work reported here was motivated by the need for flow control on SAAL links deployed in the current common channel signalling (CCS) networks. It is shown that flow-control becomes even more important in broadband networks, thereby requiring well-designed credit control schemes.

Keywords

Asynchronous transfer mode (ATM), Signalling ATM adaptation layer (SAAL), Service-specific connection oriented protocol (SSCOP), Flow-control, Buffer sizing, Transfer controlled (TFC) procedure, Signalling message handling (SMH) congestion.

1 INTRODUCTION

This paper examines the problem of link-level flow-control in an asynchronous transfer mode (ATM) signalling network. The purpose of signalling is to establish and tear-down end to end connections and to manage the network. In an ATM network, signalling information is carried on specially designated virtual channels in the control plane and is supported by the ATM adaptation layer 5 (AAL5). The ATM signalling suite of protocols is also known as SAAL (Signalling AAL), and includes a number of sublayers: CP (common part), SSCOP (service-specific, connection-oriented protocol), SSCF (service-specific coordination function), and LM (layer management). Of these, the transmission protocol, SSCOP, is the one that is of primary interest in this paper.

SSCOP provides a *credit* mechanism by which the receiver periodically grants credit to the transmitter for sending *user* protocol data units (PDUs)* Section 2 briefly describes SSCOP and the credit granting scheme. The SSCOP standard does not specify how the credit is to be updated, thereby allowing the flow-control scheme to be implementation dependent. In this paper, we shall propose two schemes for updating the credit, compare their performance, and study their parameterization under various scenarios.

In general, if the receiver grants credit for n additional PDUs, it can be viewed as a commitment on its part to accept them even if they all happen to arrive back to back. This commitment can always be fulfilled by using a large receive buffer, but doing so is undesirable from a delay perspective. On the other hand, dropping a solicited PDU (i.e., one for which credit was granted) because of lack of receive buffer is also undesirable because the link error monitor cannot distinguish between losses due to errors and losses due to deliberate rejections of PDUs. In this paper we shall discuss these issues and suggest a few receive buffer sizing schemes.

The work reported here was motivated by the need to enhance the capacity of common channel signalling (CCS) links used in the current narrow-band telecommunication networks. We first discuss this aspect briefly and then turn to broadband networks.

The currently deployed CCS network in the US uses "quasi-associated signalling", where each switch "homes" only on a pair of STPs (signalling transfer points), and the connectivity is provided via STPs only. The SCPs (signalling control points), that provide database capabilities, also "home" on a pair of STPs. Such a network architecture may face two limitations as the network size and traffic grow (a) inadequate number of physical link ports available on the STP, and (b) inadequate total capacity of STP–STP and STP–SCP link-sets. With a rapid deployment of advanced telecommunications services such as PCS and AIN, these limitations are already being faced by several network providers and have prompted the deployment of high speed links (HSLs) in the CCS network. For a variety of reasons, it was decided to use SAAL based HSLs running over physical 1.536 Mb/sec channels. (See GR-2878). However, this deployment results in two complications that are relevant from a flow-control perspective:

1. If the driver for converting a DS0 link-set to a HSL is the shortage of port capacity or future traffic growth, the conversion may cause capacity problems. For example, replacing 4 DS0 links by a single HSL not only frees up 3 ports, but also allows the traffic to come in at the rate of about 16 DS0 links.† Thus, during high usage intervals the receive end may become overwhelmed and cause excessive delays and buffer overflows.
2. Consider a STP serving a mix of HSLs and DS0 links. Since a HSL can bring much

---

*A more accurate terminology for user PDUs is sequenced data PDUs; we use user PDUs here since signalling does not use unnumbered data capability of SSCOP.

†Because of much higher padding overhead of SAAL compared to MTP2, one 1.5 Mb/sec SAAL link is equivalent to about 16 DS0 MTP2 links for current messages sizes (Kant 1995a).

more traffic than a DS0 link, a focussed overload coming from the HSL may result in *signal message handling* (SMH) congestion, i.e., congestion at level 3 of the STP. If SMH congestion control is *not* implemented, this situation may result in excessive delays, timeouts, and customer retries. Otherwise, the STP will send TFCs (transfer controlled messages) to all its traffic sources, which is also undesirable since a global action is initiated in response to overloading of a single link.

Both problems occur essentially because of coarse granularity in selecting a link speed — the link speed can be either 56 Kb/sec or 1.5 Mb/sec with no intermediate values allowed. In a broadband network, intermediate values *are* allowed; in fact, since the signalling channels in a broadband network are merely virtual circuits, it is easy to change their bandwidth as needed. Yet, it is unlikely that this flexibility will actually be used. The problem is that a number of system parameters (e.g., buffer sizes, congestion thresholds, PDU pickup policy from level 2, etc.) are tied to link speed, and it is expensive to keep track of them or change them via an operations interface. Moreover, given that a 1.5 Mb/sec signalling channel occupies 1 percent or less of the ATM "pipe" capacity, there is little to be gained by using a lower rate. Thus the problems mentioned above do not go away in broadband networks.

The large disparity between the bandwidth of the signalling channel and the ATM pipe brings another complication in broadband networks: Over short periods (defined by the burst size parameter of the policing function), the signalling PDUs can arrive at a much higher rate than the peak rate of the signalling channel.[‡] In the absence of any flow-control, such bursts could easily overwhelm processing resources at SAAL and MTP3 levels, since those processors would certainly not be engineered for any more than the peak rate of the signalling channel. This makes the flow-control even more important in broadband networks.

2 SAAL PROTOCOL AND FLOW-CONTROL MECHANISM

SSCOP belongs to the class of ARQ (automatic repeat request) algorithms with selective retransmission of errored messages along with periodic polling of the receiver by the transmitter. The full protocol may be found in (Quinn 1993). Here, we shall describe only its most essential aspects.

In SSCOP, each message, or protocol data unit (PDU), carries a sequence number (or *seqno* for short) that is used for detecting missing PDUs and for delivering them in proper order. The transmitter maintains a counter VT(S) for assigning these sequence numbers and another counter VT(A) for keeping track of the seqno up to which all PDUs have been correctly acknowledged by the receive end. The receiver maintains two main counters, VR(R) and VR(H). VR(R) indicates that all PDUs with a lower seqno have already been received correctly. VR(H) indicates that the highest seqno seen thus far is VR(H)-1. That is, VR(H)>VR(R) indicates that there is a gap in the seqnos (i.e., certain PDUs were lost). Whenever a new gap develops, the receiver sets VR(H) to 1 plus the

[‡]This assumes that the signalling traffic is not shaped to look like a constant bit-rate (CBR) traffic.

highest seqno seen thus far, and alerts the transmitter by sending a USTAT (unsolicited status) message. In response, the missing PDU is retransmitted. All correctly received PDUs are placed in a *receive buffer* by SSCOP. Of these, the PDUs with a seqno less than VR(R) (i.e., in-sequence PDUs) become available for pick up by the MTP3 layer.

The transmitter periodically sends a POLL message to the receiver to enquire its status. In reply, the receiver sends a STAT (status or solicited status) message, which contains a list of all currently existing gaps. All missing PDUs are retransmitted in response to a STAT. The poll/stat mechanism is essential for the protocol in that the USTAT is dispatched only to report the first loss of a given PDU. Poll/stats also serve the purpose of "I am alive" signal between the peers. Unlike user PDUs, POLLs, STATs, and USTATs are never retransmitted. Both STATs and USTATs bring back VR(R), so that VT(A) can be advanced.

For flow control, the SSCOP receiver maintains a counter called VR(MR), which gives the maximum seqno up to which the receiver will accept the arriving PDUs. The transmitter maintains a corresponding counter called VT(MS). Every STAT and USTAT message conveys the current VR(MR) value to the transmitter so that VT(MS) counter can be kept in sync with VR(MR). The difference $VT(MS) - VT(S)$ gives the currently available *credit* to the transmitter. If VT(MS)-VT(S)=0, the transmitter ceases transmission of *new* user PDUs. The credit granting is done by updating VR(MR), but SSCOP does not specify how or when this should be done.

When VT(S) reaches VT(MS), the transmitter starts a timer T_{nc} (with a default value of 1.5 seconds). If T_{nc} expires before the no-credit situation is remedied, SSCF takes the link out of service. In the event of a credit-rollback (i.e., when the credit is taken away for an already transmitted PDU), arriving PDUs may fall outside the credit window and will be dropped. The proposed SAAL error monitor (Kant 1995c) ignores any retransmissions while the no-credit situation is in effect. In extreme situations, this may cause it to miss out on some real error phenomenon. It follows that the credit control scheme should not lead to long or frequent zero credit periods.

3 CREDIT ASSIGNMENT SCHEMES

In this section we propose two simple schemes for advancing the VR(MR) counter (and thereby granting the credit). However, before doing so, let us motivate an important requirement that any credit granting scheme must satisfy. Toward this end, note that the receive buffer holds both in-sequence and out of sequence PDUs. Typically, the in-sequence PDUs are picked up by level 3 using a periodic circular scan of all level 2s controlled by it. Thus, in the event of link errors or slow pickup by level 3, the receive buffer may become full. If the transmitter is assigned more credit than the available buffer space, more PDUs will continue to arrive and will be dropped. The resulting retransmissions will be considered as "errors" by the SAAL error monitor. Worse yet, if the PDU with seqno of VR(R) is dropped while the receive buffer is occupied by out-of-sequence PDUs, a deadlock will result since out-of-sequence PDUs cannot be delivered to level 3. To avoid these problems, *assigned credit should never exceed the available buffer space.* The two credit allocation schemes considered here are as follows:

(a) **Moving Credit:** In this scheme, whenever a PDU is picked up by level 3 from the receive buffer, VR(MR) is incremented by one. VR(MR) is initialized to an appropriate value, say N_m, which is chosen with the objective of keeping the link busy.
(b) **Fixed Credit:** This scheme involves two parameters, say N'_f and b_f. Here, VR(MR) is set to $VR(R) + N'_f$ at the time of sending every b_fth stat. VR(MR) is initialized to N'_f. N'_f is computed from considerations of the maximum rate at which the traffic can be accepted by the receiver.

In both schemes, VT(MS) can be initialized to ∞, since the first STAT will reset VT(MS) to the initial value of VR(MR). Note that the assignment VT(MS)←VR(MR) happens for every STAT even in scheme (b); thus, the loss of a POLL or STAT has similar consequences for both mechanisms. The following list compares and contrasts these two mechanisms in other respects.

1. In scheme (a), updated VR(MR) is communicated to the transmitter by every stat, whereas in scheme (b), this happens on every b_fth STAT only. Thus, with $b_f > 1$, scheme (a) needs to grant credit for traffic generated over a smaller interval, and thereby requires a smaller credit and buffer size.
2. If a user PDU is corrupted by a random error, it will be reported almost immediately via a ustat. In scheme (a), this USTAT will increase the credit by the number of PDUs that were correctly received before the error. This results in a larger room for out of sequence PDUs before the corrupted PDU is re-received correctly, without requiring additional buffer space. That is, with $b_f = 1$, scheme (a) has a slight edge over scheme (b) in error situations.
3. Scheme (a) can automatically ensure that the granted credit never exceeds the available buffer space, since if VR(MR) is initialized to the receive buffer size (in messages), $VR(MR) - VR(R)$ always gives the number of available message slots in the receive buffer. In contrast, scheme (b) must explicitly adjust N'_f to achieve this property. In particular, if N_f is the desired credit amount without considering the receive buffer occupancy, one needs to set $N'_f = \max(N_f - O, 0)$ where O is the number of occupied message slots.[§]
4. In scheme (a), if an adequate amount of credit is granted initially, that credit is maintained continuously so long as level 3 keeps up with the PDU arrivals. That is, the scheme does not provide any flow control under normal conditions. In contrast, scheme (b) explicitly grants a fixed credit periodically, and thus can effect flow control even under normal conditions.
5. Since scheme (a) provides no flow control until the receive buffer is almost full, it results in an almost on-off control that becomes more sluggish as the receive buffer size increases. In contrast, scheme (b) begins to control traffic whenever it exceeds a desired threshold, independent of the receive buffer size.
6. As shown in section 4, the credit requirements must be somewhat overestimated to account of jitter in STAT arrivals and link propagation delays. This "error" becomes

[§]Thus, the parameter N_f is really the one that this paper is attempting to estimate, whereas N'_f is a local variable that is computed for updating VR(MR).

relatively less important as the credit allocation interval increases. In this sense, scheme (b) can achieve a more accurate flow control.

Item (4.) above points to a fundamental difference between the two schemes: Suppose that VR(MR) is initialized to a higher value than the desired credit amount. Then, under scheme (a), credit window plus the number of in-sequence PDUs in the receive buffer will remain invariant, thereby retaining the excess credit. On the other hand, scheme (b) advances VR(MR) relative to the current VR(R); consequently, any unused credit during an allocation interval will not propagate into the next interval, and each interval will only get the desired credit. (This property justifies the name "fixed credit" for scheme (b).)

The next section addresses the problem of estimating the parameters of these schemes (N_m for the moving credit scheme, and N_f, b_f for the fixed credit scheme). Although the determination of N_m (or N_f) directly yields the required receive buffer size in messages, specifying the size in bytes is a bit more involved and is also addressed.

4 CREDIT REQUIREMENTS AND BUFFER SIZING

Although the SSCOP standard does not contain any specific guidelines for credit allocation, it does provide a default size for the credit window (Appendix IV in (Quinn 1993)). This default size is chosen very generously so that the credit is almost never a limiting factor. This may result in large receive buffer size requirements and consequent delays. The analysis in this section comes up with smaller, and more realistic, numbers.

As discussed in section 1, level 2 and 3 processors may be engineered assuming that the signalling link carries less than 1 Erlang load. Thus, it is necessary to examine the parameterization problem without and with the processing capacity limitations. Sections 4.1 and 4.2 study these two cases. Section 4.3 then examines some issues in sizing the receive buffer.

4.1 Basic Credit Requirements

We start with situations where there are no link errors and estimate the minimum credit needed to ensure that the transmitter can indefinitely transmit PDUs at the full link capacity. It is clear that regardless of the scheme for updating VR(MR), the updated value is reflected on the transmit side (by the assignment of VT(MS)=VR(MR)) only when a STAT arrives. Thus, it suffices to compute credit requirements only for the fixed credit scheme, since the moving credit scheme becomes a special case with $b_f = 1$. The basic idea is that when a STAT arrives and updates VT(MS), the new value of VT(MS) must be large enough to allow the transmitter to keep going until the next credit granting instant.

Let T_{poll} denote the polling interval, δ_s a suitable measure of jitter in the arrival time of successive STATs. For example, one may choose δ_s as twice the standard deviation of T_a, the interarrival time of STATs on the transmit side. Assuming, for simplicity, that the queuing/processing delays experienced by successive POLLs (and STATs) are iid (independent, identically distributed), the jitter in the interarrival time of successive b_fth STATs is $\sqrt{b_f}\delta_s$. Thus, the transmitter should have enough credit to keep transmitting for the $b_f T_{poll} + \sqrt{b_f}\delta_s$ seconds. Let r_{pdu} denote the maximum rate at which user PDUs can be transmitted, and $CT_{\min}$ the required minimum credit. Then

$$CT_{\min} = r_{pdu}(b_f T_{poll} + \sqrt{b_f}\delta_s) \quad (1)$$

Let τ_u denote the end to end transit delay of POLLs and STATs. Because of this delay, the transmitter and receiver don't have the same view of the SSCOP state. Let T_u denote the maximum number of PDUs that can be transmitted during the time τ_u. Then, at the time the receiver sends a stat, the transmitter could already be ahead by T_u, i.e., VT(S) may already be $VR(R)+T_u$. By the time the STAT reaches the transmit end, VT(S) could have advanced by another T_u. Thus, if a credit of CR PDUs is granted by the receiver when sending the stat, it will result in VT(MS)=VR(R)+CR on the transmit side. Since $VT(S) = VR(R) + 2T_u$ when the STAT arrives on the transmit side, the actual credit granted to the transmitter is only $VT(MS) - VT(S) = CR - 2T_u$. It follows that the minimum credit on the receive side, $CR_{\min}$, should be set to $CT_{\min} + 2T_u$. That is,

$$CR_{\min} = r_{pdu}(b_f T_{poll} + \sqrt{b_f}\delta_s + 2\tau_u) \quad (2)$$

Let us now consider the estimation of the variables used in the above equation. Let n_{cell} denote the average PDU size in cells at the ATM layer, and r_{cell} the link speed in cells/sec. Then r_{pdu} can be estimated as

$$r_{pdu} = [r_{cell} - 2/T_{poll}]/n_{cells} \quad (3)$$

For a 1.5 Mb/sec SAAL link with $T_{poll} = 0.1$ sec, $r_{pdu} \approx 3600/n_{cells}$. The parameter τ_u includes one-way link propagation delay (t_p), plus processing, queuing, and transmission time of management messages (t_u). That is, $\tau_u = t_p + t_u$.

Because of frequent physical link rearrangements, it is undesirable to use the actual link length in estimating t_p. In fact, current systems do not have any built-in mechanism to keep track of actual link lengths. If the chosen t_p is larger than the actual, excess credit will be allocated to short links. Similarly, if the chosen t_p is smaller than actual, long links may be starved of credit. Of these two choices, the first one appears preferable since both credit control schemes will cut-down credit if the receive buffer fills up. Thus, we propose estimating t_p using, say, 90 percentile link length for the network of interest. For example, in a regional network, a benchmark length of 2000 miles will give $t_p = 20$ ms (using 0.01 ms/mile rule).

Estimation of t_u and δ_s requires computing delays experienced by management messages. The SSCOP specification suggests priority queuing for transmission with highest priority accorded to management messages. This is desirable from a credit allocation perspective, since one doesn't need as much credit over-allocation to allow for statistical fluctuations. However, some implementations may not use priority queuing. Similarly, the delay suffered by the PDU at lower layers is implementation dependent. For example, if the ATM layer takes cells from CP in large batches rather than individually, it may add significantly to the delays. In general, analytic estimation of t_u or δ_s is intractable, and one may have to use a reasonable upper bound determined from the architectural details. (However, (Kant 1995c) provides an approximate method for estimating SSCOP level delays.)

Let us now consider the impact of link errors on the credit requirement. A single link error (which results in a USTAT being generated almost immediately) results in about $2\tau_u$ seconds delay before the retransmitted PDU will arrive again. During this time, VR(R) does not change, which can reduce the granted credit by as much as $2\tau_u r_{pdu}$ because of the reception of out-of-sequence PDUs. Thus to ensure that a single link error does not result in a zero-credit (or dead) period, it is necessary that $CT > 2\tau_u r_{pdu}$. Assuming that δ_s is small, this condition, along with equation (1) implies that $b_f T_{poll} > 2\tau_u$. With moving credit scheme, this condition may not be satisfied if the link is too long or the management messages experience significant queuing delays. The fixed credit scheme with $b_f > 1$ may be preferable in this case.

Links often suffer from severe error bursts such that during the error burst almost all PDUs are lost or corrupted. The SAAL error monitor is designed to ride over error bursts of length up to $t_b = 0.4$ secs. Thus, following the error burst, the separation between VR(R) and VR(H) could be $t_b r_{pdu}$, and grow to $(t_b + 2\tau_u) r_{pdu}$ until the first retransmission arrives. The utility of giving the transmitter enough credit to continue transmission over this entire period is questionable, particularly, since the credit granting STATs would themselves get lost. However, if the user PDUs are long, it is possible that polls/stats get through, but the user PDUs do not. The credit requirements in this case are:

$$CR_{\max} = (t_b + 4\tau_u) r_{pdu} \tag{4}$$

where the additional $2\tau_u$ factor results from the difference between CT and CR parameters. If the credit is set according to this equation, the condition $CT > 2\tau_u r_{pdu}$ implies that $t_b > 0$; i.e., the dead periods cannot occur in case of single errors.

This analysis shows that in the absence of any processor capacity limitations, the parameters N_m and N_f can be set somewhere between the limits $CR_{\min}$ and $CR_{\max}$. A larger credit requires more buffer space and results in longer delays, but does not constrain the transmitter as much under error conditions. Given that link errors occur rarely, it appears that the disadvantages of choosing a credit larger than $CR_{\min}$ outweight its advantages.

Let us now examine the choice of the allocation interval b_f for the fixed credit scheme. Clearly, it is really the duration $b_f T_{poll}$ that is of interest here. The default value of T_{poll} is 100 ms, and it has been shown in (Kant 1995b) that this value of T_{poll} is adequate even at very high link speeds. Thus, we can regard T_{poll} as a constant and regard b_f as the parameter of interest.

As b_f increases, less control is exercised on the arriving traffic over short periods. For example, with $b_f = 10$, enough credit must be granted to last for 1 second, and the receiver will be obligated to accept the incoming traffic during this time even if the conditions change and require lowering the incoming traffic rate. A large b_f also requires a large receive buffer to hold the traffic in the event of a level 3 congestion. However, a large b_f has some advantages as well. First, a larger b_f results in less reduction in credit due to a single error; in fact, for very long links, a large b_f may be necessary to satisfy the condition $b_f T_{poll} > 2\tau_u$. Second, a large b_f results in more accurate credit allocation. To see this, note that the δ_s term in equation (2) can be regarded as "error", since it merely accounts for statistical fluctuations in the arrival times of successive STATs. The same is true of the t_p and t_u terms, since, as discussed above, they need to be estimated from upper bounds rather than actual delays. Thus, the worst-case relative error, denoted $\eta_{\max}$, is given by:

$$\eta_{\max} = \frac{\delta_s \sqrt{b_f} + 2\tau_u}{b_f T_{poll}} \quad (5)$$

It is clear that $\eta_{\max}$ decreases as b_f increases, which means that credit allocation falls more in line with what it is designed for. This aspect becomes even more important if the credit must be limited due to processor capacity limitations.

In the broadband environment, the ATM policing function will ensure that the long term traffic rate on the signalling channel does not exceed its peak bandwidth; however, short-term traffic bursts will still be allowed. For example, the leaky-bucket policing mechanism has allowable burst length as one of its parameters. It may be reasonable to allow the bursts to last several hundred milliseconds for the signalling channel. However, this implies that the credit allocation duration should also be at least a few hundred milliseconds long.

From the above considerations, a b_f in the range 3-5 seems appropriate. With $\tau_u = 0.025$ secs and $\delta_s = T_{poll}/10$, the maximum error $\eta_{\max}$ is 22.4% for $b_f = 3$, and 14.5% for $b_f = 5$.

4.2 Impact of Level 3 Capacity on Credit Allocation

If the level 3 processor cannot handle the full engineered rate of the SAAL link, the allocated credit N_f must be less than $CR_{\min}$ for the fixed credit scheme. (The moving credit scheme, as stated in section 3 does not consider processing ability and thus N_m is not affected.) Let r'_{pdu} denote the actual traffic rate that level 3 can handle. Then the required credit N_f can be computed by simply substituting r'_{pdu} for r_{pdu} in equation (2). That is,

$$N_f = r'_{pdu}(b_f T_{poll} + \sqrt{b_f}\delta_s + 2\tau_u) \quad (6)$$

The appropriate technique for estimating r'_{pdu} depends on the precise details of the situation involved. Following are two simple examples illustrating how to determine r'_{pdu} for a new SAAL link:

1. Suppose that the signalling needs or processor capacity limitations indicate that only a fraction ξ of the full bandwidth need be supported. Then, $r'_{pdu} = \xi r_{pdu}$.
2. Suppose that the average occupancy of the level 3 processor without the new link is ρ, which is less than the desired engineered load ρ_e. Then, $r'_{pdu} = (\rho_e - \rho)/s_{l3}$, where s_{l3} is the average processing time of a message by level 3.

In telecommunication networks, nodes often employ active redundancy for enhanced reliability. For example, with quasi-associated signalling, STP–STP links are engineered for only 0.2 Erlang traffic under normal conditions, so that a double failure only results in 0.8 Erlang load. In such an environment, the level 3 (and other) processors will also be engineered in a similar way. Thus, in estimating r'_{pdu}, we need to specify the operating conditions of the nodes on either end of a SAAL link. If the credit allocation is done based on normal conditions, the granted credit will be inadequate under failure scenarios. On the other hand, if the credit allocation assumes a double-failure scenario, the credit granted under normal conditions will be 4-times the expected value which will allow a substantial surge in traffic before the flow-control goes into effect. Of these, the latter method seems more appropriate as it does not restrict the traffic unnecessarily. A third option is to vary the granted credit appropriately as a side effect of TFR/TFP/TFA procedures.

With limited level 3 capacity, the flow-control proposed here will prevent level 3 from going into SMH congestion; however, the transmitting end of the flow-controlled link will perhaps go into congestion as a result. Consequently, congestion control procedure will be triggered and reduce the traffic going over this link. As discussed in section 1, this is a better response than letting SMH congestion occur. Finally, neither of the proposed schemes causes a credit rollback, and within reasonable limits, they are unlikely to lead to long enough dead periods to let T_{nc} timer expire. Thus, the schemes should work well even for $r'_{pdu} << r_{pdu}$.

4.3 Impact of Message Size and Receive Buffer Sizing

SSCOP requires that the credit be specified in PDUs. However, equation (3) involves n_{cell}, the average PDU size in units of cells. Estimating n_{cells}, in turn, requires detailed knowledge of message lengths so that the overhead of padding and headers could be computed correctly (Kant 1995a). Since actual message sizes could be different than assumed, the credit allocation may be inaccurate. In particular, if the assumed message size is less (more) than actual, r_{pdu}, and hence the the granted credit is overestimated (underestimated). Of these two cases, credit overestimation is perhaps preferable as it avoids unnecessary flow control. Since a fill-up of the receive buffer will eventually cut-down the credit, the flow control will still take effect eventually, although somewhat belatedly.

Depending on the implementation details, the receive buffer may be sized either in terms of messages or bytes. With message sizing, the receive buffer size can be chosen to be equal to the credit amount (N_m or N_f, depending on the scheme). Unfortunately, message sizing necessarily implies that the buffer space will be allocated dynamically depending on the sizes of the messages to be held. In contrast, byte sizing allows static allocation of buffer space and is usually preferred. With byte sizing, there are two issues to consider:

1. Determining buffer size to be allocated (needed either at design time or when the signalling channel is set up.)
2. Determining available buffer space in terms of messages whenever VR(MR) is to be updated (needed if the receive buffer holds some in-sequence messages at this time).

Let us start with issue (1). Suppose that there are no capacity limitations at level 3. In this case, equation (2) could be used for credit allocation for both schemes. The same equation could also be used for buffer sizing, provided that we consider r_{pdu} in the units of bytes/sec. In particular, a 1.5 Mb/sec SAAL link can carry at most $3600 \times 48 = 172$ Kbytes/sec of user traffic. Thus, irrespective of the quality of estimates of δ_s and t_u, or the prevailing message sizes, the receive buffer size computed using $r_{pdu} = 172$ Kbytes/sec in equation (2) cannot fall short so long as no backlog develops in the receive buffer. Now, if the backlog does develop, both credit schemes will reduce allocated credit by the number of backloged PDUs. (In the moving credit scheme, this adjustment occurs automatically, whereas in the fixed credit scheme, the credit must be explicitly reduced.) However, if the messages transmitted during periods of backlog are longer, the link may carry more than the allowed number of bytes. To illustrate this, suppose that during one credit allocation interval, the link can carry 100 average sized messages. Now, if the receive buffer already contains 90 average sized messages, the allocated credit will be only 10. Because of their small number, it is possible that these 10 messages have a significantly higher average size. They will still be carried by the link; however, the receive buffer will run out of space for them. This problem can be addressed by a slight increase in the buffer size. In fact, since the credit allocation and buffer sizing already assumes rather extreme situation with respect to end-to-end transit delays, an increase may not even be necessary.

If level 3 capacity limitation does exist, we can use a similar approach; i.e., compute the buffer size using equation (6) with r'_{pdu} converted from messages to bytes by using some average message size, say s_m. Let s_{ma} denote the prevailing average message size in the network. Clearly, we want $s_m > s_{ma}$ to avoid buffer overflows. However, s_{ma} may increase over time but the growth may be difficult to predict. Thus, if level 3 capacity is not significantly lower than the link bandwidth, computing buffer size without regard to level 3 capacity is still the best approach. Otherwise, an automatic adjustment of credit and/or buffer size is needed in order to avoid the need for intervention from operation systems. This involves keeping track of current average message size s_{ma}, and must be implemented as an enhancement to SSCOP, since the current SSCOP standard does not keep this information.

Let us now consider issue (2). The easiest approach is to work with message units only. That is, we initialize the available buffer to the credit amount (N_f or N_m, depending on the scheme), and measure available space in the units of messages. This will work so long as the receive buffer is sized conservatively.

5 CONCLUSIONS

The paper proposed two credit control schemes for SAAL links and studied their parameterization. It is clear from this discussion that both schemes can provide effective flow control. The moving credit scheme is somewhat simpler to implement, but does not provide the rate control provided by the fixed credit scheme. Instead, as level 3 starts to fall behind, the PDUs accumulate in the receive buffer until the receive buffer is full and the credit drops to zero. The fixed credit scheme (with $b_f > 1$) also allows longer traffic bursts in a broadband environment and a smoother control over traffic if the allowed traffic is much less than the link capacity; however, this is at the cost of increased receive buffer space. Overall, it appears that the fixed credit scheme is somewhat superior.

One issue not addressed in this paper is the interaction of congestion control and flow control mechanisms. With the national congestion control option, the congestion control mechanism will selectively throttle messages based on their congestion priority. Thus, if some messages are much longer than others, the average message size may change drastically during congestion periods and cause problems of buffer overflow, overcontrol, or undercontrol of traffic. This may force a dynamic calculation of average message size for flow-control purposes, as suggested above.

ACKNOWLEDGMENTS

I am grateful to J.R. Dobbins for his help in weeding out errors in the design and analysis of the credit control schemes presented in this paper. Discussions with and comments from G. Ganguli and A. Jacob were also helpful in designing the schemes.

6 REFERENCES

Quinn, S. (ed.) (1993), "BISDN - ATM Adaptation Layer - Service Specific Connection Oriented Protocol", TD PL/11-20C Rev1, 1993.

GR-2878, "Generic Requirements for CCS Nodes Supporting ATM High-Speed Signalling Links", GR-2878-CORE, Bellcore, Oct 1995.

Kant, K. (1995a), "Comparison of achievable throughputs of SAAL and MTP2 Protocols", unpublished memo.

Kant, K. (1995b), "Analysis of Delay Performance of ATM Signaling Links", Proc of INFOCOM 95, Boston, MA, April 1995.

Kant, K. (1995c), "An Error Monitoring Algorithm for ATM Signaling Links", in *Data Communications and their Performance*, edited by S. Fadida and R. Onvural. (Proc. of sixth Intlċonfȯn the Performance of Computer Networks, Oct 1995, Istanbul), pp 367-381.

16

A Traffic Management Framework for ATM Networks

Jon W. Mark and Jing-Fei Ren
University of Waterloo
Department of Electrical and Computer Engineering
Waterloo, ON, Canada, N2L 3G1
email: jwmark@bbcr.uwaterloo.ca

Abstract

A traffic management framework, which couples the usage parameter control (UPC) and connection admission control (CAC) functions in an ATM network, is described. The method incorporates an open-loop rate-reservation control mechanism to guarantee cell loss performance for high priority traffic and a feedback adaptive control mechanism for efficient utilization of the available network resources. A novel traffic control strategy, called a credit-based controller (CBC), is introduced to mechanize the dual goal of UPC and CAC. The CBC allows a user to send cells as untagged and tagged, or as untagged only. Untagged cells are protected in the backbone network nodes, whereas tagged cells are selectively discarded upon the onset of congestion. The amount of tagged traffic a user is allowed to send is governed by a CBC parameter, α, which is adaptively adjusted according to the traffic load in the network.

Keywords

Traffic management, credit-based control, connection admission control, usage parameter control, feedback parameter adjustment.

1 INTRODUCTION

The high link speed, the diverse quality of service requirements and the widely different traffic characteristics render traffic management in ATM networks a relatively complex task. For example, at the link speed of 600 Mb/s, a 53 byte ATM cell needs to be transmitted in about 0.7 μs. Cell processing schemes used in ATM networks must, therefore, operate at a comparable speed. This fast processing speed can only be achieved by simplifying the processing schemes. Also, the large propagation delay-bandwidth product in high speed networks render reactive control impractical. For this reason, preventive rate-based control strategies are likely solutions for cell-level congestion control in ATM networks. Network control has to be administered at the call and cell levels using Connection Admission Control (CAC) and Usage Parameter

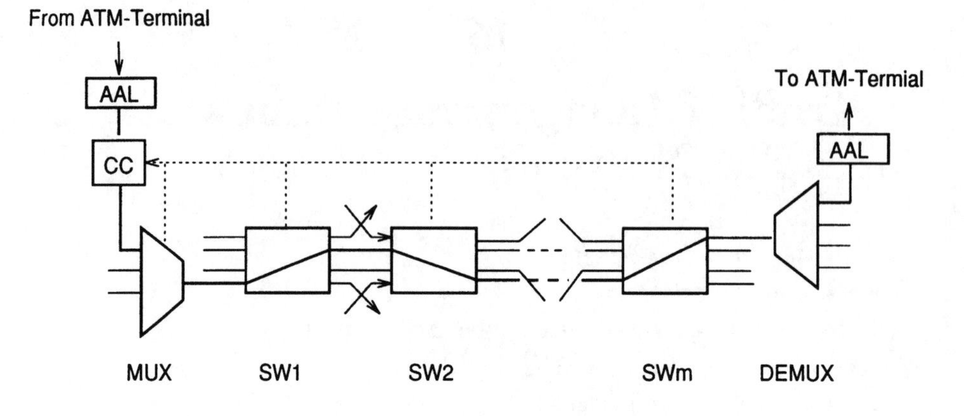

Figure 1 An ATM Network Partition

Coontrol (UPC), respectively. CAC restricts the number of on-going calls admitted into the network, while UPC ensures that the on-going calls obey the connection establishment agreement.

The network must provide satisfactory Quality of Service (QoS) for different connections. These requirements are usually measured in terms of end-to-end cell loss rates, cell delay and delay variation. An effective congestion control scheme should be able to provide both guaranteed QoS and best effort QoS for users in order to achieve an economical and efficient use of the network resources. It is also desirable to allow a user to make cell loss rate and cell delay tradeoffs.

Broadband traffic is in general unpredictable. Because of this, it is not always possible to completely characterize, e.g., a VBR source, by using just a few traffic parameters. Moreover, the derivation of flexible and robust control strategies that incorporate both traffic parameters and QoS requirements is not a simple task. In the formulation of a flow/congestion control framework, we must strive for (i) simplicity, (ii) efficiency, (iii) flexibility, and (iv) robustness.

Following the above design principle, we propose a congestion control structure consisting of an open-loop rate-based control mechanism to offer cell loss rate guarantee and a closed-loop adaptive control mechanism to provide best effort service. This combination enhances network resource utilization.

The proposed control structure offers the following salient features:

- Simplicity in implementation.
- Efficiency in network resource utilization.
- Flexibility in allowing effective tradeoff between cell loss and cell delay performance.
- Robustness in that noncompliant users cannot interfere with guaranteed services.

An ATM network is a mesh connection of multiplexers and switches. For the purpose of investigating a framework for traffic management, we consider a network partition consisting of a tandem connection of boundary and internal nodes (see Fig 1).

Traffic sources with large peak rates (say, larger than 1/10 of the ATM link capacity), high burstiness (ratio of mean to peak rate smaller than, say, 1:10) and long burst lengths (say, larger than a few hundred cells) require large buffer spaces. It is generally believed that the transmission and control of this type of sources require much more dedicated network resources [Lau and Li 1993, Roberts 1991]. For sources with long bursts, resource reservation on a burst basis appears to be attractive [Doshi and Heffes 1991, Turner 1992]. In our work, we assume bursty sources, but with mean to peak rate ratios greater than 1:10.

We assume that the different input flows at a network access point are statistically independent. One implication of this assumption is that the network cannot assume that individual users will act in the best interest of the network or other users. For the connection-oriented services to be efficient, we assume that the holding time of a connection is much longer than the setup time, which is roughly equal to a round-trip end-to-end propagation delay.

The traffic management structure is comprised of a set of control functions: (i) a connection admission control (CAC) performed at an access point based on the information sent from all the nodes along the route, (ii) a usage parameter control (UPC) performed at the access point, (iii) an adaptive adjustment of the UPC parameters at the access point, based on feedback control indications from the output buffers of all the relevant nodes, and (iv) measurements at the ingress and egress ports of switch nodes supporting the connection.

The paper is organized as follows: In Section 2, we briefly describe the credit-based controller (CBC) for traffic regulation. Connection admission control, buffer dimensioning and usage parameter control based on the credit-based controller are discussed in Section 3. The information provided in sections 2 and 3 completely defines the open-loop control. In Section 4, the issues on adaptive adjustment of the weighting parameter of the UPC and the selective cell discarding mechanism are discussed. In Section 5, we compare the performance of the proposed control structure with those of related work reported in the literature. Finally, concluding remarks are given in Section 6.

2 CREDIT-BASED CONTROLLER

The core element in the proposed traffic management structure is the credit-based controller described in [Ren and Mark 1995]. The CBC, consisting of a data buffer and a credit counter, is a UPC device whose parameters also serve as the basis for connection admission control. Cells waiting to be sent are stored in the data buffer. Cells can be sent untagged by consuming positive credits, if any, or tagged by borrowing credits when the credit level is too low. A source (user) has the option of sending only untagged cells (if cell loss guarantee is desired) or untagged and tagged cells (if best effort). Credits, which represent permits for cell transmission, are accumulated at the rate λ, up to a maximum value H. In general, the sending of an untagged cell consumes l units of credit where l is proportional to the cell length. For fixed size cells, as in ATM networks, $l = 1$.

Let λ be the arrival rate and p be the peak cell rate. The minimum spacing between two successive cells sent by the CBC is $1/p$ ($p > \lambda$). When credits are consumed faster than it is accumulating, the credit level will eventually diminish to below 1 unit. At this time, the user can choose one of two options: (i) delay sending until enough credits have accumulated and then send as untagged cells or (ii) send as tagged cells immediately if borrowing is still permitted. Sending a tagged cell will consume α, $(0 \leq \alpha \leq 1)$, credits. The tagged cells may be discarded in the backbone network when congestion is imminent. The value of the parameter α is adjusted adaptively according to the traffic load (perhaps measured) along the route of the end-to-end connection. When the traffic load is low, α will be set to a small value and only a small amount of credit is consumed by sending a tagged cell. In contrast, if congestion is imminent, α will be adjusted progressively to approach 1. The credit level will eventually be driven to the negative limit L

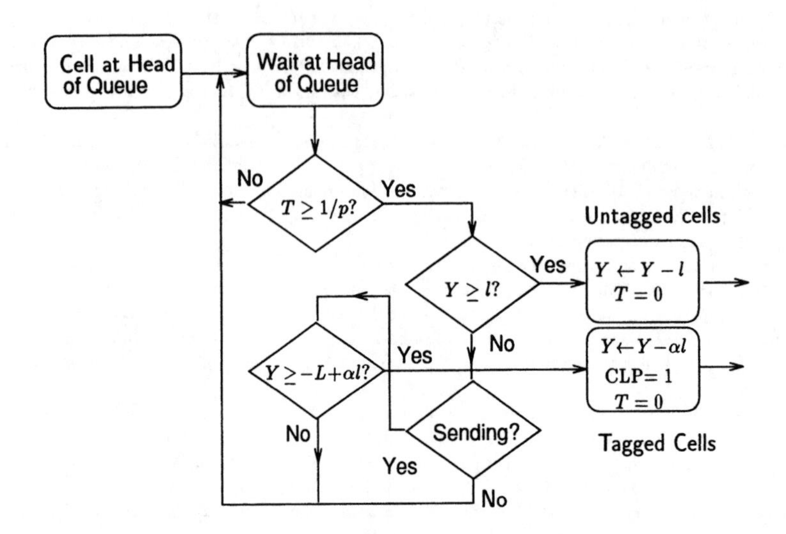

Figure 2 A Flow Chart of the CBC Operation

($|L|$ is the maximum borrowing power) if the user continues to send tagged cells. When the credit level reaches the negative limit L, sending tagged cells is suspended until the credit level becomes greater than $L + \alpha$ again.

At connection establishment time, the counter value X is set to 0. The CBC then operates as follows:

1. **(data buffer non-empty)** If $1 \leq X \leq H$, the cell at the head of the queue is transmitted every $1/p$ slots; if $L + \alpha \leq X < 1$, depending on the option the user chooses, either the cell at the head of the queue is tagged and transmitted every $1/p$ slots or the cell is delayed and transmitted as an untagged cell later when $X \geq 1$.
2. **(data buffer empty)** If the elapsed time T since the last cell transmission is at least $1/p$ slots, the arriving cell is considered for transmission as in step 1; otherwise, it waits until $T \geq 1/p$ slots.
3. **(update of X)** Whenever a cell is considered for transmission, X is updated as $\min(X + \lambda T, H)$. Following the transmission of an untagged cell, X is decreased by 1; following the transmission of a tagged cell, X is decreased by α.
4. **(negative limit)** If a user persists in sending tagged cells when $\alpha > \lambda/p$, it will eventually drive the credit level to L. When this happens, a tagged cell can only be sent when $X \geq L + \alpha$. This effectively constrains the long term maximum transmission rate of the tagged cells to $\min(\lambda/\alpha, p)$.

A flow chart portraying the operation of the CBC is shown in Fig. 2.

Remark 1: If a user chooses the delay option when the credit level becomes zero, the CBC emulates the behaviour of a buffered leaky bucket combined with peak-rate regulation. The burstiest traffic pattern of the untagged traffic transmitted from the CBC is a periodic *on/off* process with burst length

$$b = H/(p - \lambda), \tag{1}$$

silence period

$$s = H/\lambda \tag{2}$$

and peak rate p. By properly dimensioning the buffers and performing admission control, the cell loss/delay performance of the untagged cells can be guaranteed, as will be shown later.

Remark 2: The long time average rates of the untagged traffic and the total traffic (including the untagged and tagged cells) are bounded by λ and $\min(p, \lambda/\alpha)$, respectively. The peak-rate of the total traffic can never exceed p. The burst length of the untagged cells is bounded by b. Moreover, the bursts of a full length b are interleaved by silent periods (with respect to the untagged cells) whose lengths are at least s. In particular, when $\alpha = 1$, if the send option is chosen, the burstiest pattern of the total traffic is again a periodic *on/off* process with burst length

$$\hat{b} = (H - L)/(p - \lambda), \tag{3}$$

silence period

$$\hat{s} = (H - L)/\lambda \tag{4}$$

and peak rate p. Compared with the case where the delay option is chosen, the two burstiest traffic patterns have the same average rate λ and the same burstiness factor $d = \lambda/p$. The only difference is that the burst and silence lengths are extended by a factor of $(H - L)/H$. If we consider the untagged traffic alone, it is a periodic *on/off* process with burst length b, peak rate p and silence period length $\hat{b} + \hat{s} - b$.

Remark 3: When the credit level is less than 1, the CBC gives the user the flexibility to choose either the delay option or the send option. When a user chooses the send option, the credit level may move towards the negative limit. The value of α determines the speed of the movement. As the values of α changes from 1 to 0, we obtain a spectrum of control strategies. At one end of the spectrum is $\alpha = 1$ where sending a tagged cell consumes the same amount of credit as an untagged one. To restore the credit level so that cells can be sent without being tagged, the CBC needs to suspend sending for some time interval. However, by choosing the send option, the user can get earlier access to the network. The send option can be particularly useful for delay sensitive services. When α becomes smaller than 1, a cell stream with average rate larger than λ can be sent to the network through the CBC. As α decreases, the constraint becomes loose. At the other end of the spectrum is $\alpha = 0$, where sending tagged cells does not have any effect on the credit level. The only limit on sending tagged cells is the peak-rate constraint p. Therefore, the weighting parameter α provides us a tuning device between the system constraints and the service flexibility. In our control framework, α is adaptively adjusted based on the measurement/estimation of the network loads to enhance network efficiency. Users can choose the delay option when the services are loss sensitive and α is large, or choose the send option if services are delay sensitive and α is small. In general, when α is not very large, tagged cells are very likely to get through the network without being dropped. By exploiting the opportunity of sending tagged cells, users are able to send bursts of VBR traffic quickly and economically.

3 ADMISSION CONTROL, BUFFER SIZING AND USAGE PARAMETER CONTROL

Because of the diversity and uncertainty associated with integrated services, traffic modeling is a relatively difficult problem. Restrictions imposed on the number of the parameters permissible in the traffic descriptor makes traffic characterization even more ambiguous. Thus, real-time Connection Admission Control (CAC) and Usage Parameter Control (UPC) based on traffic models may not be viable. In order to avoid the ambiguity in traffic characterization, we choose three parameters λ, H and p in the CBC as the basis for contract negotiation between the users and the network. Given the values of these three parameters, the worst traffic pattern of the untagged traffic can be easily identified. When a connection request is accepted, the network is obligated to satisfy Quality of Service (QoS) requirements for compliant connections. Excess loads arising from noncompliancy are regulated by the CBC. Since the worst traffic pattern is solely determined by the parameters λ, H and p, uncertainty and ambiguity of the traffic characteristics do not impose adverse effect on CAC or UPC.

The worst traffic pattern of the untagged cells outputted from the CBC is a periodic *on/off* stream which can be completely characterized by the triple (p, λ, b), where b is the burst length defined earlier. To explicitly specify the amount of information contained in a burst, we use $b' = pb$ to represent the number of cells in a burst.

When a user requests a connection, it estimates the three parameters (λ, p, H) for the CBC based on performance requirements, cost considerations and a knowledge of its traffic characteristics. The three parameters are then submitted to the CAC unit and are converted to the parameter triple (λ, p, b'). The CAC then chooses a route for the connection and sends a connection_request to nodes along the route to ensure that the suggested parameters are acceptable by all the nodes. A decision at a node is made by taking into consideration the CBC parameter values for all existing connections and the new connection_request as well as the nodal resources. If a decision to accept is made, the connection_request is forwarded to the next node along the route and the nodal resource for the new request is reserved. The process is repeated until the connection_request reaches the destination node where a connection_ack is sent back to the user. If the connection_request is denied at any node, a connection_deny message is directly sent back to the user and all the reserved network resource for this request is released. The CAC may suggest a set of new parameter values for the user. The user can renegotiate the parameter values with the CAC or wait and resubmit a new connection request later.

Next, We derive the conditions for connection acceptance. First, we consider only the untagged traffic. We assume that all users will take the delay option since this requires the most network resources. If a user claims it will take the send option, the procedure to be described later can be modified accordingly to accept more calls. To simplify the CAC procedure, we assume that the connection acceptance decision is made solely based on the cell loss performance. The cell delay performance can be bounded by the buffer size under an FCFS discipline. Consider the fluid flow model for a multiplexer at the access point with buffer size B and link capacity c. The inputs to the multiplexer are a number of independent periodic *on/off* fluid flows, with the ith flow characterized by (λ_i, p_i, b'_i). In [Bensaou, Guibert and Roberts 1990], a fluid-flow queue fed by the superposition of multiple *on/off* sources with generally distributed *on* and *off* periods has been analyzed using the Beneš result. An upper bound of the cell loss rate in the buffer is derived. The results are applied to our periodic *on/off* sources case. The accuracy of the approximation

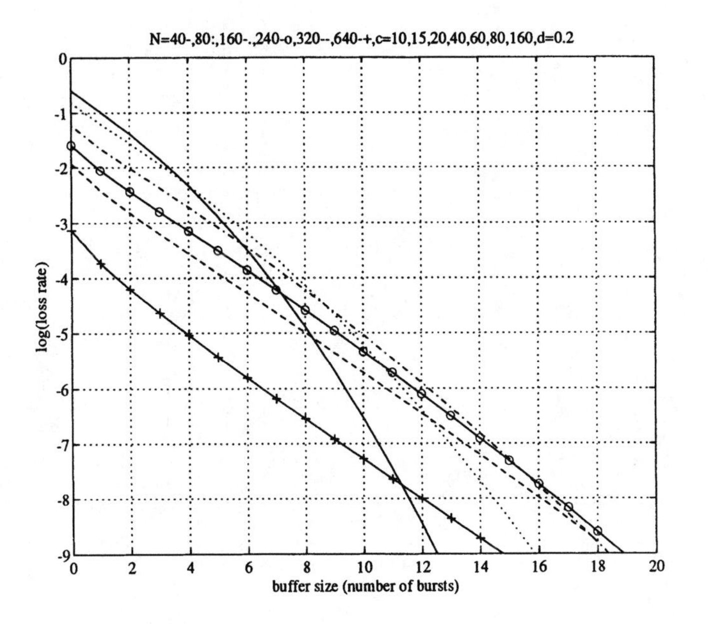

Figure 3 Cell Loss Rate vs. Buffer Requirement I

to this particular case has been examined in [Roberts, Bensaou and Canetti 1993] and is found to be very good. However, the computation includes a numerical integration which is rather CPU intensive and apparently cannot be performed in real time for CAC. Fortunately, we can do numerical computations off-line and deduce guidelines for CAC and buffer dimensioning.

In the sequel we examine two numerical examples. To simplify computation, only homogeneous source cases are considered. In the first example (Fig. 3), the traffic intensity is fixed at 0.8, and the burst factor d, given by $d = \lambda/p$, is fixed at $d = 0.2$. The channel capacity c is measured by the number of sources it can accommodate under peak rate assignment. Thus, $c = 10$ indicates that the peak rate of any source is 1/10 of the channel capacity. For $c = 10$, $N = 40$ sources can be multiplexed onto the output link. For a link speed of 600 Mb/s, this is interpreted as multiplexing 40 sources, each of which has a peak rate of 60 Mb/s, an average rate of 12 Mb/s, a burst length of b' cells per burst and a silence length of $4b$. It is noted that the buffer size is measured in terms of the number of bursts with length b'. A larger c actually means multiplexing more sources with a lower peak rate. We plot the relationship between the cell loss and the buffer size requirement in Fig. 3 for different combinations of c and N. The solid line corresponds to the case of $N = 40$ and $c = 10$. To achieve a loss rate lower than 10^{-8}, a buffer size of about $12b'$ is

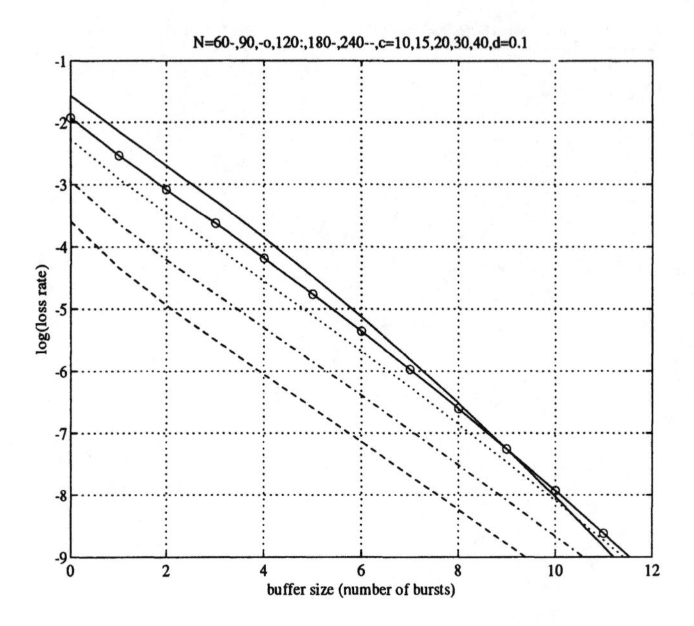

Figure 4 Cell Loss Rate vs. Buffer Requirement II

required. As N increases and c decreases, the buffer requirement increases first. It attains the maximum of about $17b'$ at $N = 240$ and $c = 60$. This corresponds to multiplexing 240 sources each of which has a peak rate of 10 Mb/s and an average rate of 2 Mb/s on a 600 Mb/s ATM link. The buffer requirement then decreases as N increases further. This observation is in keeping with our intuition that, for a given b', the buffer size requirement for the periodic *on/off* sources will be bounded by Nb'. When N increases starting from $N = 1$, the buffer requirement increases initially. However, for sufficiently large N, the superposition of N sources tends to be smoothed out because of the law of large numbers, which requires a smaller buffer size. Thus, the buffer requirement attains the maximum at moderate value of N. It is also noted that, in general, the buffer size requirement is not very sensitive to the different combinations of c and N for modest c's (N's). For small and large values of N, the buffer requirement decreases.

A second example being considered represents the case of burstier sources (with burstiness factor $d = 0.1$) but smaller utilization, $\rho = 0.6$ (see Fig. 4). For a loss rate lower than 10^{-8}, the buffer requirement attains a maximum of $10b'$ at $N = 90$ and $c = 15$. This represents the case of multiplexing $N = 90$ sources, each of which has peak rate 40 Mb/s, average rate 4 Mb/s, burst length b' and silence length $9b$. It is noted that in a fluid flow model, only the cell loss at the burst level is accounted for, while the micro dynamics inside the bursts are ignored. Considering the cell loss at the cell level, the buffer size should be augmented by about a few tens of cell spaces, a buffer size required in an $M/D/1$ queue with the same utilization. Again, we observe that the buffer size requirement is not very sensitive to the different combinations of c and N except for very large c's (N's).

Buffer dimensioning and CAC are closely related. We can either dimension the buffer size and perform CAC based on the buffer size or determine the CAC procedures and choose a suitable buffer size. Suppose we take the former approach, i.e., we fix the buffer size B based on the maximum queueing delay at a node (which is bounded by the buffer size under an FCFS discipline). Suppose user i, with parameters (λ_i, p_i, b_i'), requires a connection from a network operating on the conditions specified in Fig. 4. CAC at the multiplexer can be performed by checking the following conditions:

- the peak-rate p_i is at most 1/10 of the link capacity c: $10p_i \leq c$,
- the burst length (in cells) is at most 1/10 of the buffer capacity B: $10b_i' \leq B$,
- the total utilization (including the new connection) is lower than 0.6: $\sum_i \lambda_i/c = \rho \leq 0.6$,
- the burst factor d_i is larger than or equal to 0.1: $d_i = \lambda_i/p_i \geq 0.1$.

If all the above conditions are satisfied, the connection can be accepted by the multiplexer at the access point. Otherwise, it is rejected. It is very easy for the CAC unit to check these conditions in real-time. Because each output buffer of an internal switch node can be viewed as a multiplexer, CAC in an intermediate node can be performed in the same way by using the CBC parameters. In an intermediate node, the CAC outlined here is somewhat conservative, as the smoothing effect of the queueing on the sources has been ignored. To utilize the network resources more efficiently, as suggested in [Ren, Mark and Wong 1994], an approach of characterizing the smoothing effect needs to be developed for the periodic *on/off* sources. A connection is considered accepted only when it is accepted by all the nodes along its route. If the connections are accepted following this procedure and the selective cell discard mechanism (discussed later) works properly, the loss rate of untagged cells will be guaranteed.

Returning to the numerical examples, suppose the buffer size of a multiplexer/switch node is 1,800 cells. For the first example, each source can send a burst up to 100 untagged cells at peak rate at utilization 0.8; while in the second example, a source can send a burst of 180 untagged cells at utilization 0.6. If a user wishes to send a longer burst with guarantee, it can still do so by requesting a λ larger than the source average rate.

Because CAC is performed by using the parameters of the CBC, usage parameter control of untagged cells can be performed efficiently by restricting the input traffic to conform to the traffic pattern determined by the CBC parameters in real-time.

The control of tagged (low loss priority) traffic is realized by properly setting L for the CBC. In conjunction with the selective cell discarding, the impact of tagged cells on untagged cells can be made minimum. The selection of L should strive for an effective tradeoff between flexibility and controllability. For simplicity, we consider the case $L = -H$. With this value of L, if a user chooses the send option, it can double the burst length to $2b'$ when $\alpha = 1$, which signifies a potential congestion in the network. The silence period is doubled accordingly (see (3) and (4)). If all the users send traffic following the burstiest traffic pattern, the loss rate over all (untagged and tagged) cells can be evaluated from Figs. 3 and 4. Because the burst length is doubled, the buffer size should be scaled by 1/2. For the first example, the maximum total loss rate 3.5×10^{-5} is attained when $N = 160$ and $c = 20$ for a buffer size $9b'$ (here b' is the burst length of the total traffic in cells). The untagged cell loss rate is guaranteed to be smaller than 10^{-8} by the selective cell discarding mechanism at each node. Because half of the cells in a burst are tagged ones, by ignoring the loss of untagged cells, the loss rate of tagged cells is approximately 7×10^{-5}. Under the parameter setting $L = -H$, a rather flexible transmission is possible for all users.

For instance, a user who has underestimated its network resource requirement at the connection setup stage can send a substantially larger/burstier traffic to the network if the network is not heavily loaded (when α is small). The only performance degradation caused by this user to the untagged traffic from other users is a slight increase in the queueing delay. Also, for cost consideration, a user may reserve less network resources than it really needs to take advantage of the low cost transmission of tagged cells. The network adaptively adjusts the parameter α for different connections so that the network resources unused by the untagged cells can be fairly shared by all the existing connections. If a stricter loss rate is required for the tagged cells, the absolute value of the negative limit can be set smaller.

4 ADAPTIVE ADJUSTMENT OF WEIGHTING PARAMETER AND SELECTIVE CELL DISCARDING

The performance of untagged cells can be guaranteed by the CAC and the UPC by considering the worst case traffic pattern. Under normal operating conditions, users probably will generate traffic which is substantially better than the worst traffic pattern. Also, it is very unlikely that the network will always operate in a high utilization for the untagged cells alone. These two factors in general will lead to underutilization of the network resources. Because of the unpredicability of the user behaviour, any guaranteed service relying on sending tagged cells appears impossible. However, the network resources unused by the untagged cells provide an economical vehicle for applications which can tolerate certain degree of cell loss. If CAC is performed following the example in Fig. 3 and the untagged cells actually constitute a utilization of 50%, to attain a total utilization of 70%, the utilization contributed by the tagged cells should be 20%. In addition, to prevent users from overbooking the network resources when a great uncertainty exists in the source traffic characteristics, it is necessary to allow the users to use more network resources than they claimed at the connection setup stage, at the risk of a potentially higher cell loss rate. The objective of adaptively adjusting the weighting parameter α is to allow users to have enough network resources when the network traffic load is low and receive a fair share when the traffic load is high.

We argue that any sudden surge of the network traffic load occurs only when relatively large connections (with large rates and long burst lengths) are accepted into the network. There are several reasons that support this argument: First, the traffic from a single user only constitutes a small fraction of the total traffic; a short surge which is relatively large with respect to the traffic source is small compared with the high link capacity and is likely to be absorbed by well dimensioned buffers. Because all the users behave independently, it is very unlikely that several short surges from different sources will overlap. Secondly, the higher layer flow control protocols should limit a long life surge from entering the network. Finally, the CBCs further regulate the input traffic: The peak-rate of the total traffic is restricted to p and so long as α is greater than λ, sending untagged cells at peak rate p will push the credit count to the negative limit. This will not only reduce the rate of sending tagged cells, but also have a negative effect on sending untagged cells. Based on these arguments, adaptive adjustment of α can be targeted on the congestion from medium term to long term.

Suppose the individual output buffers of switches and multiplexers are equipped with adaptive controllers, each with the responsibility to measure/monitor the local traffic behaviour. The measured parameters are then filtered and converted to control information for adjusting α. A simple approach is to use a multiplicative increase/additive decrease algorithm tailored from the additive increase and multiplicative decrease algorithm in [Chiu and Jain 1989] to update α. In particular, an adaptive learning mechanism [Narendra and Thathackar 1989] could be attractive in this application. The updating of α is essential only when a connection is accepted or an existing connection is terminated when the traffic load is high. Because the holding time of a connection is much longer than the propagation delay and the sudden surge caused by other factors is negligible, the negative effect of propagation delay on the adaptive control is limited. Furthermore, because acknowledgment of acceptance of a new connection takes one end-to-end propagation delay, controllers usually have enough time to send the updated α to most users sharing the same output buffer. Therefore, the CBCs can react quickly to sudden surges caused by accepting new connections. Oscillations seem to be unavoidable due to the long propagation delay in a wide area network [Fendick, Rodrigues and Weiss 1992]. However, because adjusting the values of α virtually does not affect the performance of untagged cells, the control structure is rather robust to oscillations in the adjustments of α. Alternatively, feedback signals for adjusting α can be sent from the local controllers periodically. As indicated in [Faber, Landweber and Mukherjee 1992], the update period should be at least a round-trip propagation delay, which is in the range of 50ms for a continent-wide ATM network, to ensure that the effect of the last update is seen by the controllers.

Because of the presence of CAC and UPC, sending untagged cells alone will not cause network congestion. However, the sending of tagged cells is only loosely controlled. A temporary overload may occur inside the network. When congestion occurs at a node due to excessive tagged cells, it becomes necessary to selectively discard tagged cells to protect untagged cells. A threshold-style space priority mechanism can be adopted at the multiplexers and the output buffers of intermediate switch nodes. The major advantage of the mechanism is simplicity and good protection of untagged cells [Bala, Cidon and Sohraby 1990, Elwalid and Mitra 1992]. For example, the mechanism may discard tagged cells when the buffer occupancy exceeds the threshold B_1 ($B_1 < B$). All incoming cells are accepted if the buffer occupancy is less than B_1 and untagged cells are discarded only if the buffer is full. Service to cells already in the buffer is provided on an FCFS basis. Hence, resequencing is not necessary. The threshold B_1 controls the tradeoff between the delay of untagged cells and the loss rate of tagged cells. By suitably choosing the threshold B_1, the mechanism can provide a good protection for untagged cells while achieving a reasonably low loss rate for tagged cells.

By instituting different limits for borrowing, the CBC can handle more than one class of tagged cells. However, this can lead to multiple loss priorities. The selective cell discarding algorithm can be easily modified to handle multiple priorities. The tradeoff is between the increased fairness and the increased complexity.

5 COMPARISON WITH RELATED WORK

The work described here has been motivated by related works reported in the literature. The tagging and selectively discarding of excess cells have been proposed, e.g., in [Eckberg, Luan and Lucantoni 1990]. Using the UPC parameters as the basis for contract negotiation is proposed in [Eckberg, Luan and Lucantoni 1990]. The idea is further explored in [Kvols and Blaabjerg 1992, Roberts, bensaou and Canetti 1993], where numerical results of the periodic *on/off* queues are obtained by using the approach developed in [Bensaou, Guibert and Roberts 1990]. Nevertheless, our proposed traffic management framework is unique. In this section, we compare our work with the leaky-bucket and equivalent bandwidth approaches.

5.1 Leaky-bucket and its variants

The leaky-bucket mechanism is a very popular rate/burst control scheme. As mentioned before, the CBC is similar to a leaky-bucket if a user only chooses the delay option. There are numerous variants of leaky-buckets. One of the generalized leaky-bucket approaches considered in the literature tags cells when the token pool is empty (corresponding to the zero credit in the CBC) and the data buffer is full [Elwalid and Mitra 1991]. However, the CBC is different from the generalized leaky-bucket in that it lets the tagged cells consume credit in a controlled manner — it allows the credit level to go negative but requires a positive credit level before the user is allowed to send untagged cells again. Thus, the CBC provides a more flexible way for users to tradeoff between the delay performance and the loss performance while giving a better regulation on both the tagged cells and the untagged cells. The worst case total traffic pattern outputted from the CBC is easily predicted. Moreover, the parameter α in the CBC serves as a tuning parameter between the system controllability and the service flexibility. By adaptively adjusting α to match the traffic load inside the network, an efficient tradeoff could be achieved. These properties are not shared by the leaky-bucket and its variants.

5.2 Equivalent bandwidth approach

An approach of congestion control has been developed by [Guérin and Gün 1992]. in [18] based on the concept of effective bandwidth [Guérin, Ahmadi and Naghshinch 1991]. UPC is performed by leaky-buckets and peak-rate regulators. One major advantage of the equivalent bandwidth is that the equivalent bandwidth of Markov modulated *on/off* sources can be simply computed from a few traffic descriptor parameters and the equivalent bandwidth of multiple sources is additive. Thus, CAC based on the equivalent bandwidth can be performed in real-time. However, estimates of the equivalent bandwidth makes use of Markovian property in the source traffic. In particular, the concept cannot be straightforwardly extended to the periodic *on/off* process. Furthermore, because of the tradeoff between access delay and efficiency, the equivalent bandwidth of a source traffic is not easily controlled by a UPC device. Therefore, when users try to fully use the network resource regulated by the UPC, a significant increase of the cell loss rate is observed in the multiplexer (see examples in Fig. 6 and Fig. 7 of [Guérin and Gün 1992]). In contrast, the CAC proposed here is solely based on the parameters of the UPC device. This completely eliminates the uncertainty and ambiguity in the source traffic. In the proposed control structure, the UPC can strictly limit the output traffic to stay in the range specified by the UPC parameters. Thus, well-behaving users are protected. The excess cells sent by misbehaving users are either blocked outside of the network by UPCs or are tagged as low priority cells. We feel that the decision to fully use the network resources reserved by the UPC belongs to the user. A disadvantage in CAC based on the UPC parameters is a lower utilization if the users do not make full use of their reserved network resources. This problem is overcome by allowing other users to send tagged cells.

6 CONCLUDING REMARKS

We have proposed a new end-to-end traffic management framework for an ATM network. The core element in the control structure is the credit-based controller (CBC) for implementing the UPC and the use of the CBC parameters as the basis for decision-making by the CAC. The scheme has potential to satisfy all four traffic management goals: simplicity, flexibility, efficiency and robustness.

Further work will be directed toward understanding the performance of the CBC and the dynamics of adaptive feedback control.

ACKNOWLEDGEMENT

for Telecommunications Research under the NCE program of the Government of Canada and in part by a grant from the Information Technology Research Centre under the Centres of Excellence program of the Province of Ontario.

REFERENCES

Lau, W.-L. and Li, S.-Q. (1993) Traffic analysis in large-scale high-speed integrated networks: Validation of nodal decomposition approach. *Proc. of IEEE INFOCOM '93*, 1320-1329.

Roberts, J.W. (1991) Variable-bit-rate traffic control in B-ISDN. *IEEE Communications Magazine,* **29**(9), 50-56.

Doshi, B.T. and Heffes, H. (1991) Performance of an in-call buffer-window reservation/allocation scheme for long file transfers. IEEE Journal on Selected Areas in Communications, **9**(7), 1013-1023.

Turner, J.S. (1992) Managing bandwidth in ATM networks with bursty traffic. *IEEE Networks*, 50-58.

Ren, J.-F. and Mark, J.W. (1995) Design and analysis of a credit-based controller for congestion control in B-ISDN ATM networks. *Proc. of IEEE INFOCOM '95*, 40-48.

Bensaou, B., Guibert, J and Roberts, J.W. (1990) Fluid queueing models for superpostion of bursty sources in ATM multiplexers. *Proc. of ITC 7th Specialists Seminar*, Morristown, NJ.

Roberts, J.W., Bensaou, B. and Canetti, Y. (1993) A traffic control framework for high speed data transmission. *Proc. of Modelling and Performance Evaluation of ATM Technology*, 6.1.1-6.1.22, La Martinique, French Carribean.

Ren, J.-F., Mark, J.W. and Wong, J.W. (1994) End-to-end performance in ATM networks. *IEEE ICC '94 Conf. Record*, 996-1002, New Orleans, LA.

Chiu, D.-M. and Jain, R. (1989) Analysis of the increase and decrease algorithms for congestion avoidance in computer networks. *Computer Networks and ISDN Systems*, **17**, 1-14.

Narendra, K. and Thathachar, M. (1989) **Learning Automata: An Introduction**. Prentice-Hall, Inc.

Fendick, K.W., Rodrigues, M.A. and Weiss, A. (1992) Analysis of a rate-based control strategy with delayed feedback. *ACM SIGCOMM '92*, 136-148.

Faber, T., Landweber, L.H. and Mukherjee, A. (1992) Dynamic time windows: Packet admission control with feedback.*ACM SIGCOM '92*, 124-135.

Bala, K., Cidon, I. and Sohraby, K. (1990) Congestion control for high speed packet switched networks. *Proc. of IEEE INFOCOM '90*, 344-349.

Elwalid, A.I. and Mitra, D. (1992) Fluid models for the analysis and design of statistical multiplexing with loss priorities on multiple classes of bursty traffic. *Proc. IEEE INFOCOM '92*, 415-425.

Eckberg, J.A.E., Luan, D.T. and Lucantoni, D.M. (1990) An approach to controlling congestion in ATM networks. *International Journal of Digital and Analog Communication Systems*, **3**, 199-209.

Kvols, K. and Blaabjerg, S. (1992) Bounds and approximations for the periodic on/off queue with applications to ATM control. *Proc. of IEEE INFOCOM '92*, 487-494.

Elwalid, A.I. and Mitra, D. (1991) Analysis and design of rate-based congestion control of high speed networks,: Stochastic fluid models, access regulation. *Queueing Systems*,**9**, 29-64.

Guérin, R. and Gün, L. (1992) A unified approach to bandwidth allocation and access control in fast packet-switched networks. *Proc. of IEEE INFOCOM '92*, 1-12.

Guérin, R., Ahmadi, H. and Naghshinch, M. (1991) Equivalent capacity and its application to bandwidth allocation in high-speed networks. *IEEE Journal on Selected Areas in Communications*, **9**(7), 968-981.

17

Congestion Control in a Constrained-Worst-Case Framework *

S. Valaee and M. A. Kaplan
INRS-Télécommunications
16, Place du Commerce
Verdun, Quebec
CANADA H3E 1H6

Abstract

We present an approach to worst-case reactive congestion control in ATM-based broadband networks. The effective capacity of a virtual circuit in such a network can vary dynamically because of traffic fluctuations, call rerouting or equipment failures; we propose a single mechanism by which to ensure that the traffic offered to the circuit remains matched to capacity, whatever the source of the variation. The approach is non-statistical. It applies to circuits subject to Leaky Bucket regulation at the network interface, and to networks implementing a GPS-like service schedule at the nodal output buffers. The key element is an algorithm which modulates the parameters of the Leaky Bucket on the basis of measured values of instantaneous capacity, and in such a way as to avoid cell loss, to the extent possible, inside the network.

1 INTRODUCTION

Statistical traffic models are of undisputed importance for telecommunication network design. Nonetheless, it can happen that source statistics are unavailable, uncertain or variable, in which cases an alternative, worst-case approach to design may be useful. We are interested in using the theory developed in (Parekh (1993, 1994)) as a possible basis for Constrained Worst-Case Design. The principal features of that theory are that the data streams entering the network are subject to Leaky Bucket (Turner (1986)) regulation at the network boundary, that access to any transmission line in the network is allocated to the various flows contending for it acording to some non-preempting variant (Demers (1989), Golestani (1994), Parekh (1993)) of a Generalized Processor Sharing (GPS) schedule, and that network performance is measured by the maximal values of backlogs and delays relative to the most adverse set of inputs permitted by the regulators.

We attempt here to formulate a strategy for reactive congestion control in the framework of constrained worst-case design. The network model includes Leaky Bucket regulators at every traffic source and GPS schedulers at every link. The essential feature of the model is that the link capacities are timevarying. Such time variation can be interpreted as resulting from timevarying commitments of resources to higher priority or rate-constrained traffic, from faults or rerouting, or simply from changes in the number or character of the sources being served, such changes being reflected, in the case of GPS-based service, by changes in the rate guaranteed to each connection. The Leaky Bucket regulator associated

*This research was supported by a grant from the Canadian Institute for Telecommunications Research under the NCE program of the Government of Canada.

with a particular flow is viewed as defining the gateway to the network; the congestion controller, reacting to changes in network operating conditions, exercises its function at the cell-level by modulating the width of the gateway — that is, by tuning the Leaky Bucket parameters; the goal is to ensure, to the extant possible in the presence of faults or overload, that the network operates within advertised limits on cell loss and delay. Our design philosophy is that quality-of-service guarantees in terms of cell loss and delay are sacrosanct, and that the Leaky Bucket parameters of individual sessions may be modified, if necessary, to accommodate them. If at some time a particular data stream is incompatible with the instantaneous values of its Leaky Bucket parameters, then cells may be lost or delayed — but the losses and delays in question will occur at the boundary of the network, rather than in the interior. To the extent that individual sources are capable of buffering their own data and reacting to changes in the width of their gateway to the network, the control mechanism we have in mind can be viewed as traffic shaping at the network boundary.

The problem, then, is to provide a uniform mechanism by which variations in available capacity are reflected in the cell admission control strategy. The input to that mechanism is a capacity estimate; the ensuing action, following the principle that the traffic in each session is characterized solely by the corresponding Leaky Bucket, is to recompute the Leaky Bucket parameters. The control is to be end-to-end, meaning that control decisions are communicated (subject to the intervening signal propagation delays) directly to the traffic sources. We deal first with the case that the network consists of a single multiplexor, and then with a multi-node version of the model.

2 THE SINGLE NODE CASE WITHOUT PROPAGATION DELAY

The network is a single multiplexor with output capacity which has some nominal value C, and which may vary according to a time function $C(\cdot)$ whose value at the present time t is determined by online measurement. The traffic sources are designated S_i, $i \in \mathcal{S}$. The communication path between $\mathcal{S}_i$ and the multiplexor involves a signal propagation delay of D_i seconds, which in the present section we take to be zero. The scheduling of transmissions on the multiplexor output line is approximated by a GPS strategy with parameters ϕ_i, $i \in \mathcal{S}$.

There are three kinds of sources. A rate-constrained source (representing the CBR class) is characterized exclusively by its peak rate, which is also the token generation rate in the associated Leaky Bucket, and the rate at which its cells are served in the multiplexor. Delay-constrained sources are characterized by their Leaky Bucket parameters and by a threshold representing the maximum acceptable cell delay. Buffer-constrained sources (approximating ABR) come with a constraint on backlog in place of one on delay. We use $\mathcal{S}_R$, $\mathcal{S}_D$ and $\mathcal{S}_B$, respectively, for the index sets of rate-constrained, delay-constrained and buffer-constrained sources.

When the available capacity is C, Sesion (i) has Leaky Bucket parameters ρ_i (the token generation rate) and σ_i (the size of the token pool). In the event that C is changing, these nominal values are replaced by controlled values $\rho_i(t)$, $\sigma_i(t)$, reflecting the results of service rate measurements. It can be useful as well to control the instantaneous value of $\ell_i(t)$, the pointer which records at each time instant the number of tokens currently available in the token pool, and the values of the GPS parameters ϕ_i. The objective of control is to transfer unavoidable congestion from the network interior to the network boundary.

The $\phi_i(\cdot)$ are assumed normalized so as to sum to unity for every t. Define functions β_i, δ_i by

$$\rho_i(t) = \beta_i(\gamma(t))\rho_i, \tag{1}$$
$$\sigma_i(t) = \delta_i(\gamma(t))\sigma_i, \tag{2}$$

where

$$\gamma(t) = \frac{C(t)}{C} \tag{3}$$

is the normalized link capacity. We require that

$$\sum_{i \in \mathcal{S}} \rho_i(t) \leq C(t) \tag{4}$$

for all t.

The control of a particular session depends on whether the corresponding traffic is rate-constrained, delay-constrained or buffer-constrained. In the case of rate-constrained sources, we attempt (where possible) to scale the ϕ_i so that ρ_i remains constant and equal to the received service rate. For delay-constrained sources, we attempt to scale σ_i, ρ_i so as to respect the delay constraint. The approach to buffer-constrained traffic is to scale σ_i, ρ_i and ϕ_i so as to avoid buffer overflow. In the event of sufficiently dramatic loss of capacity, these goals may not be possible and existing commitments to service may have to be renegotiated.

Thus, the control parameters for $i \in \mathcal{S}_{\mathrm{R}}$ are

$$\phi_i(t) = \frac{\phi_i}{\gamma(t)}, \tag{5}$$
$$\rho_i(t) = \rho_i, \tag{6}$$
$$\sigma_i(t) = \sigma_i; \tag{7}$$

these values correspond to $\beta_i(\gamma(t)) = \delta_i(\gamma(t)) = 1, i \in \mathcal{S}_{\mathrm{R}}$, in (1) and (2). This ensures that the guaranteed rate for the rate-constrained traffic is insensitive to changes in C. The token pool size is 1 cell. The congested node first attempts to satisfy the requirements of the rate-constrained sessions, then reallocates the remaining bandwidth among delay-constrained and buffer-constrained sources.

The basis for the analysis of worst-case backlog and worst-case delay in GPS networks with Leaky Bucket regulation is the Universal Service Curve (USC) introduced in (Parekh (1993)). For any particular configuration of the system — that is, given C and all the ρ_i, σ_i and ϕ_i — there is a corresponding USC. Changes in C and in the parameters of the various sessions will produce changes in the USC. We write $v(t, \cdot)$ for the USC corresponding to the system parameters in force at time t. It is easy to check that the first line segment of $v(t, \cdot)$ has slope

$$\mu_1(t) = C(t). \tag{8}$$

The assignment

$$\begin{aligned}\phi_i(t) &= \phi_i, &(9)\\ \rho_i(t) &= \gamma(t)\rho_i, &(10)\\ \sigma_i(t) &= \gamma(t)\sigma_i, &(11)\end{aligned}$$

thus ensures that the maximum delay for the delay-constrained Session (i) does not exceed the prescribed threshold. Indeed, the USC, and the load lines $(\frac{\sigma_i + \rho_i t}{\phi_i}), i \in \mathcal{S}_\mathrm{D}$, are scaled equally. The control in this case is said to be *proportional* in that the token pool sizes and token generation rates of the delay-constrained sessions are modified by the same factor. For the proportional control, $\beta_i(\gamma(t)) = \delta_i(\gamma(t)) = \gamma(t), i \in \mathcal{S}_\mathrm{D}$.

The remaining step is to devise a control suitable for buffer-constrained sources. Since the proportion of available bandwidth allocated to rate-constrained traffic has increased above the nominal operating value, it can be expected that

$$\rho_i(t) \le \gamma(t)\rho_i, \qquad i \in \mathcal{S}_\mathrm{B}. \tag{12}$$

We are interested in a controller that modifies the token generation rate of the buffer-constrained sources by the same factor. We assume for simplicity that the system is designed so that (4) holds with equality. Defining $\rho_\mathrm{R} = \sum_{i \in \mathcal{S}_\mathrm{R}} \rho_i, \ \ \rho_\mathrm{B} = \sum_{i \in \mathcal{S}_\mathrm{B}} \rho_i, \ \ \phi_\mathrm{R} = \sum_{i \in \mathcal{S}_\mathrm{R}} \phi_i, \ \ \phi_\mathrm{B} = \sum_{i \in \mathcal{S}_\mathrm{B}} \phi_i$ and

$$\begin{aligned}\gamma'(t) &= 1 + \Big(1 - \frac{1}{\gamma(t)}\Big)\frac{\phi_\mathrm{R}}{\phi_\mathrm{B}}, &(13)\\ \gamma''(t) &= \gamma(t) + \big(\gamma(t) - 1\big)\frac{\rho_\mathrm{R}}{\rho_\mathrm{B}}, &(14)\end{aligned}$$

we set

$$\begin{aligned}\phi_i(t) &= \gamma'(t)\phi_i, &(15)\\ \rho_i(t) &= \gamma''(t)\rho_i. &(16)\end{aligned}$$

Because the slope of the load line for a buffer-constrained source can be larger or smaller than $\gamma(t)$, the depletion time for such a session under the new operating conditions can be smaller or larger than before. In the case that it is smaller, the new USC lies above the old one and the constraints on maximum buffer size are respected. If, on the other hand, the new USC is below the one devised for nominal operating conditions, then certain constraints may be violated. We propose to choose the token pool sizes so that the buffer depletion times in the new setting are the same as before; that is, so that

$$\sigma_i(t) = \frac{\phi_i(t)}{\phi_i}\sigma_i + \frac{\phi_i(t)}{\phi_i}\rho_i T_i - \rho_i(t)T_i \tag{17}$$

where T_i is the abscissa at which the Session (i) load line intersects the USC. We require also that the maximum backlog should not exceed the prescribed level. These considerations suggest setting

$$\sigma_i(t) = \min\Big\{\mathcal{B}_i - \rho_i(t)T_{[i]} + \phi_i(t)v(t,T_{[i]}), \gamma'(t)\sigma_i + (\gamma'(t) - \gamma''(t))\rho_i T_i\Big\}, \qquad i \in \mathcal{S}_\mathrm{B}. \quad (18)$$

Transitory effects caused by the changes in parameters can be controlled by adjusting the pointer values $\ell_i(\cdot)$. It can be shown that if $A_i(t_1,t_2)$ is the total volume of Session (i) traffic actually entering the network in the time interval $[t_1,t_2]$, then

$$A_i(t_1,t_2) \le \ell_i(t_1) + \int_{t_1}^{t_2} \rho_i(\tau)\, d\tau, \quad (19)$$

meaning $\ell_i(t)$ determines the maximum burst size for the traffic of Session i at time t. This suggests that repositioning the $\ell_i(t)$ (for $i \in \mathcal{S}_\mathrm{D} \cup \mathcal{S}_\mathrm{B}$) might help to reduce cell discard inside the network.

Let $\mathcal{B}_i(t)$ be the maximum backlog (allowed to depend on time) in an all-greedy version of our system, and let $q_i(t)$ be the number of Session (i) cells backlogged in the node at time t. Then

$$\ell_i(t) + \rho_i(t)T_{[i]} - \phi_i(t)v(t,T_{[i]}) = \mathcal{B}_i(t) - q_i(t). \quad (20)$$

The foregoing definition of $\sigma_i(t)$, and the fact that $\ell_i(t)$ cannot be negative, lead to

$$\ell_i(t) = \Big\{0, \sigma_i(t) - q_i(t)\Big\}. \quad (21)$$

This gives the largest $\ell_i(t)$ for the smallest cell loss inside the network; any other choice of $\ell_i(t)$ will will either increase cell loss or decrease throughput.

In summary, for rate-constrained sessions there is no feedback control signal; the GPS parameters are selected (where possible) so as to respect the rate constraints. For delay-constrained sources,

$$\begin{aligned}
\phi_i(t) &= \phi_i, && (22)\\
\rho_i(t) &= \gamma(t)\rho_i, && (23)\\
\sigma_i(t) &= \gamma(t)\sigma_i, && (24)\\
\ell_i(t) &= \max\Big\{0, \sigma_i(t) - q_i(t)\Big\}, && (25)
\end{aligned}$$

while for buffer-constrained sources

$$\begin{aligned}
\phi_i(t) &= \gamma'(t)\phi_i, && (26)\\
\rho_i(t) &= \gamma''(t)\rho_i, && (27)\\
\sigma_i(t) &= \min\Big\{\mathcal{B}_i - \rho_i(t)T_{[i]} + \phi_i(t)v(t,T_{[i]}), \gamma'(t)\sigma_i + \big(\gamma'(t) - \gamma''(t)\big)\rho_i T_i\Big\}, && (28)\\
\ell_i(t) &= \max\Big\{0, \sigma_i(t) - q_i(t)\Big\}. && (29)
\end{aligned}$$

Because the effects of propagation delay have so far been ignored, the reaction to capacity changes is instantaneous.

3 THE SINGLE NODE CASE WITH PROPAGATION DELAY

Now assume that there is a propagation delay D_i between the i-th source and the node. As before, the variables subject to control are the GPS parameters $\phi_i(t)$ (implemented in the node) and the parameters $\big(\rho_i(t), \sigma_i(t), \ell_i(t)\big)$ fed back to the source. The corresponding feedback signal is received at the source after D_i seconds. The newly regulated traffic reaches the node after another D_i seconds. The effect of the control signal sent to Session (i) is thus felt at the node after $2D_i$ seconds, while application of the new GPS parameters is immediate. The GPS parameter satisfies

$$\phi_i(t) = \begin{cases} \dfrac{\phi_i}{\gamma(t)}, & \text{for } i \in \mathcal{S}_\mathrm{R}, \\ \phi_i, & \text{for } i \in \mathcal{S}_\mathrm{D}, \\ \gamma'(t)\phi_i, & \text{for } i \in \mathcal{S}_\mathrm{B}. \end{cases} \tag{30}$$

Because the feedback control signal depends on the GPS schedule, we require prediction of the GPS parameters $2D_i$ seconds ahead. The prediction is used only for the feedback calculation; the actual schedule is as defined above.

One approach to predicting the GPS parameter for a rate-constrained session is to use

$$\phi_i(t) = \frac{\phi_i}{\gamma(t+2D_i)}; \qquad i \in \mathcal{S}_\mathrm{R} \tag{31}$$

this indicates how much bandwidth will have been used for transmitting rate-constrained sources. If a precise prediction of the future values of the link capacity is available, then (31) assures that each rate-constrained session receives enough bandwidth to send at its nominal bit rate.

Similarly, the proportional controller for a delay-constrained session is described by

$$\phi_i(t) = \phi_i, \tag{32}$$

$$\rho_i(t) = \gamma(t+2D_i)\rho_i, \tag{33}$$

$$\sigma_i(t) = \gamma(t+2D_i)\sigma_i, \tag{34}$$

$$\ell_i(t) = \max\big\{0, \sigma_i(t) - q_i(t+2D_i)\big\}, \tag{35}$$

where $q_i(t+2D_i)$ is the Session (i) buffer occupancy at time $t+2D_i$. The buffer backlog in the node at time $t+2D_i$ is computed from

$$q_i(t+2D_i) = q_i(t) + A_i(t, t+2D_i) - S_i(t, t+2D_i) \tag{36}$$

where $A_i(t, t+2D_i)$ is the input traffic of Session (i) to the queue in the time interval $[t, t+2D_i]$ and $S_i(t, t+2D_i)$ is the output traffic exiting the queue; they are represented by

$$A_i(t, t+2D_i) = \int_t^{t+2D_i} a_i(\tau - D_i)\, d\tau, \tag{37}$$

$$S_i(t, t+2D_i) = \int_t^{t+2D_i} u_i(\tau)\, d\tau, \tag{38}$$

where $a_i(t)$ is the instantaneous rate of the source departing the Leaky Bucket regulator and $u_i(t)$ is the service rate of the traffic of Session (i).

In designing the feedback control signal, we choose a conservative approach, assuming that the buffer is continually backlogged in the interval $[t, t+2D_i]$ and that each source receives, at most, its guaranteed bandwidth. This corresponds to the case that all sources are active and the buffers of delay-constrained and buffer-constrained sessions do not underflow in the time interval $[t, t+2D_i]$. Thus, the output traffic is

$$S_i(t, t+2D_i) = \int_t^{t+2D_i} \phi_i C(\tau)\, d\tau. \tag{39}$$

The control signal for a delay-constrained source is then given by

$$\phi_i(t) = \phi_i, \tag{40}$$
$$\rho_i(t) = \gamma(t+2D_i)\rho_i, \tag{41}$$
$$\sigma_i(t) = \gamma(t+2D_i)\sigma_i, \tag{42}$$
$$\ell_i(t) = \max\Big\{0, \sigma_i(t) - q_i(t) - \int_t^{t+2D_i} a_i(\tau - D_i)\, d\tau + \int_t^{t+2D_i} \phi_i C(\tau)\, d\tau\Big\}. \tag{43}$$

Similarly, the control signal for a buffer-constrained session is

$$\phi_i(t) = \gamma'(t+2D_i)\phi_i \tag{44}$$
$$\rho_i(t) = \gamma''(t+2D_i)\rho_i, \tag{45}$$
$$\sigma_i(t) = \min\Big\{y_i(t), z_i(t)\Big\}, \tag{46}$$
$$\ell_i(t) = \max\Big\{0, \sigma_i(t) - q_i(t) - \int_t^{t+2D_i} a_i(\tau - D_i)\, d\tau + \int_t^{t+2D_i} \phi_i \gamma'(\tau) C(\tau)\, d\tau\Big\}, \tag{47}$$

where $\phi_i(t)$ and $\rho_i(t)$ in (46) are given in (44) and (45), and

$$y_i(t) = B_i - \rho_i(t) T_{[i]} + \phi_i(t) v(t, T_{[i]}), \tag{48}$$
$$z_i(t) = \frac{\phi_i(t)}{\phi_i}\sigma_i + \Big(\frac{\phi_i(t)}{\phi_i}\rho_i - \rho_i(t)\Big) T_i. \tag{49}$$

To compute the control signal, $a_i(t - D_i)$ and $C(t)$ are to be determined in the time interval $[t, t+2D_i]$. A predictor of the link capacity is applied to ascertain $C(\tau)$, $\tau \in [t, t+2D_i]$. To estimate the input traffic, one of the following alternative approaches might be used:

(1) The input rate $a_i(\tau), \tau \in [t, t+2D_i]$ might be predicted from its past on the basis of a stochastic model. This might result in a computationally expensive method since $a_i(\tau)$ should be predicted for all $\tau \in [t, t+2D_i]$. For a B-ISDN switch, operating at high speeds, this approach may not be practically attractive. Furthermore, the objective here is to introduce a method that does not rely on source modelling.

(2) The quantity $\int_t^{t+2D_i} a_i(\tau - D_i)\, d\tau$ can be approximated by its maximum (greedy) value. This will tend to make $\ell_i(t) = 0$, an option that we are trying to improve upon.

(3) The message rate $a_i(t)$ can be approximated by its average value. This might prove useful in cases where $2D_i$ is large compared to the fluctuations of $a_i(t)$. Since $a_i(t)$ is controlled by a Leaky Bucket with a parsimonious token generation rate $\rho_i(t-D_i)$, and assuming that the source traffic is the most adverse permitted by the regulator, we have

$$\bar{a}_i(t) = \rho_i(t - D_i) \tag{50}$$

where at each time t $\bar{a}_i(t)$ is the average traffic when the source is greedy and bounded by a Leaky Bucket with token generation rate $\rho_i(t - D_i)$.

We have adopted the last approach, which is consistent with the fluid-flow description of a slow random Markov walk. Consider the Markov walk (36). If the value of $A_i(t, t+2D_i) - S_i(t, t+2D_i)$ is small compared to the buffer size, or if the tail of its distribution is small, then (36) can be approximated by the (deterministic) finite-difference equation formed by substituting the terms in the brackets with their expected values (Meerkov (1972), Lim (1991), Benmohamed (1993)). In such a case, it can be shown that the trajectories of (36) and those of its deterministic finite-difference equation are close in probability. Substituting $\bar{a}_i(t)$ in place of $a_i(t)$ gives

$$\int_t^{t+2D_i} a_i(\tau - D_i)\, d\tau = \int_t^{t+2D_i} \bar{a}_i(\tau - D_i)\, d\tau = \int_{t-2D_i}^{t} \rho_i(\tau)\, d\tau. \tag{51}$$

This result is then used in (43) and (47) to determine $\ell_i(t)$. Note that this ccontrol does not completely confine cell loss to the network boundary. It does represent a compromise that provides a feasible solution.

For the proportional assignment of GPS parameters in which

$$\rho_i(t) = \phi_i(t + 2D_i)C(t + 2D_i) \tag{52}$$

(the token generation rate being equal to the GPS guaranteed rate), we have

$$\int_t^{t+2D_i} a_i(\tau - D_i)\, d\tau = \int_{t-2D_i}^{t} \rho_i(\tau)\, d\tau = \int_t^{t+2D_i} \phi_i(\tau)C(\tau)\, d\tau. \tag{53}$$

In this case, (43) and (47) simplify to

$$\ell_i(t) = \max\Big\{0, \sigma_i(t) - q_i(t)\Big\}, \tag{54}$$

and the control signal simply depends on the value of $\gamma(t + 2D_i)$. A predictor can be used to estimate $\gamma(t + 2D_i)$.

4 THE MULTI-NODE CASE

A virtual circuit in an ATM network usually extends over several links. Each link will be represented here by the subscript of its input node; for instance, the link from N_k to N_l is denoted L_k, with capacity C^k.

Suppose that the capacity of link L_l is changing. The traffic on the virtual circuits passing through L_l should be controlled to prevent overflow of the buffers in N_l. To that end, a control signal is generated in N_l and transmitted backwards to the upstream nodes, each of which, in turn, may modify it according to local operating conditions before pasing it on. The process continues until the control signals arrive at the Leaky Buckets for which they are intended; each Leaky Bucket adopts the new parameter setting and transmits an acknowledgement cell reporting the new parameter setting to all the nodes along its path. This information can be used by downstream nodes for efficient management of bandwidth and buffer space.

Assume that $\mathcal{S}^l$ sources contend for the bandwidth $C^l(t)$ and that a control signal is generated at time instant t^l, the superscript l indicating that time is measured relative to N_l. As before, priority is given to rate-constrained flows, and and the remaining bandwidth divided among delay-constrained and buffer-constrained sessions. We do not be concerned with rate-constrained traffic, assuming simply that the corresponding GPS parameters are modified so that the rate-constraints are satisfied.

Delay-constrained traffic

The proportional control signal generated at N_l for $i \in \mathcal{S}^l_{\mathrm{D}}$ is given by

$$\begin{aligned}
\phi^l_i(t^l) &= \phi^l_i, && (55)\\
\rho^l_i(t^l) &= \gamma^l(t^l + 2D^l_i)\rho_i, && (56)\\
\sigma^l_i(t^l) &= \gamma^l(t^l + 2D^l_i)\sigma_i, && (57)\\
\ell^l_i(t^l) &= \max\Big\{0, \sigma^l_i(t^l) - q^l_i(t^l + 2D^l_i)\Big\}, && (58)
\end{aligned}$$

where D^l_i is the propagation delay between Source (i) and node N_l; the superscripts indicate that the control signal is generated in N_l. The signal is then sent back, along its virtual circuit, to an upstream node N_k, which modifies it according to the available bandwidth on Link L_k.

Let the control signal for the token generation rate of a session connected to N_k be given by $\rho^k_i(t^k)$ where t^k is measured at N_k. In the absence of a control signal from N_l, this control signal is given by $\gamma^k(t^k + 2D^k_i)\rho_i$. When a control signal from N_l is present, it should be compared to $\gamma^k(t^k + 2D^k_i)\rho_i$ and the smaller of the two is passed on to an upstream node. Take, for example, the case in which

$$\rho^l_i(t^l) \geq \gamma^k(t^k + 2D^k_i)\rho_i. \qquad (59)$$

In this case, L_k is the bottleneck link and the control signal is

$$\begin{aligned}
\phi^k_i(t^k) &= \phi^k_i, && (60)\\
\rho^k_i(t^k) &= \gamma^k(t^k + 2D^k_i)\rho_i, && (61)\\
\sigma^k_i(t^k) &= \gamma^k(t^k + 2D^k_i)\sigma_i, && (62)\\
\ell^k_i(t^k) &= \min\Big\{\ell^l_i(t^l), \max\Big\{0, \sigma^k_i(t^k) - q^k_i(t^k + 2D^k_i)\Big\}\Big\}. && (63)
\end{aligned}$$

Now consider the case

$$\rho_i^l(t^l) < \gamma^k(t^k + 2D_i^k)\rho_i. \tag{64}$$

This signifies that N_l is the bottleneck node and that the token generation rate should be the one computed by N_l. In fact, the control signal is given by

$$\begin{aligned}
\phi_i^k(t^k) &= \phi_i, &(65)\\
\rho_i^k(t^k) &= \rho_i^l(t^l), &(66)\\
\sigma_i^k(t^k) &= \sigma_i^l(t^l), &(67)\\
\ell_i^k(t^k) &= \min\Big\{\ell_i^l(t^l), \max\big\{0, \sigma_i^k(t^k) - q_i^k(t^k + 2D_i^k)\big\}\Big\}. &(68)
\end{aligned}$$

In this case, the token generation rate fed back to the source is smaller than that for a proportional control based on $C^k(t)$; this suggests that the intersection of the load line with the USC at N_k is smaller than under nominal operating conditions, so that the delay constraints are satisfied.

The two cases can be combined in the form

$$\begin{aligned}
\phi_i^k(t^k) &= \phi_i^k, &(69)\\
\rho_i^k(t^k) &= \min\Big\{\gamma^k(t^k + 2D_i^k), \gamma^l(t^l + 2D_i^l)\Big\}, \rho_i, &(70)\\
\sigma_i^k(t^k) &= \min\Big\{\gamma^k(t^k + 2D_i^k), \gamma^l(t^l + 2D_i^l)\Big\}\sigma_i, &(71)\\
\ell_i^k(t^k) &= \min\Big\{\ell_i^l(t^l), \max\big\{0, \sigma_i^k(t^k) - q_i^k(t^k + 2D_i^k)\big\}\Big\}. &(72)
\end{aligned}$$

Buffer-constrained traffic

For a buffer-constrained source, the control signal generated by L_l is given by

$$\begin{aligned}
\phi_i^l(t^l) &= \gamma'^l(t^l + 2D_i^l)\phi_i^l, &(73)\\
\rho_i^l(t^l) &= \gamma''^l(t^l + 2D_i^l)\rho_i, &(74)\\
\sigma_i^l(t^l) &= \min\Big\{y_i^l(t^l), z_i^l(t^l)\Big\}, &(75)\\
\ell_i^l(t^l) &= \max\Big\{0, \sigma_i^l(t^l) - q_i^l(t^l + 2D_i^l)\Big\}, &(76)
\end{aligned}$$

where

$$\begin{aligned}
y_i^l(t^l) &= B_i^l - \rho_i^l(t^l)T_{[i]}^l + \phi_i^l(t^l)v^l(t, T_{[i]}^l), &(77)\\
z_i^l(t^l) &= \frac{\phi_i^l(t^l)}{\phi_i^l}\sigma_i + \Big(\frac{\phi_i^l(t^l)}{\phi_i^l}\rho_i - \rho_i^l(t^l)\Big)T_i^l. &(78)
\end{aligned}$$

At N_k, $\rho_i^l(t^l)$ is compared to $\gamma^k(t^k + 2D_i^k)\rho_i$ and the smaller of the two is passed on to the upstream node. We consider two cases:

If $\rho_i^l(t^l) \geq \gamma^k(t^k + 2D_i^k)\rho_i$, then the control signal for an upstream node is

$$\phi_i^k(t^k) = \gamma'^k(t^k + 2D_i^k)\phi_i^k, \tag{79}$$

$$\rho_i^k(t^k) = \gamma''^k(t^k + 2D_i^k)\rho_i, \tag{80}$$
$$\sigma_i^k(t^k) = \min\{\sigma_i^l(t^l), y_i^k(t^k), z_i^k(t^k)\}, \tag{81}$$
$$\ell_i^k(t^k) = \min\{\ell_i^l(t^l), \max\{0, \sigma_i^k(t^k) - q_i^k(t^k + 2D_i^k)\}\}. \tag{82}$$

On the other hand, if $\rho_i^l(t^l) < \gamma^k(t^k + 2D_i^k)\rho_i$, the control signal for an upstream node is

$$\phi_i^k(t^k) = \gamma'^k(t^k + 2D_i^k)\phi_i^k, \tag{83}$$
$$\rho_i^k(t^k) = \rho_i^l(t^l) = \gamma''^l(t^l + 2D_i^l)\rho_i, \tag{84}$$
$$\sigma_i^k(t^k) = \min\{\sigma_i^l(t^l, y_i^k(t^k), z_i^k(t^k)\}, \tag{85}$$
$$\ell_i^k(t^k) = \min\{\ell_i^l(t^l), \max\{0, \sigma_i^k(t^k) - q_i^k(t^k + 2D_i^k)\}\}. \tag{86}$$

The control signals in the two cases above can be combined as follows:

$$\phi_i^k(t^k) = \gamma'^k(t^k + 2D_i^k)\phi_i^k, \tag{87}$$
$$\rho_i^k(t^k) = \min\{\gamma''^l(t^l + 2D_i^l)\gamma''^k(t^k + 2D_i^k)\}\rho_i, \tag{88}$$
$$\sigma_i^k(t^k) = \min\{\sigma_i^l(t^l), y_i^k(t^k), z_i^k(t^k)\}, \tag{89}$$
$$\ell_i^k(t^k) = \min\{\ell_i^l(t^l), \max\{0, \sigma_i^k(t^k) - q_i^k(t^k + 2D_i^k)\}\}. \tag{90}$$

This process continues till the control signal arrives at the source.

5 EXAMPLE

The network model contains two access nodes connected to a remote switch. There are six sources, of which two are rate-constrained and the others, delay-constrained. Each access node serves one rate-constrained and two delay-constrained sources. The rate-constrained sources have a constant bit rate of 5 Mbs (Mega bit per second). The delay-constrained flows are generated by Poisson processes with an average rate of 5 Mbs; the nominal values of the corresponding Leaky Bucket parameters are $\rho = 2.5$ Mbs and $\sigma = 12.5$ Kb (25 cells). The capacity of the links connecting the sources to their access nodes and the link between each access node and the multiplexing node are fixed at 10 Mbs. The total traffic is multiplexed on a link with a nominal capacity 20 Mbs. The capacity of this link varies in the interval $[12.5, 20]$ Mbs. The transition times are exponentially distributed with the average period 5 ms. The access links are 100 km long and the links connecting the access nodes to the multiplexer are 500 km long. The maximum buffer size in all nodes is 100 kb. A control signal is generated and send back to the upstream nodes when the capacity of the multiplexing link changes — the multiplexing node is the bottleneck.The cell loss at the nodes and the throughput of the system were averaged over 20 independent trials. The simulation ran one second for each trial.

In the absence of feedback control, 5513.40 cells were lost and 14 518.35 cells delivered. The proposed feedback control reduced the cell loss to 0.95 cells and the throughput (marginally) to 14 505.60 cells. To study the importance of the pointer in the feedback controller, we simulated the same example with $\ell(t)$ in (70) always reset to zero; the cell loss fell to zero and the throughput, to 14 152.75 cells. Note that simply setting the pointer to zero reduces throughput.

6 SUMMARY

We introduced a constrained-worst-case, reactive congestion control technique for ATM-based telecommunication networks. The technique distinguishes three broad classes of traffic sources. It acts by modulating the parameters of the Leaky Bucket regulators attached to the various sessions, with the objective of transferring (to the extent possible) the effects of congestion from the interior of the network to the boundary. A traffic predictor is included to accommodate signal propagation delays between sources and the bottleneck nodes.

7 REFERENCES

Benmohamed, L. and Meerkov, S.M. (1993) Feedback control of congestion in packet switching networks: the case of a single congested node. IEEE/ACM Transactions on Networking,1, 693-708.

Demers, A., Keshav, S. and Shenker,S. (1989) Analysis and simulation of a fair queuing algorithm. ACM SIGCOMM, 19, Volume 4, 1-12.

Golestani, S.J. (1994) A self-clocked fair queueing scheme for broadband applications. IEEE Infocom'94, 636-646.

Lim, J.T., Meerkov, S.M. and Zeng, T. (1991) Simplified description of slow-in-the-average Markov walks. Journal of Mathematical Analysis and Applications, 476-486.

Meerkov, S.M. (1972) Simplified description of slow Markov walks — Part I. Automation and Remote Control, 33, 404-414.

Parekh, A.K. and Gallager, R.G. (1993) A generalized processor sharing approach to flow control in integrated services networks: the single node case. IEEE/ACM Transactions on Networking, 1, 344-357.

Parekh, A.K. and Gallager, R.G. (1994) A generalized processor sharing approach to flow control in integrated services networks: the multiple node case. IEEE/ACM Transactions on Networking, 1, 137-150.

Turner, J.S. (1986) New directions in communications (or which way to the Information Age?). IEEE Communications Magazine, 24, 8-15.

18

Performance Analysis of a Fuzzy System in the Policing of Packetized Voice Sources

Giuseppe Ficili and Daniela Panno
Istituto di Informatica e Telecomunicazioni, Facoltà di Ingegneria
V.le A. Doria 6, 95125 Catania - ITALY
phone: +39 95 339449 - fax: +39 95 338280
e-mail: {gficili, dpanno}@iit.unict.it

Abstract

In this paper we analyze the performance of a policing mechanism based on Fuzzy Logic for the control of packetized voice sources. The results obtained show an excellent selectivity, close to that of an ideal policer, and a responsiveness, assessed by the combined measures of reaction time and rise time, which is decidedly better than that of the Leaky Bucket. Analysis by simulation also shows that the fuzzy policer is efficient in the combined control of two traffic descriptors, the Sustainable Cell Rate and the long-term Average Cell Rate.

Keywords

ATM, Congestion Control, Policing, Fuzzy Logic.

1 INTRODUCTION

One of the most critical functions in the management of high-speed networks using the ATM technique is that of "policing" [(Hong, 1991); (Habib, 1991); (Fratta, 1992); (Onvural, 1994)]. Also referred to as Usage Parameter Control (UPC), it has the task of ensuring that each traffic source stays within the parameter values negotiated during the set-up phase of the call.

In order to define an efficient policing mechanism, a first issue is identifying the traffic parameters which best characterize the behaviour of a source. The difficulty lies in the fact that the sources to be characterized have different statistical properties as they range from video to data services, and it is necessary to define parameters that can be monitored during the call [(Berger, 1991); (Berger, 1994); (Andrade, 1994)]. A traffic parameter contributing to a source traffic descriptor should be understandable by the user, of significant use in resource allocation, and enforceable by the network provider through the UPC. Ordinarily, UPC is performed for each traffic parameter in a source traffic descriptor. For example, if the source traffic descriptor consists of the peak bit rate and the mean bit rate, UPC is necessary for both of them.

Another key issue is defining a traffic enforcement mechanism which will be efficient in coping with the conflicting requirements of ideal flow enforcement: high selectivity with respect to the traffic monitored (that is, the capability of detecting any illegal traffic situation and transparency for connections that respect the parameter values negotiated, on whose cells no policing action need be taken); and high responsiveness, that is, low response time to parameter violations.

In literature several mechanisms such as the Leaky Bucket (LB) and window mechanisms have been proposed to police such parameters as Peak Cell Rate (PCR) and Average Cell Rate (ACR) seen as the average calculated over the whole duration of the connection. Policing of the peak rate is generally not complex and can be achieved, for example by using a cell spacer [(Guillemin, 1992)] or other mechanisms [(Buttò, 1991)]. Enforcement of the mean rate is more problematic, since short-term statistical fluctuations of the source traffic are admissible as long as the source respects the average value negotiated, λ_n, in the long term. An extensive review of traditional mechanisms can be found in [(Rathgeb, 1991); (Dittmann, 1991)]; for none of them has it been possible to achieve a satisfactory tradeoff between the above-mentioned conflicting requirements.

This difficulty in compiling the control know-how may be due to non linearity, to time variant behaviour of the system, or to the fact that the measurements available have poor quality. In order to overcome the limits that traditional mechanisms seem to display, in [(Catania, 1995)] a new policing mechanism based on fuzzy logic was proposed.

Fuzzy logic [(Zadeh, 1965); (Munakata, 1994); (Zadeh, 1994)] is an important tool for formalizing processes of approximate reasoning in which the knowledge base can be acquired from human expert. To reproduce the concepts expressed in natural language, fuzzy logic replace true and false with continuous membership values ranging from zero to one. This allows the processing of linguistic concepts such as "small", "big", "low", "high" or "approximately", which can be expressed in the fuzzy inferential rules that describe the control algorithm. Thanks to this inherent features, fuzzy logic proves to be efficient in controlling of real time processes which are too complex to be representative by exact mathematical models.

Exploiting the advantages offered by fuzzy logic in [(Catania, 1995)] we formalized the control actions of the policer translating the know-how of an expert in the field into fuzzy rules. The fuzzy policer obtained was analyzed using the same bursty sources as those examined by Rathgeb (1991) and compared with conventional mechanisms such as the LB and the Exponential Weighted Moving Average (EWMA). The results obtained showed that fuzzy logic, when applied to policing, is extremely promising on account of both the performance obtainable and the simplicity of the control algorithm.

In this paper we focus on a dual aim. The first is to analyze the behaviour of the Fuzzy Policer (FP) in controlling a real source, namely a packetized voice source, evaluating its selectivity and dynamic response. The latter is evaluated by means of a combined measure of reaction time and rise time. The second aim is to see whether the FP is capable of controlling not only the ACR but also the Sustainable Cell Rate (SCR) defined by ITU in [(ITU-TS, 1995)] and the ATM Forum in [(ATM FORUM, 1995)]. Some of the main issues linked to use of the ACR as a traffic descriptor parameter are discussed and the need is shown for combined control of ACR and SCR. We then present a simulation analysis from which it emerges that the FP is extremely flexible in the combined control of ACR and SCR, unlike mechanisms such as the LB, guaranteeing an optimal level of performance.

The paper is organized as follows: in Sect. 2 we define the fuzzy policing mechanism model and evaluate its performance in the control of packetized voice sources. In Sect. 3 we discuss the efficacy of the ACR as a descriptor of source behaviour and propose the FP for the combined control of ACR and SCR. Finally in Sect. 4 some conclusions are drawn.

2 POLICING USING FUZZY LOGIC

Fuzzy logic is based on the concepts of linguistic variables and fuzzy sets. A fuzzy set in a Universe of Discourse is characterized by a membership function μ_f which assumes values in the interval [0,1]. A fuzzy set F is represented as a set of ordered pairs, each made up of a generic element $u \in U$ and its degree of membership $\mu_f(u)$.

A linguistic variable x in a Universe of Discourse is characterized by a set $W(x)=(W_{1x}, \ldots, W_{nx})$ and a set $M(x)=(M_{1x},\ldots,M_{nx})$, where W(x) is the term-set, i.e. the set of names the linguistic variable x can assume, and W_{ix} is a fuzzy set whose membership function is M_{ix}. If, for instance, x indicates a temperature, W(x) could be the set W(x)=(Low, Medium, High), each element of which is associated with a membership function.

The rules governing a fuzzy system are often written using linguistic expressions which formalize the empirical rules by means of which a human operator is able to describe the process in question using his own experience. If x and y are taken to be two linguistic variables, fuzzy logic allows these variables to be related by means of *fuzzy conditional rules* of the following type:

'IF (x is A) THEN (y is B)'

where *(x is A)* is the *premise* of the rule, while *(y is B)* is the *conclusion*. This rule makes it possible to deduce, using specific inferential methodologies, a fuzzy set for y for each input value of x, whether it is associated with a fuzzy set or assumes a numerical value (*crisp*).

2.1 Model of the System

Our policer is a window-based control mechanism in which the maximum number, N_i, of cells that can be accepted in the i-th window of size T_F is dynamically updated by inference rules based on fuzzy logic. We have defined a control mechanism that aims to make a VBR source respect the Average Cell Rate negotiated value, λ_n, i.e. to ensure that on average the source transmits N cells per window with $N=T_F \cdot \lambda_n$.

As pointed out in the introduction, a policing mechanism has to allow for short-term fluctuations, as long as the source respects the negotiated parameters over the long term, and also has to be able to recognize a violation immediately.

The philosophy on which our mechanism is based is one of granting credit to a source which in the past has respected the parameters negotiated by increasing its control threshold, N_i, as long as it perseveres with non-violating behaviour. Vice versa, if the behaviour of the source is violating or risky, the mechanism reduces its credit by decreasing the threshold value.

The parameters describing the behaviour of the source and the policing control variables are made up of linguistic variables and fuzzy sets, while control action is expressed by a set of fuzzy conditional rules which reflect the cognitive processes that an expert in the field would apply.

The source descriptor parameters used are: the average number of cell arrivals per window since the start of the connection, A_{oi}, and the number of cell arrivals in the last window, A_i. The first gives an indication of the long-term trend of the source; the second indicates its current behaviour. A third parameter, the value of N_i in the last window, was also introduced to indicate the current degree of control (degree of permissiveness) the mechanism has over the source. These parameters are the three fuzzy policer inputs.

The output chosen was the linguistic variable ΔN_{i+1} which represents the threshold variation to be made in the next window.

The model of the fuzzy system, comprising the control rules and the term sets of the variables with their related fuzzy sets, was obtained through a tuning process which started from a set of initial insight considerations and progressively modified the parameters of the system until it reached a level of performance considered to be adequate. During this tuning phase we used a stochastic model, with different parameter values, and also some real voice samples.

For the input variable a term-set with a cardinality of three (Low, Medium, and High) was derived, while for the output variable the term-set has a cardinality of seven (Negative Big, Negative Medium, Negative Small, Zero, Positive Small, Positive Medium, Positive Big). In particular, the Universe of Discourse of A_i and A_{oi} ranges from 0 to the maximum number of cells that can arrive in a window (T_F/t_c), where t_c is the cell interarrival time during a burst; the Universe of Discourse of N_i ranges from 0 to N_{i_max} which indicates the upper bound value fixed for the N_i variable; finally, the Universe of Discourse of ΔN_{i+1} ranges from -N/4 to N/4. We observe that the Universe of Discourse for all the variables is defined parametrically as a function of N, t_c and T_F. This allows us to use the same model for sources with different statistical properties.

The following are the fuzzy conditional rules which implement the control action of the fuzzy policer.

1. *If (A_{oi} is low) and (N_i is high) and (A_i is low) then (ΔN_{i+1} is positive big).*
2. *If (A_{oi} is low) and (N_i is high) and (A_i is medium) then (ΔN_{i+1} is positive small).*
3. *If (A_{oi} is low) and (N_i is high) and (A_i is high) then (ΔN_{i+1} is zero).*
4. *If (A_{oi} is medium) and (N_i is medium) and (A_i is low) then (ΔN_{i+1} is positive big).*
5. *If (A_{oi} is medium) and (N_i is medium) and (A_i is medium) then (ΔN_{i+1} is positive small).*
6. *If (A_{oi} is medium) and (N_i is medium) and (A_i is high) then (ΔN_{i+1} is zero).*
7. *If (A_{oi} is medium) and (N_i is high) and (A_i is low) then (ΔN_{i+1} is positive big).*
8. *If (A_{oi} is medium) and (N_i is high) and (A_i is medium) then (ΔN_{i+1} is zero).*
9. *If (A_{oi} is medium) and (N_i is high) and (A_i is high) then (ΔN_{i+1} is negative big).*
10. *If (A_{oi} is high) and (N_i is low) and (A_i is low) then (ΔN_{i+1} is positive big).*
11. *If (A_{oi} is high) and (N_i is low) and (A_i is medium) then (ΔN_{i+1} is positive medium).*
12. *If (A_{oi} is high) and (N_i is low) and (A_i is high) then (ΔN_{i+1} is positive small).*
13. *If (A_{oi} is high) and (N_i is medium) and (A_i is low) then (ΔN_{i+1} is positive big).*
14. *If (A_{oi} is high) and (N_i is medium) and (A_i is medium) then (ΔN_{i+1} is positive medium).*
15. *If (A_{oi} is high) and (N_i is medium) and (A_i is high) then (ΔN_{i+1} is zero).*

16. *If (A_{oi} is high) and (N_i is high) and (A_i is low) then (ΔN_{i+1} is negative small).*
17. *If (A_{oi} is high) and (N_i is high) and (A_i is medium) then (ΔN_{i+1} is negative medium).*
18. *If (A_{oi} is high) and (N_i is high) and (A_i is high) then (ΔN_{i+1} is negative big).*

In order to ensure transparency towards a respectful source, a suitable initial credit has to be given to a source. For this reason N_1 is set to a value greater than N.

2.2 Performance evaluation in policing of packet voice sources

In a previous paper [(Catania, 1995)] we showed that the FP offers performance levels which are decidedly better than those obtainable with the LB and EWMA, recognized as better than other conventional mechanisms [(Rathgeb, 1991)]. We considered a bursty source that is widely accepted as the worst-case traffic pattern for policing and refers to the bursty source characteristics studied by Rathgeb, (1991) which, however, are not representative of any real source.

In this section we assess the efficiency of the FP proposed in policing a real bursty source, namely a packetized voice source. We assume that the number of cells per burst is geometrically distributed with a mean of E[x] = 29 cells; the duration of the idle phase is exponentially distributed with a mean of E[s] = 650 ms; and the intercell time during a burst is t_c = 12 ms. So, the cell arrival rate negotiated is λ_n = 29 cell/s.

We will compare the performance of our FP with that of some LBs of various sizes.

The LB [(Turner, 1986)] is based on the concept of pseudoqueue and consists of a counter which is increased on the arrival of cells and decreased, if positive, at a constant frequency λ_e. When the counter exceeds a pre-established threshold, Q, the cells are detected as excessive and the policing action agreed on is taken. In enforcing the mean cell rate negotiated, λ_n, it has emerged from the analysis in [(Buttò, 1991), (Rathgeb, 1991), (Monteiro, 1990)] that in order to achieve greater flexibility in Q and reduce the probability of false alarms, it is necessary to introduce an overdimensioning factor C (C > 1) between the negotiated cell rate, λ_n, and that which is really policed; it follows that $\lambda_e = C \cdot \lambda_n$. On the other hand, as we will demonstrate, this artifice reduces the capacity to detect violation over a long term.

The parameters for the policing mechanisms considered for comparison are set to enforce the ACR negotiated, λ_n = 29 cell/s, guaranteeing a given false alarm probability of $P_{FA} = 10^{-7}$:

LB_1 with Q=1660 and C=1.1
LB_2 with Q= 487 and C=1.4
LB_3 with Q= 85 and C=2.3
FP with T_F =3 sec, $N_1 = 4 \cdot N$ and $N_{i\_max} = 9 \cdot N$, where $N = T_F \cdot \lambda_n$ = 87 cells.

Let us focus our attention on selectivity. As a performance measure of this requirement we consider the probability that the policing mechanism detect a cell as excessive, P_d. The ideal behaviour would be that P_d is zero with the mean cell rate up to the nominal one (the false alarm probability is zero), and $P_d = (\sigma - 1)/\sigma$ for $\sigma > 1$, where σ is the long-term actual mean cell rate of the source normalized to the negotiated mean cell rate.

In order to obtain the curve P_d versus σ, we assume that a variation in the average cell rate is due to a change in the average number of cells per burst, while the average silence time is assumed to be constant. As shown in Figure 1, our fuzzy policer presents a negligible false alarm probability ($\sigma < 1$) and, in the case of violating sources ($\sigma > 1$), a probability of detection of violation very close to ideal and certainly much greater than that of the other mechanisms.

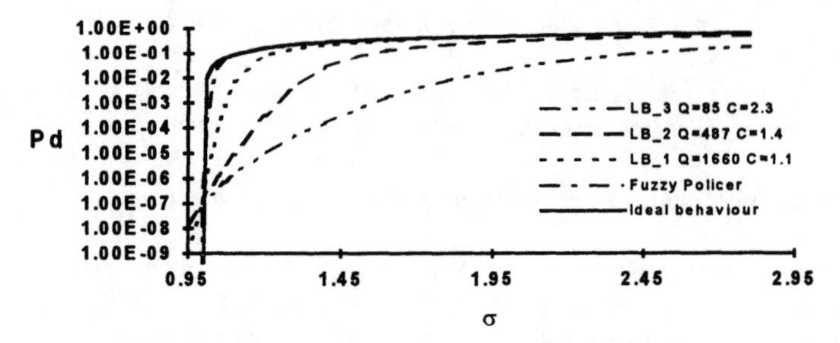

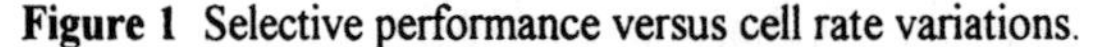

Figure 1 Selective performance versus cell rate variations.

Another important requirement for a policing mechanism is responsiveness. In connection with this, we introduce two performance measures: reaction time and rise time. We define the reaction time of a policing mechanism as the time required to begin the excessive cell detection of a violating source, and the rise time as the interval between the reaction time instant and that at which the system reaches 90% of its steady-state P_d value (i.e. 90% of its selectivity).

Figure 2 shows the reaction time versus σ. As can be seen, for LBs whose C is less than the σ value considered, the reaction time assumes very high values. Indeed, in certain cases with a particular traffic flow evolution the reaction time is infinite, i.e. violation is never detected.

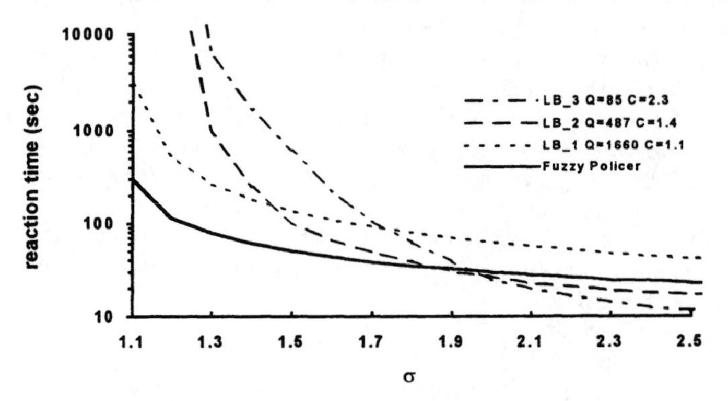

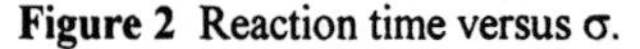

Figure 2 Reaction time versus σ.

From Figures 1 and 2, a comparison of LBs shows that the system which has the best selectivity, that is, the best behaviour towards long-term violations (namely LB_1), is the worst as far as responsiveness is concerned; and vice versa. The selectivity performance of LB can be further improved if C is chosen close to unity, i.e. if the depletion rate of the pseudoqueue is close to the rate negotiated. However, to guarantee the same value for the P_{FA}, the choice to reduce C involves excessive Q values, thus drastically reducing the speed at which the mechanism detects violations.

This problem is not encountered with our fuzzy policer. In fact, a trend very close to the ideal curve in the steady state corresponds to decidedly better dynamics than those of the other mechanisms.

The reaction time measure alone may be considered to be insufficient for correct evaluation of the dynamic behaviour of the policing method. We therefore took another figure of merit into consideration: rise time. Figure 3 shows that with values of $\sigma < 2$ the FP shows much better behaviour in terms of rise time. In addition, although LB_2 and LB_3 present a lower rise time than the FP with high values of σ, as can be seen in Figure 1 in the steady state these mechanisms reach a much lower P_d value than the FP.

To conclude, the FP succeeds where traditional mechanisms, the LB in particular, fail: in controlling the Average Cell Rate in the long term, achieving an excellent tradeoff between selectivity and responsiveness.

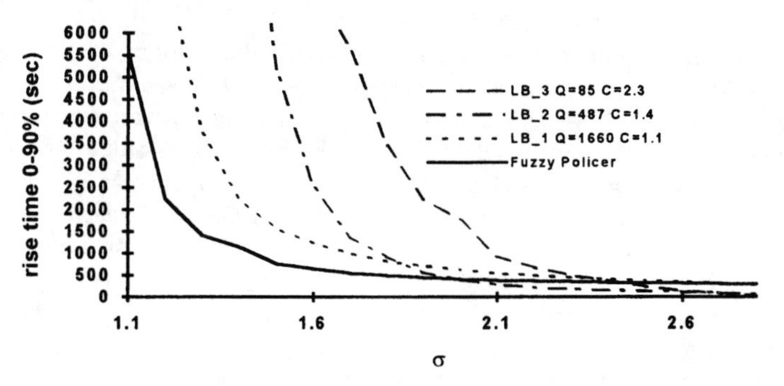

Figure 3 Rise time versus σ.

3 ISSUES ON THE CONTROL OF THE AVERAGE CELL RATE

The analysis presented in the previous section compares the capacity of the two methods to police the Average Cell Rate parameter, evaluated on the whole duration of the connection (ACR). Although this parameter has often been indicated as a traffic descriptor for several policing methods [(Rathgeb, 1991)], there are certain cases, not taken into consideration in the previous section, in which use of this parameter conflicts with the aim of preventing network congestion.

Let us consider a source S for which a value, λ_n, has been negotiated for the ACR and let us assume that S transmits for a long period of time at an average cell rate which is much higher than that negotiated, but which in all, over the whole connection, respects the value λ_n.

In this case, the policing action has to treat the cells generated by the source as non-excessive. However, none of the policers presented in literature, including the FP, is capable of guaranteeing such a result. After a time interval depending on their various characteristics, each of them would begin to consider the cells sent by the source as violating, thus generating a false alarm probability. In the specific case of the LB, this time interval depends on the tolerance offered by the depth of the bucket, while in the case of the FP it is linked to the value of the credit the source is granted.

There would therefore seem to exist cases in which the control action is not always transparent towards a 'compliant' source that respects the parameter values negotiated, i.e. it does not always guarantee control without unfairly penalizing the traffic of a compliant source. There is, however, a basic consideration which has to be made in connection with the use of the ACR as the traffic descriptor parameter.

Let us assume that it is possible to perform ideal policing in long-term control of the ACR, and that in the traffic conditions outlined above its behaviour is 'transparent' to S. This may contrast with the aim of exploiting network resources to the full. If, in fact, a generic source is allowed to transmit for long periods of time at an average cell rate which is much higher than the λ_n, the probability that several sources will behave in the same way in the same time interval increases, with a consequent risk of congestion. Hence to avoid congestion, the number of connections accepted in the network will have to be limited, with a consequent reduction in the efficiency of use of the network's resources.

This problem shows how use of the ACR alone as a traffic descriptor parameter may create a conflict between the aim of achieving correct policing of S and that of preventing congestion. In other words, whereas on the one hand the selectivity requirement requires the source S to be treated as non-violating, on the other the prevention of congestion requires S to be prevented from transmitting for too long a time at a cell rate higher than λ_n. This conflict can be solved by introducing another parameter besides the ACR, the Sustainable Cell Rate (SCR), which expresses the maximum average bit rate at which S can transmit in an interval of T without being considered to be violating. This new parameter can be seen as an upper bound imposed on the maximum credit the source can be granted so as to prevent it from transmitting for too long a time at an average bit rate equal to SCR, higher than λ_n.

The combined policing of the two parameters has a dual purpose:

- optimizing the allocation policy in that the period of time during which the source can transmit at a bit rate of SCR is limited;
- eliminating the probability of false alarms as, in the traffic conditions considered for S at the beginning of this section, it is now legitimate to consider cells violating the parameter SCR as excessive.

At present, the Peak Cell Rate and the Sustainable Cell Rate are the only parameters standardized by ITU [(ITU-TS, 1995)] and the ATM Forum [(ATM FORUM, 1995)] as a Variable Bit Rate source traffic descriptor. Along with the SCR, Intrinsic Burst Tolerance (IBT) is also provided for, which is a parameter necessary to determine the Maximum Burst Size that may be transmitted at the PCR and the time interval considered.

The formal definition of SCR together with the Intrinsic Burst Tolerance (IBT) parameter, uses a virtual scheduling algorithm (which is equivalent to a continuous state leaky bucket algorithm) referred to as the Generic Cell Rate Algorithm (GCRA).

Although the SCR is enforceable by the network provider through the GCRA, it cannot by itself cope with the objective of traffic control in using the multiplexing gain to establish VBR connections economically.

Let us consider the following example. Let A be an application of retrieval of digital image information stored at a remote location, supported by an ATM connection with a source traffic descriptor containing PCR and SCR. Let us assume that one of the QoS requirements is that the delay between the time epoch a new image is requested and the epoch when the new image is fully displayed be less than d_1 seconds. Let us also assume that the time interval between requests for a new image is more than d_2 seconds. If M is the number of cells needed to encode the image, the user could choose the PCR to be M/d_1 cell/s, the SCR to be M/d_2 cells/s, and the IBT to be $(d_2 - d_1)$ seconds.

Choice of the SCR does not, however, provide any information as to the average rate at which a new image is requested. Without this information, the network provider would have to allocate a bandwidth capable of guaranteeing the maximum rate for requests for new images, with a consequent reduction in the statistical multiplexing gain.

It is evident that if the user declares the average new image request rate to be kM/d_2, with $0<k<1$, the network provider could use this information in the CAC policy to optimize the use of network resources. This example shows that the parameters PCR and SCR alone do not ensure efficient control of resources. If, however, the long-term average (ACR) is added, the CAC function can count on more complete information about the statistical behaviour of the source and can thus optimize the allocation control strategies.

The SCR and the ACR are therefore a pair of traffic descriptors which allow the behaviour of the source to be characterized better and in an unambiguous way. They are, in fact, understandable by the user, who knows that not only the ACR but also the limit imposed by SCR has to be respected; at the same time they are of significant use in resource allocation.

At this point there remains the open issue of implementing a policing mechanism which will be efficient in the simultaneous control of SCR and ACR.

3.1 Fuzzy policer for the combined control of SCR and ACR

Let us see whether the FP we propose is capable of policing the two traffic descriptors, SCR and ACR.

First of all, let us analyze whether there exist one or more FP parameters which will allow control of SCR. It is clear that one of these could be N_{i_max}, that is, the upper bound value of the threshold. It represents, in fact, the maximum credit a source can subsequently spend.

Table 1 σ_{SCR} values enforced by the FP for different pairs (N_{i_max}, T)

	$T=2\,T_F$	$T=3\,T_F$	$T=5\,T_F$	$T=10\,T_F$
$N_{i\_max}=2N$	1.90	1.78	1.57	1.35
$N_{i\_max}=3N$	2.33	2.09	1.86	1.59
$N_{i\_max}=4N$	2.41	2.13	1.91	1.63
$N_{i\_max}=5N$	2.43	2.16	1.94	1.65

By simulation we evaluated the FP control, with various N_{i_max} values, of packetized voice sources which do not violate the ACR but have heavy traffic activity for periods of varying length. Table 1 gives, for some values of N_{i_max}, the SCR values normalized with respect to λ_n (σ_{SCR}). The numerical results show that with a fixed N_{i_max} value the control action is more or less restrictive in terms of the maximum traffic intensity allowed in T. It is important to note that, with the same N_{i_max} value, the value of SCR in T, decreases as T increases, as is to be desired.

It is natural to wonder whether the FP, since it has to combine the value assigned to the parameter N_{i_max} with control of the SCR negotiated, is still capable of controlling the ACR, and if so, with what level of performance.

The answer can be found by analyzing the effect of N_{i_max} on policing the ACR. High N_{i_max} values determine a degradation in the dynamic response of the system since, if the source changes its behaviour and starts violating, a long transition period will be required before the violation is detected. Low N_{i_max} values, on the other hand, limit the maximum fluctuation the source traffic is allowed, with the risk of raising the false alarm probability. This risk, however, disappears with combined control of the ACR and SCR, because when the source presents long periods of intense traffic the violations detected, as explained previously, are no longer to be considered as false alarms but as violations of the negotiated SCR value.

From the simulation analysis performed it emerged that even high SCR values, close to PCR, are controllable with a much lower N_{i_max} than the value of 9N used for policing of ACR alone, as shown in Sect. 3.2.

This is to the advantage of responsiveness because it improves as N_{i_max} decreases. Considering that with N_{i_max} =9N the responsiveness of the FP is much higher than that of the LB, it would be interesting to calculate the minimum time interval, T_{min}, during which the source is allowed an SCR value equal to PCR. Let us suppose that N_i in the i-th window reaches the value of N_{i_max} =9N. In the subsequent windows the source can transmit at the peak rate as long as N_i remains greater than the maximum number of cells the source can generate in T_F (T_F/t_c, which is about 3N). Considering that in each window T_F N_i can be decreased at most by N/4 (see the Universe of Discourse for the ΔN_{i+1} output variable), the resulting T_{min} is $24 \cdot T_F$; this is equivalent to 72 secs! This example shows that the range of SCR values that can be controlled with $N_{i\_max} \leq 9N$, i.e. without the responsiveness undergoing any penalization, is extremely wide.

From this analysis it emerges that when fuzzy logic is applied to policing it allows the often conflicting requirements linked to combined control of the two traffic descriptors to be reconciled in a single framework which cannot be achieved by using conventional mechanisms. If we consider the LB as provided for by the ITU, for example, when sized to control the SCR it is not suitable for control of the ACR. With a LB of this size, in fact, a source could legitimately transmit at the SCR value for the whole connection. A possibility would be to use two parallel LBs [(Fratta, 1992)], one sized for control of the SCR and the other for control of the long-term ACR. The Double Leaky Buckets operation is an OR operation: a cell is detected as excessive whenever one of the buckets is in overflow. In such a way, although control of the SCR is guaranteed, the same cannot be said for the long-term average. We have, in fact, demonstrated that the LB fails as the selectivity of the mechanism is in conflict with its responsiveness.

4 CONCLUSIVE REMARKS

In this paper we have presented an analysis to evaluate the behaviour of a policing mechanism based on fuzzy logic. It is a window control mechanism in which the number of cells that can be accepted per window is dynamically updated in accordance with the degree of compliance of the source with the negotiated parameter.

The parameters describing both the source behaviour and the policing control actions are expressed by linguistic variables and fuzzy sets. The control strategy is described through a set of fuzzy inferences which emulate the knowledge base that is typical of human expertise.

The behaviour of the FP has been evaluated in terms of selectivity and dynamic response when it is used to control the ACR of a real packetized voice source. The results obtained show an excellent selectivity, close to that of an ideal policer, and a responsiveness, assessed by the combined measures of reaction time and rise time, which is decidedly better than that of the Leaky Bucket.

An analysis has also been made to assess the FP's capacity for simultaneous control of two traffic descriptors, the ACR and the SCR. As simulation tests have shown, the FP is capable of simultaneously controlling the SCR, simply acting on the value established for N_{i_max}, and the ACR negotiated. It is important to point out that this is achieved at a zero cost in terms of performance. Modifying the value assumed by the parameter N_{i_max} is not in conflict with the other requirements of the mechanism. The mechanism's responsiveness in controlling the ACR is not, in fact, penalized; indeed it is enhanced as the maximum credit a source can be assigned is reduced.

In conclusion, when fuzzy logic is applied to policing it allows the often conflicting requirements linked to combined control of the two traffic descriptors to be reconciled in a single framework which cannot be achieved by using conventional mechanisms.

5 REFERENCES

Andrade, J. (1994) ATM source traffic descriptor based on the peak, mean and second moment of the cell rate, *Proc. of the 14th ITC*, Antibes, June 1994.

ATM FORUM, (1995) Traffic management specification, version 4, April 1995.

Berger, A. and Eckberg, A. (1991) A B-ISDN/ATM traffic descriptor, and its use in traffic and congestion controls, *Proc. GLOBECOM '91*, Phoenix, December 1991.

Berger, A. (1994) Desiderable properties of traffic descriptors for ATM connections in a broadband ISDN, *Proc. of the 14th ITC*, Antibes, June 1994.

Buttò, M. Cavallero, E. and Tonietti, A. (1991) Effectiveness of the leaky bucket policing mechanism in ATM networks, *IEEE Journal on Selected Areas in Communications*, Vol. 9, No. 3, April 1991.

Catania, V. Ficili, G. Palazzo, S. and Panno, D. (1995) A fuzzy expert system for usage parameter control in ATM networks, *Proc. GLOBECOM '95*, Singapore, November 1995.

Dittmann, L. Jacobsen, S.B. and Moth, K. (1991) Flow enforcement algorithms for ATM networks, *IEEE Journal on Selected Areas in Communications*, Vol. 9, No. 3, pp. 343-350, April 1991.

Fratta, L. Musumeci, L. Gallassi, G. and Verri. L. (1992) Congestion control strategies in ATM networks, *European Transaction on Telecommunications*, Vol. 3, No. 2, pp. 183-193, Mar.-Apr. 1992.

Guillemin, F. Boyer, P. Dupuis, A. and Romoeuf, L. (1992) Peak rate enforcement in ATM networks, *Proc. INFOCOM '92*, Firenze, May 1992.

Habib, I.W. and Saadawi, T.N. (1991) Controlling flow and avoiding congestion in broadband networks, *IEEE Communications Magazine*, Vol. 29, No. 10, October 1991.

Hong, D. and Suda, T. (1991) Congestion control and prevention in ATM networks, *IEEE Network Magazine*, Vol. 5, No. 4, pp. 10-16, July 1991.

ITU - TS (1995) Recommendation I.371, Paris, March 1995.

Monteiro, J. Gerla, M. and Fratta, L. (1990) Leaky bucket input rate control in ATM networks, *Proc. ICCC 90*, New Delhi, Oct. 1990.

Munakata, T. and Jani, Y. (1994) Fuzzy systems: an overview, *Communications of the ACM*, Vol. 37, No. 3 pp. 69-76, March 1994.

Onvural, R.O. (1994) Asynchronous Transfer Mode networks: performance issues, *Artech House* 1994.

Rathgeb, E. (1991) Modeling and performance comparison of policing mechanisms for ATM networks, *IEEE Journal on Selected Areas in Communications*, Vol. 9, No. 3, pp. 325-334, April 1991.

Turner, J.S. (1986) New directions in communications (or which way to the information age?), *IEEE Communications Magazine*, Vol. 24, No. 10, pp. 8-15, Oct. 1986.

Zadeh, L.A. (1965) Fuzzy sets, *Inform. Contr.*, Vol. 8, pp. 338-353, 1965.

Zadeh, L.A. (1994) Fuzzy logic, neural networks, and soft computing,. *Communications of the ACM*, Vol. 37, No. 3 pp. 69-76, March 1994.

19

Virtual Network Concept and Its Applications for Resource Management in ATM Based Networks *

Zbigniew Dziong, Yijun Xiong and Lorne G. Mason
INRS-Telecommunications
16, place du Commerce, Verdun, Quebec H3E 1H6, Canada
e-mail: dziong@inrs-telecom.uquebec.ca

Abstract

In the paper we describe a framework for resource management and traffic control which is based on the virtual network concept. In this context the virtual networks are used as a tool for customization of networks management functions and for virtual separation of network resources. We identify three main categories of virtual network applications (service, user and management oriented virtual networks). We discuss generic and application oriented problems, which have to be solved to take full advantage of the proposed framework. We also study the relation between the virtual network concept and the virtual path concept in the context of bandwidth management. The study indicates that using virtual paths for bandwidth management involves some inherent contradictions.

Keywords

ATM networks, resource allocation, virtual networks, virtual paths.

1 INTRODUCTION

The integration of all services into one uniform transport layer is seen as a major advantage of the ATM standard. Nevertheless, this integration also creates several new problems. In particular a broad range of services, traffic characteristics, time scales and performance constraints, which are integrated into one transport system, causes the *resource management and traffic control* issues to become very complex and difficult. One can compare the situation to traffic control on a highway where the supersonic jet traffic is integrated with the personal car traffic. That is why, in many cases, the resource management and traffic control is trying to re-introduce some kind of separation in order to make the problem manageable. In this paper we argue that the virtual network concept can serve ideally for this purpose since it can provide a vehicle for two types of separation. The first is *separation of management functions* in order to allow customization to particular needs of some services and user groups. The second is *virtual separation of resources* in order to simplify the resource management functions and provide grade of service (GoS - connection layer performance metrics) guarantees for some services and user groups. The aim of this work is to propose a coherent framework for resource management and traffic control which takes full advantage of the virtual network concept.

*The research was supported by a grant from the Canadian Institute for Telecommunications Research under the NCE program of the Government of Canada.

The notion of virtual networks is not new and was used in several papers (e.g. Mason (1987,1990), Walters (1992), Atkinson (1992), Dziong (1993,1994), Wernik (1993), Fotedax (1995), Dupuy (1995)) for particular applications such as virtual private networks or virtual networks associated with different qualities of service. In this paper we generalize the virtual network concept by showing that a common generic definition of virtual networks can be applied to all potential applications. There are five main objectives of this paper. First, to introduce a precise and coherent definition of the virtual network notion in order to avoid any vagueness and misinterpretations. Second, we identify the possible applications of the virtual network concept which would improve network management and utilization. The third objective is to specify the generic problems which are common to all virtual network applications. The fourth objective is to identify problems specific to particular applications. Finally we discuss what should be the relation between the virtual network concept and the virtual path concept defined by the standard bodies.

We start from the generic virtual network definition which is based on a virtual network link concept (Section 2). Then we discuss the potential applications of the virtual network (VN) concept which can be divided into three categories: service oriented VN, user oriented VN and management oriented VN (Section 3). The relation between virtual networks and virtual paths is described in Section 4. In particular we argue that the virtual path connections should not be used for bandwidth management purpose but rather for routing and switching simplification. The general architecture of the network bandwidth management system based on the virtual network concept as well as interaction of the virtual network with other traffic layers (cell, connection and physical network layers) are presented in Section 5. Finally in Section 6 we identify the generic and application oriented problems, which have to be solved to take full advantage of the proposed framework.

2 VN DEFINITION

A virtual network is defined by a set of network nodes and a set of virtual network links connecting the nodes. The virtual network is referred to by a virtual network identifier, VNI. The virtual network link (VNL) defines a path (consisting of one or more physical links) between two VN nodes and is referred to by a virtual network link identifier, VNLI. Examples of different VN topologies are given in Figure 1. Connections (VCCs or VPCs) associated with a particular VN can be established only on VNLs belonging to the VN.

Several virtual networks can co-exist in a physical network. They can constitute independent entities but in some cases a virtual network can be nested in another virtual network. In the latter case a VCC can belong to two or more virtual networks at the same time. In the example of Figure 1, VN3 can be nested in VN2. This means that the connections associated with VN3 belong also to VN2. At the same time there can be some connections associated with VN2 which do not belong to VN3 although they are carried on the same sequence of physical links as connections from VN3.

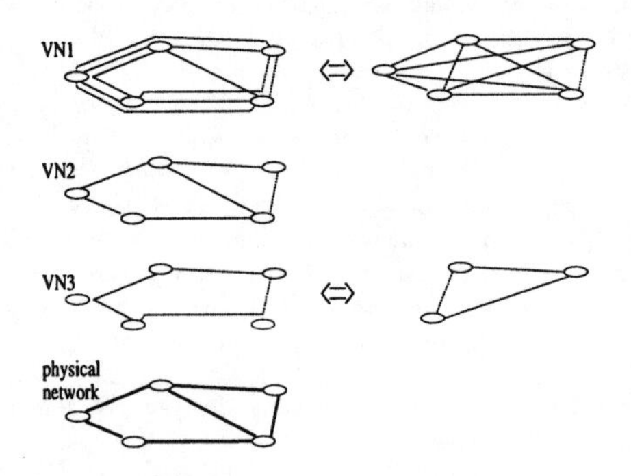

Figure 1 Examples of virtual network topology.

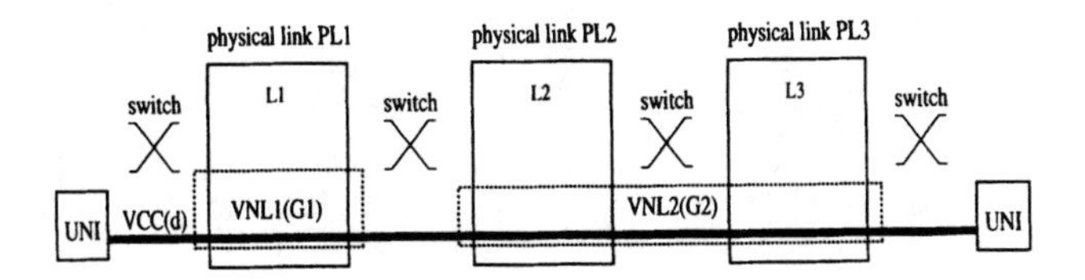

Figure 2 Allocation of resources to VN.

A set of resources can be allocated to the virtual network. In particular this set can include bandwidth and a set of resource management objects. It is important to emphasize that in this paper the notion of bandwidth is used in the sense of a qualified bandwidth. The qualified bandwidth includes the scheduling algorithm and buffer allocation and is characterized by quality if service (QoS - cell layer performance metrics) constraints for given traffic characteristics of connections using this bandwidth. In the following we concentrate on the application of the virtual network concept for bandwidth management purposes.

Figure 2 illustrates the logical bandwidth allocation to virtual network links along a particular VC connection which requires bandwidth d (denoted as VCC(d)). The connection is established on two VNLs: VNL1 and VNL2. Each of these VNLs is allocated a certain bandwidth (G1 and G2 respectively) from the physical link bandwidths, L1, L2, L3 (G1 from L1 and G2 from L2 and L3). The connection can be established if bandwidth d can be reserved on each of the VNLs.

3 POSSIBLE VN APPLICATIONS

In general the VN applications can be divided into three categories: service, user and management oriented.

Service oriented virtual networks are created to separate management functions specialized for different services (e.g. real-time vs data services) and/or to simplify the QoS management (each QoS class is served by a separate virtual network). Allocation of bandwidth to service oriented VNs aims at providing sufficient GoS and fairness for different services. Moreover, bandwidth allocation to QoS virtual networks can increase bandwidth utilization and simplify bandwidth allocation to connections as will be discussed in Section 6

User oriented virtual networks are created for some group of users who have specific requirements (e.g. guaranteed throughput, customized control algorithms, bandwidth management under the user control, increased security and reliability, "group" tariff, etc). The two most likely applications are private networks and multi-point connections. Note that in most cases the set of VN nodes will include only a subset of nodes to which the users are connected.

Management oriented virtual networks are created to facilitate some of the management functions (not associated with particular service or user group). The first application is connected with fault management and is called back-up virtual network. The bandwidth allocated to back-up VN should provide that in case of failure of network components (e.g. a link or node), all (or a given fraction of) connections affected by a failure can be restored in the back-up VN. The second application (henceforth referred to as CAC VN) is aimed at simplification of bandwidth reservation procedure during the connection setup. In particular, if all connections are routed via VNLs which connect directly origin and destination nodes, the connection admission procedure has to ask only the VNL bandwidth manager at the connection origin node for the required bandwidth (no need for bandwidth reservation in the transit nodes).

In the following we discuss interrelation between different types of VNs which is also illustrated in Figure 3:

- *Service oriented VN vs. management oriented VN:* In general, management oriented networks can divide bandwidth allocated to the backbone QoS VNL into several VNLs. Backbone QoS VN is defined as a VN where each VNL corresponds to one physical link.
  - The back-up networks are created for each QoS virtual network. It means that the bandwidth allocated to the backbone QoS VN is divided between primary QoS VN and back-up VN.

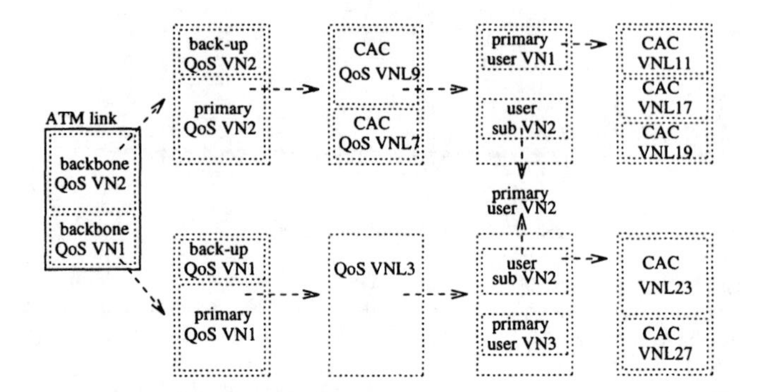

Figure 3 Interrelations between virtual networks.

- Application of CAC virtual network may cause the bandwidth allocated to the primary QoS VNL to be divided among several CAC VNLs (in general this case is not likely).

- *Service oriented VN vs. user oriented VN:* In general user oriented networks are nested in QoS virtual networks. In other words the user oriented VNLs are treated as connections within QoS VNLs. If a user oriented network requires different QoS connection classes it can be realized as:
  - one VN within the most stringent QoS virtual network.
  - a group of virtual subnetworks each of them nested in a corresponding QoS virtual network.
- *User oriented VN vs. management oriented VN:*
  - Since the user oriented VN is treated by its service oriented VN as a connection, there is no separate back-up network for user oriented VN.
  - Application of CAC virtual network may cause the bandwidth allocated to the primary user oriented VNL to be divided among several CAC VNLs

4 RELATION BETWEEN VNL AND VPC

There is a substantial literature on using virtual path connections for bandwidth management purposes by allocating bandwidth to a VPC. The main argument used by proponents of this approach is that the connection set-up is simplified since the first node of the VPC can reserve bandwidth for the new connection along the entire path. We argue that in general this approach has several drawbacks which significantly overweight the aforementioned advantage. To illustrate the problem we compare possible relations between VNL, VPC and routing paths (RP). To facilitate the comparison a network example is depicted in Fig.4. In this example we assume three QoS VNs. In the following we outline the main features of the three possible applications of VPC:

- *VPC=RP, bandwidth management by VNL:*
  - High bandwidth utilization (VNLs are optimized to provide this feature).
  - Simple bandwidth management on the virtual network layer (small number of variables).
  - Admission of a new VC connection does not require any changes in the routing tables of the transit nodes.
  - Admission of a new VC connection requires bandwidth reservation in the VNL's bandwidth managers located in the transit nodes of the path.
- *VPC=RP, bandwidth management by VPC:* - Low bandwidth utilization due to division of the link bandwidth into a large number of VPCs using the link (see Figure 4).
  - Complex bandwidth management (large number of interdependent variables).
  - Admission of a new VC connection does not require any changes in the routing tables of the transit nodes.

- Admission of a new VC connection does not require requires bandwidth reservation in the VNL's bandwidth managers located in the transit nodes of the path.

● *VPC=VNL, bandwidth management by VPC=VNL:*
- High bandwidth utilization (VNLs are optimized to provide this feature).
- Simple bandwidth management (small number of variables).
- Admission of a new VC connection requires changes in the routing tables of the transient nodes between concatenated node-to-node VPCs (VC switches required).
- Admission of a new VC connection requires bandwidth reservation in the VNL's bandwidth managers located in the transit nodes of the path.

Routing paths (RP) on link 2-5

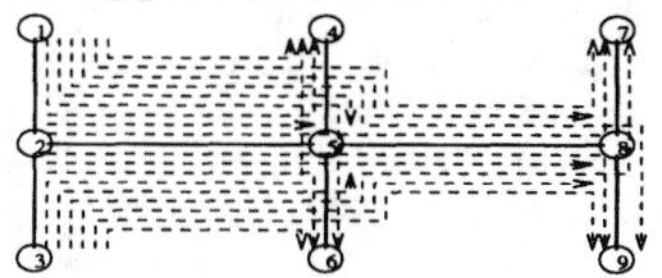

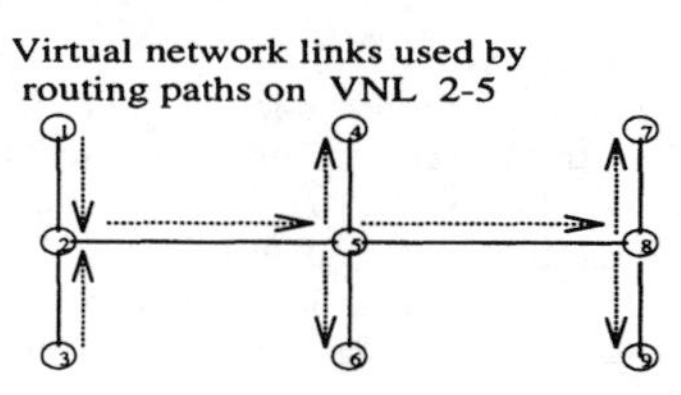

Possible applications of VPC:

VPC=RP, BM by VNL
VPC=RP, BM by VPC
VPC=VNL, BM by VNL

Example:
three QoS categories,
150 Mbps links,
symmetrical traffic matrix,

Result:
VPC=RP, BM by VPC
- 2.7 Mbps per VPC

VPC=RP, BM by VNL
- 50.0 Mbps per VNL

Figure 4 Analysis of possible VPC applications within the VN framework.

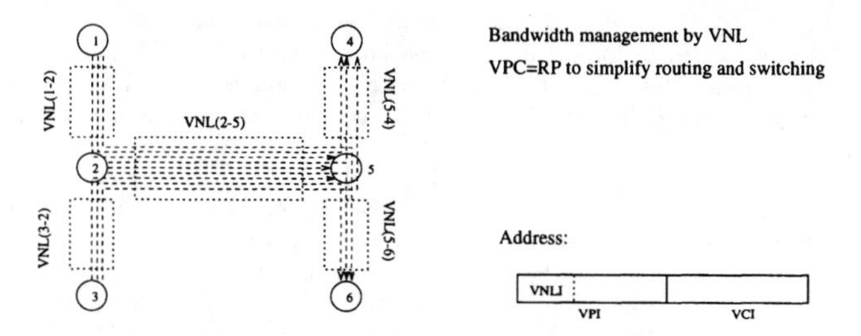

Figure 5 Decomposition of VN and VP functions.

The comparison of the main features shows that using VPCs for bandwidth management involves some contradictions. First, when used for call set-up simplification, the bandwidth utilization drops significantly due to the lack of statistical multiplexing between the VPCs on the connection level. Second, when used to optimize bandwidth utilization, the connection set-up becomes complex since the routing tables in the nodes between VPCs constituting the routing path have to be updated. To avoid these inherent contradictions we propose separation of bandwidth management function from VPC implementation. In this case the virtual paths should be used to simplify routing and switching (VPC=RP) only while the virtual networks should be used to optimize resource utilization. Observe that the processing and signaling cost of reserving bandwidth at VNL bandwidth managers for a new connection can become negligible if the resource management (RM) cells are used for this purpose on the pre-established routing paths (Dziong, 1995). This follows from the fact that the signaling traffic generated by ABT or ABR resource management cells will be significantly higher than the one generated by connection set-up RM cells. Other problems with application of VPC to bandwidth management are also indicated in (Dziong, 1991).

It should be noted that for particular network cost function assumptions and/or technical constraints, the optimization procedure can give VNL=VPC=RP for all VNLs.

Concerning implementation of the VNL identifier it can be defined by a part of VPI field with the highest weight as shown in Figure 4 (alternatively VNLI can be a general function of VPI or it may be unrelated directly to VPI).

5 BANDWIDTH MANAGEMENT ARCHITECTURE BASED ON VN CONCEPT.

The bandwidth management algorithms are distributed over three layers: connection layer, virtual network layer and physical layer. Interaction between these layers is discussed in Section 5.1. Besides distribution over different layers, these algorithms can be implemented in different locations and in different ways (e.g. centralized vs. distributed). These additional dimensions are illustrated in Figure 6 where a general architecture of the bandwidth management objects is presented. In the following the main functions of these objects are described:

- **Connection admission control and routing manager, CAC&RM**
  The main function of this object is to provide UNI-CAC objects with CAC and routing information which would facilitate establishment of the connections within the VN. In particular CAC&RM is responsible for evaluation of equivalent bandwidth functions used to map the connection's declared parameters into the bandwidth which should be reserved for the connections. Moreover CAC&RM should design a set of alternative paths for each OD pair (these paths can be implemented in the form of VPCs). Another important CAC&RM function is to provide the VN bandwidth manager with connection performance metrics which can be used to adapt the VN topology and bandwidth allocation. CAC&RM can be implemented in a centralized or distributed (over UNI-CAC and VNL-BM) fashion.
- **User-network interface connection admission control, UNI-CAC**
  The user-network interface CAC object is responsible for bandwidth allocation to connections originating at this interface. Based on the routing recommendation from CAC&RM, the UNI-CAC asks the bandwidth managers of VNLs constituting the chosen path to reserve bandwidth d required by the connection. If there is not enough bandwidth on this path, another path can be tried or the connection can be rejected.
- **Virtual network bandwidth manager, VN-BM**
  The main function of the VN-BM object is to update the VN topology and bandwidth allocation based on the performance measures from the connection layer. When there is a need to increase the bandwidth allocation, the VN-BM asks the network bandwidth manager to reserve this bandwidth. If there is not enough free bandwidth, the network bandwidth manager can realize only a part of the demand or reject the demand. The virtual network manager can be implemented in a centralized or distributed (over VNL-BM) fashion.
- **Virtual network link bandwidth manager, VNL-BM**
  The main function of the VNL-BM is link connection admission control. In particular VNL-BM decides whether a connection demand (VCC or VPC or another VNL) should be granted the required bandwidth. Several fixed or state dependent policies can be considered (e.g. complete sharing, coordinate convex, trunk reservation, dynamic trunk reservation, shadow price). Optionally, the VNL-BM can ask directly the link bandwidth managers to increase the bandwidth allocated to the VNL in the case of a high connection rejection rate.
- **Physical network bandwidth manager, PN-BM**
  There are two main functions of the network BM object. The first is to update the physical network topology and bandwidth allocation based on the performance measures from the VN layer. The second is to allocate bandwidth to virtual networks. The second function has to take into account fairness criteria when there is not enough resources to satisfy all VN demands.
- **Physical link bandwidth manager, PL-BM**
  The PL-BM can be seen as a part of the PN-BM. Its function is to control admission of new VNs and to update bandwidth allocation of the existing VNs. Several different policies can be applied.

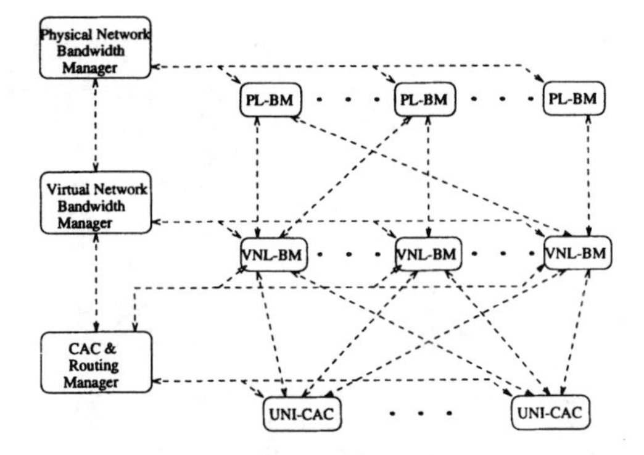

Figure 6 Bandwidth management architecture based on VN concept.

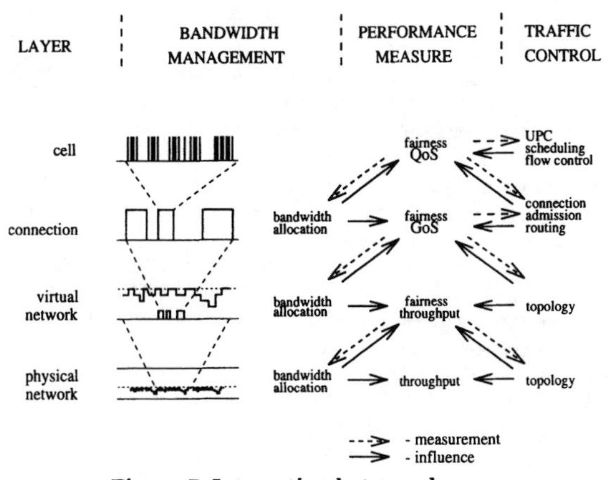

Figure 7 Interaction between layers.

5.1 Interaction between layers

The interaction between different layers is illustrated in Figure 7. The critical part of this interaction is measurement which allows one to adapt resource management algorithms to varying traffic conditions. In the following we specify the main measurement functions which are associated with the bandwidth management on the virtual network layer (directly or indirectly).

- *Cell layer measurements for connection layer:* The connection layer can use measurements on the cell layer to adapt bandwidth allocation to connections. This adaptation can have two objectives. The first aims at adapting bandwidth allocation to existing connections due to the source declaration errors and nonstationarity. The second objective is to adjust the bandwidth allocation function if some systematic allocation error is observed.
- *Connection layer measurements for virtual network and connection layers:* In general the values measured on the connection layer correspond to the GoS metrics (e.g. connection rejection rates) or to connection flow distribution in the virtual network. These values can be used by CAC and routing algorithms (connection layer) to improve bandwidth utilization and/or to provide fair access for all

users (according to fairness criteria). The virtual network layer can use the measured values to adapt the VN topology and bandwidth allocation.

- *Virtual network layer measurements for physical network layer:* The virtual network layer measurement metrics provide statistical information about the VN bandwidth allocation demands. The physical layer network uses these metrics to adapt the physical network topology and bandwidth allocation. These measurements can also serve to provide fair access to resources for all VNs (according to fairness criteria).

6 GENERIC AND APPLICATION ORIENTED PROBLEMS

We start by describing generic problems which are common to all basic applications of the virtual network concept. We identify three main generic problems associated with virtual network set up, virtual network update and back-up virtual network. Two main application oriented problems are described in Section 6.1 and Section 6.2.

The issue of virtual network set up is illustrated in Figure 8. The key element of the VN set-up procedure is an optimization procedure which designs the VN topology, allocation of bandwidth to VNLs and routing algorithm parameters. The optimization procedure is fed by the offered traffic matrix, GoS constraints, bandwidth cost functions, route set-up cost, bandwidth reservation cost during the set-up procedure and routing policy. Note that this formulation also includes design of CAC VN since the cost of bandwidth reservation and route set-up is taken into account. The result of the optimization procedure constitutes a VN demand for resource allocation which is considered by higher level bandwidth manager (physical network or virtual network in case of nested virtual networks). The demand can be accepted if there are enough free resources or rejected in case of resource shortage. In some cases of resource shortage the demand can be realized partially. In this case the optimization procedure can be applied one more time to take into account the resource limitations.

The issue of virtual network update is illustrated in Figure 9. The objective of the VN update is to correct the original VN set-up design and to adapt the VN design to the changes in traffic profiles. The structure of the VN update procedure is similar to the set-up procedure. The difference is that the algorithm is fed by measurements which are associated with offered traffic matrix, GoS metrics and bandwidth utilization metrics. This information is used to update the VN topology, allocation of bandwidth to VNLs and routing algorithm parameters. As in the case of VN set-up demand, the update demand can be accepted, rejected or realized partially.

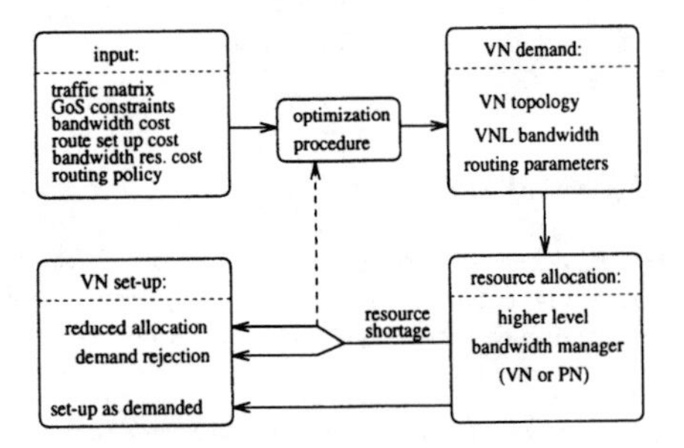

Figure 8 Virtual network set-up.

The back-up virtual networks are created to enable all (or a fraction) of the carried connections to be restored in case of failure of network components (e.g. a link of node). The restoration can be implemented on a link, fragment or path basis, as depicted in Figure 10, where fragment is a portion of the path. The path restoration seems to be most attractive in the ATM based networks. The design of the back-up VN has to provide the topology of the back-up virtual paths and bandwidth allocation to back-up VNLs. The backup virtual network usually has the same network topology as the physical network. Note that the

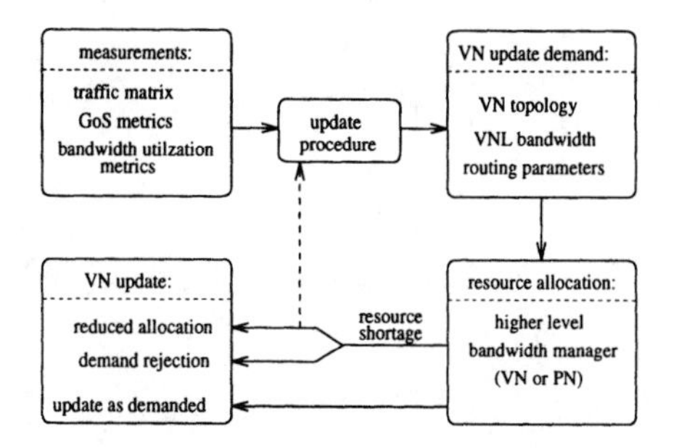

Figure 9 Virtual network update.

bandwidth allocated to the back-up VNL is not a sum of bandwidth required by all back-up VPs using this VNL since only a part of these VPs will be activated by failure of one link or node. That is why the back-up VN design is different from the design of the primary network. To illustrate this issue let us consider the case where two back-up VPs are asking the same back-up VNL bandwidth manager for equal bandwidth allocation increase. It may happen that the demand of one back-up VP is rejected while the other is accepted. To provide network reliability cost-effectively, an important objective of the back-up VN design is to minimize the bandwidth allocation to back-up VNLs while satisfying the restoration requirements.

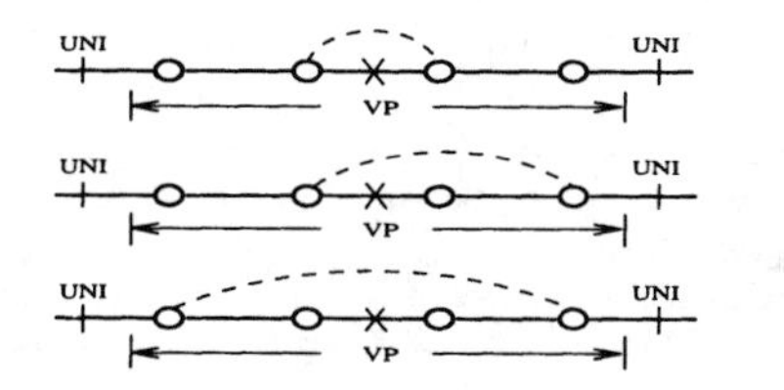

Figure 10 Link, fragment and path restorations.

6.1 Resource allocation in service oriented VN

A major application of service oriented virtual networks is to provide specialized treatment to different traffic classes. In particular (Classes A, B, C and D) as identified in the B-ISDN reference model, or (CBR, VBR, ABR, UBR) by ATM-Forum, as well as different QoS levels within those classes provide a natural classification for defining service oriented virtual networks. A fundamental question that arises is the manner in which these different virtual networks, which support different traffic and QoS classes, are managed as a function of the scheduling and flow control procedures at the cell level. In particular one should explore the bandwidth management options based on priority queuing as well as GPS-like scheduling (fair-queuing). While priority queuing is simpler to implement than "fair queuing" the later has certain advantages, among them flexibility, which is very desirable for a future-proof network.

Consider an example with two distinct QoS classes in the network. Each of these classes can be associated with the QoS virtual network. To provide high bandwidth utilization a buffer can be allocated to each of these virtual networks. Assume that the higher VN index, the more stringent the QoS requirements. The service priorities are implemented by a scheduler which can be realized as non-preemptive multi-priority system or a fair queuing system.

Three basic approaches to bandwidth allocation to virtual networks are presented in Figure 11. To illustrate the main features of these approaches, the figure includes the exact admissible region (continuous line) and linear admissible regions corresponding to each of the approaches (dotted lines). These admis-

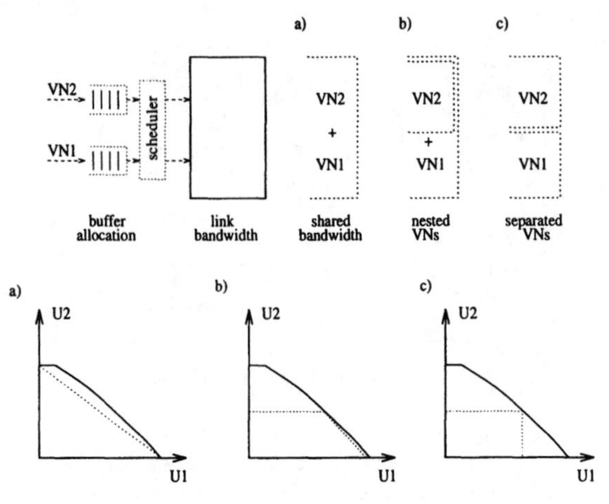

Figure 11 Resource allocation in service oriented VN.

sible regions are evaluated for a non-preemptive multi-priority system and statistical QoS constraints (Dziong, 1993). Nevertheless similar features can be expected for other scheduling algorithms and deterministic QoS constraints. In the following we indicate the main features of the three resource allocation schemes:

- All virtual networks share the same pool of link bandwidth Figure 11(a):
  - full statistical multiplexing between virtual networks (connection layer),
  - linear admissible regions are less efficient,
  - requires additional tools for GoS fairness between VNs.
- Nested virtual networks Figure 11(b) (the bandwidth allocated to a particular priority can be used by lower priority connections):
  - limited statistical multiplexing between virtual networks (connection layer),
  - linear admissible regions are efficient in the inclined region,
  - limited GoS fairness between VNs (tools to protect high priority traffic access against low priority traffic are required),
  - a scheme for fast bandwidth allocation to low priority VN is required for controllable traffic services (flow control problem).
- Each virtual network is allocated separate bandwidth Figure 11(c):
  - no statistical multiplexing between virtual networks (connection layer),
  - linear admissible regions are efficient only for the cases where the ratio of service traffic levels is similar to the one in the designed operating point,
  - GoS fairness between VNs is provided.

The choice of a particular resource allocation scheme depends on the services and the design objectives. In the following we identify the main problem categories which should be resolved to take advantage of the service oriented VNs:

- Evaluation of connection admissible regions as a function of bandwidth allocation to virtual networks and scheduling algorithm.
  - Statistical QoS guarantees (e.g. Dziong, 1993).
  - Deterministic QoS guarantees (fair queuing e.g. Noiseux, 1995).
- Fast bandwidth allocation to controllable traffic VNs (flow control).
  - Rate-based schemes.
  - Credit-based schemes.

- Providing GoS fairness for shared bandwidth and nested VN cases (fair CAC policies).
- Bandwidth allocation to virtual networks when demands exceed link capacity (fairness issue).

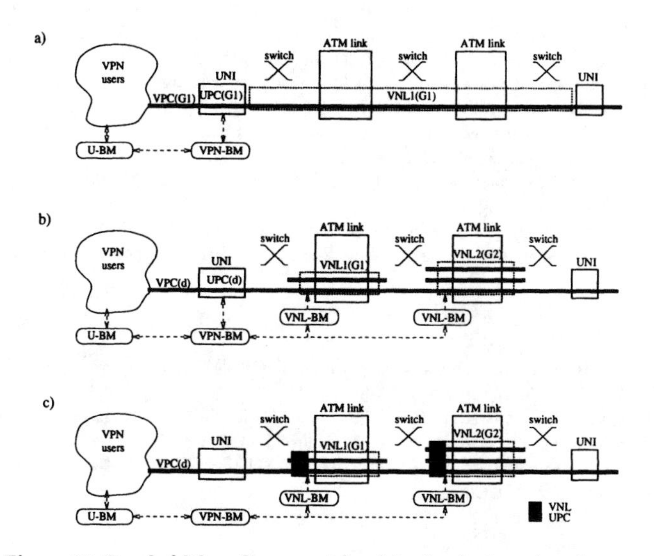

Figure 12 Bandwidth enforcement in virtual private networks.

6.2 Bandwidth enforcement in virtual private networks

In the case of service oriented VNs and user oriented VNs created for multi-point connections, the enforcement of bandwidth allocation to VCCs is under the network manager responsibility and is realized by UPC mechanism located at UNI. Thus the enforcement of bandwidth allocation to VNLs can be done logically by the connection admission procedure at VNL-BM.

Concerning virtual private networks, bandwidth allocation to VCCs and connection admission procedure can be under the user responsibility. In this case the virtual private network bandwidth manager has to provide enforcement of bandwidth allocation to VNLs independently of the user actions. Below we indicate three basic possibilities:

- Enforcement at UNI, VPC=RP=VNL, Figure 12(a):
  Here the virtual network links are realized as end-to-end VPC. Thus the bandwidth allocated to VNL can be enforced by a UPC algorithm applied to VPC at UNI.
- Enforcement at UNI, VPC=RP≠VNL, Figure 12(b):
  This option requires that the user bandwidth manager (U-BM) asks the virtual network bandwidth manager for the bandwidth allocation to the end-to-end VPCs. This bandwidth allocation can be updated on a call-by-call basis and can take into account statistical multiplexing (cell layer) between the VPCs using the same VNL. For each update the VN-BM verifies whether the sum of bandwidth required by VPCs does not exceed the VNL capacities. If this condition is fulfilled the bandwidth required by VPC is enforced by a UPC algorithm at the UNI.
- Enforcement at switch output ports, VPC=RP≠VNL, Figure 12(c).
  Bandwidth allocated to VNL is enforced by means of UPC mechanisms (e.g. fair queuing) installed in the VN node switch output port originating VNL.

7 CONCLUSIONS

In the paper we described a framework for resource management and traffic control which uses virtual networks as a tool for customization of network management functions and for virtual separation of network resources. The proposed generic definition of the virtual network is based on the virtual network link concept. We have shown that this definition fits many potential applications which can be divided into service, user and management oriented virtual network categories.

In the context of the virtual network application for bandwidth management there is an evident overlap with the concept of bandwidth allocation to virtual paths investigated in many publications. The presented study of the relation between the two concepts indicates that using virtual paths for bandwidth management involves some inherent contradictions. To avoid these drawbacks we propose separation of the bandwidth management function from VPC implementation. In this case the virtual paths are used to simplify routing and switching only.

In the last part of the paper we have identified generic and application oriented problems, which need to be solved to take full advantage of the proposed framework. These problems are currently under study.

REFERENCES

Anderson J., Doshi B., Dravida S. and Harshavardhana P. (1994) "Fast Restoration of ATM Networks", IEEE J. Select. Areas. Commun., Vol.12, No.1, pp. 128-138.

Atkinson D.J., Anido G.J. and Bradlow H.S. (1992) "B-ISDN Traffic Characterization and Management", 7th Australian Teletraffic Research Seminar, Mannum, South Australia.

Crocetti P., Fratta L., Gallassi G. and Gerla M. (1994) "ATM Virtual Private Networks: alternatives and performance comparisons" ICC'94.

Dupuy F., Nilsson G. and Inoue Y. (1995) "The TINA Consortium: Towards Networking Telecommunications Information Services", ISS'95, Berlin.

Dziong Z., Liao K-Q., Mason L.G. and Tetreault N. (1991) "Bandwidth Management in ATM Networks", Proc. of ITC13, Copenhagen.

Dziong Z., Liao K-Q. and Mason L.G. (1993) "Effective Bandwidth Allocation and Buffer Dimensioning in ATM Based Networks With Priorities", Computer Networks and ISDN-Systems, Vol. 25, 1065-78.

Dziong Z., Montanuy O. and Mason L.G. (1994) "Adaptive Bandwidth Management in ATM Networks ", International Journal of Communication Systems (John Wiley & Sons), VOL.7, 295-306.

Dziong Z., Juda M. and Mason L.G. (1995) "A Framework for Bandwidth Management in ATM Networks - Aggregate Equivalent Bandwidth Estimation Approach", submitted to IEEE/ACM Transactions on Networking.

Fotedax S., Gerla M., Crocetti P. and Fratta L., (1995) "ATM Virtual Private Networks", Communications of the ACM, Vol.38, No 2.

Kawamura R., Sato K-I and Tokizawa I. (1994) "Self-Healing ATM Networks Based on Virtual Path Concept", IEEE J. Select. Areas. Commun., Vol.12, No.1, pp. 120-127.

Letourneau E. and Mason L.G. (1995) "Comparison Study of Credit-Based and Rate-Based ABR Control Scheme" submitted to Workshop on ATM Traffic Management, Paris.

Mason L.G. and Gu X.D. (1986) "Learning Automata Models for Adaptive Flow Control in Packet Switching Networks", in *Adaptive and Learning Systems — Theory and Applications*, K.S. Narendra, New-York:Plenum Press, pp. 213–227.

Mason L.G. (1987) "Virtual Network Services and Architectures", INRS-Telecom. report 87-24.

Mason L.G., Dziong Z., Liao K.-Q. and Tetreault N. (1990) "Control Architectures and Procedures for B-ISDN", Proc. of ITC Specialist Seminar, Morristown, USA.

Murakami K. and Kim H. (1995) "Joint Optimization of Capacity and Flow Assignment for Self-Healing ATM Networks", IEEE ICC'95, pp. 216-220.

Noiseux L. (1995) "Stratégies d'allocation des ressources dans les réseaux ATM", Master's Thesis, INRS-Télécommunications, Université du Québec.

Walters M.S. and Ahmed N. (1992)"Broadband Virtual Private Networks and Their Evolution", ISS'92.

Wernik M., Pretty R. and Smith D. (1993) "Evolution of Broadband Network Services - A North American Perspective" 1993

20

Some Traffic Aspects in Virtual Private Networks over ATM

Fabrice Guillemin and Isabelle Hamchaoui
France Télécom, Centre National d'Etudes des Télécommunications
2 avenue Pierre Marzin, 22307 Lannion, France

Abstract

The emergence of broadband technologies such as ATM gives rise to new developments in the domain of Virtual Private Networks (VPNs). In this paper, a target Broadband VPN (B-VPN) service is proposed, which combines the costs savings of classic VPN services and the powerful features of private corporate networks (service integration, statistical multiplexing, etc.), enhanced by the use of ATM. To reach this B-VPN service, some architectural issues related to the network supporting the B-VPN service are discussed. The traffic implications of architectural choices are then examined and from this discussion, some basic principles for a VPN architecture are drawn. Finally, target architectures are proposed.

Keywords
Virtual Private Network, Statistical multiplexing, Traffic control, Quality of service.

1. INTRODUCTION

Although VPN and corporate network services may appear nowadays as to be in competition, the emergence of broadband technologies such as ATM allows new possibilities to be foreseen for encompassing both types of services. This convergence is further motivated by two growing needs expressed by users. On the one hand, users of voice VPNs based on shared network resources wish for a better level of service integration and flexibility. On the other hand, users of private corporate networks expect to reduce their costs through resource sharing and greater flexibility of their network.

The challenge for B-VPN services is to satisfy the needs of both types of customers, and possibly, to meet new requirements. In particular, while still providing for more classic functions such as voice services and PBX interconnection, B-VPN services should realise LAN interconnection and open the door to new high speed multimedia services (e.g., video).

The basic feature of a VPN service is that it relies on a set of resources offered by a network operator to a customer with multiple sites spread over a wide area. These resources together with access equipment constitute the customer network, which may be compared to a real private network. The difference, however, is that the resources constituting the customer network are taken from a pool of resources shared between various users of the general infrastructure. The

VPN customer has then the possibility to handle freely his end-to-end connections on his customer network, for instance by statistically sharing the bandwidth of his network between the different end-to-end connections.

In Section 2 of this paper, a target Broadband VPN (B-VPN) service intended to meet both voice VPN and private network customer expectations is discussed along with the benefits for a network operator to offer such a service. Recognising that the B-VPN service should be offered via a customer network, referred to as VPN in the following and whose architecture is derived from that of private corporate networks, the different components of a VPN, namely virtual trunks, VPN nodes, and access equipment are introduced in Section 3. The traffic management issues related to some basic architectural choices are analysed in the subsequent section. Some possible VPN architectures consistent with the previous analysis are described in Section 5. The concluding section presents directions for future work.

2. DESIGNING A BROADBAND VPN SERVICE

As mentioned above, the design of a B-VPN service has to take into account the expectations of both voice VPN and private network customers. As a matter of fact, the main requirements of those customers are currently related to four major items. Specifically, a target B-VPN service should offer **user-friendly operation** of customer network and **cost savings**, while offering the same level of **Quality of Service** (QoS) and **security** as in real private networks. Taking into account such considerations, a minimum set of functions may be considered as a definition for a basic B-VPN service.

Global bandwidth allocation

A VPN service requires the prior design of a VPN to support the various end-to-end connections of the VPN customer. Specifically, this consists in allocating a certain amount of network resources, notably a global bandwidth, to a particular VPN customer. This global bandwidth should be managed as dynamically as possible and may be evaluated, in a first step, by taking into account the volume of traffic between customer sites.

Free handling of end-to-end connections

Several end-to-end services should be offered to the customer in the context of the B-VPN service (voice, video, data transmission, etc.), so that all usual applications of a private corporate network could be available. Service integration notably increases the statistical gain, since traffic is no longer partitioned on a service basis. The operation of the corresponding end-to-end connections should remain under the control of the customer, exactly as if the VPN were a real private network. In particular, if the network operator is involved in the operation of end-to-end connections, the response time of the network should be very short, so that the user is unaware of the network operator's intervention.

Statistical multiplexing of end-to-end connections

To optimise the utilisation of the VPN and since the various applications supported by a VPN may have different characteristics in terms of bandwidth as well as QoS, it would be highly desirable to perform statistical multiplexing of end-to-end connections. For instance, the end-to-end connections of a VPN may use different ATM transfer capabilities (I.371, 1995). Moreover, in the context of a VPN over ATM, statistical multiplexing should be applied not only to the end-to-end connections within the VPN (dynamic bandwidth sharing of the global bandwidth) but also between several VPNs (dynamic adjustment of the global bandwidth).

Enhanced management

A VPN service has to provide the user with the usual management features of a private network. Basic functions, such as supervision of end-to-end connections and homogeneous representation of the VPN, should be provided. In particular, the homogeneous representation of the VPN should allow the user to ask for precise modifications of the VPN (bandwidth, QoS, topology, etc.). In addition, this basic set of functions may be enhanced by more sophisticated options. For example, a B-VPN service may provide for private numbering, which is essential once voice is concerned, and Closed User Group (CUG) facilities.

3. HIGH LEVEL ARCHITECTURE ISSUES

To offer the target B-VPN service outlined in the previous section, two possibilities may be envisaged. The first one is based on an overlay network and consists in introducing special equipment dedicated to the VPN service on the top of the ATM infrastructure. This solution may be attractive if a large number of customers subscribe to the VPN service. But it presents the major disadvantage of being a dedicated VPN solution, that is, reusing the VPN overlay network for other services may be difficult.

The second possibility is an integrated solution. The components of the VPN are network elements of the general ATM infrastructure, except perhaps the service multiplexers at network edges. Although attractive, this solution may raise some technical difficulties due to the increase of complexity in network elements (e.g., specific traffic control functions, management facilities, etc.). However, the capabilities of ATM are fully exploited and the costs are consequently reduced. This is why only the second solution will be considered in the following. Note that the architecture of a VPN should actually be derived from that of a private corporate network, since the target features of a B-VPN service are very close to those of a private corporate network (see Figure 1). Such a VPN is composed of virtual trunks, VPN nodes, and the access equipment.

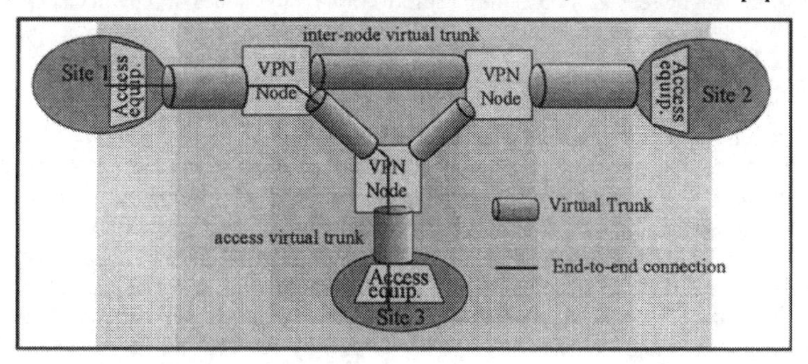

Figure 1. Architecture of a VPN.

Virtual trunk

The virtual trunk is the logical entity used to describe the transport of information between two VPN network elements. Specifically, cells belonging to end-to-end connections, namely Virtual Channel Connections (VCs), are transported over virtual trunks. The bandwidth of a virtual trunk is the amount of bandwidth allocated to the VPN customer between two VPN elements. From a functional point of view, this means that the virtual trunk, similarly to the Virtual Path Connection (VP), is at a higher level than the VC.

In ATM networks, a virtual trunk may be composed of a part of a VP, a VP, or several VPs, as shown in Figure 2. In the first case, only a subset of the VCs within the VP are used for the VPN, other VCs may be used by other VPNs or for other services. However, a VPN relying on VPs seems easier to operate, since the VP is a standard ATM entity commonly handled in current ATM infrastructures (e.g., the European ATM Pilot). As a consequence, only this solution will be examined in the following. The virtual trunk bandwidth is equal in this case to the aggregate bandwidth of the VPs constituting the virtual trunk.

The concept of virtual trunks is introduced only for convenience, notably to associate with a unique entity the resources allocated to the VPN at the access or between two nodes (access virtual trunks and inter-node virtual trunks). At any time, the bandwidth allocated to the virtual trunk should not be exceeded by the aggregate traffic of all customer's end-to-end connections.

Figure 2. Examples of virtual trunks.

VPN nodes

VPN nodes may switch or cross-connect the end-to-end connections supported by the virtual trunks. These nodes are the originating or terminating points of virtual trunks and represent basically the splitting points of VPN traffic. VPN node functions are supported by usual network nodes (VP cross connects, VC switches, VP/VC cross connects, etc.), possibly enhanced with specific traffic control and management functions. Note that, for each VPN, only a limited number of physical nodes will be seen as VPN nodes. Moreover, the VPN nodes of a particular VPN may be different from those of another VPN.

Access equipment

The main function supported by access equipment is to multiplex several connections on the VPN access trunk. In addition, such equipment may host the mechanisms required for the control of individual connections, especially in the case of end-to-end control via private signalling and/or when the end-to-end connections of the VPN are statistically multiplexed. Finally, the access equipment may implement interworking functions so that non-native ATM services may be offered by the VPN service. The access equipment is typically composed of a service multiplexer or an access switch. Unlike the architecture discussed by Walters (1991), it is assumed in this paper that the access equipment belongs to the VPN and is operated by the VPN service provider. This is probably the best way of further providing optimised outsourcing facilities (i.e., management of the customer network by the service provider).

4. TRAFFIC CONSIDERATIONS

4.1. Controlling End-to-end Connections

Each ATM end-to-end connection should be established between the user and the access equipment via the negotiation of a traffic contract as specified in I.371 (1995). The connection is subsequently policed at the input of the access equipment according to the declared traffic parameters. For a non-native ATM service, some service parameters are declared at connection establishment. The connection is then policed at the service level via specific actions and service parameters are mapped onto ATM layer parameters via an Inter Working Unit.

4.2. Controlling the Virtual Trunks of a VPN

At the establishment of a VPN, each virtual trunk is assigned a global bandwidth. To protect other customers (in particular users of other VPNs) from excess traffic on a particular VPN, the bandwidth of each virtual trunk should be controlled by the network. Two cases have to be considered.

Access Virtual Trunk

The bandwidth of the access virtual trunk is implicitly controlled by the Connection Admission Control (CAC) function implemented in the access equipment. Indeed, when multiplexing all the end-to-end connections onto the access virtual trunk, the access equipment may use a buffering capability and a CAC algorithm to prevent any buffer overflow. The traffic control functions performed on individual connections (e.g., policing at access equipment input) together with the CAC further limit the load so that the bandwidth allocated to the access virtual trunk is not exceeded.

Internode Virtual Trunks

For the same reasons as for access virtual trunks, internode virtual trunks have to be controlled within the ATM network. In fact, congestion problems arise in the case of a "Y" structure, where the peak bandwidth of the output virtual trunk is less than the aggregate peak bandwidth of the input virtual trunks, and where the network does not take into account the bandwidth limitation of the virtual trunks at the establishment or the activation of the end-to-end connections. In this context, buffer overflow may occur at the output of the VPN node, possibly leading to QoS degradation for all the connections multiplexed in the output buffer. As suggested by Walters (1991), a special function, referred to as output policing, should be implemented at the output of the VPN node.

However, the suitability of such a solution should be questioned from a VPN user view point. Indeed, while the service is designed to allow free handling of the individual end-to-end connections, VPN users expose themselves to sudden QoS degradation of their communications, without any prior indication from the network because output policing mechanisms freeze out excess traffic with respect to virtual trunk capacities. In fact, it seems to be more desirable to provide the user with complementary mechanisms in order to prevent such situations. Two possibilities may be envisaged for this purpose :

1. The instantaneous cell rate of a virtual trunk is not known by VPN nodes. In this case, VPN nodes are aware only of the allocated peak bandwidth of the virtual trunk and has to implement output policing. The end-to-end connections are freely handled by the VPN users without any prior authorisation by the VPN nodes. To avoid any QoS degradation due to output policing, end-to-end signalling may be used between the different VPN sites before the establishment of a new end-to-end connection to check whether the requested resources are available. Thus, each access equipment of the VPN has an image of the traffic configuration within the VPN. Nevertheless, the overhead due to the exchange of such messages between VPN sites may be significant and the response time of such a mechanism may be very large, leading to poor performance. Moreover, deadlock situations may occur if the traffic configuration is rapidly changing and such a mechanism is not adapted to elastic data applications, whose transmission rates are continually changing.
2. The instantaneous cell rate of a virtual trunk is known by the VPN node. Since it is difficult to estimate the peak rate of an ATM connection via observation, the best solution is that the access equipment involved in the establishment or the modification of an end-to-end connection indicates the traffic parameters of this connection to the VPN nodes. Those VPN nodes check whether the peak cell rate increase is possible, given that the global bandwidth of each inter-node virtual trunk should not be exceeded. From a theoretical point of view, three possibilities may be envisaged to indicate the peak cell rate of a connection in an ATM network, namely network management, signalling (Q.2963, 1994), or Resource Management (RM) procedures. In particular, the ATM Block Transfer (ABT) capability based on RM procedures is perfectly adapted to the in-call renegotiation of the peak cell rate of a connection.

Note that in the case where the instantaneous cell rate of a virtual trunk is known by the VPN node, the need for output policing disappears, since the network can prevent any overflow of the global bandwidth of a virtual trunk.

4.3. Quality of Service of End-to-End Connections

Impact of Traffic Shaping on the End-to-end CDV of VCs within a VP

In earlier studies, it has been recognised (e.g., Boyer *et al.*, 1992) that traffic shaping is unavoidable for efficient usage of the network. The shaping function may actually be performed at the output of the access equipment (Boyer and Servel, 1995) or by the transit network, either as an associated function of the UPC or within the network node via the implementation of special queuing disciplines such as Fair Queuing, see Golestani (1994), Parekh and Gallager (1993), and Roberts (1994). The impact of traffic shaping by the network has to be carefully examined from a QoS point of view namely cell dispersion (Guillemin and Monin, 1992) affecting the end-to-end connections. Cell dispersion is also referred to as end-to-end Cell Delay Variation within the standardisation bodies (I.371, 1995).

As a matter of fact, when several end-to-end connections (typically VCs), in particular real time connections (supporting for example voice communications), are multiplexed onto a VP, traffic shaping may result in a significant increase in cell dispersion on the end-to-end connections. This is due to the fact that the only entity known by the network is the VP and cell spacing can only be performed at the VP level with no regard to the VCs.

To prevent problems due to traffic shaping by the network, two solutions may be envisaged.

1. Some VPs should be dedicated to real time end-to-end connections and contain only this type of connections; other VPs should contain non delay-sensitive connections (e.g., supporting data applications). To avoid the potential inefficiency of this partitioning, dynamic bandwidth sharing mechanisms, such as resource management procedures, may be used to share a global bandwidth (typically the peak bandwidth of a virtual trunk of a VPN) between the different VPs. Note that the cells of the different real-time end-to-end connections within a VP should be interleaved in such a way that cell spacing should not adversely disturb the time structure of a VC. Moreover, such a VP may have stringent requirements on the end-to-end CDV through the network (pseudo-synchronous VP).
2. The second solution is to take into account the shaping function within the access equipment as in Boyer (1995). In this case, cells of real-time and non real-time end-to-end connections should be interleaved in such a way that cell dispersion objectives are met for real-time connections, given that the VPs supporting the different end-to-end connections may be altered by CDV within the network, which may perform traffic shaping (in particular cell spacing). This solution is the most attractive since the number of VPs to be managed is minimum, but may entail difficulties for building up a VP within both access equipment and VPN nodes if VPs are terminated within the network.

Terminating a VP within the network

Consider several VCs multiplexed onto an end-to-end VP, which is then supported by a VP-based network. The only entity known by the network is the VP, which is offered a QoS in terms of CLR and cell dispersion (say, γ_{VP}). With regard to cell dispersion, it can easily be shown that the magnitude of cell dispersion affecting a particular VC within the VP through the network is equal to γ_{VP}. Taking into account the prior multiplexing of VCs onto the VP, resulting in a cell dispersion of magnitude γ_{mux} for the VC considered, the end-to-end CDV affecting the VC is equal to $\gamma_{VP}+\gamma_{mux}$. It thus appears that the VC multiplexing scheme plays a critical role with regard to cell dispersion for the VCs of a VP. In particular, depending on the VC multiplexing scheme, γ_{mux} may take large values, incompatible with cell dispersion requirements for real time connections. In any case, the value of γ_{mux} should be carefully evaluated for any multiplexing scheme.

Let us now considered the reference configuration depicted in Figure 3, where VPs are terminated within the network. The previous discussion applies for each segment of the VPs. In particular, the end-to-end CDV on a VC is increased by the quantity γ_{switch}, which is the magnitude of cell dispersion introduced by the VC switch. In particular, γ_{switch} may be of the same order of magnitude as γ_{mux} and should not take too large a value in order to be compatible with the end-to-end CDV requirements for real-time connections. This emphasises the need to evaluate carefully the increase in cell dispersion when multiplexing several real-time connections on a given VP.

Moreover, if the VC switch is operated by a network operator different from the one operating the multiplexer, each incoming VC has to be policed at the VC switch access. The traffic parameters to be controlled (e.g., the peak cell rate by taking into account the associated CDV tolerance τ for a given VC) should be determined. The CDV tolerance τ may depend on the CDV affecting the VP, the CDV experienced in the multiplexer, and the initial CDV tolerance τ_0 declared in the traffic contract negotiated between the user and the operator of the multiplexer. The initial CDV tolerance and the CDV altering the VP are known through the traffic contracts. Moreover, the CDV introduced by the multiplexer may be evaluated or at least upper-bounded. Consequently, at the set up of a VC, the access equipment should replace the CDV tolerance in the traffic with another CDV tolerance computed on the basis of the original CDV τ_0, the CDV tolerance of the VP, and the CDV introduced by the multiplexer.

It thus turns out that terminating a VP within the network requires the development of new tools in terms of signalling and further investigations with regard to traffic and congestion control, which have not so far been addressed within the standardisation bodies.

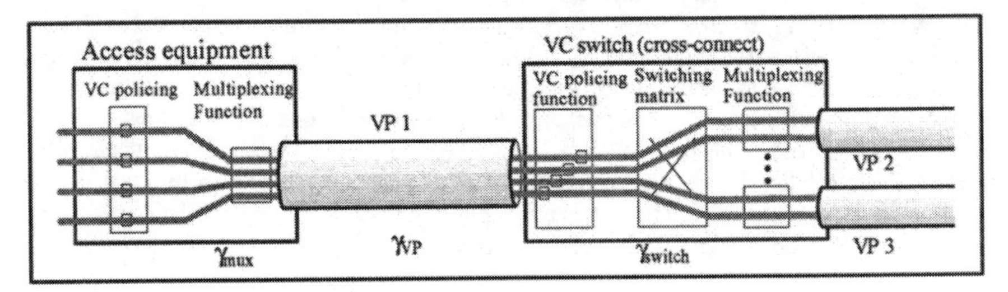

Figure 3. Terminating a VP within the network.

5. POSSIBLE ARCHITECTURES OF A VPN

In view of the above discussion, two architectures are proposed for supporting a VPN. The first one relies on end-to-end cross-connected VPs. In the second proposal, VPs are terminated within the network and VCs are individually switched (or cross-connected).

5.1. VP-based architecture (Access Switching)

General description

Virtual trunks are composed of several VPs (a VP per destination site) and VPN nodes are VP cross-connects (see Figure 4). End-to-end connections (VCs) are controlled at the input of access equipment and multiplexed on the relevant VP depending on their destination. End-to-end connections are not further seen by VPN nodes. If N is the number of sites of the VPN customer, each access virtual trunk is then composed of (N-1) bidirectional VPs (fully meshed architecture). N (N-1) bidirectional VPs are thus required to connect N sites. These VPs are said to be end-to-end in the sense that they are not terminated within the network (VP trails).

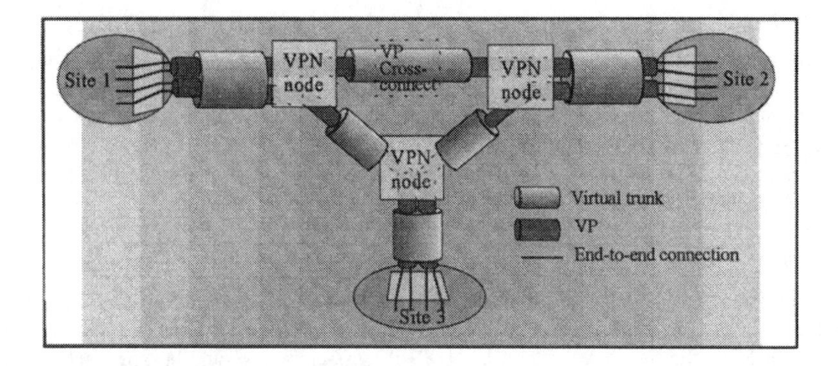

Figure 4. VPN based on end-to-end VPs.

Bandwidth Sharing

The different VCs within a given VP share the transmission capacity of the VP. Dynamic bandwidth sharing may be achieved at the connection level through the set-up and release of end-to-end connections, and for a given configuration of end-to-end connections, via statistical multiplexing procedures, for instance by using the ATM Block Transfer (ABT), Available Bit Rate (ABR), or Statistical Bit Rate (SBR) transfer capabilities. Specifically, ABR and ABT procedures consist in dynamically negotiating the traffic parameters of a connection. The set-up and release of end-to-end connections require specific procedures (e.g., signalling facilities).

Since the number of VPs per virtual trunk in this type of architecture may potentially take large values, it seems highly desirable to share dynamically virtual trunk bandwidth between VPs. As a matter of fact, static bandwidth allocation may lead to significant waste of network resources and to prohibitive cost for the customer. Such a dynamic bandwidth sharing mechanism may rely on RM procedures at the VP level or Network Management procedures. The former are faster than the latter and are well-adapted to relatively unpredictable traffic fluctuations between VPN sites. Network Management procedures may be used when traffic fluctuations are known in advance (e.g., periodic traffic fluctuations with a calendar).

Connection Admission Control

At the establishment (or modification) of an end-to-end connection (namely a VC), a traffic contract is negotiated explicitly or implicitly between the VPN user and the service provider. This connection is subject to CAC in the sense that:

1. the access equipment checks whether it can support this new connection, namely that the peak bandwidth of the access virtual trunk is not exceeded and that the peak bandwidth of the relevant VP (i.e., the VP between the originating and corresponding sites) is not exceeded. These actions require a CAC local to the access equipment.
2. possibly, the peak cell rate of the VP intended to support the new end-to-end connection may have to be renegotiated, given that the allocated bandwidth of each virtual trunk containing the end-to-end VP should not be exceeded. This may be achieved via a bandwidth sharing procedure (RM or Management) and amounts to performing a global CAC at the VPN level.

Advantages

This architecture may support the target B-VPN service outlined in the previous section and meets the requirements stated above. Specifically, a QoS may be guaranteed for each end-to-end connection and optimal bandwidth sharing between connections may be achieved. Only minor enhancements of standard network elements are needed. Moreover, cell dispersion on end-to-end connections is limited, since the VPs are end-to-end and the end-to-end connections pass through only one VC multiplexing stage.

Limitations

The major drawback of the VP-based architecture is the need for a large number of VP identifiers, which may be critical in a large network. Since the support of RM procedures by cross-connects may be needed for efficient operation of a VPN, a VP cross-connect should take into account the virtual trunk level when operating the network through RM procedures. Indeed, it should check that the sum of the peak cell rates of the VPs managed by an RM procedure does not exceed the value of the peak bandwidth allocated to the virtual trunk. This principle slightly increases the complexity of the bandwidth management within the cross-connects.

5.2. VP/VC-based Architecture

General description

A virtual trunk is composed of one VP and a VPN node is a VC switch or a VC cross-connect (see Figure 5). The access equipment controls and multiplexes end-to-end connections (VCs), regardless of their destination, onto the access virtual trunk, which is composed of a single VP. This VP-virtual trunk is terminated at the input of a VPN node. As shown in Figure 6, each VC is then independently switched or cross-connected toward different output VPs (inter-node virtual trunks).

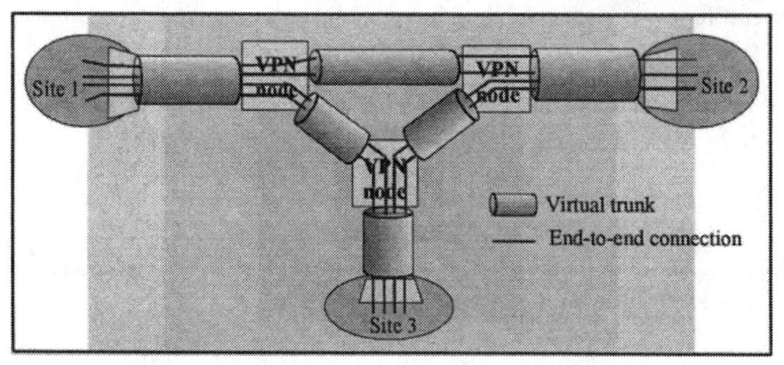

Figure 5. VP/VC based VPN.

Bandwidth Sharing

As mentioned in the previous section, dynamic sharing may be performed through set-up and release of end-to-end connections and via statistical multiplexing procedures. Concerning the set-up and release of end-to-end connections, two possibilities may be envisaged. An end-to-end connection may be established and released on-demand by using network public signalling, thus reflecting the customer's time-dependent needs in terms of connections and resources. Another possibility consists in pre-establishing (by means of Network Management or Signalling) a pool of VCs and operating them through resource management procedures. While not used, the pre-established VCs are dormant (i.e., assigned a zero bit rate). When needed, a VC is activated via an RM procedure, which allocates to the VC the requested bandwidth.

Connection Admission Control

In the on-demand signalling solution, a mechanism should check whether the establishment of a new connection is possible, namely that the bandwidth of each virtual trunk (i.e., a VP) along this connection is not exceeded. This mechanism does not require any enhancement of standard network elements, since any standard switch or cross-connect should perform such a CAC function. In the pre-established VC case, such a verification has to be performed at each activation of a dormant connection. This action is part of standard RM procedures (e.g., ABT) and does not require any enhancement of the switches.

Limitations

The major drawback of the pre-established VC solution is that a huge number of VC identifiers (VCIs) should be reserved for each customer. Approximately, for each VPN user, a VC should be pre-established between this particular user and each of his potential destinations.

The on-demand signalling option requires the implementation of signalling capabilities within the access equipment and the availability of VC switching facilities. In this case, special management functions are required within a VC switch in order to recognise a VC as a VPN end-to-end connection and to route it to the appropriate virtual trunk. As a matter of fact, for reliability purposes in a public network, there are always several routes between two sites. If no additional functions are implemented within the switches (such as recognising that a VC belongs to a VPN), then it is quite possible to set up end-to-end connections outside the virtual trunks, which may entail complex network-dimensioning problems. Actually, a VC belonging to a VPN should be routed along virtual trunks in order to consume the bandwidth reserved for the VPN. An easy solution to this problem consists in reserving some VCIs for the VCs belonging to a VPN, but this solution, however, presents the same drawback as the pre-establised configuration, that is the reservation of a large number of VCIs per VPN. Other solutions may be envisaged, but at the expense of an increased complexity of the switches and establishment procedures (e.g., use of a data base).

In conclusion, it turns out that in the signalling solution, enhanced signalling facilities are requested for establishing or modifying end-to-end connections. Furthermore, the set-up delay may be inappropriate in the case of connectionless traffic. The pre-established solution is relatively simple in terms of network element enhancements and seems quite easy to operate, but it consumes a large number of VCI values.

6. CONCLUSION

A possible solution to offer the target Broadband VPN service described in this paper is to choose a VPN architecture integrated into the general ATM infrastructure, possibly by introducing additional network elements such as service multiplexers at network edges. From traffic and QoS considerations, two VPN architectures have been introduced, one relying on end-to-end VPs and the other one on VPs terminated within the network and switched VCs.

Some traffic and congestion control functions should be thoroughly described and analysed, for instance RM procedures to activate or deactivate end-to-end connections, multiplexing schemes in access equipment to meet delay requirements for real-time applications, etc. Furthermore, for a given architecture, a management system should developed. These issues will be addressed in further investigations.

In fact, the final choice of a VPN architecture depends on the evolution of the general ATM architecture, in particular the location of VC switching, given that the traffic issues raised in Section 4 may apply not only to VPN but also to any ATM network. Notably, the increase of cell dispersion in multiplexing stages (switches or multiplexers) should be kept rather low. The feasibility of multiplexing schemes compliant with stringent CDV requirements may play a crucial role in the choice of the general architecture for future ATM networks.

7. REFERENCES

Boyer, P. and Servel, M. (1995) A Spacer-multiplexer for Public UNIs, in *Proc. ISS'95*, Berlin.

Boyer, P. Guillemin, F. Servel, M. and Coudreuse J.P. (1992) Spacing cells protects and enhances utilisation of ATM network links. *IEEE Network Magazine*, 38-49.

Golestani, J. (1994) A self-clocked fair queueing scheme for broadband applications, in *Proc. Infocom'94*.

Guillemin, F. and Monin, W. (1992) Management of Cell Delay Variation in ATM networks, in *Proc. Globecom'92*, Orlando.

ITU-T Recommendation I.371 (1995) Traffic control and congestion control in B-ISDN, Geneva.

ITU-T Recommendation Q.2963 (1994) Digital Subscriber Signalling System number 2: point-to-point call connection modification, Geneva.

Parekh, A. and Gallager, R. (1993) Generalised processor sharing approach to flow control integrated services networks - the multiple nodes case, in *Proc. Infocom'93*.

Roberts, J. (1994) Virtual Spacing for flexible traffic control. *Int'l J. Commun. Systems*, 307-318.

Walters, S. (1991) A new direction for broadband ISDN. *IEEE Commun. Magazine*, 39-42.

21

Statistical Resource Allocation and Pricing in Broadband Communication Networks

Hongbin Ji and Joseph Y. Hui*
Rutgers, The State University of New Jersey
Department of Electrical and Computer Engineering
Piscataway, New Jersey 08855-0909, USA

Abstract
This paper presents resource management for broadband communication networks from an economic point of view.† To be specific, we consider bandwidth allocation and buffer dimensioning. A pricing scheme is also proposed to be dependent on the amount of allocated bandwidth and buffer.

First, we exploit the utility function to represent satisfaction level of a user who requests a certain type of connection service from the network. Then we address the problem of how to allocate bandwidth and buffer capacity of one network component into virtual paths (VPs). On the one hand, the problem is formulated as a non-cooperative K-person game. Two cases, namely unconstrained and constrained, are considered. On the other hand, the problem may be formulated as a cooperative game. The objective is to find an optimal resource allocation such that the total utility of one network component will be maximized. Alternatively, the objective is to maximize the minimum utility among VPs. It is shown that these two game problems may have the same solution for unconstrained case.

Keywords
Asynchronous Transfer Mode (ATM) networks, utility function, resource management, pricing, Nash equilibrium.

1 INTRODUCTION

Broadband communication networks (e.g. ATM networks) are under development for multimedia applications such as data, audio, image and video transmission. Extended services and multiple quality-of-service (QoS) will be provided. The capability to reserve bandwidth and dynamically set up calls make ATM an ideal basis for the support of multimedia applications as pointed out by McDysan and Spohn (1994).

---

†The author is now at AT&T Bell Laboratories, Room 3H-309, Holmdel, NJ 07733-3030.

Bandwidth and buffer are the major resources in broadband networks. When a certain type of connection request arrives, a virtual circuit (VC) is set up through the network. The VC carries the call from its origin to destination on one physical connection. Multiple VCs may be integrated to use a virtual path (VP). How to effectively allocate the resources of a physical ATM trunk among the VPs is crucial for managing QoS. A good resource allocation scheme may avoid congestion and entail efficient utilization of resources.

People usually approach this problem from an engineering point of view. For example, Hui (1988) addressed resource allocation for broadband networks through measuring congestion at different levels, i.e., the cell, burst and call levels. An algorithm was proposed for multilayer bandwidth allocation emulating some functions of virtual circuit setup, fast circuit switching, and cell switching. Dziong, Choquette, Liao and Mason (1990) investigated effective bandwidth allocation in an ATM link. The result was used to construct some simple admission control strategy. A routing scheme was given based on residual effective bandwidth. Resource management in broadband ISDN was considered by Burgin (1990) based on VPs. A cost-benefit analysis was presented to determine the conditions under which capacity should be reserved on VPs and the benefits obtained by dynamically updating this reservation. Eckberg, Luan and Lucantoni (1990) studied bandwidth management for congestion control in broadband packet networks. It was noted that bandwidth management can be thought of as applying a throughput-burstiness filter to packet flows entering the network. They focused on characterizing this filter in teletraffic terms.

Hui, Gursoy, Moayeri and Yates (1991) proposed a layered broadband switching architecture with physical or virtual path configurations. A graph framework was introduced to describe network layers of network design, path configurations, dynamic call routing, burst switching and ATM cell switching. These hierarchical layers of switching are performed at decreasing time scales. A layered notion of equivalent bandwidth for satisfying layered grade of service parameters was introduced for making connections at these time-scales.

Monteiro and Gerla (1994) presented bandwidth allocation in ATM networks at different levels and in different stages. At physical level, ATM topology can be dynamically reconfigured by adding or removing trunks between ATM switches. The bandwidth allocation in this level is made possible by the SONET synchronous transfer mode (STM) infrastructure equipped with digital cross connection systems (DCSs). At ATM level, bandwidth can be allocated to individual VCs and VPs. They found that it was convenient to organize the VPs in a connectionless overlay network. This introduced connectionless server bandwidth allocation. Gun, Kulkarni and Narayanan (1994) looked at the problem of bandwidth allocation and access control in high-speed networks. They formulated the design problem so as to minimize the allocated bandwidth subject to service guarantees and stability conditions for the input and output buffers.

In this paper, we consider resource allocation from an economic point of view. A utility function is defined for each class of connections. The utility depends on allocated resources and the number of connections of the same class. As more bandwidth and buffer are allocated, QoS will be improved. Thus users will be more satisfied. When QoS is upgraded to a certain level, users will become indifferent to any further QoS improvement. Thus, we assume that utility is an increasing concave function of allocated bandwidth and buffer.

MacKie-Mason and Varian (1994) defined utility based on allocated bandwidth and the network congestion level. Ji, Hui and Karasan (1996) developed an economic model in a more general way, i.e. users' utility is based on the QoS. Cocchi, Shenker, Estrin and Zhang (1993) specified utility as a linear function of delay and cell loss for different applications such as ftp, mail, telnet, etc. They also studied the role of pricing policies in multiple service class networks. By simulations, they found that it was possible to set the prices so that users of every application type were more satisfied with the combined cost and performance of a network with service-class sensitive prices.

In this paper, we also study pricing scheme. The price is charged to each VP (instead of each connection) according to the allocated resources. This pricing scheme is so simple that it will not cause much extra cost when it is implemented in practical ATM networks. Ji, Hui and Karasan (1996) developed a pricing scheme which charges each connection according to its externalities on the other connections. Pricing a network service in a competitive environment was studied by Liau, Lutton, and Kouatchon (1994). Murphy, Murphy and Posner (1994) presented distributed pricing for embedded ATM networks. A tariff structure, which encourages the cooperative sharing of information between users and the network, was proposed by Kelly (1994) for high speed multiservice networks. Sairamesh, Ferguson and Yemini (1995) discussed the economy and formulated economic based problems for allocating resources.

The rest of this paper is organized as follows. In Section 2, we develop system and economic model for pursuing resource allocation. Some reasonable assumption is also made. Then in Section 3, we address non-cooperative resource allocation. There exists Nash equilibrium under the assumption. An algorithm is proposed to find the non-cooperative optimal resource allocation. In Section 4, we present cooperative resource allocation. The problem is formulated as maximization of total utility in one ATM component or maximization of the minimum utility of the VPs in the ATM component. An algorithm is also proposed to search for cooperative optimal resource allocation. Last section concludes the paper and gives direction for future work.

2 SYSTEM AND ECONOMIC MODELS

Since a broadband network trunk has very large amount of bandwidth and buffer capacity, it is possible to support hundreds of connections simultaneously. Each connection has distinct characteristics and QoS requirements. Without loss of generality, we make the following assumption.

Assumption 1 *The carried connections are generically categorized into K classes according to their characteristic and QoS requirement.*

It is suggested in ATM Forum that traffic is classified into constant bit rate (CBR), variable bit rate (VBR), available bit rate (ABR) or unspecified bit rate (UBR). CBR traffic may include real-time video applications, VBR traffic consists of real-time and non-real-time video and data services, ABR traffic is for data service and UBR for any kind of non-real-time services.

Furthermore, we make another assumption:

Assumption 2 *A physical trunk is correspondingly decomposed into K virtual paths (VPs), and each VP just supports one class of connections exclusively.*

Assumption 2 was also made by Bolla, et al, (1993) and Mishra and Tripathi (1993), Ji (1995), Ji, Hui and Karasan (1996). The major advantage of such decomposition lies in twofold. First, interferences among different classes of traffic are avoided. Thus, QOS management becomes flexible and easy. For example, bursty data traffic would cause significant cell delay variance (CDV) for real-time video traffic if they were multiplexed into the same VP. Second, the benefits of statistical multiplexing are still maintained under the decomposition since the VCs which carry the same class connections are multiplexed into the corresponding VP.

When a call arrives, a virtual circuit (VC) will be established over its correspondingly VP. Each virtual path has dedicated bandwidth and buffer, which are dynamically assigned by the network. It is associated with one type of QoS. So, multiple grades of services are supported through the K virtual paths.

Different VPs might implement distinct queueing disciplines like FCFS, LCFS, etc. The discipline implemented at one VP should be appropriate for the carried class of traffic. And each class of traffic is separated into its corresponding VP.

Let C_i and B_i denote the bandwidth and buffer capacity allocated for the i-th VP, respectively, and N_i denote the number of connections currently established in the i-th VP. These N_i connections are statistically multiplexed into the VP. And each connection has an associated virtual circuit identifier (VCI).

A wide range of applications with various preferences are envisioned to use ATM networks. Users' preferences are represented by utility function, which is defined as a function of allocated resources and the number of connections. Formally,

Definition 1 *User's utility function, denoted by $u_i(C_i, B_i, N_i)$, is a function of the allocated resources, C_i, B_i and the number of connections N_i, for $i = 1, 2, ..., K$.*

We observe that users will be more satisfied when QoS is improved as more resources are allocated. When QoS is upgraded to a certain level, users will become indifferent to any further QoS improvement. According to this observation, we make the following reasonable assumptions.

Assumption 3 *Utility u_i is differentiable and monotonically increasing concave function of C_i, B_i, for $i = 1, 2, ..., K$.*

The above assumption will guarantee the existence of optimal resource allocation. It should be noted that the number of connections may change once a connection is admitted or terminated. The time scale of changing N_i may be a few milliseconds. However, resource allocation should be implemented in a larger time scales, for instance, one second. The reason is that change of resource allocation is costly. It might also cause instability and chaos if resources were re-allocated frequently.

3 NON-COOPERATIVE RESOURCE ALLOCATION

Without loss of generality, we consider one ATM component. The idea generated in this section can be extended into the whole ATM networks.

By collecting historical data, we may obtain statistics on the number of i-th class connections N_i for $i = 1, 2, ..., K$. Let $P_{N_i}(l_i)$ denote the probability when the number of i-th class connections is equal to l_i. The problem is how to allocate the available resources of one ATM component among the K VPs such that the resources will be efficiently utilized.

We know that the total utility of the i-th VP is given by $N_i u_i(C_i, B_i, N_i)$, for $i = 1, 2, ..., K$. The expected value is given by $E[N_i u_i(C_i, B_i, N_i)]$. If u_i is assumed to be a concave function of N_i (Ji, Hui and Karasan, 1996), from Jensen's inequality, we have:

$$E[N_i u_i(C_i, B_i, N_i)] \leq E[N_i] u_i(C_i, B_i, E[N_i]), \ \forall i \tag{1}$$

A simple resource allocation may try to optimize the upper bound in the above equation (Ji, 1995; Ji, Hui and Karasan, 1996). In this paper, our objective is to maximize the expected total utility.

3.1 Unconstrained Resource Allocation

Before multimedia applications become popular, resources in the broadband ATM networks seem relatively plentiful. In this case, the constraints of available resources can be neglected.

Problem Statement

Let p_j and q_j denote the usage cost for unit bandwidth and unit buffer, respectively, for $j = 1, 2, ..., K$. Note that the prices p_j and q_j are specified by network service providers. They may be the same for all j if the providers don't want to distinguish services cost. In general, p_j and $q_j, \forall j$ are different due to distinct services cost of different classes of connections. How to appropriately specify the prices will be addressed in Section 3.1.

The expected total utility of the j-th VP is given by $E[N_j u_j(C_j, B_j, N_j)]$ while the cost charged by network service provider is given by $p_j C_j + q_j B_j$. The revenues for the j-th class of users, denoted by J_j, is given by the difference between expected utility and service charges, i.e. $J_j = E[N_j u_j(C_j, B_j, N_j)] - p_j C_j - q_j B_j$. Each class of users request an optimal amount of resources such that their expected revenues are maximized, i.e.

$$\max_{\{C_j; B_j\}} J_j = E[N_j u_j(C_j, B_j, N_j)] - p_j C_j - q_j B_j \tag{2}$$

for $j = 1, 2, ..., K$. Since $P_{N_j}(l_j)$ represents the probability when the number of j-th class of connections is l_j, straightforwardly we have:

$$J_j = \sum_{l_j=1}^{L_j} P_{N_j}(l_j) l_j u_j(C_j, B_j, l_j) - p_j C_j - q_j B_j \tag{3}$$

where L_j denotes the maximum number of connections which can be supported on j-th VP.

From Assumptions 3, we know that J_j is a concave function of C_j, B_j. Therefore, there exists a optimal solution, which is called non-cooperative (or Nash) equilibrium (Nash,

1950). The optimal solution, denoted by C_j^*, B_j^*, for $j = 1, 2, ..., K$, should satisfy:

$$\sum_{l_j=1}^{L_j} P_{N_j}(l_j) l_j \left. \frac{\partial u_j(C_j, B_j, l_j)}{\partial C_j} \right|_{C_j^*, B_j^*} - p_j = 0$$

$$\sum_{l_j=1}^{L_j} P_{N_j}(l_j) l_j \left. \frac{\partial u_j(C_j, B_j, l_j)}{\partial B_j} \right|_{C_j^*, B_j^*} - q_j = 0 \quad (4)$$

For each VP, we have two unknowns and two equations. Then we may represent the solution by:

$$C_j^* = C_j^*(p_j, q_j)$$
$$B_j^* = B_j^*(p_j, q_j) \quad (5)$$

for $j = 1, 2, ..., K$. It should be noted that there is no conflict among different classes of users. This is because the available resources are assumed to be plentiful (or unlimited). Also note that Nash equilibrium is unique in this case.

Algorithm

We develop an algorithm to find the optimal resource allocation. It should be noted that the algorithm is applicable to each VP.

Algorithm 1 *Find a non-cooperative optimal resource allocation:*
Step 1. Set an appropriate initial resources C_j^0, B_j^0, the desired precision of convergence ϵ, and iteration number $t = 0$.
Step 2. Find C^{t+1}, B^{t+1} such that

$$\sum_{l_j=1}^{L_j} P_{N_j}(l_j) l_j \left. \frac{\partial u_j(C_j, B_j, l_j)}{\partial C_j} \right|_{C_j^{t+1}, B_j^t} = p_j \quad (6)$$

and

$$\sum_{l_j=1}^{L_j} P_{N_j}(l_j) l_j \left. \frac{\partial u_j(C_j, B_j, l_j)}{\partial B_j} \right|_{C_j^t, B_j^{t+1}} = q_j \quad (7)$$

Step 3. If $|C_j^{t+1} - C_j^t| < \epsilon$ and $|B_j^{t+1} - B_j^t| < \epsilon$, stop. Otherwise, let $t = t + 1$ and go back to step 2.

Pricing Schemes

Roughly speaking, if the prices p_j, q_j are set too small, then the j-th class of users can always increase their revenues by choosing larger amount of resources. It implies that users will choose infinite amount of resources. In other words, resources will be abused if the prices are too low. On the other hand, if the prices are set too high, users can not get any revenues no matter how much resources they may request. In other words, users can not afford service charges. In both of these cases, there will be no solution to the problem of (2). Therefore, prices should be appropriately set such that resources will be efficiently utilized.

We can see that necessary conditions for good price lie on:

$$\lim_{C_i \to \infty} \sum_{l_i=1}^{L_i} P_{N_i}(l_i) l_i \frac{\partial u_i(C_i, B_i, l_i)}{\partial C_i} < p_i < \lim_{C_i \to 0} \sum_{l_i=1}^{L_i} P_{N_i}(l_i) l_i \frac{\partial u_i(C_i, B_i, l_i)}{\partial C_i}$$
$$\lim_{B_i \to \infty} \sum_{l_i=1}^{L_i} P_{N_i}(l_i) l_i \frac{\partial u_i(C_i, B_i, l_i)}{\partial B_i} < q_i < \lim_{B_i \to 0} \sum_{l_i=1}^{L_i} P_{N_i}(l_i) l_i \frac{\partial u_i(C_i, B_i, l_i)}{\partial B_i} \quad (8)$$

If the prices do not satisfy the left hand inequalities, users will abuse resources. On the other hand, if the prices do not satisfy the right hand inequalities, users will not request any resources because the prices seem too expensive.

Note that users' utility may saturate as resources are expanded continuously. It implies that:

$$\lim_{C_i \to \infty} \sum_{l_i=1}^{L_i} P_{N_i}(l_i) l_i \frac{\partial u_i(C_i, B_i, l_i)}{\partial C_i} = 0$$
$$\lim_{B_i \to \infty} \sum_{l_i=1}^{L_i} P_{N_i}(l_i) l_i \frac{\partial u_i(C_i, B_i, l_i)}{\partial B_i} = 0 \quad (9)$$

Then the left hand inequalities of equation (8) become trivial since the prices must be positive.

From network service providers' point of view, their objective is to find optimal prices such that their profits will be maximized, i.e.

$$\max_{\{p_i; q_i\}} \sum_{j=1}^{K} p_j C_j^*(p_j, q_j) + q_j B_j^*(p_j, q_j) \quad (10)$$

Then a necessary condition for optimal pricing is:

$$C_i^* + p_i^* \left.\frac{\partial C_i^*}{\partial p_i}\right|_{p_i^*, q_i^*} + q_i^* \left.\frac{\partial B_i^*}{\partial p_i}\right|_{p_i^*, q_i^*} = 0$$
$$B_i^* + p_i^* \left.\frac{\partial C_i^*}{\partial q_i}\right|_{p_i^*, q_i^*} + q_i^* \left.\frac{\partial B_i^*}{\partial q_i}\right|_{p_i^*, q_i^*} = 0 \quad (11)$$

for $i = 1, 2, ..., K$.

3.2 Constrained Resource Allocation

When multimedia applications become widespread, the resources in broadband ATM networks will become scarce. In this case, we may formulate the resource allocation as an K-person game problem, i.e.

$$\max_{\{C_j; B_j\}} J_j = \sum_{l_j=1}^{L_j} P_{N_j}(l_j) l_j u_j(C_j, B_j, l_j) - p_j C_j - q_j B_j, \quad \forall j \tag{12}$$

subject to the constraint of available bandwidth in the ATM component, denoted by C:

$$\sum_{j=1}^{K} C_j \leq C \tag{13}$$

and to the constraint of available buffer in the ATM component, denoted by B:

$$\sum_{j=1}^{K} B_j \leq B \tag{14}$$

If the resources are so plentiful that the following condition is satisfied

$$\sum_{j=1}^{K} C_j^* \leq C$$
$$\sum_{j=1}^{K} B_j^* \leq B \tag{15}$$

where C_j^*, B_j^* are solutions to equation (4), then the constraints (13) and (14) can be relaxed. Thus, the solution of constrained resource allocation is the same as that of unconstrained case.

Suppose the network resources are so scarce that the condition (15) is violated. In this case, the optimal solution, denoted by C_j^c, B_j^c, to problem of (12) becomes:

$$C_j^c = \min[C_j^*, C - \sum_{i=1, i \neq j}^{K} C_i^c]$$
$$B_j^c = \min[B_j^*, B - \sum_{i=1, i \neq j}^{K} B_i^c] \tag{16}$$

for $j = 1, 2, ..., K$. Note that the set $\{C_1^c, C_2^c, ..., C_K^c; B_1^c, B_2^c, ..., B_K^c\}$ is a non-cooperative equilibrium (or Nash equilibrium) (Nash, 1950). It should also be noted that the Nash equilibrium is not unique in this case. In next section, we shall consider resource allocation in cooperative way.

4 COOPERATIVE RESOURCE ALLOCATION

In this section, we consider resource allocation cooperatively. First, we formulate resource allocation as a problem of maximizing total utility of the ATM component. Second, we formulate it as max-min problem which may guarantee fairness among different kinds of applications.

4.1 Total Utility Maximization

Let us formulate the problem of maximizing total utility of one ATM component. Then we give an algorithm to find the solution.

Problem Statement

The problem is formulated as

$$\max_{\{C_i,B_i\}} \sum_{j=1}^{K} \left[\sum_{l_j=1}^{L_j} P_{N_j}(l_j) l_j u_j(C_j, B_j, l_j) - p_j C_j - q_j B_j \right] \tag{17}$$

subject to the constraints (13) and (14).

If the resources are plentiful, i.e., inequality (15) is satisfied, then the solution of above maximization problem, which is called cooperative equilibrium, is the same as non-cooperative equilibrium.

If inequality (15) is not met, then the optimal solution to (17) must be at the boundary of the convex set specified by (13) and (14). Using Lagrange multiplier method, we have

$$\begin{aligned}
&\sum_{l_i=1}^{L_i} P_{N_i}(l_i) l_i \frac{\partial u_i(C_i, B_i, l_i)}{\partial C_i} - p_i - \lambda_1 = 0 \\
&\sum_{l_i=1}^{L_i} P_{N_i}(l_i) l_i \frac{\partial u_i(C_i, B_i, l_i)}{\partial B_i} - q_i - \lambda_2 = 0
\end{aligned} \tag{18}$$

for $i = 1, 2, ..., K$ and where λ_1, λ_2 are Lagrange multipliers. It should be noted that λ_1, λ_2 may represent investment cost of bandwidth and buffer, respectively (Ji, 1995, Ji, Hui and Karasan, 1996).

Algorithm

From equation (18), we have

$$\begin{aligned}
&\sum_{l_i=1}^{L_i} P_{N_i}(l_i) l_i \frac{\partial u_i(C_i, B_i, l_i)}{\partial C_i} - p_i = \sum_{l_j=1}^{L_j} P_{N_j}(l_j) l_j \frac{\partial u_j(C_j, B_j, l_j)}{\partial C_j} - p_j \\
&\sum_{l_i=1}^{L_i} P_{N_i}(l_i) l_i \frac{\partial u_i(C_i, B_i, l_i)}{\partial B_i} - q_i = \sum_{l_j=1}^{L_i} P_{N_j}(l_j) l_j \frac{\partial u_j(C_j, B_j, l_j)}{\partial B_j} - q_j
\end{aligned} \tag{19}$$

for $i, j = 1, 2, ..., K$.

We propose the following algorithm to obtain the cooperative optimal resource allocation.

Algorithm 2 *Searching Cooperative Optimal Resource Allocation*
Step 1. *Set an appropriate initial resource allocation for the first VP,* C_1^0, B_1^0, *a positive initial scaling factors* $\alpha^0 < 1, \beta^0 < 1$, *and the desired precision of convergence* ϵ. *Set iteration index* $t = 0$.
Step 2. *Measure the partial derivatives* $\partial u_1/\partial C_1$ *and* $\partial u_1/\partial B_1$ *at the point* (C_1^t, B_1^t, N_1). *Then calculate* $\partial u_i/\partial C_i, \partial u_i/\partial B_i$ *according to the equation (19) for* $i = 2, 3, ..., K$. *Then obtain* C_i^{t+1}, B_i^{t+1} *for* $i = 2, 3, ..., K$.
Step 3. *If* $|\sum_{i=1}^K C_i^{t+1} - C| \leq \epsilon$ *and* $|\sum_{i=1}^K B_i^{t+1} - B| \leq \epsilon$, *then stop. Otherwise,*

Step 3.1. *If* $|\sum_{i=1}^K C_i^{t+1} - C| > \epsilon$,
If $\sum_{i=1}^K C^{t+1} - C$ *are positive*
If $\sum_{i=1}^K C^{t+1} - C$ *has the same sign as* $\sum_{i=1}^K C^t - C$,
double the scaling factor, i.e. $\alpha^{t+1} = 2\alpha^t$;
set $C_1^{t+1} = \alpha^{t+1} C_1^t$, *t=t+1 and go to step 3.2.*
Otherwise, reduce the scaling factoring by 2, i.e. $\alpha^{t+1} = \alpha^t/2$;
set $C_1^{t+1} = \alpha^{t+1} C_1^t$, *t=t+1 and go to step 3.2.*
Otherwise,
If $\sum_{i=1}^K C^{t+1} - C$ *has the same sign as* $\sum_{i=1}^K C^t - C$,
double the scaling factor, i.e. $\alpha^{t+1} = 2\alpha^t$;
set $C_1^{t+1} = \frac{C_1^t}{\alpha^{t+1}}$, *t=t+1 and go to step 3.2.*
Otherwise, reduce the scaling factoring by 2, i.e. $\alpha^{t+1} = \alpha^t/2$;
set $C_1^{t+1} = \frac{C_1^t}{\alpha^{t+1}}$, *t=t+1 and go to step 3.2.*

Step 3.2. *If* $|\sum_{i=1}^K B_i^{t+1} - B| > \epsilon$,
If $\sum_{i=1}^K B^{t+1} - B$ *are positive*
If $\sum_{i=1}^K B^{t+1} - B$ *has the same sign as* $\sum_{i=1}^K B^t - B$,
double the scaling factor, i.e. $\beta^{t+1} = 2\beta^t$;
set $B_1^{t+1} = \alpha^{t+1} B_1^t$, *t=t+1 and go to step 2.*
Otherwise, reduce the scaling factoring by 2, i.e. $\beta^{t+1} = \beta^t/2$;
set $B_1^{t+1} = \beta^{t+1} B_1^t$, *t=t+1 and go to step 2.*
Otherwise,
If $\sum_{i=1}^K B^{t+1} - B$ *has the same sign as* $\sum_{i=1}^K B^t - B$,
double the scaling factor, i.e. $\beta^{t+1} = 2\beta^t$;
set $B_1^{t+1} = \frac{B_1^t}{\beta^{t+1}}$, *t=t+1 and go to step 2.*
Otherwise, reduce the scaling factoring by 2, i.e. $\beta^{t+1} = \beta^t/2$;
set $B_1^{t+1} = \frac{B_1^t}{\beta^{t+1}}$, *t=t+1 and go to step 2.*

In Step 2, since u_i is a monotonically increasing function of C_i, B_i, we can easily obtain C_i, B_i once the information of the partial derivatives $\partial u_i/\partial C_i, \partial u_i/\partial B_i$ is available.

4.2 Max-Min Optimization

Alternatively, we may formulate the cooperative resource allocation as a max-min optimization problem:

$$\max_{\{C_i;B_i\}} \min_j \sum_{l_j=1}^{L_j} P_{N_j}(l_j) l_j u_j(C_j, B_j, l_j) \tag{20}$$

subject to the constraints (13) and (14).

If the resources are plentiful, i.e., inequality (15) is satisfied, then solution of above problem is the same as the unconstrained case.

If the resources are scarce, i.e., inequality (15) is not met, then solution of max-min optimization, denoted by C_i^m, B_i^m, will occur at the boundary of the convex set specified by the constraints (13) and (14). The optimal solution will satisfy:

$$C_j^m = \min[C_j^*, C - \sum_{i=1, i\neq j}^{K} C_i^m]$$
$$B_j^m = \min[B_j^*, B - \sum_{i=1, i\neq j}^{K} B_i^m] \tag{21}$$

for $j = 1, 2, ..., K$. Note that above form is similar to equation (16). However, in this max-min optimization, fairness is achieved for those VPs whose $C_i^m \neq C_i^*, B_i^m \neq B_i^*; C_j^m \neq C_j^*, B_j^m \neq B_j^*$ where $i, j = 1, 2, ..., K$, i.e.

$$\sum_{l_j=1}^{L_j} P_{N_j}(l_j) l_j u_j(C_j^m, B_j^m, l_j) - p_j C_j^m - q_j B_j^m = \sum_{l_i=1}^{L_i} P_{N_i}(l_i) l_i u_i(C_i^m, B_i^m, l_i) - p_i C_i^m - q_i B_i^m \tag{22}$$

It is important to maintain fairness among different classes of multimedia applications.

5 CONCLUSION

In this paper, we study resource allocation in broadband communication networks. Each connection is associated with a utility function which depends on the allocated bandwidth, buffer capacity and the number of same class connections. Then the resource allocation is formulated as a non-cooperative and/or cooperative game problem. Both unconstrained and constrained cases are considered.

A simple pricing scheme is proposed and discussed. Future work may be concerned with pursuit of dynamic resource allocation. Time may be divided into relatively large intervals. Resources are optimally allocated at the start of an interval according to current number of connections. The network service providers may give a price at the start of an interval according to current usage of the resources. This price is fixed during that interval. As time goes on and the number of users is changing, resources will correspondingly be re-allocated. The producer will also change the pricing scheme such that the resources are efficiently utilized.

REFERENCES

Bolla, R., Danovaro, F., Davoli, F. and Marchese, M. (1993) An integrated dynamic resource allocation scheme for ATM networks, *Proceedings of IEEE INFOCOM'93*, San Francisco, CA.

Burgin, J. (1990) Broadband ISDN resource management. *Computer Networks and ISDN Systems*, **20**,323–331.

Cocchi, R. and Shenker, S., Estrin, D. and Zhang, L. (1993) Pricing in computer networks: motivation, formulation, and example. *IEEE/ACM Trans. on Networking*,**1**, no. 6.

Dziong, Z., Choquette, J., Liao, K. Q. and Mason, L. (1990) Admission control and routing in ATM networks. *Computer Networks and ISDN Systems*, **20**, 189–196.

Eckberg, A. E., Luan, D. T. and Lucantoni, D. M. (1990) Bandwidth management: a congestion control strategy for broadband packet networks–characterizing the throughput-burstiness filter. *Computer networks and ISDN Systems*, **20**, 415–423.

Gun, L., Kulkarni, V. G. and Narayanan, A. (1994) Bandwidth allocation and access control in high-speed networks, *Annals of Operations Research*, **49**, 161–183.

Hui, J.Y. (1988) Resource allocation for broadband networks. *IEEE J. Selected Areas in Commun.*, **6**.

Hui, J. Y., Gursoy, M. B., Moayeri, N. and Yates, R. (1991) A layered broadband switching architecture with physical or virtual path configurations. *IEEE Journal on Selected Areas of Communication*, **9**, no. 9.

Ji, H. (1995), Resource allocation, admission control, pricing and routing in multimedia ATM networks. *Proceedings of IEEE Conference on Asia Pacific Communication Conference (APCC)*, Jakarta, Indonesia, Nov. 27-29.

Ji, H., Hui, J. Y. and Karasan, E. GoS-based pricing and resource allocation for multimedia broadband networks. *Proceedings of IEEE INFOCOM'96*, San Francisco, CA, March 24-28.

Kelly, F.P. (1994) On tariffs, policing and admission control for multiservice networks *Operations Research Letters.*, **15**, no. 9.

Liau, B., Lutton, J. L. and Kouatchou, J. (1994) Pricing a network service in a competitive environment. *Proc. of 14 ITC,* Antibes Juan-les-Pins, France, June.

MacKie-Mason, J. K. and H. Varian (1994), Pricing the Internet, *Public Access to the Internet,* Kahin, B. and Keller, J. (Eds.), Prentice-Hall, Englewood Cliffs, New Jersey.

McDysan, D. E. and Spohn, D. L. (1994) ATM: Theory and Application, McGraw-Hill, New York.

Mishra, P. P. and Tripathi, S. K. (1993) Dynamic bandwidth allocation in high speed integrated service networks, *Proceedings of IEEE INFOCOM'93*, San Francisco, CA.

Monteiro, J. A. S. and Gerla, M. (1994) Bandwidth allocation in ATM networks. *Annals of Operations Research*, **49**, 25–50.

Murphy, J., Murphy, L. and Posner, E. C. (1994) Distributed pricing for embedded ATM Networks, *Proc. of 14 ITC*, Antibes Juan-les-Pins, France, June.

Nash, J. (1950) Equilibrium points in n-person games, *Proc. Nat. Acad. Sci. U.S.A.*, **36**.

Sairamesh, J., Ferguson, D. F. and Yemini, Y. (1995) An approach to pricing, optimal allocation and quality of service provisioning in high-speed packet networks, *Proceedings of IEEE INFOCOM'95*, Boston, MA, April.

22
Degradation Effect of Cell Loss on Speech Quality Over ATM Networks

Mohamed M. Meky and Tarek N. Saadawi

The City University of New York

New York N. Y. 10031

Tel: (212) 650-7263 Fax: (212) 650-8249

mmeki and eetns@ee-mail.engr.ccny.cuny.edu

Abstract

As recommended, broadband ISDN is expected to carry all the telecommunications services provided in the future, including real time services such as telephony, videoconferencing, and videotelephony. An ATM based network will introduce some impairments not experienced in synchronous networks, such as cell delay variation (jitter) and cell loss. For these real-time services, if a cell is corrupted or lost, retransmission is not possible and so degradation of the signal may occur. In this paper, we study the impact of cell loss on speech quality over ATM networks. Moreover, we compare the results between two different cell loss's replacement techniques: stuffing silent samples and inserting the previous information in the lost cell. Study shows that the second replacement techniques produces better result when compared with the first one. The study also shows that up to 10% of speech cells can be lost over ATM networks while keeping the speech quality over MOS (Mean Opinion Score) of 3.2 for some speech coders. Understanding of the impact of cell loss on speech quality over ATM networks is important for the proper design of network algorithms such as routing, flow control, and management techniques.

1 Introduction

Broadband Integrated Service Digital Network (B-ISDN) will transport diverse classes of traffic such as data, voice, image, and video. ATM (Asynchronous Transfer Mode) is being standardized as the transport mechanism to integrate such services in a single network (Pryker, 1993). These services are likely to have a wide range of traffic characteristics, performance, and quality of service (QOS) requirements (Gibert, 1991). ATM poses some problems when applied to transmission of real-time sources such as speech (Kondo, 1993). Among the central problems in the support of real-time applications (voice, video) with ATM networks are the existence of delay jitter (Cidon, 1994) and cell loss. Designers of speech coders and networks need to work separately and together to heighten our understanding of QOS as perceived by the user (Wolf, 1991). The need for a pre-connection quality of service for statistically multiplexed connections must be assessed (Gibert, 1991).

In this paper, our objective is to understand the impact of cell loss on speech quality over ATM networks. Understanding of that impact is important for the proper design of network algorithms such as routing, flow control, and management techniques. The management techniques achieve the objective of maintaining the QOS of the ATM layer by managing the number of connections that are accepted and assigning prioritizing to control the jitter and cell loss tolerances. In emerging technology, the user expects a minimum guaranteed value of QOS regardless of traffic intensity, service variety, or network imperfections (Jayant 1993). A careful definition of the user requirements would also greatly assist in the design of future telecommunication systems, services (Wolf, 1991), and audio applications (Clark, 1992).

An objective measure of perceived speech quality is used to study the degradation effect of the transmission of speech over ATM networks. The validity of the proposed measure technique has been checked and has been found that it is highly correlated to human responses across a wide range of quality levels and for a wide range of speech processing, transmission, and transport technologies (Meky, 1996). In that algorithm, we emulate several known features of perceptual processing of speech sounds by human ear (including critical-band masking, equal loudness, and the intensity-loudness power law operations) to map the speech power spectrum into auditory power spectrum (Bark domain). Then, we use the auditory power spectrum in calculating the Bark spectral distance per band (BSDB) between the input and the output speech signals. Finally, we use the abductive network, that evolved from neural network, statistical modeling, and artificial intelligence concepts, to estimate the speech quality from the BSDB.

The study of the impact of cell loss on speech quality over ATM networks, for different cell loss distributions and speech coding algorithms, shows that:

• Up to 10 % of speech cells can be lost over ATM networks while keeping the speech quality over MOS (Mean Opinion Score) of 3.2 for some coders such as LDCELP at 16 kbps, ADPCM at 32 kbps, GSM at 13 kbps, and CS-CELP at 8 kbps;

• Replacement of the lost cell by the previous successfully received cell achieve better speech quality than insertion silent samples in the lost cell;

• Speech quality over ATM networks doesn't strongly depend on the cell loss distribution, but it mainly depends on the value of the cell loss rate.

2 Calculation of the Bark Spectral Distance per Band (BSDB) Parameters

In the proposed technique we would like to emulate several known features of perceptual processing of speech sounds by the human ear to map the speech power spectrum, P(f), into auditory power spectrum, B(z), which is assumed to represent the information conveyed by the auditory nerve to the brain. This section introduce the signal processing operations needed to calculate the BSDB parameters.

2.1 Preprocessing

As a first step, the input and output speech signals are aligned in time. The relative delays are determined by cross correlating the input and output speech envelops. To avoid system gain effects, the records are corrected to have equivalent average power in the speech periods. The speech frame of 10 msec is weighted by Hamming window and the consecutive frames overlapped by 50 %. If in a given frame the signal was found to fall below a threshold power level, the contribution of that frame (silent period) to the average distortion was set to zero. By computing the magnitude square FFT spectrum, the frame's power spectrum P(f) is calculated and followed by several stages.

2.2 Perceptual model

The perceptual model emulates the perceptual processing of speech sounds by human ear. A block diagram of the perceptual model is shown in Figure 1.

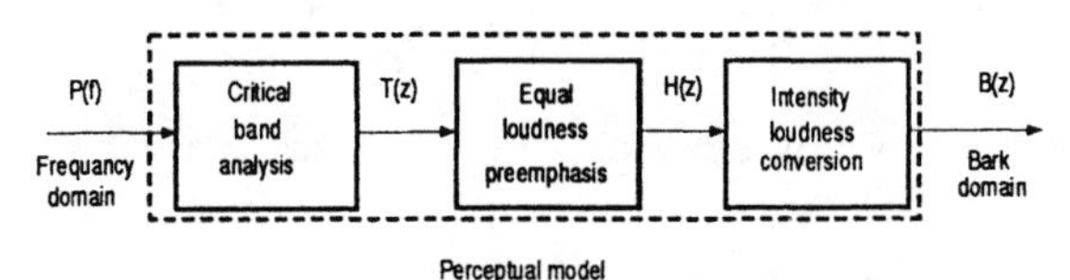

Figure 1 Block diagram of the perceptual model.

2.2.1 Critical band analysis

The procedure of converting Hz to Bark follows the established view of auditory perception in psychoacoustics, which holds that the frequency-to-place transformation along the basilar membrane of the inner ear is in terms of critical bands whose bandwidth is one Bark (Bladon, 1981). Thus as a first step, the power spectrum P(f) is warped along its frequency axis, f, into the bark frequency, z, to obtain what is called "critical-band density" (Schroeder, 1979), P(z), via the relation (Fourcin, 1977):

$$f = 600\, sinh\,(z/6) \tag{1}$$

where f is the frequency in Hz. The neural excitation pattern,T(z) which models the auditory nerve response to vowel sounds (Bladon, 1981), is calculated by convolving the critical-band density, P(z), with the critical-band masking curve, ψ(z) (Hermansky, 1990). The excitation pattern, T(z), is sampled in approximately 1–Bark intervals. Typically, 17 spectral samples of T(z) are used to cover the 0–15.575 bark (0–4 kHz) analysis bandwidth (with sample 1 equal to sample 2 and sample 16 equal to sample 1 (Hermansky, 1990)).

2.2.2 Equal-loudness preemphasis

In this stage, the threshold of hearing, the nonlinear and frequency-dependent response of the ear to intensity differences are taken into account. This is calculated by multiplying the samples of the excitation pattern T[z(f)] by the simulated equal-loudness curve, E(f) (Hermansky, 1990):

$$H(z(f)) = E(f)\,T(z(f)) \tag{2}$$

The function E(f) is an approximation to the nonequal sensitivity of human hearing at different frequencies (Robinson, 1956) and simulates the sensitivity of hearing at about the 40–dB level.

2.2.3 Intensity-loudness power law

As a last stage, the samples of the auditory power spectrum B(z) is given by applying cubic-root amplitude compression of H(z) (Zwicker, 1990):

$$B(z) = [H(z)]^{0.33} \tag{3}$$

This operation simulates the nonlinear relation between sound intensity and perceived loudness.

In summary, the perceptual model takes into account the human ear's nonlinear transformations of frequency and amplitude, together with important aspects of its frequency analysis and masking behavior in response to complex steady-state sounds.

For telephony application, thirteen samples (bands) of the auditory spectrum B(z) are used to cover the spectrum from 300–3400 Hz

2.3 Bark Spectral Distance per Band (BSDB)

The auditory spectrum B(z) reflects the ear's nonlinear transformations of frequency and amplitude, together with aspects of its frequency analysis and spectral integration properties in response to complex sounds (Gersho, 1992). For each band, i, the square Euclidean distance between the auditory spectrum of the input and the output is given by:

$$dis\left[B_x^l(i)\,,\,B_y^l(i)\right] = \left[B_x^l(i)\,-\,B_y^l(i)\right]^2 \tag{4}$$

where $B_x^l(i)$ *and* $B_y^l(i)$ are the auditory spectrum samples of frame, l, of the input and output speech respectively. We define the Bark spectral distance per band, BSDB(i) as:

$$BSDB(i) = \frac{\sum_{i=1}^{N} dis\left[B_x^l(i)\,,B_y^l(i)\right]}{\sum_{l=1}^{N}\sum_{i=1}^{b}\left[B_x^l(i)\right]^2} \tag{5}$$

where N is the number of frames in the utterance while b is the number of bands. Figure 2 illustrates the basic transformations used in obtaining BSDB(i).

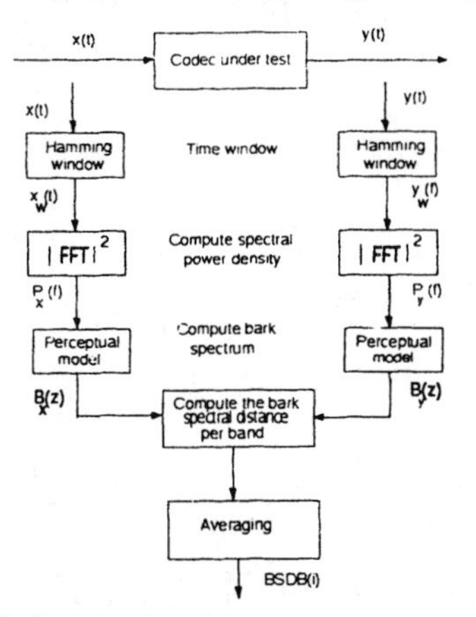

Figure 2 Basic transformations used in obtaining BSDB(i).

3 Speech Quality Evaluation System

Our evaluation system first undergoes a training phase which selects connectivity and adjusts the summation weighting functions of the abductive network (Barron, 1984), (Hess, 1987) that used to map the BSDB into the predicted MOS. The output speech sentences that processed by four different speech coders with the input speech sentences are used to prepare the learning data. The learning data base set contains 48 BSDB vectors, each of 13 elements (cover a spectrum from 300 to 3400 Hz that used in telephony), with their desired speech quality scores, each of 11 elements that match the output layer's size.

During the learning phase of our evaluation system, the actual output speech quality scores is compared to the desired scores and the errors between the actual and desired scores are then used to determine the best network structure, element types, coefficients, and connectivity that minimize the predicted square error (PSE).

The validity of the proposed evaluation system is proved by comparing the average predicted MOS obtained from our technique to those obtained from the subjective test. Figure 3 and Figure 4 present the actual MOS and the predicted one for the mixed speakers for two different evaluation systems (each uses different learning data set). The symbol r, that appears in the figures, is the correlation coefficient between the actual and predicted MOS values while the symbol s is the standard deviation of the prediction error.

It is clear that our evaluation system is robust in evaluating the MOS ratings. For example, the actual MOS for coders is54 (bit rate = 7.95 Kbps), fs1016 (bit rate = 4.8 Kbps), and g728b (bit rate = 16 Kbps) are 3.49, 3.03, and 2.31 respectively and the predicted MOS for them, using evaluation system 1, are 3.465, 3.058, and 2.4 respectively and their predicted MOS scores using the second evaluation system are 3.38, 2.988, and 2.35 respectively. These results explain that the proposed technique successfully predicts speech quality that are highly correlated to human responses across a wide range of quality levels and coding algorithms. More results and details can be found in (Meky, 1996).

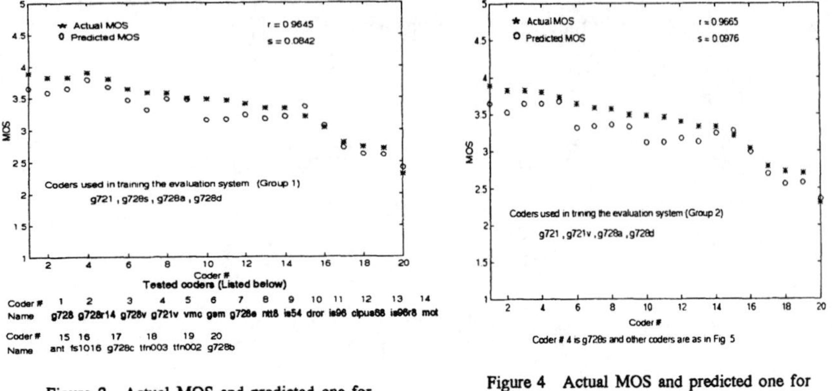

Figure 3 Actual MOS and predicted one for mixed speakers using evaluation system 1.

Figure 4 Actual MOS and predicted one for mixed speakers using evaluation system 2.

4 Impact of Cell Loss on Speech Quality

Each cell generated by a speech source is routed to the destination via a sequence of intermediate nodes. Cells may be rejected at the intermediate nodes because of buffer overflow or if the delay of that cell goes behind a predefined upper delay limit used in reconstruction of the speech cells. When a cell is lost, the receiver coder needs to deal with the resulting discontinuity in the output signal in some way. Stuffing zero samples in the lost slot, or replacing the information from a previous unlost slot are two simple possibilities (Gould, 1993).

For certain loss-rate distribution (uniform, binomial, and Poisson), and speech signal, we define the number of the losing cells and replace the lost cells either by a silent samples or by the samples of the previous cell. We use the abductive network to predict the speech quality of speech files (24 files) for certain bit-rate (coder algorithm) by feeding the trained abductive network with the BSDB between the corrupted output speech (output of ATM network) and the input speech signal (input of ATM network).

4.1 Numerical Result

In this section, we illustrate the impact of cell loss-rate (up to 10%) on the speech quality over ATM networks. Moreover, we compare the results between two replacement techniques: stuffing silent samples and inserting the previous information in the lost cell.

4.1.1 Replacement of the lost cell by silent samples

Figures 5–7 show the relation between speech quality and cell loss for speech bit rate of 128, 32, 16, 13, and 8 kbps for different cell loss distribution: uniform, binomial, and Poisson. These figures depict how the predicted MOS scores (obtained from the corresponding BSDB values) vary with cell loss rate when the lost cell was replaced by silent samples.

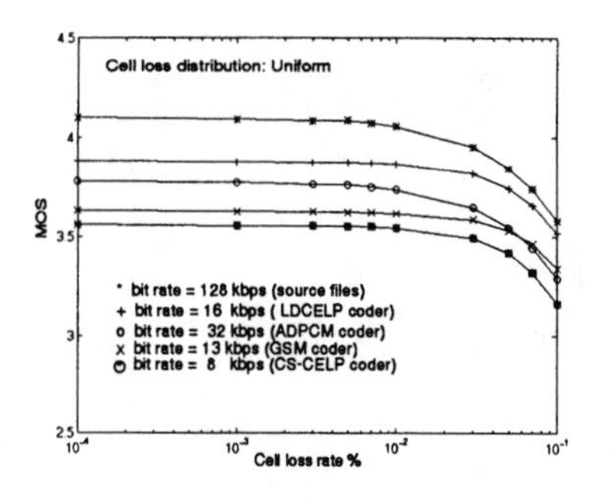

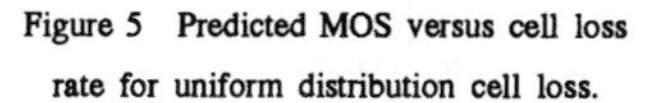

Figure 5 Predicted MOS versus cell loss rate for uniform distribution cell loss.

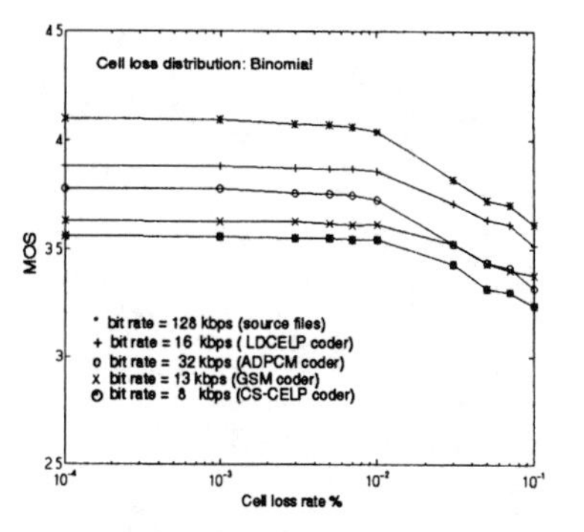

Figure 6 Predicted MOS versus cell loss rate for binomial distribution cell loss.

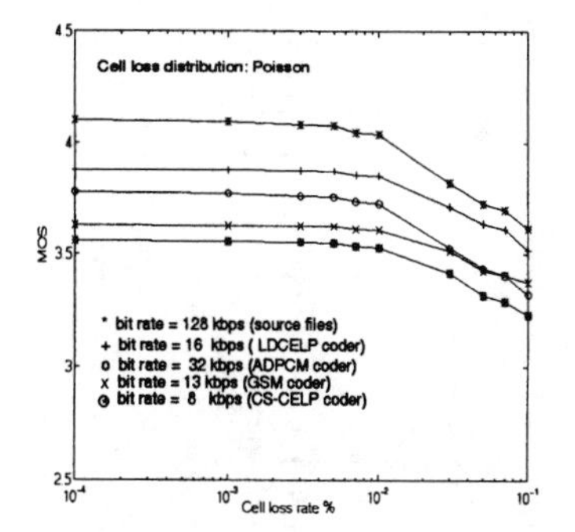

Figure 7 Predicted MOS versus cell loss rate for Poisson distribution cell loss.

Figures 5–7 show, as expected, that the speech quality decreases with increase of the cell loss rate. For example, speech quality at 10^{-4} (which we consider as zero cell loss) are 4.1, 3.77, 3.88, 3.63, and 3.56 for the source files, ADPCM coder, LDCELP coder, GSM coder, and CS-CELP coder respectively while the corresponding MOS values at 10% cell loss, for uniform cell loss distribution are 3.58, 3.29, 3.52, 3.34, and 3.16 respectively. Figures 5–7 show that with 10% cell loss rate, which is assumed to be a worst case in private ATM networks (Kondo, 1993), the quality is kept above 3.2 for bit rate ≥ 8 kbps. To study the degradation behavior for each coder algorithm, we calculate the degradation in speech quality, taking MOS at zero cell loss as a reference point, versus the cell loss as shown in Figures 8–10 for the three cell loss distributions.

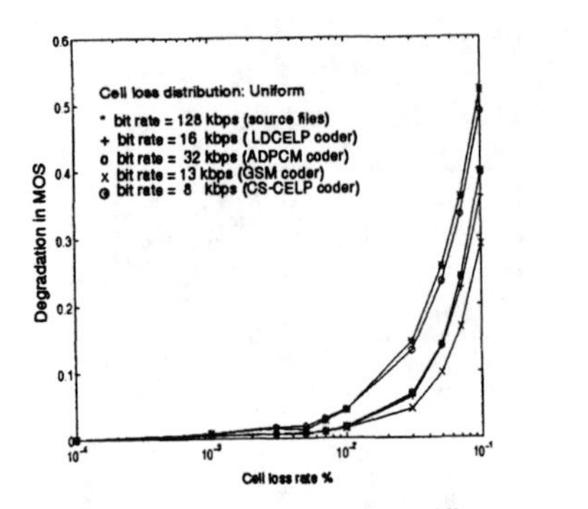

Figure 8 Degradation of speech quality versus cell loss: Uniform distribution.

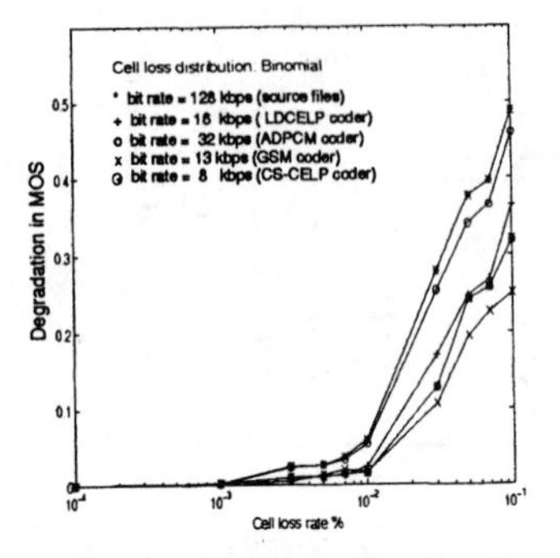

Figure 9 Degradation of speech quality versus cell loss: Binomial distribution.

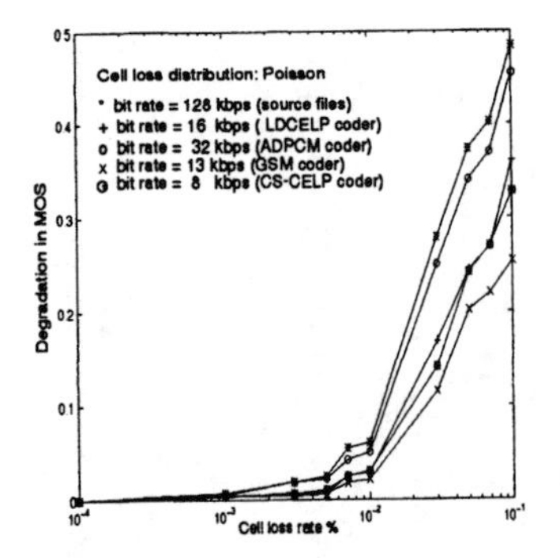

Figure 10 Degradation of speech quality versus cell loss: Poisson distribution.

Figures 8–10 shows that the degradation rate increases with the increase of coding bit rate for bit rate 128, 32, 16, 8 kbps and the lowest degradation rate is for GSM coder at 13 kbps. We believe that when the bit rate is high, the speech utterances are clear and the user can easily perceive the degradation effect, while for the lower bit rate, the speech utterances are not so clear so that the user can't easily distinguish the degradation effect for small variation in cell loss. A careful study of figures 5–10 shows that the degradation behavior of different bit rate coders are the same for the three cell loss distribution assumptions (uniform, binomial, and Poisson). Figures 11–13 depict the variation of MOS versus the cell loss for certain bit rate coder under different loss distributions. Figures 11–13 illustrate that the speech quality doesn't strongly depend on the cell loss distribution, but it mainly depends on the value of the cell loss rate. Thus, we can normalize the results for the different cell loss distributions as described in Figures 14–15.

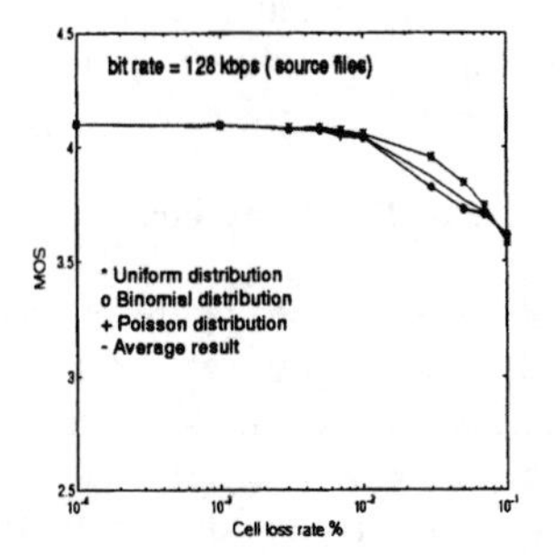

Figure 11 Degradation of speech quality of 128 kbps bit rate versus cell loss for different cell loss distributions.

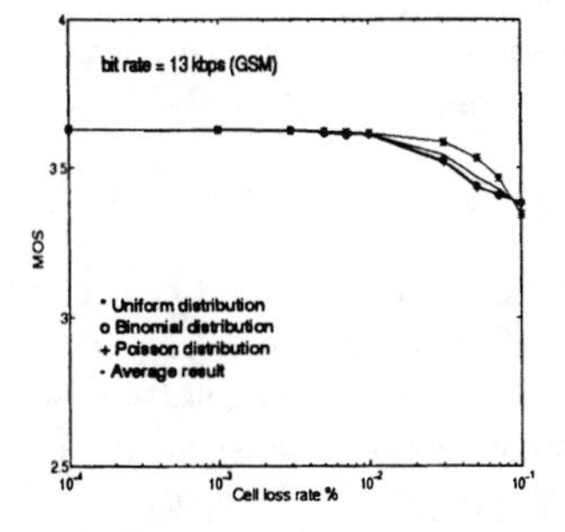

Figure 13 Degradation of speech quality of 13 kbps bit rate versus cell loss for different cell loss distributions.

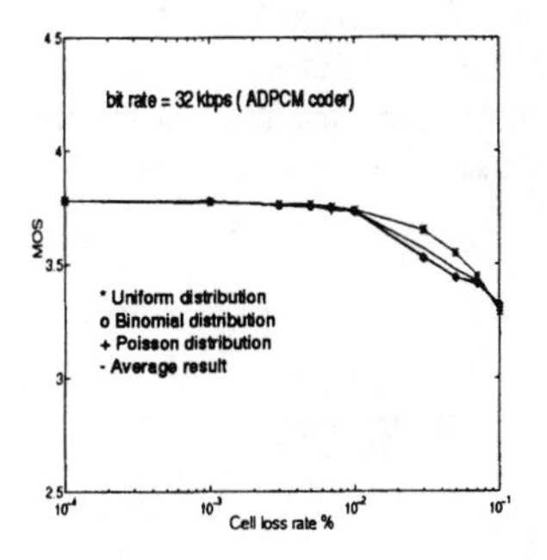

Figure 12 Degradation of speech quality of 32 kbps bit rate versus cell loss for different cell loss distributions.

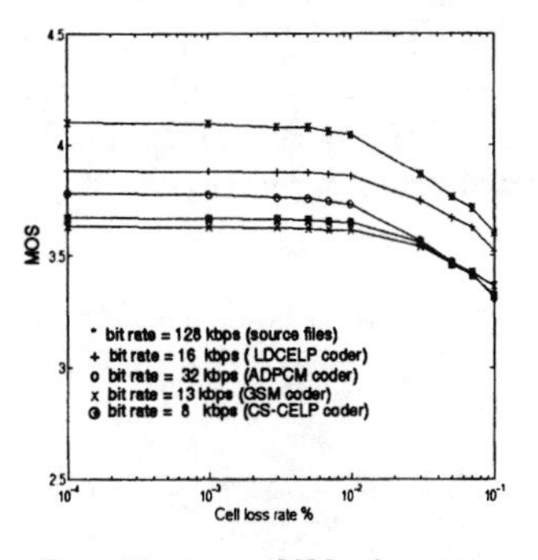

Figure 14 Average MOS values versus cell loss for different bit rate coding.

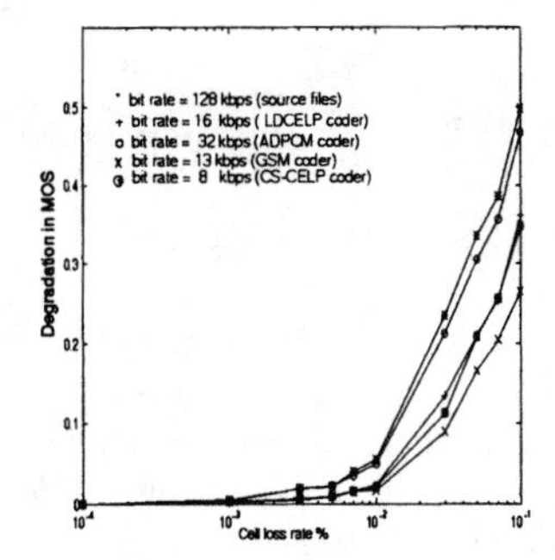

Figure 15 Average degradation in the speech quality versus cell loss for different bit rate coding.

The results from Figures 14–15 is consistent with the previous results (Figures 8–10) and can be used to study the degradation effect of cell loss on the speech quality for different coders.

4.1.2 Replacement of the lost cell by the previous successfully received one

For the second replacement technique, in which the lost cell is replaced by the previous successfully received one, we repeated the previous study (done in the first replacement technique) and we got the same behavior results as for the first replacement technique. But the second algorithm shows improvement in the speech quality with respect to the first one. For example, we choose a 32 kbps bit rate to compare the speech quality differences in case of using the two replacement techniques. This comparison is depicted in Figure 16. It is clear that for the same cell loss, second replacement technique gives a higher speech quality. For example at 10% cell loss, MOS value for the first replacement technique is 3.31 while it is 3.6 for the second replacement technique. Instead of introducing the improvement effect of the second replacement technique for every coder algorithm, we plot the improvement effect of the second replacement technique, for all coders, with cell loss variation as shown in Figure 17. In summary, Figure 17 demonstrates that the second replacement technique produce better results when compared with the first one.

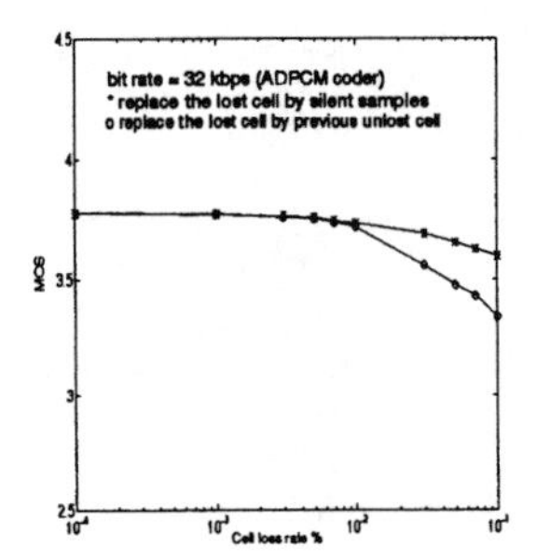

Figure 16 Average MOS for a 32 kbps bit rate for the two replacement techniques.

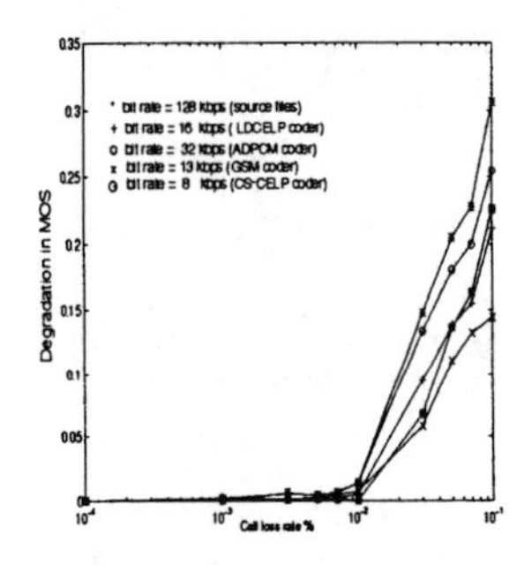

Figure 17 Improvement effect of using the second replacement technique rather than using the first one.

5 Conclusion and Further Work

This paper has presented a discussion of the issues involved in predicting the degradation impact of cell loss on speech quality over ATM network. From speech coder designing point of view, for given speech coding algorithms, the proposed technique can be used to predict the quality performance of speech coding algorithms due to cell loss impairments that will be introduced by ATM networks. Prediction the speech quality over ATM network help in designing the speech coders and controlling their electrical parameters to maintain certain speech quality.

From network design point of view, the proposed technique can be used as a tool to predict the performance of speech reconstruction algorithms, that deal with the cell loss problem, and select the reliable method. Also degradation information produced by the proposed technique can be used to aid in designing of the management, congestion control protocols and assignment rules that allow the meeting of connection performance standards and the achieving of certain quality of service (QOS) requirements. The study also shows that up to 10% of speech cells can be lost while keeping the speech quality over MOS (Mean Opinion Score) of 3.2 for some coders such as LDCELP at 16 kbps, ADPCM at 32 kbps, GSM at 13 kbps, and CS-CELP at 8 kbps.

Major research is still necessary in this area to predict the degradation impact of jitter on speech quality over ATM network.

References

[1] Barron, A. R. (1984) Predicted Squared Error: A Criterion for Automatic Model Selection: Self-Organizing Methods in Modeling, edited by S. J. Farlow, Marcel-Dekker, Inc. , New York.

[2] Bladon, R. A. W. , and Lindblom, B. (1981) Modeling vowel perception: The Journal of the Acoustical Society of America , vol. 69, pp. 1414–1422.

[3] Cidon, I. , Khamisy, A. , and Sidi, M. (1994) Dispersed Messages in Discrete-Time Queues: Delay, Jitter and Threshold Crossing: Proceedings of IEEE INFCOM'94, Toronto, Ontario , Canada, pp. 218–223, June 12–16.

[4] Clark, D. D. , Shenker, S. , and Shang, L. (1992) Supporting real-time applications in an integrated services packet network: Architecture and mechanism: Proceedings of ACM Sigcomm'92, Baltimore, MD, Augest, pp. 14–26.

[5] Fourcin, A. (1977) Speech processing by man and machine-Group report," Recognition of Complex Acoustic Signals: Dahlem Workshops, Berlin, Germany.

[6] Gersho, A. , and Wang. S. (1992) An objective measure for predicting subjective quality of speech coders: IEEE Journal on Selected Areas in Communications, vol. 10 SAC-10, pp. 819–829.

[7] Gilbert, H. , Magd, A. , and Phung, V. (1991) Developing a cohesive Traffic Management Strategy for ATM Networks: IEEE Communications Magazine, October, pp. 36–45.

[8] Gould, K. W. , et. al. (1993) Robust Speech Coding for the Indoor Wireless Channel: AT&T Technical Journal , October–November.

[9] Hermansky, H. (1990) Perceptual linear predictive (PLP) analysis of speech: Journal of the Acoustic Socity of America , vol. 87, April, pp. 1738–1752.

[10] Hess, P. (1987) Neural Network Approach to Problem Dealing with Uncertainty: Proceedings of the 3th Annual Aerospace Application of Artificial Intelligence Conference (AAAIC), pp. 89–100.

[11] Jayant, N. (1993) High Quality Networking of Audio-Visual Information: IEEE Communications Magazine, pp. 84–95.

[12] Kondo, K. , and Ohno, M. (1993) Packet Speech Transmission on ATM Networks using a variable Rate Embedded ADPCM Coding Scheme: IEEE Transactions on Communications, vol. E76B, no. 4, April.

[13] Meky, M. , and Saadawi, T. N. (1996) A Perceptually-Based Objective Measure for Speech Coders using Abductive Network: Proceedings of ICASSP: IEEE International Conference on Acoustics, Speech, and Signal Processing, Atlanta, Georgia, May 7–10.

[14] Pryker, M. (1993) Asynchronous Transfer Mode: Solution for Broadband ISDN: New York, Ellis Horwood, second edition.

[15] Robinson, D. W. , and Dadson, R. S (1956) A redetermination of the equal-loudness relationships for pure tones: Journal of the Applied Physics. vol. 7, pp. 166–181.

[16] Schroeder, M. R. , Atal, B. S. , and Hall, J. L. (1979) Objective measure of certain speech signal degradations based on masking properties of human auditory perception: Frontiers of Speech Communication, New York: Academic.

[17] Wolf, S. , Dvorak, C. A. , Kubichek, R. F. , South, C. R. , Schaphost, R. A. , and Voran, S. D. (1991) How will we rate Telecommunications system performance?: IEEE Communications Magazine, October, pp. 23–29.

[18] Zwicker E. , and Fastl H. (1990) Psychoacoustics: Facts and Models: Springer-Verlag, Berlin, 1990.

23

Shuffle vs. Kautz/De Bruijn Logical Topologies for Multihop Networks: a Throughput Comparison

*F. Bernabei *, V. De Simone *, L. Gratta *, M. Listanti #*

** Fondazione Ugo Bordoni - via B. Castiglione 59, 00142 Roma (Italy); e-mail:{bernabei, laura}@fub.it*
 # INFOCOM Dept., University "La Sapienza"; via Eudossiana 18, 00184 Roma (Italy); e-mail: marco@infocom.ing.uniroma1.it

Abstract
This paper deals with the analysis of the throughput performance of various logical topologies for Multihop Networks. In particular, ShuffleNets, De Bruijn graphs and Kautz graphs are analyzed. For the comparison, routing algorithms adopting minimum path length are considered. A hot-spot traffic scenario is adopted, modeling the presence of a centralized network resource to which a quota of the internal traffic is directed or originated from. The analysis is carried out by varying the traffic unbalance degree, from a uniform traffic distribution to a completely unbalanced one (all the traffic is concentrated in the hot-spot node).

For ShuffleNets, simple analytical expressions of the actual throughput limits are utilized. In the case of De Bruijn and Kautz graphs, instead, a lower bound of the throughput is utilized, which coincides with the actual throughput in a wide range of values of the network size.

The results obtained show that Shuffle and Kautz graphs always outperform De Bruijn topologies. Moreover, ShuffleNets present a further advantage on the other topologies; in fact, since the nodes are topologically equivalent, the placement of the hot-spot node does not affect the throughput performance.

Keywords
Multihop lightwave networks, Shuffle Networks, De Bruijn graphs, Kautz graphs

1. INTRODUCTION.

Multihop Networks (MNs) [1,2] represent an attractive solution to implement optical infrastructure interconnecting a large set of stations distributed in a local-metropolitan area.

In MNs the *physical topology* must be distinguished from the *logical topology*. The former consists in the physical medium connecting the network nodes, and is usually a star or bus; the latter is overlaid on the physical one, for example using a number of different wavelengths each between a pair of stations. In this way, each node is connected, by means of dedicated logical links, to a set of other nodes; so, a packet must be forwarded for multiple hops to reach its destination if origin and destination nodes are not directly connected in the logical topology.

The logical topology can be usefully represented by means of a graph. The indegree and the outdegree of the graph are represented by the number of logical links incoming to and outgoing from a node, respectively. The graph diameter determines the minimum number of hops needed, in the worst case, to go from to a node to another. In the class of graphs with N nodes and maximum outdegree p the minimum diameter is of $O(log_pN)$. A logical topology should be designed aiming at: i) simplifying the routing algorithm; ii) reducing, as much as possible, the number of hops to reach a destination node; iii) avoiding bottlenecks in the network in order to increase the maximum throughput. In this paper, three logical topologies are described (Shuffle, Kautz and De Bruijn topologies) and their throughput performance are compared.

The Shuffle logical topology is based on Perfect Shuffle graphs. Many papers dealt with ShuffleNets (SNs) [1,3,4,5,6,7]. They showed the highly efficient use of the communication channel under uniform traffic, whereas they outlined a remarkable throughput decrease in case of non-uniform load, partially limited by the use of an adaptive packet routing scheme.

Logical topologies based on De Bruijn and Kautz graphs have been proposed and studied in recent papers [8,9]. The comparison with the SNs is basically focused on topological properties and is limited to a uniform traffic scenario.

In [8] it is stated that, for the same maximum degree and average number of hops, logical topologies based on De Bruijn graphs can support a much larger number of nodes than ShuffleNets. Networks based on De Bruijn graphs have higher throughput and lower average delay. In [9] a comparison between De Bruijn and Kautz graphs is discussed. The main outcomes are that logical topologies based on Kautz graphs can accommodate a greater number of nodes than that given by a De Bruijn graph with the same degree and diameter. Moreover, the average queueing delay in a Kautz graph network gives better results than in a network based on De Bruijn graphs.

In this paper, a more complete analysis of the throughput performance of the three topologies is carried out, with reference to modulated hot-spot traffic configurations. Over a background given by a uniform traffic component, a hot spot traffic models the presence of a centralized network resource (e.g. a file server, a mass storage equipment, etc.) which a quota of the internal traffic is directed to or is originated from. As a particular case, if the hot spot component vanishes, a uniform traffic configuration is obtained. A high hot-spot traffic quota leads to critical conditions for MN performance, since it involves high loads on links incoming to and outgoing from the hot spot. So, a worsening of throughput performance is expected with respect to the uniform case. The study aims at evaluating the capacity of various logical topologies to mitigate the throughput degradation as the hot spot traffic quota increases.

Simple analytical expressions of the actual throughput limits, derived in [7], are used for Shuffle Networks, whereas for the other topologies an enumeration approach is applied.

Obtained results show that the performance of all the three topologies are strongly influenced by the traffic unbalance degree. Moreover, De Bruijn graphs always perform worst than Kautz graphs and ShuffleNets. However, for high values of traffic unbalance and network size the differences among the three topologies tend to vanish, and they fundamentally depend only on the network size and on the traffic unbalance value. Lastly, ShuffleNets have the exclusive advantage that, since all nodes are topologically equivalent, there is no impact of the hot-spot node placement on the throughput performance.

The paper is organized as follows. In Sec. 2 the logical topologies are described; in Sec. 3, the adopted traffic configuration is explained. In Sec. 4 the procedures employed for the performance evaluation are reported, while in Sec. 5 a performance comparison is carried out.

2. LOGICAL TOPOLOGIES FOR MULTIHOP NETWORKS

In this section, Shuffle, Kautz and De Bruijn logical topologies are described. Let $\mathcal{N}$ and $\mathcal{L}$ be the set of nodes and the set of links of a generic MN, respectively. Any node is able to detect p input frequencies and to transmit over p output frequencies. Whichever the logical network topology is, it results: $|\mathcal{L}|=p\,|\mathcal{N}|$, wherein $|\mathcal{S}|$ denotes the cardinality of the set $\mathcal{S}$.

2.1 Shuffle topology

A (p,k) ShuffleNet (SN) [2] is composed of $|\mathcal{N}|=N^S=kp^k$ nodes, logically arranged in k columns. Two adjacent columns are connected by means of a p-shuffle; the kth column is connected to the first one, in such a manner that the overall connectivity graph is wrapped around a cylinder.

A generic node $\chi\in\mathcal{N}$ is identified by the couple $[c_x, \mathbf{x}]$, where c_x ($0\leq c_x\leq k-1$) is the *column index* and $\mathbf{x}$ represents the *row index*, i.e. the string $<x_k,x_{k-1},...,x_1>$ being the p-ary representation of the row index of the node χ. Moreover, $[c_x, \mathbf{x}, x_0]$ represents the node χ outgoing link, identified by the routing digit x_0 processed by the node itself. Each value of the digit x_0 corresponds to a specific output link. The node $\chi= [c_x,<x_k,x_{k-1},...,x_1>]$ is connected to the p nodes $[c_{x+1},<x_{k-1},x_{k-2},...,x_1,x_0>]$ obtained assigning the p possible values to x_0. Figure 1 shows the logical topology of a (2,2) SN.

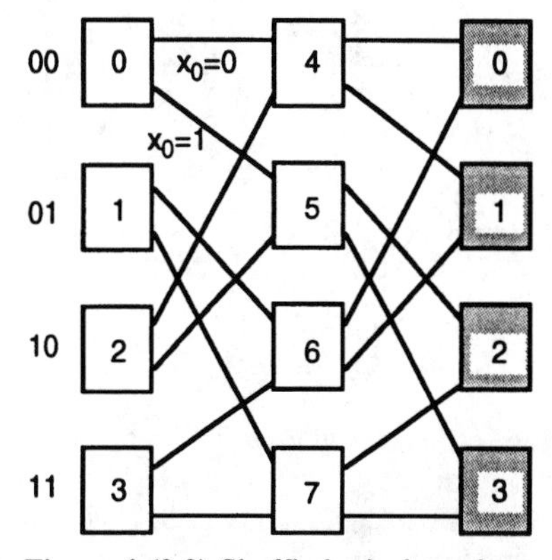

Figure 1 (2,2) Shuffle logical topology.

2.2 Kautz topology

A (p,k) *Kautz Network* (KN) [9] is composed of $|\mathcal{N}|=N^K=p^k+p^{k-1}$. The generic node $\chi\in\mathcal{N}$ is identified by a string $\mathbf{x}=<x_k, x_{k-1}, ..., x_1>$ of k $(p+1)$-ary digits, called *node index*. Each link is identified by the couple $[\mathbf{x}, x_0]$, being x_0 the routing digit processed by the node itself. The Figure 2 represents a (2,2) KN. The generic node $<x_k, x_{k-1}, ..., x_1>$ is connected to the p nodes $<x_{k-1}, x_{k-2}, ..., x_1, x_0>$ obtained by assigning the p possible values to x_0 such that $x_0\neq x_1$. It is to be noted that this constraint implies that there is no link connecting a node to itself, and there is no node such that: $x_i = x_{i+1}$, $\forall\ i\in[0,k-1]$.

2.3 De Bruijn topology.

A (p,k) *De Bruijn Network* (DN) [8] is composed of $|\mathcal{N}|=N^D=p^k$. The node index of a generic node $\chi\in\mathcal{N}$ is identified by a string $\mathbf{x}=<x_k, x_{k-1}, ..., x_1>$ of k p-ary digits. Each link is identified by the couple $[\mathbf{x}, x_0]$, being x_0 the routing digit processed by the node itself. The Figure 3 represents a (2,3) DN. The generic node $<x_k, x_{k-1}, ..., x_1>$ is connected to the p nodes $<x_{k-1}, x_{k-2}, ..., x_1, x_0>$ obtained by assigning the p possible values to x_0. It is to be noted that, if all the digits of the node index are equal, there is one link connecting the node to itself. As it will be shown in the following, the presence of these links impairs the throughput performance of DNs as compared to KNs.

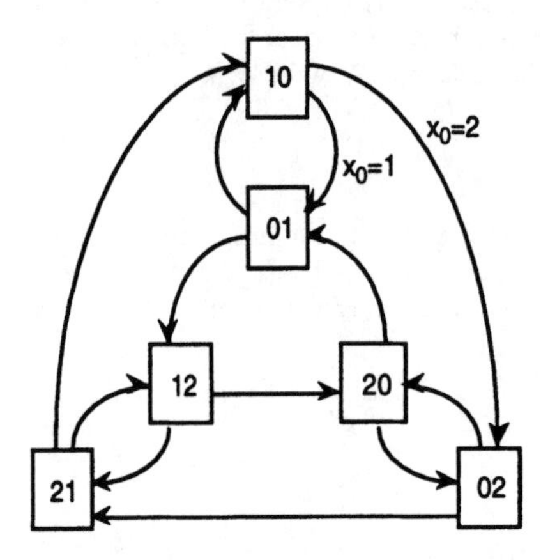

Figure 2 (2,2) Kautz logical topology.

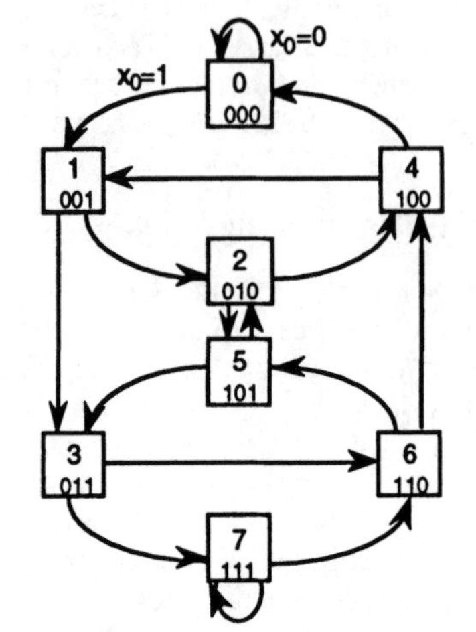

Figure 3 (2,3) De Bruijn logical topology.

3. TRAFFIC SCENARIO.

In this section, the adopted traffic scenario is described, and the performance measures used in the comparison of the logical topologies are introduced. In the following, all the traffic measures are intended as normalized with respect to the link capacity, supposed equal for all the network links. Let:

a_{oxy} the normalized mean traffic generated by the traffic relation (x,y); it is worth noting that, the traffic components relevant to the traffic relations originating from and terminating at the same node are not considered, therefore $a_{oxx} = 0, \forall x$;

a_{cxy} the normalized mean carried traffic relevant to the traffic relation (x,y);

A_{ox} the overall normalized mean traffic offered by a node $x \in \mathcal{N}$ i.e. $A_{ox} = \Sigma_{y \in \mathcal{N}} a_{oxy}$;

A_{cx} the normalized throughput of a node $x \in \mathcal{N}$ defined as the sum of the normalized mean carried traffic relevant to all the traffic relations originating from the node i.e. $A_{cx} = \Sigma_{y \in \mathcal{N}} a_{cxy}$;

A_o be the overall normalized mean load offered to the network, i.e. $A_o = \Sigma_{x \in \mathcal{N}} A_{ox}$;

A_c be the normalized total traffic carried by the network, i.e. $A_c = \Sigma_{x \in \mathcal{N}} A_{cx}$.

The evaluation of the maximum network throughput is here carried out by considering traffic configurations resulting from the superposition of different percentages of uniform and hot-spot traffic patterns. A specific traffic configuration is characterized by: a) a percentage, α, $(0 \leq \alpha \leq 1)$, called *hot-spot factor*, of A_o associated with the traffic relations originating from and destined to a given node (that is, the *hot-spot*); b) the remaining load, $(1-\alpha)A_o$, is uniformly distributed among all the traffic relations.

Let the node z be the hot-spot; by considering that the number of traffic relations originating from or terminating at z is equal to $2\cdot(N-1)$, we have:

$$a_{oxy}(\alpha) = \begin{cases} \frac{F}{N-1} & x \neq z \text{ and } y \neq z \\ \frac{F}{N-1} + \frac{\alpha A_o}{2(N-1)} & x = z \text{ or } y = z \end{cases} \qquad A_{ox}(\alpha) = \begin{cases} F + \frac{\alpha A_o}{2(N-1)} & x \neq z \\ F + \frac{\alpha A_o}{2} & x = z \end{cases}$$

wherein $F = \frac{(1-\alpha)A_o}{N}$.

It is to be noted that, $\alpha=0$ corresponds to the uniform traffic case. In the following, for the sake of brevity, a traffic configuration with $\alpha\neq0$ will be called "α-hot-spot".

The α-hot-spot traffic configuration models the presence of a centralized network resource, corresponding to the hot-spot node (e.g. a file server, a mass storage equipment, etc.), to which a quota of the internal traffic is directed or is originated from. The higher α is, the more the traffic configuration represents a critical case. This is due to the fact that it involves high loads on incoming and outgoing hot-spot node links. Therefore, a worsening of throughput performance can be expected as α increases.

If a value of α is fixed, the normalized maximum throughput, $T_{max}(\alpha)$, is defined as the maximum value of traffic carried by the network (normalized with respect to the number of nodes N) provided that the characteristics of the carried traffic be identical with those of the offered one, i.e.

$$T_{max}(\alpha) = \left\{ \frac{1}{N} \max_{A_o}[A_c] \mid \frac{a_{cxy}}{a_{czw}} = \frac{a_{oxy}}{a_{ozw}}, \forall x, y, z, w \right\}$$

In the case of uniform load, $T_{max}(\alpha=0)$ represents the maximum value of the offered traffic such that the traffic carried by the network be equally distributed over all the traffic relations.

It is worth noting that in some cases $T_{max}(\alpha)$ does not correspond to the actual maximum possible value of network throughput. In fact, if the network links are not uniformly loaded, a node could generate, at least in principle, more traffic over those paths crossing non-congested network zones. That would lead to a higher throughput, but the distribution of the carried traffic on the various traffic relations would not correspond to the hypothesized characteristics of the offered traffic. In this sense, $T_{max}(\alpha)$ corresponds to the maximum throughput arising from a perfectly "fair" network operation with respect to the various traffic relations.

According to the hypotheses on the traffic model, the evaluation of $T_{max}(\alpha)$ can be carried out by identifying the link carrying the maximum load and by counting the number of paths belonging to the uniform and hot-spot traffic components crossing that link.

Let $\ell \in \mathcal{L}$ be a generic network link. Let:

N^x the number of nodes of a generic topology (x = S, K, D);

$N_h(\ell)$ the number of paths belonging to the hot-spot traffic component passing through ℓ;

$N_u(\ell)$ the number of paths belonging to the uniform traffic component passing through ℓ;

ℓ_{max} the most loaded link(s) in the network;

$a(\ell)$ the mean traffic carried by the link ℓ

It results:

$$a(\ell_{max}) = \max_{\ell \in \mathcal{L}} \left\{ N_u(\ell) \frac{(1-\alpha) A_o}{N^x (N^x-1)} + N_h(\ell) \frac{\alpha A_o}{2 (N^x-1)} \right\} \quad (1)$$

Since $a(\ell_{max}) \leq 1$, and normalizing with respect to N^x, it follows that:

$$T^x_{max}(\alpha) = \frac{2(N^x-1)}{2(1-\alpha)N_u(\ell_{max}) + \alpha N^x N_h(\ell_{max})} \quad (2)$$

4. MAXIMUM THROUGHPUT EVALUATION.

A sequence of links and nodes connecting two nodes χ and y defines a path. The number of hops of a path identifies the *path length*. In order to identify a path of length h a string of $k+h$ digits can be used. A schematic description of the general path format is given in Figure 4. The first k digits correspond to the index (row index for SNs, node index for KNs and DNs) of origin node χ, while the rightmost h digits identify the routing digits. The digit r_j is processed at the jth hop. A rightward sliding window of width k identifies, step by step, the index of all the crossed nodes from the origin to the destination. In particular, the last k digits identify the index of the destination node y. Routing algorithms for Multihop Networks can be classified according to the criterion adopted to fix the number and the value of the routing digits $<r_1,r_2,\ldots,r_h>$.

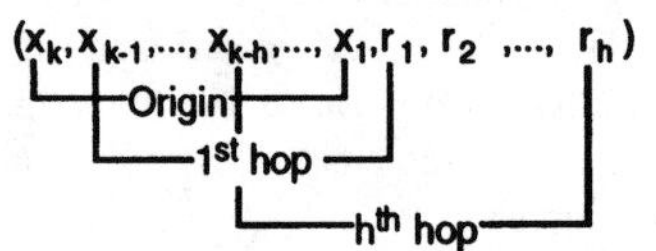

Figure 4 General path format.

The performance comparison of the three logical topologies is carried out by uniquely considering minimum path length algorithms. In order to better understand the comparison, it is to be noted that, in the case of SNs, since the minimum length path is not unique, the multipath capability can be exploited. This does not hold for Kautz and De Bruijn, since there is a single minimum length path.

In the following, a brief description of the minimum length routing algorithms adopted for the three logical topologies will be given. At this aim, we give some definitions.

Let $\mathbf{x} = <x_k,x_{k-1},\ldots,x_1>$ be a string of k digits. We denote as $\lambda_i(\mathbf{x})$, $i \leq k$, the i least significant digits of $\mathbf{x}$, i.e., $\lambda_i(\mathbf{x}) = <x_i,x_{i-1},\ldots,x_1>$ and with $\mu_i(\mathbf{x})$, $i \leq k$, the i most significant digits of $\mathbf{x}$, i.e., $\mu_i(\mathbf{x}) = <x_k,x_{k-1},\ldots,x_{k-i+1}>$. Moreover, δ_k represents a string composed of k *don't care* digit. Lastly, $\mathbf{x}\cdot\mathbf{y}$ represents the juxtaposition of the two strings $\mathbf{x}$ and $\mathbf{y}$.

4.1 Maximum throughput in SNs

Let w be the number of columns from the origin to the destination node; it results:

$$w = \begin{cases} (k + c_y - c_x) \bmod k & \text{if } c_y \neq c_x \\ k & \text{if } c_y = c_x \end{cases}$$

wherein c_x and c_y are the column indexes of nodes χ and y, respectively.

The minimum path length is equal to either w or $w+k$, and therefore its range is between 1 and $2k$-1. Let $\mathcal{M}(\chi,y)$ be the set of minimum length paths for the traffic relation (χ,y). If $h=w$ the minimum length path is unique ($|\mathcal{M}(\chi,y)| = 1$) and is unambiguously determined by the string $\mathbf{x}\cdot\lambda_w(\mathbf{y})$, otherwise there exist p^w minimum length paths ($|\mathcal{M}(\chi,y)| = p^w$), which are obtained by fixing arbitrarily the first w routing digits of the path $\mathbf{x}\cdot\delta_w\cdot\mathbf{y}$.

In [2] a fixed minimum length routing algorithm, called in the following *Fixed Routing on ShuffleNets* (FR), is proposed. According to this algorithm, if $\mu_{k-w}(\mathbf{y}) = \lambda_{k-w}(\mathbf{x})$ the routing path is $\mathbf{x}\cdot\lambda_w(\mathbf{y})$; otherwise it is $\mathbf{x}\cdot\lambda_w(\mathbf{y})\cdot\mathbf{y}$. This algorithm identifies, for each traffic relation, a unique route, and therefore it does not exploit the multipath capability of SNs. More details on the properties of FR can be found in [7].

Closed formulae for the evaluation of the maximum throughput for FR ($T_{max}^{FR}(\alpha)$) are demonstrated in [7]; here we only recall the fundamental results of that analysis. Given a (p,k) SN operating according to FR and loaded by an α-hot spot traffic, with $0 \leq \alpha \leq 1$, $N_h(\ell_{max})$ and $N_u(\ell_{max})$, are given by:

$$\begin{cases} N_h(\ell_{max})=4 & N_u(\ell_{max})=8 & \text{if } k=p=2 \\ N_h(\ell_{max})=(k-1)p^k & N_u(\ell_{max})=\dfrac{p^{k+1}[p^3(k^2-k)-p(3k^2-k)+2k^2]-2k+2kp}{2\cdot(1-p)^2} & \text{otherwise} \end{cases} \tag{3}$$

Therefore $T_{max}^{FR}(\alpha)$ can be evaluated by substituting eq.(3) in eq.(2).

Throughput advantages can be obtained if the multipath capability is exploited, still keeping the minimum length constraint. The maximum throughput for minimum path length algorithms can be found by solving a linear optimization problem (*Globally Optimized Minimum length Routing*, GOMR). For network sizes greater than a few tens of nodes, the solution of the problem becomes unfeasible. However, a tight upper bound for the maximum network throughput ($U_{GOMR}(\alpha)$) is given by [7]:

$$U_{GOMR}(\alpha) = \frac{1}{\max\left[C_1(k,\alpha),\ C_2(k,\alpha)\right]} \tag{4}$$

where

$$C_1(k,\alpha) = \frac{\alpha N(N-1)(1-p)^2 + 2(1-\alpha)[k(p^{k+2}-p^{k+1}-(1-p)^2)+p^2-p^{k+2}+(1-p)^2kp^k]}{2\,p\,(1-p)^2\,(N-1)}$$

$$C_2(k,\alpha) = \frac{3k^2p^k(1-p)+kp^k(1+p)-2k}{2\,p\,(1-p)\,(N-1)}$$

The upper bound provided by eq. (4) is very close to the actual value of the maximum throughput, except for low values of α for the (2,2) SN. For any other values of p and k the upper bound practically coincides with the maximum, as it arises from the analysis of Figure 5.

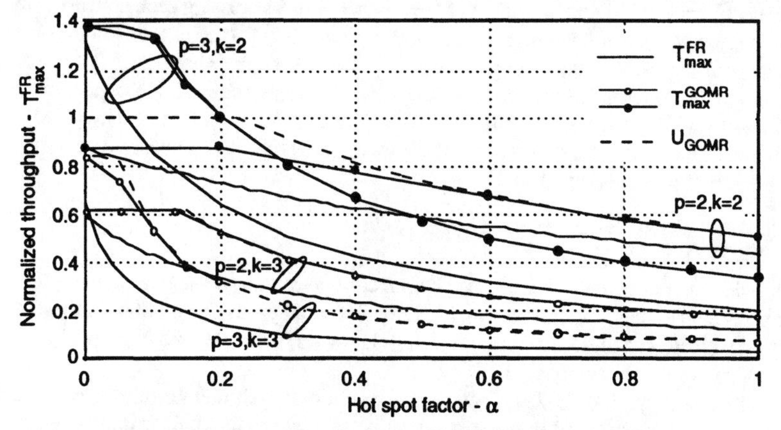

Figure 5 Throughput limits for ShuffleNets, p = 2 and 3.

This figure plots $T_{max}^{FR}(\alpha)$, $T_{max}^{GOMR}(\alpha)$ and $U_{GOMR}(\alpha)$ vs. α for k = 2,3, for p = 2 and 3, respectively. In the following, $U_{GOMR}(\alpha)$ will be employed in the performance comparison with DNs and KNs, except in the case of the (2,2) SN, where $T_{max}^{GOMR}(\alpha)$ will be reported.

From the figure it is evident that the node throughput considerably decreases as α increases; however, for low values of α GOMR is able to achieve a throughput close to the value relevant to the uniform traffic case. It must be observed that, in the SN case, the hot-spot node can be placed in any network node without affecting the throughput performance, since all the nodes are topologically equivalent [7]. This is not true for the other network topologies, as it will be shown in the following.

4.2 Maximum throughput in KNs and DNs

The routing algorithm for KNs and DNs is based on the same principles; what follows applies to both topologies. Given a (p,k) KN or DN, for each traffic relation $(\mathcal{X},\mathcal{Y})$ there is a unique minimum length path ($|\mathcal{M}(\mathcal{X},\mathcal{Y})| = 1$). The maximum number of hops of a minimum length path is equal to k; therefore, the length s of the string representing the path is such that $k+1<s<2k$. The minimum length path can be found according to the ***Minimum Length routing on Kautz graphs*** (MLK) (***Minimum Length routing on De Bruijn graphs*** (MLD)) consisting in the following two steps:

For each traffic relation $(\mathcal{X},\mathcal{Y})$,

- find the largest integer $0 \le i \le k$ such that $\lambda_i(\mathbf{x}) = \mu_i(\mathbf{y})$
- the path from $\mathcal{X}$ to $\mathcal{Y}$ is identified by $\mathbf{x} \bullet \lambda_{k-i}(\mathbf{y})$.

It is to be noted that, taking into account the routing algorithm operation, all the links (nodes) whose index can be obtained by a transliteration of a same link (node) string are topologically equivalent [(1)]. This means that, in the case of links, they are loaded by the same number of paths while, for the nodes, this means that they have the same configuration of incoming and outgoing links.

A detailed description of the above routing algorithm for KNs and DNs and its properties can be found in [9] and [8], respectively.

In order to evaluate the throughput performance of MLK (MLD), $N_u(\ell_{max})$ and $N_h(\ell_{max})$ must be evaluated. That would imply the count of the number of paths crossing each network link.

As for the uniform component, a useful upper bound for $N_u(\ell_{max})$, that holds for both topologies, can be easily found. The generic path passing through $\ell = <l_k, l_{k-1}, \ldots, l_1, l_0>$ can be written as $\delta_a \bullet \ell \bullet \delta_b$, wherein $0 \le a+b \le k-1$. All the possible origin nodes $\mathcal{A}$ of the path are such that $\mathcal{A} = \delta_a \bullet \mu_{k-a}(\ell)$, while the possible destination nodes $\mathcal{B}$ are such that $\mathcal{B} = \lambda_{k-b}(\ell) \bullet \delta_b$. An upper bound of the maximum number of paths passing through $\forall \ell \in \mathcal{L}$ is obtained by assigning all the possible values to δ_a and δ_b, i.e.

$$N_{UP} = \sum_{a=0}^{k-1} \sum_{b=0}^{k-1-a} p^{(a+b)} = \frac{kp^k(p-1)-p^k+1}{(p-1)^2} \tag{5}$$

N_{UP} provides an upper bound for the exact number of paths crossing a generic link ℓ. In fact, shorter paths connecting $\mathcal{A}$ and $\mathcal{B}$ not passing through ℓ could exist. The number of these paths must be taken into account to determine $N_u(\ell_{max})$. However, in most cases $N_{UP}=N_u(\ell_{max})$; this occurs in the following conditions:

- for Kautz topology, if $k \le 2p-2$; in fact, there exists a link such that no shorter path can be found among those counted in N_{UP}; in the string relevant to this link at most one digit is repeated, and this digit appears neither in l_k nor in l_0. For example, the string <0121314> (and any its transliteration) represents the most loaded link in a (4,6) KN.

(1) A transliteration of the string $<x_k, x_{k-1}, \ldots, x_1>$ is any string such that at least one digit is substituted, and the equality and inequality relations are kept. For example, in a (2,2) KN the links <010>, <101>, <020>, <202>, <121> and <212> are all transliterations of the same link.

- for De Bruijn topology, the condition is $k \leq 2p-4$; in fact, in addition to the above statements, the constraint that two occurrences of the repeated digit cannot be adjacent in the string must be taken into account. This constraint is automatically satisfied in the case of KNs. The string <0121314> represents the most loaded link in a (5,6) DN.

If the above conditions are not satisfied $N_u(\ell)$ is here evaluated by enumeration.

As far as $N_h(\ell)$ is concerned, a closed form expression is not presently available; so its evaluation was carried out by enumeration.

Figures 6 and 7 plot the maximum throughput of MLK and MLD vs. α, respectively. It is to be noted that, in both cases, the network performance depends on the position of the hot-spot node. In fact, the number of paths crossing the links outgoing from a node varies with the node index, i.e., the network nodes are not topologically equivalent. Consequently, for each value of α, there exist a best and a worst position for the hot-spot node; therefore, in the figures the throughput performance is represented by a region delimited by the curves corresponding to the best and worst placement of the hot-spot. It is to be noted that the curves relevant to SN, KN and DN cannot be quantitatively compared, since they are relevant to networks of different size.

Figure 6 plots $T_{max}^{MLK}(\alpha)$ vs. α for k=2,3 and for p=2,3. It should be noted that, when k=2 (whichever p is), all the node indexes can be derived from the index of a node by means of a transliteration, and therefore they are all topologically equivalent. Consequently, the hot-spot placement has no impact on the throughput.

If k=3, in the two cases reported here, the best hot-spot position is the same for any value of α, and the throughput decrease caused by a wrong hot-spot placement is not sensitive.

For higher values of k it has been verified that the best hot-spot position varies with α, although the relevant curves still have the same qualitative behaviour.

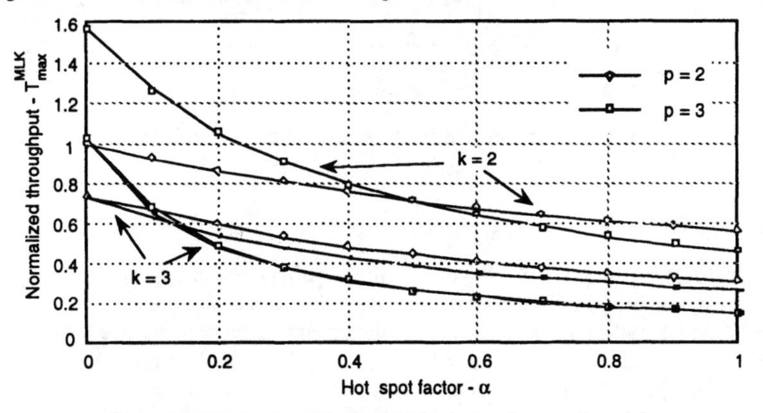

Figure 6 Throughput limits for Kautz graphs, p = 2 and 3.

Figure 7 plots $T_{max}^{MLD}(\alpha)$ vs. α for k = 2,3 and for p = 2,3. In the DN case, it resulted that in most cases the best hot-spot position varies with α.

For example, in the (2,3) case for low values of α ($\alpha \leq 0.3$) the best position for the hot-spot is in the node <000> (or equivalently in <111>). That can be justified by considering that the links incoming and outgoing in these nodes are those less loaded by the uniform traffic. Consequently, the addition of the hot-spot traffic component on these links does not cause a more stringent bottleneck to be introduced in the network; conversely, the load on the bottleneck links (<1001> and <0110>) decreases, since it is mostly determined by the uniform traffic component. This causes a growth of the maximum throughput for low values of α. As the traffic unbalance further increases, the best hot-spot position moves to <010> (or <101>). In fact, since the nodes <000> and <111> have only one incoming and one outgoing link (the recirculating links don't carry traffic) for high values of traffic unbalance these links would become very stringent bottlenecks. The new position of the hot-spot node causes the bottleneck links to be <0101> and <1010> for any further increase of α. The same considerations can be carried out for p=3. The case (2,2) is somehow degenerate since the uniform traffic component is the same on all the links (excluding <000> and <111>); consequently, the best hot-spot position is always the same (<01> or <10>), and the maximum throughput decreases with α. As it can be noted, in the case of DN an inconvenient placement of the hot-spot node can lead to a sensitive worsening of the network performance.

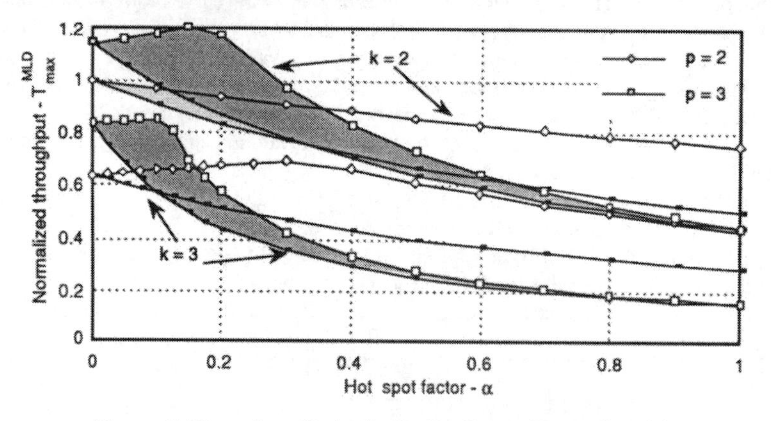

Figure 7 Throughput limits for De Bruijn graphs, p = 2 and 3.

5. PERFORMANCE COMPARISON

In the following, the performance of the three logical topologies, with respect to the maximum throughput they can carry in a hot-spot traffic configuration, will be analyzed.

Figures 8 and 9 plot the maximum normalized throughput vs. the network size N for SN, DN and KN, for $\alpha = 0$, 0.2 and 1 and for $p = 2$ and 4, respectively. In the case of SN, both $T_{max}^{FR}(\alpha)$ and $U_{GOMR}(\alpha)$ are plotted. The former represents the basic performance of SNs, while the latter represents the best obtainable performance for these networks with minimum length routing algorithms. Although the curve of $U_{GOMR}(\alpha)$ cannot be actually reached unless a globally optimized algorithm is used, the performance of locally optimized adaptive algorithms can be very close to it [7].

If a value of $\alpha \neq 0$ is fixed, the curves relevant to $T_{max}^{FR}(\alpha)$ for SNs tend toward 0 as $2/\alpha N$, independently of p [7]. That can be explained by observing that, from eq. (3), $N_h(\ell_{max})$ is equal to $(k-1)p^k$, while the hot-spot component on all the other hot-spot outgoing links is equal to $kp^k-1-(k-1)p^k = p^k-1$. Consequently, the remaining p-1 links outgoing from the hot-spot have a negligible effect on the throughput limit.

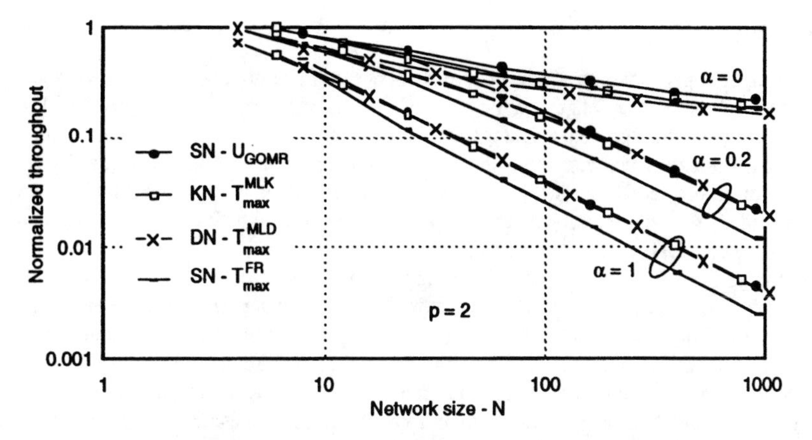

Figure 8 Normalized throughput various values of α, $p = 2$.

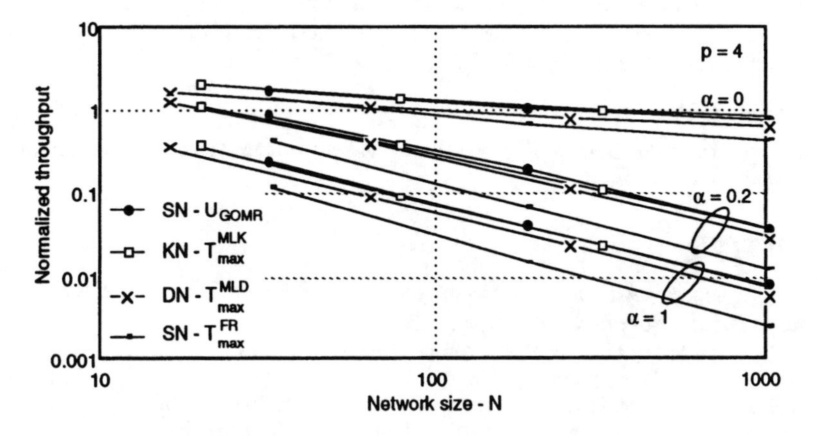

Figure 9 Normalized throughput various values of α, $p = 4$.

Conversely, from eq. (4) it results that, when the term $C_1(k,\alpha)$ dominates, the curves relevant to $U_{GOMR}(\alpha)$ tend to 0 as $2p/\alpha N$. Consequently, they are shifted upwards with respect to those relevant to $T_{max}^{FR}(\alpha)$. By comparing $C_1(k,\alpha)$ and $C_2(k,\alpha)$ it can be seen that, for increasing values of k and p, $C_1(k,\alpha)$ is greater than $C_2(k,\alpha)$ even for very small α. Conversely, for $\alpha=0$ it is always $C_1(k,\alpha) < C_2(k,\alpha)$. The considerable difference between $U_{GOMR}(\alpha)$ and $T_{max}^{FR}(\alpha)$ for $\alpha \neq 0$ can be intuitively explained by considering that, for increasing network size, FR is not able to reroute through alternative paths the traffic component which is not direct to the hot-spot node. Moreover, the asymptotic behaviour of the curves relevant to FR does not depend on p.

From the graph it results that also in the case of DNs and KNs the maximum throughput has the same behaviour for increasing values of N, i.e., it tends to 0 with decay rate $1/\alpha N$. In all the cases, the performance of MLK and MLD is between that of FR and GOMR; in most cases, $T_{max}^{MLK}(\alpha)$ practically coincides with $U_{GOMR}(\alpha)$.

6. CONCLUSIONS

The throughput performance of Multihop Networks loaded by hot-spot traffic and operating according to minimum path length routing strategies has been analyzed. Three logical topologies based on Perfect Shuffle graphs, Kautz graphs and De Bruijn graphs have been considered.

Analytical expressions of maximum throughput values have been utilized for ShuffleNets; whereas, for Kautz and De Bruijn networks, a lower bound has been found which, in most cases, provides the exact throughput value. When this is not true, a path enumeration technique for the throughput evaluation has been used.

From the analysis it turned out that the throughput performance of the three topologies is qualitatively equivalent. More precisely, ShuffleNets and Kautz networks always slightly outperform De Bruijn graphs, while they have throughput performance practically coincident.

Generally speaking, the network performance depends fundamentally on the number of network nodes, whereas the impact of the particular choice of p and k is negligible for all the topologies. The only distinctive element among these topologies consists in the higher flexibility of ShuffleNets, since in these networks the hot-spot can be placed indifferently in each node.

Further work is now proceeding to find closed form expressions for the maximum throughput for Kautz and De Bruijn topologies. Moreover, impact on adaptive routing schemes on Kautz and De Bruijn networks have to be analyzed.

ACKNOWLEDGEMENT

Work carried out in the framework of the agreement between the Italian PT Administration and the Fondazione Ugo Bordoni.

REFERENCES

[1] A. S. Acampora: "A Multichannel Multihop Local Lightwave Network". *Globecom'87*, Tokyo (Japan), November 1987, paper 37.5.

[2] A. S. Acampora, M. J. Karol: "An Overview of Lightwave Packet Networks". *IEEE Network*, January 1989, pp. 29-41.

[3] M. Eisenberg, N. Mehravari: "Performance of the Multichannel Multihop Lightwave Network under non uniform traffic". *IEEE JSAC*, Vol. 6, N° 7, August 1988.

[4] M. J. Karol, S. Z. Shaikh: "A simple adaptive routing scheme for congestion control in ShuffleNet Multihop Lightwave Networks", *IEEE JSAC*, Vol. 9, N° 7, September 1991, pp. 1040-1051.

[5] A. S. Acampora, S. I. A. Shah: "Multihop lightwave networks: a comparison of store-and-forward and hot-potato routing", *IEEE Transactions on Communications*, Vol. 40, N° 6, June 1992, pp. 1082-1090.

[6] M. Kadoch, A.K. Elhakeem: "Adaptive Routing for Lightwave Shufflenet Under Unbalanced Loads". *Globecom'91*, San Diego (USA), November 1991, paper 52.7.

[7] F. Bernabei, L. Gratta, M. Listanti: "Throughput Analysis of Multihop ShuffleNets in a Hot-Spot Traffic Scenario: Impact of Routing Strategies". *Computer Networks and ISDN Systems*, Vol. 28, N° 6, April 1996.

[8] K. N. Sivarajan, R. Ramaswami: "Lightwave Networks Based on De Bruijn Graphs". *IEEE/ACM Transactions on Networking*, Vol. 2, N° 1, February 1994, pp. 70-79.

[9] G. Panchapakesan, A. Sengupta: "On Multihop Optical Network Topology Using Kautz Digraphs". *Infocom 95*, Toronto (Canada), April 1995, paper 6a.1.1, pp. 675-682.

24

Service Specific Connection Oriented Protocol (SSCOP) with no credit limitation

Gérard Hébuterne
Institut National des Télécommunications,
gerard.hebuterne@int-evry.fr

Wei Monin
France Telecom/CNET
wei.monin@lannion.cnet.fr

Abstract

An error correction procedure is currently being defined by ITU-T under the name of "Service Specific Connection Oriented Protocol". It is to be used in the B-ISDN in conjunction with AAL5 in order to provide an error-free data link layer. We study the performances of the procedure in terms of end-to-end delay. An analytical model is presented, which allows to address various traffic engineering aspects for realistic loss probability figures, and especially the influence of the Poll period.

Keywords: Error correction, Selective retransmission, SSCOP, B-ISDN.

1 Introduction

According to ITU-T Recommendation I.121 [4], the future Broadband ISDN is to be based upon ATM, which has thus to be seen as an universal and powerful transport network. To achieve this goal, and to attain its maximum efficiency, the ATM layer only provides basic transfer functions. Especially, no error correction is performed at this level. This is because such a function is not of "universal" nature: real time services must not rely on error correction, for instance.

However, there does exist services which strongly rely on error-free connections. Signaling is one of such services, and ITU-T has undertaken the definition of the "Service Specific Connection Oriented Procedure" (SSCOP). SSCOP is to be used as Service Specific Convergence Sublayer of the ATM Application Layer type 5 (AAL5). Although SSCOP has been devised first for signaling needs, its use is by no means restricted and any data service based upon the AAL5 may rely on it.
In order to operate an error-free data connection, various engineering rules have to be devised. Quantiles of end-to-end delay are probably the most important Traffic-related QOS figures – allowing to define response time of signaling functions; also, their value are necessary for a correct dimensioning of timeouts, etc. A preliminary simulation study [3] allowed to characterize the protocol behaviour for high (and rather unrealistic) loss figures.
In this paper, we give an analytical derivation for the probability distribution of the end-to-end transfer delay for the Selective Retransmission SSCOP. The expression allows to estimate the quantiles of the delay, as a function of loss probability, of the round-trip delay and of the POLL period, for loss figures corresponding to what can be expected from nowadays networks.

2 Protocol Description

The description given here conforms with the ITU-T references of SG XI, May 1993 [5], both as SDL or textual parts. Only the data transfer part is of concern.
SSCOP ensures reliable data transfer between users of the ATM Application Layer (AAL). When the connection is established, the Sender may send data as long as it has credits available, which correspond to available buffer capacity at the Receiver's end. In case of lost data, the Receiver requests retransmission to the Sender, according to a selective retransmission scheme (see e.g. [1]).
The protocol makes use of three types of Protocol Data Units (PDUs): Sequences Data PDUs (SD PDU), Poll PDUs and Stat PDUs. SD PDUs, refered to as data packets or frames, are variable-length packets which carry user's data, and are identified by their sequence number. Poll PDUs are periodically generated by the Sender to request information about the state of the Receiver, and especially about possibly lost SD PDUs. Stat PDUs are either "sollicited" (sent by the Receiver on receipt of a Poll PDU) or "unsollicited" (sent by the Receiver on detection of a loss, by detecting a gap in sequence numbers).

2.1 Data Transfer using SSCOP

1. Upon arrival of a data block, a SD PDU is generated and buffered. As soon as the Sender is available, the SD PDU is sent. At the same time, it is copied in a retransmission buffer, along with the current value of the Poll period.

2. The Sender periodically generates and sends a Poll PDU. Each Poll PDU is numbered and carries the number of the last sent PDU.

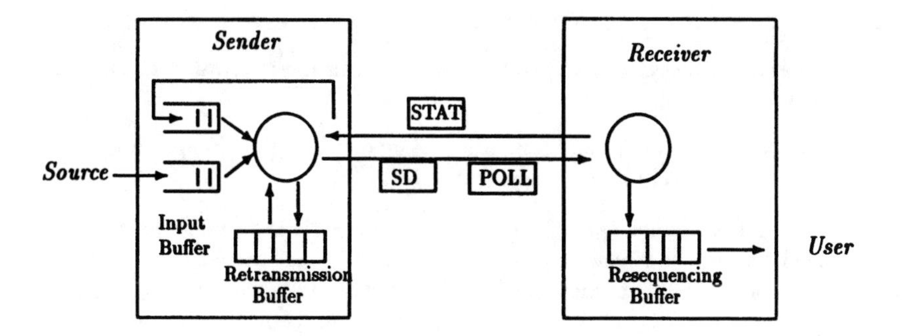

Figure 1: The elementary data link

3. Upon receipt of a POLL PDU, the Receiver sends back to the Sender a "Sollicited STAT PDU", which acknowledges correctly received frames and tells the Sender a list of missing ones.

4. Correctly received SD PDUs are immediately delivered to the user, provided all SD PDUs of lower number have been already received. If this is not the case, the SD PDU is stored until missing PDUs are correctly received.

   If the number of a received SD PDU is strictly greater than the expected one, the Receiver detects loss. It then sends back to the Sender a "unsollicited STAT PDU", reporting the loss.

5. Upon receipt of a STAT PDU, the Sender erases all correctly received SD PDUs. If missing SD PDUs are reported, the Sender retransmits them immediately if the STAT PDU is unsollicited. In case of a "sollicited STA PDU", it compares the number of the POLL PDU given in the STAT with the one stored in the buffer along with the SD PDU. Only if this last is lower will the retransmission be triggered. This avoids redundant retransmissions while allowing for detection of multiple losses (loss of a resent frame).

   Each time a SD PDU is retransmitted, the Poll number associated with it in the retransmission buffer is updated.

Remark: The "Flow Control" Procedure

The receiving end transmits in the STAT PDUs the amount of credits the Sender is allowed to use. This corresponds to the memory available in the resequencing buffer, and can serve two different purposes: either the memory is of limited and variable size (a common memory dynamically shared among different "Receivers", for instance), or the end user may control the speed at which the Sender is working.

In what follows, we assume no restriction due to credit management.

2.2 Parameters, Measures of Performance and Assumptions

The following variables denote the parameters used in the study, and the measures of performance:

- λ is the arrival rate of new frames in the input queue (number of frames per time unit). The arrival process is assumed to be Poisson.
- T^{prop} is the *propagation delay* (time needed for a bit to travel from the Sender to the Receiver). It depends on the distance between end users and on the underlying network (e.g. ATM connections probably give rise to variable propagation delays). One must also take into account processing and switching delays.
- T^{RT} is the *Round-Trip delay*: the time needed for a single bit to travel to the Receiver and to come back to the Sender. If processing delays are neglected, this corresponds to twice the propagation delay. The correspondance between delays and distances has to be made for each particular network
- T^{poll} is the time interval between two successive Poll PDUs. They are assumed to be sent with a fixed periodicity.
- p is the probability for a frame to be lost or errored. It is assumed independent of its length, for simplicity. It is also independent of the distance.
- The *End-to-End Transfer Delay* is defined as the time elapsed between the instant the first bit of the frame enters the input queue, and the instant the frame is made available to the User.
- The *End-to-End Additional Delay* (additional delay for short) is defined as the Transfer Delay decreased by the propagation delay and by the emission time; it accounts for any supplementary waiting a frame incurs due to retransmissions (of itself, or of previous frames, since the frames are to be delivered in sequence).

See figure 2 for the definition of delays: the transfer delay is denoted as w' and the additional delay as w.

3 The End-to-End Delays

3.1 Derivation of the End-to-End Additional Delay

For the frame numbered n, let w'_n be the transfer delay: see Figure 2. We assume that frames in sequence are immediately read by the user, so that:

- if the frame is sent in a "no-error period" (e.g. frame #1 in the figure), w'_n is simply the sum of the emission time and the propagation time;
- if the frame is sent in an "error period" (e.g. frame #3), w'_n ends as soon as the error is corrected: in Figure 2, a "batch" is observed, composed with frames #2, 3, etc. when frame #2 is received.

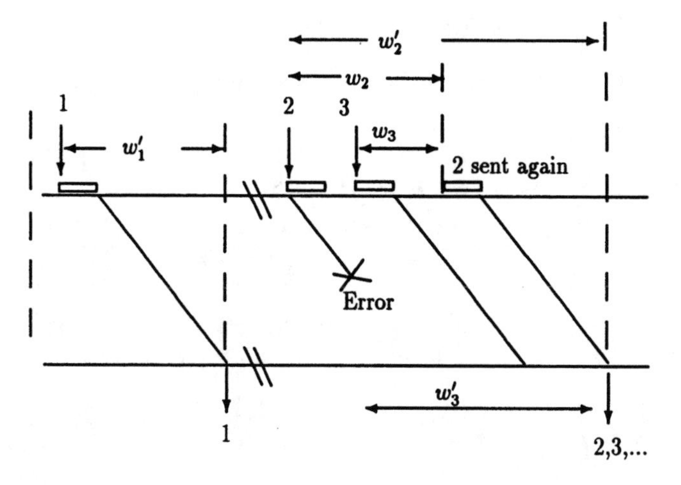

Figure 2: End-to-end transfer delay, End-to-end additional delay

The analysis considers the additional delay, defined as $w_n = w'_n - T^{prop} - s_n$, with s_n representing the emission time and T^{prop} the propagation delay.
Let us define the *successfull sending* of a frame as the event: beginning of the last emission of the frame. Note that even if the frame is sent successfully it may have to wait in the resequencing queue. The "equivalent" service time of a frame, x_n, is defined as the time interval between the first sending and the successfull sending. x_n represents the contribution of the frame to the overall queueing.
Let t_n be the time elapsed between sending of $n-1$-th and n-th frame:

- if frame #n is such that $x_n = 0$ (it incurs no loss), then $w_n = w_{n-1} - t_n$, or 0 if the expression is negative.

- if frame #n is lost, and if x_n is smaller than $w_{n-1} - t_n$, then $w_n = w_{n-1} - t_n$ - that is, the frame #n is corrected before the preceding error is corrected, and it remains in the queue due to this previous error (this is the case for instance if the previous error needs more than one reemission to be corrected).

- if frame #n is in error, and if x_n is larger than $w_{n-1} - t_n$, then $w_n = x_n$ (this is the case for Frame #2 in the Figure).

Finally, one has the following relation between the w_n, t_n and x_n's:

$$w_n = \max\{x_n, w_{n-1} - t_n\} \tag{1}$$

From the recurrence relation, one can derive an equation for $W(t)$, the probability distribution function of w. Let:

- $W_n(t) = P\{w_n \leq t\}$, and $W(t) = \lim_{n\to\infty} W_n(t)$;

- $F_n(t) = P\{x_n \leq t\}$, and $F(t) = \lim_{n\to\infty} F_n(t)$;
- t_n is distributed according to a Poisson process with rate λ.

$$\begin{aligned} Pr\{w_n \leq t\} &= Pr\{x_{n-1} \leq t \underline{\text{ and }} w_{n-1} - t_n \leq t\} \\ &= Pr\{x_{n-1} \leq t\}.Pr\{w_{n-1} \leq t_n + t\} \\ &= Pr\{x_{n-1} \leq t\} \int_u Pr\{w_{n-1} \leq t + u\}.Pr\{t_n \in [u, u + du[\} \end{aligned}$$

The second equation follows from independence assumptions. The third equation is obtained by conditioning on u, the value of t_n. Assuming the existence of a stationary limit:

$$W(t) = F(t).\int_u \lambda e^{-\lambda u} W(t+u)dx \tag{2}$$

The equation is solved by putting first $Q(t) = e^{-\lambda t}W(t)$, yielding

$$Q(t) = \lambda F(t) \int_{u \geq t} Q(u)du$$

One then introduce the function $H(x) = \int_{u \geq x} Q(u)du$, that is $Q(x) = -H'(x)$; the equation becomes then $H'(x) = -\lambda F(x).H(x)$, the solution of which is immediate.
The normalisation condition ($W(\infty) = 1$) gives finally:

$$\begin{aligned} W(x) &= F(x)e^{-\lambda G(x)} \\ \text{where} \quad & G(x) = \int_x^\infty [1 - F(u)]du \end{aligned} \tag{3}$$

Actually, Equation 3 gives only the "transmission part" of the delay. Frames may be delayed in the input buffer, due to other frames being sent. In the case where no credit management can block the Sender, one may assume independence between the two components of the end-to-end delay; the delay in the Input Buffer is then estimated as the sojourn time in a M/GI/1 queue (the service time being the transmission time of the frame). Preliminary simulation studies have shown [3] this delay to be negligible, and we omit it in the following. Note that with a 10 Mbit/s link rate, it takes 100μs to send a 1 kbit-packet.
The independence assumption implies also that the total delay is the *convolution* of the two distributions.

3.2 The components of the equivalent service time

The End-to-End Waiting Time is null if the packet is successfully received at the first sending. Figure 3 shows the components of the delay in case of a single error and in case of multiple errors.

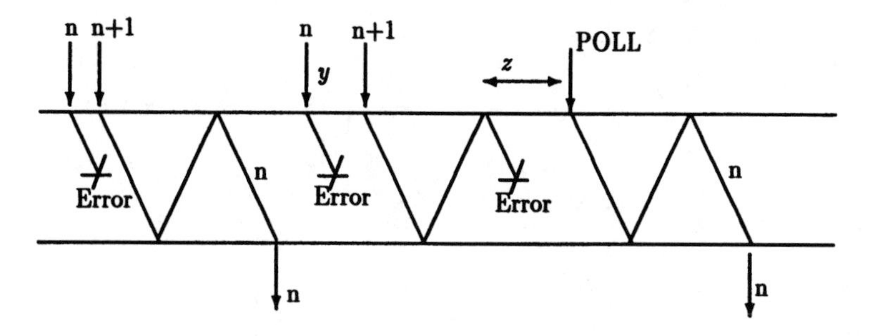

Figure 3: End-to-end delay : one or two successive errors

Assume the packet numbered n is lost. The next packet $n+1$ to be sent, or the next POLL PDU, allows the receiver to detect the loss. Either the U-Stat (as in the left part of Figure 3) or the S-Stat reports the event to the sender and the missing packet is then sent again.
Let us denote as y the random variable which measures the time elapsed between the packet sending and the next successfull sending which signals the previous loss (either the data or POLL). One sees that y is the minimum of the two variables: remaining time until next POLL, and time to the next successfull STAT. In fact, the next packet or the STAT could be lost, too. In this case, packet $n+2$ detects the event, etc. It is possible to show that this is equivalent to assume that the interdepartures between successfully sent packets is exponentially distributed with parameter $\lambda(1-p)$ instead of λ. The distribution of y is easily obtained, taking into account the independence between data packets and POLL PDUs.
In this single-retransmission configuration, the waiting time is the sum $y+T^{\mathrm{RT}}$.
Assume now that the second sending of packet n is lost too. In this case, only the POLL PDU which follows the second sending allows the detection of the loss. The detection scheme conforms with the right hand part of Figure 3. The additional waiting time becomes $y+2T^{\mathrm{RT}}+z$, where z stands for the delay between the S-STAT PDU is received and the next POLL is sent.
Should a third error occur, the next POLL detects it, with this time an additional waiting time of $y+2T^{\mathrm{RT}}+z+\tau$, where τ stands for the delay between the S-STAT and the next POLL. The general case (although rather theoretical) follows easily.
From these relations, the distribution of the variable X can be derived. The complete expression is not given here. In fact, the actual estimation of the distribution depends on the relative values of T^{RT}, T^{prop}, etc. See [2] for more details about the actual calculation process.

4 Application to Traffic Engineering

The following results show the influence of the control variables. Of special concern is the value to be given to T^{poll}.

4.1 Influence of the Loss Probability

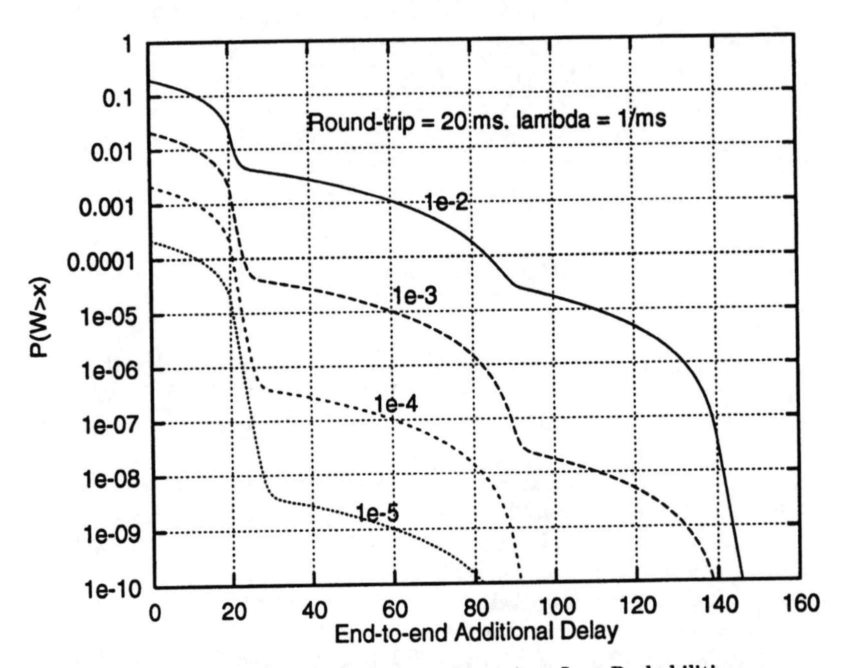

Figure 4: Delay distribution, for various Loss Probabilities

The first curves give the distribution of the end-to-end additional delay: probability that the delay is larger than t, for various packet loss ratios. As expected, the higher the probability, the longer the additional delay.
Figure 4 is drawn for a round-trip delay of 20 ms, and for a POLL time of 50 ms. The arrival rate is 1/ms.
Each "knee" of the curves corresponds to losses. When a data packet is lost, an "error period" begins, which ends when the packet is successfully retransmitted. All packets sent during the error period are delayed (Remind a PDU is delayed if it is errored and has to be resent, or if a preceding PDU has been lost and must be sent again.
PDUs involved in a single-error period have a delay responsible of the first knee, while PDUs sent in a 2-errors period give rise to the second knee, etc. The "knees" appear such due to the log scale on the vertical axis. Were the scale be linear, each of the domains would approximately look like a piece of straight line.
In a first approximation, the limits of the domains are as follows: the k-th error (i.e., the domain corresponding to waiting times due to frames incurring exactly k consecutive transmission errors) produces waiting time probability distributions ranging from $\lambda T^{\mathrm{RT}} p^k$ to $\lambda T^{\mathrm{RT}} p^{k+1}$. Recall that λ is the frame arrival rate in the Sender, T^{RT} is the round trip delay (approximately 2 propagation delays) and p the frame loss probability.

4.2 Influence of POLL periodicity

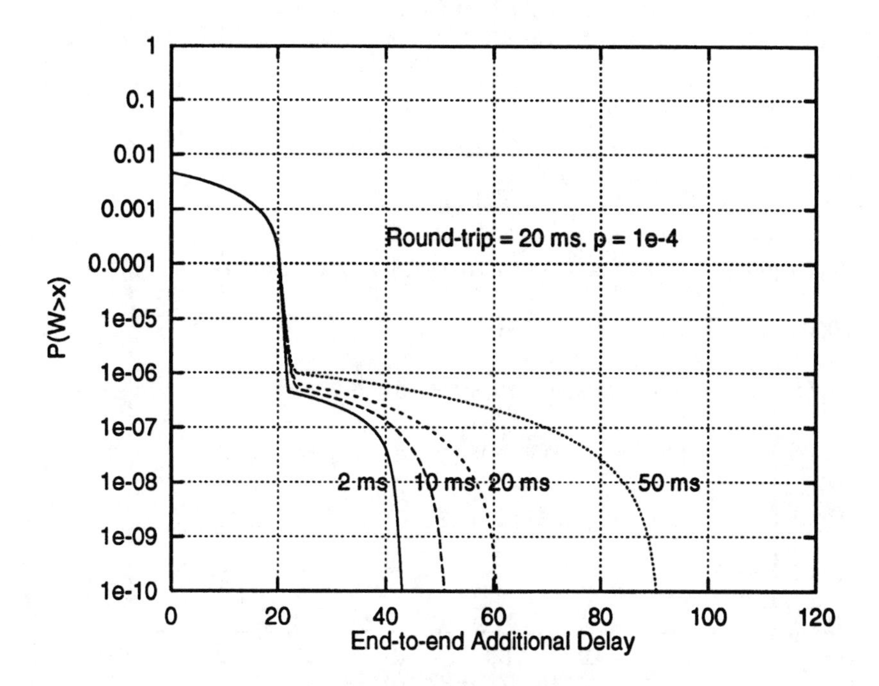

Figure 5: Delay distribution, for various POLL timers

The POLL period has no effect on PDUs involved in a single error: their additional delay is around T^{RT} (see Section 3.2). On the other hand, packets incurring two retransmissions (or more) have an additional delay which can be as large as $2T^{\mathrm{RT}} + T^{\mathrm{poll}}$, as shown on Figure 3. This can be verified on Figure 5.
The abscisses of the knees are around T^{RT} for the first one, and $2T^{\mathrm{RT}} + T^{\mathrm{poll}}$ for the following one (at least, as long as λ is large enough, see the following Figure).

4.3 Influence of the connection rate

Figure 6 illustrates the effect of the PDU rate. For low data rates, the additional delay increases. For instance, the delay at which the first knee ends is around 20 ms for $\lambda = 1$ or 3 /ms, while it goes to around 30 for 0.5/ms, and to 70 ms for lower rates.
The point is in the delay to detect the PDU error. Usually, the next packet detects the gap in numbering sequence. However, if the input rate is too low, the next POLL is likely to be sent first, and the detection delay increases up to T^{poll}, so that the first knee ends at $T^{\mathrm{RT}} + T^{\mathrm{poll}}$. This is what happens here with $\lambda = 0.02$/ms. At the same time, the level of each knee decreases, since each packet is much less likely to be involved in an error recovery period (since less packets are sent). In the limit, the additional delay a packet

incurs is only due to its own losses, so that the probability levels at which the successive steps begin are around p^k instead of $\lambda p^k T^{RT}$ as for higher rates.
Note that with 1 kbit-packets, a link rate of 0.02/ms corresponds to a 20 kbit/s connection.

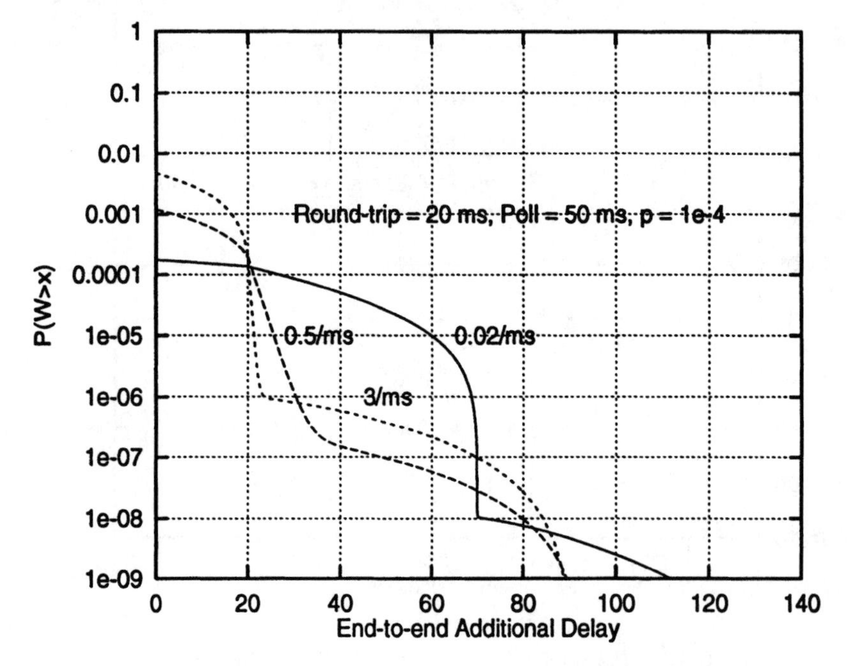

Figure 6: Delay distribution, for various packet arrival rates

5 Conclusions

We have presented an analytical derivation of the end-to-end delay to be expected with the Selective Repeat scheme used in the SSCOP. The analysis is summarized Equation (3) which can be easily programmed and which allows a study of the protocol for realistic figures.
The results show the influence of the main traffic and control parameters: the round-trip delay, the PDU error ratio, the packet load, the poll period. Figure 7 summarizes the typical behavior of the probability distribution function of the End-to-end additional delay: successive "steps" correspond to packets involved in 1, 2, etc. retransmissions. The actual distribution lays below the envelope. The results show that for typical loss levels, the additional delay, i.e. the part of the delay which is related with error recovery, is less than $\lambda p T^{RT}$. For most cases, this figure is less than 1%. This may question the utility of the dimensioning criterion presently proposed by Rec. E733. Concerning the poll period, its value is of little importance as long as loss figures remain moderate and the load is not too low. For data links with low utilisation, T^{poll} would have to be shortened.

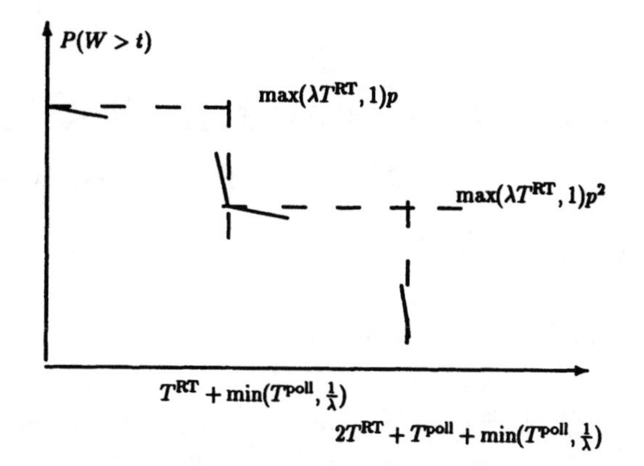

Figure 7: Schematic representation of the shape of Additional Delay

References

[1] D. Bertsekas, R. Gallager: *Data Networks.* Prentice Hall, second edition 1992.

[2] G. Hébuterne: *Service Specific Connection Oriented Protocol: delay calculation (no credit limitation.* INT Technical Report, october 1995.

[3] G. Hébuterne, W. Monin: *Preliminary performance of the SSCOP.* 14 th International Teletraffic Congress, Antibes Juan les Pins, June 1994.

[4] ITU-T Recommendation I.121: *Broadband aspects of ISDN.*

[5] ITU-T Draft Recommendation Q.2110: *Service Specific Connection Oriented Protocol (SSCOP).*

25
Window based estimation of the intensity of a doubly stochastic Poisson process

J. Virtamo, S. Aalto and D. Down
VTT Information Technology
P.O.Box 1202, FIN-02044 VTT, Finland. Telephone: +358 0 456 5612.
Fax: +358 0 455 0115. email: jorma.virtamo@vtt.fi

Abstract
Tasks related to advanced management of telecommunications networks, such as optimal dynamic routing or dynamic allocation of bandwidth for virtual paths in an ATM network, call for a method to estimate the current call arrival intensities. Such an estimate must be based on the observed history of the call arrival instants. In this paper, we study window type estimators with the purpose of giving a clear idea of their accuracy. In particular, the simple straight case and exponential windows are considered and applied to an example where the correlation function of the intensity process is of the exponential type – a system for which the exponential window, in fact, is the optimal window. It turns out that the accuracy is controlled by a single dimensionless traffic parameter. The achievable accuracy is little affected by the shape of the window function. It also turns out that a Bayesian approach, applied in an earlier study, gives only a slight improvement over the window based estimator. More important instead is the proper choice of the size of the window. In order to find the optimal size, one has to know some basic characteristics of the intensity process. We briefly address the problem how these can be inferred from the long term statistics of the counting process.

Keywords
estimation, call arrival intensity, window, kernel, doubly stochastic Poisson process

1 INTRODUCTION

Many of the advanced management schemes of telecommunications networks assume that the current call arrival intensity λ of any given traffic stream is known. Such information is needed e.g. for optimal dynamic routing in telephone networks (Krishnan, 1990, Mitra et al., 1993) or dynamic forward-looking allocation of bandwidth for virtual paths in an ATM network (Bruni et al., 1994). In practice, of course, call arrival intensities are not given, and as a separate problem one must estimate their values on the basis of the available information. Typically one can assume that the arrival process constitutes a doubly stochastic Poisson process, i.e. a Poisson process where the intensity λ_t is also a stochastic process. What is observed is the point process of call arrivals (counting process N_t) and the task is then to estimate λ_t based on these observations. Such a real-time traffic estimation problem is essentially a filtering problem (Warfield et al., 1994).

In (Aalto and Virtamo, 1995) we developed such an algorithm for the case where the nature of the process λ_t was fully known in advance. In particular, a Bayesian estimator was developed with the assumption that λ_t is an Ornstein-Uhlenbeck process. This estimator exploits to a maximum degree all the available information and in that sense is an optimal estimator. Though the algorithm presented was not complex, one might desire an even simpler algorithm. Moreover, the particular process model does not have general validity, and it would be desirable to have a scheme which does not depend on specific assumptions on the underlying process.

The most straightforward estimators are based on a sliding window, i.e. on a count of past arrivals within a window divided by its length, possibly with different weights in different parts of the window. In this paper, our aim is to give a clear idea on the accuracy of window estimators. The problem belongs to the area of linear filtering theory and has been subject to earlier analysis (Snyder, 1975).

In section 2 we rederive a formula for the mean squared error of the estimator. This turns out to depend only on the correlation function of the process λ_t (in addition to the mean and variance of the stationary distribution of λ_t) but not on other dynamic characteristics of the process. Two different window types are considered in section 3: a straight case window and an exponential window. The latter is the most attractive one from the algorithmic point of view. Besides the standard window estimator, we consider an improved estimator which exploits the prior knowledge of the mean of λ_t, using a linear combination of this prior value and the window "measurement" as the actual estimator. The optimal mixture of these two components is derived. This improved estimator is indeed found to increase the accuracy, especially in those cases where the accuracy of a pure window estimator is poor. In all the cases the achievable accuracy is controlled by a single traffic parameter.

In section 4 we consider the problem of determining the optimal window function. This satisfies a known integral equation (Snyder, 1975, Van Trees, 1971). We give a general solution to this equation in terms of Fourier transforms. For an exponential correlation function the optimal window function itself turns out to be of exponential type. In section 5, the problem of inferring the characteristics of the underlying intensity process from the observed Poisson counting process is briefly addressed. The same dimensionless parameter which governs the accuracy of the estimator also determines the accuracy of the long term measurements. Finally, numerical results are presented and discussed in section 6.

2 ANALYSIS

The intensity process λ_t is assumed to be stationary with mean Λ and variance Σ^2. Its covariance function is denoted by $\Sigma^2(t)$ and can be expressed in terms of the correlation function $\rho(t)$,

$$\mathrm{Cov}[\lambda_t, \lambda_u] = \Sigma^2(t-u) = \Sigma^2 \rho(t-u). \tag{1}$$

The covariance is symmetric in its variables. Therefore, the correlation function is a symmetric function, i.e. $\rho(t) = \rho(-t)$.

Let N_t be the counting process of the arrivals. A window based estimator X for the arrival intensity λ_0 at time 0 is

$$X = \int_0^\infty W(t)\mathrm{d}N_t = \sum_i W(t_i), \tag{2}$$

where $W(t)$ is the window function such that $\int_0^\infty W(t)\mathrm{d}t = 1$ and the t_i are the arrival instants. (For notational simplicity we use here reversed time). X itself is, of course, a stochastic variable. Its mean is

$$\mathrm{E}[X] = \int_0^\infty W(t)\mathrm{E}[\mathrm{d}N_t] = \int_0^\infty W(t)\Lambda \mathrm{d}t = \Lambda.$$

In choosing the window function one has to balance between two trends: if the function is concentrated around time 0 the estimator is based on a few recent arrivals and can be very noisy; on the other hand broadening the window too much means that the estimator is based on outdated information.

The estimator can be improved by taking it to be of more general linear form

$$Y = aX + a'. \tag{3}$$

Let us consider the error Δ of the estimator

$$\Delta = Y - \lambda_0 = aX + (a' - \lambda_0). \tag{4}$$

We wish to choose the window function $W(t)$ and parameters a and a' such that the mean squared error

$$\mathrm{E}[\Delta^2] = \mathrm{D}^2[\Delta] + \mathrm{E}[\Delta]^2$$

is minimized. Since the variance of Δ in (4) is not affected by the constant a', the optimal choice of a' makes the estimator unbiased, i.e.

$$\mathrm{E}[\Delta] = (a-1)\Lambda + a' = 0$$

from which

$$a' = (1-a)\Lambda.$$

With this value we can write

$$Y = a(X - \Lambda) + \Lambda, \qquad \Delta = aX - \lambda_0 + (1-a)\Lambda, \tag{5}$$

i.e. the estimator is Λ plus a correction term where the deviation of the "measured" value X from Λ is multiplied by a constant a. The optimal choice of a obviously depends on the confidence of the measurement. With full confidence we have $a = 1$ and $Y = X$.

Now, because $\mathrm{E}[\Delta] = 0$, we have using (5)

$$\mathrm{E}[\Delta^2] = \mathrm{D}^2[\Delta] = \mathrm{D}^2[aX - \lambda_0] = a^2\mathrm{D}^2[X] - 2a\,\mathrm{Cov}[X, \lambda_0] + \mathrm{D}^2[\lambda_0].$$

Let us calculate each of the terms. First,

$$\mathrm{D}^2[X] = \int_0^\infty W^2(t)\,\mathrm{D}^2[\mathrm{d}N_t] + \int_0^\infty \int_0^\infty W(t)W(u)\,\mathrm{Cov}[\mathrm{d}N_t, \mathrm{d}N_u].$$

Here, because $\mathrm{d}N_t^2 = \mathrm{d}N_t$,

$$\mathrm{D}^2[\mathrm{d}N_t] = \mathrm{E}[\mathrm{d}N_t] - \mathrm{E}^2[\mathrm{d}N_t] = \Lambda\mathrm{d}t - (\Lambda\mathrm{d}t)^2 \to \Lambda\mathrm{d}t,$$

and

$$\mathrm{Cov}[\mathrm{d}N_t, \mathrm{d}N_u] = \mathrm{Cov}[\lambda_t, \lambda_u]\,\mathrm{d}t\mathrm{d}u.$$

Thus,

$$\mathrm{D}^2[X] = \Lambda\int_0^\infty W^2(t)\,\mathrm{d}t + \int_0^\infty \int_0^\infty W(t)W(u)\mathrm{Cov}[\lambda_t, \lambda_u]\,\mathrm{d}t\mathrm{d}u.$$

Second,

$$\mathrm{Cov}[X, \lambda_0] = \int_0^\infty W(t)\mathrm{Cov}[dN_t, \lambda_0] = \int_0^\infty W(t)\mathrm{Cov}[\lambda_0, \lambda_t]\,\mathrm{d}t.$$

Third, $\mathrm{D}^2[\lambda_0] = \Sigma^2$. By collecting the results together and using the correlation function (1), we have

$$\frac{\mathrm{E}[\Delta^2]}{\Sigma^2} = 1 + a^2(\beta I_1 + I_2) - 2aI_3, \tag{6}$$

where $\beta = \Lambda/\Sigma^2$ and I_1, I_2 and I_3 stand for the integrals

$$I_1 = \int_0^\infty W^2(t)\,\mathrm{d}t, \quad I_2 = \int_0^\infty V(t)\rho(t)\,\mathrm{d}t, \quad I_3 = \int_0^\infty W(t)\rho(t)\,\mathrm{d}t, \tag{7}$$

and

$$V(t) = 2\int_0^\infty W(u)W(u+t)\mathrm{d}t.$$

The standard window estimator X is obtained by setting $a = 1$

$$\frac{\mathrm{E}[\Delta_X^2]}{\Sigma^2} = 1 + \beta I_1 + I_2 - 2I_3. \tag{8}$$

A better estimator is obtained by minimizing (6) with respect to a,

$$\frac{\mathrm{E}[\Delta_Y^2]}{\Sigma^2} = 1 - \frac{I_3^2}{\beta I_1 + I_2}, \qquad \text{with} \quad a^* = \frac{I_3}{\beta I_1 + I_2}. \tag{9}$$

Note that the results depend on Λ and Σ only through the single parameter $\beta = \Lambda/\Sigma^2$.

3 EXAMPLES

In this section we consider the problems where the *form* of the window function $W(t)$ is given but it contains one free parameter, T, which characterizes the temporal extent of the window (loosely speaking the window size) and a further minimization is performed with respect to this parameter. In particular, the window function is assumed to be either a straight case window or an exponential window.

3.1 Straight case window

In this case the window function is

$$W(t) = \frac{1}{T}U(t/T)$$

where $U(x)$ is the unit step function, i.e. $U(x) = 1$ for $0 \leq x < 1$ and 0 otherwise. The window estimator (2) then is

$$X = \frac{1}{T}\sum_i 1_{t_i<T},$$

i.e. the count of arrivals in the window divided by the length of the window. This is algorithmically a simple formula. However, one has to keep a list of the arrival times, and an entry in the list cannot be deleted until time T has elapsed since the corresponding arrival time. From the standard window estimator X one then obtains the optimal estimator using (5).

With this window function, the integrals (7) become

$$I_1 = \frac{1}{T}, \quad I_2 = \frac{2}{T}\int_0^T (1 - t/T)\rho(t)\,\mathrm{d}t, \quad I_3 = \frac{1}{T}\int_0^T \rho(t)\,\mathrm{d}t.$$

This is as far as one can get in the general form. To proceed, one has to specify the correlation function $\rho(t)$. As an example we consider the following correlation function,

$$\rho(t) = e^{-|t/\tau|}, \tag{10}$$

where τ is the correlation time. Such an exponential form is valid for e.g. the Ornstein-Uhlenbeck process considered in (Aalto and Virtamo, 1995). By denoting $x = T/\tau$ we get with this correlation function

$$I_2 = \frac{2}{x^2}(e^{-x} - 1 + x), \qquad I_3 = \frac{1}{x}(1 - e^{-x}).$$

Substitution into (8) and (9) gives the mean squared error of the standard and the optimal estimators X and Y,

$$\frac{\mathrm{E}[\Delta_X^2]}{\Sigma^2} = 1 - 2 \cdot \frac{1 - bx - (1+x)e^{-x}}{x^2}, \qquad \frac{\mathrm{E}[\Delta_Y^2]}{\Sigma^2} = 1 - \frac{1}{2} \cdot \frac{(1-e^{-x})^2}{bx + e^{-x} - 1 + x}, \tag{11}$$

with the optimum obtained at

$$a^* = \frac{x}{2} \cdot \frac{1 - e^{-x}}{bx + e^{-x} - 1 + x}.$$

In these formulae b is the dimensionless parameter

$$b = \frac{\Lambda}{2\Sigma^2 \tau} = \frac{1}{2} \cdot \frac{1}{(\Sigma/\Lambda)^2} \cdot \frac{1}{\Lambda \tau}. \tag{12}$$

The expressions in (11) are still to be minimized with respect to the parameter x, i.e. the window size. This cannot be done in analytic form: zero points of the derivatives have to be found by numerical methods. Nevertheless, the result of the minimization is solely a function of b, i.e. this parameter alone determines the achievable accuracy of the estimator.

3.2 Exponential window

Now the window function is

$$W(t) = \frac{1}{T} e^{-t/T}. \tag{13}$$

This is particularly attractive from the point of view of algorithmic simplicity. From eqs. (2) and (13) we have

$$X = \frac{1}{T} \sum_i e^{-t_i/T}.$$

Thus, by advancing the time by an interval Δt without arrivals, X is multiplied by a factor $\exp(-\Delta t/T)$ (i.e. by a constant for fixed intervals), and a new arrival at time 0 increments its value by $1/T$. There is no need to record the individual arrival times. As before, the optimal estimator is calculated from (5) with X given by the above algorithm.

With this window function the integrals are

$$I_1 = \frac{1}{2T}, \qquad I_2 = I_3 = \frac{1}{T} \int_0^\infty e^{-t/T} \rho(t) \mathrm{d}t,$$

and in the specific case of the exponential correlation function (10) we get (with $x = T/\tau$)

$$I_2 = I_3 = \frac{1}{1+x}.$$

The mean squared errors (8) and (9) of the standard and the optimal estimators are

$$\frac{\mathrm{E}[\Delta_X^2]}{\Sigma^2} = 1 + \frac{b}{x} - \frac{1}{1+x}, \qquad \frac{\mathrm{E}[\Delta_Y^2]}{\Sigma^2} = 1 - \frac{1}{1+x}\cdot\frac{x}{x+b(1+x)}, \tag{14}$$

where the parameter b is given by (12) and the optimum is obtained at $a^* = x/(x+b(1+x))$.

As expressions in (14) are rational functions of x the minimization with respect to this parameter can be done analytically giving the optimal window size for the standard estimator X and for the optimal estimator Y

$$x_X^* = \frac{\sqrt{b}}{1-\sqrt{b}}, \qquad x_Y^* = \sqrt{\frac{b}{1+b}}. \tag{15}$$

Finally, substituting these optimal window sizes into (14) gives the mean squared errors,

$$\frac{\mathrm{E}[\Delta_X^2]}{\Sigma^2} = 1-(1-\sqrt{b})^2, \qquad \frac{\mathrm{E}[\Delta_Y^2]}{\Sigma^2} = 1 - \frac{1}{(\sqrt{b}+\sqrt{1+b})^2}, \tag{16}$$

with the optimal a being

$$a^* = 1 - \sqrt{\frac{b}{1+b}}. \tag{17}$$

In the next section we will show that for the exponential correlation function the exponential window estimator Y with the optimal parameter values x_Y^* of (15) and a^* of (17) indeed is the optimal estimator also with respect to the form of the window function.

4 OPTIMAL WINDOW

Thus far we have considered window estimators with a given form of the window function. In order to find the truly optimal window estimator the form of the window should also be subject to optimization. For the present purposes it is most convenient to absorb even the coefficient a in (3) into the definiton of the window function, i.e. here the window function $W(t)$ is not normed but both its shape and norm are simultaneously optimized. The optimal a is then regained just as the norm of the optimal window function, and the best value for the constant a' in (3) is as before $(1-a)\Lambda$. It can be shown (Snyder, 1975) that the optimal window function $W(t)$ satisfies the following integral equation

$$\beta W(t) + \int_0^\infty W(u)\rho(t-u)\mathrm{d}u = \rho(t), \qquad t \geq 0. \tag{18}$$

This integral equation arises naturally in the study of linear filters for the detection of signals in the presence of additive noise, see for example (Van Trees, 1971). In fact, it is described in (Snyder, 1975) how determining the optimal window may be recast as such a detection problem. There it is also discussed how such problems may be treated with various tools from linear filtering theory, including a Kalman-Bucy approach. Here we will simply indicate how the solution may be obtained directly by transform methods. Let us define the Fourier transform $f^*(k)$ of a function of time $f(t)$

$$f^*(k) = \int_{-\infty}^{\infty} e^{-ikt} f(t) dt.$$

Define

$$f_+(t) = 1_{t \geq 0} f(t), \qquad f_-(t) = 1_{t<0} f(t),$$

and denote

$$S(t) = \int_0^{\infty} W(u)\rho(t-u) du = \int_{-\infty}^{\infty} W_+(u)\rho(t-u) du = W_+ \otimes \rho$$

with $\otimes$ representing the convolution operation. Equation (18) may now be rewritten as

$$\beta W_+(t) + S_+(t) = \rho_+(t), \qquad \forall t. \tag{19}$$

Adding $S_-(t)$ to both sides of (19) and taking Fourier transforms yields

$$W_+^*(k) = \frac{\rho_+^*(k) + S_-^*(k)}{\beta + \rho^*(k)}. \tag{20}$$

It can be seen that the only singularities of $W_+^*(k)$ in the upper half plane are the roots of $\beta + \rho^*(k) = 0$. The fact that $\rho^*(k) = \rho_+^*(k) + \rho_+^*(-k)$ implies that these roots appear in conjugate pairs. Let us denote them by $\pm i\kappa_j$ with $\kappa_j > 0$.

We can now formally write the inverse transform as

$$W(t) = i \sum_j \text{res} \left\{ \frac{\rho_+^*(k) + S_-^*(k)}{\beta + \rho^*(k)} \right\}_{k=i\kappa_j} e^{-\kappa_j t}. \tag{21}$$

The function $S_-^*(k)$ may be determined from the condition that expression (20) as a whole represents the Fourier transform of a "+ function" and thus is analytic in the lower half plane. The function $S_-^*(k)$ must be such that the numerator has roots at $k = -i\kappa_j$ in order to cancel the roots of the denominator in the lower half plane.

Let us again consider the example of the exponential correlation function $\rho(t) = e^{-|t/\tau|}$. We use τ as the unit of time and, as before, denote $b = \beta/2\tau$. Then we have

$$\rho_+^*(k) = \frac{1}{1+ik}, \qquad \rho^*(k) = \frac{2}{1+k^2}, \qquad \kappa = \sqrt{1 + \frac{1}{b}},$$

where we have dropped the subscript as there is only one pair of roots $\pm i\kappa$. We can easily calculate that

$$S_-^*(k) = \frac{C}{1-ik}, \qquad C = \int_0^{\infty} dt W_+(t) e^{-t}.$$

Then, from (21),

$$W(t) = i \cdot \operatorname{res}\left\{\frac{1 - ik + C(1 + ik)}{2b(k - i\kappa)(k + i\kappa)}\right\}_{k=i\kappa} e^{-\kappa t}.$$

The constant C can now be determined by requiring that the numerator has a root at $-i\kappa$ from which it follows that $C = (\kappa - 1)/(\kappa + 1)$, and $W(t) = a\kappa e^{-\kappa t}$, where $W(t)$ is written in the form of a constant times a normed function with

$$a = 1 - \frac{1}{\kappa} = 1 - \sqrt{\frac{b}{1+b}}. \tag{22}$$

Thus we have regained the optimal solution Y of section 3.2 with an exponential window: an exponential window is the optimal window for the case of an exponential correlation function. Finally, we can easily check that the constant C is defined consistently, i.e. the substitution of $W(t)$ into the defining integral of C yields $(\kappa - 1)/(\kappa + 1)$.

5 INFERRING THE CHARACTERISTICS OF THE INTENSITY PROCESS FROM THE COUNTING STATISTICS

All the results presented above were based on the assumption that the basic characteristics of the intensity process were known, viz. its mean Λ, variance Σ^2 and the covariance function $\Sigma^2(t)$. Information on these can be obtained only indirectly via the long term statistics of the counting process. An estimate for the mean Λ is of course provided by the mean count per time interval $\lim_{t\to\infty} N_t/t$. The covariance function can also, in principle, be obtained straightforwardly by determining the counting statistics $\mathrm{E}[\mathrm{d}N_u \mathrm{d}N_v]$ of two non-overlapping intervals $\mathrm{d}u$ and $\mathrm{d}v$, with

$$\mathrm{E}[\mathrm{d}N_u \mathrm{d}N_v] \approx (\mathrm{Cov}[\lambda_u, \lambda_v] + \Lambda^2)\mathrm{d}u\mathrm{d}v.$$

The problem with this is that in order to make the points of time u and v precise, $\mathrm{d}u$ and $\mathrm{d}v$ should be chosen small. This implies a poor statistics or very long measurement periods to get a reliable expectation in the above formula.

A better approach is to determine experimentally $\mathrm{D}^2[N_t]$ as a function of t and derive the covariance function from this function. To this end let us denote

$$n_t = \int_0^t \lambda_t \mathrm{d}t.$$

Given the value of n_t the variable N_t is Poisson distributed with mean n_t. Therefore we have

$$\mathrm{E}[N_t] = \mathrm{E}[\mathrm{E}[N_t \mid n_t]] = \mathrm{E}[n_t] = \Lambda t, \qquad \mathrm{D}^2[N_t] = \mathrm{E}[n_t] + \mathrm{D}^2[n_t].$$

From the definition of n_t we can easily deduce that

$$\mathrm{D}^2[n_t] = \int_0^t \mathrm{d}u \int_0^t \mathrm{d}v\ \Sigma^2(u - v) = 2\int_0^t \mathrm{d}u \int_0^u \mathrm{d}v\ \Sigma^2(v)$$

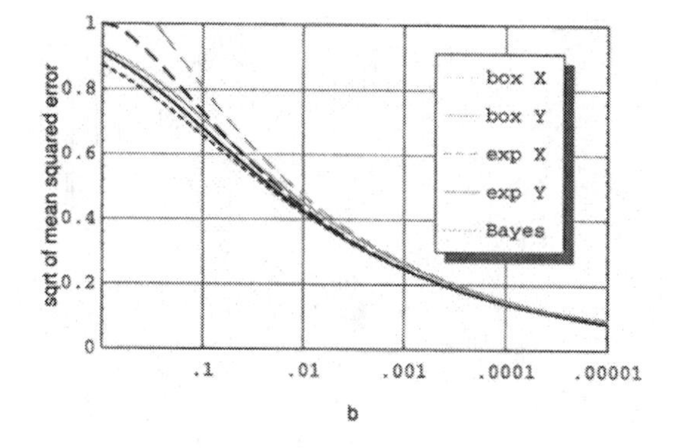

Figure 1 Accuracy of the estimators.

and thus

$$D^2[N_t] = \Lambda t + 2\int_0^t du \int_0^u dv\, \Sigma^2(v). \tag{23}$$

Differentiating, we then have

$$\frac{d}{dt}D^2[N_t] = \Lambda + 2\int_0^t \Sigma^2(u)du, \qquad \frac{d^2}{dt^2}D^2[N_t] = 2\Sigma^2(t).$$

The latter relation can be used for determining the covariance function. However, a numerical evaluation of the second derivative is involved which is susceptible to noise. This is a reflection of the same problem we alluded to in discussing the direct measurement of the covariance function. A more stable procedure would entail choosing a parameterized functional form for $\Sigma^2(t)$ and determining the parameters by a global fit to (23).

6 NUMERICAL RESULTS AND DISCUSSION

Numerical results of the accuracy $\sqrt{E[\Delta^2]}/\Sigma$ of different estimators as a function of b are shown in Figure 1. These were obtained from (11) in the case of a straight case window, with the additional numerical optimization with respect to the window size x, and from (16) in the case of an exponential window. For comparison, the accuracy of a Bayesian estimator (for which an approximation was derived in (Aalto and Virtamo, 1995)) is also shown. The latter is based on full information about the process model of λ_t, which here is assumed to obey the Ornstein-Uhlenbeck process (exponential correlation function). The accuracy of the Bayesian estimator is not solely a function of b but depends separately on $\Lambda\tau$ and Σ/Λ. Here we have plotted the curve for the case where Σ/Λ is kept fixed to 0.3 and $b = 1/(2(\Sigma/\Lambda)^2(\Lambda\tau))$ is controlled by the value of $\Lambda\tau$.

First we note that the Bayesian estimator is the best one, as it should by construction. Second, the estimators with an exponential window function are more accurate than their straight case window counterparts. For the case of the estimator Y this is obvious because

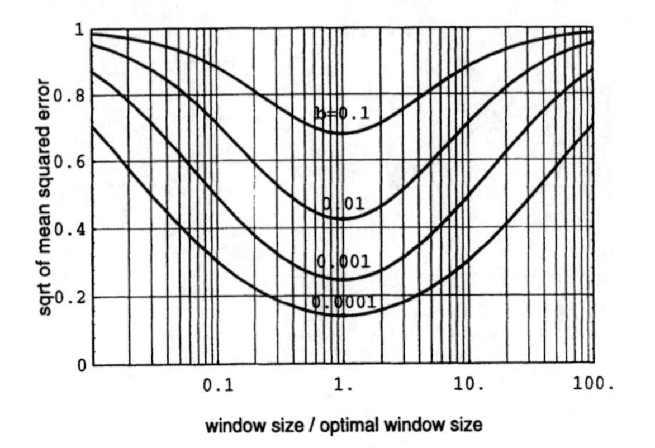

Figure 2 Sensitivity of the accuracy of the estimator Y with respect to the window size.

we have shown that the exponential window is the optimal window. For very small b, the accuracies of the estimators differ only depending on the window function (X and Y are identical). An exponential window function gives an estimator which is asymptotically as good as the Bayesian scheme. For values of b closer to 1, the more important factor is whether one uses the simple window estimator or the improved one with a mixture of the mean value Λ and the window estimator.

In general, the differences between different estimators are relatively small. The main differences appear in the range of large values of b where the accuracies of all the estimators are poor. When b is small the estimators are more accurate and essentially the same for all the schemes. However, it should be noted that these results were obtained with the *optimal* window size. With a wrong window size the accuracy can deteriorate. This is shown in Figure 2 where the accuracy of the estimator Y for an exponential window, given by (14), is plotted against the ratio of the window size to the optimal window size for different values of b. An order of magnitude error in the window size typically doubles the standard deviation of the estimator.

A somewhat surprising finding is that the achievable accuracy for the estimator does not change dramatically over a wide range of values of b, i.e. even though b changes by five orders of magnitude the standard deviation of the estimator changes only by a factor of ten. By local observations of the doubly stochastic Poisson process it is hard to increase the accuracy over the estimate which is provided by the mean intensity determined from long term statistics.

Returning to the motivation of our work by the needs of dynamic resource management schemes, it should be noted that an estimate of the traffic intensity contains inherent inaccuracies. At best, the accuracy is determined by the mean squared error such as given by (16). However, errors in the long term measurements of the basic parameters can lead to the choice of a non-optimal window size, and consequently the estimation error may be larger. For the sake of the robustness of a management scheme relying on the knowledge of the call arrival intensities, the applied algorithms should explicitly take into account the uncertainty involved in the estimator.

REFERENCES

Aalto, S. and Virtamo, J.T. (1995) Real-time estimation of call arrival intensities, in the proceedings of the seminar *ATM hot topics on traffic and performance: from RACE to ACTS*, Milano.

Bruni, C., D'Andrea, P., Mocci, U. and Scoglio, C. (1994) Optimal capacity assignment of virtual paths in ATM networks, *Globecom'94*, San Francisco, pp. 207-11.

Krishnan, K.R. (1990) Markov decision algorithms for dynamic routing. *IEEE Communications Magazine,* **28**, 66-9.

Mitra, D., Gibbens, R.J. and Huang, B.D. (1993) State-dependent routing on symmetric loss networks with trunk reservations – I, *IEEE Transactions on Communications,* **41**, 400-11.

Snyder, D.L. (1975) *Random Point Processes.* John Wiley & Sons, New York.

Van Trees, H.L. (1971) *Detection, Estimation, and Modulation Theory: Part III.* John Wiley & Sons, New York.

Warfield, R., Chan, S., Konheim, A. and Guillaume, A. (1994) Real-time traffic estimation in ATM networks, *The Fundamental Role of Teletraffic in the Evolution of Telecommunications Networks* (ed. J. Laboutelle and J.W. Roberts), the 14th International Teletraffic Congress – ITC14. Elsevier, Amsterdam, pp. 907-16.

26

Efficient Estimation of Cell Loss Probability in ATM multiplexers with a Fuzzy Logic System*

Shirley T. C. Lam, Brahim Bensaou and Danny H. K. Tsang
Hong Kong University of Science and Technology
Department of Electrical and Electronic Engineering
Clear Water Bay, Kowloon, Hong Kong.
E-mail: {lamtc, eebben, eetsang}@ee.ust.hk
fax: (852)-2358-1485

Abstract

An important parameter in ATM-based network design and management is the cell loss probability in ATM multiplexers. However, it depends on many unknown and unpredictable traffic parameters such as burst length distribution, mean rate, etc. In this paper, we propose a simple and robust fuzzy-based algorithm to predict the cell loss probability in large-sized systems based only on both a small amount of information from small-sized systems, and the asymptotic behavior for very large systems. Numerical results show that the value predicted by this algorithm is quite accurate. The application of the proposed algorithm to call admission control is also presented.

Keywords

ATM, cell loss probability, fuzzy logic, call admission control.

1 INTRODUCTION

An ATM-based Broadband Integrated Services Digital Network (B-ISDN) is a single high speed network designed to support all known and unknown services with different quality of service (QoS) requirements. It transfers the information through fixed size packets called cells. The network takes advantage of the statistical behavior of the sources with different traffic characteristics to efficiently share transmission resources through statistical multiplexing. An ATM multiplexer consists of a buffer of size K (cells) and a single output link with a transmission capacity of C cells per second. When the input rate is greater than the transmission capacity and the buffer is full, cell loss occurs. Cell loss has a harmful effect on the QoS. Therefore, an accurate estimation of the cell loss probability

*Supported by HongKong Telecom Institute of Information Technology grant HKTIIT93/94. EG01

(CLR[†]) not only gives a good estimation of this QoS but also provides an important parameter needed for network design and management (e.g., buffer dimensioning, congestion control, call admission control and routing).

Many approaches for evaluating the CLR at the so-called burst-scale congestion for multiplexers loaded with a superposition of on/off sources have been proposed in the literature. The first approach approximates the actual arrival process by fluid flow (FF) (see Anick et al. (1982) and Bensaou et al. (1994)). In this approximation, the fluctuation of cell arrival rates can accurately be represented by assuming that the information arrives in a continuous flow rather than in discrete cells. The CLR is accurately approximated by the overflow probability which is obtained by solving an adequate eigensystem (Anick et al. (1982)). Another approach, very similar to the fluid flow method, approximates the actual arrival process by a Markov modulated deterministic process (MMDP) in which cells arrive according to a deterministic renewal process whose rate is controlled by a Markov process as discussed by Yang and Tsang (1995). In this approach, the exact cell loss probability is obtained by solving a set of linear equations through the Gauss-Seidel algorithm.

These approaches are efficient in predicting the CLR in an ATM multiplexer. However, when the system size becomes large, computation complexity increases in the FF approximation and memory problems arise in the MMDP approximation. It is also shown in Bensaou et al. (1994) that the CLR depends on many unknown and unpredictable traffic parameters (e.g., burst length distribution,...). To avoid these problems, the goal is to derive a model-independent algorithm to predict the CLR in large-sized systems by relying only on information from some small-sized systems.

In this spirit, the Global Rational Approximation (GRA) algorithm is proposed by Yang et al. (1995) to approximate the CLR. In this method, the CLR as a function of buffer size (or of number of users) is approximated by a rational function, $R(x) = P_m(x)/Q_n(x)$ where P_m and Q_n are polynomial functions of degree m and n respectively. The coefficients of P_m and Q_n are determined by solving the system of linear equations: $R(x_i) = \Pr(x_i), i = 1, \cdots, m+n$; where the pairs $(\Pr(x_i); x_i)$ represents the small-sized system information (e.g., (CLR; buffer size), (CLR; number of users)).

Even though the GRA approximation is efficient and accurate in many cases, it still has some major problems:

- the accuracy of the approximation is closely related to the degrees m and n, which in turn determine the number of pairs $(R(x_i); x_i)$ required (i.e., $m+n$ pairs). When the degrees are not large enough, the accuracy of GRA can be poor;
- computation time can sometimes be quite long since the fitting process is based on an iterative algorithm to determine the suitable values of m and n for the required accuracy. In addition, depending on the values of the pairs $(R(x_i); x_i)$, the system of equations to solve sometimes can encounter singularity problems, which make the fitting process difficult. This problem occurs mainly when the known values of CLR do not fit on a smooth curve. This fact restricts the practical use of GRA when the known values of CLR are obtained from measurements that normally contain a small amount of error.

[†]in the sequel we use the generic term cell loss probability or CLR to designate either the cell loss probability, the overflow probability or the cell loss ratio

Fuzzy logic technique has been proposed to efficiently solve several ATM problems (e.g. Chang and Cheng (1994) and Cheung et al. (1994)). Thus, in view of the shortcomings encountered in the above mentioned approaches, we propose to use an adaptive fuzzy system to efficiently predict the CLR in large-sized systems based on only a small amount of information from small-sized systems. This information can be obtained by either real-time measurement when the traffic characteristics are unknown or from any analytic model when the traffic characteristics are known. This predicted CLR is then refined, by taking advantage of the knowledge of the asymptotic behavior of the performance curve (CLR). The advantages of using our proposed fuzzy logic based algorithm include robustness, simplicity in implementation, computational efficiency, and good accuracy.

In Section 2, we present a brief introduction to an adaptive fuzzy system and then describe our proposed fuzzy approximation (FA) algorithm. In Section 3, we give some numerical results to validate our algorithm. The FA algorithm is used to predict the CLR as a function of buffer size. The results are compared to those obtained from other alternative methods. In Section 4, we show through a simple example how this algorithm can be used to develop a call admission control solution in ATM networks. The algorithm is used to predict the CLR as a function of the number of connected users for both homogeneous and heterogeneous traffic. We finally draw our conclusion in Section 5.

2 FUZZY APPROXIMATION

In this section, we briefly introduce an adaptive fuzzy system and show how it can be used to represent an unknown mathematical model. We will then show how this system is applied to predict the CLR in a large-sized system by taking information from both the small-sized systems and the asymptotic behavior of the performance curve for the large-sized system.

2.1 Fuzzy Rule-based System

Suppose we have a system which is almost impossible to be represented accurately by any analytical model. And suppose we know N sets of input-output pairs $(\mathbf{X}_0^j; y_0^j)$, $\mathbf{X}_0^j = (x_{01}^j, \dots, x_{0n}^j)^T \in \mathbf{R}^n, j = 1, 2, \dots, N$, where N is a small number. A general method is provided by Wang (1994) to design a fuzzy system which can match all the N input-output pairs to any given accuracy (i.e., for any given $\epsilon > 0$, we require that $|f(\mathbf{X}_0^j) - y_0^j| < \epsilon$ for all $j = 1, 2, \dots, N$). This optimal fuzzy system is constructed as

$$f(\mathbf{X}) = \frac{\sum_{j=1}^{N} y_0^j \prod_{i=1}^{n} \mu_{\mathbf{A}_i^j}(x_i)}{\sum_{j=1}^{N} \prod_{i=1}^{n} \mu_{\mathbf{A}_i^j}(x_i)}, \tag{1}$$

where $\mathbf{A}_i^j$ are fuzzy sets defined by their membership functions $\mu_{\mathbf{A}_i^j}, i = 1, \dots, n$ and $j = 1 \dots, N$.

The above fuzzy system is constructed from N IF-THEN rules which have the following form:

$$\text{IF } x_1 \text{ is } \mathbf{A}_1^j \text{ and ... and } x_n \text{ is } \mathbf{A}_n^j \text{ THEN } y \text{ is } \mathbf{B}^j, \quad (2)$$

where $\mathbf{A}_i^j$ and $\mathbf{B}^j$ are fuzzy sets in $\mathbf{R}$, and $\mathbf{X} = (x_1, x_2, ..., x_n)^T \in \mathbf{R}^n$ and $y \in \mathbf{R}$ are respectively the input and output variables of the fuzzy system. The membership function of $\mathbf{A}_i^j$ is defined as $\mu_{\mathbf{A}_i^j}(x_i) = \exp\{-(x_i - x_{0i}^j)^2/(\sigma_i^j)^2\}$ where σ_i^j is used to control the accuracy of the matching and the center of $\mathbf{B}^j$ equals to y_0^j. By appropriately choosing the parameters σ_i^j, the fuzzy system in (1) can match all the N input-output pairs to any given accuracy ϵ.

As mentioned by Wang (1994), the fuzzy system defined in (1) is exactly the same as the probabilistic general regression formula.

2.2 Application to CLR estimation

The fuzzy approximation (FA) we propose to estimate the cell loss probability in ATM multiplexers is based on the above fuzzy system. We view the CLR as an unknown function of a variable which can be the multiplexer's buffer size, the number of connected users, or the service capacity.

The algorithm we propose to predict the CLR does not assume anything about the traffic parameters but requires only the knowledge of

- the CLR of the multiplexer when the system size is small (e.g., small buffer size). In this case, the CLR is relatively large and can be calculated or measured quickly and easily;
- the asymptotic behavior of the CLR when the system size is very large (e.g., infinitely large buffer size or number of users).

Let I be the number of known values from the small-sized system. Let $P(m)$ be the cell loss probability when the system is in state S_m, where S_m can be the buffer size B_m or the number of users N_m. The set of states (S_m) should define a linearly increasing function of the form: $S_m = m\delta + S_0$, where $\delta > 0$ is the step size and S_0 is the initial state. Denote by k the index of the current state S_k.

Based on the above informations, the FA algorithm is constructed as follows:

Step 1: Input-Output Pairs Definition

The $(i-2)$ sets of input-output pairs are chosen as follows:

$$\begin{aligned} &\{(P(k-2), P(k-1)); P(k) - P(k-1)\}, \\ &\{(P(k-3), P(k-2)); P(k-1) - P(k-2)\}, \\ &\vdots \\ &\{(P(k-(I-1)), P(k-(I-2))); P(k-(I-3)) - P(k-(I-2))\}, \end{aligned} \quad (3)$$

where $(P(k-(j+1)), P(k-j))$ is the input vector and $P(k-(j-1)) - P(k-j)$ is the output for $j = 1, \ldots, I-2$.

By taking advantage of the monotonicity of the parameterized curve, representing the

CLR as function of the buffer size or the number of users, the fuzzy system can be constructed from only a few input-output pairs. The inputs are represented by the CLR values from the previous and the current states and the output is the difference between the next and the current CLR values. The input-output pairs are updated in real time.

Step 2: Rule Base Generation

For each input-output pair, one fuzzy rule base is generated according to (2). The rules in the fuzzy rule base become

$$\begin{array}{l} \text{IF } P(k-1) \text{ is } \mathbf{A}_1^1 \text{ and } P(k) \text{ is } \mathbf{A}_2^1 \text{ THEN } (P(k+1)-P(k)) \text{ is } \mathbf{B}^1, \\ \text{IF } P(k-1) \text{ is } \mathbf{A}_1^2 \text{ and } P(k) \text{ is } \mathbf{A}_2^2 \text{ THEN } (P(k+1)-P(k)) \text{ is } \mathbf{B}^2, \\ \vdots \\ \text{IF } P(k-1) \text{ is } \mathbf{A}_1^{I-2} \text{ and } P(k) \text{ is } \mathbf{A}_2^{I-2} \text{ THEN } (P(k+1)-P(k)) \text{ is } \mathbf{B}^{I-2}, \end{array} \quad (4)$$

where the membership functions of the input fuzzy sets are given by

$$\begin{cases} \mu_{\mathbf{A}_1^j}(P(k-1)) = \exp\left\{-\dfrac{(P(k-1)-P(k-j-1))^2}{(\sigma_1^j)^2}\right\}, \\ \mu_{\mathbf{A}_2^j}(P(k)) = \exp\left\{-\dfrac{(P(k)-P(k-j))^2}{(\sigma_2^j)^2}\right\}, \end{cases} \quad j = 1, \ldots, I-2, \quad (5)$$

with

$$\begin{cases} \sigma_1^j = \dfrac{\max_j(P(k-j-1)) - \min_j(P(k-j-1))}{I-2}, \\ \sigma_2^j = \dfrac{\max_j(P(k-j)) - \min_j(P(k-j))}{I-2}, \end{cases} \quad j = 1, \ldots, I-2, \quad (6)$$

and the center of the fuzzy set $\mathbf{B}^j$ is $P(k-j+1) - P(k-j)$, $j = 1, \ldots, i-2$.

Note that σ_1^j and σ_2^j, $j = 1, \ldots, i-2$, are free parameters that determine the accuracy of the approximation. The above choice of σ_1^j and σ_2^j, $j = 1, \ldots, I-2$, makes the membership functions uniformly cover the range of the inputs vectors.

Step 3: Fuzzy system construction

The fuzzy system given in (1), with $n = 2$ and $N = I-2$, is chosen to estimate the increment (or decrement) of the CLR. The output of the fuzzy system will be added to (or subtracted from) $P(k)$ to estimate $P(k+1)$. This predicted value of $P(k+1)$ is fed back into the fuzzy system as an input to estimate $P(k+2)$ using $P(m)$, $m = k-N, ldots, k+1$ and the process continues until we get the CLR for the desired size of the system.

Step 4: Use of the asymptotic information

Generally, the FA algorithm provides an accurate prediction when the number of states to predict is not too large. Nevertheless, predicting CLR values of the order of 10^{-10} requires a large number of states, this makes the accuracy of the prediction poor because of the accumulation of small errors. In order to improve the estimation, we have to take into account the asymptotic behavior of the CLR as the system size S_m becomes infinitely large.

When S_m is the multiplexer's buffer size (B_m), we know that the CLR behaves like (Anick et al. (1982))

$$P(m) \sim L_0 \exp\{-\eta B_m\}, \quad \text{as } B_m \to \infty, \tag{7}$$

where η is given in Anick et al. (1982). In other words, we have $\log P(m+1) - \log P(m) \sim -\eta(B_{m+1} - B_m) = -\eta\delta$, for $B_m \to \infty$, where δ is the step increment in buffer size.

When S_m is the number of users (N_m), we know that

$$P(m+1) - P(m) \sim 0, \quad \text{whenever the system load } \geq 1. \tag{8}$$

By taking the advantage of this additional knowledge, the predicted CLR value becomes

$$P(k+1) = (1 - \lambda(k))P_1(k+1) + \lambda(k)P_2(k+1), \tag{9}$$

where $\lambda(m)$ is a nonlinear increasing function of S_m taking its values in $[0,1]$, $P_1(k+1)$ is the predicted value in Step 3 above and $P_2(k+1)$ is the predicted value based on the above asymptotic behavior. Finally, we feed this predicted value to the fuzzy system in Step 1 to predict the next value.

The function $\lambda(m)$ we use is given by

$$\lambda(m) = \begin{cases} \exp\left(-\dfrac{(S_m - S_\infty)^2}{\sigma^2}\right) & \text{if } S_m \leq S_\infty \\ 1 & \text{elsewhere,} \end{cases} \tag{10}$$

where S_∞ is any state at which the asymptotic behavior begins to hold and σ is a parameter controlling the accuracy of the fuzzy approximation. The details of determining S_∞ and σ will be discussed later in Section 3.3.

3 NUMERICAL RESULTS

The implementation of the algorithm described above proves straightforward, and the CPU time needed to predict the CLR is quite small. To illustrate the accuracy and the robustness of the FA algorithm, we present in the following a sample of figures representing the CLR as a function of the buffer size. On this figures, our results are compared to the other alternative methods presented in the literature. The model we use to derive the "exact" results is based on the FF approximation with different burst length distributions as discussed in Bensaou et al. (1994). The characteristics of the input traffic are depicted in Table 1.

3.1 Validation of the FA approximation

To show the accuracy and the effectiveness of the FA algorithm, we use the minimum necessary set of input/output pairs. In Figure 1, this number is equal to five ($I = 5$). This means that the fuzzy rule base contains three IF-THEN rules. The figure shows the overflow probability as a function of the buffer size for voice calls (peak rate = 64

Table 1 Traffic Characteristics used.

Traffic class	Peak Rate (Mb/s)	Mean Rate(Mb/s)	Burst length (cells)
Data	10	1	339
Image	2	0.087	2604
Voice	0.064	0.022	58

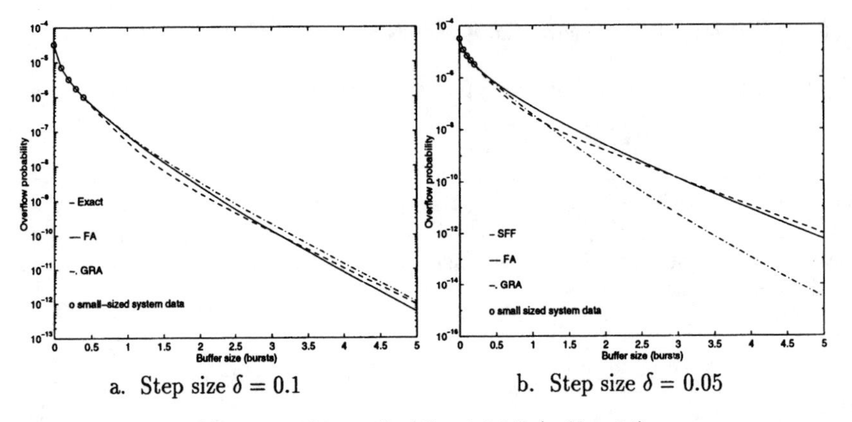

a. Step size $\delta = 0.1$ b. Step size $\delta = 0.05$

Figure 1 Voice calls (C = 1.5 Mb/s, N = 34).

kb/s, mean rate = 22 kb/s, mean burst duration = 350 ms). The system load is 0.5 and the number of multiplexed sources is equal to 34. In Figure 1.a and Figure 1.b, the five input/output pairs are obtained from the FF approximation with a step size (δ) of 0.1 and 0.05 (mean burst length) respectively, beginning at 0. As shown in the figures, our fuzzy algorithm gives a very good approximation compared to the exact results and to those provided by GRA. The CPU time (10ms) needed by the FA algorithm is, however, 10 times smaller than that (100ms) needed by the GRA algorithm and about 20 times less than that (200ms) needed by the exact algorithm. We mention however that the CPU time needed by the FF approach is not representative of the CPU time needed by analytic models. The above value is obtained from the only known closed-form result. For the other algorithmic solutions, this can be much larger.

3.2 Independence from a model

The fuzzy approach used to predict the CLR is to some extent independent of any analytical modelling of the actual input process. To illustrate this independence, in this section we show two kinds of figures.

The first kind (Figure 2.a) shows the overflow probability against the buffer size for generally distributed burst lengths, as discussed by Bensaou et al. (1994). To simplify the exact model, the number of sources is assumed to be infinitely large. The burst arrival

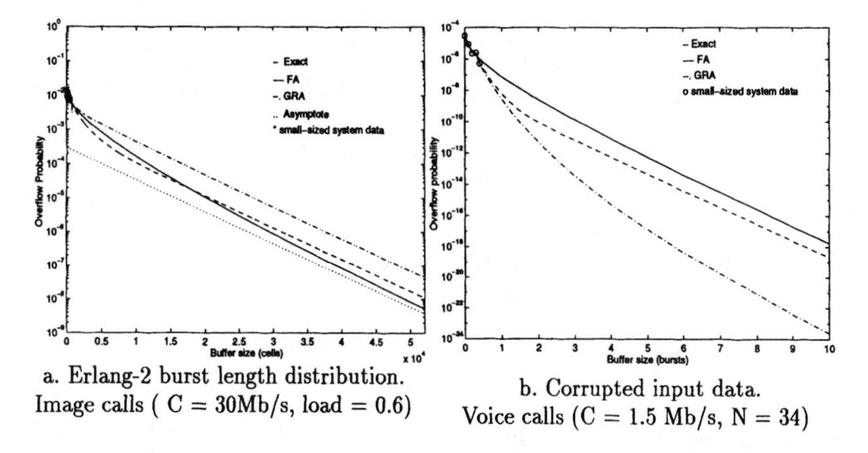

a. Erlang-2 burst length distribution. Image calls (C = 30Mb/s, load = 0.6)

b. Corrupted input data. Voice calls (C = 1.5 Mb/s, N = 34)

Figure 2 Independence of the FA approximation from the model.

process thus becomes Poissonian and the individual mean rate is infinitely small. This assumption does not restrict the generality of our algorithm. In Figure 2.a the burst lengths are 2-phase Erlang distributed with a mean value of 2604 cells and the peak rate is equal to $2Mbits/s$. As shown by the curves, the FA algorithm gives a very reasonable approximation. The error is always within one order of magnitude. More numerical examples with other burst length distributions have shown the same accuracy. This shows that our approach is independent of any assumption on the traffic parameters such as the burst length distributions, etc.

The second kind (Figure 2.b) shows the prediction of the overflow probability from a measured set of input/output pairs with some random measurement errors. To simulate this measured set, we take the exact (FF) curve for a superposition of voice sources from Figure 1.a, and introduce a small perturbation to the first 5 points by adding (or subtracting) a value ranging from 0% to 50% respectively to (or from) these points. The smaller is the overflow probability, the larger is the introduced error. In other words, the first point has no error while the last point can have relative error as large as 50%. This simulates the situation of on-line measurement of real traffic when the measurement time is not very large. Figure 2.b shows the accuracy of the FA algorithm in predicting the overflow probability. It is clear that the refinement based on the knowledge of the asymptotic slope is very valuable in this kind of situations.

These two figures have shown the robustness and the independence of our algorithm from any specific traffic model. Moreover, the algorithm can be easily extended to the case of a superposition of heterogeneous classes of traffic without increasing the complexity. This makes it very appealing for real-time applications.

3.3 Determination of S_∞ and σ

We would like to draw the attention of the reader to the fact that the performance of the FA approximation is largely determined by the choice of the values of the parameters S_∞ and σ. In this section, the state S_m has been chosen to be the buffer size B_m. The value for $S_\infty(= B_\infty)$ must be chosen such that for a buffer of size B_∞ the asymptotic behavior holds. A heuristic approach is to set B_∞ equal to ten times the mean burst length when the traffic intensity is light (< 0.85), while a value of a few times the mean burst lengths is sufficient in heavy traffic conditions. In the figures above, the value of B_∞ was equal to 10 mean burst lengths. In addition, extensive calculations have shown that a value of σ equal to 60% of the value of B_∞ is quite reasonable. Furthermore, we can combine our fuzzy system with other adaptive learning algorithm such as neural networks to determine the optimal values of B_∞ and σ. Therefore, one can first train the fuzzy system with off-line simulation and then use it with on-line traffic measurement.

4 APPLICATION TO CALL ADMISSION CONTROL

One of the most important procedures to control congestion in the network is the admission control process, which restricts the number of calls within the network to provide and ensure QoS guarantees to all users in progress. To allow the network to decide quickly whether a new call can be accepted or not, a fast algorithm is required to estimate either the CLR or the required bandwidth. Most of the proposed call admission control (CAC) algorithms (e.g., Guérin et al. (1991)) are based on the effective bandwidth (the asymptotic upper bound or the stationary Gaussian approximation) without taking into account of statistical multiplexing among sources. However, both algorithms overestimate the required aggregate bandwidth, resulting in an under-utilization of expensive network resources such as bandwidth and memory.

In this section, we present a simple admission control algorithm based on our FA algorithm. This CAC algorithm should just be taken as a simple example to illustrate the application of the FA algorithm to real ATM problems. The algorithm we present below is effective only when the traffic model can be known. It considers the CLR, as a function of the number of users. A more practically interesting measurement-based CAC algorithm using our FA algorithm to predict the required aggregate bandwidth is presented in the work of Chu et al. (1996).

4.1 Homogeneous traffic sources

Without loss of generality, we consider an ATM multiplexer with a high speed channel serving homogeneous sources and a buffer of finite size to bound the delay requirements. To provide QoS guarantees such as the cell loss probability (e.g., CLR $< \epsilon$), we need to be able to estimate the CLR with different number of users. Let $P(N)$ denote the CLR when the number of users is N. When N is small, we can use any proposed analytic model (such as FF, MMDP, etc.) to calculate the values of the CLR. Nevertheless, the computational complexity increases dramatically for large number of sources. As discussed in Section 2, we can easily apply our FA algorithm to predict the CLR as a function of the number of users. Therefore, we can implement our CAC according to the following steps.

1. We first obtain four initial values of the CLR $\{P(N_0), P(N_0+1), P(N_0+2)$ and $P(N_0+3)\}$ with some analytic model, where N_0 is chosen large enough to ensure that the CLR is not equal to zero. For instance, the number N_0 can be chosen such that $N_0 - 1$ is the maximum number of admissible users with peak bandwidth allocation.

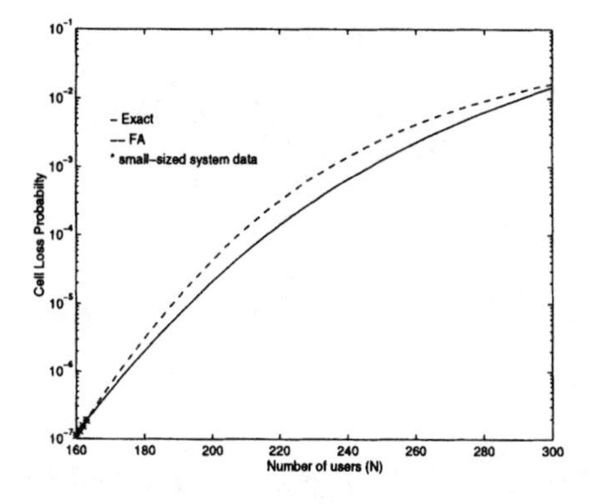

Figure 3 Cell loss probability vs. number of users (Data source, C = 350 Mb/s).

Table 2 Mixed traffic sources.

Traffic class	Peak Rate (Mb/s)	Mean Rate(Mb/s)	Burst duration (s)
Class 1	0.1	0.01	0.35
Class 2	0.2	0.02	0.175

2. Using $\{P(N_0), P(N_0+1), P(N_0+2)$ and $P(N_0+3)\}$, the FA algorithm is used to predict $P(N_0+4)$ and then $P(N_0+k)$ for $k > 4$ successively. The asymptotic behavior is given in (8).
3. We obtain the curve of CLR as a function of the number of users from Step 2. When a new user requests a connection, the call admission controller will decide to accept the new call if the resulting CLR is smaller than ϵ or to reject it if the value is greater.

In order to test the performance of the FA algorithm, Figure 3 shows the FA prediction of the CLR against the number of users. The exact model we used is based on the MMDP approximation developed by Yang and Tsang (1995). The sources are data sources with traffic characteristics given in Table 1. The service capacity is equal to $350Mbits/s$ and the buffer can accomodate upto 50 cells. From the figure, the accuracy of this algorithm is much better than the results obtained from the effective bandwidth approach using the asymptotic FF upper bound (see Guérin (1991)). We omit this latter curve from the figure, the CLR being always close to 1. The CPU time needed is independent of the number of sources and is linearly dependent on the number of initial values.

4.2 Two heterogeneous classes of traffic sources

The simplicity of this algorithm and its speed make it very easy to extend to the heterogeneous case, for which the functions to estimate become multidimensional. To illustrate this, Figure 4 shows the relative error for the mix of two generic classes of traffic, each one contributing 50% of the total load. The channel capacity is $1.5Mb/s$ and the buffer can accommodate up to 2358 cells. The traffic characteristics are given in Table 2.

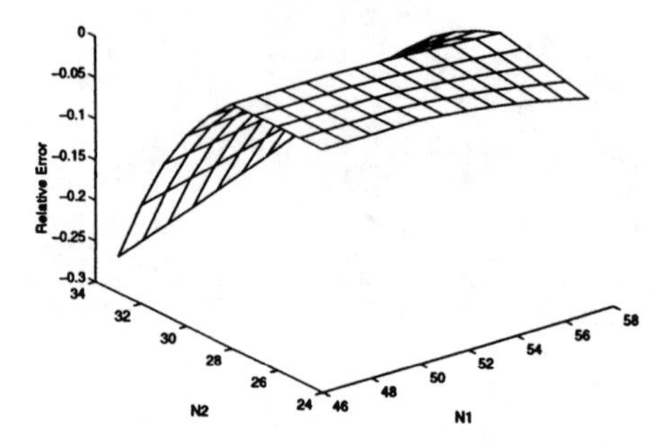

Figure 4 Prediction error of the aggregate overflow probability for 2 classes.

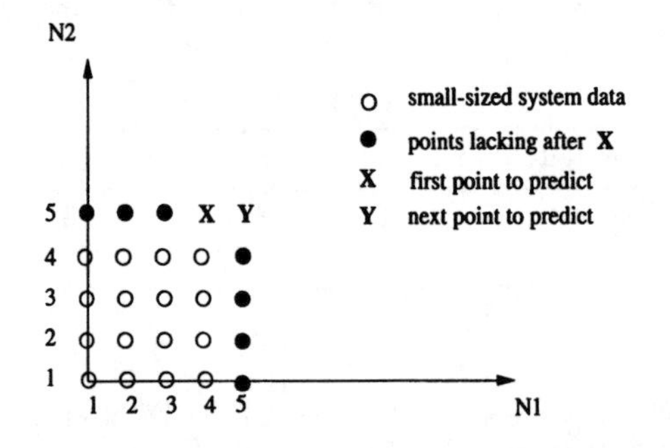

Figure 5 Prediction path for 2 classes of traffic sources.

In this case, the FA algorithm relies on the knowledge of 16 initial points (the 16 empty circles in Figure 5). From these values and knowing that the CLR converges to 1 when the number of sources is very large, we can predict the CLR surface. Since the predicted surface and the exact one are too close, we show the relative prediction error instead. The figure shows that the relative error is reasonably small (less than 30%). Extensive numerical tests have shown that the accuracy of prediction is independent of the followed path: fix the number of class 1 sources and predict towards the increasing number of class 2 sources, or vice versa. The complexity in this heterogeneous case is unfortunately dependent on the number of sources in $M - 1$ classes, where M is the total number of traffic classes. As shown in Figure 5, assume the first user to arrive is from class 2. By using the four points in column 4, the first point to predict in Figure 5 is then **X**. Assume the next user to request a connection is from class 1, from Figure 5 we see that in either direction (row or column), there is a lack of points to predict the point **Y**. We therefore need to predict at most 4 points preceding **Y** in either direction (3 points in row 5 or 4 points in column 5). Nevertheless, since the algorithm is intended for on-line use for which only one point is predicted when a new connection arrives, the prediction of the future lacking points can be done off-line after a call is accepted. This dependence on the number of sources then becomes unimportant.

The values of N_∞ and σ are very important. To obtain tight predictions, we have noticed from numerical tests that the suitable value of N_∞ is equal to 10 times the ratio of the channel capacity to the mean rate, and σ is approximately 48% of the value of N_∞.

5 CONCLUSIONS

A new approximation algorithm is proposed to estimate the CLR for large sized systems by using a small amount of information from small-sized systems. In this algorithm, we use an adaptive fuzzy system to predict the CLR of the actual system. This prediction combined with the knowledge of the asymptotic behavior allows us to obtain a good approximation of the CLR.

To illustrate the efficiency and the practical utility of our algorithm, we have proposed one possible implementation of a CAC mechanism using the predicted CLR from the proposed FA algorithm. The numerical results have shown that the FA-based CAC algorithm outperforms the traditional approaches such as effective bandwidth without requiring excessively more CPU time than these approaches.

Currently, we are extending our work to other types of traffic such as self-similar traffic. Since our approach is independent of any traffic model, the FA algorithm is expected to predict accurately the CLR for this kind of traffic sources as well. The extension of the FA algorithm towards CAC in more complex networks composed of multiple multiplexing stages is also under examination.

REFERENCES

Anick, D., Mitra, D., and Sondhi, M. (1982). Stochastic theory of a data handling system with multiple sources. *Bell Systems Technical Journal*, 61(8):1871–1894.

Bensaou, B., Guibert, J., Roberts, J. W., and Simonian, A. (1994). Performance of an ATM multiplexer queue in the fluid approximation using the Beneš approach. *Annals of Operations Research*, 48.

Chang, C. J. and Cheng, R. G. (1994). Traffic control in an ATM network using fuzzy set theory. In *IEEE Infocom'94*, pages 1200–1207.

Cheung, K. F., Tsang, D. H. K., Cheng, C. C., and Liu, C. W. (1994). Fuzzy logic based ATM policing. In *IEEE ICCS'94*, pages 535–539.

Chu, H.-W., Bensaou, B., Lam, S. T. C., and Tsang, D. H. K. (1996). Call admission control in ATM using fuzzy logic. Submitted.

Guérin, R., Ahmadi, H., and Naghshineh, M. (1991). Equivalent capacity and its application to bandwidth allocation in high-speed networks. *IEEE JSAC*, 9(7):968–981.

Wang, L. X. (1994). *Adaptive fuzzy system and control: design and stability analysis.* Prentice Hall, Inc.

Yang, H., Towsley, D., and Gong, W. (1995). Efficient calculation of cell loss in ATM multiplexers. In *IEEE Globecom'95*, pages 1226–1230.

Yang, T. and Tsang, D. H. K. (1995). A novel approach to estimating the cell loss probability in an ATM multiplexer loaded with homogeneous sources. *IEEE Transactions on Communications*, 43(1):117–126.

PART TWO

Broadband Technologies and Network Design

27

A TDMA Based Access Control Scheme for an ATM Passive Optical Tree Network

Frans J.M. Panken
University of Nijmegen, Computing Science Institute
Toernooiveld 1, 6525 ED Nijmegen, The Netherlands
Tel. +31 24 3652450; E-mail: fransp@cs.kun.nl

Abstract
This paper proposes a MAC protocol for a broadband access facility, using an ATM Passive Optical Network with a tree structure. Access to a shared medium is controlled by means of a request/permit mechanism and allows a flexible way of access for services which are sensitive for (variation in) delay and for Available Bit Rate & Connection Less services. In order to guarantee a fast reaction time on changing traffic characteristics and a flexible use of available data rate, Request Access Blocks are introduced. Quantitative simulation studies show that the traffic distortion caused by the proposed MAC protocol is very small. An additional complex traffic shaper is therefore not necessary. Moreover, the protocol still provides a good performance with respect to transfer delay.

Keywords
Protocol Design and Analysis, Multiple access, Medium Access Control Protocol, TDMA, APON, Performance, Tree Network.

1 INTRODUCTION

So far, research regarding access to broadband networks mainly addresses the needs for big business customers, located in traffic intense metropolitan areas. This is natural, since these users need and can afford a dedicated fiber to have access towards B-ISDN. On the other hand, small business customers and residential users who want access to B-ISDN often neither need nor can afford the full offered data rate. In order to offer these users services from the B-ISDN, a flexible way of sharing resources is required, with the aim to concentrate traffic and to reduce costs. The ATM Passive Optical Network (APON) technology is a good candidate for such an access network, leading to the set up as depicted in Figure 1. This figure shows that the *Optical Line Terminator* (OLT) is located at the root of the tree and the fiber is splitted into several branches which on their turn contain an *Optical Network Unit* (ONU), with connections towards customer(s).

calculates the required electronic delay for each ONU in order to know how long an ONU should wait before transmitting a cell. This gives each ONU the same perception of time, but is transparent for the MAC protocol to which the distance from OLT to each ONU appears to be the same. The way ranging can be realized is beyond the scope of this paper. For general issues with respect to ranging, we refer to d'Ascoli (1994).

2 THE PROPOSED MAC PROTOCOL

The main function of the MAC protocol is to avoid collisions between upstream ATM cells, originating from the different users. Moreover, a MAC protocol should aim at a good:

- *Efficiency :* the overhead introduced by the MAC protocol should be low.
- *Performance :* the delay and the delay variation introduced by the MAC protocol should be kept within certain bounds, particularly for services whose delay requirements were guaranteed during connection set up.
- *Priority :* cells originating from non-delay sensitive services (e.g. ABR and CL services) must be handled in such a way that the presence of these services does not have a large impact on the performance of services which are sensitive for (variation in) delay.
- *Fairness :* one customer should not be subject to more delay than another.
- *Robustness :* when an error occurs (especially when it is introduced by the medium), the MAC protocol should be able to recover from this error.

As described in Section 1, a maximum of 6 terminations are grouped into one ONU which saves all cells originating from these terminations in two separate buffers: the sensitive buffer and the non-sensitive buffer. An ONU consequently advertises its transmission requirements through requests (see Section 2.1) which are sent in a RAB to the controller, located at the OLT. These requests contain information about the state of the queues at the ONU. Using these requests, together with parameters agreed on during call set up, the MAC protocol distributes dynamically the available upstream data rate among the ONUs by means of the bandwidth allocation algorithm (BAA, see Section 2.3). The ONUs are then informed about the allocated transmission capacity by means of permits. When an ONU receives a permit, it is allowed to send one cell from the appropriate buffer in the upstream direction.
The request/permit mechanism allows us to assign the protocol to the class of *reservation* based protocols, with centralized control (see Kurose,Schwartz and Yemini (1984)).

2.1 Requests

In order to be able to declare the need of transmission capacity, two *Request Counters* (of 5 bits each) are assumed to be present at each ONU. A request counter counts the number of arrivals per ONU-buffer. Each time a new cell enters one of the ONU-buffers, the corresponding counter is incremented by one. The need for an empty slot for each of the buffers is passed on to the MAC controller by sending *Request Access Blocks* (RABs).
An RAB consists of a collection of requests originating from all 16 ONUs and is sent in the upstream direction every twenty slots. Figure 2 shows the upstream information structure. During two successive RABs a maximum of 5 cells can be generated by each termination, such that a total of $6 \cdot 5 = 30$ cells

may arrive at an ONU during this period. This requires an information field of 5 bits. However, these arrivals may originate from either solely the non-sensitive buffer or solely the sensitive buffer. Therefore, the contribution to an RAB of each ONU needs a field of $2 \cdot 5 = 10$ bits in an RAB in order to inform the controller about all new arrived cells. Every twenty slots the MAC controller issues a permit which forces all ONUs to form an RAB. Each ONU knows its position at the PON and consequently sends after a fixed period of time the value of its request counters (with a maximum of 31), such that an RAB is formed. At the same time, the value of the request counters is decreased with this value.

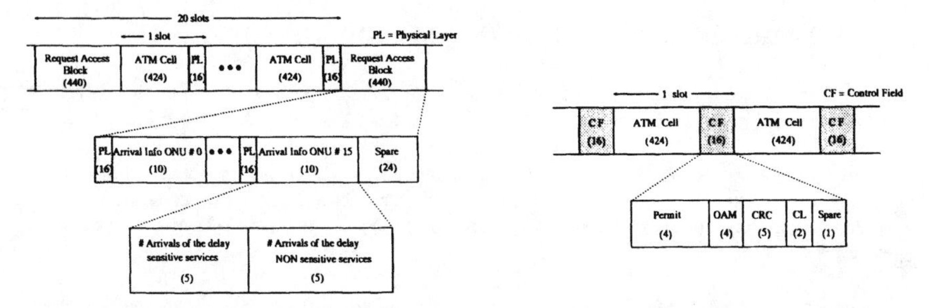

Figure 2 The upstream transmission format. All values given in bits.

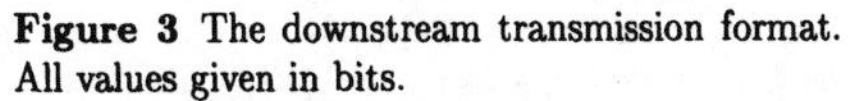

Figure 3 The downstream transmission format. All values given in bits.

Besides the 10 bits field containing information about new arrivals, the contribution of each ONU to one RAB also consists of a preamble of 16 bits. This preamble includes a gap (≈ 6 bits) and bits which are required for fine ranging, bit/byte synchronization and correction for drifts due to temperature changes. It also includes 3 bits which are used for tuning. This is necessary since due to distance equalization, the optical intensity of each ONU at the OLT may differ such that the transceiver in the OLT needs to tune for this. For the same reason such a preamble is attached to each upstream ATM cell. Time is therefore divided into slots of $424 + 16 = 440$ bits. Since there are 16 ONUs, each RAB occupies $16 \cdot (10 + 16) = 416$ bits. To make synchronization easier, an RAB simply takes the position of an ATM cell, enforcing a 24 bit spare field (see Figure 2).

2.2 Permits

The distribution of data rate is handled by the *Bandwidth Allocation Algorithm* (see next subsection), whose output consists of *permits*. A permit is sent downstream and contains the address of an ONU. As soon as an ONU receives a permit, it is allowed to send one ATM cell in upstream direction.
The permit makes part of a *Control Field* which has a length of 2 bytes. We refer to Figure 3 for the downstream transmission format. This control field travels together with a downstream ATM cell from OLT to termination. Because of the broadcast nature of the PON, the destination address of the ATM cell and the permit do not need to be the same. When at the OLT no ATM cell was scheduled for downstream transmission, the control field (containing the permit) can just take the place of an ATM cell.
Three kind of permits can be distinguished:

1. Permits for cells in the sensitive queue.

2. Permits for cells in the non-sensitive queue.
3. Permits to contribute to a Request Access Block.

We decided to let all ONUs wait for a permit to send an RAB (instead of an autonomous decision at the ONUs) to reduce the complexity at the side of an ONUs and for making synchronization easier. A *Permit Distinguish* (PD) field of two bits is added to the downstream control field in order to indicate the functionality of the permit (see Figure 3).

2.3 The Bandwidth Allocation Algorithm (BAA)

This algorithm distributes the available data rate of the shared medium among all terminations and is the *engine* of the MAC protocol. The main characteristics of the BAA of this MAC protocol are:

1. Approximation of a *VP-based UPC device.* By doing this, we concentrate any possibly miscellaneous traffic behaviour to one ONU and therefore supply a protection mechanism for the access network.
2. Use spare transmission capacity of the shared access medium to grant access to cells originating from non-sensitive services.

Upon reception of an RAB, the OLT has a clear overview of the number of cells arrived in both the sensitive buffer and in the non-sensitive buffer of all ONUs during a period of 20 slots. Generating permits for these requests can be realized very easily, since the sequence of ONUs within an RAB is fixed, such that each 10 bit information field within an RAB corresponds with a pointer to an ONU-address. However, if permits are generated according to this fixed sequence, an unfairness between the ONUs is introduced. To accomplish a fair access mechanism, it would be desirable to consider all requests in one RAB to the same collection. Consequently one request can be removed randomly from this collection and a permit for the corresponding ONU can be generated. This process may continue until no requests are left. However, since the number of requests carried within one RAB is not known in advance, this solution is difficult to implement in hardware. Furthermore, it is also desirable to have a mechanism that *protects* the access network.

The solution of fairness and protection is solved at the same time, by making use of 4 sets of k buffers, all located in the OLT. Let us first consider the protection mechanism in case of delay sensitive requests. This mechanism decides whether the permit is assigned to the first set or the second set of k buffers, by verifying if the number of requests corresponds with the parameters agreed on during connection set up. In other words, the MAC controller checks if the terminations, connected to an ONU, are requesting a data rate that agrees with their contracts. By enforcing this, we *expand* the basic functionality of the MAC protocol. Such an enlargement of the MACs functionality was also proposed in Casals, García and Blondia (1993), by making use of a *bundle spacer* (see also García,Blondia,Casals and Panken (1994)). We propose to protect the access network by making use of an algorithm that approximates the well-known *killing window* UPC mechanism, based on the sum of all connections which are shared by one ONU. We shall describe shortly how this can be implemented, by considering the parameter X, defined by

$$X = X_{old} + Nquantum \cdot (Time - Last) \cdot Alloc. \quad (1)$$

In this equation, *Nquantum* is the number of elementary quantum in the link bit rate (measure of *granularity*). *Alloc* is the number of quantum allocated to the policed connections which share an ONU. *Alloc* can easily be determined by $\lceil$ (Nquantum / nett link capacity) $\cdot \sum_i$ peak bit rate$_i \rceil$, where we sum over all VPI/VCI combinations which share an ONU. Finally, there is a parameter *Window*, representing the discarding threshold. Three situations can occur:

1. $X < 0$. When this occurs the BAA creates a permit and assigns it for the first set of k buffers. The value X_{old} is reset to zero and *Last* is set to *Time*.
2. $0 \leq X \leq Window$. When this occurs, the BAA creates a permit and assigns it for the first set of k buffers. X_{old} is updated and *Last* is set to *Time*.
3. $X > Window$. For this situation the request is considered to be *non compliant*. The BAA consequently creates a permit and assigns it for the second set of k buffers. X_{old} is not updated and *Last* is set to *Time*.

This same procedure can be carried out for requests of delay unsensitive cells, by simply substituting the first set of k buffers by the third set and the second set of k buffers by the fourth set.
After all requests are served, we start serving queue i (where i is randomly chosen out the interval [1,k]) and forward the permits from this queue to a queue Q1. When all permits in buffer i are served, we repeat this procedure with buffer $i + 1 \pmod{k}$ until all k buffers of the first set are empty. We repeat this procedure for sets 2,3 and 4, whose permits are forwarded to queues Q2, Q3 and Q4 respectively.
Clearly, this procedure makes the access scheme less unfair. It is not a very complex procedure, with respect to the required amount of additional hardware (except for the sets of buffers, of course). Both the fairness and the costs of the access mechanism are an increasing function in k, such that there is a trade-off between costs and fairness. In Section 3 we show the impact of the number of buffers k on the mean delay, as experienced by cells in each of the ONU buffers.

All permits waiting in queues Q1, Q2, Q3 and Q4 are *served*, i.e. we put them in a control field which is attached to a downstream ATM cell, in the following way. Whenever queue Q1 contains permits, this queue is served first. Only when queue Q1 is empty, queue Q2 is served. When queue Q1 and Q2 are empty, queue Q3 is served. Finally, queue Q4 is served whenever queue Q1, Q2 and queue Q3 are empty. By choosing an appropriate queue length of queues Q2 and Q4, this scheme provides a very efficient protection mechanism for the access network. The robustness scheme of the protocol (see Section 2.4) eliminates the possibility that cells keep waiting in the ONU buffer forever in case a permit is lost due to buffer overflow from queue Qi (i=1..4).

In general, X_{old}, *Last, Nquantum, Alloc* and *Window* can be defined for each ONU. However, in order to reduce the amount of memory to store all these parameters, only *Alloc, Last* and X_{old} are recommended to be defined for each ONU. The parameters *Nquantum* and *Window* are recommended to be chosen in a restricted set of 4 pair, each pair based on one value of *Nquantum* and one value of *Window*.

2.4 Robustness of the MAC protocol

As described in Subsection 2.1, the MAC protocol generates permits according to the information received from RABs. However, the loss of a request or a permit due to bit errors on the medium, leads to the situation that cells remain in the buffer of the ONU for a long time, ruining the low CDV behaviour. In order to recover from this situation, one additional *robustness counter* is necessary for each of the ONU buffers. These counters are used in the following way to recover from errors:
Whenever either a cell is sent in upstream direction or a cell enters an empty buffer, the corresponding robustness counter is set to MAX-VALUE. MAX-VALUE equals the time between two successive RABs (i.e. 20 slots) increased with the round trip delay, expressed in slots of the shared medium. When during a slot no ATM cell from a buffer is sent in the upstream direction, the corresponding robustness counter

is decremented by one. As soon as a robustness counter reaches zero, the maximum delay due to both the round trip delay and the time required to wait until an RAB was sent in upstream direction, is reached. When this occurs, the *critical period* starts. An additional time due to buffering in the OLT queues may be the cause of the fact that no permit has arrived yet. However, when during the critical period an ONU spots either a permit for a cell in a non-sensitive buffer (addressed to any ONU) or an empty permit (i.e. a permit with no ONU address), an error must have occurred. An error for non-sensitive queue is ascertained as soon as the ONU observes an empty permit while its robustness counter is zero and the non-sensitive queue is not empty.

As soon as an error is detected, both the robustness counter of this queue is rest to MAX-VALUE and the appropriate request counter is increased with one. Consequently, after a while, another request will be issued by this ONU.

This scheme allows the MAC protocol to recover from errors in a fast and efficient manner. Since we assumed a 5-bits request counter and a maximum of 30 cells at one ONU can arrive between two RABs, the last bit of this counter guarantees the possibility to correct for errors under each traffic condition.

Since an appropriate distance between ONU and OLT is $\approx$ 10 km, an 8 bits robustness counter for both cell buffers is sufficient. Finally, we remark that the round trip delay is already available, since the ranging procedure requires it.

2.5 Efficiency of the MAC protocol

In the PON we divided time into slots of 440 bits. The assumed overhead of 16 bits per ATM cell which is inherited from the physical layer, corresponds very well with the standardized overhead to transport ATM cells such as SOH and POH in case of SDH and F1, F2 and F3 flows in case of cell based transmission. So, by using an APON slot of 440 bits, the transmission rate of 155 Mbit/s can be used while still preserving 149.76 Mbit/s for ATM cells.

However, the introduction of RABs in the upstream direction requires an additional reservation of 440 bits, every 20 slots. This reduces the actual capacity for ATM cells by 100%/20 = 5 %, to about 142.27 Mbit/s on a standardized 149.76 Mbit/s interface. Since it would be very unlikely that a connection with such a large capacity would be established, this is not much of a problem. However, the CAC mechanism has to take this reduced capacity into account (i.e. change the maximum data rate from 149.76 Mbit/s to 142.27 Mbit/s).

3 PERFORMANCE EVALUATION

Since the MAC protocol controls the information flow at the entrance of a B-ISDN network, it has a large impact on the overall performance of the system. In particular, the profile of the traffic as it enters the network is highly influenced by the MAC protocol. Therefore, a performance evaluation of the proposed MAC protocol is carried out in this section.

In order to study the impact of the presence of the proposed MAC protocol, we show the complementary distribution function (i.e. $\Pr\{X > x\}$) of two important performance measures: the transfer delay and the CDV.

The *Transfer Delay* of the access network is defined as the time difference between the sending of a cell at the T_b-Interface and the receiving of this cell at the U_{1b}-Interface.

An accurate performance characterization of CDV is a network performance parameter, known as *1-Point-*

CDV. The 1-Point-CDV describes the variation of the arrival times pattern with respect to the negotiated peak cell rate. It is measured by observing successive upstream cell arrivals at the U_{1B}-Interface and only considers cell clumping, i.e. the effect of cell interarrival distances which are shorter than the reciprocal of the peak cell rate. The characterization of CDV by means of 1-Point-CDV was given in CCITT (1992) and is recommended for CDV assessment by ITU-T.

All presented results in this section are obtained from simulation. In order to make a comparison between certain aspects easier, several related curves are shown in one figure. To prevent that figures become disordered, confidence intervals are omitted. We remark that the 95 % confidence intervals did not show a deviation larger than 10 %, for values larger than 10^{-3}.

To obtain the results presented in Figures 4- 8, we simulated an access network where at each ONU three delay sensitive CBR cell stream of 1, 4 and 10 Mbit/s respectively were offered. In addition, a total of 3 ABR cell streams with the bit rates 1, 4 and 10 Mbit/s were offered at ONU 5,12 and 15 respectively. This accomplishes a load of 0.4. In order to study the impact of the presence of the MAC protocol under a load of 0.8, three delay sensitive CBR cell streams with a bit rate of 5 Mbit/s were added to each ONU. From the two curves in Figures 5 - 12 which represent the performance of connections sharing the same ONU, the one with the lowest quantiles corresponds with a load of 0.4. For each load condition the 1 Mbit/s and the 10 Mbit/s connections were *tagged* during successive simulations and their performance was studied.

Fairness

In order to try to make the delay for each connection of the access network the same, 4 sets of k buffers were proposed in Subsection 2.3. As explained in this subsection, the fairness of the MAC protocol increases with k.

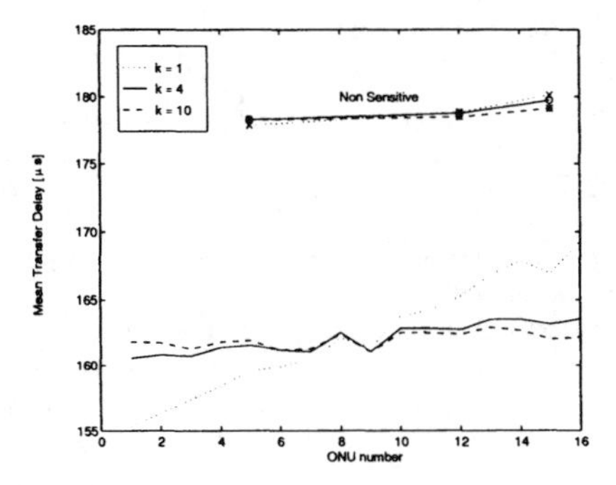

Figure 4 Mean transfer delay as function of the ONU number, for different values of k.

Figure 4 shows the mean delay as function of the ONU number (consider only integer values) under a load of 0.8. We observe that $k = 4$ buffers already shows a rather horizontal curve, corresponding with a fair MAC protocol. For this reason, $k = 4$ buffers were chosen to obtain the rest of the presented results.

Transfer Delay and CDV

From Figures 5 - 8 we conclude that, compared to either centralized MAC protocols or basic equipment in an ATM network such as switches, the proposed access scheme guarantees low CDV and both bounded and acceptable transfer delays for CBR traffic streams at all ONUs. Furthermore, we observe that the performance seen by a connection does not depend very much on its bit rate, as occurs e.g. for some of

the MAC protocols analyzed in Chapter 3.6 of Killat (1996). The load of the access network appears to be an important parameter for those connections which are **not** sensitive for (variations in) delay.
Figures 9 and 10 show that the above statements are still valid when we increase the bit rate of a tagged connection above the bit rate which corresponds with the situation that interarrival times between successive cells are smaller than the time between two successive RABs.

In Figures 11 and 12 we study the performance in case of VBR input traffic. For VBR traffic we used an *on-off* source, where cells are generated according to an underlying Markov chain which alternates between two states: "on" and "off". When the Markov chain is in the on-state, cells are generated according to a CBR pattern, whereas during the off-state no cells are emitted. Both the number of cells generated during the on-period and the sojourn time in the off-state are chosen to be geometrically distributed. In this configuration, temporary overload situations may occur, causing longer tails.
We observe that the priority scheme for sensitive services becomes effective when the load of the access network is high or when the input traffic is more variable.

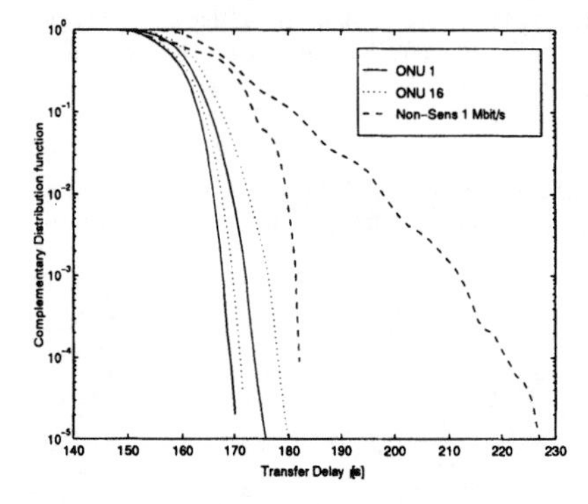

Figure 5 Complementary distribution of the transfer delay of a 1 Mbit/s CBR connection, under the condition that the access network is loaded up to 0.4 and 0.8 respectively.

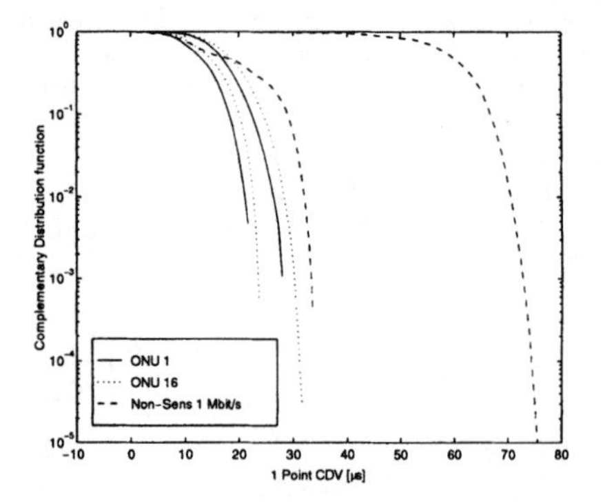

Figure 6 Complementary distribution of the 1 point CDV of a 1 Mbit/s CBR connection under the condition that the access network is loaded up to 0.4 and 0.8 respectively.

4 CONCLUSIONS

In this paper we propose a MAC protocol for an ATM Passive optical tree network. We show how terminations must be grouped into one Optical Network Unit (ONU) and how the ONUs can declare their needs for transmission capacity to a centrally situated controller. Besides this, a robustness scheme and the bandwidth allocation algorithm of the protocol are explained in detail. We showed how the bandwidth allocation algorithm ensures a protection mechanism for the access network, expanding the functionality of the access control scheme.
The MAC protocol can be used for both fiber to the home and fiber to the kerb and allows ABR services in such a way that it can not harm the performance of delay sensitive services. In order to make a make

a statement with respect to the performance of the protocol, simulation results are presented. These results show that the proposed protocol can supply a relative fair access scheme for all users and ensures a bounded delay and very little traffic distortion (CDV), especially for delay sensitive services.

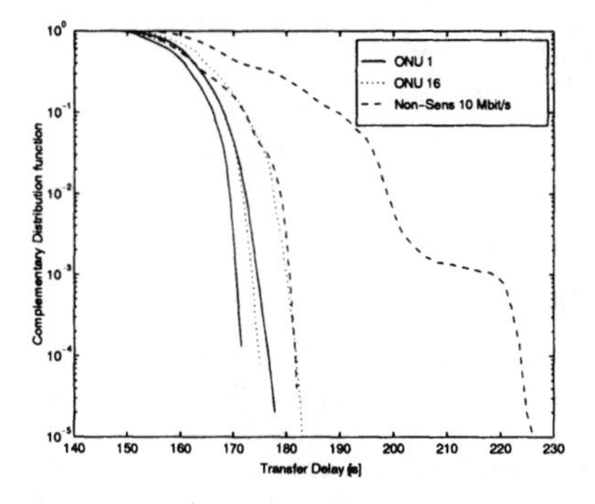

Figure 7 Complementary distribution of the transfer delay of a 10 Mbit/s CBR connection under the condition that the access network is loaded up to 0.4 and 0.8 respectively.

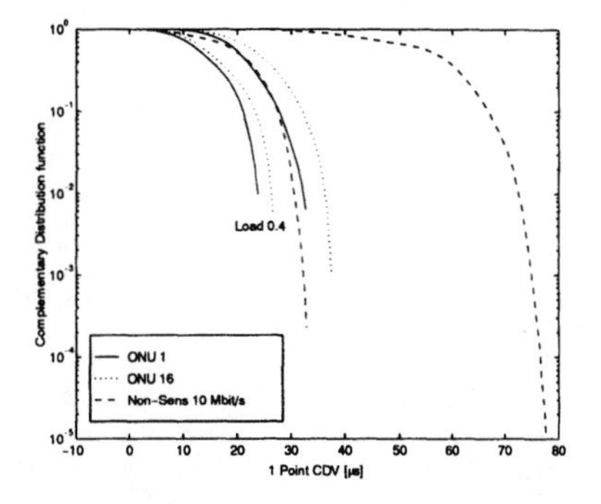

Figure 8 Complementary distribution of the 1 point CDV of a 10 Mbit/s connection under the condition that the access network is loaded up to 0.4 and 0.8 respectively.

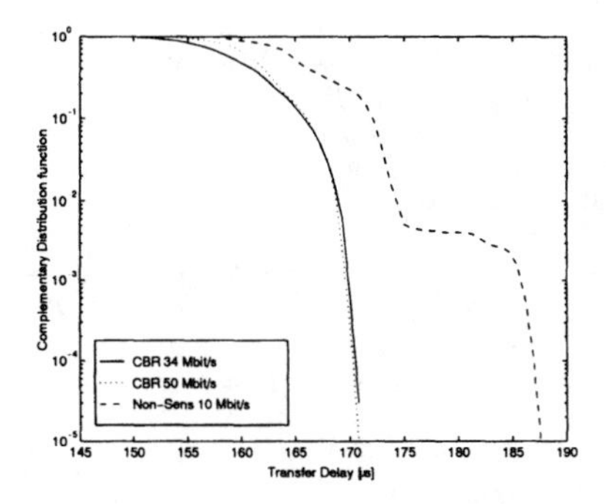

Figure 9 Complementary distribution of the transfer delay when the access network is loaded with CBR traffic of bit rates 34 Mbit/s and 50 Mbit/s. There is one delay sensitive connection and the total load equals 0.8.

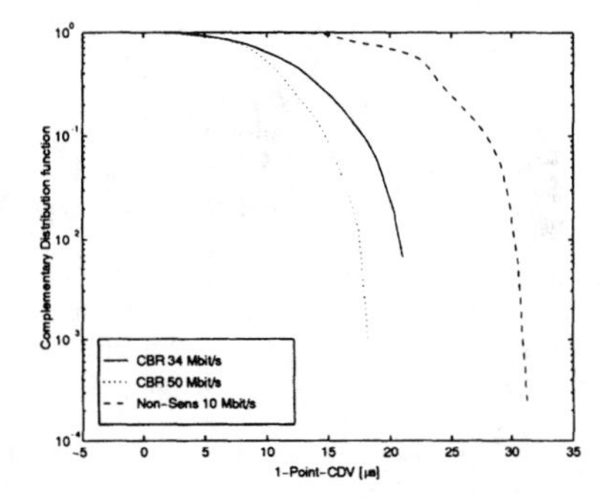

Figure 10 Complementary distribution of the 1 Point CDV when the access network is loaded with CBR traffic of bit rates 34 Mbit/s and 50 Mbit/s. There is one delay sensitive connection and the total load equals 0.8.

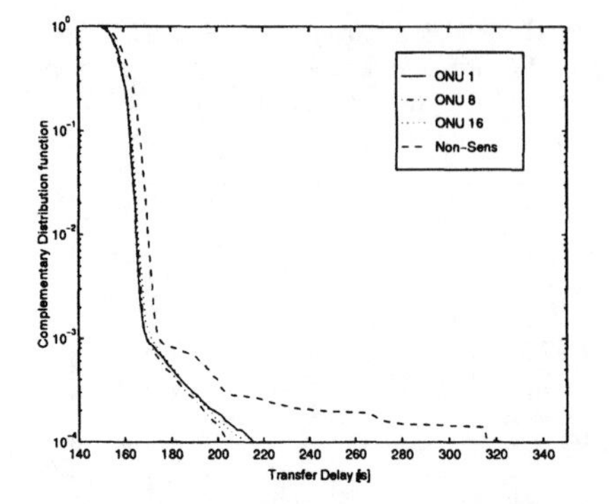

Figure 11 Complementary distribution of the transfer delay when the access network is symmetrically loaded with a total of 45 VBR sources with a peak bit rate of 50 Mbit/s, a mean bit rate of 5 Mbit/s and a burst length of 100 cells.

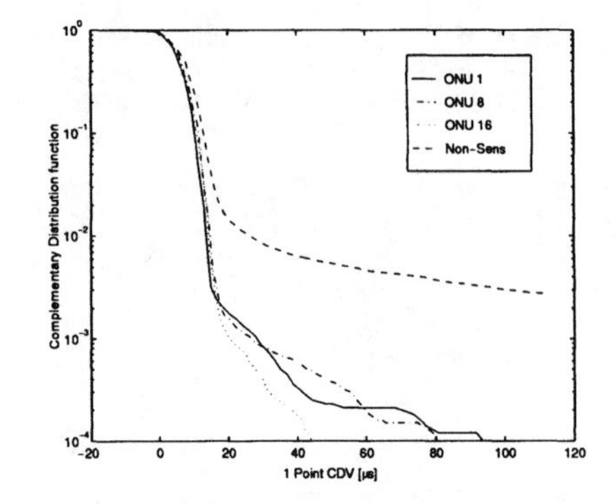

Figure 12 Complementary distribution of the 1 Point CDV when the access network is symmetrically loaded with a total of 45 VBR sources with a peak bit rate of 50 Mbit/s, a mean bit rate of 5 Mbit/s and a burst length of 100 cells.

REFERENCES

Angelopoulos, J.D. and Venieris, I.S. (1994) A Distributed FIFO Spacer/Multiplexer for Access to Tree APONs. *Proceedings of SUPERCOMM/ICC Conference* New Orleans, Louisiana, May 2.

Angelopoulos, J.D. and Venieris, I.S. and Stassinopoulos, G.I. (1993) A TDMA Based Access Control Scheme for APON's. *IEEE Journal Lightwave Technology*, **vol. 11, no. 5/6**, 1095-1103.

Blondia, C. and Verri, L. and Toniatti, T. and Casals, O. and García, J. and Angelopoulos, J.D. and Venieris, I. (1994). Performance of Shared Medium Access Protocols for ATM Traffic Concentration. *European Transactions on Telecommunication*, Special Issue on Teletraffic Research for B-ISDN in the RACE program **Vol. 5, No. 2**, 219-226.

Casals, O and García, J. and Blondia, C. (1993) A Medium Access Control Protocol for an ATM Access Network. *Proceedings of the Fifth International Conference on Data Communication Systems and their Performance, High Speed Networks.*, Eds. H.Perros and Y.Viniotis North Carolina, USA.

CCITT (1992) Draft Recommendation I.35B. B-ISDN ATM Layer Cell transfer Performance. Geneva, June 1992.

d'Ascoli, L. et al. (1994) Out-of-Band Ranging Method for ATM over PON Access Systems. *ECOC, Firenze, Italy.*

García, J. and Blondia, C. and Casals, O and Panken, F. (1994) The Bundle-Spacer: A Cost Effective Alternative for Traffic Shaping in ATM Networks. *Proceedings of IFIP Conference on Local And Metropolitan Communication Systems, Kyoto.*, 177-196.

Killat, U (1996) Access to B-ISDN via PONs. To be published by Wiley.

Kurose, J.F. and Schwartz, M. and Yemini, Y. (1984) Multiple-access protocols and time-constrained communication. *COMPUT. SERV.*, **16**, 1, 43-70.

Van der Plas, G. (1993) APON: An ATM-based FITL system. *EFOC & N'93*, 91-95.

28

The Super-PON concept and its technical challenges

Denis J.G. Mestdagh and Claire M. Martin
Alcatel Corporate Research Center, Francis Wellesplein 1, B-2018 Antwerp, Belgium
Phone : (32 / 3) 240 9070 - Fax. : (32 / 3) 240 9932
E-mail : martinc@btmaa.bel.alcatel.be

Abstract

This paper describes the architecture and the technical challenges for the design of a large-split and long-range Passive Optical Network (PON) intended for Fiber-To-The-Home (FTTH) deployment. The system, called Super-PON, can support a splitting factor of 1024 (or more but with a more complex and hence costly implementation) and a range of 100 km. The overall network capacity is 2.5 Gbps TDM downstream and 300 Mbps ATM-based TDMA upstream. As compared with already designed PON systems, the proposed Super-PON requires the introduction of optical amplifiers along the fiber paths. The presence of these optical amplifiers brings new technical challenges especially for burst mode amplification in the upstream direction of transmission. Analysis and potential solutions to solve these problems while allowing further capacity upgrades using WDM/WDMA techniques are described in this paper along with a discussion of network survivability issues.

Keywords

PON, TDMA, ATM, optical amplification.

1 INTRODUCTION

Passive Optical Networks (PONs) using optical splitters to share the overall system cost among subscribers were recognized about 7 years ago as an economical way forward for Fiber-In-The-Loop (FITL) (Stern, 1989). At that time, the need to modernise the copper-based access network for narrowband services triggered the development of the so-called TPON systems (Telephony over PON) which were specifically aimed at minimizing the cost of fiber technology for telephony and narrowband services (Stern 1989, McGregor 1989). The encouraging results demonstrated by several field trials using these TPON systems in the early 1990's have stimulated the development of PON systems capable of supporting ATM-based broadband services. The development of these ATM-PON systems was both technology and market driven due to the fact that much of the present growth in telecommunication networks is for the business sector where the demand is for high-speed data transfer in addition to telephony and narrowband-ISDN.

Typically, the capacity of ATM-PON ranges between 50 Mbps and 622 Mbps with a splitting factor of 16 or 32 over lengths not longer than 20 km (Ballance 1990, Mestdagh 1991, Ishikura 1991, du Chaffaut 1993). The recent results of an ATM-PON field trial that has been carried out in the Bermuda's islands can be found in (Van der Plas, 1995). The economical viability of these current ATM-PON systems has been demonstrated for business users as well as for a cluster of residential users located in buildings (Fiber-To-The-Building, FTTB). However, with the current cost of optoelectronic devices ATM-PONs for residential users living in separate houses can only be justified economically when deployment is done in a Fiber-To-The-Curb (FTTC) configuration where fiber terminates on a street cabinet from which Asymmetric/Very-high speed Digital Subscriber Line (ADSL/VDSL) technology can be used to provide broadband services, like video on demand, over the embedded twisted-wire pair network primiraly used for the Plain Old Telephone Service (POTS) service. Upgrades of these hybrid fiber/copper networks towards FTTH for residential users can be accomplished in several ways, one of which being the replacement of active remote electronics in the curb site by optical splitters. The resulting large-split PON-based FTTH system, later referred to as Super-PON, shares costs among a much larger number of subscribers than conventional ATM-PON systems do and may thus become an attractive techno-economic solution in the medium terms for the provisionning of existing and future anticipated services for residential users. Besides this upgrade strategy for local access network, Super-PONs may also be an attractive solution for FTTH deployment in so-called green fields where the wired telecommunication infrastructure is not yet in place. In this case, preliminary studies have shown that due to the required/expected switching node consolidation, the access network will have to cover a much longer range (say, 100 km) than the 20 km of conventional ATM-PON systems.

Taking into account the evolution of FTTC systems towards FTTH systems as well as the deployment of FTTH in green fields, Super-PONs will have to support a high capacity over a large-split (say, of the order of 1000 or more) and a large-range (say, up to 100 km) fiber-based access network. These requirements bring new technical challenges that are mainly associated with multiple access techniques for upstream transmission over the shared PON medium, the use of optical amplifiers along the optical paths to overcome the large network losses, and the need to provide some sort of network survivability mechanisms in order to fulfil the requirements of service availability. These challenges and their potential solutions are discussed in this paper which is organized as follows. The next Section presents the general architecture of Super-PONs and will briefly discuss the trade-offs that exist between the required user capacity and the maximum splitting factor. This section also details the network power budget and will serve to introduce the many parameters that will be used in the subsequent section. Section 3 discusses the technical challenges and presents alternative solutions to overcome them. Emphasis will be given to the analysis of amplified burst-mode upstream transmission. Finally, Section 4 provides a summary and final remarks.

2 THE SUPER-PON ARCHITECTURE

Figure 1 depicts the general architecture of a Super-PON access network system. A single-fiber cable plant is proposed with wavelength multiplexing in the 1.3 μm and 1.5 μm windows for, respectively, up- and downstream transmissions. As an alternative, a two-fiber system could be designed as well, depending on the level of allowable crosstalk introduced by WDM devices in the single-fiber solution.

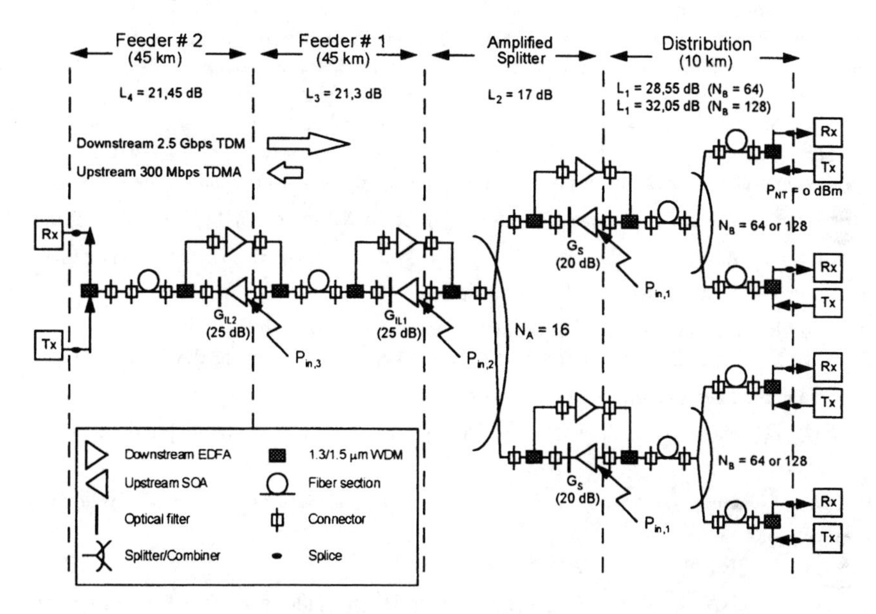

Figure 1 The general architecture of a Super-PON showing the placement of optical amplifiers, connectors/splices, WDM devices for the separation of downstream (at 1.5 μm) and upstream (at 1.3 μm) transmissions and optical amplifiers. The power losses of the different network sections are also indicated along with the gain of optical amplifiers.

Due to the targetted high splitting factor, $N = N_A . N_B$ (see Figure 1 for the definition of N_A, and N_B), high bit rates have to be supported over Super-PONs.

In addition, the choice of multiplexing/multiple-access schemes over Super-PON is of prime concern for an optimum design keeping in mind current technological maturity and eventual upgrading scenarios according to forecast advances in opto-electronic devices. A comparison of multiple access schemes and their potential domains of applications can be found in (Mestdagh, 1995). A 2.5 Gbps downstream Time Division Multiplexing (TDM) and a 300 Mbps upstream packet-based Time Division Multiple Access (TDMA) scheme is proposed for the first generation of Super-PONs. This will allow to later introduce Wavelength Division Multiplexing (WDM) and Wavelength Division Multiple Access (WDMA) schemes as an overlay to TDM/TDMA when the demand for higher network capacity will arise.

The high network losses experienced by the optical signals in the Super-PON requires the introduction of optical amplifiers for both downstream and upstream directions of transmission. The architecture of the Super-PON system that will be analyzed in this paper is depicted in Figure 2. The access network system has been divided into four sections, namely the distribution section, the amplified splitter/combiner section, the first feeder section, and the second feeder section. This will facilitate the calculation of the optical signal power at the input of each amplifier as a function of the losses and gains experienced by the signal in the previous network sections.

The network contains two in-line optical amplifiers, one in each feeder section and 16 optical amplifiers in the input/output branches of the amplified splitter/combiner section. Two different values for the splitting factor of the splitter/combiner located in the distribution section will be considered, N_B=64 and N_B=128, while N_A will be fixed at 16. As such, we will focus our analysis on Super-PONs having either an overall splitting factor N of 1024 or 2048.

Each section is characterised by a total loss $L_{T,i}$ (i=1,2,3,4) equal to the sum of the link losses L_i, the splice losses L_{si}, the connector losses L_{ci}, the WDM losses $L_{WDM,i}$ and the splitter losses $L_{Spl,i}$. The gain of the optical amplifiers will be denoted G_s for the amplified splitter/combiner and $G_{IL,i}$ for the feeder sections.

The five different losses (expressed in dB units) can be calculated as follows:

. $L_i = l_i.\alpha(\lambda)$, where l_i [in km] and $\alpha(\lambda)$ [in dB/km] are respectively the fiber length of section i and the fiber attenuation at the wavelength λ;

. $L_{si} = N_{si} . L_s^{'}$, where N_{si} is the number of splices in section i and $L_s^{'}$ [in dB] is the loss per splice;

. $L_{ci} = N_{ci} . L_c^{'}$, where N_{ci} is the number of connectors in section i and $L_c^{'}$ [in dB] is the loss per connector;

. $L_{WDM,i}$ = 0.5 dB is the loss per WDM device;

. $L_{Spl,i} = 3{,}5. Log_2 N_{A,B}$, where $N_{A,B}$ is the splitting factor of the splitter/combiner in section i [in dB], with $N_A = 16$ in the amplified splitter section and $N_B = 64$ or 128 in the distribution section (the factor 3.5 is taken as a conservative rule).

Table 1 summarises the numerical values of the losses and amplifier gains of the four sections for the upstream direction of transmission operating at a wavelength of 1.31μm. It is seen that the end-to-end upstream network losses are 88.3dB for N=1024 and 91.8dB for N=2048. In order to compensate for these high losses, we will assume $G_s = 20dB$ and $G_{IL,1} = G_{IL,2} = G_{IL} = 25dB$. Such gains are readily achievable with standard commercially available optical amplifiers.

Essentially, it is the presence of optical amplifiers along the fiber path that imposes new technical challenges in the design of TDM/TDMA Super-PONs. These challenges are described in the next section to which we turn now.

Table 1 Network power budget for upstream transmission at 1.31 μm.

		Feeder # 2	Feeder # 1	Amplified splitter	Distribution
	l_l	45 km	45 km		10 km
	α	0,36 dB/km	0,36 dB/km		0,36 dB/km
L_l		16,2 dB	16,2 dB		3,6 dB
	N_{sl}	15	14		3
	L'_s	0,15 dB	0,15 dB		0,15 dB
L_{sl}		2,25 dB	2,1 dB		0,45 dB
	N_{cl}	4	4	4	5
	L'_c	0,5 dB	0,5 dB	0,5 dB	0,5 dB
L_{cl}		2 dB	2 dB	2 dB	2,5 dB
	$N_{WDM,l}$	2	2	2	2
	L'_{WDM}	0,5 dB	0,5 dB	0,5 dB	0,5 dB
$L_{WDM,l}$		1 dB	1 dB	1 dB	1 dB
	$N_{A,B}$			16	64 or 128
$L_{spl,l}$				14 dB	21 or 24,5 dB
$L_{T,l}$		21,45 dB	21,3 dB	17 dB	28,55 or 32,05 dB
G_l		25 dB	25 dB	20 dB	

3 TECHNICAL CHALLENGES

As indicated, one of the major problems for bidirectional transmission over Super-PONs is associated with the use of optical amplifiers along the fiber path. For downstream transmission, Erbium-doped Fiber Amplifiers (EDFAs) operating in the 1.5 μm window can be used to broadcast the TDM data stream to all network terminations. In fact, a PON network comprising a cascade of 4 x 3 x 7^6 x 4 x 7 splittings, corresponding to 39 530 064 potential users, has already been experimentally demonstrated for the broadcast of 384 digital video channels over 12 optical carriers each supporting 2.2 Gbps (Hill, 1990).

Upstream TDMA transmission presents much more challenging problems. Two main issues can be identified: the accumulation of amplified spontaneous emission noise (ASE) of the optical amplifiers, and the time response of optical amplifiers with μsec bursts of data. The former issue becomes particularly detrimental when several optical amplifiers (OAs) are placed in parallel since each amplifier generates its own ASE noise that accumulates at the output of the amplified power combiner (i.e., noise funneling effect). In order to quantify this effect, let us first consider Figure 2 which represents a 1:16 amplified power combiner comprising 16 optical amplifiers (for upstream amplification) of which only n ($n \leq 16$) are biased to their operating point while the other (16-n) OAs are set in the OFF state. With the TDMA MAC protocol, only one of the n biased OAs actually amplifies the input burst data signal while the other (n-1) biased OAs only generate ASE noise.
The Signal-to-Noise Ratio (*SNR*) degradation due to the ASE noise funneling effect can be evaluated as follows. Suppose the amplifier gain is G_s and that each biased amplifier generates identical (but independent) ASE power. Then, the output of the useful amplifier that effectively amplifies the burst of data is given by $G_s.P_{in}$, where P_{in} is the signal power at the input of the amplifier (P_{in} is assumed to be low enough so that OAs don't saturate). Therefore, assuming an ideal photodetector with unity quantum efficiency, the photocurrent generated by the amplified optical signal is given by :

$$i_{sig} = \frac{e}{h\nu}.G_s.P_{in} \tag{1}$$

where e is the electronic charge, h is the Planck's constant and ν is the optical carrier frequency.

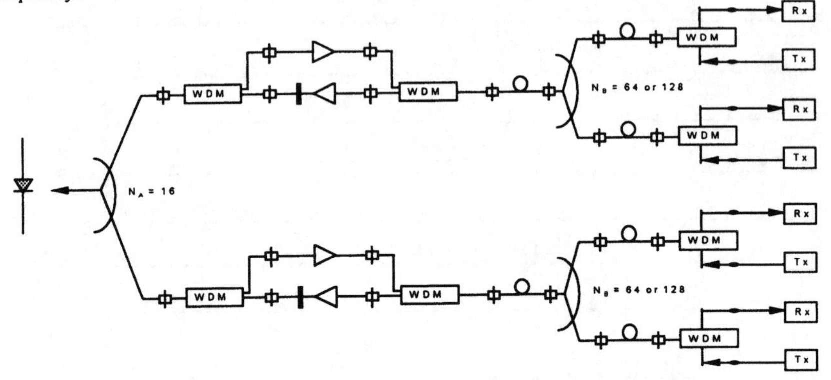

Figure 2 The Super-PON configuration without in-line amplifiers (i.e., short feeder length).

Among the distinct noise contributions due to ASE components, the two most dominant will be considered; that is the signal-ASE and the ASE-ASE beat noises. This assumption is justified provided the gains of OAs are sufficiently high (say, higher than 15 dB) so that shot noises as well as thermal receiver noise become negligible as compared with the beat noises. With a single OA, the electrical power of the beat noises can readily be expressed as follows (Mestdagh, 1995):

$$P_{sig-ASE} = \frac{4e^2}{h\nu}.n_{sp}.(G_s - 1).G_s.P_{in}.B_e \tag{2}$$

$$P_{ASE-ASE} = 2.e^2.n_{sp}^2.(G_s - 1)^2.\Delta\nu_{opt}.B_e \tag{3}$$

where $P_{sig-ASE}$ and $P_{ASE-ASE}$ are respectively the power of the beat between the signal and the ASE and the power of the beat between the ASE components themselves, n_{sp} is the spontaneous emission factor of the OA ($n_{sp} = 2$, which is equivalent to a noise figure of 6 dB, will be assumed throughout the present analysis), $\Delta\nu_{opt}$ is the OA's optical bandwidth ($\Delta\nu_{opt} = \frac{c.\Delta\lambda_{opt}}{\lambda^2}$ where c is the speed of light), and B_e is the electrical bandwidth of the receiver.

With the Super-PON configuration depicted in Figure 2, expressions (2) and (3) must be modified to take into acount the additional beatings between the signal and the ASE noises generated by the other (n-1) biased OAs as well as the beatings between ASEs from all possible combinations in pair of the n biased amplifiers. This gives:

$$P'_{sig-ASE} = n . P_{sig-ASE} \tag{4}$$

$$P'_{ASE-ASE} = \frac{n(n+1)}{2} . P_{ASE-ASE} \tag{5}$$

where $P_{sig-ASE}$ and $P_{ASE-ASE}$ are given by Eqs.(2) and (3), respectively.
Using Eqs.(1), (4) and (5), the *SNR* at the receiver is given by :

$$SNR = \frac{(G_s . P_{in})^2}{2h\nu . n_{sp} . (G_s - 1) . B_e . \left[2.n.G_s . P_{in} + \frac{n(n+1)}{2} . n_{sp} . (G_s - 1) . h\nu . \Delta\nu_{opt} \right]} \tag{6}$$

where $P_{in} = L_1 . P_{NT}$ (in what follows we will assume $P_{NT} = 0dBm$).

Assuming large gains (e.g., $G_s \geq 15$ dB) and $n_{sp} = 2$, Eq.(6) reduces to:

$$SNR = \frac{P_{in}^2}{4.h\nu . B_e . \left[2.n.P_{in} + n(n+1) . h\nu . \Delta\nu_{opt} \right]} \tag{7}$$

and is therefore independent of the gain G_s.

Eq.(7) shows that the *SNR* increases with P_{in} (i.e., when N_B decreases) while it decreases as n increases. In order to achieve a bit error rate $BER \leq 10^{-9}$ with binary On-Off Keying (OOK), the required *SNR* must satisfy $SNR \geq 15.6\ dB$. Figure 3 plots the maximum value of n as a function of N_B assuming a 3dB margin above the minimum required *SNR* (i.e., we imposed the condition $SNR \geq 18.6\ dB$ in Eq.(7)) and $\Delta\lambda_{opt} = 10\ nm$. The maximum number of connected users, $N = N_A . N_B$, is also shown on the same plot. It is seen that as long as $N_B \leq 64$ each of the 16 OAs can be continously biased (i.e., even if there is no upstream packet at their input) without degrading the *SNR* below 18.6dB. For $N_B > 64$, the maximum allowed number of biased OAs monotonically decreases with N_B. In other words, this means that if $N_B > 64$, then the OAs in the 1:16 amplified power combiner must be able to be switched on and off to minimize the *SNR* degradation due to the accumulation and beatings generated by ASE noises. Since the duration of upstream ATM bursts is of the order of a μsec (53 bytes (ATM+overhead) at 300 Mbps), fiber-doped optical amplifiers with long time response of the order of 10 msec are excluded from consideration so that only semiconductor-based optical amplifiers (SOAs) with time response of the order of a few nsec can be used. Each SOA in the amplified power combiner must be controlled so that it can be quickly set in the ON state (i.e., biased at their operating point) when a TDMA upstream packet arrives at its input. At the same time, all other SOAs are set in the OFF state. Although this solution would increase N up to about 2000 as it can be seen from Figure 3, it would be much more attractive to restrict N_B

to a maximum of 64 (providing N=1024) since then the SOAs must not be switched ON and OFF according to the MAC protocol. Therefore, in order to simplify the design and control of the amplified power combiner, it is proposed to dimension the Super-PON system with a maximum splitting factor of 1024 with N_A=16 and N_B=64. Notice that even in this case, SOAs are the only useful candidates for use in upstream direction of transmission thanks to their very short response time.

Eq.(7) shows also that the *SNR* can be improved by incorporating an optical filter with bandwidth $\Delta\nu_{opt}$ behind each SOA. Figure 4 plots the *SNR* calculated from Eq.(7) as a function of n for various values of $\Delta\lambda_{opt}$ ($\Delta\nu_{opt} = c.\Delta\lambda_{opt} / \lambda^2$). It is clear that *SNR* increases as $\Delta\lambda_{opt}$ decreases. However, care must be taken in order to not restrict by too much the usable signal wavelength window. Indeed, this may have serious consequences not only on the cost of the subscriber equipment that would need very precise emitting wavelength control, but also on the possibility to later upgrade the network capacity by using WDMA techniques. $\Delta\lambda_{opt} = 10\ nm$ seems to provide the best compromise between first installation equipment cost and capability of capacity upgrade by WDMA.

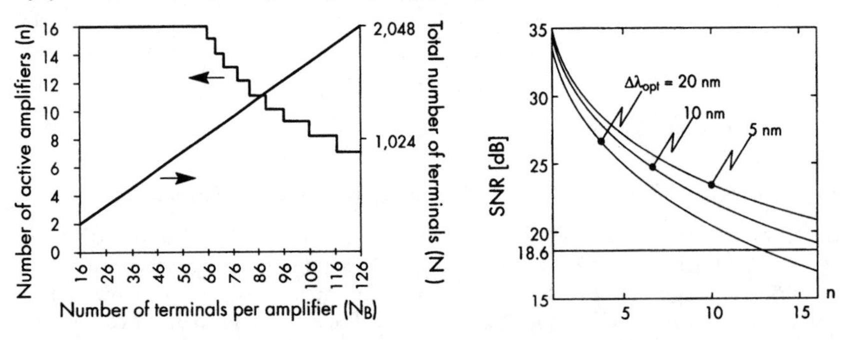

Figure 3 The maximum value of biased amplifiers in the amplified power combiner as a function of N_B assuming $SNR \geq 18.6dB$ and $\Delta\lambda_{opt} = 10nm$.

Figure 4 *SNRs* in front of the optical detector shown in Figure 3 as a function of n for various values of $\Delta\lambda_{opt} = 5$ nm, 10nm, and 20nm. N_B=64 so that $N = N_A . N_B = 1024$.

Let us now turn our attention to the case of a long feeder Super-PON as depicted in Figure 1. As compared with Figure 2, this long feeder Super-PON contains two additional OAs for in-line amplification. These two OAs will introduce their own ASE noise that will mix with the amplified signal and ASEs from the splitter/combiner amplifiers to create many beat noises.
The generated beat noises can be grouped into two main categories: (a) signal-*ASE* beat noises; between the signal and the *ASE* from the n OAs in the power combiner (ASE_s) and between the signal and the *ASE* from the first (ASE_{IL1}) and second (ASE_{IL2}) in-line OAs, and (b) the *ASE-ASE* beat noises from all combinations in pair. Following the same analysis as that outlined in (Mestdagh, 1995), and using the parameters defined in Figure 1, the signal-ASEs and ASEs-ASEs beat noise electrical powers are readily obtained as :

$$P_{sig-ASE_s} = n.\frac{4e^2}{h\nu}.n_{sp}.G_{IL}^3.(G_s-1).L_2L_3L_4^2.P_{in,3}.B_e \quad (8)$$

$$P_{sig-ASE_{IL1}} = \frac{4e^2}{h\nu}.n_{sp}.G_{IL}^2.(G_{IL}-1).L_3L_4^2.P_{in,3}.B_e \quad (9)$$

$$P_{sig-ASE_{IL2}} = \frac{4e^2}{h\nu}.n_{sp}.G_{IL}.(G_{IL}-1).L_4^2.P_{in,3}.B_e \quad (10)$$

$$P_{ASE_s-ASE_s} = \frac{n(n+1)}{2}.2e^2.n_{sp}^2.G_{IL}^4.(G_s-1)^2.L_2^2L_3^2L_4^2.\Delta\nu_{opt}.B_e \quad (11)$$

$$P_{ASE_{IL1}-ASE_{IL1}} = 2e^2.n_{sp}^2.G_{IL}^2.(G_{IL}-1)^2.L_3^2L_4^2.\Delta\nu_{opt}.B_e \quad (12)$$

$$P_{ASE_{IL2}-ASE_{IL2}} = 2e^2.n_{sp}^2.(G_{IL}-1)^2.L_4^2.\Delta\nu_{opt}.B_e \quad (13)$$

$$P_{ASE_s-ASE_{IL1}} = n.2e^2.n_{sp}^2.G_{IL}^3.(G_{IL}-1).(G_s-1).L_2L_3^2L_4^2.\Delta\nu_{opt}.B_e \quad (14)$$

$$P_{ASE_s-ASE_{IL2}} = n.2e^2.n_{sp}^2.G_{IL}^2.(G_{IL}-1).(G_s-1).L_2L_3L_4^2.\Delta\nu_{opt}.B_e \quad (15)$$

$$P_{ASE_{IL1}-ASE_{IL2}} = 2e^2.n_{sp}^2.G_{IL}.(G_{IL}-1)^2.L_3L_4^2.\Delta\nu_{opt}.B_e \quad (16)$$

where $P_{in,3} = P_{NT}.L_1.L_2.L_3.G_s.G_{IL}$ is the optical signal power at the input of the second in-line optical amplifier.
After straightforward algebraic manipulations, we obtain :

$$SNR = \frac{P_{in,3}^2}{2.h\nu.n_{sp}.B\left[F_1(n,P_{in,3},G_s,G_L,L_2,L_3)+F_2(n,\Delta\nu_{opt},G_s,G_{IL},L_2,L_3)\right]} \quad (17)$$

where $F_1(n,P_{in,3},G_s,G_{IL},L_2,L_3) = 2.(n.G_sG_{IL}.L_2L_3+G_{IL}.L_3+1).P_{in,3}$ (18)

and $F_2(n,\Delta\nu_{opt},G_s,G_{IL},L_2,L_3) = n_{sp}.h\nu.\Delta\nu_{opt}.[\frac{n(n+1)}{2}.G_s^2G_{IL}^2.L_2^2L_3^2$

$+n.G_sG_{IL}.L_2L_3.(G_{IL}.L_3+1)+G_{IL}^2.L_3^2+G_{IL}.L_3+1]$ (19)

Figure 5 plots the *SNR* as a function of *n* for three distinct values of the optical filter bandwidth placed just behind each OA.
When n=1, $SNR \cong 32$ dB and is almost independent of the filter bandwidth (for the range of $\Delta\lambda_{opt}$ considered here). The introduction of the two in-line OAs therefore degrades the *SNR* by about 3dB as can be seen by comparison with Figure 4 which provides $SNR \cong 35$ dB when n=1. As *n* increases, the degradation in *SNR* due to the two additional in-line OAs becomes relatively less significant. Even for n=16 (i.e., all OAs in the amplified power combiner are biased and generate independent ASE noises) the *SNR* can still be maintained above 18.6dB (actually $SNR = 18.7$ dB). Notice that when the two in-line OAs are absent, $SNR = 19$ dB for n=16 and $\Delta\lambda_{opt} = 10nm$ (Figure 4) so that the degradation due to the in-line OAs is only 0.3dB for n=16. In essence, this is because the dominant beat noises are those stemming from

beatings of the n ASEs from the OAs in the power combiner with the signal and among themselves. Beatings with ASEs from the two in-line OAs are comparatively negligible. In conclusion, upstream burst-mode transmission over a 1024-split Super-PON covering a range of 100 km can be achieved by the use of SOAs without requiring on-off switching capability of the SOAs.

The situation becomes much more complex if N_B has to be increased up to 128 in order to obtain an overall splitting factor of 2048. Indeed, as shown in Figure 6, n must satisfy $n \leq 6$ in order to have $SNR \geq 18.6\,\text{dB}$ with $\Delta\lambda_{opt} = 10nm$. This means that the SOAs in the power combiner must be switched on and off according to packet arrivals. This requires a complex control system related to the MAC protocol that manages the packet flow through the network. Notice finally that this complex control may be avoided by reducing $\Delta\lambda_{opt}$. However, as shown in Figure 6, $\Delta\lambda_{opt}$ must be reduced down to $\Delta\lambda_{opt} \leq 0.5nm$ in order to achieve $SNR \geq 18.6\,\text{dB}$ when n=16. Clearly, this is unacceptable since there will be no space left for network capacity upgrade by WDMA.

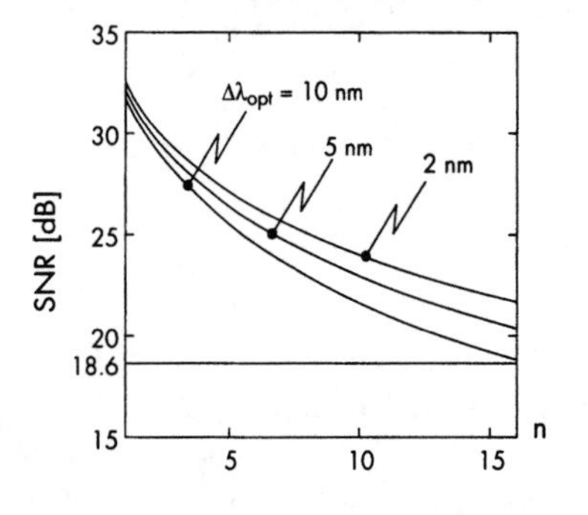

Figure 5 : *SNRs* for the long feeder Super-PON as a function of n for various values of $\Delta\lambda_{opt}$ =2nm, 5nm, and 10nm. N_B=64 so that $N = N_A . N_B = 1024$.

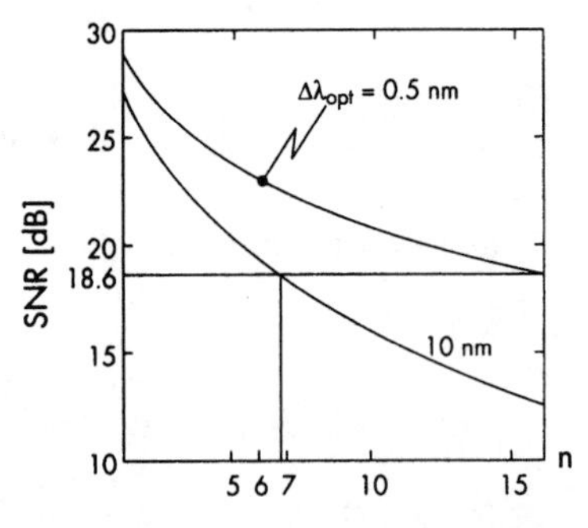

Figure 6 : *SNRs* for the long feeder Super-PON as a function of n with $\Delta\lambda_{opt}$ =0.5nm and 10nm. N_B=128 so that $N = N_A . N_B$ =2048.

Network survivability - A key issue for the viability of Super-PONs is to assure sufficient robustness against failures of network elements that are shared by a large number of subscribers. This is especially the case for the line card at the central office, the feeder, and the amplified splitter/combiner. Therefore, redundancy of these network elements appears to be mandatory. A possible configuration of a survivable Super-PON network is shown in Figure 7. It is seen that, in addition to duplication of these network elements, the duplicated feeder should be installed along a diverse route in order to avoid network breakdown due to cable cuts. Moreover, it might also be required to provide diverse routing of the sub-feeder of the drop section that serves up to 64 or 128 subscribers (not shown in the figure).

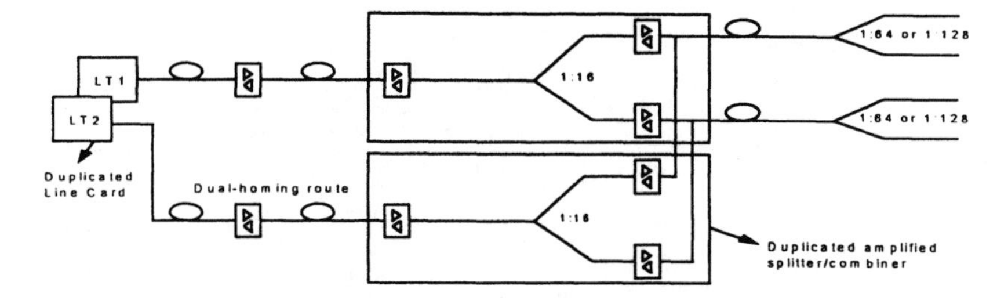

Figure 7 One possible configuration to improve the access network reliability by duplication and diverse routing of high group failure network elements.

Capacity upgrades - To upgrade the capacity of the initial TDM/TDMA Super-PON, additional channels could be added by the introduction of WDM/WDMA techniques. The challenge is to define an upgrading strategy that has no impact on the subscriber equipment in order to minimize the installation costs of the upgrade. One possible downstream upgrade scenario that does not require any change or replacement of already installed subscriber equipment is depicted in Figure 8.

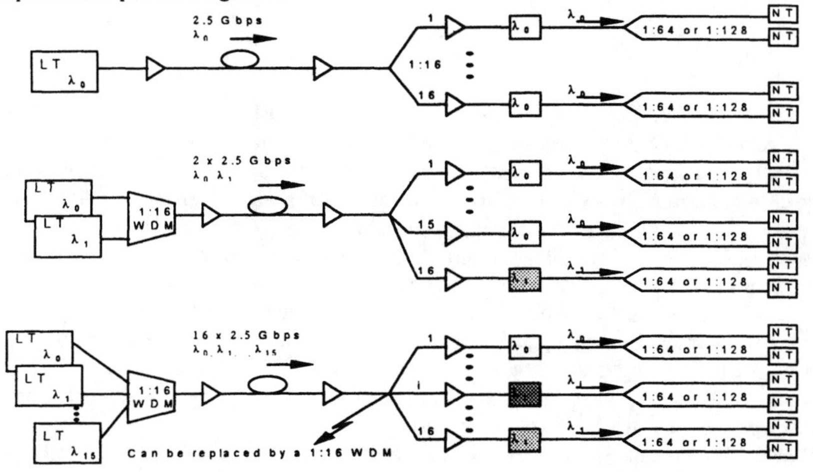

Figure 8 An upgrade scenario for downstream Super-PON capacity by using WDM technology without modification of the subscriber equipment.

The 1:16 amplified splitter initially contains passband optical filters centered around the transmitter wavelength λ_0. Downstream capacity upgrade is achieved by transmitting separate 2.5 Gbps TDM channels on distinct wavelengths within the 1.5 μm window. Up to 16 wavelength channels can be added gradually by replacing the λ_0-centered passband filter by another passband filter centered around one of the wavelengths not yet in use for the upgrade. Eventually, when all 16 WDM downstream channels are in operation, then the 1:16 splitter can be replaced by an almost lossless WDM device resulting in an improved overall network power budget. Upgrading of the upstream capacity can be accomplished by wavelength conversion at the amplified splitter as shown in Figure 9.

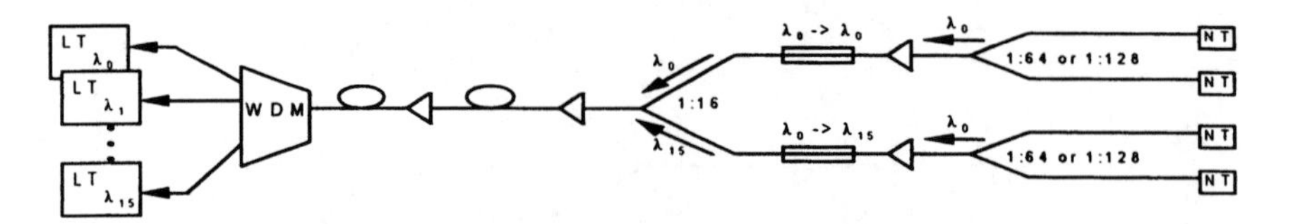

Figure 9 An upgrade scenario for upstream Super-PON capacity by using wavelength convertors within the amplified power splitter/combiner.

Although wavelength convertors are still at the preliminary laboratory stage, encouraging results have recently been achieved. A review of the state-of-the-art alternatives for wavelength conversion can be found in (Masetti, 1995). A very promising device is the so-called "clamped-gain SOA" which can be monolithically integrated to provide both wavelength-conversion and optical amplification (Soulage, 1995). In addition, these devices have a much improved performance regarding crosstalk than conventional SOAs.

4 CONCLUSIONS

The new concept of Super-PON for application in the local access network has been introduced and its technical challenges discussed. One of the major challenging problems with Super-PONs is associated with the use of optical amplifiers and packet-based TDMA for upstream transmission. It has been shown by an analytical analysis that an overall splitting factor of up to 1024 can be sustained without the need to provide switching capability for the necessary semiconductor optical amplifiers (SOAs) placed in parallel within the power splitter/combiner. However, if the target splitting factor must be increased up to 2048 or more, then the SOAs in the power splitter/combiner must be switched on and off according to the TDMA MAC protocol unless optical filters with a passband less than 0.5nm are inserted behind each amplifier (forbidding capacity upgrade by WDMA). Therefore, in order to simplify the design and control of Super-PONs, and hence improving its reliability, it is advised to limit the maximum splitting factor to 1024 at the most.

The Super-PON concept is currently investigated within the PLANET Consortium that is partly funded by the Commission of the European Communities in the framework of the ACTS programme. Results of the studies and developments will be presented during the conference.

5 REFERENCES

Ballance, J.W. *et al.* (1990) ATM access through a passive optical network *IEE Elect. Lett.*, vol.26, n.9, 558-60.

Du Chaffaut, G. *et al.* (1993) ATM-PON: une famille de systèmes optiques pour la distribution - l'exemple SAMPAN *L'Echo des Recherches*, n.154, 28-38.

Hill. A.M. *et al.* (1990) 39.5 million-way WDM broadcast network employing two-stage of erbium-doped fiber amplifiers *IEE Elect. Lett.*, vol.26, n.22, 1882-3.

Ishikura, A. *et al.* (1991) A cell-based multipoint ATM transmission system for passive double star access networks *Third IEEE Workshop on Local Optical Networks*, Tokyo, Japan, paper G.3.4.1-10.

Masetti, F. *et al.* (1995) ATMOS (ATM Optical Switching): results and conclusions from the RACE R2039 project *Proc. of the 21st European Conf. Opt. Commun.*, ECOC'95, Brussels, Belgium, paper n.243.

McGregor, I.M. *et al.* (1989) Implementation of a TDM optical network for subscriber loop applications *IEEE J. Light. Techn.*, vol. LT-7, n.11, 172-8.

Mestdagh, D.J.G. *et al.* (1991) ATM local access over passive optical networks *Third IEEE Workshop on Local Optical Networks*, Tokyo, Japan, paper G.3.1-8.

Mestdagh, D.J.G. (1995) *Fundamentals of Multiaccess Optical Fiber Networks*, Artech House Publishers, Boston (ISBN-0-89006-666-3).

Stern, J.R. (1989) Passive Optical Networks for telephony applications and beyond *IEE Electronics Letters*, vol.23, n.24, 1255-7.

Soulage, G. *et al.* (1995) Clamped-gain SOA gates as multiwavelength space switches *Techn. Digest OFC'95*, San Diego (CA), paper Tu.D.1, 9-10.

Van der Plas, G. *et al.* (1995) Demonstration of an ATM-based Passive Optical Network in the FTTH trial on Bermuda, *Globecom'95*, Singapore.

Zaganiaris, A. *et al.* (1995) Etudes technico-économiques des réseaux d'accès optiques dans des projets européens, *L'Echo des Recherches*, n.160, 3-14.

29

Satellite full mesh ATM LAN: interconnection, satellite access scheme, signaling and performance

Bernard Perrin, Centre Suisse d'Electronique et de Microtechnique (CSEM) Maladière 71, 2007 Neuchâtel, Switzerland,
Phone: +41 38 205 263, Fax: +41 38 205 720,
email: bernard.perrin@csemne.ch

Roberto Donadio, European Space Agency (ESA)
PO Box 299, 2200 AG Noordwijk, The Netherland,
Phone: +31 71 565 3134, Fax: +31 71 565 4598,
email: rdonadio@t.estec.esa.nl

Abstract

This paper presents the main results of a study which investigated the opportunities to use the geostationary satellite for the interconnection of ATM LAN's in a meshed configuration. The strategical aspects in terms of user requirements, infrastructure evolution and ATM equipment availability are first analyzed. Both short and long term scenarios are considered in this paper. For the short term, a configuration involving only minor modifications to existing equipment is proposed while for the long term a more efficient configuration is outlined. Several satellite access schemes are compared and one scheme is selected respectively for the short term and the long term scenario. The modifications to ATM equipment available today for the satellite gateway are evaluated together with the issues related to the set-up of a connection. The elements of the system which affect the overall performance are analyzed.

Keyword

Satellite, ATM, LAN interconnection, connection set-up, performance.

1 INTRODUCTION

1.1 Scope of the project

This document reports the results of the study project entitled *Broadband communication over satellite* performed by CSEM for the European Space Agency (ESA). Eutelsat, Swiss Telecom PTT and LOGICA (UK) have collaborated in this project which had a larger scope than the one presented in this paper: a first part has been dedicated to the Synchronous Digital Hierarchy (SDH) and the second part to the Asynchronous Transport Mode (ATM). Only the results of the latter are reported here.

The project objectives were:

- to determine if and how satellites can play a role for ATM LAN interconnection,
- to identify suitable system concepts and architectures,
- to check the performance of the identified solutions by means of simulations.

The scope of the study was limited to geostationary orbit satellites (GEO) and to the consideration of a limited number of nodes in the network. Only transparent 'bent-pipe' satellite payloads were considered. On-board switching payloads were therefore not part of this activity. The focus was on defining a configuration which could be set up in the short term, reusing existing equipment with minimum development. A network node, as shown in figure 1, consists of an earth station, a bank of modulators and demodulators, a switch and the terrestrial interface.

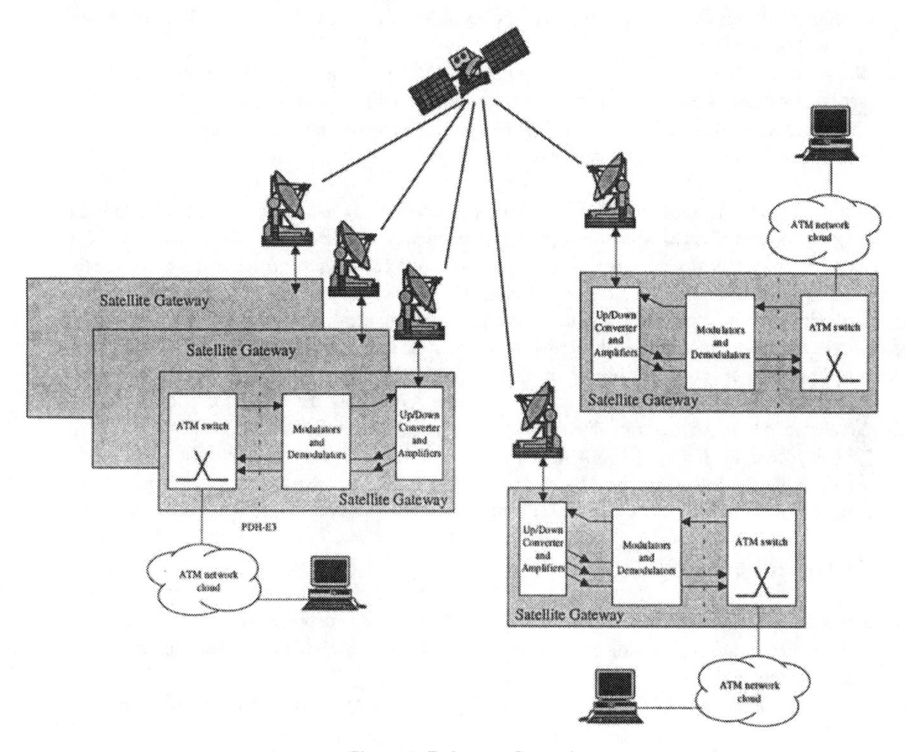

Figure 1: Reference Scenario

1.2 Background

Over the last years, massive investments have been devoted to the development of terrestrial broadband networks based on fiber optic transmissions and on the use of cell switching techniques. ATM is first being implemented in the local area, creating local ATM islands which need to be interconnected by ATM trunks. A ubiquitous fiber-based broadband infrastructure cannot be made available in the short term. On the other hand, the advantages deriving from the use of satellites capable of interconnecting these ATM broadband islands are apparent: easier set-up of the network infrastructure, rapid reconfiguration of the network, inherent possibility to implement multicasting and wide area coverage potential. It is therefore expected that satellite systems can play a major role in the evolving broadband networks.

Among the user requirements, one can enumerate:

- Video-conferencing: today, the time spent traveling for business or technical meetings is consuming a lot of resources. There is no doubt (or it is at least highly probable) that in the future, information, voice and images will travel instead of people.
- Remote demonstration: it is not always possible or desirable to move people and machines to show something on a computer. Sending part or all of the display from one machine to a remote location, including the voice and image of the people initiating the demo will be much used as soon as it will be possible.
- Real-time database access and manipulation: access to the same piece of information from the various locations of one company or from a consortium is highly desirable.
- Collaborative work, enabling the members of a project team to share and revise information simultaneously available on their displays.

The satellite has an inherent capability to distribute information, which, combined with the good switching capabilities of ATM, allows an efficient connection of remote sites.

A number of technical challenges have to be considered for broadband networking via satellite. Geostationary satellites introduce a roundtrip delay of 270 ms. The overall performance can fall below acceptable limits when errors occur and cells have to be retransmitted. Not being physically possible to decrease the end-to-end delay under the roundtrip delay, some care has to be taken in order to maximize the throughput. The effects of the bursty errors of coded satellite links on the ATM protocol has been extensively studied and is discussed in [1]. It has been shown that the optimized choice of coding schemes allows to have a quasi error-free link for a high percentage of the time.

Moreover, due to the full mesh configuration we are working with, the data broadcast by the satellite from an Earth Station (ES) is received by all the remote ESs, even those which are not supposed to get them. As we will show later on, this makes the signaling protocol more complex.

1.3 Organization

In this paper, two of the scenarios considered in the mentioned study are presented:

- a short term option only relying on equipment available today, which can be used with only small modifications
- a long term option providing a more efficient system, but for which extra development is needed.

Section 2 summarizes the results of an ATM market survey which was part of the study, where three areas were covered: the user requirements, the Telecom Operators' (TO) plans and the ATM equipment availability.

In section 3, an analysis of the terrestrial and satellite tariffs is done to extract a range of opportunity where satellite solutions are attractive from an economical point of view.

Various existing satellite access schemes are compared. Section 4 treats this subject for the short term option.

From an end-user point of view, the establishment of a connection with a colleague or a machine located in a remote site should be as transparent as possible. Section 5 highlights the issues and proposes outline solutions.

Section 6 selects one configuration for the short term and develops an outline design for the long term, where two options are evoked: the encapsulated and the integrated signaling.

Some system elements play a key role to optimize the overall performance. They are listed in section 7. Section 8 concludes this paper.

2 ATM MARKET SURVEY

In the course of the study several corporate organizations and telecom operators have been interviewed. Some of the former expressed the need for ATM communications over satellite. These are above all organizations which operate in areas where high rate terrestrial links are unavailable. Among the latter, trials have been undertaken, to validate ATM over satellite. Satellites are seen as a possible means to avoid transit fees with direct satellite connections to distant countries.

3 RANGE OF OPPORTUNITY FOR THE SATELLITES

In order to be able to make a meaningful comparison between satellite and terrestrial networks offering ATM, one needs to compare similar services offered by the terrestrial and satellite approaches. This comparison has been done on tariffs rather than on costs, which would imply the calculation of full telecommunications costs such as infrastructure costs, operating costs and profit margins.

3.1 Short term

The results of a tariff analysis carried out by the Swiss PTT are: for short (from Switzerland to the bordering countries) and medium distances (within western Europe), satellites are not price competitive. They could only play a role in this area when an immediate need of connection arises and no cable connection exists within the expected bandwidth. For long distances, satellites are currently very interesting from an economic point of view. The tariff comparison is done for point-to-point links; in a full mesh configuration, by dividing the satellite transponder rent (main costs contributing to the tariffs) over all the ES users, the satellites become competitive even for medium distances.

From this analysis, a range of opportunity has been found: a full mesh satellite network with a limited number of nodes will be viable for medium and long distances.

3.2 Long term

The evolution of tariffs in the long term is very unpredictable: the market deregulation in 1998 will probably decrease the communication tariffs at least for international traffic. This change in tariffs is also expected to apply to satellite capacity and satellites may become more competitive for this reason. Configurations where the users have to pay for the satellite link even when they are not using it are not acceptable. Some form of demand-assignment of the available capacity is necessary.

4 SATELLITE ACCESS SCHEME

This section presents candidate access schemes. Although CDMA provides security and a good resistance to fading, there is no existing equipment now on the market. Hence, we are not considering it. We explicitly excluded any kind of topology leading to a double hop, such as the star topology. In such a topology any kind of data exchange between two remote ESs has to go through the hub station. This introduce a delay of 0.54s (same for the return). This is not compatible with the time constraints of voice and video services over ATM.

4.1 Frequency Division Multiple Access (FDMA)

IDR (Intermediate Data Rate) is a well known standard [2] and IDR equipment is widely used to transmit data rates from 2 to 45 Mbit/s, mostly in point-to-point configurations. Nevertheless, a multi-destination option is considered in the standard.

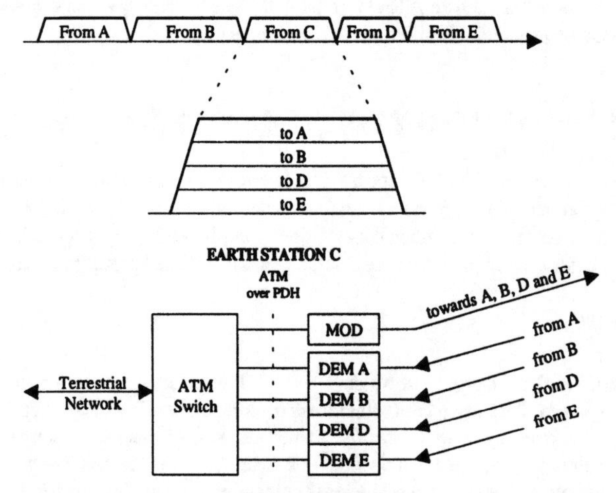

Figure 1: IDR

Within IDR the multiplexing of the available bandwidth is achieved by allocating different frequency bands to each ES (see Figure 1). Each ES sends data to the others (up link) and receive on a separate frequency (each requiring a dedicated demodulator) from every other ES as shown. The standard allows up to five stations. This scenario is the point-to-multipoint extension of the configuration demonstrated by Eutelsat in a series of field trials in mid-'94 [1], aiming at validating ATM over satellite at 34 Mbit/s using standard IDR RF modems in a point-to-point configuration.

In this figure, 4 demodulators are placed in the ES, each one assigned to the carrier frequency A, B, D or E. The frequency bandwidth allocation cannot be modified easily with IDR since it is adjusted in hardware at the input filters. The maximum throughput capacity on a link is given with its bandwidth.

This scenario shows a network node composed of an ATM switch, an IDR modulator and 4 IDR demodulators. For the terrestrial network interface several configurations could be used:

- ATM cells encapsulated into a STS-1(SONET)/STM-1(SDH) frame (OC-3c interface);
- ATM cells encapsulated into a 34 Mbps PDH frame (E-3 interface).

The connections between the ATM switch and the modulator/demodulators can either be done through an E-3 interface (34 Mbps PDH) or a dedicated 'Low rate ATM'.

Standards for mapping ATM cells over PDH exist for 2 Mbit/s and 34 Mbit/s only [3][4]. IDR modems may support also any other intermediate bit rate. In order to take advantage of such bit rates, dedicated interfaces should be defined.

As indicated in [2], IDR carriers have a framing period of 125 μs for transporting 12 bits of overhead and 256 bits of payload at 2 Mbit/s. This overhead is relatively lower for the higher rates (up to 44.736 Mbit/s).

Each ES transmits data destined to all the other ESs on one frequency. The receiving ESs will receive multiple frequencies and will retain only the relevant ATM cells, ignore those destined to another station. Usually, switches hold a mis-routed cells counter. In current switch implementations, alarms are generated when a given threshold is violated. A mechanism should be devised to inhibit the generation of these alarms.

4.2 Time Division Multiple Access (TDMA)

Within TDMA the multiplexing is achieved by allocating different time slots to each ES (see Figure 2). The number of cross-connected stations is theoretically limited by the transmission bit rate via satellite and the incoming information rate at each ES.

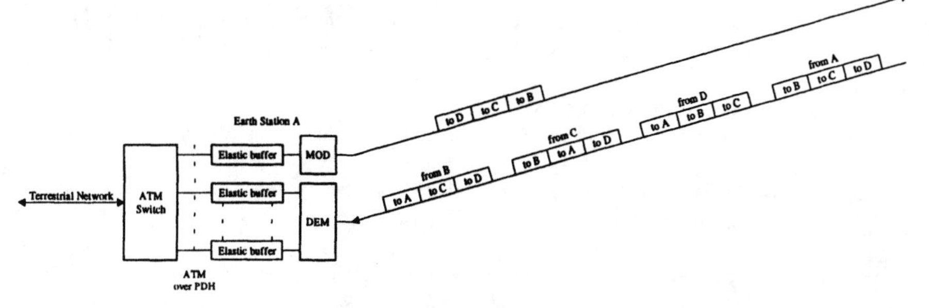

Figure 2: TDMA

For example, a conventional total bandwidth of 120 Mbit/s is used per transponder with the EUTELSAT II satellite. The picture above shows a single transponder scenario. Transponders can also work in parallel.

Since each ES has a given allocated timeslot to transmit, synchronization of all the ESs is necessary. Due to the geographical distribution of the ESs, the notion of a timeslot in term of absolute time is different for each one: two ESs transmitting exactly at the same time during a given window will not necessarily provoke a conflict when the two data packets arrive at the satellite. Moreover, the distance from the satellite to the ESs is changing all the time due to the satellite movement. The additional overhead necessary to ensure synchronization adds a lot of complexity and makes TDMA more expensive than IDR. However, for a large number of ESs (>5), TDMA becomes interesting.

One advantage of TDMA is that there are no inter-modulation problems since the transponder only has to deal with one carrier at a time.

The information bit rate allocation for each transmitting ES can be changed by adapting the time slot size for each ES. However, it is not possible to change the burst plan allocation dynamically. Using EUTELSAT II, it takes few hours to do it.

This scenario shows a network node, including an ATM switch, a TDMA modulator and a TDMA demodulator. Terrestrial interface aspects are identical to the FDMA case.

The connections between the ATM switch and the elastic buffers and between the elastic buffers and the modulator/demodulator can either be done through an E-3 interface (34 Mbps PDH) or a dedicated 'Low rate ATM'.

The cells sent by the switch need to be stored in an elastic buffer before the modulator can re-transmit them: the modulator can send data only during a given window (its allocated timeslot). This buffer was not needed in the previous scenario since the IDR modem is always transparently carrying the incoming data flow. Idle cells should be generated as soon as the elastic buffer is empty.

So, an elastic buffer is also needed at the terrestrial station, in order to reassemble the ATM cells. The buffer is also needed to store all the cells which are received in one burst and to transmit it 'continuously' to the cross-connect in order to meet the cell delay variation requirements.

5 SIGNALING AND CALL CONTROL

In this section, we assume that the satellite links are established and the connections between the satellite gateways are transparent.

5.1 Permanent and Switched Virtual Circuit

ATM is a connection oriented technology: before data can flow over the network from the source to the destination, a connection should first be established. The connection establishment can be based on two kind of solutions:

1. Permanent Virtual Circuit (PVC) if a circuit is established manually by a network operator. It is considered permanent since it is present after a connection is made and until the operator changes it.
2. Switched Virtual Circuit (SVC) if a circuit is dynamically allocated for each new connection using ATM signaling, and is present only for the time of the connection.

The ATM Forum has defined a common user signaling protocol. This is reflected in the ITU-T recommendation Q.2931. The absence of a stable standard has led to the definition of proprietary implementations of the access signaling protocol. For example, the FORE switches considered in this study use a proprietary protocol named SPANS (Simple Protocol for ATM Network Signaling) to create a SVC. This solution has the drawback to work only with FORE equipment and does not allow to connect FORE equipment with equipment from another vendor.

Since the purpose of the study was to propose a configuration which could be set-up in the short term, therefore reusing existing equipment with minor developments, the SPANS signaling protocol was taken as a reference and is shortly introduced in the next section. Nevertheless, the considerations related to the user signaling protocol are general and may apply to any proprietary implementation.

5.2 The Simple Protocol for ATM Network Signaling (SPANS)

This protocol relies on the ATM layer with a bi-directional PVC (VPI=0, VCI=15). All links used by participants in the signaling protocol must declare this PVC. The adaptation layer could either be AAL3/4 or AAL5. The transport layer does not exist. So it is assumed that the signaling application will retransmit a lost message and suppress a duplicate one.

The presentation layer is encoding/decoding messages from/to the application layer, using the External Data Representation (XDR) [5], an abstract notation used as the input of the publicly available *rpcgen* tool [6].

The application layer is based on the well known Remote Procedure Call (RPC)[7] and is materialized by a client/server structure program. It uses a set of messages like:

- **open_request**: request for opening a connection (providing arguments like AAL type, preferred VPI/VCI, desired bandwidth, minimum bandwidth);
- **open_response**: response to the above request, providing arguments like chosen VPI/VCI, allocated bandwidth and result of the request (ok or not ok);
- **multi_request**: contains the same set of parameters as open_request but here concerns a multicast channel;

.....

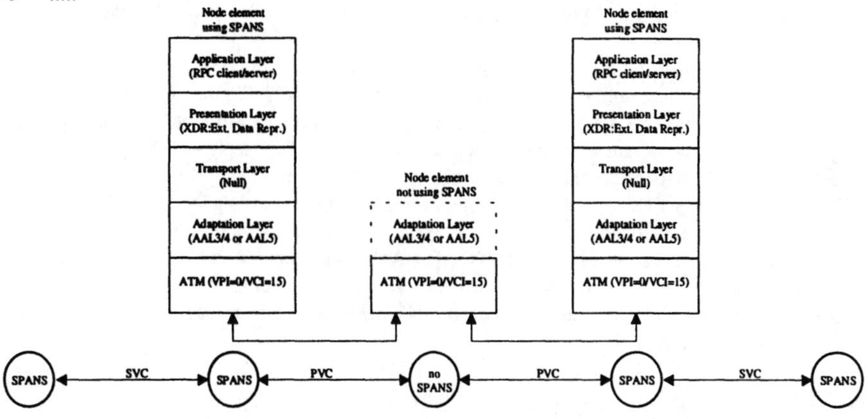

Figure 3: SVC over PVC protocol stack

This protocol can run transparently over an ATM node which doesn't support it. In this case an existing PVC will connect the SVCs and make the links transparent as represented in Figure 3. However, the following pre-requisite exists to do that: two PVCs (VPI=0/VCI=15) should be created (one per direction) in order to carry this out of band signaling transparently over the non-compatible node.

5.3 PVC based configuration

The ATM switch at the satellite gateway of ES_x could be programmed with through paths having VPI=X0Y.

Then, the ES_2 switch will discard all the ATM cells having a VPI other than 0, 201 and 203 to 205 from the satellite. VPI=0 can be reserved for multicast, where the VCI could be used to specify which ES should receive the cells, in a 1 among n notation:

VCI=0x006= B0000 0000 0110 will be used to reach ES_3 and ES_2.
VCI=0x01F=B0000 0001 1111 will be used to reach all the 5 ESs.

The problem of a such configuration is its rigidity: a workstation has to be directly attached to the satellite gateway or through PVCs.

5.4 SVC over PVC based configuration

A more flexible solution is to use SVCs over PVCs with the protocol stack illustrated in Figure 3. Then, one through path has to be created to broadcast the signaling requests to the nearest ATM switch running SPANS and 4 through paths to merge the incoming traffic from the satellite.

An *open_request* call, asking to open a connection, from the nearest switch connected to the ES will be carried over the satellite and received by the four remote stations. Since each one has an entry in its through path table corresponding to 0/15, the call will be forwarded to the next switches connected to B1 ports of the remote ESs. Suppose they are enabling SPANS, **then only one of them** (this one able to reach the targeted host) will send a positive response with *open_response*. The three other one will send a negative acknowledgement.

A regular implementation of SPANS will consider the first reply and ignore the next one. If the first acknowledge is the one accepting the connection and returning the VPI/VCI, then everything is fine. Otherwise, the connection will not be established, sending to the workstation operator a message like: *Sorry, there is no way to reach xx*. As nothing ensures that the first reply will be positive, the establishment of SVCs through the existing PVCs will not work correctly.

This problem does not exist when the satellite acts as a simple point-to-point link. This is very specific to the special configuration under study we have in this satellite full mesh network.

A single *open_request* call will be sent from the last ATM switch connected to the ATM gateway. This switch should be prepared to receive several *open_response* messages from the in port associated with the out port from where the request was sent. It should know how many responses are expected, so how many ESs are involved in this full mesh link.

5.5 SVC based configuration

In this case, we are considering the SVC establishment over the satellite link. Several PVCs will have to be created for the out-of-band signaling. For example, we can easily imagine to keep VPI=0 and use VCI=X0Y for an out-of-band signaling from ES_Y to ES_X. The signaling protocol should listen to all these channels: a request coming with VPI=0/VCI=X0Y (on any port) should be acknowledged with VPI=0/VCI=Y0X on port A1.

This configuration implies the need to modify the existing signaling software. If we are looking at the SPANS again, software changes will have to be done only concerning the out-band signaling. The amount of modifications appear to be very reasonable and focused on a lower layer.

6 SELECTED CONFIGURATIONS

6.1 Short term selected configuration

The two following tables summarize the motivation for the selection of the configuration:

Satellite access scheme comparison

TDMA Advantages	IDR (FDMA) Advantages
Is not limited to a given number of ESs	Simpler, cheaper and can reach data rates high enough in order to be interesting (up to 44736 kbit/s)

Signaling comparison

SVCs over PVCs Advantages	SVCs only Advantages
A single out-band signaling channel is uniformly used	Only the gateway has to deal with modifications for the satellite
Traffic between ES is allocated by the network operator (if the total bandwidth shouldn't be shared in a first came in first serve scheme).	Changes are concentrated on a single point
	Multicast will not need special development
	Bandwidth can be shared by any ES, so it will be more efficiently used.

This led to the choice of IDR for the satellite access scheme and SVCs only for the signaling.

6.2 Long term scenario: Reservation TDMA

The access schemes previously listed have the drawback not to be able to allocate bandwidth dynamically. The presented long term configuration is based on bandwidth on demand. Studies in this area have already been done, but for lower bandwidth (up to 2-8 Mbps), like in the FODA-TDMA project [8]. These access schemes principles could be re-used for the developments of this scenario, just scaling the bandwidth in order to fit it for the ATM based application requirements.

The main difference between TDMA described in paragraph 4.2 and reservation TDMA comes from the TDMA burst plan. In 4.2, the burst plan is fixed and can eventually be changed by an operator. The objective of the reservation TDMA is to give the opportunity to any ES to seize a free timeslot dynamically.

From this point, there are two main options:

- encapsulate all the ATM traffic together (signaling and data) and use the existing system with minor modifications. Opening/closing of virtual circuits must be performed by higher level protocols;
- integrate the ATM signaling into the control sub-frame and reference burst structure.

Comparison

ATM traffic encapsulation Advantages	ATM signaling integration Advantages
Cheaper development	Only the TDMA controller is concerned by the changes
Is not bound to ATM	Short connection setup delay
The number of potential ESs is not limited	The ATM 'bandwidth on demand' capabilities are used

The advantages and disadvantages of both approaches are very well balanced, and the choice is not obvious. The second argument of the first option is partially true: at least, the physical interface will have to be changed to move to other technologies. Moreover, as noted, the burst plan should be optimized for ATM if we want to avoid a per frame queuing buffer development. We believe that the ATM signaling integration is a better candidate and it surely presents long term advantages.

7 SYSTEM ELEMENTS AFFECTING PERFORMANCE

Each element involved in the end-to-end communication will have an influence in the overall performance. The switch and the RF modem are considered here:

7.1 The ATM switch

The following characteristics are considered:

- Contentionless time division switching fabric: contrarily to the time division switching, this switching fabric type is non-blocking, even when multi-cast (the most probable case where a congestion could occur) is used.
- Distributed switch control architecture: contrarily to the centralized switch control architecture, this architecture type can better integrate the switch configuration changes of the satellite gateway which can be made locally.
- Extendable output buffers: cell loss has to be avoided as much as possible, since the affected packets will have to be re-transmitted, introducing a delay greater than twice the roundtrip delay.
- Bandwidth management capabilities[9]: for the same reason mentioned in the previous item, choose smart output buffers offering:
  1. A queue for each VC.
  2. Multiple classes of services through dynamic buffer allocation.
  3. Packet level discard: Early Packet Discard (EPD) and Partial Packet Discard (PPD).

7.2 The RF modem

The use of suitable Forward Error Correction (FEC) is essential to keep the Bit Error Rate (BER) within limits which do not impair the performance of the ATM link ($BER<10^{-10}$). The suggested technique is to use a Reed Solomon in conjunction with a Viterbi encoder/decoder. The satellite link presents a fiber-like performance most of the time with a high percentage of availability. This is achieved at the expense of a moderate bandwidth increase.

8 CONCLUSIONS

A five node fully meshed satellite network based on multidestination IDR was selected for implementation in the short term. The ATM switch residing at the satellite gateway will need to run a modified implementation of the signaling protocol in order to support end-to-end SVCs. The modification appears very reasonable and is located in the lower layers of the signaling protocol stack. This configuration can be commercially interesting in a corporate networking environment with the need to exchange large amounts of data. Satellite links can be designed in order to be highly reliable ($BER < 10^{-10}$), hence making data retransmission very improbable and the roundtrip delay a secondary drawback.

A configuration for the long term has been outlined using a demand assignment reservation TDMA satellite access scheme, where the ATM signaling is integrated with the TDMA signaling. The TDMA controller will require an adapted ATM signaling implementation with the benefit that no other network node has to be modified.

Finally, we would like to acknowledge Gaye Cassidy (LOGICA), Rudolph Brönnimann (Swiss-PTT) and Jonathan Castro (CSEM) for their valuable work in this project.

9 REFERENCES

[1] Transmission of framed ATM cell streams over satellite: A field experiment, Stefano Agnelli, Paolo Mosca, ICC 95.

[2] The INTELSAT earth station Standard 'IESS-308' for Intermediate Data Rate Digital Carrier.

[3] G.704 Synchronous frame structures used at primary and secondary hierarchical levels.

[4] G.832 Transport of SDH elements on PDH networks: Frame and multiplexing structures.

[5] External Data Representation Standard: Protocol Specification, Network Programming, SunOS 4.0 Volume 10, SUN MicroSystems, 9 May 1988.

[6] The *rpcgen* programming guide, Network Programming, SunOS 4.0 Volume 10, SUN MicroSystems, 9 May 1988.

[7] Remote Procedure Call Programming Guide, Network Programming, SunOS 4.0 Volume 10, SUN MicroSystems, 9 May 1988.

[8] The FODA-TDMA Satellite Access Scheme: Presentation, Study of the system, and Results, Nedo Celandroni and Erina Ferro, IEEE Transactions on communications, Vol 39, No 12, December 1991.

[9] ForeThought Bandwidth Management, White Paper, Version 1.0, 4/95.

30

Fair Prioritized Scheduling in an Input-Buffered Switch

Carsten Lund, Steven Phillips, Nick Reingold
AT&T Bell Laboratories
600 Mountain Avenue, Murray Hill, NJ 07974-0636

Abstract

The rapid growth of inter-networking and the popularity of ATM have resulted in a need for high-speed low-cost network components. This paper presents a new algorithm, Fair Arbitrated Round Robin (FARR), for scheduling the crossbar of a high-speed input-buffered switch. FARR respects virtual circuit (VC) priorities and has per-VC fairness properties that have previously only been achieved in output-buffered switches. Input-buffering is more cost-effective than output-buffering at high speeds, due to much more lenient memory speed requirements. Simulations are presented using a variety of work loads, traffic types and switch sizes. The simulations demonstrate the performance benefit of FARR over previous input-buffered switch algorithms and show that FARR performs similarly to Fair Prioritized Round Robin running on an output-buffered switch.

Keywords

ATM, input-buffer switch, per-virtual circuit fairness

1 INTRODUCTION

In recent years there has been a rapid rise of interest in high-speed cell-based networking. A number of factors have been responsible, including political support for the "information super-highway" in the United States, the high bandwidth requirements of future video-on-demand services, the convergence of technology trends enabling high-speed communication, most notably fiber-optic technology. The burgeoning demand for network bandwidth requires the development of new high-speed switches. This paper describes a new algorithm for scheduling a high-speed input-buffered switch, Fair Arbitrated Round Robin (FARR), that has desirable properties previously only obtained in output-buffered switches. The benefits of these properties, per-virtual circuit fairness and respect for different virtual circuit priorities, are demonstrated in extensive simulations.

It is generally recognized that fair prioritized round robin scheduling (PRR) provides fair resolution when there is conflict for resources (Demers *et al.* [1990], Fraser & Morgan [1984], Hahne [1986], Katavenis [1987], and Nagle [1987]). Fair scheduling has many advantages over first-come-first-served service, for example enabling performance guarantees for high-priority traffic and providing policing of sources. Some flow control methods require fair round robin-like scheduling (Keshav [1991]), and it has been shown that fair scheduling can improve the stability of ensembles of adaptive rate-controlling sources (Shenker [1990]). However, round robin scheduling is applicable only in an output-buffered or shared-memory switch, and such switches running at multi-gigabit speeds are extremely expensive to build, since exotic memory technology must be used to attain the necessary buffer bandwidth. Input-buffering is cheaper at very high speeds, as the buffers need only run at line speed, rather than the total switch speed, but no previous input-buffered switch algorithms have performed fair scheduling.

Our scheduling algorithm, FARR, emulates the behaviour of PRR in an *input-buffered* switch. FARR respects cell priorities, enabling voice and video cells to be transmitted with very small delay (well under a millisecond), while bandwidth is shared fairly between different VCs at the same priority level.

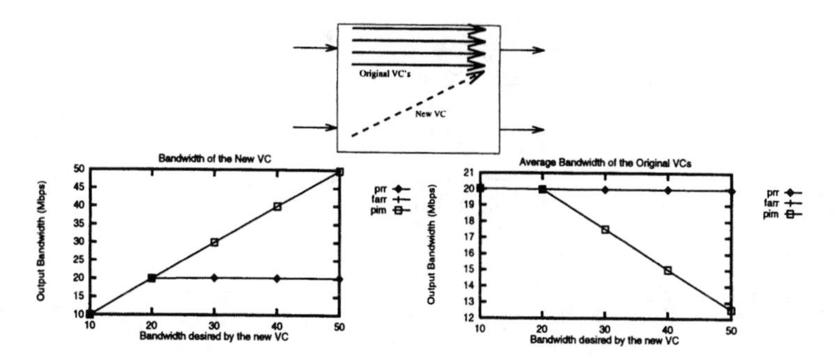

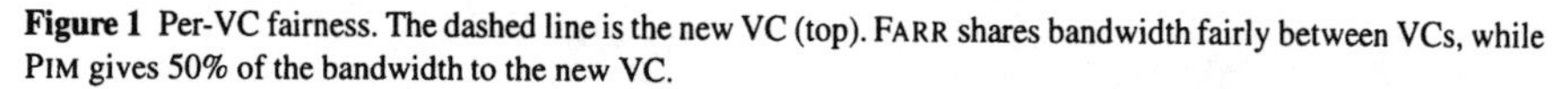

Figure 1 Per-VC fairness. The dashed line is the new VC (top). FARR shares bandwidth fairly between VCs, while PIM gives 50% of the bandwidth to the new VC.

To demonstrate the feasibility of an input-buffered switch using FARR, we are designing a switch that makes cost-effective use of current technology to achieve a total throughput of 40 gigabits per second (Gbps), with configurations from 32×32 at link speeds of 1.2 Gbps through to 256×256 at link speeds of 155 Mbps.

The paper is organised as follows: Section 2 discusses the importance of per-VC fairness. The algorithm FARR is presented in Section 3. Related work is described in Section 4. The implemention and cost of FARR are discussed in Section 5, while Section 6 gives the results of extensive simulations comparing FARR to other input-buffered switch schedulers and to PRR. Finally our findings are summarized in Section 7.

2 PER-VIRTUAL CIRCUIT FAIRNESS VERSUS PER-LINK FAIRNESS

Previous studies of scheduling in an input-buffered switch have addressed the issue of per-link fairness, or fairly allocating switch capacity among the different input and output links (Anderson *et al.* [1993], McKeown & Anderson [1994]). However, *per-VC* fairness is more important than per-link fairness.

For example, consider a 2×2 switch with 100 Mbit links and 4 VCs running from input 1 to output 1, each using 20 Mbps of bandwidth. Let us investigate the effect of introducing a new VC, from input 2 to output 1, whose desired bandwidth varies between 10 Mbps and 50 Mbps. Figure 1 shows the effect on the bandwidth given to the original circuits. We see that FARR, which has per-VC fairness, limits the bandwidth given to the new circuit to 20 Mbps, so bandwidth allocation is fair. On the other hand, the algorithm PIM (described in Section 4.1 below), which on this workload exhibits per-link fairness,* allows the new circuit to receive up to 50 Mbps of bandwidth, which is 4 times the bandwidth given to each original circuit.

Since PIM tries to share resources fairly between the links, the VCs using more heavily loaded links get starved. Thus a switch scheduling algorithm that allocates bandwidth fairly among the links may be very *unfair* in allocating bandwidth between VCs. However FARR has no such problem. This effect is also evident in simulations with more realistic traffic.

The notion of fairness that FARR exhibits has been formalized as *max-min fairness* (Ramakrishnan & Jain [1990]). If there is only 1 priority class, max-min fairness can be described as follows. At the most heavily loaded link, all VCs should receive bandwidth equal to the minimum of b and the bandwidth desired by the VC, where b is set such that all the bandwidth of the link is used up. Bandwidth should be shared in the same way at less loaded links, subject to the limits set by more heavily loaded links.

*In general PIM does not even exhibit per-link fairness, see McKeown & Anderson [1994].

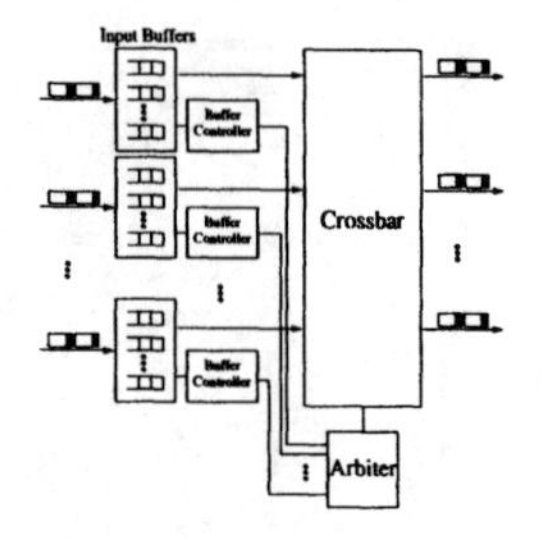

Figure 2 An Input-Buffered Switch with Centralized Arbiter

3 THE ALGORITHM FARR

This section presents the algorithm FARR for scheduling an input-buffered switch. Before presenting the algorithm in Section 3.3, we will first specify the requirements of a input-buffered switch-scheduling algorithm (Section 3.1), and then describe Fair Prioritized Round Robin (PRR), the output-buffered switch scheduler on which FARR is based (Section 3.2).

3.1 Input Buffering and Matching

It has been traditional to regard input-queued switches as impractical because of head-of-line blocking at the input queues. The solution is to use more sophisticated buffers at the inputs, which we refer to as *input-buffering*. This approach has been finding increasing support (see for example Anderson *et al.* [1993]). What is then needed is a fast and effective algorithm for choosing which cell to send from each input buffer during each cell time. A switch that uses a centralized arbiter to implement the scheduling algorithm is pictured in Figure 2. The arbiter needs to do the following task:

> Each cell time, the arbiter must select a set of cells to enter the crossbar. Each cell must reside in a different input buffer, and each cell must be destined for a different output.

We see that the arbiter must repeatedly solve a *bipartite matching* problem. However, finding a *maximum* matching in each cell time is both infeasible, since even the best algorithms are too slow, and undesirable, as it can cause starvation of some inputs and outputs. In the context of ATM, it is necessary to have per-VC fairness and respect for VC priorities.

3.2 Fair Prioritized Round Robin (PRR)

Although a number of variations on Prioritized Round Robin have been developed (Fraser & Morgan [1984], Hahne [1986], Katavenis [1987], Nagle [1987], and Demers *et al.* [1990]), most are approximations of the following algorithm, which we call "Ideal Round Robin":

> Whenever an output link becomes free, a cell is selected to be sent from the corresponding output buffer as follows: from among the VCs of the highest priority level that have cells in the buffer, choose the VC which has sent a cell least recently. Send the oldest cell from that VC.

If we define the *timestamp* of a VC to be the last time it sent a cell from the switch, and the *extended timestamp* to be the pair (VC priority, timestamp), then Ideal Round Robin can be described as follows:

When an output link becomes free, select the VC with smallest extended timestamp among VCs with cells buffered at that output, and send its oldest cell.

Comparing timestamps of a large collection of VCs is infeasible, so Prioritized Round Robin is typically implemented using service queues, in which case it is called Fair Prioritized Round Robin (PRR). All VCs that have cells in the buffer are kept in service queues, one per priority level. When the link becomes free, the VC at the head of the highest priority non-empty service queue sends a cell and is removed from the service queue. If the VC has more cells queued, it is reinserted at the tail of the service queue. When a cell arrives in a VC that is not currently in the service queue, the VC is inserted at the tail of the queue.

PRR differs slightly from Ideal Round Robin: when a VC that does not currently have cells buffered receives a cell, PRR will send that cell only after sending a cell from each VC that is currently in the same service queue, even if the new cell's VC has an older timestamp than them. This is because the new VC is inserted behind them in the service list. In our simulations below, the input-buffered algorithms are compared against PRR, which is implemented using service queues.

3.3 The Algorithm FARR

The main idea of FARR is to use timestamps to obtain Round Robin information across the entire switch, while keeping service lists *only at the inputs*. Most of the work is done at the input buffers, where speed requirements are less severe: each input buffer maintains a service list for each priority level for each output.

Each VC has a timestamp which is the time when that VC last sent a cell across the crossbar (as in the definition of Ideal Round Robin in the previous section). The time is just a global counter that is incremented each cell time.

Algorithm FARR

1. (Preselection) Each input preselects the VC at the head of the highest priority non-empty service queue for each output for which it has buffered cells.
2. (Request) Each input sends to the arbiter the extended timestamp of each preselected VC that was not preselected in the previous cell time. Each extended timestamp is a *request* to send a cell from some input to some output.
3. (Arbitration) Initially matching is empty. Repeat r times:
   (a) If an unmatched output has any requests from unmatched inputs, it grants the request with smallest extended timestamp.
   (b) If an unmatched input receives any grants, it accepts the grant with smallest extended timestamp.
   (c) Accepted grants are added to the matching.

Because the timestamp of a preselected VC is sent to the arbiter only if the VC was not preselected in the previous cell time, extended timestamps of preselected VCs must be stored in the arbiter. The extended timestamps form a matrix, with at most one entry per pair of input and output links.

It has been suggested that not all priority levels need round robin service, for example we might wish to use first in first out (FIFO) scheduling for voice traffic in the input buffer controllers (see Morgan [1994]). In this case we should associate timestamps with voice cells in the FIFO queue, recording the time that each cell arrived, rather than with VCs. This would make FARR emulate FIFO queueing across the input buffers at the voice priority level, while still providing round-robin service at other priority levels. The arbiter need not even know which service discipline is used at the input buffers for each priority level.

4 RELATED WORK

Karol *et al.* [1987] showed that under a simple uniform traffic model, if each input buffer is a single FIFO queue, then the queues saturate at a utilization approaching $2-\sqrt{2} \approx 0.586$ for large switches due to head-of-line blocking. Much worse behavior is possible for other traffic models (Li [1988]).

These findings have been taken as proof of the unsuitability of input-buffered switches. On the other hand, output-buffered switches become prohibitively expensive at high link speeds, since the switch fabric and output buffers need to operate at n times link speed, in an $n \times n$ switch. An input-buffered switch is cheaper to build, since the buffers need only run at link speed.

Many switch designers have explored ways to avoid having buffers working at n times line speed (where n is the number of input links to the switch). Input-buffering is an approach that has been extensively studied. A number of authors have studied contention mechanisms that have reduced the impact of head-of-line blocking and raised the throughput close to 100% under the uniform traffic model, see for example Karol *et al.* [1992], Obara & Yasushi [1989], and Tamir & Frazier [1988].

Recently a switch design has been proposed that eliminates head-of-line blocking by keeping in each input buffer a separate queue for each output buffer, as pictured in Figure 2. This architecture was used with the Parallel Iterative Matching (PIM) algorithm by Anderson *et al.* [1993] and with Iterative Round Robin Matching with Slip (SLIP) by McKeown & Anderson [1994], McKeown *et al.* [1993]. These algorithms, described more fully below, are the closest in spirit to FARR, and we use them as yardsticks for measuring FARR's performance.

A number of other architectures work around the memory speed problems of output-buffering. The Gauss switch de Vries [1990] broadcasts each cell to all output modules. Each output buffer receives the cells in n small buffers (one per input link), which are drained at 4 times line speed into a single larger buffer. The Sunshine switch Giacopelli *et al.* [1991] has a fabric that allows 2-4 cells to arrive at an output buffer per cell time, while cells that do not reach their respective output are recirculated. The Knockout switch (Yeh *et al.* [1987], and generalized by Eng *et al.* [1992]) has an interconnection network that can route cells to output buffers at a constant factor times line speed.

None of these alternative switch architectures addresses the basic topic of this paper: achieving per-VC fairness in an input-buffered switch.

4.1 Parallel Iterative Matching (PIM) and Iterative Round Robin Matching with Slip (SLIP)

Parallel Iterative Matching (PIM) is the randomized algorithm implemented in Digital's Autonet 2 (AN2) switch. For a complete description, see Anderson *et al.* [1993]. Parallel iterative matching does not use VC priorities, nor does it provide fairness or performance guarantees. The basic algorithm is as follows.

Algorithm PIM Repeat r times:

1. Each unmatched input sends a request to every output for which it has a buffered cell.
2. Each unmatched output randomly selects one of its requests, and sends a grant to that input.
3. Each input that receives some grants selects one at random, and accepts that grant.
4. Accepted grants are added to the matching.

The biggest difficulty in implementing PIM is sufficiently fast generation of pseudo-random numbers for randomly selecting between requests. Anderson *et al.* describe extensions to PIM for real-time performance guarantees, including reserving bandwidth for preallocated frames, and *statistical matching*, in which the random selections are biased rather than fair in order to allocate bandwidth unevenly.

Iterative Round Robin Matching with Slip (SLIP) eliminates the use of randomness, instead resolving conflicts with the aid of rotating priorities (see McKeown & Anderson [1994] and McKeown *et al.* [1993]). Simulations have shown that SLIP and PIM have similar performance on simple Bernoulli traffic models, while SLIP is easier to implement than PIM.

The rotating priority is implemented by a counter a_i associated with each input and a counter g_j associated with each output. Lines 2 and 3 of Algorithm PIM are replaced by:

2′ Each unmatched output j that receives some requests grants the request from the input i that is closest to g_j in cyclic order (i.e. $g_j - i \bmod n$ is minimized).
3′ Each input i that receives any grants accepts the grant that is closest to a_i in cyclic order. If the grant is from j, the assignments $a_i = (j + 1 \bmod n)$ and $g_j = (i + 1 \bmod n)$ are made.

A second input-buffered algorithm described by McKeown & Anderson [1994] is LRU, where in Step 2, each output selects the request from the input that least recently sent that output a cell, and similarly for Step 3. McKeown and Anderson conclude that LRU is much more complex to implement than SLIP, with no performance gain, so we do not consider it further.

5 IMPLEMENTATION AND COST OF FARR

At most 2 extended timestamps need be sent from each input during each cell time: the first to replace the extended timestamp of a cell that was sent in the last cell time, and the second if a cell arrives on the input link and is immediately preselected in Step 1 of the algorithm. (If multicast is supported in the switch, more timestamps may need to be sent from each input — some implementations require 3 rather than 2.) Thus the number of bits entering the arbiter each cell time is $2n(\log n + \tau)$, where τ is the size of an extended timestamp. In practice, τ is between 8 and 10, so for a switch with 32 input and output links running at 1.2 Gbps, less than 1000 bits must enter the arbiter per cell time of 353 ns.

Each of FARR's input buffers is essentially performing the task of a shared-memory switch running PRR. So why should FARR be simpler or cheaper to build than a single shared-memory switch? The answer is that the price of a shared-memory switch does not scale linearly with throughput. Using current technology it is cost-effective to build an ATM switch running PRR with total throughput of up to 1-4Gbps, using only DRAM or SRAM buffers and a VLSI buffer management chip. Therefore an input-buffered switch running FARR makes fair prioritized switching cost effective at a much higher total throughput than is possible in a shared memory switch.

An input-buffered switch running FARR has a very natural pipeline, consisting of three steps:

1. Contention resolution to determine the set of cells to be sent.
2. Setting up the crossbar to send the cells.
3. Transmitting the cells.

Each of these steps can therefore use a full cell time to complete. In addition, much of the queue manipulation at the input buffers can be done in parallel with contention resolution.

FARR's contention resolution is designed for efficient parallel implementation – for each of steps 2, 3(a), 3(b) and 3(c), the step can be executed for each input (or output) in parallel. The basic operation of steps 3(a) and 3(b) involves finding the minimum of n τ-bit numbers, and can be performed in combinational logic with $c \log n \log \tau$ gate delays, where c is a small constant, for example using a tree of fast comparators.

The number of timestamps that need be compared is small, since only the timestamp at the head of a service list is used in arbitration. The number of rounds of matching (r) is flexible, and should be maximized within the timing and cost constraints of the implementation. Beyond a certain point further rounds are unnecessary, when no more pairs can be added to the matching, i.e. a maximal matching has been found. In our simulations we use $r = 4$, and present evidence showing that using 3 rounds is as good as continuing until a maximal matching is found.

We believe that a single-chip VLSI implementation of the arbiter is feasible for a 32×32 switch with 1.2 Gbps links, using current VLSI technology.

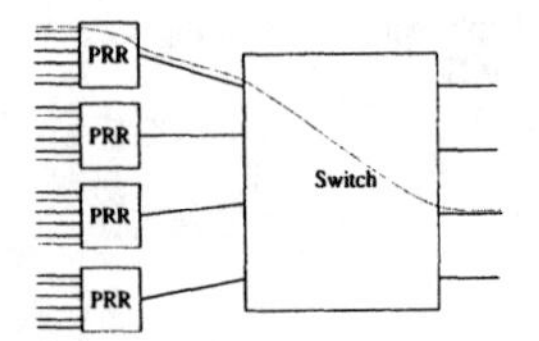

Figure 3 The general workload model. The dotted line denotes a VC entering input 1 and leaving output 3.

6 EMPIRICAL EVALUATION

To evaluate the performance of our switch algorithm FARR, we compared it in simulations with PIM, SLIP, and PRR under a variety of work loads, traffic types and switch sizes. The algorithms SLIP and PIM behaved very similarly under simulation, so the comparisons below use mainly PIM. We have not simulated algorithms that use FIFO input buffers, due to the detrimental effect of head-of-line blocking.

The comparison with PRR is somewhat unfair, since PRR requires a more expensive output-buffered switch. Nevertheless, FARR is designed to emulate the performance of PRR. We are thus evaluating how close we can get to output-buffered performance in an input-buffered switch.

6.1 Workload Modeling

A source of frustration in the design of components for ATM networks is the lack of good workload models for very high bandwidth multimedia networks. Indeed, there are as yet no large high-speed networks from which empirical data could be gathered, and furthermore, as tools and applications for multimedia become available, network workloads are likely to evolve. A case in point is the World Wide Web, which has in a short time become a significant influence on traffic carried by the Internet backbone.

Given these uncertainties, one can hope only to make a reasonable estimate of the properties and relative proportions of the different traffic types. For this estimation we defer to the work of Morgan [1994], who suggests that future ATM network may use 10% of the bandwidth for voice traffic, 40% for video traffic, and the remaining bandwidth for data traffic. He suggests that voice traffic should be at the highest priority level, followed by video traffic, with data traffic having lowest priority.

Our models of the sources for these basic traffic types are as follows. A *voice* source generates constant bit rate traffic at a rate of 64 Kbps, while a *Bernoulli* source generates Bernoulli traffic at a rate of 1 Mbps. A *video* source generates cells using the trace of MPEG2 frame sizes for a full length feature film. Each source starts transmitting from a random point in the film. For each frame, the video source sends sufficient cells to contain the frame, and spreads the cells out over the time till the next frame. Lastly, for a data source we use a *packet train* model, in which bursts of cells are interspersed with large gaps. The length of a burst is a geometric random variable with mean 12, while the inter-burst gap is chosen uniformly up to some limit, where the limit is such that the average bandwidth is 1 Mbps.

The simulations we perform are on a single ATM switch. Because different VCs feeding into the same input can produce cells at the same time, we need to multiplex these VCs. In a real network this multiplexing is done by earlier switches in the network. In our simulations we place a PRR scheduler between each input and the VCs feeding into it (see Figure 3). The PRR scheduler simulates the interleaving of VCs by earlier switches in a real network.

It is important to note that when we measure cell delays in the switch, we do not include delays inside the PRR filters at the inputs. This is because we are interested in the delay induced by the switch scheduling algorithm inside a switch, and the delays at other points in the network are not of interest here.

In addition to the three basic traffic types, one simulation involves adding a *file transfer* after the other VCs have been set up. A file transfer uses a Bernoulli source and absorbs all the available bandwidth. Its bandwidth is thus the minimum of the remaining bandwidth on the input and output links it uses.

Our workload models are derived by selecting a set of VCs each having one of the above traffic source types. We use the following models:

`vvd` Our basic workload model, with 10% of the bandwidth voice, 40% video and 50% data.

`server` A model with non-uniform work load, as might arise in a client-server environment. The first 4 input and output links are connected to *servers*, while the remaining links are connected to clients. The load on client-client connections is only 10% of the load on server-server or server-client connections. This is implemented by first constructing an instance of the vvd model, then randomly dropping 90% of the client-client VCs. A similar model was used in Anderson *et al.* [1993].

`Bernoulli` A model with 100% Bernoulli traffic. This gives a Bernoulli traffic model for which the load is approximately uniform across the input and output links.

`ft` After constructing an instance of the vvd model, an additional VC of type File Transfer is added with enough bandwidth to saturate some switch link.

To generate an instance of the `vvd` or `Bernoulli` model, we set up a large number of VCs, with traffic types chosen according to the above distributions. The input link and the output link for each VC is chosen at random among all the links, subject to the condition that the VC does not enter and leave the same link, and does not increase the bandwidth through any incoming or outgoing link above 95%. VCs are added until the average link load reaches a threshold percentage p. The parameter p, a measure of the switch utilization, is varied in most of the simulations to determine the behaviour of the algorithms under varying switch loads. Notice that the actual load will vary from link to link.

6.2 Simulations

In most of the simulations we use a 16×16 switch. Although we expect that an input-buffered architecture is more cost-effective than output-buffering only when the total switch bandwidth is multiple gigabits per second (with current technology), there are difficulties in simulating such high bandwidths. In particular, VCs at reasonably high bandwidths of around 1 Mbps would only produce a cell every thousand switch cell times. Thus a simulation would have to run for a very long time to give statistically significant results for such VCs, and for VCs running at voice bandwidths the problem is much more severe. Therefore our simulations use link speeds of 155 Mbps.

In our simulations we varied the workload model, the switch scheduling algorithm, the switch size, and the utilization parameter p. In each simulation, we measured the average and maximum queue sizes, as well as the average and maximum delay of each traffic type. We ran each simulation for 100000 cell times in order to overcome any initial transient. (We determined that 100000 cell times was sufficient by considering runs of 20000, 50000, 100000, 200000 and 500000 cell times. The transient behaviour from the start of the simulation had ended well before 100000 cell times.) To obtain as much statistical significance as possible, we ran each simulation at least 125 times, or until every 5% confidence interval had width at most 10% of the average for the statistic, whichever came later. Whenever we measured the average of a parameter, for example the delay on voice cells, we report the average over all cells in all runs. Similarly, when we report the maximum value of a parameter, we mean the maximum over all cells in all runs. When the maximum values are plotted, they do not always produce smooth curves, since the values are determined by rare events.

It is important to note that running a simulation 125 times over is not the same as running the same simulation for 125 times as long. The reason is that we set up a random configuration of VCs at the start of a simulation, and these VCs stay constant during the run. By running the experiment many times over, we obtain values that are representative of all configurations of VCs, rather than just a single configuration.

In the simulations presented below, whenever the model is not mentioned it should be understood to be `vvd`.

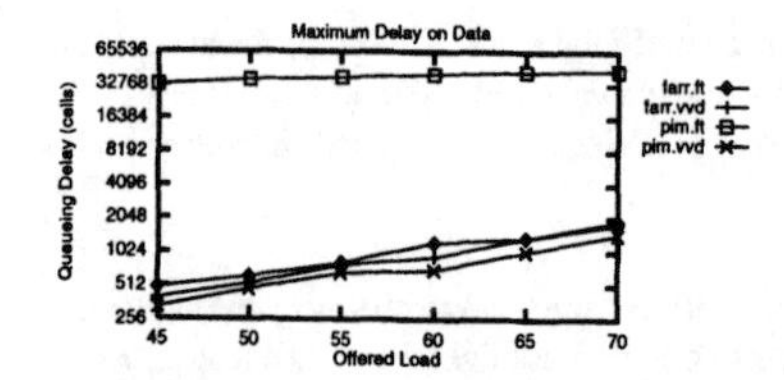

Figure 4 Effect of adding a file transfer to the vvd model. Maximum delay on other VCs is unaffected for FARR but greatly increased for PIM.

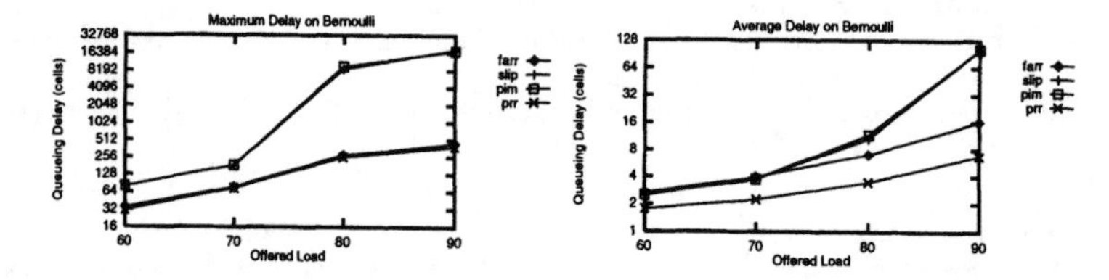

Figure 5 FARR outperforms other input-buffered algorithms on pure Bernoulli traffic. The first graph shows fairness of FARR, the second graph shows lower delays under high utilization. The traffic model is `Bernoulli`.

Similarly the default switch size is 16 x 16, and we mention the switch size only when deviating from the default. We use a logarithmic scale when presenting average or maximum queueing delays or queue sizes.

Per-VC Fairness The first simulation studies the effect that high-bandwidth VCs have on lower bandwidth VCs. In Figure 4 we show two algorithms (PIM and FARR) and two workload models (`vvd` and `ft`). The only difference between the workload models is that `ft` contains a single file transfer circuit that was added after forming a `vvd` configuration of VCs. Figure 4 shows the average delay on data cells from VCs other than the file transfer for PIM (the line `pim_ft`) and FARR (the line `farr_ft`). When we compare with average delays in the `vvd` model, we see that the file transfer has almost no effect on the maximum queueing delay of other data cells if we use FARR as the switch scheduling algorithm. This demonstrates that FARR insulates VCs from the behaviour of other VCs, as is the case for PRR.

However, the file transfer causes PIM to greatly increase the queueing delay of some data cells in previously existing VCs. This is true also of other scheduling algorithms for input-buffered switches.

Performance on Bernoulli traffic In the second simulation, we investigate the performance of FARR compared to PIM, SLIP and PRR on the `Bernoulli` workload model. The reasons for presenting data derived from this source are twofold: this source is quite close to the uniform Bernoulli workload model, which has often been used as a simple model for analyzing the performance of switch algorithms, and secondly, it demonstrates the importance of per-VC fairness. Our `Bernoulli` workload differs from the uniform Bernoulli workload in two ways: we model VCs, and the load varies somewhat from link to link due to the random selection of VCs.

The first graph in Figure 5 shows that FARR simulates PRR very closely but PIM and SLIP both have maximum cell delays of up to a factor of 32 larger. The reason for this is that different inputs have different numbers of VCs. Algorithms such as SLIP and PIM that don't exhibit per-circuit fairness will create large cell delays on the more loaded inputs. This is similar to the effect described in Section 2.

The second graph in Figure 5 shows that FARR, PIM and SLIP all achieve the same average cell delay for offered loads up to about 80%. However, above 80%, FARR starts achieving lower average cell delays than the other input-buffered algorithms. A possible explanation for this is that FARR favors more heavily loaded links, so there are more often cells available to be transmitted from the less loaded links.

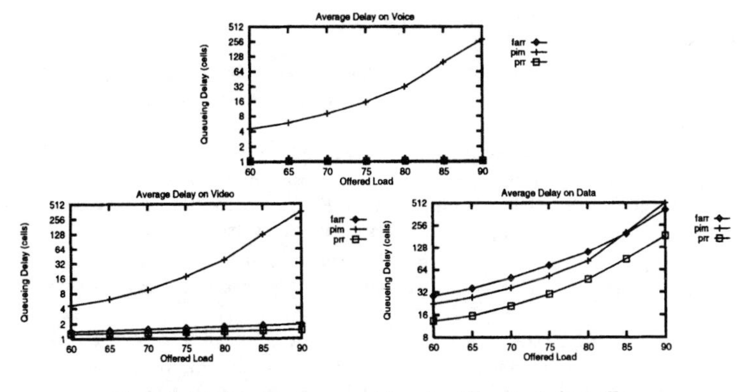

Figure 6 VC Priorities. Average delays for different priority traffic.

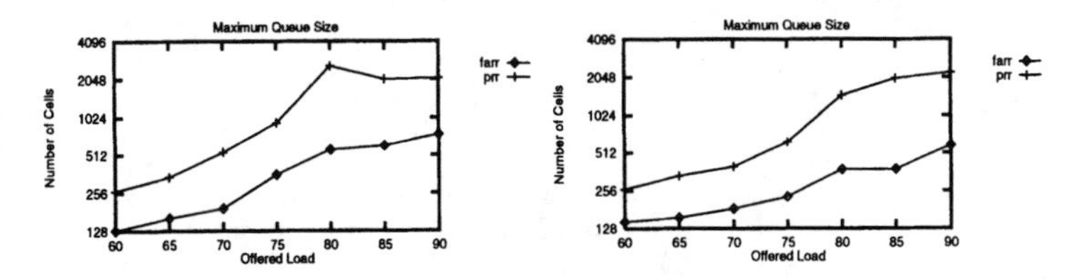

Figure 7 Input queueing may give smaller maximum queues. FARR and PRR are shown on the vvd model (left) and the server model (right).

Priorities for Different Traffic Types The next simulation shows the importance of using multiple priority levels in an ATM switch (see Figure 6). Both FARR and PRR respect cell priorities, and therefore have much smaller average queueing delay for voice and video cells than for data cells. In contrast, PIM does not use multiple priority levels, so voice and video cells are subject to large delays under high traffic loads.

Input buffering may yield smaller maximum queue size The maximum queue size is an important measure of a switch scheduling algorithm, because an algorithm that has a larger maximum queue size requires larger buffers to avoid dropping cells. In the case that cell loss is tolerable at low levels, the maximum queue size is an indication of the buffer size needed to achieved a low enough level of cell loss probability.

In many of our simulations we have noticed that the maximum queue size for input-buffered switches is smaller than that of an output-buffered switch.† There is a good reason for this: when there is "hot spot" contention at an output link in an output-buffered switch, the cells building up in that buffer are being received from a number of inputs. In an input-buffered switch, those cells remain at the inputs, and each input buffer holds a smaller number of cells than the output buffer at the overloaded link.

Figure 7 shows that the maximum queue is significantly less for FARR than for PRR under two workload models, the vvd model and the server model.

It is important to note that this phenomenon may not always occur. If for a particular workload the bottleneck of an input-buffered switch is the switch fabric, rather than an output link, then output-buffering may achieve smaller

† By an *output-buffered switch* we mean a switch with a separate buffer at each output link. The argument made here does not hold for a shared-memory switch, in which all output links shared a common buffer.

maximum queue size for that workload. This is a question that should be revisited once more detailed models of ATM network traffic are available.

Scaling with Switch Size Since the advantages of input-buffering increase with larger switches, we studied how FARR scales to larger switch sizes. Simulations show that the average and maximum delay remains almost constant across different switch sizes.

Crossbar Speedup We considered the effect of running the crossbar faster than the incoming and outgoing links. Although speeding up the crossbar cannot increase throughput or decrease queueing delays beyond that of an output-buffered switch, our simulations show that a crossbar speedup of 10% provides a distinct improvement in the performance of FARR, while a speedup of 20% gives performance which is extremely close to that of PRR.

The Size of the Timestamps The size of the timestamp is a factor in the cost of implementing FARR: smaller timestamps require less storage and simpler arbiter logic. Our simulations suggest that 8 bit timestamps are sufficient to achieve fairness in the vvd model.

Number of Rounds Needed by FARR We investigated the effect of varying r, the number of rounds used by FARR. Increasing r will tend to increase throughput, while decreasing r makes implementing FARR easier and faster.

Our simulations show that 1 round is inadequate, and that 3 rounds provides a substantial improvement over 2 rounds. With 3 rounds the performance is very close to continuing FARR for as many rounds as are needed to produce a maximal matching.

7 DISCUSSION

We have presented a new algorithm, Fair Arbitrated Round Robin (FARR), for scheduling the crossbar of a high-speed input-buffered switch. Input-buffering is more cost-effective than output-buffering at the high speeds demanded by the rapid growth of internetworking, multimedia and ATM. Output buffering has the advantages of prioritization and per-VC fairness, when used in conjunction with Fair Prioritized Round Robin (PRR) scheduling. FARR gives the advantages of both approaches, providing the cost benefit of an input buffered switch, while respecting VC priorities and demonstrating per-VC fairness properties that have previously only been achieved in output-buffered switches. FARR is designed for efficient parallel implementation, and we believe that a single-chip VLSI implementation of the arbiter is possible for a 32×32 switch with 1.2 Gbps links.

We have performed simulations using a variety of work loads, traffic types and switch sizes, measuring throughput, maximum and average queue sizes, and maximum and average queueing delays for each traffic type. The simulations demonstrate the performance benefit of FARR over previous input-buffered switch algorithms such as Parallel Iterative Matching (PIM) and Iterative Round Robin Matching with Slip (SLIP). Furthermore, FARR achieves similar performance to Fair Prioritized Round Robin (PRR) running on an output-buffered switch, and if the crossbar is run 20% faster than line speed, the performance of FARR becomes almost identical to that of PRR.

Acknowledgments We would like to thank Alan Berenbaum, Joe Condon, Hemant Kanakia, S. Keshav and Rae McLellan for many helpful discussions and for good advice.

REFERENCES

THOMAS E. ANDERSON, SUSAN S. OWICKI, JAMES B. SAXE, AND CHARLES P. THACKER, High-speed switch scheduling for local-area networks. *ACM Transactions on Computer Systems* **11**(4) (1993), 319–352.

A. DEMERS, S. KESHAV, AND S. SHENKER, Analysis and simulation of a fair queueing algorithm. *Journal of Internetworking Research and Experience* (1990), 3–26. also Proc. ACM SigComm, Sept. 1989, pp 1-12.

K. Y. ENG, M. J. KAROL, AND Y. S. YEH, A growable packet (ATM) switch architecture: Design principles and applications. *IEEE trans. on commun.* **COM-40, 2** (1992), 423–430.

A. G. FRASER AND S. P. MORGAN, Queueing and framing disciplines for a mixture of data traffic types. *AT&T Bell Laboratories Technical Journal* **63**(4) (1984), 1061–1087.

J. N. GIACOPELLI, W. D. SINCOSKIE, AND M. LITTLEWOOD, Sunshine: A high-performance self-routing broadband packet switch architecture. *IEEE Journal on Selected Areas in Communications, October 1991* **SAC-9, 8** (1991), 1289–1298.

E.L. HAHNE, *Round Robin Scheduling for Fair Flow Control in Data Communication Networks.* Laboratory for Information and Decision Systems, Massachusetts Institute of Technology, Cambridge, MA 02139, 1986. Thesis LIDS-TH-1631.

M. KAROL, M. HLUCHYJ, AND S. MORGAN, Input versus output queueing on a space-division packet switch. *IEEE Trans. Commun.* **12** (1987), 1347–1356.

M. KAROL, K. ENG, AND H. OBARA, Improving the performance of input-queued ATM packet switches. In *Proceedings of INFOCOM,* 1992, 110–115.

M.G.H. KATAVENIS, Fast switching and fair control of congested flow in broadband networks. *IEEE JSAC* **SAC-5**(8) (1987).

S. KESHAV, A control-theoretic approach to flow control. In *Proc. ACM SigComm 1991,* 1991.

S.-Y. LI, Theory of periodic contention and its application to packet switching. In *Proceedings of INFOCOM,* 1988, 320–325.

NICK MCKEOWN AND THOMAS E. ANDERSON, A quantitative comparison of scheduling algorithms for input-queued switches. *Submitted for publication* (1994).

NICK MCKEOWN, PRAVIN VARAIYA, AND JEAN WALRAND, Scheduling cells in an input-queued switch. *Electronics Letters* (1993).

S. P. MORGAN, Engineering guidelines for an integrated services ATM network. AT&T Bell Laboratories, Technical Memorandum 11270-940601-02TM, 1994.

J. NAGLE, On packet switches with infinite storage. *IEEE Trans. on Communications* **COM-35** (1987), 435–438.

H. OBARA AND T. YASUSHI, An efficient contention resolution algorithm for input queueing ATM cross-connect switches. *Int. J. Digital Analog Cabled Syst.* **2**(4) (1989), 261–267.

K. K. RAMAKRISHNAN AND RAJ JAIN, A binary feedback scheme for congestion avoidance in computer networks. *ACM Transactions on Computer Systems* **8**(2) (1990), 158–181.

S. SHENKER, A theoretical analysis of feedback flow control. In *Proc. ACM SIGCOMM '90; (Special Issue Computer Communication Review),* 1990, 156–165.

Y. TAMIR AND G. FRAZIER, High-performance multi-queue buffers for VLSI communication switches. In *Proceedings of the 15th Annual Symposium on Computer Architecture,* 1988, 343–354.

R. J. F. DE VRIES, Gauss: a simple high performance switch architecture for ATM. In *Proc. ACM SIGCOMM '90; (Special Issue Computer Communication Review),* 1990, 126–134. Published as Proc. ACM SIGCOMM '90; (Special Issue Computer Communication Review), volume 20, number 4.

Y. S. YEH, M. G. HLUCHYJ, AND A. S. ACAMPORA, The knockout switch: A simple, modular architecture for high performance packet switching. *Proc. ISS'87, March 1987 and IEEE JSAC-5, 8, 1987, 1274-1283* (1987), 801–808.

31

Fair and Flexible Contention Resolution for Input Buffered ATM Switches Based on LAN Medium Access Control Protocols

Andreas Kirstädter, TU München, Germany
andreas@lkn.e–technik.tu–muenchen.de

Abstract:

Only switches with input buffers offer the possibility to handle effectively the large and full–rate bursts that arise from the transport of data traffic and the burst accumulation within large ATM networks. Efficient contention resolution mechanisms are necessary to prevent output blocking in these input buffered switch architectures and to allow a fair and waste–free utilization of the switch.
This paper first reviews the different types of existing contention resolution mechanisms and shows their limits concerning fairness, scalability, and the possibility to support switching of multicast and prioritized cell streams. Then a new approach is presented that achieves an absolutely fair and efficient contention resolution on the cell level by using modified LAN medium access control (MAC) protocols. The requirements that a MAC protocol has to accomplish are investigated and the excellent performance (using an adapted version of the CRMA–II MAC protocol) of the proposed approach is shown. Finally native extensions toward the integration of multicast and prioritized traffic are given and it is demonstrated how this architecture can easily be scaled up to coordinate switches with throughputs of several Terabits per second.

Keywords
ATM Switch, Contention Resolution, Input Buffering, LAN, MAC Protocols

1 Introduction

During the last few years the application scenarios of ATM have changed dramatically. At the beginning the primary intention of ATM was the bandwidth–effective replacement of existing synchronous multiplexing hierarchies. By now it seems however that the biggest part of ATM traffic will be data traffic, i.e. ATM is going to be used increasingly for extending, replacing, and connecting the so called legacy LANs (802.x, FDDI).

The consequences of this development on switching architectures and their buffering strategies are no more neglectable: the switching and multiplexing of classic voice and video channels requires no big buffering efforts. So the switch concepts of the first years of the ATM life cycle focused on methods for reducing the connectivity expenses within the switch core (Ahmadi, H. and Denzel, W., 1989). And the performance analysts of that era used uncorrelated (i.e. non–bursty, bernoulli) traffic sources for exercising their models. But with the deployment of available bit rate (ABR) services for data traffic at large bandwidth delay products buffering strategies for ATM switches become increasingly important. Huge cell bursts will have to be managed by the switches during the reaction gaps of the data sources. And quite low cell loss guaranties have to be met in order not to disturb higher layer data communication protocols (ATM Forum, 1994). Corresponding simulation models have to use bursty traffic sources. At the same time future ATM switches will be confronted with the tasks of video distribution and handling different services classes so that multiple levels of priority and multicast traffic will have to be supported.

After reviewing buffering and buffer management strategies proposed in the literature shortly in the next section this paper presents a new buffer management scheme based on LAN medium access protocols that accomplishes all the requirements mentioned above. Two well known high speed LAN MAC protocols (CRMA–II and DQDB) are investigated concerning their suitability for the coordination of input buffered ATM switches. The resulting fairness is demonstrated and the delay throughput performance is compared to that of an ideal (output buffered) switch and a rate based control approach. Then a straightforward extension is shown that allows the coordination of an arbitrary and varying number of priority and multicast classes. Finally implementation, feasibility, and scalability aspects are considered.

2 ATM Buffering and Contention Resolution Strategies

ATM switches have to deal increasingly with highly bursty and asymmetric client–server data traffic and have to absorb large bursts during the reaction gaps of ABR controlled end systems (within networks with large bandwidth delay products). While it was shown that the best throughput delay performance can be achieved by pure output buffering (Hluchyj, M. G. and Karol, M. J., 1988) this strategy is simply not applicable to large numbers of switch ports and asymmetric load scenarios (Simcoe, R. J. and Pei, T.–B., 1995): in this case a relatively small speed–up of the output buffers (in the order of 4 to 8, s. e.g. Karol, M. J. et al., 1987) of the input bit rate, that was found to be sufficient in the case of symmetric load scenarios with bernoulli sources, no longer results in sufficiently low levels of cell loss so that the loss now becomes highly dependent on the applied load pattern. Shared memory buffering used by most current generation ATM switches (Garcia–Haro, J. and Jajszczyk, A., 1994) suffers not only from the fact that the necessary speed of the common buffer limits the maximum number of ports per switching module. The size of the common buffer is also severely limited since it has to be implemented on the (full custom) switch ASIC. Thus the only feasible buffering strategy is to use large input buffers that can easily be implemented at line speed.

Another problem arises from the need to coordinate the input buffers, i.e. to determine during each T_{cell} which of the contending input buffers is allowed to send a cell to a certain output. The main requirements for the contention resolution mechanism are:

- fair arbitration between contending inputs;
- waste–free operation (the max. switch throughput has to be impaired as less as possible);
- mechanisms for handling multicast and prioritized traffic;
- low implementation costs;
- independence from the special implementation of the switch core;
- scalability (i.e. a large maximum number of ports that can be coordinated).

Several contention resolution principles have been proposed in the past:

2.1 Usage of a Central Scheduler

Often the decision which input buffer may send a cell to which output port is taken by a central scheduling engine (s. figure 1, Obara, H., 1991, Matsunaga, H. and Uematsu, H., 1991). During

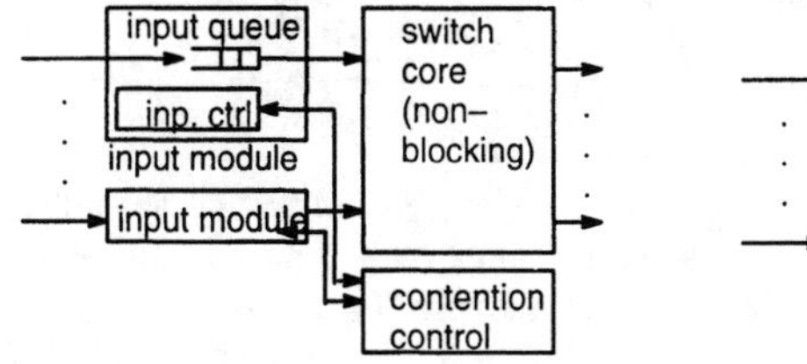

Figure 1: Centralized scheduling.

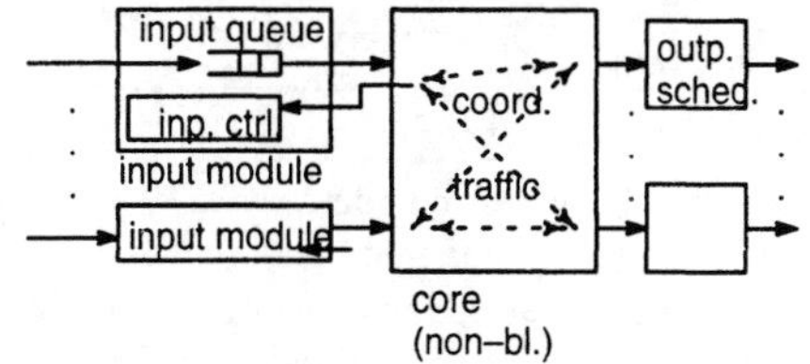

Figure 2: Output scheduling.

each T_{cell} it first receives information about all head of line (HOL) cells at all input buffers because otherwise its decision would be based on outdated information. Then it uses certain algorithms (e.g. round robin or neural network based etc.) to determine the winners between contending input buffers. This information is then transferred back to the input buffer control units that then at the end will send the corresponding cells into the switch kernel. Thus during the same T_{cell} a large amount of information has to be transferred and processed.

The scalability of this architecture is further limited if the input buffers are split into sub–buffers (one per switch output at each input, s. e.g. LaMaire R. O. and Serpanos, D. N., 1994) in order to prevent the HOL blocking effect present in simple FIFO input buffering solutions. Then the number of cells that have to be considered by the central scheduler rises with $o(N^2)$, N being the switch size. Also the usage of time stamps for circumventing the strict FIFO order increases the control data transfer and processing requirements (Obara, H., 1991).

2.2 Output Scheduling

A way to alleviate this problem by parallelism is the so called output scheduling (figure 2, s. e.g. Main, J. and Sarkies, K., 1995). Here a smaller decision engine is placed at each output port. All inputs having a cell ready for transmission to a certain output send a request to that output. The output then randomly selects one of the requesting inputs and notifies it by a confirm message. Some inputs may have got two or more confirm messages from different outputs. Since they can send only one cell during T_{cell} subsequent request–confirm rounds are necessary to allocate the excess confirm messages that those input returned to the outputs (a randomly chosen confirm is held back by each of the inputs).

So this parallel procedure substitutes the transmission and processing speed requirements of the central scheduling solutions above by a corresponding amount of hardware: slow but many ($o(N^2)$) lines are necessary for connecting each input buffer controller (IBC) with each output scheduler. Thus in a real systems an extra "switch" (maybe with its own contention resolution) is necessary for the transmission of the contention resolution information.

2.3 Dynamic Bandwidth Allocation (DBA)

The dynamic bandwidth allocation (DBA) (Worster, T. et al., 1995) tries to reduce the hardware expenses for contention resolution by allocating cell transmission rates instead of transmission instants for single cells to the inputs (s. figure 3). Each statistical multiplexing unit (SMU) at the input of the switch core sends a bandwidth request to the scheduler within the SMU at the corresponding destination output if one of several thresholds has been exceeded within any of its sub–buffers. The output SMU then considers this request together with the requests from other inputs and sends back an acknowledge signal indicating the cell rate that the requesting SMU can use. Request and acknowledge messages are transmitted by ATM cells through the switch core.

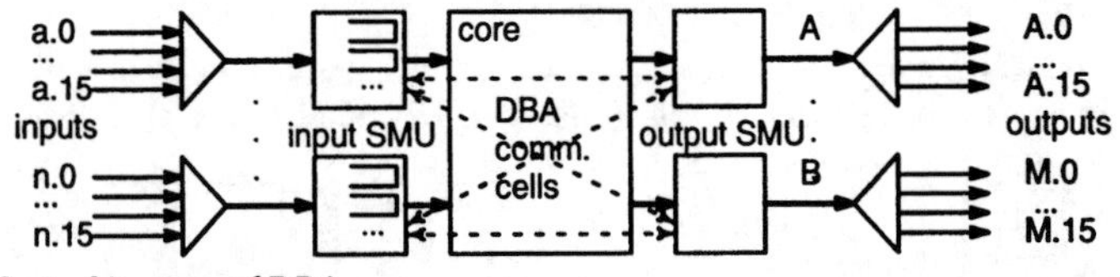

Figure 3: Architecture of DBA.

Since only cell rates are coordinated the switch core must offer high speed buffering capabilities (nearly as an output buffered core). Furthermore a low efficiency of this approach is to be expected since either a large amount of payload cells are used for the transport of coordination messages or the coordination itself will be rather slow and bandwidth wasting in the case of highly bursty traffic. These problems increase with the number of ports and if multicasts and priorities are to be considered (the MUX / DEMUX solution for increasing the number of ports shown in figure 3 implies fairness problems at the output trunks A, B, etc.).

2.4 Two– / Three–Phase–Algorithms

Another family of coordination mechanisms for input buffers, back again on the cell level, are the so called 2– or 3–phase–algorithms (s. e.g. Hui, J. Y. and Arthurs, E., 1987). But since they can only be used together with a special kind of switch core (Banyan network with a bidirectionally transmitting Batcher sorter) they are not further considered in this paper.

2.5 Ring Reservation

A very promising approach for the coordination of input buffered ATM switches is the so called ring reservation (Bingham, B. and Bussey, H., 1988). A number of bit positions each controlling

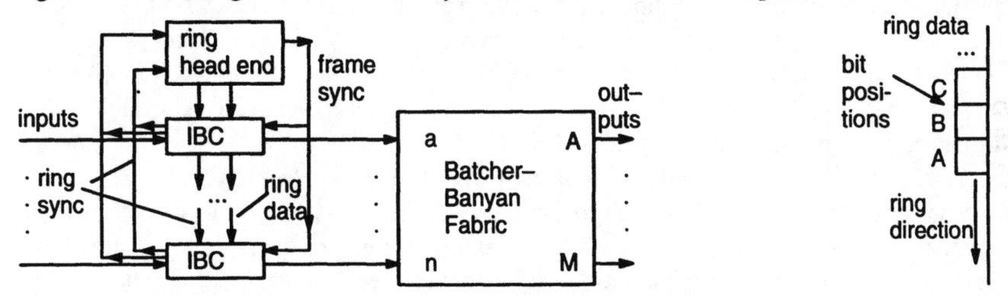

Figure 4: Ring reservation: architecture and ring data structure (control bits).

the sending of cells to a certain output (corresponding to the index of that bit) are clocked along a ring (a full rotation per T_{cell}) that connects all IBCs (s. figure 4). At the start of T_{cell} all bits have been set to "0" and an input can reserve for its HOL cell a sending opportunity for the next T_{cell} by successfully setting (from "0" to "1") the corresponding bit.

The big advantage of this solution is the strong reduction of the necessary coordination traffic and / or the corresponding hardware expenses since for the access control only a neighbor–neighbor communication in one direction between the IBCs has to be established. Splitting the input buffers internally into sub–buffers to avoid HOL blocking does not increase the necessary flow of coordination information. Neither a central controller is required nor are the core outputs involved in this communication. Further this mechanism is completely independent of the design of the core.

But so far the fairness problem arising from the sequential probing of inputs has not been solved successfully. The upstream one of two neighboring IBCs stays favored even if (as proposed in Bingham, B. and Bussey, H., 1988) the virtual origin of the bit frame is rotated (by a single IBC per T_{cell}) around the ring. Using a round robin mechanism by starting the search just behind the last serviced IBC for each output port is also not a solution to this problem. Not only the ring bandwidth would have to be doubled because now 2 rotations of the complete bit frame are required (since the sequence of the bit positions cannot be adapted) but also an unfairness in the case of coincidences (similar to that of DQDB, s. below) would occur. Furthermore up to now no satisfying native solution has been presented for the coordination of multicast and prioritized connections. Multicast cells have to be duplicated before entering the switch core (e.g. Xerox Corp., 1993) thus severely limiting the amount of multicast the switch can handle. Priorities can only be processed by rotating the bit frame a corresponding number of times per T_{cell} around the ring (each time coordinating a single priority level) thus inadequately increasing the ring bandwidth requirements.

3 Input Buffer Management Using LAN Medium Access Control Protocols

3.1 Basic Principles and Requirements

The idea behind this new solution for the buffer management problem is first to reduce the promising idea of ring reservation to its main principle: to tie the authorization for sending a cell to the successful setting of a certain bit position on a serial line. But instead of rotating the origin of the bit frame in a ring like fashion the proposed buffer management scheme uses always a fixed origin (making the ring to a bus and leading to new possibilities for an efficient handling of multicast and prioritized traffic). The fairness between contending inputs is guaranteed then by adapted medium access control protocols for bus networks from the local area networks domain.

Since the access of cells from the *N* switch inputs to each of the *M* outputs is controlled by its own medium access control this coordination system can be viewed as *M* virtual LANs (VLANs). All these VLANs exist on the same serial medium (the bit positions corresponding to the controlled outputs) and comprise *N* stations. Each input buffer controller (IBC) then comprises *M* finite state machines (FSMs) each of them controlling the transmission of cells of a single sub–buffer to its corresponding switch output.

The result is a LAN–switch hybrid system in the sense that the switched payload is still transmitted by input buffers into an ATM switch core but the coordination is done now by LANs whose bandwidth requirements have been reduced to a mere transmission of the fairness information.

Correspondingly the task of coordinating the cell stream from contending inputs can be considered to be split into three levels:

- The basic mechanism of preventing output blocking by the requirement to first set a bit before the corresponding cell can be transmitted.
- The superimposed medium access mechanism controlling the fair access of all inputs to all outputs.
- Intended interferences into the fairness mechanism to provide efficient means for the hand–ling of different kinds of traffic (multicast, priorities).

The basic requirements the MAC protocol has to accomplish for each LAN in order to be suited for the coordination of input buffered ATM switches are:

- Low processing overhead: *M* parallel MAC FSMs have to be implemented within each IBC.
- Low consumption of bandwidth for the transmission of MAC fairness information between the IBCs: each VLAN can only use a few bits per T_{cell} on the serial line.
- Slotted transmission structure: the fairness mechanism itself has to be based on the assignment of throughput in discrete quantities (cells respectively bit positions).
- The MAC protocol must be able to work on bus LANs since they show the same asymmetry of access as the IBC control line.
- The MAC protocol must allow a full utilization of the LAN since no transmission capacity on a switch output is to be wasted (even if only one of the stations has traffic to transmit).

The above requirements can only be met by MAC protocols originally designed for high–speed bus LANs. During the investigation of several of those protocols for this task it has been found that another requirement arises from the fact that the *M* VLANs are not fully independent from each other. The underlying effect is that the hardware implementations of the switch core and the input buffers severely limit the number of cells an input buffer can send into the switch core during a single T_{cell}. So called coincidence situations arise when an IBC is able to occupy more bit positions within a single T_{cell} than the maximum number of cells it can emit during the same T_{cell}. The consequence is that the MAC protocol cannot rely on the fact that an IBC is able to use all the opportunities for setting bits it has been allocated to by the MAC protocol in order to restore its fairness.

3.2 Cyclic Reservation Multiple Access (CRMA–II)

Implemented on a dual bus topology a CRMA–II (van As, H. et al., 1991) network consists of a fairness scheduler (located in one of the two bus headends) and two counterdirectional buses. The buses are used for the payload transport between the stations and for conveying fairness information between the stations and the scheduler. Fairness is controlled in cycles that are initiated by a request frame emitted by the scheduler. The stations inform the scheduler about their fairness situation by entering into the request frame the actual number of slots pending within their transmission buffers and the number of slots transmitted since reception of the last request request frame. After the reception of the request frame (looped back by the other headend) the scheduler calculates a fairness threshold from the received values of the different buffer lengths and transmission counters (several algorithms have been proposed for this task in: van As, H. and Lemppenau, W. W., 1992, and Lemppenau, W. W. et al., 1993). The scheduler then emits a confirm frame containing the threshold value. At the reception of the confirm frame each station takes the appropriate measures to restore the fairness in a distributed manner: if its transmission count was above the threshold it defers a corresponding number of times its access to empty slots. Otherwise it is allowed to access a number of reserved slots issued by the scheduler. This number of confirms is the smaller value of either the difference between the threshold and the transmission count or the buffer length that it had previously informed the scheduler about. Before re–transmitting the confirm frame down the bus the station finally subtracts the threshold value from its transmission counter. Immediately following the emission of the confirm frame the scheduler generates a number of slots that are marked as reserved and can only be used by stations having calculated a positive number of confirms. So that the sum of marked slots equals the sum of confirms of all stations. After that the scheduler starts a new fairness cycle by emitting the next request frame.

In order to apply this cyclic fairness mechanism to the coordination of the IBCs the following "infrastructure" has to be provided between the IBCs (s. figure 5):

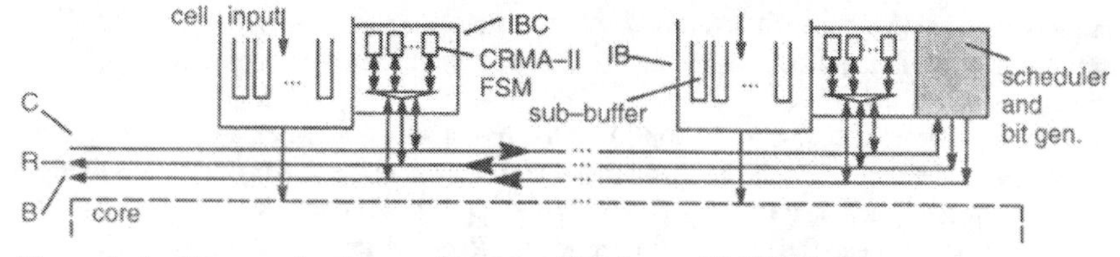

Figure 5: Architecture for the coordination of IBs by modif. CRMA–II.

- A scheduler entity generating the bit positions for all the VLANs is placed within the first IBC (before its FSMs) at the origin of the serial line B that conveys the bit positions that control the access to the outputs. For an increased reliability of the system the scheduler FSM could be present in each of the IBCs (only the first one being active).
- Each bit position on B has to be complemented by a second bit showing the reservation state of the bit position. This second bit may either be conveyed by a parallel line R (as shown in figure 5) or it could be time–multiplexed onto B.
- Finally a line C is necessary for the transmission of fairness information (transmission counters and buffer length values) from the IBCs to the scheduler. The threshold values for the different VLANs can be transmitted from the scheduler to the IBCs on the line B between the frames containing the bit positions.

  The duration T_{cycle} of the scheduling cycles of the single VLANs is mainly determined by the time T_t necessary for the transmission of the fairness information from the IBCs to the scheduler. At a switch size of 64x64 ports a T_t below 2* T_{cell} per VLAN has been found to be sufficient leading to a T_{cycle} in the order of 128*T_{ell}. If at much larger switch sizes a reduction of T_t is desired a few parallel lines can be used for C.

Also the CRMA–II protocol itself has to be adapted to the task of coordinating switch IBCs by the following modifications:

- The transmission counters are reset each cycle after notifying the scheduler about their values instead of decrementing them by the threshold value. Otherwise (depending on the difference between the loads of the single stations) the transmission counters would assume a large range of values (as in the original CRMA–II) and the resulting transmission time part (T_t) of T_{cycle} would lead to a slowed down reaction (determined by T_{cycle}) of the fairness mechanism.
- The deferment of stations with excess throughput is no longer used since the IBCs are only able to defer the access on those bit positions that have not been marked as reserved. A spatial re–use of bit positions (as with the destination release in the original LAN) does no longer exist so that during the marking phase each marked bit position also works as a implicit deferment (of the IBCs that cannot access it). The usage of deferments also becomes unreliable if not enough unmarked bit positions exist.
  Since now only confirms can be used for the equalization of throughput the scheduling threshold has to be always equal to the largest transmission counter value.
- At the beginning of the marking period each station pre–subtracts the number of calculated confirms from its transmission counter since they serve as the equalization of unfair throughput situations during the last cycle. Otherwise an IBC would be "punished" for the act of trying to get its fare share by accessing the reserved slots granted to him in this T_{cycle}.

3.3 Distributed Queue Dual Bus (DQDB, IEEE 802.6)

The distributed queue dual bus (s. e.g. Conti, M. and Lenzini, L., 1991) is usually the first MAC one comes to think of when looking for high–speed bus LANs. The fairness of the transmissions on one unidirectional bus is controlled by the emission of transmission requests in the opposite direction so that a global queue of data slots (not more than one per station) to be scheduled for transmission (by all stations) on a bus is constructed by implicitly considering the position of the stations.

Applying this fairness scheme to the control of the IBs in an ATM switch a topology similar to that of figure 5 (without the scheduler and line C) results where the line R is used this time to transfer the requests (each again a bit position) in the opposite direction. So for each of the *M* VLANs a distributed queue (DQ) is constructed for the transmission of cells to that output.

Complications arise in the case of coincidences when a IBC is in the HOL position of more than one DQ at the same time but can emit only one of the corresponding cells into the core. It has to pass the other sending opportunities to the downstream stations that were originally behind the losing station in the DQ: the DQ mechanism is severely disturbed resulting an unfair throughput behaviour. The number of places a station has to go back or advances within the DQ depends on the actual load pattern and the coincidence situation of the stations. Each single station cannot asses the actual throughput situation (i.e. its own throughput compared to that of every other station) on its own. Thus a distributed reconstitution of fairness (as in the case of the bandwidth manager in large DQDB networks) is not possible.

In the CRMA–II solution the scheduler compares each scheduling cycle the throughput values of the different IBCs on the different VLANs and is able to fully equalize throughput losses caused by coincidence situations within the regular scheduling. The result is an absolute fairness even at overload and asymmetric load situations (s. figures below).

4 Simulation Results

Each of the plots has been derived by time discrete process–oriented simulations at the cell level. Figures 6 and 7 compare the fairness behavior of DQDB and (modified) CRMA–II in the case

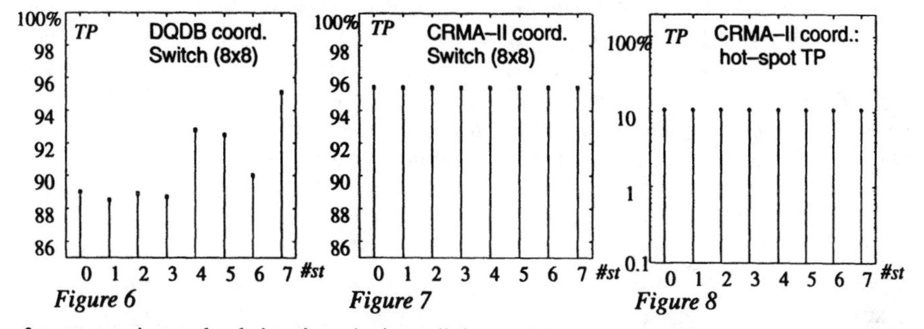

Figure 6 *Figure 7* *Figure 8*

of a symmetric overload situation: the input links are fully utilized (normalized *load* = 100%) and the destination addresses of the incoming cells are equally distributed. Figure 7 also shows that the maximum achievable throughput of the CRMA–II controlled switch gets very close to 100%. In the figures 8 and 9 the same model of the CRMA–II controlled switch has been exer-

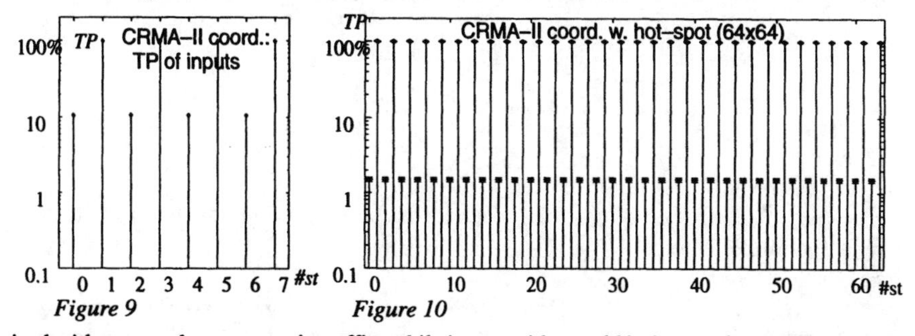

Figure 9 *Figure 10*

cised with extremely asymmetric traffic: while inputs with an odd index get the same load as before in figure 7, the cells on even inputs are directed only to the output with index 0 (the "hot–spot"). Figure 8 demonstrates the absolute fair sharing of this hot–spot output by the single inputs. Figure 9 shows the total throughput (as a sum) each input gets in this situation: the interesting fact is that the throughput values of the odd inputs are only neglectably impaired by the reactions of the fairness mechanism (the values get very close to 100%). In figure 10 a 64x64 switch under the same type of asymmetric load shows that this behavior is maintained even if much larger switches are considered.

The MAC coordination principle (for a 64x64 switch with a link bit rate of 150 Mbit/s) finally has been compared to two alternatives of the DBA method (discussed above) and an ideal (output buffered) switch of the same size. Two different types of cell sources have been used for the simulation of the CRMA–II coordinated switch: an uncorrelated type (where the destination of each cell is independent from the destination of previous cells) and an on–off type with full–rate bursts. In the latter case the mean burst size was 16 cells and the medium cell rate was adjusted by choosing the off–time accordingly. The data for the two DBA alternatives have been taken from (Worster, T. et al., 1995). The source model used in those simulations can also be considered to be uncorrelated since each of the inputs was loaded by a large number of independent small on–off sources (the peak bit rate being 10 Mbit/s at a medium bit rate of 0.1 Mbit/s and a mean burst size of 100 cells). Thus no significant increase of the probability that contiguous cells are destined to the same output will be observed compared to the case of a random destination selection on a cell by cell base.

The comparison of the throughput delay behaviour in figure 11 shows several facts:
First the performance of the MAC coordinated switch under uncorrelated traffic gets very close to that of the ideal switch. Under full–rate bursts the performance of the new contention resolution mechanism is still comparable to the better one of the DBA architectures (having a core size of 4x4). The DBA architecture with the 600 Mbit/s SMUs (and a core size of 16x16) shows the worst performance. This might at least be partially attributed to the throughput degradation resulting from the much larger amount of coordination traffic required at larger core sizes.

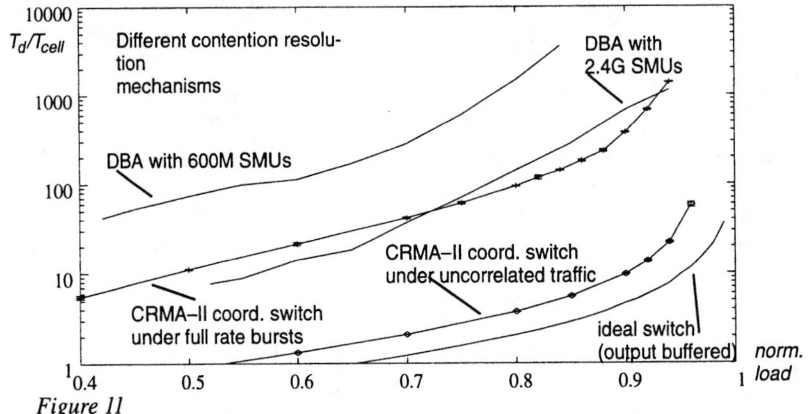

Figure 11

5 Integration of Multicast and Prioritized Cells

The above proposed MAC based control mechanism can be extended in a native way to permit differently prioritized inputs:

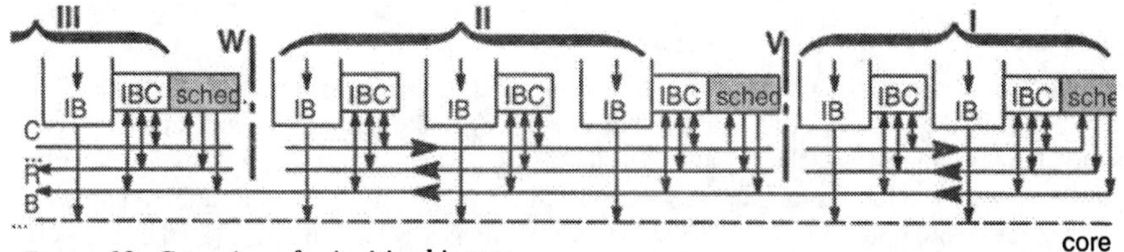

Figure 12: Grouping of prioritized inputs.

By splitting the lines R and C (at the positions W and V) and additionally activating the scheduler entities in the IBCs beneath each of those interruptions the IBCs are separated into several groups (s. figure 12). Since the R and C lines now only connect the IBCs within a single group fairness is only established between the members of this group. The line B that controls the access into the core still connects all of the IBC members of all groups. As the consequence the members of a group downstream on B can only access the bit positions left empty by the members of all upstream groups. Thus each group represents a own priority class of inputs.

This structuring principle can now be used to process correctly and with little overhead cells with an arbitrary number of priority levels that enter an ATM switch (s. figure 13).
The inputs of the switch are directly connected to the IBCs of the lowest priority group (LPIBCs). Cells of the lowest priority are completely managed by the IBCs of this group: they are stored

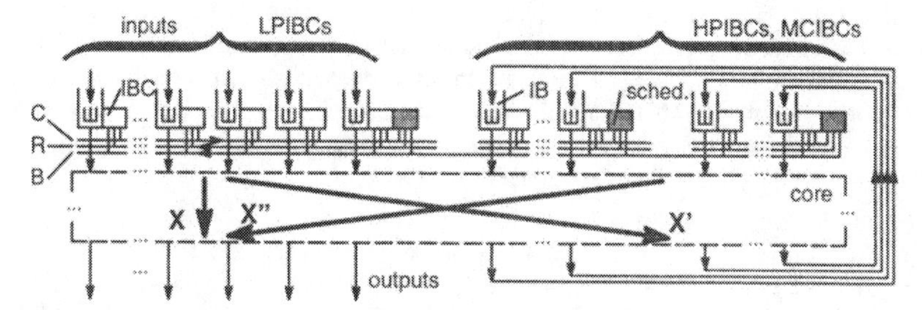

Figure 13: Flexible solution for multicast and prioritized traffic.

in sub–buffers corresponding to their destination output and are emitted into the core if the required bit positions are still found empty (path X in figure 13).

Cells of any higher priority level are processed by high–priority IBCs (HPIBCs) consisting of the members of the higher priority groups upstream. Corresponding to the expected throughput ϱ_i of cells of a certain priority class i a group of s_i IBCs is composed for the service of that class. A number of s_i specialized bit positions on the serial line control the access to the HPIBCs of that class. Any LPIBC manages a number of s_i special sub–buffers for each priority class i (i>0). A LPIBCs receiving a cell of a prioritized connection (priority level i) stores that cell in one of its s_i sub–buffers for that priority level. If the LPIBC receives a corresponding specialized bit position as empty it sets the bit on line B and is now able to transmit the cell to one of the HPIBCs of that priority class i (feed–back path X' in figure 13). The HPIBC then stores the cell in one of its sub–buffers and re–transmits it to its destination output (path X") if it receives the necessary empty bit position.

This server based principle exhibits a number of advantages compared to "traditional" solutions for the problem of coordinating multiple priorities:

- An arbitrary mixture of priority classes can be handled by the same infrastructure.
- The number of necessary HPIBCs always has an upper limit: since each of the N LPIBCs can receive and emit at most one cell per T_{cell} a maximum number of N HPIBCs (of all possible server groups together) is needed. The number of bit positions is also extended not more than by N. In the case of a symmetric switch the bit rate on the serial lines has only to be doubled to handle a nearly arbitrary number of priority classes. Traditional approaches always require an additional arbitration round per priority class.
- All the necessary HPIBCs will not need more space for their implementation than a single line card of the switch since each of them requires only a very small buffer capacity: in the case of an increasing buffer length a HPIBC can send an internal backpressure signal to all the LPIBCs at once by presetting the specialized bit position that controls the access of the LPIBCs to its own input.
- The HPIBCs can be flexibly re–partitioned into more appropriately sized groups in order to adapt the system to fluctuations of the relative throughput of the single priority classes. Only the lines C and R have to be split or re–connected at predetermined joints and some schedulers have to be activated respectively deactivated. Since just the assignment of sub–buffers to priority classes within the LPIBCs has to be adapted it is not necessary to interrupt the operation of the whole system. Only the access to the HPIBCs under reconfiguration has to be stopped for a few cell durations by presetting the corresponding specialized bit positions.

In the case of multicast (MC) connections this server principle also makes it unnecessary to have a cell duplication entity in each of the IBCs. One group of HPIBCs is assigned to the task of serving multicast cells: they become MCIBCs. Normally this will be the highest priority group but also other solutions can be selected. The LPIBCs are informed about this fact by the configuration management and treat the cells of the MC connections using the same mechanisms as with the other priority classes: they are deflected into the corresponding sub–buffers and routed on the path X' through the core to the MCIBCs. Only there the duplication takes place as far as necessary (depending on the usage of different switch cores) and the copies are emitted into the core on path X" if the corresponding bit positions are available.

6 Implementation and Scalability

Most important for the scalability of the proposed approach is the number of switch ports that can be coordinated. Within one T_{cell} the frame with the bit positions (on the lines B and R) has to be completely clocked through all IBCs in order to control the emission of cells within the following T_{cell}. Assuming conservatively that the serial lines (and the serial parts of the IBCs) can be implemented with the same speed as the serial ports of the switch core a number of around 450 clock periods (length of an ATM cell plus internal header information) are available for this task. For the coordination of a symmetric switch (N input and output ports) with a medium number of HPIBCs and MCIBCs $1.5*N$ bits have to be shifted along $1.5*N$ IBCs. So in this case the maximum switch size will be N_{max}=450/3=150.

With a single serial line N_{max} is limited by the technology for connecting the serial lines to the IBC chips. A larger N_{max} therefore can be reached if parallel bundles are used. Each line within a bundle then conveys the bits belonging to a certain part of the switch outputs. The parallel lines within the bundle can mainly be processed in parallel. The only processing step that has to evaluate all of the lines of a bundle together is the reaction to coincidences: an IBC cannot set a number of bits (on all lines together) that exceeds the number of cells its input buffer can emit during T_{cell}. Thus the maximum number of coordinated ports is no longer limited by the interconnection technology used for the implementation of the serial lines. It is only limited by the maximum delay of a small number of gates within the IBC chip. So a much larger number for N_{max} is possible.

Another interesting point is the fact that not only a special family of switch cores (with its own very specific properties like speedup or buffering strategy) can be used. A great deal of the fundamental switch functions are now already done before the cells enter the core: e.g. the sorting of the cells corresponding to destination ports and the coordination of competing cells according to their priorities. Thus the switch core has only to deliver the pure connectivity function, i.e. to offer a way from any input to one or more outputs. This can also be implemented by using simple cross–bar or backplane solutions; with the cost–effective side effect that these core topologies offer the inherent capability of duplicating the cells within the core.

Summary and Future Activities

A new approach for the resolution of output port contentions in input buffered switches been presented that is based on the usage of MAC protocols of high–speed LANs. The requirements for the applied MAC protocol have been discussed and it has been shown that a modified version of CRMA–II provides a very high throughput and a complete fairness even under very asymmetric load scenarios. The proposed architecture can be easily extended to allow an arbitrary number of priority levels together with an effective coordination of multicast cells. Thus very large switches with throughputs of several Terabits per second can now be efficiently coordinated on the cell level.

The next step in the process of further examining this approach will be a VHDL–based hardware–level simulation of the critical parts of this topology (e.g. input buffer controller and scheduler) in order to further demonstrate its feasibility with common ASIC technologies. Also the investigation of two additional MAC protocols together with advanced simulations of the multicast and priority behavior are currently under way.

References:

Ahmadi, H. and Denzel, W. (1989): A Survey of Modern High–Performance Switching Techniques, IEEE Journal on Selected Areas in Communications, Vol. 7, No. 7, pp. 1091 – 1103

ATM Forum (1994): Reliability and Performance Considerations for ABR and VBR+ to support LAN Applications, Traffic Management Subworking Group, 94–0262

Bingham, B. and Bussey, H. (1988): Reservation–Based Contention Resolution Mechanism For Batcher–Banyan Packet Switches, Electronics Letters, 23 June 88, pp. 772 – 773

Conti, M. and Lenzini, L. (1991): A Methodological Approach to an Extensive Analysis of DQDB Performance and Fairness, IEEE J. on Sel. Areas in Comm. Vol 9 No 1, pp. 76 – 87

Garcia–Haro, J. and Jajszczyk, A. (1994): ATM Shared–Memory Switching Architectures, IEEE Network July / August 1994, pp. 18 – 26

Hluchyj, M. G. and Karol, M. J. (1988): Queueing in High–Performance Packet Switching, IEEE Journal on Selected Areas in Communications Vol. 6 No. 9, pp. 1587 – 1597

Hui, J. Y. and Arthurs, E. (1987): A Broadband Packet Switch for Integrated Transport, IEEE Journal on Selected Areas in Communications Vol. 5 No. 8, pp. 1264 – 1273

Karol, M. J. et al. (1987): Input Versus Output Queueing on a Space–Division Packet Switch, IEEE Transactions on Communications Vol. 35

LaMaire R. O. and Serpanos, D. N. (1994): Two–Dimensional Round–Robin Schedulers for Packet Switches with Multiple Input Queues, IEEE/ACM Trans. on Netw., pp. 471 – 482

Lemppenau, W. W. et al. (1993): ATM Implementation of the CRMA–II Dual Ring LAN and MAN, Proceedings of EFOC&N 93, The Hague, June 30 – July 2, 1993

Main, J. and Sarkies, K. (1995): Cell Scheduling Using Status Arrays in Input Buffered ATM Switches, Proceedings of the First IEEE Worksh. on Broadb. Switching Syst., Poznan, Polen

Matsunaga, H. and Uematsu, H. (1991): A 1.5 Gb/s 8x8 Cross–Connect Switch Using a Time Reservation Algorithm, IEEE Journal on Selected Areas in Communications, pp. 1308 – 1317

Obara, H. (1991): "Optimum Architecture for Input Queueing ATM Switches, Electronics Letters, 28th March 1991, pp. 555 – 557

Obara, H. and Hamazumi, Y. (1992): Parallel Contention Resolution Control for Input Queueing ATM Switches, Electronics Letters, 23rd April 1992, pp. 838 – 839

Simcoe, R. J. and Pei, T.–B. (1995): Perspectives on ATM Switch Architecture and the Influence of Traffic Pattern Assumptions on Switch Design, Computer Comm. Review, pp. 93 – 105

van As, H. et al. (1991): CRMA–II: A Gbit/s MAC Protocol for Ring and Bus Networks with Immediate Access Capability, Proceedings of EFOC/LAN 91, London, June 19–21, 1991

van As, H. and Lemppenau, W. W. (1992): Performance of CRMA–II: A Reservation–Based Fair Media Access Protocol for Gbit/s LANs and MANs, Proceedings of EFOC/LAN 92, Paris, June 22–24, 1992

Worster, T. et al. (1995): Buffering and Flow Control for Statistical Multiplexing in an ATM Switch, Proceeding of the ISS 95, Berlin, Germany, April 1995, Paper no. 488

Xerox Corp. (1993): A Switching Network, European Patent Application, Publication Number: 0 571 166 A2, 24 Nov 1993.

32

Throughput Analysis of Multiple Input-Queuing in ATM Switches†

Christos Kolias
Department of Computer Science
University of California at Los Angeles, Los Angeles, CA 90024, U.S.A.
email: kolias@cs.ucla.edu, tel.: +1 310 825-1563, fax : +1 310 825-2273

Leonard Kleinrock
Department of Computer Science
University of California at Los Angeles, Los Angeles, CA 90024, U.S.A.
email: lk@cs.ucla.edu, tel.: +1 310 825-2543, fax : +1 310 825-2273

Abstract

In this paper we investigate various schemes of input-queueing ATM switching systems. We first give a brief description of the *Odd-Even* switch model which is followed by an approximate analysis for evaluating its throughput. We then introduce an extension of the Odd-Even model which employs a *Multiple Input-Queueing* strategy, where an input port is expanded into m queues. In fact, we consider two policies as far as arbitration among the input queues is concerned and we show how thoughput can increase as m gets larger. We also comment on the special case where $m = N$, for an $N \times N$ switch, and show that the achieved throughput is actually 100%. We call this last scheme *Virtual Output-Queueing.* The models under examination assume a uniform output destination distribution and a Bernoulli process for the cell arrivals.

Keywords

Throughput, Odd-Even, Multiple Input-Queueing, ATM Switches.

1 INTRODUCTION

As we witness the proliferation of the ATM technology, it becomes quite evident the critical role switches are required to play to support the objectives of ATM networks. Switches, as an integral part of an ATM network not only support fast cell-relaying but also provide for network management, such as connection admission control, flow and congestion control, resource allocation, multicasting and other functions. This has allowed for various switch

†Supported by ARPA/CSTO under contract DABT63-93-C-0055, *The Distributed Supercomputer Supernet - A Multi Service Optical Intelligent Network.*

architectures to emerge and be proposed as standards. As a matter of fact, we cannot single out a particular switch design as a superior one, since each switching system is suitable, in terms of functionality, according to the specific network requirements. ATM switching systems have been covered quite extensively in the literature (Chen, 1991), (Onvural, 1994) and different schemes have been studied and characterized in terms of various aspects, such as location of buffers, switch fabric design, switching (i.e. time vs. space division), routing and congestion control (i.e. leaky bucket) and buffer management.

If cells are stored between the input ports and the switch fabric then the switching architecture is characterized as an input-queueing. On the other hand, if buffering of cells occurs between the switch fabric and the output trunks then the switch is classified as an output-queueing. A third option exists, where cell buffers are located inside the switch fabric (thus shared among the input and output ports), in which case this queueing discipline is described as internal or central queueing. Hybrid switch architectures exist adopting a combination or extension (i.e. recirculation) of these buffering strategies with the analogous cost in terms if complexity and management.

Input-buffered switches exhibit a blocking phenomenon known as *head-of-line* (HOL). HOL blocking occurs when more than one cell attempts to access the same output port. Also, since during the period of a time slot only one cell can be switched to the requested output, the remaining cells destined to the same output port can either remain buffered at their HOL positions (and try again in the next slot), thus blocking the cells queued behind them, or they can be simply discarded from the input queues, so no blocking occurs. As a result, HOL blocking can severely affect the throughput (which we define here as the utilization of the output ports averaged over the number of ports) of an input-queueing switch and normally large buffers are needed. For an ordinary input-buffered switch throughput is found to be $2 - \sqrt{2} \approx 0.586$ (Karol, 1987). In contrast, in output-buffering, each output is assigned a dedicated FIFO buffer where cells contending for the same output are all switched, within the same time slot, to the corresponding FIFO (but only one can be transmitted on the output link). Therefore, with output-queueing the ideal throughput of 100% can be actually achieved.

In this paper we demonstrate a few alternatives to ease the HOL blocking and consequently improve the switch throughput. The simulation results presented refer to the maximum attainable throughput (we consider a saturation point) for different values of N (switch size). The analytical results show the asymptotic ($N \to \infty$) throughput behaviour of the switch under investigation. Switches are modeled as multi-queue, multi-server, time-slotted queueing systems. Queues at the inputs are FIFOs and assumed of infinite size. In clarifying some of the terminology we will be occasionally using thoughout the paper, by *new* cells we mean any incoming cells that arrive at the input ports, while by *fresh* cells we describe these cells that move up to the HOL position (they were either queued or, if the FIFO was empty, they are new).

As for the organization of the paper : section 2 presents a rather concise description of the Odd-Even model and continues with an approximation analysis of the model. In section 3 we extend and generalize the Odd-Even model by introducing the Multiple Input-Queueing switch, which we study under two different arbitration policies. Finally, section 4 gives some concluding remarks.

*We would like to thank Larry Roberts of *Connectware Inc.* for bringing this model to our attention.

2 THE ODD-EVEN MODEL

2.1 Switch Description

In this subsection we briefly present the Odd-Even switch (a more thorough treatment can be found in (Kolias,1996)). We consider an $N \times N$ crossbar, nonblocking, input-buffered switch, where the innovation is that each input is split (expanded) into two FIFO queues which we call *Odd* and *Even* for this type of dual input queueing. By convention, we refer to outputs numbered as 2,4,6,... as even while to 1,3,5.... as odd ports. An incoming cell destined to an odd (even) numbered output port joins the odd (even) queue of the input port to which it was fed and waits for its turn to be switched to its output. Queued cells move up to the head of these queues that switched their HOL cells, so that they can get involved in the next output contention resolution cycles. Contention resolution for the HOL cells takes place in two consecutive contention resolution rounds. Arbitration during the first round involves the HOL cells at the even input queues. In the second round cells at the HOL of the odd input queues contend for the odd output addresses. However, those input ports whose even queues could not access an output port in the first round, because they either lost the contention (and therefore were blocked) or simply did not have any HOL cells present, are allowed to participate in a contention among their odd queues in the subsequent second round. In this fashion, an input port always gets a chance to route a cell either from an even or an odd queue, but not from both, within the same time slot. For fairness, if the even FIFOs were tried first in a time slot the odd ones are polled in the next one and vice versa.

Traffic intensity is equal for every input link (same traffic load is applied to each of the N input ports). The output address of an incoming cell is assigned randomly among the N output ports with equal probability. Both these assumptions characterize our system as a homogeneous one.

As far as the selection policy is concerned, the longest waiting in the HOL position cell is chosen among those contending for the same output, by the arbitration controller (when there is a tie the lowest numbered input queue is the winner). However, in the analysis that follows, we implicitly assume that the winning HOL cell is chosen randomly, since the selection policy has no bearing on the switch throughput (for this homogeneous system).

2.2 Approximate Analysis

Because of the various service dependencies introduced by the arbitration policy, where we assumed that even slots (during which even FIFOs can first transmit) are interchanged with odd slots (odd FIFOs can first transmit), an exact analysis of the switch's behaviour becomes intractable in terms of calculating the stochastic quantities involved. Therefore, we propose an approach where we analyze an approximate model, which is based on assuming a slightly modified arbitration policy. More specifically, we assume that all time slots are characterized as even, which means that the even FIFOs are given some priority in transmitting their HOL cells. The intuition behind this approach is that unclaimed output ports will still get the opportunity to receive (serve) HOL cells, given that output addresses are uniformly distributed. Due to this modified service policy we expect that the throughput achieved by the even output ports will be higher than the one for the odd outputs.

We make the following distinction among the N input ports: we call an input port *available* (for the second round) if no HOL cell was switched from its even FIFO (either there was not one or it lost the contention). Conversely, it is *unavailable* if the corresponding even FIFO switched a cell through one of the even output ports. We denote the throughputs achieved by the even and the odd output ports, as γ_E and γ_O respectively.

We first derive an exact expression for γ_E and then approximate γ_O, where, as we will see, the latter actually depends on γ_E. Our approach is similar to the one described in (Hui, 1987). We focus on an even numbered output j (the "tagged" output) and we denote by N_j^E the number of HOL cells at the even queues destined for output j. In steady state, γ_E can be defined as :

$$\gamma_E = \lambda_E \frac{2}{N} \sum_{j \text{ is } Even} E[\epsilon(N_j^E)] = E[\epsilon(N_j^E)], \tag{1}$$

where λ_E is the arrival rate of new cells at the even queues, assuming a non-saturation situation and $\epsilon(x)$ is an indicator function ($\epsilon(x) = 0$ if $x \leq 0$, $\epsilon(x) = 1$ if $x > 0$). Note that the even HOL cells can be switched from any of the N (even) input queues to any of $N/2$ even output ports.

Let also N_b^E denote the total number of HOL cells blocked (i.e. lost the contention) at the even queues after the first contention round. Then :

$$N_b^E = \sum_{j \text{ is } Even} N_j^E - \sum_{j \text{ is } Even} \epsilon(N_j^E). \tag{2}$$

Taking now expectations in (2) and combining it with (1), we get, by symmetry :

$$\lambda_E = \gamma_E = E[N_j^E] - \frac{2E[N_b^E]}{N}. \tag{3}$$

Let ρ_E be the steady-state probability that there is a fresh even cell at the HOL position of an even queue given that the HOL cell has departed and M_E denote the number of released even HOL positions (those that had a cell switched) at the end of the first contention round, then clearly :

$$M_E = N - N_b^E. \tag{4}$$

The flow conservation relationship describing the system for the even queues states :

$$E[M_E]\rho_E = \frac{N}{2}\lambda_E, \tag{5}$$

where intuitively $\frac{E[M_E]\rho_E}{N/2} = \lambda_E$ is simply the expected fraction of the busy even servers (output ports), which defines the throughput of the even outputs. By taking expectations on both sides in (4) and using (5) we get :

$$\frac{E[N_b^E]}{N} = 1 - \frac{\lambda_E}{2\rho_E}. \tag{6}$$

Let now K_j^E denote the number of HOL cells destined to output j, in the next time slot, then :

$$K_j^E = N_j^E - \epsilon(N_j^E) + A_j^E, \tag{7}$$

where A_j^E is the number of fresh HOL cells destined for output j. Forming the expectations on both sides of (7) and considering the steady-state case we have :

$$E[K_j^E] = E[N_j^E]. \tag{8}$$

Taking into account (1) and (8), (7) yields[†] :

$$E[A_j^E] = E[\epsilon(N_j^E)] = \lambda_E. \tag{9}$$

In finding $E[N_j^E]$ we first square (7) as follows :

$$(K_j^E)^2 = (N_j^E)^2 + \epsilon(N_j^E) + (A_j^E)^2 - 2N_j^E + 2N_j^E A_j^E - 2\epsilon(N_j^E)A_j^E,$$

where by taking expectations and then using (8) and (9) we further get :

$$E[N_j^E] = E[A_j^E] + \frac{E[A_j^E(A_j^E - 1)]}{2(1 - E[A_j^E])}. \tag{10}$$

In determining $E[A_j^E(A_j^E - 1)]$, as in (Hui, 1987), we can argue that A_j^E becomes Poisson(λ_E) as $N \to \infty$[§], thus we have :

$$E[A_j^E(A_j^E - 1)] = \lambda_E^2, \tag{11}$$

and $E[N_j^E]$ can be directly expressed as :

$$E[N_j^E] = \lambda_E + \frac{\lambda_E^2}{2(1 - \lambda_E)}. \tag{12}$$

Applying (6) and (12) in (3), we further obtain :

$$\lambda_E \;=\; \lambda_E + \frac{\lambda_E^2}{2(1 - \lambda_E)} - 2(1 - \frac{\lambda_E}{2\rho_E}),, \tag{13}$$

which leads to the following equation :

$$(2 - \rho_E)\lambda_E^2 - 2(2\rho_E + 1)\lambda_E + 4\rho_E = 0. \tag{14}$$

Since we are interested in maximizing throughput, then by letting $\rho_E = 1$ (i.e. all even HOL

[†] Also derived from first principles about flow conservation.

[§] Note that $N \to \infty$, practically means a switch with sufficiently large N, i.e. close or larger than 100.

positions are occupied), we get :

$$\lambda_E = 3 - \sqrt{5} \approx 0.764, \tag{15}$$

which, again, represents the throughput achieved by the even output ports.

It now remains to obtain the throughput of the odd output ports. Because not all the input ports are available for the second contention round it is clear that the throughput of the odd output ports can be potentially limited due to that unavailability factor. Let δ be the fraction of the N input ports that are available for the second arbitration round, then δN is the expected number of odd queues that are allowed to switch a HOL cell to an odd destination port.

Setting up the same equations as we did for evaluating λ_E, where i is now the "tagged" odd output, we have :

$$\gamma_O = \lambda_O = \frac{2}{N} \sum_{i \text{ is } Odd} E[\epsilon(N_i^O)] = E[\epsilon(N_i^O)], \tag{16}$$

or

$$\gamma_O = \lambda_O = E[N_i^O] - \frac{2E[N_b^O]}{N}, \tag{17}$$

where λ_O is the input arrival rate at the odd queues and N_i^O is the number of available odd HOL cells destined to (odd) output i and N_b^O is the number of blocked odd HOL cells. Then, by letting M_O be the number of odd ports that became free at the end of the second contention round, we have :

$$E[M_O] = \delta N - E[N_b^O]. \tag{18}$$

The flow conservation for the odd outputs is expressed as follows :

$$E[M_O]\rho_O = \frac{N}{2}\lambda_O. \tag{19}$$

Then, by taking expectations in (18) and using (19) we have :

$$\frac{E[N_b^O]}{N} = \delta - \frac{\lambda_O}{2\rho_O}. \tag{20}$$

If K_i^O is the r.v. for the number of available odd HOL cells destined for output i in the next time slot and A_i^O is the number of fresh odd HOL cells then, in general, we cannot claim that :

$$K_i^O = N_i^O - \epsilon(N_i^O) + A_i^O, \tag{21}$$

since queues that were available may become unavailable in the next slot (after the even queues arbitration round) while any of those that were unavailable may become available.

However,

$$\sum_{i \text{ is odd}} K_i^O = \sum_{i \text{ is odd}} N_i^O - \sum_{i \text{ is odd}} \epsilon(N_i^O) + \sum_{i \text{ is Odd}} A_i^O + R - B \tag{22}$$

where R denotes the number of input ports that become available in the next time slot while B represents the number of input ports that become unavailable (and which may include any fresh HOL cells). At steady state the expected number of unavailable input ports should remain the same, therefore $E[R] = E[B]$. Now, taking expectations in (22) we have :

$$\sum_{i \text{ is Odd}} E[K_i^O] = \sum_{i \text{ is Odd}} E[N_i^O] - \sum_{i \text{ is Odd}} E[\epsilon(N_i^O)] + \sum_{i \text{ is Odd}} E[A_i^O], \tag{23}$$

and we can make the assumption that, by symmetry :

$$E[K_i^O] = E[N_i^O] - E[\epsilon(N_i^O)] + E[A_i^O], \tag{24}$$

and since $E[K_i^O] = E[N_i^O]$, (24) then implies :

$$E[A_i^O] = E[\epsilon(N_i^O)] = \lambda_O. \tag{25}$$

Proceeding as before, we can approximate the distribution of A_i^O by a Poisson(λ_o) distribution as $N \to \infty$ and conclude that :

$$E[N_i^O] = \lambda_O + \frac{\lambda_O^2}{2(1-\lambda_O)}. \tag{26}$$

By substituting (20) and (26) in (17) we get :

$$\lambda_O = \lambda_O + \frac{\lambda_O^2}{2(1-\lambda_O)} - 2(\delta - \frac{\lambda_O}{2\rho_O}) \tag{27}$$

which, for $\rho_O = 1$, yields :

$$\lambda_O = 1 + 2\delta - \sqrt{4\delta^2 + 1}. \tag{28}$$

Now, since δ denotes the fraction of those input ports that are available at the end of the first arbitration round, then we can express it, using (6), as

$$\delta = 1 - \frac{\lambda_E}{2\rho_E}. \tag{29}$$

Then from (28) and (29) and for $\rho_E = 1$ we finally get $\lambda_O \approx 0.646$, which is the throughput for the odd output ports.

Note, that in (28) for $\delta = 1$ (i.e. all input ports are available at the beginning of the second round) $\lambda_O = 3 - \sqrt{5}$ and for $\delta = 0$, obviously, $\lambda_O = 0$. Also, for $\delta = \frac{1}{2}$, ($N/2$ input ports are available), we get $\lambda_O = 2 - \sqrt{2}$, which of course agrees with previous analysis (Karol, 1987) (i.e. consider an $\frac{N}{2} \times \frac{N}{2}$ switch).

We also observe that $1 + 2\delta - \sqrt{4\delta^2 + 1}$ is an increasing function of δ and from (13) we actually have $\delta = \frac{\lambda_E^2}{4(1-\lambda_E)}$ which is maximized when λ_E is maximized, namely for $\rho_E = 1$, which yields $\lambda_E = 3 - \sqrt{5}$. Thus, $\rho_E = \rho_O = 1$ yields the maximum achievable throughput for both the even and the odd outputs and therefore for the switch.

Since we now have the throughput for both the even and the odd output ports, we can finally find, as an approximation, the throughput for the whole system, namely the Odd-Even switch, by normalizing λ_E and λ_O, as follows :

$$\gamma_s = \frac{\lambda_O + \lambda_E}{2} \approx 0.705. \tag{30}$$

We notice that $\gamma_s \approx 0.705$ is in a very good agreement (within 1% range) to the simulation result of 0.713 for the Poisson, uniform traffic case (Kolias,1996).

3 MULTIPLE INPUT-QUEUEING SWITCHES

In this section we present an extension of the Odd-Even model. Assuming an $N \times N$ single-stage, input-buffered switch, instead of having only two FIFOs (odd and even) per input port, we allow m queues per input port, where $m \leq N$ (the switch size). It is essential to emphasize here the distinction between an input FIFO queue and an input port. As in the Odd-Even model, output ports are partitioned into m groups (it is not particularly important how we allocate the output ports to the groups, as long as we assume uniformity of the output addresses). That means each of the m queues is associated with one or more (depending on N) output ports, so that an incoming cell joins a particular queue according to its output port destination. We realize that the larger the m is, the more complex the system becomes, in terms of implementation of the arbitration scheme. There is no extra cost induced regarding the additional buffers, since each input port buffer (of size i.e. b) can be throught as being partitioned into m smaller buffers (of size b/m).

We study this *Multiple Input-Queueing* system in two different versions with respect to the arbitration policies.

3.1 Policy A

The arbitration process, under this policy, is a direct extension of the one mentioned in the Odd-Even scheme and takes place in m consecutive contention rounds (within the same time slot). Therefore, it can be considered as a generalization of the Odd-Even scheme. At the i-th round, only FIFOs that are available (whose input ports have not switched cells during the previous i-1 rounds) can compete for output ports. This kind of policy can limit the switch throughput, since even if there are HOL cells destined to a particular output, that output might stay idle. However, we notice from the simulation results (figure 1) that as m grows to infinity, the asymptotic throughput approaches 1 very quickly (always under the assumption of saturation at the inputs).

In fact, for $m = N$, that is, the number of FIFOs per input port is equal to the total number of outputs, throughput is 1. Basically, in this form of input-queueing each input port has a dedicated buffer for each of the outputs. Therefore, the achieved throughput is

100% since, effectively, there is no HOL blocking. This type of buffering has exactly the same result as the pure output buffering scheme, where there is no blocking whatsoever and cells are switched immediately to their corresponding output buffers. We call this type of input queueing, where essentially output buffering is emulated by buffers at the inputs, *Virtual Output-Queueing.*

Figure 1 illustrates the simulation behaviour of this general model, for different values of m, namely for m=1 (the ordinary input-buffered switch, (Karol, 1987)), m=2 (the Odd-Even switch), and m=5, 10, 20, 50 and 100. We see that for $m > 10$ we obtain diminishing gains in throughput and most notably when $m = 50$ is doubled.

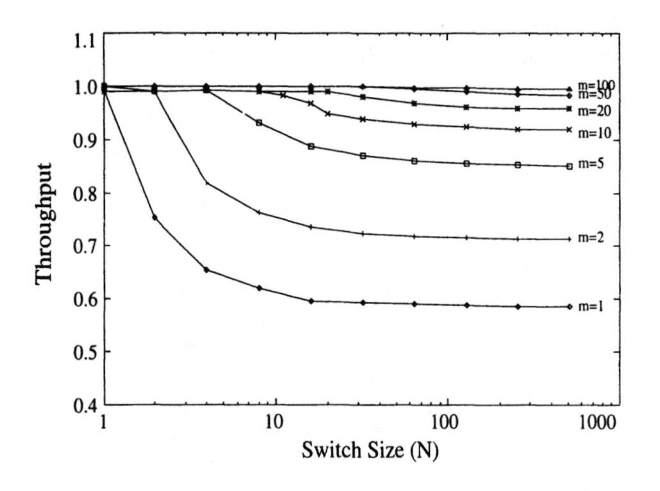

Figure 1 Throughput of a Multiple Input-Queueing Switch under policy A (simulation).

3.2 Policy B

Here we study a variation of policy A, where we allow all N input FIFOs (not only from those input ports that did not transmit a cell during the previous rounds) to participate in each of the arbitration rounds. Actually since we do not distinguish between available and unavailable input ports (all input ports are available), we can assume that the contention for the outputs is completed in just one round, where all mN input FIFOs can participiate. In other words, we can view this system as an $mN \times N$ switch where the N inputs are expanded into mN and a new cell joins a particular FIFO queue, depending on the cell's output address. Under this policy the mN queues are considered to be indepedent.

Let us now concentrate on analytically obtaining the switch throughput for this type of multiple input queueing. As in subsection 2.2, let γ be the total throughput of the switch and N_j denote the number of HOL cells that are destined to a "tagged" output j. Then, at steady-state, we have :

$$\gamma = \lambda = \frac{1}{N}\sum_{j=1}^{N} E[\epsilon(N_j)] = E[\epsilon(N_j)], \tag{31}$$

where λ is the probability that a new cell arrives at an input port at the begining of a time slot.

If N_b is defined as the number of all HOL cells that became blocked at the end of a slot (after the arbitration phase is over), then clearly :

$$N_b = \sum_{j=1}^{N} N_j - \sum_{j=1}^{N} \epsilon(N_j). \tag{32}$$

Taking expectations on both sides in (32) and using (31) we can then rewrite throughput as follows :

$$\lambda = \gamma = E[N_j] - \frac{E[N_b]}{N}. \tag{33}$$

Recall that, if M represents the number of unblocked FIFOs then :

$$E[M] = mN - E[N_b]. \tag{34}$$

The flow conservation rule implies :

$$E[M]\rho = N\lambda, \tag{35}$$

where ρ is again the steady-state probability that there is a fresh cell occupies the HOL position following the departure of the HOLL cell. Taking into account (34) and (35), (33) can be modified as :

$$\lambda = E[N_j] - (m - \frac{\lambda}{\rho}). \tag{36}$$

If K_j denotes the number of HOL cells destined to the j-th output port during the next time slot and by letting A_j be the number of the "fresh" HOL cells destined to j, then :

$$K_j = N_j - \epsilon(N_j) + A_j. \tag{37}$$

Taking expectations in Eq. (37) then, since at steady-state $E[K_j] = E[N_j]$, we get :

$$\lambda = E[\epsilon(N_j)] = E[A_j]. \tag{38}$$

By squaring Eq. (37) and taking expectations we finally obtain :

$$E[N_j] = E[A_j] + \frac{E[A_j(A_j - 1)]}{2(1 - E[A_j])}. \tag{39}$$

As we mentioned before, as $N \to \infty$, A_j becomes Poisson(λ), thus :

$$E[A_j(A_j - 1)] = \lambda^2, \tag{40}$$

then (39) becomes :

$$E[N_j] = \lambda + \frac{\lambda^2}{2(1-\lambda)}. \tag{41}$$

By substituting (41) in (36) we get :

$$\lambda = \lambda + \frac{\lambda^2}{2(1-\lambda)} - (m - \frac{\lambda}{\rho}), \tag{42}$$

where for $\rho = 1$, solving (42) we find :

$$\lambda = m + 1 - \sqrt{1 + m^2}, \tag{43}$$

which is the maximum throughput of the m-FIFO multiple input queueing switch.

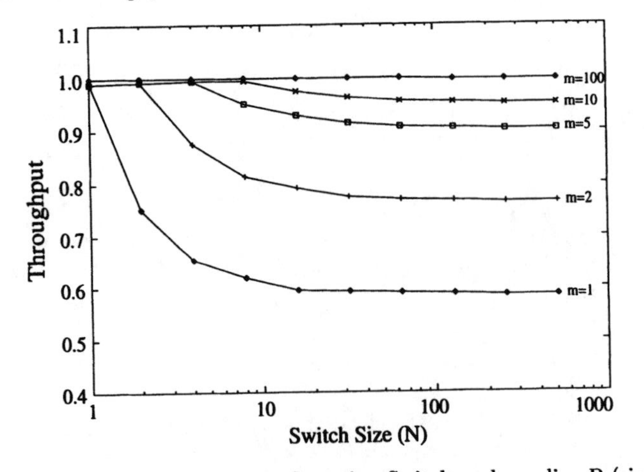

Figure 2 Throughput of a Multiple Input-Queueing Switch under policy B (simulation).

Figure 2 shows simulation results for switches with m =1,2,5,10 and 100. Applying the result of (43) for these values of m the consistency between the theoretical and the simulation results becomes obvious.

Note that for $m = 1$, (43) yields $\lambda = 2 - \sqrt{2}$, while for $m = 2$, $\lambda = 3 - \sqrt{5}$ (see (15)). Also, in the limit, as $m \to \infty$, (43) yields $\lambda = 1$. Let's give the physical explanation of this result: as $m \to \infty$ then also $N \to \infty$ which practically means that there are as many FIFOs per input port as the total number of output ports, namely $m = N$. For $m = N$, policies A and B have ultimately the same effect, in terms of throughput, and under a saturation situation; so our remarks in the previous subsection are reinforced here.

4 CONCLUSION

It is apparent that buffering (location, size, contention resolution) in an ATM switching system is of great importance to the switch designer, in terms of complexity, efficiency and certainly cost (i.e. hardware implementation). In this paper we were concerned with the performance efficiency of input-buffered switches, acknowledging that the other factors are significant too. In that respect, we studied various schemes using multiple input-queueing as their buffering strategy by analyzing their performance in terms of throughput and demonstrated how throughput can be improved.

5 REFERENCES

Chen, T. and Liu, S. (1995) ATM Switching Systems. Artech House, Norwood, MA. *

Hluchyj, M. G. and Karol, M. J. (1988) Queueing in High-Performance Packet Switching. *IEEE Journal on Selected Areas in Communications*, **6**(9), pp. 1587-97.

Hui, J. Y. and Arthurs, E. (1987) A Broadband Packet Switch for Integrated Transport. *IEEE Journal on Selected Areas in Communications*, **SAC-5**(8), pp. 1264-73.

Karol, M.J., Hluchyj, M. G. and Morgan, S.P. (1987) Input versus Output Queueing on a Space-Division Packet Switch. *IEEE Transactions on Communications*, **COM-35**(12), pp. 1347-56.

Kolias, C. and Kleinrock, L. (1996) Throughput Performance of the Odd-Even Input Queueing ATM Switch. Accepted at the *International Conference on Communications (ICC) '96.*

Kleinrock, L. (1975) Queueing Systems, Volume 1: Theory. Wiley-Interscience, New York.

Onvural, R. (1994) Asynchronous Transfer Mode Networks: Performance Issues. Artech House, Norwood, MA.

33

Architectural and Performance Aspects of OFDM Nodes for the Future Transport Network

R.Osborne[1], M.N. Huber[1], B. Edmaier[2]
[1] Siemens AG, ÖN ME 11, Hofmannstr. 51, D-81359 Munich, Germany,
phone: +49-89-722-23667, fax.: +49-89-722-27787,
e-mail: manfred.huber@oenzl.siemens.de
[2] Technical University of Munich

Abstract
For the transport of the ever increasing traffic volume, an enhancement of the existing transport network will be required. For this purpose optical cross-connects and add/drop multiplexers based on the optical frequency division multiplexing (OFDM) principle will be added to the existing transport network. In this paper different OFDM node architectures are evaluated with respect to the influence of the size of the switching network on attenuation and cross-talk, the type of optical frequency converter and suitability for multicast connections, and other factors. Various scenarios for the interconnection of these optical cross-connects with the existing electrical cross-connects are discussed. For the individual optical cross-connect architectures and the interconnection scenarios, a study on the blocking probabilities for new optical connection requests was made, taking into account a suitable connection admission control mechanism and different channel hunting strategies. The performance evaluation showed that a savings of optical frequency converters in an optical cross-connect is possible compared to the nonblocking case, without deteriorating the performance significantly. However, other relevant topics such as connection management or implementation of the switching network become more complex.

Keywords: Optical cross-connect, system architecture, performance evaluation, connection acceptance control, channel hunting strategy

1 INTRODUCTION

Currently the traffic volume for narrowband service is still increasing. This increase will be strongly intensified by the evolving broadband communication. This requires a communication network which provides multigigabit transmission and cross-connects with very high throughputs. The existing transport network which consists of SDH (synchronous digital hierarchy) and ATM (asynchronous transfer mode) cross-connects (CCs) must therefore be enhanced. For this purpose new optical cross-connects (OCCs) with optical frequency division multiplexing (OFDM) can be deployed (Figure 1). This enhanced network provides many new features and benefits as discussed in (Brackett, Derr, Fioretti, Sato). In the core area, a meshed network with OCCs and electrical cross-connects (ECCs) will be deployed. In localized areas, ring networks may be sufficient which provide access to the core network. The network is under control of an appropriate network management.

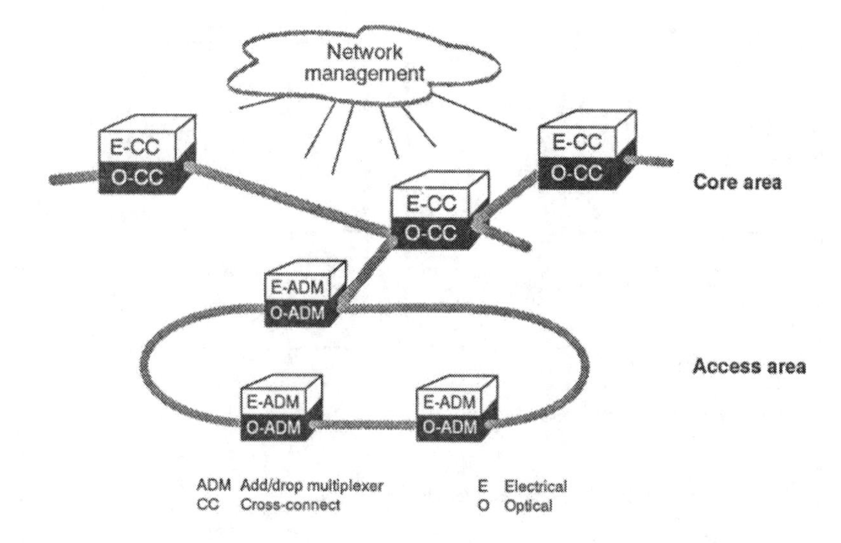

Figure 1 The future transport network.

The OCCs provide an efficient cross-connecting of multigigabit streams (e.g. 2.5 Gbit/s and beyond) which are carried by optical frequency channels. These multigigabit streams will carry several SDH containers or ATM connections, however, optical cross-connecting is only performed for the individual optical channels. Therefore, only the optical frequency has to be evaluated; there is no need for processing the information stream carried by that optical frequency channel. Splitting such a multigigabit stream into smaller streams and cross-connecting of small and moderate bit rate streams are performed in the SDH or ATM CCs. Therefore, even in the future, SDH/ATM CCs will be required and will be operated together with the OCCs in a complementary manner. This new solution will be more economically attractive compared to a network deploying only ECCs.

In Section 2 different architectures of OFDM nodes are presented. Section 3 discusses the very important aspect of interconnecting OCCs with ECCs. Section 4 deals with the performance evaluation of the different node architectures discussed in Section 2 and different scenarios of Section 3.

2 OFDM NODES

This part deals with the basic architecture of OFDM nodes required for the information transfer. Other architectural aspects such as management or operation and maintenance (Derr) have to be taken into account for a real system implementation but are not discussed here.

Figure 2a depicts the simplest form of an OFDM node. At the input the demultiplexer splits the optical signal of the input fiber into its individual optical channels. At the output the multiplexer combines the optical channels into the resulting output signal.

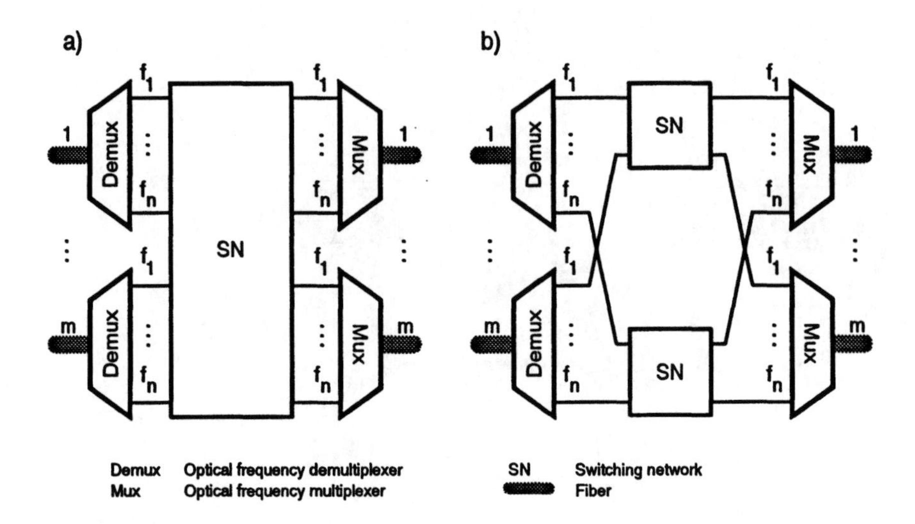

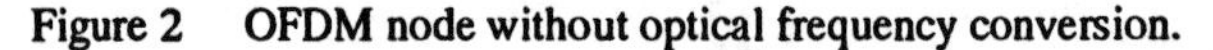
Figure 2 OFDM node without optical frequency conversion.

The switching network is able to connect any input channel to any output channel. However, for this node type the switching network can be simplified because an input channel at optical frequency f_i can only be connected to an output channel at the same frequency.

Figure 2b shows the simplified architecture for the same node size (*m* fibers, each carrying *n* optical channels). Table 1 compares the switching networks of both architectures.

If the switching network is implemented using free-space optics, e.g. (Astarte), cross-talk and attenuation are independent of the switching network size. With all other technologies Figure 2b has benefits in terms of attenuation and cross-talk because these values deteriorates with growing switching network size.

In such an OFDM node, all output channels of a multicast connection must employ the same optical frequency which results in higher blocking.

Both architectures shown in Figure 2 suffer from the lack of optical frequency conversion which results in higher blocking probabilities. To enhance these nodes with optical frequency conversion, the architecture in Figure 2a has definite advantages. The optical frequency converter can be simply integrated by connecting it between the switching network and the multiplexer or between the demultiplexer and the switching network. For the system shown in Figure 2b the enhancement requires modifications to the switching network.

Table 1: Comparision of OFDM nodes without optical frequency conversion

	Figure 2a	Figure 2b
Number of switching networks	1	n
Size of switching networks	nm x nm	m x m
Total number of crosspoints	n^2m^2	nm^2

Figure 3 depicts an OFDM node where an optical frequency converter is available for each optical channel. Only this node provides the internal non-blocking property and therefore, in the strict sense, only this node can be called a cross-connect. However, in the literature the nodes of the optical transport network are called optical cross-connects even if they do not satisfy the internal non-blocking property. For the rest of this paper we will use this definition also.

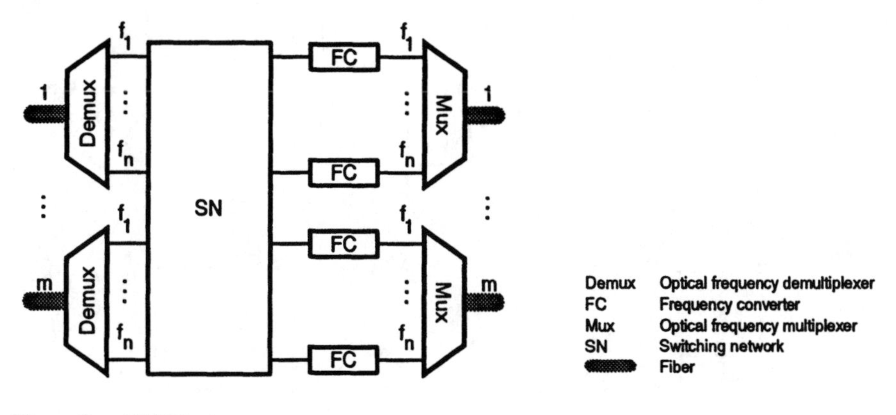

Figure 3 OFDM cross-connect.

In the OCC shown in Figure 3 the optical frequency converison is located between the switching network and the multiplexer. Thus, the optical frequency converter requires only a fixed output frequency. The converter is assumed to have a broadband optical input which accepts an arbitrary input frequency. In principle the optical frequency conversion could be located between the demultiplexer and the space divsion network. But this configuration would have some drawbacks:

- The optical frequency converter needs selectable output frequency resulting in higher realization complexity.
- For a multicast connection all output channels would be forced to use the same optical frequency resulting in higher blocking probabilities.

The system shown in Figure 3 may be an expensive solution because an optical frequency converter for each optical channel is used. To reduce the number of optical frequency converters while still providing low blocking, sharing of optical frequency converters is possible, e.g. (Lee).

A pool of optical frequency converters can be shared either among the optical channels of the output fiber or the input fiber. A combination of both is possible, too. The statistical gain of these principles is low and therefore will not be considered further in this paper.

Figure 4 shows two different solutions for OCCs where k optical frequency converters are shared among all nm optical channels (share per node architecture). In comparison to the OCC shown in Figure 3 these OCCs require a smaller number of optical frequency converters, but the expense for space division switching increases, resulting in higher attenuation for most switching network technologies as mentioned above.

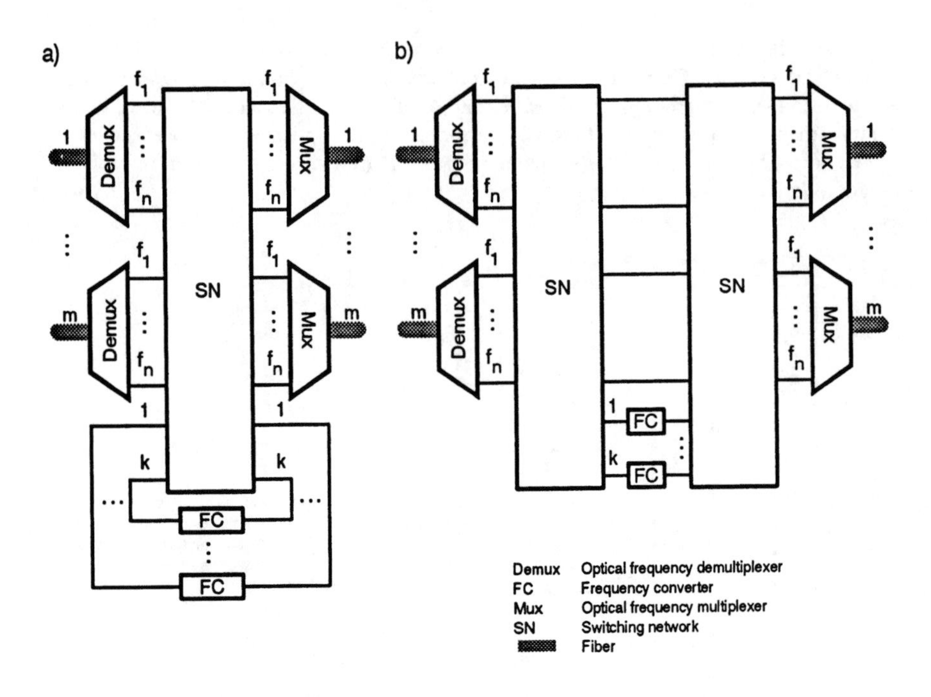

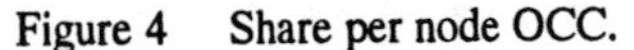
Figure 4 Share per node OCC.

Table 2 compares the switching expense for the two alternative solutions of share per node OCCs. The total number of crosspoints is lower in the system of Figure 4a. However, this system has the drawback that connections requiring optical frequency conversion have to pass through the switching network twice whereas all other channels pass through the network only once resulting in different power levels. In the system of Figure 4b all channels have to pass through two large switching networks. Additional compensation of the attenuation of the second switching network is necessary for all optical channels.

The optical frequency converters may either have variable output frequency or they may be split into *n* groups where each group consists of k/n optical frequency converters with fixed output frequency.

This share per node OCC can also support multicast connections. Multicasting must be done before optical frequency conversion (cf. OCC with one optical frequency converter per channel which is located between the demultiplexer and the switching network).

Table 2: Comparison of share per node OCCs

	Figure 4a	Figure 4b
Number of switching networks	1	2
Size of switching networks	$(nm+k) \times (nm+k)$	$nm \times (nm+k)$
Total number of crosspoints	$(nm+k)^2$	$2nm \times (nm+k)$

3 INTERCONNECTION OF OCCs WITH ECCs

In Section 2 different OCC architectures were discussed which are accessed only by OFDM links. The OFDM links connect OCCs to other OCCs but they may also be used for the connection of ECCs with the OCC. This principle requires some modifications to the ECC:

- The transmitter lasers must fulfill the requirements of the OFDM system.
- OFDM multiplexers and demultiplexers must be installed.
- The supervision of OFDM links must be supported.

Figure 5 shows an alternate solution for connecting an ECC with an OCC. Input/output lines (each line carries one channel) of an ECC are directly connected to the switching network of the OCC, e.g. (Chang). This avoids the need for OFDM multiplexers and demultiplexers in the ECC. This principle can be used for all OCC architectures discussed in Section 2 but the figure shows only the example of interconnecting an ECC with a share per node OCC.

For an OCC without wavelength conversion the lasers of the ECC transmitters have to fulfill the requirements of the OFDM system (e.g. channel allocation scheme), otherwise interconnection is impossible. The ECC inputs employ broadband optical detectors which can receive a signal at an arbitrary optical frequency.

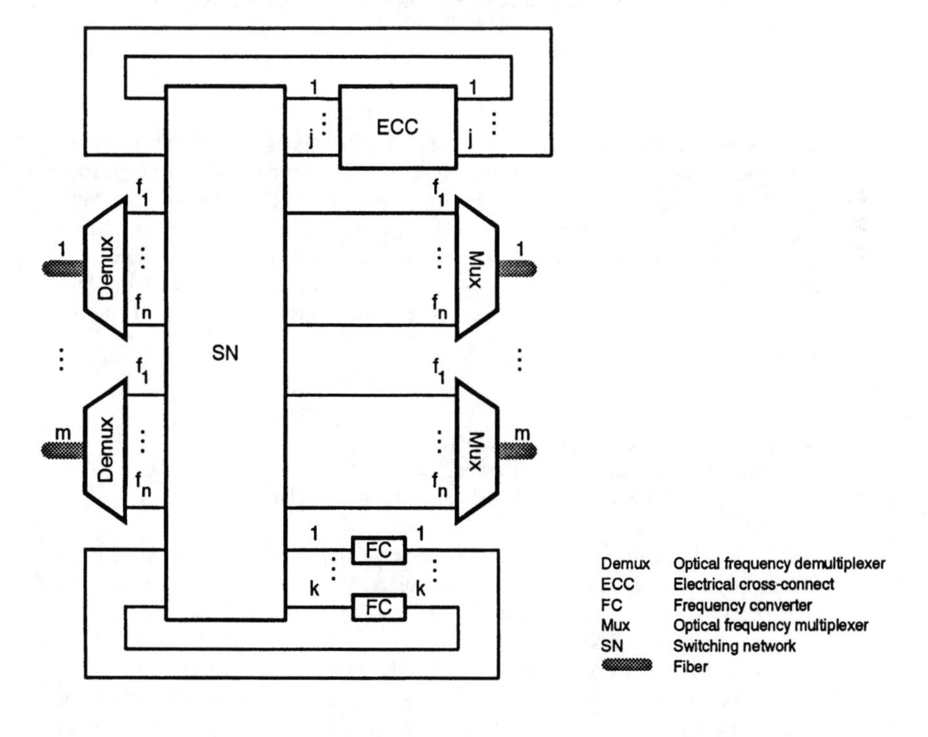

Figure 5 Example of interconnecting an OCC with an ECC.

For an OCC providing optical frequency conversion for all channels leaving the OCC on OFDM links, the lasers of the ECC do not need to conform with the channel allocation scheme of the OFDM system. Therefore, the modification efforts in the ECC can be kept low and replacement of the transmitter lasers may be unnecessary.

For the share per node OCC two possibilities exist for connecting the ECC:

1. Use of previously installed lasers which may not conform to the OFDM channel allocation scheme: Clearly the pool needs at least j suitable optical frequency converters with j = number of channels between the ECC and OCC.
2. Use of lasers satisfying the OFDM system requirements: If all lasers of the ECC have variable output frequency they can be adjusted to the required optical frequency on the outgoing fiber of the OCC and thus optical frequency conversion in the OCC can be avoided. If the lasers of the ECC have only fixed output frequency, situations may occur where optical frequency conversion in the OCC is required.

4 PERFORMANCE EVALUATION

For the performance studies, an event-by-event simulation of optical connection allocation in the OCC was performed. Set up and release of individual SDH and ATM connections were not taken into account.

4.1 Definitions

The blocking behaviour for establishing new optical connections carried by an optical frequency channel between the node input and output will be evaluated. A traffic generator producing connection requests and releases is necessary. One traffic generator per "input bundle" will be used. An input bundle may be a single optical channel, a fiber carrying several optical channels or a group of fibers. For this model it is assumed that new connection requests will only occur as long as at least one optical channel of the input bundle is not occupied. This results in a state dependent arrival process. The arrival rate is:

$$\lambda(x) = \begin{cases} \lambda = \text{const for } x = 0, ..., M\text{-}1 \\ 0 \text{ for } x = M \end{cases}$$

λ is constant and negative exponentially distributed.
M is the number of optical channels per input bundle.
x is the instantaneous number of occupied channels per input bundle

The connection holding time is also negative exponentially distributed. However, traffic generators with other characteristics can easily be implemented. The traffic generator also determines the destination of the new connection request. The destination may, for example, be a single output fiber or a bundle of fibers.

When a new connection request occurs, a check is made for at least one idle optical channel in the destination direction. If none is available, the request is rejected (output blocking). If a destination channel is available, then, in case of a non-blocking node the request will be accepted). In case of a blocking node, situations will occur where it is impossible to connect an idle channel of the input bundle to any idle channel of the output bundle due to the lack of an available optical frequency converter (internal blocking).

4.2 Connection Request Handling

This section describes the procedure for processing a new optical connection request. Two main steps are performed:

1. Searching for an idle optical channel in the destined output bundle.
2. Searching for an internal path to put through the connection from the originating input bundle to the destined output bundle.

For a node with internal blocking, the second step influences the node performance. Therefore, a proper design for this mechanism is required. The proposed solution is:

1. Are pairs of input/output channels with identical optical frequency available?
   Yes: choose one pair according to the selected channel hunting strategy and accept the request.
   No: go to step 2.
2. Are suitable optical frequency converters available which can transform the optical frequency of any idle input channel to the optical frequency of any idle output channel?
   Yes: choose an optical frequency for the input channel as well as the output channel according to the selected channel hunting strategy and accept the request.
   No: the connection request has to be rejected.

For channel hunting well-known strategies from link systems of the old telephony world (Bazlen) can be adapted. The basic mechanism is:

1. Sort all channels of a bundle in ascending order (ch_0, ..., ch_z, with z = total number of channels)
2. Use one of the possible searching strategies:
   - Ordered search, starting at a fixed point (e.g. ch_0)
   - Ordered search, starting at a random point

4.3 Results

For each incoming fiber, optical connection requests are modelled by an random source as defined in Section 4.1. Subsequently the connection request is processed according to the mechanism described in Section 4.2. Performance evaluation was only done for point-to-point connections because during the introduction phase these connections will predominate.

As an example, the blocking probability for optical connections is evaluated as a function of the number of available converters for a node with 8 input/output fibers each carrying 8 optical channels (this is a realistic assumption [der]). The output fibers are symmetrically loaded with equal connection probabilities from all input to output fibers.

Figure 6 depicts the different blocking probabilities (internal, output, total) versus the offered load per input fiber for the OCC without optical frequency conversion ($k = 0$), for OCC with frequency conversion per channel ($k = 64$) and the share per node architecture with variable optical frequency converters ($0 < k < 64$). The figure clearly indicates that the node can benefit from optical frequency conversion.

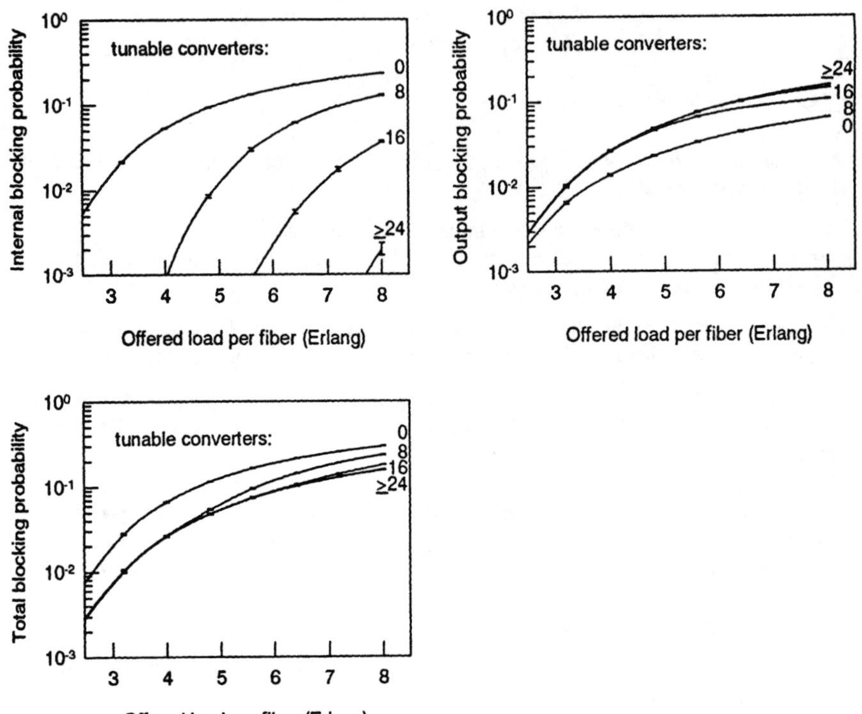

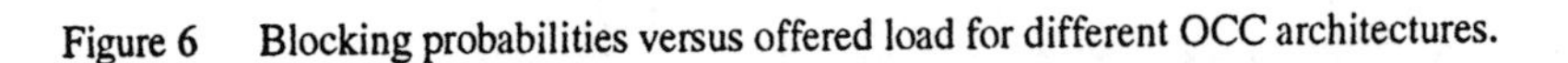

Figure 6 Blocking probabilities versus offered load for different OCC architectures.

The internal blocking probability decreases with an increasing number of optical frequency converters. Output blocking is actually seen to rise with the increasing number of converters. This is in response to the decreased internal blocking which allows a greater number of connections to be successfully made, thus raising the probability that the output channels are fully occupied. However, as expected, the total blocking probability decreases as the number of optical frequency converters is raised. For 24 or more variable optical frequency converters the internal blocking probability is very low and hence, the total blocking probability does not change significantly.

This first study showed that under symmetric traffic conditions a significant reduction in the number of optical frequency converters is possible without deteriorating the performance of the OCC over the non-blocking case. Here 24 variable instead of 64 fixed optical frequency converters are employed, but a switching network with 88 instead of 64 input and output ports is necessary.

In a second study, the share per node architecture (same size and load conditions) with fixed optical frequency converters was evaluated. Figure 7 depicts the internal blocking probability versus the offered load per input fiber. Over the shown range of loads, a node with 40 fixed converters yields approximately the same internal blocking probability as a node with 24 variable converters. Compared with an OCC with fixed optical frequency converter for each channel, the saving in optical frequency converters is not very high but the required size for the switching network increases significantly. The advantage of the share per node architecture with fixed optical frequency converters is therefore questionable.

Another study dealt with unbalanced load conditions. As an example, the connection probabilities were so chosen that connection requests on a particular optical frequency on all input fibers were directed predominantly to a particular output fiber. Compared to the symmetrically loaded system, this leads to an increased optical frequency converter demand since only one connection can be made without optical frequency conversion while any further connections must be converted. Under highly unbalanced load conditions the application of a share per node architecture may provide no benefits.

The choice of channel hunting strategy (cf. Section 4.2) has an effect on the node blocking which is particularly visible for the internal blocking probability as shown in Figure 8 for a node with 8 input/output fibers each carrying 8 optical channel and 16 variable optical frequency converters. The ordered search, starting at a fixed point provides lower internal blocking probabilities. This strategy is more successful in locating input and output channels at the same frequency at low channel numbers. Alternatively, if a conversion is needed, the search strategy assigns the lower channels first. As a result upper channels are more often available for direct connections than for a search with a random starting point.

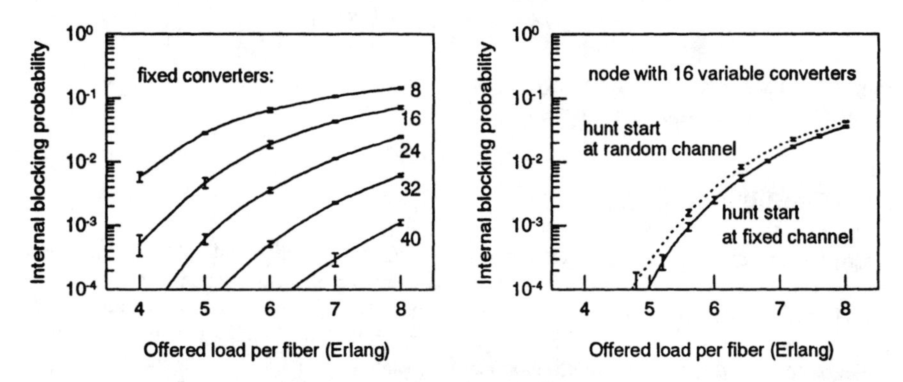

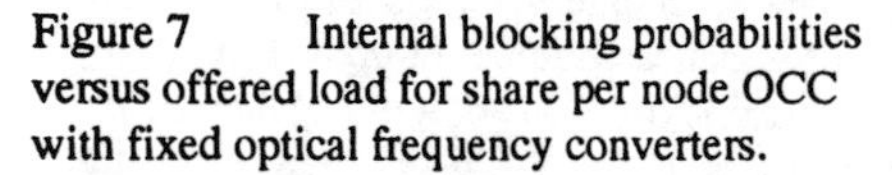
Figure 7 Internal blocking probabilities versus offered load for share per node OCC with fixed optical frequency converters.

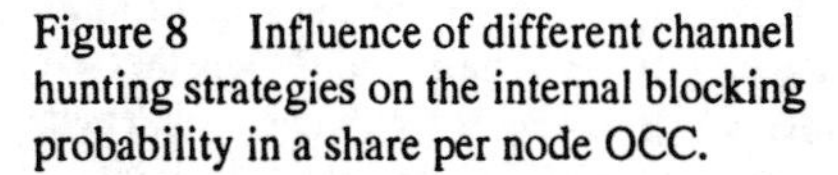
Figure 8 Influence of different channel hunting strategies on the internal blocking probability in a share per node OCC.

The last study deals with the interconnection of a share per node OCC (variable optical frequency converters) with an ECC (cf. Figure 5). In the example the OCC consists of 6 input/output fibers each carrying 8 optical channels for the interconnection with other OCCs. The ECC is connected directly to the switching network of the OCC by 16 channels (j = 16). All traffic originating from the ECC is destined for the OFDM output links. The ECC employs either lasers which can be tuned to the required optical frequency or by fixed lasers which do not conform to the frequency allocation scheme and must be frequency converted in the OCC before transmission on the OFDM link. Part of the traffic originating from the OFDM input links is destined for the ECC. The connection probabilities are chosen to have the same average traffic flow entering as leaving the ECC.

Figure 9 depicts the total blocking probability versus the offered load per channel. The node performance using tunable lasers in the ECC outputs is seen to be significantly better. The differences in the curves for a given number of variable optical frequency converters simply reflect the influence of the active ECC outputs. All of these require optical frequency conversion when fixed, non-conformant lasers are used compared to no conversion for tunable lasers with the correct optical frequency.

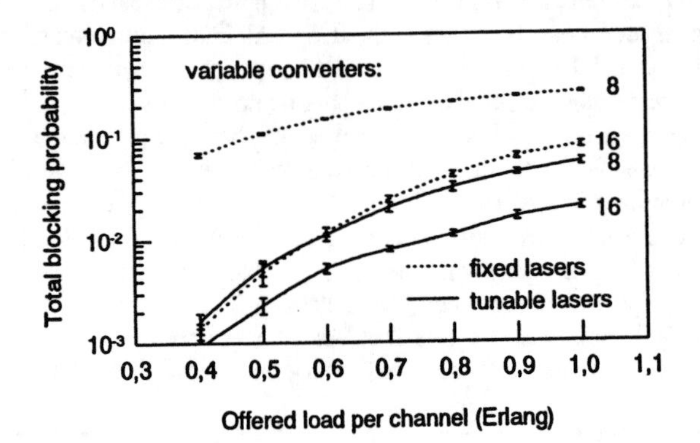

Figure 9 Total blocking probability versus offered load for different interconnection scenarios of an ECC with a share per node OCC.

5 CONCLUSION

We have presented different architectures of OCCs and different possibilities for interconnecting an OCC with an ECC. For these scenarios, the blocking probability for a new connection request was evaluated. The performance study showed that with a medium-size pool of converters with variable output frequency similar blocking behaviour can be achieved compared to a node providing optical frequency conversion for each channel.

For a final decision on the application of the share per node concept further work is necessary. It has to be examined whether the gain of a reduced number of optical frequency converters outweighs the higher implementation complexity due to the variable output frequency, the need for larger switching networks resulting in higher attenuation, the different attenuation depending whether or not an optical frequency converter is needed, and the more complex connection management and network planning.

6 REFERENCES

Astarte (1994) 'Star Switch' Product description. Astarte Fiber Networks, Inc.,

Bazlen, D., Kampe, G., Lotze, A. (1973) On the Influence of Hunting Mode and Link Wiring on the Loss of Link Systems. Proceedings of the 7th International Teletraffic Congress, Stockholm, pp. 232/1-12.

Brackett, C.A., Acampora, A.S., Schweitzer, J., Tangonan, G., Smith, M.T., Lennon, W., Wang, K.C., Hobbs, R.H. (1993) A Scalable Multiwavelength Multihop Optical Network: A Proposal for Research on All-Optical Netwoks. IEEE Journal of Lightwave Technology, vol. 11, no. 5/6, pp. 736 - 753.

Chang, G.K., Iqbal, M.Z., Ellinas, G., Shirokman, H., Young, J.C., Cordell, R.R., Bracket, C.A., Schaffner, J.H., Tangonan, G.L., Pikulski, J.L. (1995) Subcarrier Multiplexing and ATM/SONET Clear Channel Transmission in a Reconfigurable Multiwavelength All-optical Network Test Bed. Technical Digest of Optical Fiber Communication, San Diego, pp. 269-270.

Derr, F., Huber, M.N., Kettler, G., Thorweihe, N. (1995) An Optical Infrastructure for Future Telecommunication Networks. IEEE Communications Magazine, vol. 33, no. 11, pp. 84-88.

Fioretti, A., Aguilar, A., Baudron, J., Leroy, G., Masetti, F., Perrier, P., Sexton, M., Sierens, C., Sotom, M. (1995) Application of Optical Transparency to the Telecommunications Core Network. Proceedings of the 15th International Switching Symposium, Berlin, pp. 67-71.

Lee, K.C., Li, V.O.K. (1993) A Wavelength Convertible Optical Network. IEEE Journal of Lightwave Technology, vol. 11, no. 5/6, pp. 962-970.

Sato, K., Okamato, S., Hadama, H. (1993) Optical Path Layer Technology to Enhance B-ISDN Performance. Proceedings of the International Conference on Communications, Geneva, pp. 1300 - 1307.

34

Analysis of Multicasting in Photonic Transport Networks

Eugenio Iannone (*), Marco Listanti (**), Roberto Sabella (***)

(*) Fondazione Ugo Bordoni, Via B. Castiglione, 59, 00142 Roma, Italy

(**) INFOCOM Dept., University of Roma "La Sapienza", Italy

(***) Ericsson Telecomunicazioni, Research & Development Division, Roma, Italy.

Abstract

The introduction of multicasting in optical transport networks is analysed and discussed. Three different optical path realisation techniques, involving the multicasting function, are examined: the Multicast Wavelength Path (MWP), the Multicast Virtual Wavelength Path (MVWP) and the Partial Multicast Virtual Wavelength Path (PVWP). A performance evaluation model is reported and results of the performance analysis are discussed. In addition several optical cross-connect architectures, allowing multicasting to be achieved, are investigated and compared. Three different types of optical switches are considered: space division, delivery and coupling, and wavelength switches. Crucial aspects as the modularity, the complexity, the costs and some transmission aspects are taken into account in the analysis.

Keywords

Wavelength Division Multiplexing, multicast, photonic networks, transport network.

1 INTRODUCTION

The layered structure of the transport network, introduced and discussed in ITU-T [1], allows the design, the development and the operation of the network to be facilitated, also permitting its smooth evolution in pace with user demands. Likewise, the introduction of the layered concept allows network layers to evolve independently through the introduction of new technology specific to each layer. Up to now, optical technologies have been employed just within the physical media layer to greatly increase the transmission capacity. The technical advances in WDM techniques suggest that their practical application is now feasible [2], making possible the introduction of optical technology into the path layer. It considerably enhances path layer capability, and provides a transmission platform that supports different transfer modes. In addition, using the wavelength routing concept it is possible to accomplish routing and switching functions on high speed data streams, so overcoming the bottleneck induced by electronic technology [2].

As a whole the introduction of the WDM optical path layer allows the node throughput to be enhanced, and improves network flexibility and robustness in case of failures. Moreover optical layer transparency (i.e. the independence of the transmission format and the bit rate) is made possible [3,4,5,6]. This means that no restrictions on electrical path transmission mode carried by optical paths is imposed, i.e. PDH, SDH, ATM, or even analog transmission modes are possible.

In the future B-ISDN, initial traffic demands will be quite limited; therefore communications networks have to be flexible and require the minimum investment for economical introduction while also supporting future growth and incremental investment as traffic demand increases. Thus when constructing optical path networks, optical cross-connect nodes (OXCs) have to be designed taking into account these requirements [7].

Two main path layer realisation approaches have been proposed: the Wavelength Path (WP) and the Virtual Wavelength Path (VWP) [4,7]. In the former each optical path is established between two nodes by the allocation of one wavelength per each path. In the VWP scheme, the wavelengths are allocated link-by-link and thus the wavelength of the optical path is converted node-by-node. This is similar to the Virtual Path Identifier (VPI) assignment principle in ATM networks; for this reason this scheme is called Virtual WP.

The transparency provided by the optical path layer agrees with the possibility of conveying signals carrying distributive services. An important example is the transport of cable television signals (CATV), in analog or digital format, from a production centre to the access nodes. Such a potentiality is very important because consents the network infrastructure to be exploited more effectively, so increasing the revenues for a given investment. This perspective suggests the investigation of the introduction of multicasting facility in the optical path layer. This means that each OXC should have the possibility of multicast one or more input channels to several output ports.

It is not possible to foresee the evolution route of the communication networks towards broadband systems, but the flexibility and transparency of the optical path layer, together with the multicasting function, will allow the network to readily satisfy the market demand.

In this paper we analyse the issue of multicasting in the optical path layer. The concept of multicast optical path is introduced, and three different realisation approaches are discussed. The first derives from WP: the input channel and all the multicasted output channels are carried by the same wavelength. The other two derive from VWP: any of the output channels can be carried by a proper wavelength, as stated in the following. A preliminary performance evaluation study is also reported, with a performance comparison among the three strategies.

Subsequently we analyse the impact of the introduction of multicast function on the OXC architectures. Different OXC schemes, adopting several optical switching methods, are considered, discussing the most important technologies issues. The different OXC schemes are then compared, considering their modularity and flexibility, their complexity and some transmissive topics. Finally conclusions and perspective are reported.

2 OPTICAL PATH REALISATION TECHNIQUES

Two basic approaches have been proposed in literature for the realisation of the optical path layer in a global area network infrastructure [4]: the *Wavelength Path* (WP) and the *Virtual Wavelength Path* (VWP).

In the WP scheme (fig. 1a), each optical path is established between two OXCs by means of the allocation of a single wavelength; the intermediate OXCs along the path perform the WP routing using the same wavelength.

In the VWP scheme (fig. 1b), the wavelengths are allocated link-by-link, so an optical path is formed by the concatenation of as many wavelengths (possibly different each other) as the number of crossed physical links is.

Although the WP scheme leads to simpler OXC architectures in which wavelength conversion is not needed, it implies a number of drawbacks [8] (i.e. complex wavelength assignment procedures, low wavelength reuse in the network, especially when fault restoration procedures are performed). Instead, the VWP scheme permits to solve these problems.

In the context of providing an optical point-to-multipoint transport service, the concept of optical path has to be generalised to include the definition of *multicast optical path*. A multicast optical path is defined as an optical path established between one originating node and a number of terminating nodes. Topologically, a multicast optical path is a tree, in which the originating node is the root and the terminating nodes are the leaves. Each node crossed by a multicast optical path has to connect an incoming channel with a set of outgoing channels, each on a different outgoing fiber. Given a multicast optical path, the number of outgoing channels that a node has to connect to an incoming fiber is called node fan-out.

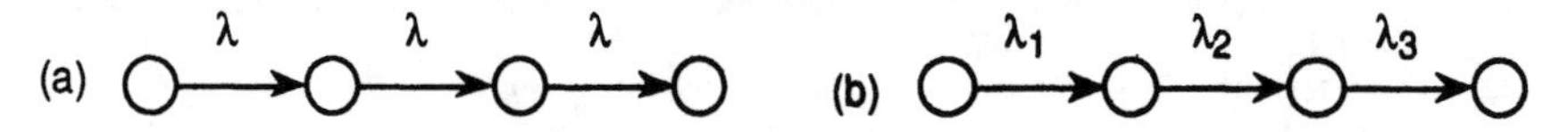

Figure 1 Optical Path Techniques: a) Wavelength Path (WP); b) Virtual Wavelength Path (VWP).

In order to cover the case of multicast optical paths, three different realisation schemes can be considered:

1) *Multicast Wavelength Path* (MWP), it is a direct extension of the concept of WP in the unicast environment; it aims at establishing a multicast optical path by utilising the same wavelength in all the links of the path; an OXC crossed by the path only performs the splitting of the incoming channel towards a set of outgoing fibers, without any wavelength conversion (Fig. 2a).
2) *Partial Multicast Virtual Wavelength Path* (PVWP), it is a limited extension of the concept of VWP; it utilises the same wavelength on all the fibers outgoing from a node, but this wavelength may be different from that utilised in the incoming link (Fig. 2b); this scheme foresees a wavelength conversion, carried out before the splitting and the switching.
3) *Multicast Virtual Wavelength Path* (MVWP), it is a general extension of the concept of VWP; with respect to the PVWP, it foresees that the wavelengths outgoing from a node of the tree may be different each other (Fig. 2c); for the application of this scheme, it is needed that nodes carry out the wavelength conversion downstream the switching stage.

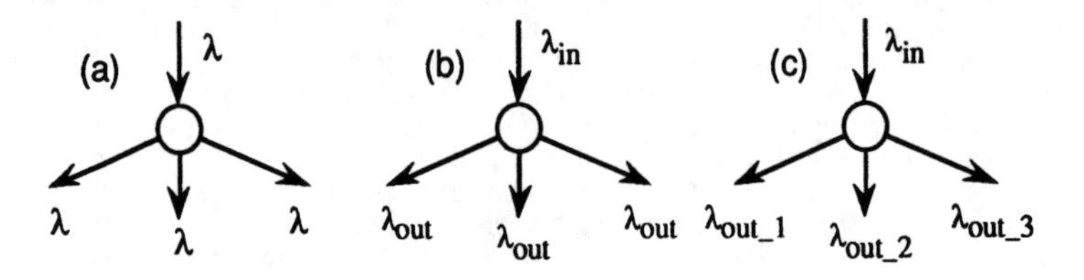

Figure 2 Multicast Optical Path Techniques. a) Multicast Wavelength Path (MWP). b) Partial multicast Virtual Wavelength Path (PVWP). c) Multicast Virtual Wavelength Path (MVWP).

As far as the applicability of the previous schemes is concerned, it is to be expected that the considerations arisen from the performance studies concerning unicast WP and VWP also hold for the comparison of MWP and MVWP schemes; moreover, PVWP should have an intermediate behaviour. A simple model to preliminary investigate the performance of the three proposed schemes is discussed in the next section.

2.1 Performance analysis

This section aims at comparing the performance of the three strategies for the handling of the multicast paths described in the previous section, i.e. MWP, PVWP and MVWP.

In some papers [8], performance are evaluated with reference to a *static scenario*, i.e. a given set of path requests has to be set up with the goal to route such requests utilising the minimum number of wavelengths. In these studies the reference figure of merit is the number of needed wavelengths. Here, as in [9], a *dynamic scenario* is assumed and only multicast path handling is considered. A single Optical Cross Connect (OXC) is considered. The paths are dynamically set up and tear down according to given input and service processes. By indicating with v the node fan-out of a multicast path, the utilised figure of merit is the probability $P_x(v)$ that a multicast path with fan-out v, can not be set up according to relevant to the strategy x (i.e. MWP, PVWP, MVWP) due to the unavailability of one or more wavelengths in the output fibers.

The analysis is carried out according to the hypothesis that the occupancies of the channels in different fibers outgoing are independent, whereas the dependence between channels belonging to the same fiber is taken into account. This *independence assumption* is the more true the lower the average value of fan-out of the multicast paths is, with respect to the number of node outgoing fibers.

In addition to the independence assumption, the analysis is also based on the following hypotheses:

a) the OXC is non blocking, i.e. it is able to connect any pair of free channels whatever the status of the switch is;
b) the path request arrival process is characterised by inter arrival times exponentially distributed with average time $1/\xi$;

c) the holding times of the multicast wavelength paths are assumed to be exponentially distributed; the average value is normalized to unit;
d) the load is uniform on both the incoming and the outgoing channels;
e) Lost Call Cleared (LCC) [10], i.e. a loss event does not modify the arrival process of the path requests.

Let:

N: number of fibers incoming to and outgoing from an OXC (including that ones coming from local transmitters, and going to local receivers);
M: number of channels per incoming and outgoing fiber;
ξ mean arrival rate of the multicast path set-up requests on each incoming wavelength;
a_v: probability that a path request with node fan-out equal to v has to be set up ($1 \le v \le N$);
K: the average value of the node fan-out of a multicast path, i.e.

$$K = \sum_{v=1}^{N} v \cdot a_v \tag{1}$$

A_O: the mean value of the offered traffic to an outgoing fiber, by recalling the above mentioned assumptions, it is easy to derive that $A_O = \xi \cdot K \cdot M$, and $0 \le A_O \le M$;
A_C: the mean value of the carried traffic on an outgoing fiber, being $0 \le A_C \le A_O$.

a) Multicast Wavelength Path (MWP).

As the wavelengths on the outgoing fibers have to be equal to that utilised on the incoming one, on the basis of the previous assumptions, it is easy to derive that the probabilities $P_{MWP}(v)$ is equal to:

$$P_{MWP}(v) = 1 - \left(1 - \frac{A_c}{M}\right)^v \tag{2}$$

A simple upper bound of $P_{MWP}(v)$ can be obtained assuming $A_C = A_O$, obtaining

$$P_{MWP}(v) \le 1 - \left(1 - \frac{A_0}{M}\right)^v = 1 - \left(1 - \frac{\xi K}{M}\right)^v \tag{3}$$

b) Multicast Virtual Wavelength Path (MVWP)

In this case there are no constraint on the wavelength to be selected on each outgoing fiber. The assumptions relevant the input and the service processes together with the independence assumptions allows the outgoing fibers to be independently modelled as M/M/M queues with no buffer space. Therefore, the probability that a single wavelength in an outgoing fiber is not available for the set-up of a multicast path is given by the Erlang-B formula [10], i.e. $B(M,A_O)$. Therefore, $P_{MVWP}(v)$ is given by

$$P_{MVWP}(v) \cong 1 - [1 - B(M, A_o)]^v \tag{4}$$

c) Multicast Virtual Wavelength Path Single Conversion (PVWP).

In this case we have to take into account the constraint that the v output wavelengths of the multicast optical path have to be equal each other.

Let us suppose that a multicast path with fan-out v has to be set-up and that the wavelength finding process follows these steps:

a) one outgoing fibers (out of the v) is randomly selected, this fiber is called *reference fiber*;
b) an unused wavelength on the reference fiber is searched; if it is not found the procedure stops and the request is lost, otherwise go to the next step;
c) the wavelength chosen in the step b) is tested on the remaining v-1 outgoing fibers; if it is free in all the fibers the request is accepted and the path is set-up, otherwise the procedure returns to the step b).

Let us suppose that in the reference fiber i ($0 \le i \le M$) wavelengths are busy, whereas M-i are free. The probability $p_O(M-i)$ that no wavelength (out of the M-i) are available in all the remaining v-1 fibers and the probability p_i that i wavelengths are busy in the reference fiber, i.e. the Erlang distribution, are given by

$$p_0(M\text{-}i) = \left[1-\left(1-\frac{A_c}{M}\right)^{v-1}\right]^{M-i} \qquad p_i = \frac{A_0^i}{i!} \Big/ \sum_{j=0}^{M} \frac{A_0^j}{j!} \qquad (1 \le i \le M) \tag{4}$$

wherein $A_c = A_0[1-B(M,A_0)]$. So, the probability $P_{PVWP}(v)$ is given by

$$P_{PVWP}(v) = \sum_{i=0}^{M} \left[1-\left(1-\frac{A_c}{M}\right)^{v-1}\right]^{M-i} \cdot p_i \qquad (1 \le v \le N) \tag{5}$$

It is to be noted that, as expected, $P_{PVWP}(1)=P_{MVWP}(1)=B(M,A_0)$.

Figure 3 plots the blocking probability P_x vs the offered traffic A_0 for the three strategies in case of an OXC with N=8 and M=8 and for a set of values of v (namely v=1,2,4). As expected, the best and the worst performance are obtained by the MVWP and MWP strategies, respectively, whereas an intermediate behaviour is presented by the PVWP. Moreover, for MVWP, the blocking probability is roughly independent of the fan-out of the path; the range of variation of $P_{MVWP}(v)$ with respect to v is less than an order of magnitude for the whole A_0 axis. This is not true for the PVWP, in fact $P_{PVWP}(v)$ rapidly reaches very high values of blocking even if the value of v is low, e.g. P_{PVWP} exceeds 10^{-2} for $v \ge 2$ and $A_0 > 0.3$.

Finally, the obtained results demonstrate that the MWP strategy is practically unusable. It leads to too high values of blocking even for very low values of A_0.

It is to be noted that, obviously, for v=1, the curves relevant to PVWP and MVWP coincide; that correspond to the implementation of the concept of Virtual Wavelength Path (VWP) in the case of unicast paths.

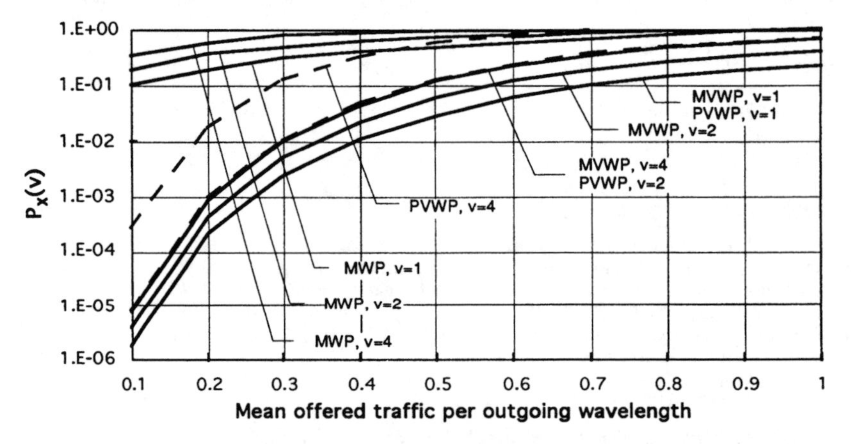

Figure 3 Blocking probabilities versus the mean value of the offered traffic to a wavelength, for the three strategies MWP, MVWP, PVWP; in case of an OXC with N=8 and M=8.

3 OPTICAL CROSS-CONNECT ARCHITECTURE

To introduce multicasting in the optical path layer, suitable Optical-Cross-Connects (OXCs) have to be devised. In this Section, some OXC architectures able to support the multicast optical path realisation strategies previously discussed will be analysed and compared taking into account node modularity, complexity, and transmissive aspects.

In order to be gradually introduced by upgrading the existing network and to respond to the rapid changes in traffic demand, the optical path layer is requested to be modular and scalable [2,6]. The path layer modularity is smoothly attained if the OXC has the possibility to add a new input/output fiber without completely changing the OXC structure. This is possible if the OXC is *link modular* [7], that is if new input/output fibers can be added without changing the OXC structure but for the addition of new components.

Moreover, during the scaling of the network, it can be needed to increase the capacity of the single WDM link. In this case, the optical path layer scalability is readily preserved if the OXC is *wavelength modular* [7]. This means that a new WDM channel can be added at each OXC input without changing the OXC structure but for the addition of new components.

In this paper we consider transparent OXC architectures in which no electro/optical conversion occurs along the signal path. We do not consider OXCs in which the signal is regenerated during the processing inside the node since, in this case, any kind of network transparency is prevented.

In a transparent network, the transmission route of the optical signal comprises all the crossed OXCs. Two main transmission impairments arise from the OXC structure: optical noise and crosstalk. These phenomena have to be somehow characterised to evaluate the transmission performances of a given OXC architecture.

Several OXC architectures have been reported in the literature based on three different types of optical switching: space division, wavelength division and delivery and coupling (D&C) switching. For each architecture we analyse the possibility to introduce the multicast function.

3.1 OXCs based on Space Switching Matrixes

The first class of OXC we consider is based on optical switching matrixes. A first architecture belonging to this class is shown in fig. 4a. At the OXC input, afterwards optical amplification, wavelength demultiplexing is accomplished through a combination of optical power splitters and tunable filters. This approach allows the order of the wavelengths at the demultiplexers output to be changed, so that any channel can be sent to any switch matrix. The space division switches actually perform the channels routing, and the combiners multiplex the channels before the output fibres. A further amplification is generally needed at the OXC outputs to compensate the losses. The set of transmitters and receivers consents to add and drop, respectively, channels in/from the traffic. For instance this scheme has been reported in [3] and practically implemented in the framework of the RACE project MWTN (RACE 2028).

This architecture does not allow the VWP scheme because it does not employ wavelength conversion. Of course, it is possible to receive a given channel through a local receiver and re-transmit it on another wavelength using a local transmitter. However this operation does not keep the transparency of the optical layer and, in any case, uses resources devoted to the add/drop function.

A second architecture is similar to the previous one, but involves all-optical wavelength converters located at the outputs of the space switching matrixes, as shown in figure 4b [11]. This scheme permits the implementation of VWP.

Both the architectures shown in fig. 4a and 4b are able to support the multicasting function. In fact let's consider the optical space switch scheme depicted in fig. 5 [12]. Each input is connected, through an optical waveguide, to every output. In any waveguide there is a gate which allows or inhibits its proper route, according to a logical command. To introduce the multicast function it is possible to grant more than one route associated to the same input; hence the signal at the input reaches more than one output. In the case of the OXC of fig. 4a the WP is to be adopted to realise a multicast optical path; conversely the OXC of fig. 4b allows the MVWP scheme to be implemented, because the wavelength conversion, following the space switching stage, allow to arbitrarily select the wavelength of each output channel.

3.2 OXCs based on Wavelength Switching

The second class of OXC we consider is based on wavelength switching. An example of OXC belonging to this class is illustrated in fig. 6 [13] while other examples are reported in [7] and in [11]. At the OXC input, the WDM combs carried by the input fibers are wavelength translated so to occupy contiguous parts of the optical spectrum. After translation, the signals are amplified and feed the inputs of an N x N M star coupler. At the star centre, all the incoming optical channels are wavelength multiplexed onto a single comb. At the star coupler outputs the tunable filters route each channel towards the proper output fiber. The wavelength converters set the channel wavelengths to a suitable value to allow multiplexing onto the output fiber. Multiplexing is performed by optical couplers and at the OXC output the signal is amplified. It is to be noted that, at the OXC input, wavelength conversion is set before amplification since wavelength converters able to shift an entire WDM comb works more efficiently with a low input signal power.

This architecture can implement the MVWP scheme since a single channel can be multicast towards a set of output fibers by an opportune tuning of the routing filters and the wavelengths of the output channels are independently selected by the wavelength converters.

3.3 OXCs based on Delivery and Coupling Switches

The third class of OXC we consider is based on D&C switches. Figures 7a and 7b report the architectures of two OXCs of this class while the scheme of the D&C switch is depicted in fig. 8 [7].

The two architectures have been contrived in order to satisfy the link modularity and the wavelength modularity requirements, respectively.

Both the architectures can adopt the PVWP scheme. Actually the D&C switch can be designed so that each 1x2 switch can operate in one of the following ways: i) inhibits both its output, ii) allows only one output (acting as a normal 1x2 switch), iii) broadcasts the input signals to both the outputs. This way it can implement the multicasting function. It is worth noting that wavelength conversion must be placed before the switch itself. Otherwise in the scheme shown in fig. 7a, contentions would occur in the output star couplers when different channels, at the same wavelength, were routed to the same output fibre. Furthermore, the OXC of fig. 8 would never work, because the D&C switches deal with channels all at the same wavelength.

This kind of cross-connect does not allow to directly perform the MVWP scheme. Actually it would be still possible to perform MVWP by putting a wavelength converter before the star coupler in the D&C switch. This way the structure of the switch would be completely flexible. Nevertheless the number of wavelength converters would increase considerably (proportional to $N^2 M$).

4 TECHNOLOGICAL ISSUES

In the architectures considered in this paper, besides mature optical components (e.g. optical transmitter and receivers, power splitters and erbium doped fiber amplifiers) there are some key optical devices which are worthy to be analysed. They are the tunable optical filters, the optical switches and the wavelength converters. Here we discuss some technological issues which concern their realisation.

4.1 Tunable Optical Filters

Two different technologies are mature for tunable optical filtering: the fiber Fabry-Perot filters [15] and acousto-optic filters [16]. The first type presents low losses, easy tunability, large tuning range (more than 30 nm) and are polarisation insensitive. However, their roll-off is quite slow. This means that the channel spacing must be great enough to not appreciably introduce crosstalk (for example 3 nm for 10 Gbit/s channels [17,18]). The second type have a large tuning range (more than 30 nm) and a good roll-off coefficient. Moreover these filters can select more frequencies at a time, being each frequency independently tunable. This property can be useful both in designing add/drop all optical multiplexers and OXC architectures based on wavelength switching [19].

A third type of filters could be employed in the future, based on InP technology [20]. However, for the time being, they are not mature for implementation. Their most important advantage is that they could be monolithically integrated with active optical devices such as lasers, semiconductor optical amplifiers and wavelength converters. The integration has the considerable advantages of removing the stabilisation and coupling problems that are present in the implementations by discrete devices and of lowering the costs.

4.2 Optical Space Switches

Basically, two types of space switching matrixes have been realised: one based on $LiNbO_3$ technology [21] and the other one based on InP technology [12].

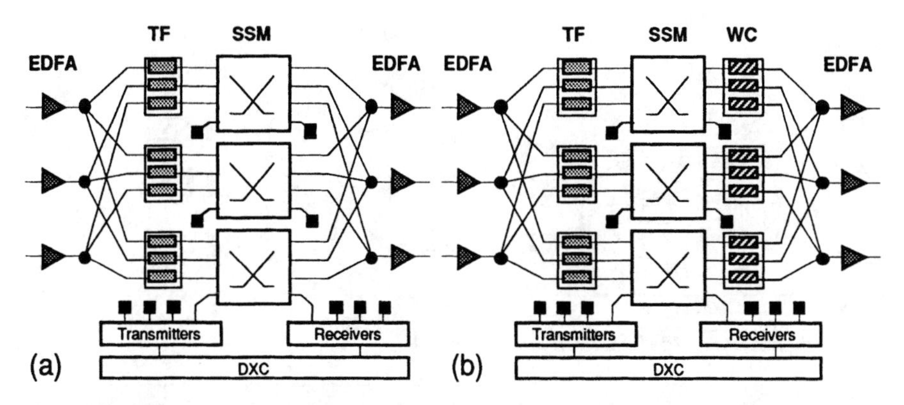

Figure 4 Optical cross-connect architecture based on space division switches: a) no wavelength conversion; b) with optical wavelength converters. EDFA: Erbium Doped Fiber Amplifier; TF: Tunable Filter; SSM: Space Switch Matrix; DXC: Digital Cross-Connect.

The latter seems more suitable for application in the OXC switching fabric. At the state of the art, it presents lower losses (for example of the order of 3 dB for a 4x4 matrix [22]) and allows the realisation of integrated optical circuits. The main difficulty in realising InP space switch matrixes is in assuring their stability, avoiding lasing inside the structure. For instance, at the state of the art, the input power of a 4x4 matrix cannot overcome a threshold of the order of -20 dB. This could be a limiting factor for the maximum feasible matrix size. In fact, relating to fig. 5, the gate element is a semiconductor optical amplifier (SOA) which allows or inhibits its optical path according to a current signal. If this current signal is low the device absorbs the radiation, while if it is high the device permits the signal to pass. In addition, if the switch dimension is (M x M) there are M SOAs at the input which provide a little amplification to partially compensate the device losses, and M SOAs at the output, which are used for equalising output signals, by adjusting the input current.

As far as the D&C switch is concerned, the same technologies used for the switching matrixes could be employed. In fact the (1:2) element could be realised both in $LiNbO_3$ or in InP technology.

4.3 All-Optical Wavelength Converters

At the time being there are not commercially available wavelength converters. Anyway there are several technological solutions reported in the literature. Among them, those based on semiconductor technology give several advantages: they can operate at high speed, present a large conversion bandwidth and can be integrated in a single chip. Moreover, they seem to be mature for a rapid engineering [23]. In fact, monolithical integration, as for instance demonstrated in [24], allows coupling and stability problems that are present when using discrete components to be avoided. More specifically two types of devices can be favourably realised to perform wavelength conversion in optical transport networks. They are the devices based on Four Wave Mixing (FWM) in SOAs [25,26] and on Cross-Phase Modulation (XPM) in SOAs put in interferometric configuration [23]. It is worth while noting that FWM devices can translate an entire WDM comb [27], while XPM do not offer this possibility. This fact influences the design of wavelength switched OXC, as better described in Section III.3.2.

In a previous work [14] we analyse the employment of such devices in an optical transparent network and show that both these solutions could be favourably employed for this type of applications. In particular, network paths covering distance of the order of 1500 and 2500 km with four channels at 2.5 and 10 Gbit/s respectively can be realised [6]. If cross phase modulation converters are adopted, the ultimate limit to transmission performance is set by fiber dispersion and Kerr effect. Moreover the dependence of the device behaviour on the input signal power and the dependence upon the transmission format impact on the design complexity.

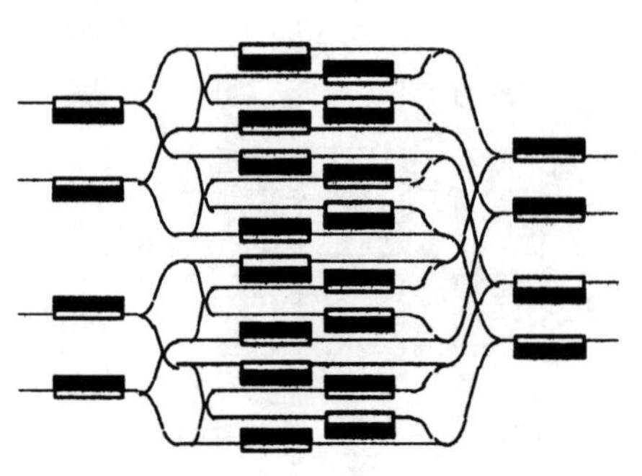

Figure 5 Scheme of an optical space division switch based on InP technology.

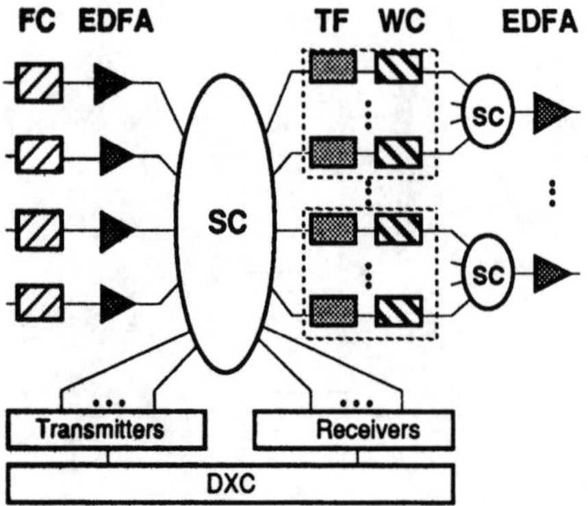

Figure 6 OXC based on wavelength switching accomplished through wavelength converters. FC: Frequency Converter for a WDM comb

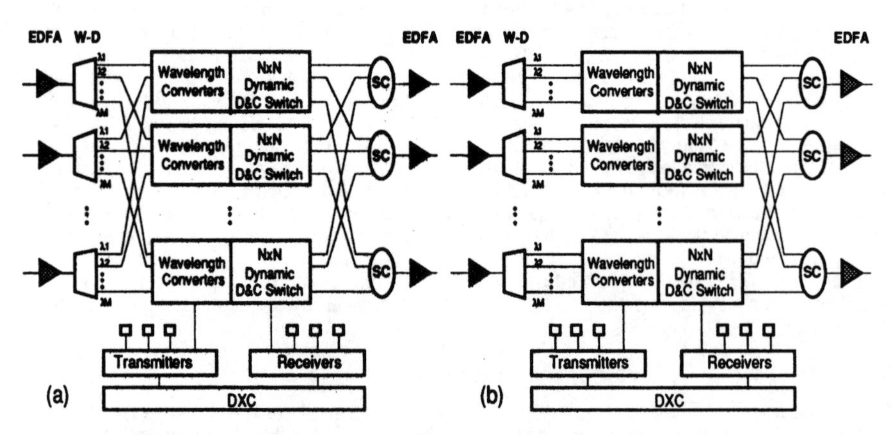

Figure 7 OXC architectures using D&C switches: a) Link modular OXC; b) Wavelength modular OXC. SC: Star Coupler; DXC: Digital Cross-Connect.

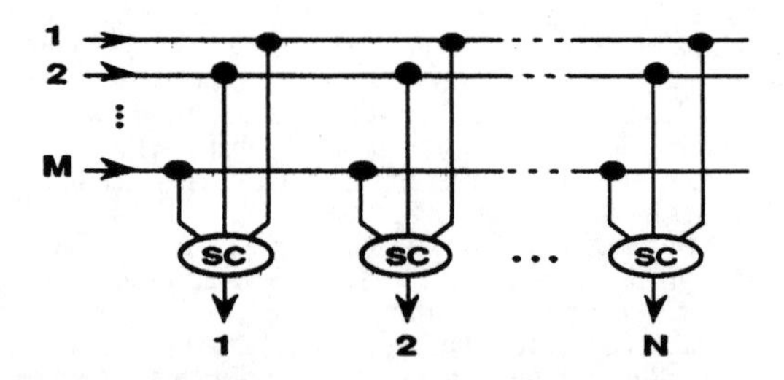

Figure 8 Scheme of MxN dimension D&C switch, used in the scheme reported in fig. 7a. In the case of the scheme shown in fig. 7b, this device has NxN dimension. SC: Star Coupler.

On the other hand, if FWM devices are adopted, tolerance with respect to signal power fluctuations and transmission format transparency are obtained [28]. Moreover, in geographical networks, the ability of FWM to compensate chromatic dispersion and Kerr effect can be very useful [29,30]. These advantages are partially counterbalanced by their higher noise factor and by the dependence of the converter efficiency on the conversion interval. As a matter of fact, this last characteristic calls for a power equalisation of the output signal.

5 ARCHITECTURES COMPARISON

To compare different OXC schemes three different criteria can be introduced: OXC flexibility and modularity, the OXC complexity and its transmission performances.

The OXC flexibility is essentially related to the ability to implement a flexible multicasting routing. Thus the OXC can be classified on the ground of the ability to implement PVWP or even MVWP. The OXC wavelength or link modularity has been already defined and is mainly related to the possibility to upgrade the OXC capacity by adding new WDM channels on the existing fibers or new input/output fibers.

To evaluate the OXC complexity, three basic elements can be individuated: space gates, wavelength converters and tunable filters. The other devices used inside the OXC, as the star couplers, the optical passive waveguides and so on, have a minor impact on the system complexity. Regarding the erbium doped fiber amplifiers, their number is the same in all the considered architectures. A useful indication can also be achieved by evaluating the overall number of semiconductor active devices (lasers and SOAs). Of course, in evaluating the utility of this figure, it is to be taken into account that semiconductor devices designed for different applications can be quite different, even under a control and cost point of view. For example, the semiconductor amplifiers used in all optical wavelength converters are quite different from those used in the switch matrixes realised in InP technology for their characteristics of size, gain and current absorption.

The OXC introduces noise due to the presence of active optical elements, e.g. optical amplifiers,. The power spectral density S can be written as

$$S = F \cdot G \cdot h \cdot \nu \qquad (6)$$

wherein G is the overall OXC gain or attenuation, h is the Plank constant and ν the optical frequency. The noise factor F, depending on OXC architecture and on the adopted technology, can be assumed as figure of merit representing the noise performances.

Concerning the crosstalk, it mainly depends on channel spacing and on the tunable filters technology. The dependence of the crosstalk on the OXC architecture is quite complex and does not critically impact on the transmission performances. Thus crosstalk will be neglected in the following comparison.

In the following, the different OXC architectures will be compared on the ground of the criteria introduced in this section. To obtain a fair comparison, the following hypothesis will be assumed for all the OXCs:

1) Space switching matrixes and D&Cs are realised in InP technology, as described in [12]; it implies that at each 1x2 switch of a D&C there are two gates realised by semiconductor amplifiers.
2) All-optical wavelength converters based on FWM in SOAs are adopted [26]; this is a realistic hypothesis for OXCs designed for multicasting since it assure the higher degree of transparency and the ability to translate a whole WDM comb. The parameters assumed for the FWM converters are those reported in [6], considering devices in which the SOAs have been designed in order to optimise frequency conversion.
3) The tunable filters that are required after each wavelength converter are accounted in the overall number of tunable filters.
4) The attenuation and noise parameters of the devices different from wavelength converters are those reported in [6]; the following typical parameters are assumed for an OXC with $G_{OXC}=1$: the channel bit rate is 2.5 Gbit/s, M=4, N=4, the power at the OXC input is -5 dBm and the channel spacing 0.5 nm.

To obtain numerical values for the noise factor of different OXCs, the analytical model reported in [11] has been used. The results of the comparison are shown in table 1. It is evident that the OXC based on wavelength switching has three main advantages: i) it can support MVWP; ii) it is both wavelength and link modular; iii) the number of active semiconductor devices increases linearly with the number N of input/output fibers, whereas in the other cases it increases quadratically. These advantages are partially counterbalanced by the need of more sophisticated devices and by a greater noise factor. As a matter of fact, the SOAs employed in the wavelength converters are more critical than those employed in the gates of space switching devices. Since even the noise factor depends mainly on the technology of wavelength converters, the possible application of the OXC based on wavelength switch is mainly related to the advances in the technology of wavelength converters.

Among the other OXC architectures, the only allowing MVWP is that based on space switching and adopting wavelength conversion. This architecture is less complex than both the architectures based on D&C and provides a comparable noise factor. However this architecture is only wavelength modular, not link modular. The only architecture not based on wavelength switching that results link modular is the second architecture exploiting D&Cs.

Finally the less complex architecture is that based on space switching without the use of wavelength converters. This architecture offers even the lowest noise factor. However these advantages are counterbalanced by the impossibility to implement effectively MVWP or PVWP.

Table 1 Comparison of OXC architectures

OXC Architecture	Space Switch-1	Space Switch-2	D&C Switch-1	D&C Switch-2	Wavelength Switch
Multicast support	WP	MVWP	PVWP	PVWP	MVWP
Link modularity	NP	NO	NO	YES	YES
Wavelength modularity	YES	YES	YES	NO	YES
Number of gates	$M \cdot N^2$	$M \cdot N^2$	$2 \cdot M \cdot N^2$	$2 \cdot M \cdot N^2$	----
Number of tunable filters	$N \cdot M$	$2 \cdot N \cdot M$	$2 \cdot N \cdot M$	$2 \cdot N \cdot M$	$N+N \cdot M$
Number of converters	—	$N \cdot M$	$N \cdot M$	$N \cdot M$	$N+N \cdot M$
Semiconductor devices	$N \cdot M \cdot (2+N)$	$NM \cdot (4+N)$	$2 \cdot N \cdot M \cdot (2+N)$	$2 \cdot N \cdot M \cdot (2+N)$	$2 \cdot N \cdot (1+M)$
Noise factor	15 dB	16 dB	17 dB	17 dB	21 dB

6 CONCLUSIONS

In this paper three different optical path strategies to implement multicasting in optical transport network have been analysed. As a result the MWP strategy seems to be practically unusable, because it leads to very high values of blocking even for very low values of the mean offered traffic. On the other hand, MVWP presents the best performance and is practically independent of the fan-out of the path. Intermediate figures are provided by PVWP strategy.

Different optical cross-connect architectures have been analysed, allowing the implementation of multicasting according to at least one out of the considered strategies. In particular, two OXC schemes allow MVWP to be achieved: one based on space division switching, the other based on wavelength switching. The comparison has revealed that the former is wavelength modular but not link modular, whereas the latter is both link and wavelength modular and employs a number of semiconductor devices which is sensitively lower with respect to the former. On the other hand, the noise factor of the latter is 21 dB regardless of the 16 dB of the former, and the number of wavelength converters is N times greater.

7 REFERENCES

[1] ITU-T Recommendation, G. 803, "Architectures of transport networks based on the synchronous digital hierarchy (SDH)", 03/93, 1-993.

[2] C.A. Brackett, "Dense wavelength division multiplexing networks, principles and applications", IEEE Journal of Selected Areas in Communications, vol. 8, no. 6, pp.948-964, 1990.

[3] G.R. Hill et al., "A transport network layer based on optical network elements", IEEE-Journal of Lightwave Technol., vol.11, no.5/6, pp. 667-679, 1993.

[4] K. Sato, S. Okamoto, H. Hadama, "Network performance and integrity enhancement with optical path layer technologies", IEEE Journal of Selected Areas in Communications, vol. 12, no. 1, pp.159-170, 1994.

[5] S. B. Alexander et. al., "A precompetitive consortium wide-band all optical networks", IEEE-Journal of Lightwave Technology, vol.11, no.5/6, pp. 714-735, 1993.
[6] R. Sabella, E. Iannone, E. Pagano, "Optical transport networks employing all-optical wavelength conversion: limits and features", in press on IEEE JSAC, special issue on "Optical Networks".
[7] A. Watanabe, S. Okamoto, K. Sato, "Optical path cross-connect node architecture with high modularity for photonic transport networks", IEICE Trans. on Communications, vol. E77-B, n. 10, pp. 1220-29, 1994.
[8] N. Nagatsu, Y. Hamazumi, K. Sato: "Optical path accomodation design applicable to large scale networks". IECE Transaction on Communications, vol. E78-B, n. 4, April 1995, pp. 597-607.
[9] I. Chlamtac, A. Ganz, G. Karmi: "Lighpath communications: an approach to high bandwidth optical WANs". IEEE Transaction on Communications, vol. 40, n. 7, July 1992, pp. 1171-1182.
[10] R. Syski: *Introduction to congestion theory in telephon systems*, North Holland, New York, 1986.
[11] E. Iannone, R. Sabella, "Performance Evaluation of an Optical Multi-Carrier Network using wavelength Converters based on FWM in Semiconductor Optical Amplifiers", IEEE-Journal of Ligthwave Technology, vol. 13, no. 2, 1995.
[12] M. Gustavsson et al., "Monolithically integrated 4x4 InGaAsP/InP laser amplifier gate switch arrays", Electronics Letters, vol. 28, pp.2223-2225, 1992.
[13] R. Sabella, E. Iannone, "A new modular optical path cross-connect", submitted to IEE Electronics Letters.
[14] E. Iannone, R. Sabella, L. de Stefano, F. Valeri, "All-Optical Wavelength Conversion in Optical Multi-Carrier Networks", in press on IEEE Transaction on Communications.
[15] P.A. Humblet, W.M. Hamdy, "Crosstalk analysis and filter optimization of single- and double-cavity Fabry-Perot filters", IEEE JSAC, vol. 8, no. 6, pp.1095-1107, 1990.
[16] G.D. Boyd and F. Heismann, "Tunable Acoustooptic reflection filter in $LiNbO_3$ without a doppler shift", IEEE Journal of Lightwave Technology, vol. 7, no. 4, 1989.
[17] G. Jacobsen, "Multichannel system design using optical preamplifiers and accounting for the effects of phase noise, amplifier noise and receiver noise", IEEE Journal of Lightwave Technology, vol.10, no. 3, pp.367-376, 1992.
[18] E. Iannone and R. Sabella, "Crosstalk in WDM optical networks adopting wavelength conversion", Proc. ECOC '95, vol. 2, pp.693-696, Brussels, Belgium, 1995.
[19] F. Wehrmann et al., "Fully packaged, integrated optical, acoustically tunable add-drop-multiplexers in $LiNbO_3$", Proc. ECIO '95, pp.487-490, 1995.
[20] O. Sahlén, "Active DBR filters for 2.5-Gb/s operation: linewidth, crosstalk, noise, and saturation properties", IEEE Journal of Lightwave Technology, vol. 10, no. 11, pp.1631-1643, 1992.
[21] P. Granestrand et al., "Pigtailed tree-structured 8x8 $LiNbO_3$ switch matrix with 112 digital optical switches, IEEE Photonics Technology Letters, vol. 6, no.1, pp.71-73, 1994.
[22] C.P. Larsen et al., "Transmission experiments on fully packaged 4x4 semiconductor optical amplifier gate switch matrix", Photonics Switching, Salt Lake City, UT, USA, March 1995.
[23] K.E. Stubkjaer et al., "Optical wavelength converters", Proceedings of ECOC '94, vol.2, pp.635-642, Firenze, Italy, 1994.
[24] M. Schilling et al., "Monolithic Mach-Zehnder interferometer based optical wavelength converter operated at 2.5 Gb/s with extinction ratio improvement and low penalty", Proceedings of ECOC '94, vol.2, pp.647-650, Firenze, Italy, 1994.
[25] A. D'Ottavi, E. Iannone, A. Mecozzi, S. Scotti, P. Spano, R. Dall'Ara, J. Eckner, G. Guekos,"Efficiency and Noise Performances of Wavelength Converters Based on FWM in Semiconductor Optical Amplifiers", IEEE Photonics Technology Letters, vol. 7, no. 4, 1995.
[26] A. D'Ottavi, E. Iannone, A. Mecozzi, S. Scotti, P. Spano, R. Dall' Ara, G. Guekos, J. Eckner, "Frequency conversion by four-wave mixing on a frequency range of 8.6 THz.", Proceedings of ECOC '94, Firenze, ITALY, September 1994, pp. 737-740.
[27] R. Schnabel et al., "Polarization insensitive frequency conversion of a 10-channel OFDM signal using four-wave-mixing in a semiconductor laser amplifier", IEEE Photonics Tech. Letters, vol. 6, n. 1, 1994.
[28] R. Ludwig and G. Raybon, "BER measurements of frequency converted signals using four-wave mixing in a semiconductor laser amplifier at 1,2.5,5,10 Gbit/s", Electronics letters, vol. 30, n. 4, pp. 338-339, 1994.
[29] M.C.Tatham, G.Sherlock, L. D. Westbrook, "Compensation of fiber chromatic dispersion by mid-way spectral inversion in a semiconductor laser amplifier", Proceedings of ECOC '93, Montreaux, Switzerland, September 12-16, paper ThP12.3, 1994.
[30] R. Sabella, E. Iannone and E. Pagano, "Impact of wavelength conversion by FWM in semiconductor amplifiers in long distance transmissions", CLEO '95 Pac. Rim, Chiba, Japan, July 1995, paper P237.

35

Integrated Analog Switch Matrix With Large Input Signal and 46dB Isolation at 1GHz

Ewa Sokolowska, Nacer Eddine Belabbes, Bozena Kaminska
NHC Communications Inc.
hfOPTEX
École Polytechnique de Montréal
Departement Génie Électrique et Génie Informatique, P.O.Box 6079, Station "Centre Ville", Montréal, PQ, H3C 3A7, Telephone: (514) 340 4270. Fax: (514) 340 4170.
email: bozena@vlsi.polymtl.ca

ABSTRACT

A complete integrated analog switch matrix allowing a large input signal and amplitude, and demonstrating outstanding performances, namely, a 46 dB feedthrough on channel at 1 GHz and a 9.5 Ω switch resistance in the ON state (R_{ON}), was developed and fabricated in a standard 1 μm QED/A TriQuint GaAs process. The main advancement of this new concept is that the drive signal opening the switch tracks the input signal with an added positive offset. This constant positive offset results in a very low R_{ON}, while keeping the switch dimensions and power dissipation acceptably small. This new concept has allowed a monolithic integration of the switches with their drive circuitry and control logic into the 13 x 13 x 2 analog switch matrix, which is a unique solution.

Keywords

Analog Switch, Design of an integrated circuit, GaAs technology.

1. INTRODUCTION

Non-blocking analog switch matrices have many applications in telecommunications, especially where the diversity of communication protocoles (e.g. ISO1, 2) must be supported Uda (1994), Kusonoki (1992), Gardiner (1989), Slobodnik (1989) and Feng (1994). In this domain where a high input signal amplitude and low insertion loss are important, an integrated solution has been lacking, although digital matrices have already appeared on the market (e.g. TQ8015-17

16 x 16 crosspoint-switch matrices from TriQuint).

Based on: Uda (1994) our new concept of the input tracking switch drivers (patent pending) leading to a low and constant R_{ON} while allowing a large input amplitude, and Kusonoki (1992) a careful choice of the technology, we have developed the first monolithically integrated analog switch matrix providing performances similar to those offered by digital matrices. This matrix was developed in collaboration with NHC Communications specialized in digital communication systems. The developed non-blocking crosspoint analog switch matrix was fabricated in a standard 1 μm QED/A GaAs process from TriQuint Semiconductors. The main reason for this choice of technology was that it supports relatively high voltages (-12V gate to drain/source breakdown) resulting in a design admitting a large input voltage swing (8 V_{p-p} for the chosen supply voltages). Furthermore, this technology is well established with a reasonable yield and offers a standard cell library, which is very useful for mixed signal designs.

The outstanding performances demonstrated by the circuit, which has been designed, such as small physical dimensions, low power dissipation, very low insertion loss and very low feedthrough has allowed us to integrate the analog switch matrix with a control logic in a 13 x 13 x 2 switch matrix. The integration of analog and digital parts resulted in a high-bandwidth integrated circuit with a low pin count.

Target performances of the designed matrix are presented in Table1.

Table 1: The main specification of Switch Matrix

switching capacity	13x13x2
analog signal swing	8 V_{p-p}
on-state resistance R_{ON}	<9.5 ohms
ΔR_{ON}	10%
frequency bandwidth	DC to over 1GHz
feedthrough on channel	<-70 dB @ 100MHz, <-45db @1GHz
switch setup time	<30 ns
dissipated power / cell	30mW
signal phase shift	0d

In this paper we describe the global architecture of the analog switch matrix, together with a new concept of the input tracking switch drivers. The excellent performances of this IC are shown as well.

2. GLOBAL ARCHITECTURE

Tight requirements for R_{ON}, feedthrough on channel and circuit surface determined the circuit architecture and limited the dimensions of the matrix. Taking into account all the performance constraints, a matrix with dimensions of 13 x 13 x 2, i.e.containing two parallel matrices 13 x 13, was designed. Note that the 8 address lines necessary for the 13 x 13 matrix allowed the maximal dimensions of 16 x 16.

The developed matrix consists of two main blocks, as shown in Figure 1. The first one is a digital block containing the control logic which decodes an address sent on the address lines (XO1-4, YO1-4), reading the input data (Din) and thus providing one internal digital control signal for each switch (CTRLi) in both 13 x 13 matrices. A full functionality of a digital control block will be described in section 2.1.

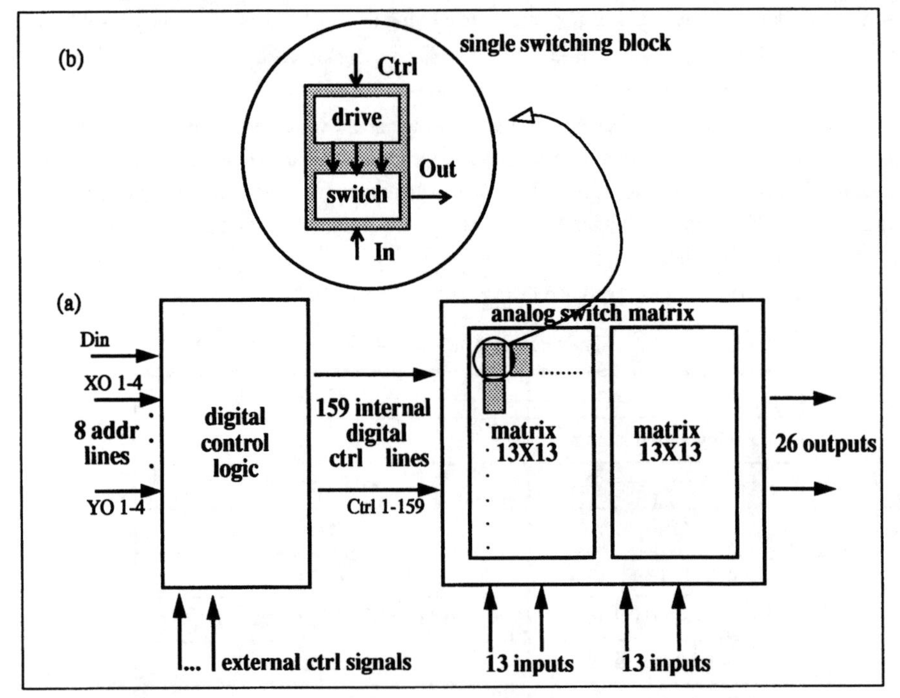

Figure 1.. Block diagram of the analog 13 x 13 x 2 switch matrix (a), showing the internal organization of the switching block (b).

The second block is an analog one , formed by a matrix of analog switches. Each of the analog switches contains a switch and drive circuitry providing the analog drive signals necessitated by the switching devices inside each switch (Figure 1 (b)). Depending on the state of the 159 control signals, the drive blocks in a corresponding switch produce the adequate drive signals. These signals are used directly to open and close the switches, as described in section 2.2. The internal architecture of the analog switch matrix is shown in Figure 2. The main matrix is divided into two identical 13 x 13 matrices having independent inputs and outputs, however they share the same control signals. Therefore, the connections effected by both matrices are parallel. The switches in each matrix are organized in a crosspoint network containing 13 rows and 13 columns of switches

Barber (1988). Therefore each input signal is propagated to an output through only one switch. This arrangement produces the lowest possible R_{ON}. Each of the switches in the matrix is controlled by a dedicated control signal, therefore a chosen input signal can be broadcasted to several outputs, or on the contrary, several inputs can be directed to a single output. In each of these 13 x 13 matrices 12 inputs and outputs are available for the external user. The 13-th input and output, although having a full functionality, are added only for the test purpose. These lines (test_lines) enable the broadcast from a chosen input to the test line or the propagation of a test signal to a chosen output.

2.1. Control logic

In a following section an abbreviated description of the functionality of the designed digital control logic, handling the read and write operations is given. To differentiate between a read of the present connection pattern of the switch and a write of the new state to an internal control line indicated by the present address, the R/W control line is provided.

During both read and write operations, the addresses sent on the 8 address lines are sequentially decoded, and corresponding input data bits sent on the Din line pass through two levels of latches. The propagation of the decoded data through the latches proceeds with the corresponding clock signals and is enabled by the chip select signal.

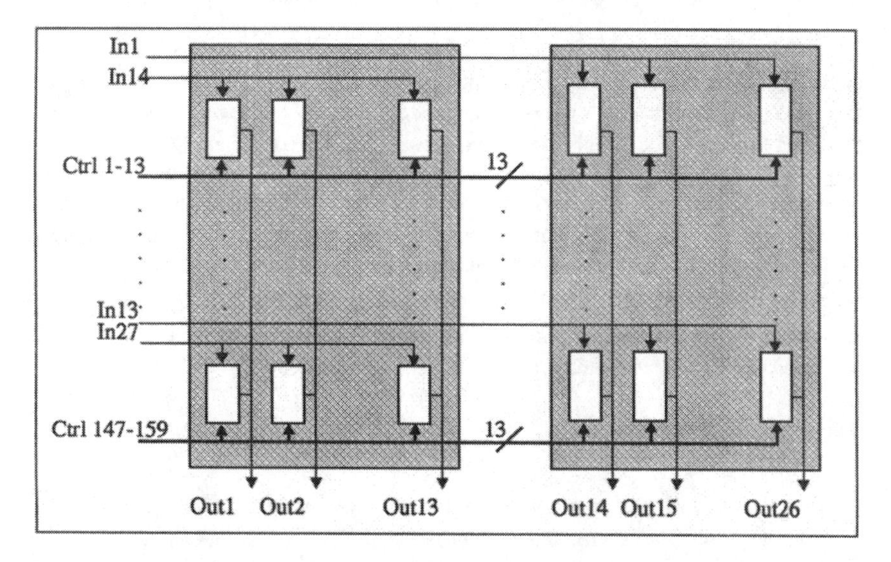

Figure 2. Internal architecture of the crosspoint switch matrix

The write operation generates the internal digital control signals CTRL1-159 used by the analog part to set the state of the switches for both 13 x 13 matrices. Each of these lines activates or deactivates one switch in each of the 13 x 13 matrices, so that the same connection pattern is present in both matrices. When the read operation is enabled, the state of the internal CTRL line indicated by the present address lines appears on output data line.

2.2. Basic switch structure

The circuit topology of one T-switch, as depicted in Figure 3, consists of two series depletion MESFETs (Q1, Q2) and one shunt depletion MESFET (Q3), and was designed to provide, simultaneously, the resistance in the on state (R_{ON}) of 9.4 ohm and isolation grater than 50 dB at 1 GHz. The main role of the transistor Q3 is to increase the isolation between input and output..

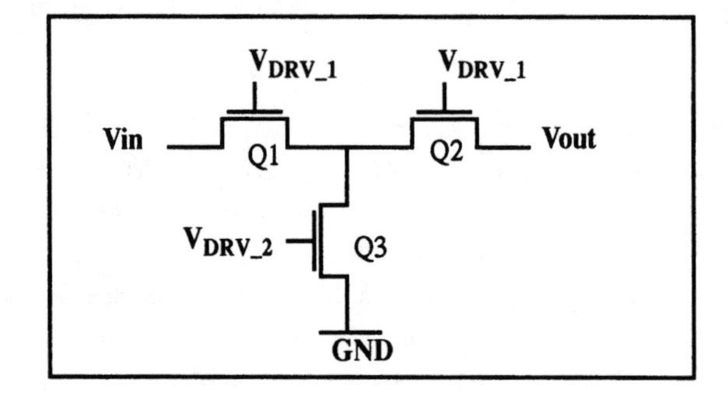

Figure 3. Configuration of a T switch

The depletion MESFETs used in the T-switch have pinch-off voltages between -0.45V and -0.75V. In order to turn the switch on, the DRV_1 signal has to be brought above V_{in} and the DRV_2 should approach V_{ee}. In this case, transistors Q1 and Q2 are on and Q3 is off, and the input signal passes through. The on resistance of the switch is determined by the resistance of two transistors connected in series. However, for a low R_{ON}, two conditions must be satisfied: the width of the transistors must be sufficient and the DRV_1 must be close to the opening of the gate diodes (about 0.6V above V_{in}), as discussed in section 3.1.

Conversely, to turn the switch off, DRV_1 must be brought at least 0.75V below Vin, and DRV_2 close to 0V. In this case, Q1 and Q2 are off and Q3 conducts, and the switch is off. To improve the isolation when the switch is off, DRV_1 should approach V_{ee}

2.3. General considerations

The major difficulty in the design of an analog switch based on GaAs MESFETs used as the transmission gates arises from the fact that the gate of the MESFET conducts a substantial current when the gate-channel diode is forward-biased (gate and source voltage approaching 0.7V). The possibility of the gate conduction in the transistors forming the switch should be avoided. Therefore, the level of a drive signal closing the switch (V_{DRV_1}) must be constrained within a range, and it is limited by

$V_{IN} + V_T < V_{DRV\_1} < V_{IN} + 0.7V$,

where V_{IN} is the analog input signal and V_T is a transistor threshold voltage. If the amplitude of the analog input signal exceeds this limit, the control voltage should track the input signal, staying within the defined limit.

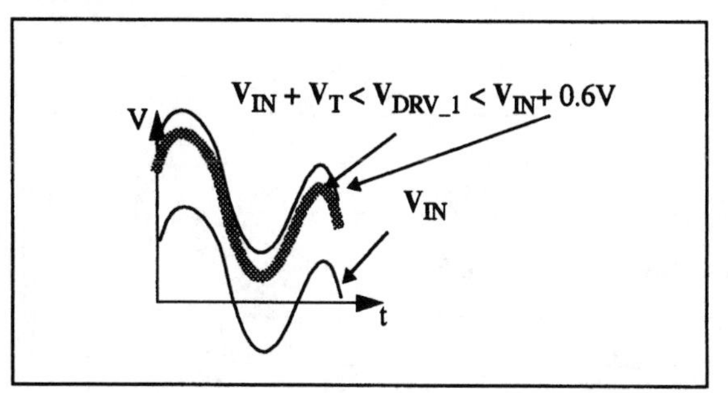

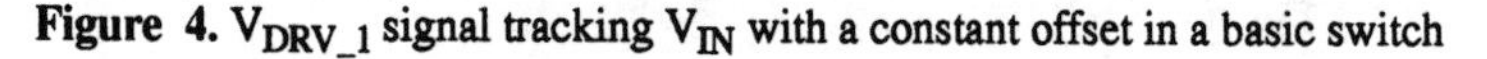

Figure 4. V_{DRV_1} signal tracking V_{IN} with a constant offset in a basic switch

For the low resistance of the closed switch R_{ON}, the control signal V_{DRV_1} should be as high as possible. To attain a very low R_{ON} of the switch while keeping the switch dimensions reasonable, it is necessary not only to track the input voltage, but also to force a positive level shift of the control voltage. In most cases it is preferable to keep the R_{ON} constant, and therefore it is necessary that V_{DRV_1} tracks the input signal V_{IN} with a constant offset (Figure 4).

Analog switch drivers with a similar feature have not yet been reported. In this paper, therefore, the single chip large signal analog switch with an up-shifted input tracking control is described.

2.4. Design of a switch-driver for large analog signal switches

In order to meet the tightest requirement - very low feedthrough - the T-switch configuration was adopted (Figure 3). In order to meet a low and constant R_{ON}, the condition $V_{IN} + V_T < V_{DRV\_1} < V_{IN} + 0.7V$ is necessary but not sufficient.

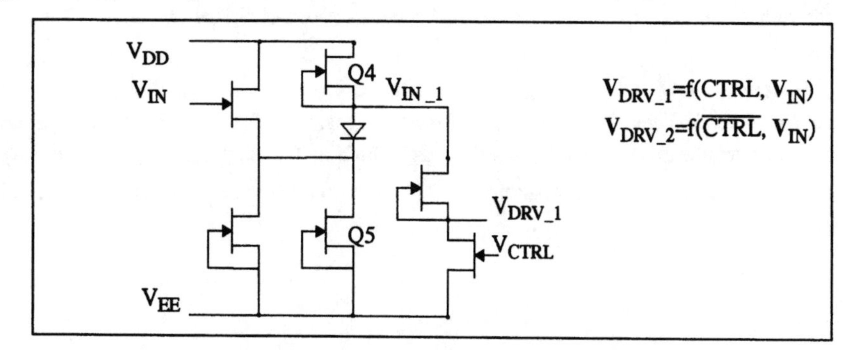

Figure 5. Input signal follower with a positive offset.

The lowest values of R_{ON} are reached with the drive signal closing the switch as high as possible without opening the gate channel diode of Q1and Q2. The switch driver therefore has to be designed in such a way that V_{DRV_1} approaches $V_{IN} + 0.7V$. This limit was set slightly lower, at $V_{IN} + 0.6V$, since the circuit should exhibit a correct behavior in all technological corners, as shown in Figure 4.

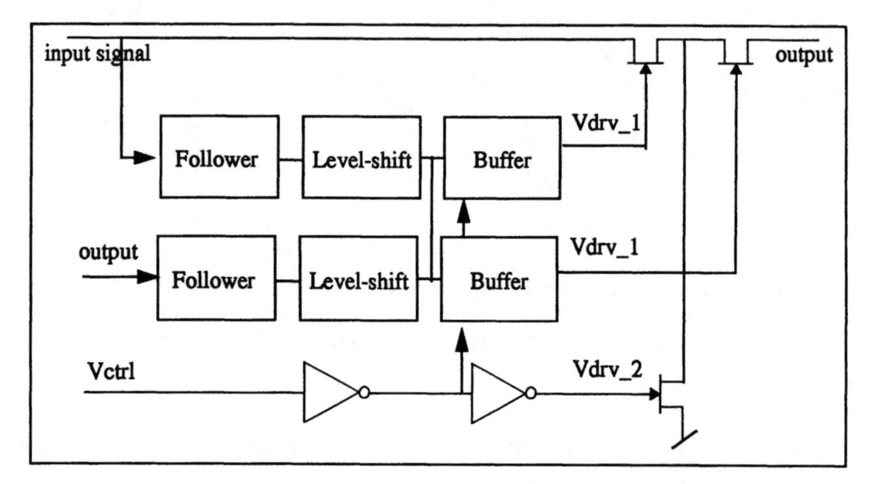

Figure 6. Block scheme of the switch drive circuitry

The circuit capable of providing the desired V_{DRV_1} is depicted in Figure 5. This circuit consists of a two-stage level-shifting circuit providing a positive level shift. The source-follower forming the first stage is used to increase the input impedance of the second stage formed by transistors Q4, Q5 and a Schottky diode. The positive level shift introduced by a forward-biased diode is maintained by the transistors Q4 and Q5. The drive signal V_{DRV_1} for the switch is provided by an inverting amplifier, whose power supply is derived from a positively shifted analog input signal V_{IN_1}, and which is driven by the control signal (V_{CTRL}). Careful ratioing of the transistor widths is necessary to meet a trade-off between the power dissipation, offset level, and attenuation

The block scheme of a complete driver for the large signal analog switch which has been designed is depicted in Figure 6, and the detailed scheme appears in Figure 7. Since the switch has to be bi-directional, the drive signal V_{DRV_1} depends on both analog input and output signals (V_{IN}, V_{OUT}). The drive signal V_{DRV_2} for transistor Q3 of the switch is provided through a cascade of two inverters

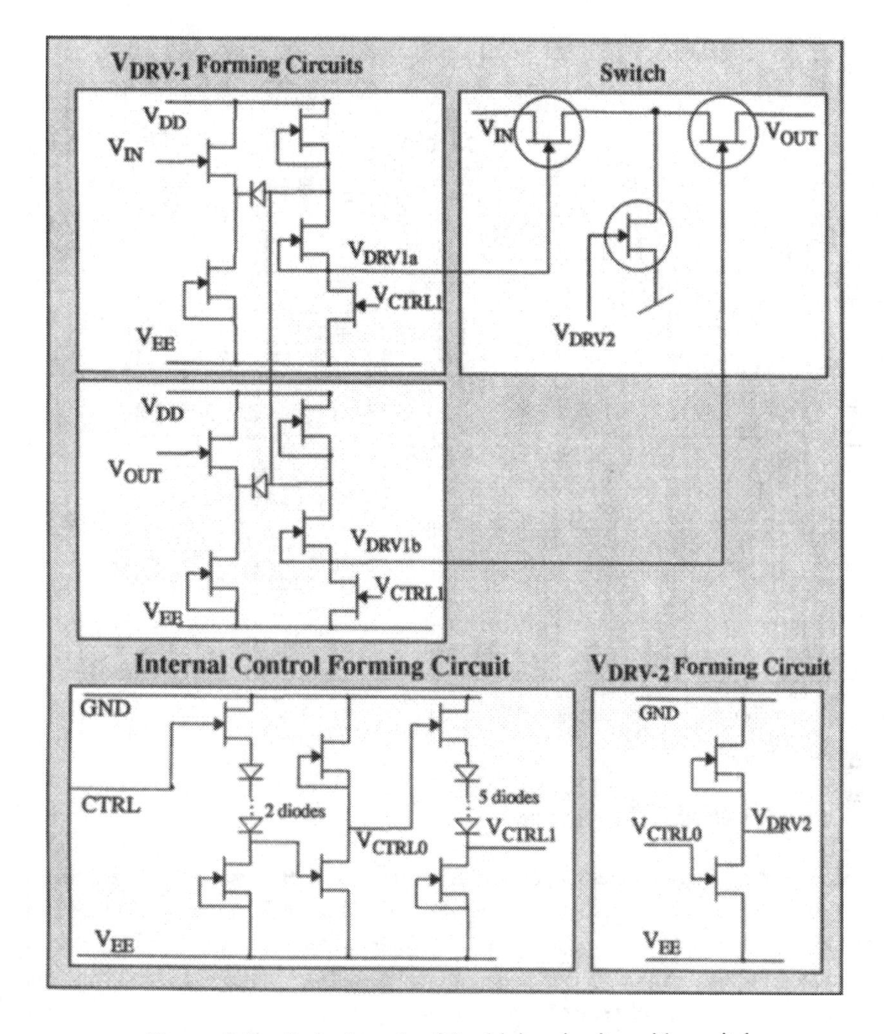

Figure 7. Detailed schematic of the driving circuitry with a switch

The layout of a complete analog switch cell containing a T-switch and its drive block is shown in Figure 8. The size of a single switch cell was reduced to 240 x 240μm, resulting in a total chip area of about $40mm^2$

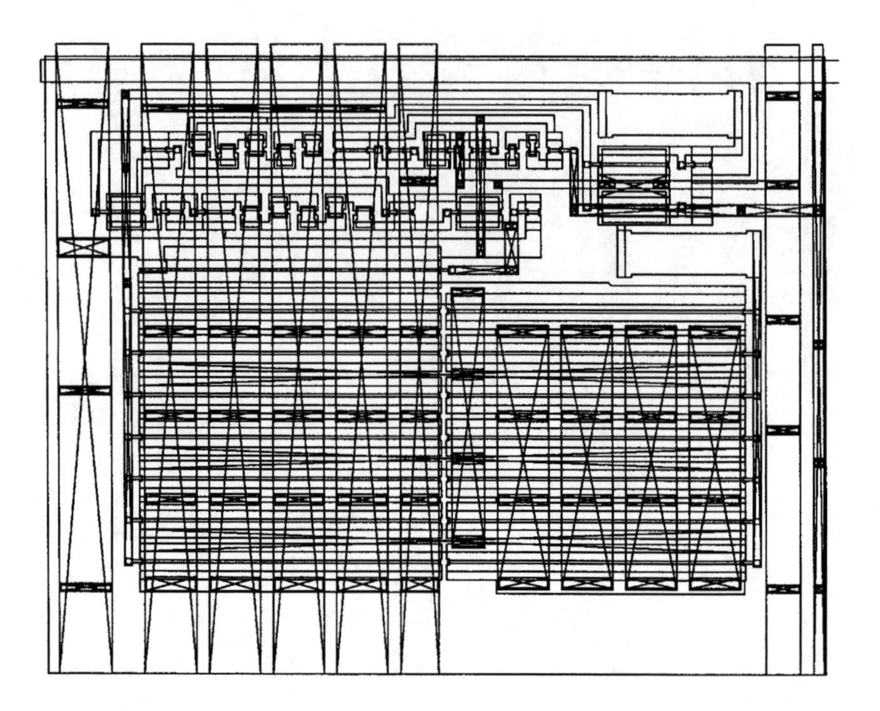

Figure 8. Layout of the analog switch cell

2.5. Characteristics of the analog switch

The frequency dependence of insertion loss and the phase shift for the closed switch are shown in Figure 9 (a), (b). All the technological corners a displayed since the switch should perform sufficiently well even in the worst case. The R_{ON} is almost insensitive to the signal DC level and remains almost invariant (within 5%) in the whole signal range. Note that up to 1GHz the insertion loss is constant and its value increase of 1dB at 4 GHz.

The frequency dependence of channel isolation for the open switch is shown in Figure 10. The switching time measured for a 100MHz full amplitude signal is less than 20ns (Figure 11). All the simulation results were obtained using PSpice.

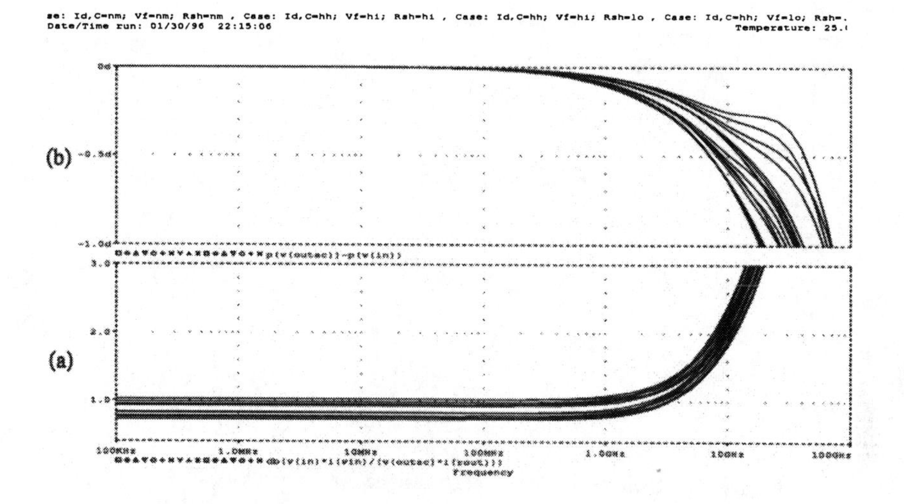

Figure 9. Frequency dependence of insertion loss (a), and signal phase shift (b) for the closed switch in all technology corners

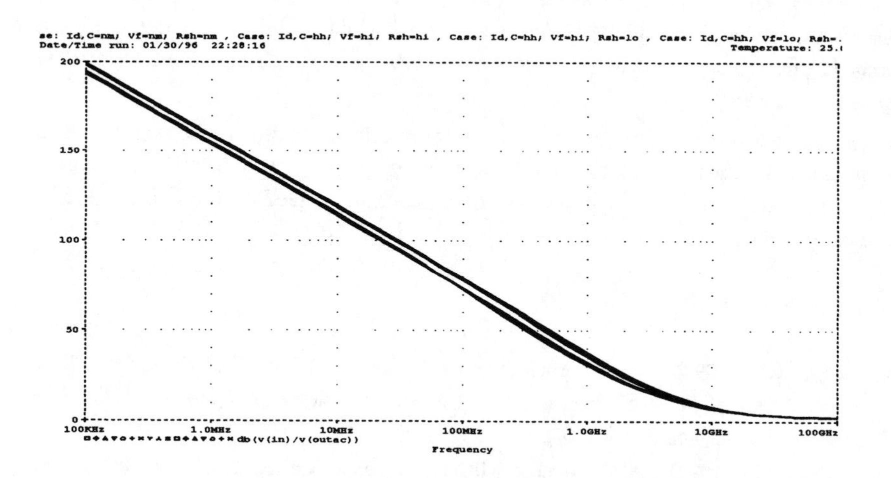

Figure 10. Frequency dependence of channel isolation for the open switch

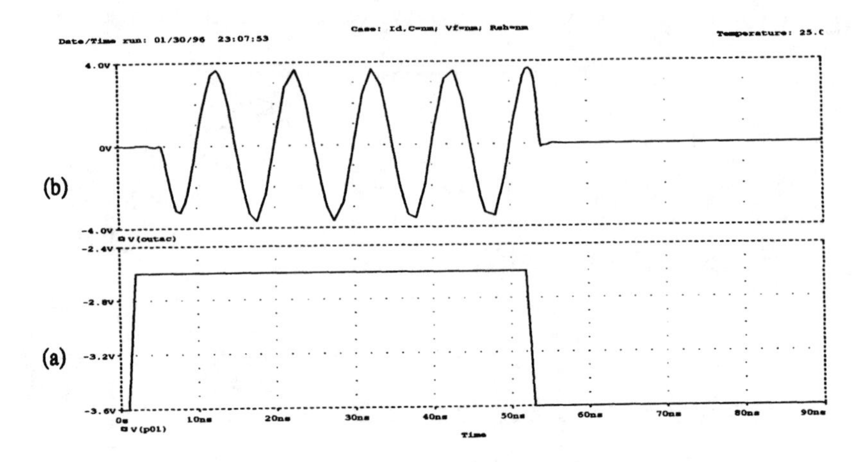

Figure 11. Switching time; (a) control signal, (b) output signal

3. CONCLUSIONS

The first fully integrated large signal analog switch matrix having the dimensions of 13 x 13 x 2 was developed based on our new concept of the input tracking switch drivers. This circuit was fabricated in a standard 1 µm QED/A GaAs process from TriQuint. The design considerations and experimental results of important parameters of the developed switch matrix were presented altogether in this paper.

The matrix cell surface and power dissipation were carefully optimized to obtain a desired high level of integration without sacrificing the circuit performances. The IC which was designed satisfies the target specifications, and is capable of operating within the DC to 3.7GHz range, with a signal of up to $8V_{p-p}$.

4. REFERENCES

Uda, H. et al., (Oct. 1994) High-Performance GaAs Switch IC's DFabricated Using MESFET's with Two Kinds of Pinch-off Voltages and a Symetrical Pattern Configuration. *IEEE J. of Solid_state Circuits*, vol. 29, No. 10, p 1262.

Kusonoki, S. et al. (1992) SPDT Switch MMIC Using E/D-mode GaAs JFET's for Personal Communications. *GaAs IC Symp. Dig.*, p 135.

Gardiner, G. J. et al. (1989) Design techniques for GaAs MESFET switches", *IEEE MTT-S Digest*. pp. 405-408.

Slobodnik, G.J. et al. (Aug. 1989) Millimeter wave GaAs switch FET modeling. *Microwave Journal*, pp. 93-104.

Feng, S. and Seitzer, D. (Apr. 1994) Improved passive model for transient prediction of GaAs E/D MESFET analogue switches. *IEEE Proc.-Circuit Devices Syst.*, vol. 141, no. 2, pp. 105-110.

Barber, F. E. et al. (1988) A 64x17 Non-Blocking Crosspoint Switch", *IEEE Int. Solid-State Circuits Conf. Dig*. pp. 112-117.

36

Current and Advanced Protocols over ATM: Evaluation, Implementation and Experiences

Sabine Kühn, Alexander Schill
Dresden University of Technology; Department of Computer Science
01062 Dresden, Germany, call: 049351/4575261
kuehn/schill@ibdr.inf.tu-dresden.de

Abstract

This paper describes recent experiences with evaluating and implementing advanced communication protocols on top of ATM. First, experiences and performance results with conventional TCP/IP over ATM based on Digital Equipment's Gigaswitch/ATM are reported. Moreover, the behaviour and performance of application-level remote procedure call protocols with mass data transfer extensions over ATM is discussed. It becomes obvious that current protocols must be tuned specifically in order to exploit ATM performance. Moreover, the lack of reservation facilities, especially in heterogeneous environments, has been identified as another significant problem; currently, no quality of service guarantees are possible.

In order to address the quality of service issues, the second part of the paper describes advanced concepts and an implementation of IPng (IP next generation) and RSVP (Resource Reservation Protocol) over ATM. Particular emphasis is put on the problem of mapping quality of service and traffic parameters in an adequate way. Moreover, the issue of address mapping from IPng onto ATM is also discussed. First implementation results in these areas are presented. The paper finally concludes with an outlook to future work in the area of resource reservation in advance.

Keywords

ATM, TCP/IP, IPng, RSVP, quality of service, resource reservation, performance measurements

1 INTRODUCTION

Recently, local ATM networks (Partridge 1994) have become widely available as products. With the current UNI 3.1 (User Network Interface) specification and the emerging UNI 4.0 specification of the ATM Forum (ATM 1993), (ATM1, ATM2 1995), interoperability can also be achieved on a broad basis. It now becomes more and more important to actually support applications on top of ATM with sufficient performance, and also in heterogeneous network environments. While it is already possible to run applications directly over AAL5, additional transport- and network-level protocols are required in heterogeneous settings, for example with Ethernet, FDDI, and ATM subnetworks being interconnected (Malamud 1992),

(Alles 1995). The typical choice of many vendors is to offer the TCP/IP protocol suite (Stevenson 1995) over ATM, based on the IP over ATM recommendation of the ATM Forum (Laubach 1994), with LAN Emulation (LANE; (LAN 1995)) as another alternative with significant limitations. Most important, compatibility of existing applications with ATM is achieved by using IP over ATM.

In this paper, we first present experiences and performance results for IP over ATM based on a local ATM network using DEC Gigaswitch/ATM. Comparisons with other standard networking technologies are also performed experimentally. It becomes obvious that current transport protocols are not ideally suited for ATM, and that tuning mechanisms and functional improvements are required. We also discuss higher-level, application-layer RPC (remote procedure call) protocols, presenting measurements of the Open Software Foundations's DCE RPC (Distributed Computing Environment). It shows that such conventional client/server interaction techniques present an even more signification bottleneck that has to be overcome.

Meanwhile, the IETF (Internet Engineering Task Force) has also specified a follow-on version of IP, the IP version 6 (or IPng - IP next generation) protocol (Hinden 1995), (Narten 1995), (Gilligan 1995). The major goal is to enhance the IP address space due to the rapid growth of the Internet and to offer additional functionality. Moreover, the resource reservation protocol RSVP (Braden 1995), (Estrin 1995), (Zhang 1993) has been developed. Such a protocol is crucial for guaranteeing quality of service (QoS) characteristics based on explicitly reserved network, memory and processing resources (Campbell 1994). It is of particular importance in heterogeneous networks with partial ATM infrastructures.

The second part of the paper therefore addresses IPng and RSVP and reports concepts and first experiences with IPng and RSVP over ATM. Particular emphasis is put on the problem of mapping QoS and traffic parameters in an adequate way. This is of specific importance as the kind of QoS specification differs significantly between RSVP and ATM so that the mapping is non-trivial. Moreover, concepts for mapping IPng addresses onto ATM addresses are also presented, and relevant implementation-level aspects are discussed. Based on this work, early experiments with these new protocols have become possible; this way, the new functionality can readily be exploited by emerging ATM applications, especially in heterogeneous network environments.

As far as it is known to the writer, only a few results in these areas have been found in the current literature as yet. In (Konstantoulakis 1995), the transfer of data over ATM using available bit rate is examined. Memory management is discussed and advanced switching concepts are proposed. However, higher-level protocols haven't been investigated yet. (Cidon 1995) presents a higher-level investigation of application-level performance in ATM networks; however, like many other similar studies, it is only based on analytical and numerical models and does not include practical measurements. While many research efforts also concentrate on QoS specification and supervision (Campbell 1994), (Vogel 1995), the actual implementation of QoS-related reservation protocols especially over ATM has hardly been addressed yet, with a few exceptions such as the ST-II work described in (Delgrossi 1994).

2 PERFORMANCE EVALUATION OF PROTOCOLS OVER ATM

In this section, we present selected performance results of conventional TCP/IP and RPC over ATM and discuss experiences and consequences. Initially, our experimental environment is briefly introduced as a general basis.

2.1 Experimental Environment

Figure. 1 shows our experimental environment. Several multimedia workstations of type DECstation 3000 AXP 700 and 300 and a server are connected with our DEC Gigaswitch/ATM via fiber optic links. ATM access is implemented by adapter cards in the workstations and by line cards in the switch. The adapter cards perform cell generation from input packets of variable size and transmit the cells using SONET/SDH frames with standard 155 Mbit/s per channel via multimode fiber. Cell assembly and disassembly is done in hardware. Only AAL5 is currently implemented.

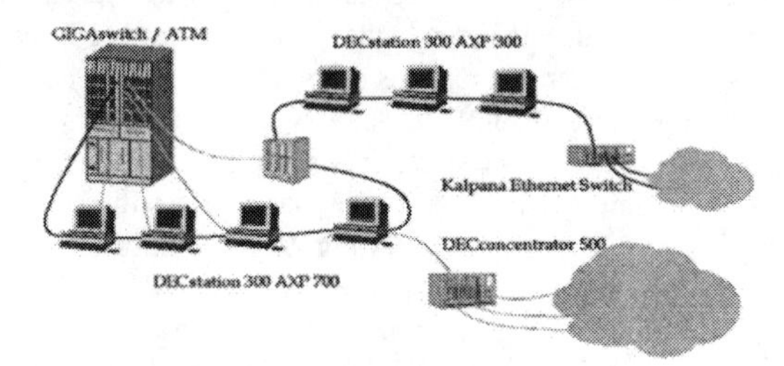

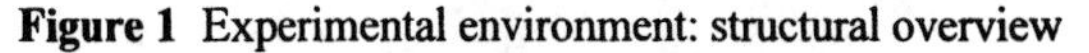

Figure 1 Experimental environment: structural overview

The switch itself offers a total performance of 10.4 Gigabit/s and is input-buffered. Possible head-of-line-blocking, a potential problem of input buffering, is reduced by a specific output port allocation algorithm (parallel iterative matching). Both PVCs (permanent virtual channels) and SVCs (switched virtual channels) are supported; signaling is based on ITU Q.93B with Q.2931 as the follow-on version. Moreover, CBR (constant bit rate), VBR (variable bit rate), and ABR (available bit rate), both with point-to-point and point-to-multipoint VCs, are basically possible. However, the driver and subsystem software does not currently support CBR yet, so that the experiments were mainly based on ABR.

The multimedia workstations are also connected via Ethernet and have access to another Ethernet switch, and also to a FDDI ring and to the Internet and via a concentrator. Each station is equipped with typical devices such as cameras, microphones, speakers etc., using MME (Multimedia Environment) as an internal software platform. In addition, the OSF Distributed Computing Environment (DCE) is installed. Over all, the installation and maintenance of our environment did not create major problems, and existing applications could easily be ported to run within this infrastructure.

2.2 TCP/IP over ATM

Within our experiments, we first evaluated the performance of TCP/IP over ATM using a tool named 'netperf'. Figure. 2 shows a major summary diagram of the results. Both the message sizes transferred via TCP/IP and the buffer sizes at the receiver's site were varied; initial experiments have already shown a strong influence of both parameters. The major target parameter was the actual throughput that could be achieved. First, transmission was unidirectional only.

The maximum throughput achieved was 135 Mbit/s with optimal parameter values. Although this equals 87% of the physical bandwidth, it also is notable that the CPU load of more than 60% was significant then, caused both by the sender and receiver applications and by protocol processing. Nevertheless, the experiment has shown that the bandwidth of ATM can really be exploited based on adequate protocol parameter setting and sufficient CPU capacity.

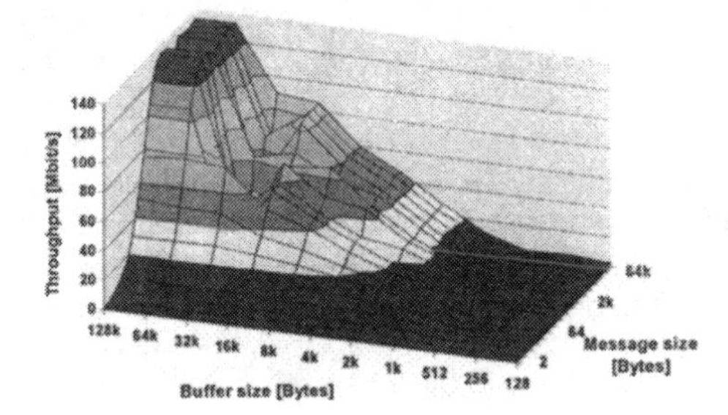

Figure 2 TCP/IP over ATM: Throughput with varying buffer and message sizes

With a buffer size of less than 64 kbyte, however, performance dropped significantly, for example down to values between 60 and 90 Mbit/s for buffers of 32 kbyte. The higher values (90 Mbit/s) could only be achieved with significant message sizes. Otherwise, the constant part of the overhead of protocol processing had increasing influence. Of course, very small messages resulted in very poor performance. Similarly, very small buffers led to very poor performance, too, due to buffer overflow with resulting loss of data and subsequent retransmissions.

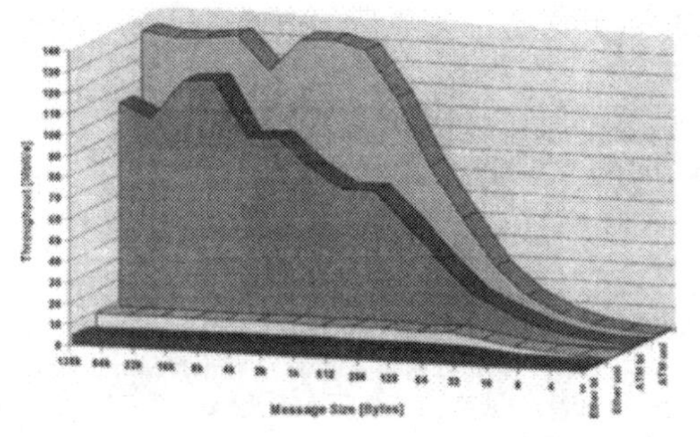

Figure 3 Throughput comparison among heterogeneous networks

Figure. 3 shows the results of further experiments with bidirectional traffic and with Ethernet to be compared with ATM, using buffer sizes of 128 kbyte. First, ATM performance drops to 110 Mbit/s and less per connection although a 155 Mbit/s VC is available for each direction. This is caused by significantly growing CPU load, now that both the sender and receiver applications and protocols have to be handled by each machine. The comparison with Ethernet shows that much lower throughput (a maximum of 8.5 Mbit/s) is achieved as expected, and that bi-directional traffic leads to even more significant reductions (down to 3 Mbit/s per connection) due to the shared medium with collisions. For FDDI, similar effects were observed, i.e. a more than 50% performance reduction per connection for bidirectional traffic. The major reason is that the FDDI bandwidth of 100 Mbit/s is to be divided among all communication partners while ATM VCs can be provided exclusively for each pair of stations, and also for each direction.

2.3 RPC over ATM

In another set of experiments, we evaluated the performance of RPC over ATM versus Ethernet. RPC is the major protocol for implementing today's client/server applications, and the DCE RPC implementation used here is now widely available and in practical use (Nicol 1993), (DCE 1994).
Performance of RPC over ATM will therefore be important for exploiting ATM performance characteristics also at the client/server application level. Basically, the major advantage of RPC as compared with socket-level communication is that RPC is much easier to use and provides a higher degree of distribution transparency. The results are summarized in Figure. 4. Over Ethernet, we achieve a throughput of 8 Mbit/s with sufficient parameter sizes.

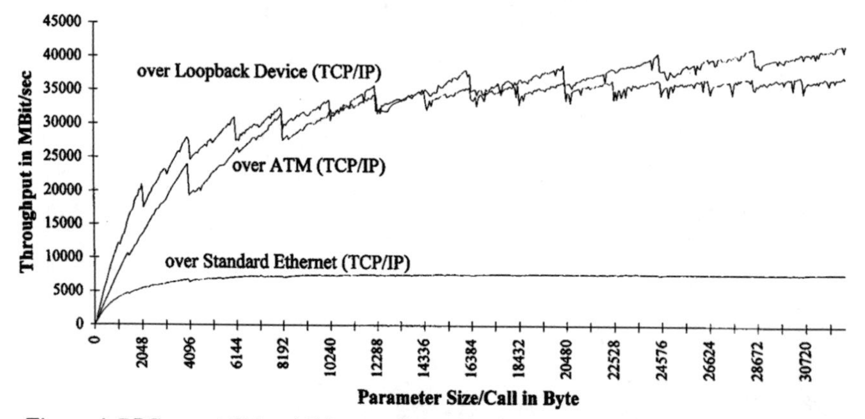

Figure 4 RPC over ATM and Ethernet: Comparison

This is a comparably good result; the RPC protocol obviously does not add too much overhead at this data rate level. With ATM, the maximum throughput, even for large parameters, was 39 Mbit/s. With a special mass data transfer mechanism of DCE RPC (so-called RPC pipes), a slightly better result of 45 Mbit/s was achieved. Nevertheless, the results were much worse than with pure TCP/IP over ATM where 135 Mbit/s were reached. There are two reasons for this.
First, the client is delayed until explicit acknowledgements are received. Even for large parameters, the time to transfer the acknowledgements and to schedule the client process upon receipt is comparably large.

Secondly, the RPC protocol does not fully exploit the MTU (message transfer unit) size of TCP/IP due to its internal implementation. Moreover, when a multiple of the MTU size is reached by the parameter size, a performance decrease is observed due to partially filled MTUs. As shown in the figure, data are blocked to form MTUs of 4096 bytes although higher MTU sizes would basically be possible as discussed above. Finally, experiments with local RPC communication based on a loopback mechanism are shown in the diagram. They illustrate that even in the local case (i.e. without ATM communication), performance is not significantly better. This also confirms that protocol processing, RPC acknowledgements and process scheduling are of major impact (rather than the actual network transfer via ATM).

2.4 Experiences

Summarizing our experiences, several recommendations can be made. First, it has been confirmed that higher-level transport and application protocols present major performance bottlenecks in ATM applications. This problem can be overcome to some degree by tuning the protocol parameters, namely message sizes and buffer sizes. However, many applications, including RPC, often do not deliver messages that are large enough to achieve satisfying performance. Moreover, buffer sizes may be limited by the hardware, especially if multiple ATM applications coexist on a system. Therefore, it can be recommended to consider the possible use of ATM already during application development, for example for designing the data transfer phases and the mechanisms to be used. Moreover, automatic adaptation of protocol parameters according to the underlying network would also be a reasonable goal. Finally, conventional client/server interaction mechanisms seem to be ill-suited for exploiting ATM facilities. It will therefore be important to integrate more efficient, basically unidirectional mass data transfer mechanisms with RPC in a semantically and syntactically uniform way. The pipe mechanism mentioned above is a first approach, but has to be tuned internally.

Moreover, under normal load conditions (with regular background load), we also observed that the quality of video transmission varied significantly due to the current network and local load conditions. This problem can only be addressed by offering CBR and VBR mechanisms with guaranteed bandwidth. However, this requires resource reservations in the end systems and in the active network components (i.e. switches, routers, bridges etc.). Important resources are bandwidth, memory or buffers, and CPU cycles.
For these reasons, we are currently working on the implementation of RSVP (resource reservation protocol) over ATM. Major problems and concepts are discussed below. This work is coupled with the implementation of IPng over ATM that is also described.

3 IPNG AND RSVP OVER ATM

The deployment of the new Internet Protocol on separate data link technologies is very important in view of a broad propagation concerning both already existent technologies such as Ethernet and new technologies such as ATM. This part of the paper describes an implemented adaptation of Internet Protocol next generation on ATM. Further, it introduces an approach to solve the problem of mapping QoS parameters required by applications onto ATM parameters to utilize the advantages of resource reservation in ATM. These considerations are especially based on the Resource ReSerVation Protocol (RSVP) as a future constituent of an Integrated Services Internet. The major characteristic of RSVP is an extended support of QoS for the Internet Protocol.

3.1 IPng - A Brief Overview

IPng is a new version of the Internet Protocol. Because IPng is an evolutionary step from the Internet Protocol (IPv4), it comprises on one hand numerous mechanisms of IP which have been expanded or kept in IPng. On the other hand, it contains new mechanisms and characteristics. The improvements of IPng in contrast to IPv4 fall primarily into the following categories:

- Header simplification and improved option support
  The IPng header is distinguished from the IP header in such a way, that some elements have been reduced and other elements are optional. The distinction of options into difference extension headers will reduce the bandwidth needed for the IPng header. Therefore, the IPng basic header size is only the twofold of the IP header even though IPng increases the IP address size from 32 to 128 bit. In addition, the time of packet processing in routers will be reduced because the extension header generally contains information concerning the endsystems. Changes in the way IPng header options are encoded allow for efficient forwarding and less stringent limits on the length of options.
- Expanded routing and addressing capabilities
  The extension of the IPng addresses provides more addressing hierarchy levels and a much greater number of addressable nodes. In conjunction with these aspects IPng offers new address formats. One of these is the Anycast Address identifying sets of nodes, whereby a packet sent to an Anycast Address is delivered to one of the nodes only.
- Security and authentication mechanism
  IPng includes additional options for the definition of extensions which provide support for authentication, data integrity, and confidentiality. These basic elements of IPng will be included in all its implementations.
- Quality of Service support
  The Flow Label and the Priority fields in the IPng header may be used by a host to identify those packets for which a special handling by IPng routers is requested, such as non-default quality of service or "real time" service. The characteristics of this special handling for the corresponding labelled packets belonging to the same flow may be conveyed by a resource reservation protocol or IPng options. This capability is important in order to support multimedia and real time applications which require some degree of consistent throughput, delay, and/or jitter.
- Transition mechanisms
  To facilitate the transfer from IP to IPng, IPng includes transition mechanisms, allowing an adoption and deployment of IPng in a highly diffuse fashion, and a direct interoperability between IP and IPng. Examples of such transition mechanisms are the dual IP layer, automatic and configured tunnelling, and the IP header translation as special case of the communication of pure IPng with IP hosts.

In conjunction with the adaptation of IPng onto different link layers, IPng hosts need to resolve or determine the neighbour link layer address which is known to reside on attached links. The neighbour discovery protocol will be applied for that, using ICMP messages for information interchange. This is generally done via multicasting the addresses of neighbouring routers, the reachability of neighbour hosts, and some additional information. Moving the address resolution up to the ICMP/IP layer makes IPng more independent from the underlying link layer. However, although the neighbour discovery protocol was designed for the deployment of different link layers, it is mainly suited for broadcast media like Ethernet.

3.2 IPng over ATM

Recently, formats and methods were specified for the transmission of IPng packets over different networks like Ethernet, FDDI and Token Ring. In the following, we outline basic concepts for an initial implementation of IPng over ATM. As a general basis, the following parameters must be derived from IPng parameters to allow support of applications via ATM networks:

- ATM address of destination,
- Quality of service and traffic parameters,
- Connection states and identifier of the ATM virtual channel.

The first three parameters represent informations which are needed for the signalling protocol to establish a virtual connection. For making reservations for dedicated flows, it is required to determine the associated virtual channel identifier. Facilitating the assignment of IPng packets to ATM VCs, the IPng packet header contains the flow label, where packets with flow label zero are sent as best effort data.
To determine the ATM destination address according to the neighbour discovery protocol, the multicast capabilities of ATM must be exploited. That is, before sending multicast messages for neighbour solicitation, an address translation of a multicast IPng address into a corresponding ATM address has to be performed. To reduce the overhead which will occur in conjunction with using a central Multicast Address Resolution Server (MARS) (Armitage 1995), the IETF currently discusses several solutions (Schulter 1995). As there are no mature results yet, address resolution for our implementation of IPng over ATM is realized based on the principle of classical IP over ATM.
The specification of QoS and traffic parameters for dedicated flows has to be performed by the application. However, with the assistance of RSVP, application-level QoS and traffic parameters can be mapped onto ATM parameters explicitly as discussed below.

3.3 RSVP

Proposed as an Internet draft, RSVP is a known constituent of the Integrated Services Internet. It provides especially real-time applications with guaranteed, predictable and controlled end-to-end performance across networks. It is a receiver-oriented protocol, which may be classified as a control protocol of the Internet Protocol. Therefore, it offers a flexible handling of heterogeneous receivers as well as an adaptation to dynamically changing multicast groups. The main task performed by RSVP is signalling of resource requirements at connection setup time. To be precise, RSVP does not actually reserve or allocate resources but rather indicates reservation requests to the underlying systems. The transmission of RSVP control information is implemented by encapsulating RSVP packets into IP/IPng or UDP packets. Reservations may be performed for both unicast and multicast connections. The basis of a reservation is a detailed description of the flow traffic and the QoS characteristics. In accordance with this fact, RSVP defines the so-called flow and filter specification. The flow specification one specifies the traffic (TSpec) using a token bucket parameter for a describing bursty traffic, and also determines the required QoS parameters (RSpec). The filter specification contains an identification of the flow for which reservations have been performed. Even though the reservation is receiver oriented, the initiator of a reservation is the sender, which informs appropriate receivers about the characteristics of the flow to be sent.

3.4 RSVP over ATM

As discussed above, reservation protocols are characterized by reservation direction, with RSVP as a receiver-oriented protocol. As opposed to that, the ATM signaling protocol is sender oriented. Moreover, using soft states to manage the state information in nodes (endsystem, router) facilitate the renegotiation of resource reservations within RSVP. Accordingly to hard states used in ATM there is no possibility to change the specified QoS parameters after call setup (Q.93b, Q.2931). In contrast to the RSVP model the QoS setup time is combined with connection establishment. These different concepts make an integration of RSVP and ATM more difficult In addition to the general functionality differences of RSVP and ATM which are shown in Table 1, RSVP supports sessions of several senders within a multicast group. ATM does not support such kind of behaviour.

In order to map RSVP onto ATM there are two possibilities. As RSVP only reserves unidirectionally, it seems to be possible to establish an ATM connection for sending the RSVP reserve message from the receiver to the sender (Borden 1995). This would be possible as ATM is able to reserve in both directions. However, this simulation of a receiver oriented establishment of a virtual channel is only applicable to unicast VCs and is in conflict with the ATM multicast environment which only allows unidirectional allocation (ATM 1993). With respect to the enhancement in UNI 4.0 (ATM1 1995) of a Leaf Initiated Join ability in point-to-multipoint connections a receiver oriented approach supporting RSVP flows is thinkable.

From our point of view there will be problems with the interactions between RSVP and ATM in ATM-nodes; so changing the RSVP specification (e.g. the traffic control interface) will be necessary.

We propose another way to avoid considerable changes in RSVP; receiving a reservation message from the downstream host, the appropriate router or host establishes an ATM connection to the downstream hop according to the reservation information. On this basis, it will also be possible to establish ATM point-to-multipoint connections, at least to homogeneous receivers. ATM presumes homogeneous receivers even in case of heterogeneous RSVP-reservations, therefore routers have to reserve according to the highest reservation requirements. Reserving VC's between routers in an ATM network which depends on address resolution (ARP) seems to be inefficient. However, a realized NHRP over ATM allows to establish ATM shortcuts without any changes of RSVP. So an extensive modification of RSVP to realize ATM shortcuts in combination with ARP will be unnecessary.

The considerable differences between the service classes of the Integrated Services IP and ATM also require detailed analysis of mapping service classes as well as traffic and quality of service parameters. Translating such kinds of parameters is an additional service for the layer-to-layer communication during the call establishment phase. After the ATM link layer gets its own QoS and traffic parameters through negotiation and translation, the admission control is performed to control resource availability. The description of data streams by an application should be enabled in such a way that the reservation at the ATM layer complies with the actual requirements. Therefore, overestimating the resource needs of applications due to an improper mapping has to be avoided to achieve an efficient bandwidth utilization.

Considering the traffic description parameters of RSVP and ATM, ATM allows both variable bit rate (burst, peak) and constant bit rate for transmission of non bursty traffic as opposed to RSVP. The traffic parameters of ATM accordingly comprise Maximum Burst Size, Sustainable Cell Rate and Peak Cell Rate. The mapping of the particular RSVP traffic parameters is discussed in conjunction with an assignment of the service classes (Table 2).

Obviously, it would be possible to transmit all IP flows with requested guaranteed service over constant bit rate channels, however this would be related with an inefficient utilization of resources e.g. for transmission of bursty traffic. In such cases, transmission via a variable bit rate channel is more convenient.

Moreover, it is apparent that the IP/RSVP service classes are not in direct relation to the service classes of ATM service classes. The basis of this consideration are the service classes

Table 1 RSVP and ATM service classes with respect to QoS and Traffic parameters

ATM		
QoS Class		
CBR	QoS	*CLR*
		CDV[1]
		Max CTD[2]
	Traffic	*PCR*
VBR-RT	QoS	*CLR*
		CDV
		Max CTD
	Traffic	*PCR*
		SCR
		MBS
VBR-NRT	QoS	*CLR*
		Mean CTD
	Traffic	*PCR*
		SCR
		MBS
ABR	QoS	*CLR*
	Traffic	*PCR*
		MCR[3]
UBR	Traffic	*PCR*

RSVP	
Service Class	
Guaranteed Service	n/a
	n/a
	n/a
	Average Token Rate (RSpec)
	n/a
	n/a
	n/a
	n/a
	Average Token Rate (R of RSpec)
	Token Bucket Depth (TSpec)
Predictive Service	n/a
	Delay as part of RSpec
	n/a
	Average Token Rate (TSpec)
	Token Bucket Depth (TSpec)
Controlled Delay	n/a
	n/a
	Average Token Rate (TSpec)
Best Effort	No declarations are necessary.

of ATM specified in the UNI 4.0 draft of the ATM Forum in order to outline mapping problems that have to be recognized in the future.

For mapping the guaranteed service class only two ATM classes are possible: the CBR and VBR real time due to guaranteed bandwidth and end-to-end delays. In conjunction with the token bucket depth, the assignment/differentiation of the appropriate ATM service class may be performed. That means, a token bucket depth with a value equal to one deals with a constant data stream. Considering the CBR service class, peak cell rate is the only traffic parameter which actually has to be specified. However, it can directly be derived from the RSVP average token rate, which will be calculated based on end-to-end delay.

Variable bit streams combined with guaranteed service can be mapped onto the VBR real time class, expecting a token bucket depth value greater than one. Parameters of RSpec and TSpec could be translated onto sustainable cell rate and maximum burst size. PCR as a necessary traffic parameter of the VBR-RT service is currently not derivable from RSVP parameters.

Basically, RSVP doesn't define any QoS parameters for the guaranteed service class yet.

Considering the predictive service of the Integrated Services Internet, it provides a very high probability that the end-to-end delay experienced by packets in a flow will not exceed a known limit. The VBR non real time class is especially suitable for this service class because

[1] Cell Delay Variation - CDV

[2] Cell Transfer Delay - CTD

[3] Mean Cell Rate - MCR

it expects a bounded end-to-end delay, too. The mapping of traffic parameters corresponds to that of VBR real time with exception of end-to-end delay as a component of RSpec.
The controlled delay service imposes relatively low requirements on network components and does not provide any quantified level of assurance about packet delays. Instead, it merely promises to avoid overloads by turning excess traffic away. A comparison of both the controlled delay service and Available Bit Rate shows that the two classes support a best effort service in assistance with congestion control. As opposed to RSVP, ATM does not provide any end-to-end delay specification. Currently, only the average token rate may be mapped onto the mean cell rate. Pure best effort service without any requirements on network resources may be mapped onto unspecified bit rate which offers the same characteristics.
In summary, a full translation or derivation of QoS parameters is impossible e.g. the cell loss rate (CLR) can generally not be mapped from RSVP parameters onto ATM parameters. As the traffic parameters of RSVP only consist of token bucket depth and average token rate, an accurate mapping or interpretation of traffic parameters is even more difficult. Introducing a peak rate as part of TSpec may facilitate the mapping onto ATM parameters. A problem which has to be taken into consideration is the additional overhead. Applications have to consider overhead calculations of the transport and network layers. An introduction of the minimum packet size as a component of TSpec would facilitate overhead calculation within the ATM layers in order to reserve the requested bandwidth of applications while considering the actual overhead.

3.5 IPng over ATM: An Implementation

IPng over ATM has been implemented within our group based on the concepts discussed above. According to the prototypical implementation of IPng and the development stage of ATM, it has first been realized for transmission of IPng packets over ATM using best effort service. This implies that packets are multiplexed over an ABR virtual channel, as reservations based on CBR channels are not yet supported by the ATM subsystem.
The adaptation of IPng to ATM is realized as a kernel module, which is implemented as a convergence module between the network layer and the ATM subsystem. To realize an automatic resolution of the IPng addresses into the corresponding ATM addresses, we currently still use the address resolution protocol. According to the principle of classical IP over ATM, the differentiation of the ARP and IPng packets which are transmitted over the same channel is performed by encapsulating both packets with LLC/SNAP. The basis for such an address resolution is an adapted ARP which is able to handle the 16 byte IPng addresses.
In the future, the LLC encapsulation will not be necessary when using the neighbour discovery protocol. This will reduce the appropriate overhead and should resolve the problem of single point of failure which the ATM ARP server represents. The adaptation of IPng onto ATM will be an evolutionary process. With the possibility to reserve CBR in one of the next releases, the analyses concerning the adaptation of RSVP onto ATM will be integrated at the implementation level. Concerning reservation for IPng flows, mechanisms for demultiplexing of IPng packets have to be considered. The following strategies will be the basis for mapping the IPng flows onto ATM channels:

- One-to-one mapping allows the transmission of an IPng flow via a dedicated virtual channel.
- Allowing also a many-to-one mapping, more than one IPng flow can be assigned to one virtual channel. This will be meaningful for flows with the same QoS service requirements.

- All-to-one mapping is currently used because no reservations are possible. In this way, all packets are delivered via ABR using best effort with flow control.

Testing this implementation of IPng over ATM, no noticeable performance losses were recognized in spite of the double sized IPng header. Such tests were be performed using the known network program 'ttcp' for IP and 'ttcpv6' for IPng to get an overview about the performance differences of both protocols. In summary, our experiences were rather positive, so that this implementation will be used as a reference work for further integrating the remaining concepts discussed in this paper, i.e. flexible address mapping, and integration of the full RSVP mechanisms.

4 CONCLUSIONS AND FUTURE WORK

This paper has presented performance measurements of TCP/IP and RPC over ATM, and has described the concepts and implementation of RSVP and IPng over ATM. It became obvious that current protocols have to be tuned explicitly in order to provide satisfying performance over an ATM network. Moreover, conventional client/server interaction mechanisms will have to be reengineered for being used efficiently over ATM environments. Applications will also have to explicitly consider the use of ATM, for example for engineering their data transfer phases to achieve maximum performance. New reservation protocols, namely RSVP, outline interesting directions towards guaranteed quality of service. We have shown that a mapping onto ATM is possible but that also new problems arise concerning adequate mapping of QoS parameters. Additional work, both at the conceptual level and at the standardization level, will be required in this area.

In the near future, we plan to do further experiments with different applications. We also completed a video transfer tool with explicit scaling mechanisms that can be used in this area. Concerning implementation, we are currently working towards the completion of the full RSVP/IPng/TCP protocol stack in cooperation with Digital Equipment. We also started a longer-term project on resource reservation in advance. The major goal is to augment the "immediate" reservation mechanisms of RSVP with advance scheduling and planning techniques. As an example, it shall be possible to schedule a videoconference with several partners for a given time. The system shall then calculate and virtually reserve the required resources for that time, however without immediately blocking them. This way, a future lack of resources during actual reservation can be avoided. Many specific problems have to be solved, for example the cooperation of advance and immediate reservation protocols, the management of application priorities, effects of "overbooking" resources, and advance reservation for weakly-specified scenarios, for example with dynamic integration of new partners. Concepts in these areas will be validated by implementation and concrete applications.

Acknowledgements

We would like to thank Digital Equipment Corporation (European Applied Research Center Karlsruhe) for extensively supporting the work described in this paper within the European External Research Program. Moreover, we would like to thank all students and colleagues who contributed to the presented results, namely Frank Breiter, Samer Habib, Sascha Kümmel, and Andreas Wentland.

5 REFERENCES

Alles, A. ATM Internetworking, Cisco Systems Inc.

Armitage, (1995) Using the MARS to support IP Unicast over ATM. draft-armitage-ipatm-mars-unicast-01;Work in Progress, Bellcore

ATM (1993) ATM User-Network Interface Specification Version 3.0, ATM-Forum: Specification, Prentice Hall

ATM1 (1995) ATM-Forum: Technical Committ; ATM Forum 94-1018R5: UNI Signaling 4.0, ATM Forum Specifikation, Work in Progress

ATM2 (1995) ATM-Forum: Technical Committ; ATM Forum 95-0013R6: Traffic Managment Specification Version 4.0, ATM Forum Specifikation, Work in Progress

Borden, M.; Crawley, E., (1995) Integration of Real-time Services in an IP-ATM Network Architecture, RFC 1821, Informational, August

Braden, R.; Zhang, L. Resource ReSerVation Protocol (RSVP) --Version 1 Functional Specification, Work in Progress

Campbell, A., Coulson, G., Hutchison, D. (1994) A Quality of Service Architecture; Computer Communication Review, Vol. 1, No. 2, pp. 6-27

Cidon, L., Guerin, R., Khamisy, A. (1995) An Investigation of Application Level Performance in ATM Networks; IEEE Infocom, Boston, pp. 845-852

Distributed Computing Environment An Overview; Open Software Foundation

Delgrossi, L., Herrtwich, R.G., Hoffmann, F. (1994) An Implementation of ST-II for the Heidelberg Transport System; Internetworking: Research and Experience, Vol. 5, pp. 43-69

Estrin, D.; Shenker, S. Routing Support for RSVP, Work in Progress

Gilligan, R.; Nordmark, E. Transition Mechanism for IPn6 Hosts and Routers, Work in Progress

Hinden, R.; Deering, S. (1995) IP Version 6 Addressing Architecture,Work in Progress

Konstantoulakis, G., Stassinopoulos, G. (1995) Transfer of Data over ATM Networks Using ABR; IEEE Symposium on Computers and Communications, Egypt, pp. 2-8

LAN (1995) LAN Emulation over ATM Specification-Version 1, ATM Forum

Laubach, M. (1994) Classical IP and ARP over ATM; RFC 1577

Malamud, C. Stacks (1992) Interoperability in Today's Computer Networks, Prentice-Hall, Englewood Cliffs, N. J.

Narten, T.; Nordmark, E. (1995) Neighbor Discovery for IP Version 6 (IPv6), Work in Progress

Nicol, J.R., Wilkes, C.T., Manola, F.A.(1993) Object-Orientation in Heterogeneous Distributed Computing Systems; IEEE Computer, Vol. 26, No. 6, pp. 57-67

Partridge, C. (1994) Gigabit Networking, Addison Wesley Publishing Company

Schulter, P. (1995) A Framework for IPv6 Over ATM. draft-schulte-ipv6atm-framework-00; Work in Progress, DEC

Stevenson, R. ;Wright, G. (1995) TCP/IP Illustrated (Volume 2) The Implementation, Addison Wesley Publishing Company

Vogel, A., Kerherve, B., Bochmann, G., Gecsei, J. (1995) Distributed Multimedia and QoS: A Survey; IEEE Multimedia, Vol. 2, No. 2, pp. 10-19

Zhang, L., Deering, S., Estrin, D., Shenker, S., Zappala, D.(1993) RSVP: A New Resource Reservation Protocol; IEEE Network, pp. 8-18

Multi connection TCP mechanism for high performance transport in an ATM Network

Masami Ishikura, Yoshihiro Ito, Katsuyuki Yamazaki, Tohru Asami
KDD R & D Laboratories
2-1-15 Ohara Kamifukuoka-shi, Saitama 356, JAPAN
Telephone: +81-492-78-7892   *Fax: +81-492-78-7510*
e-mail: ishikura@lab.kdd.co.jp

Abstract

This paper proposes a new transport mechanism, multi-connection TCP, for high-speed and long delay networks. With to the appearance of ATM network services, wide area data communication networks(WANs) are becoming faster and more world wide oriented. In this situation, the performance of window-based, flow-controlled and connection-oriented transport protocols, such as TCP, is degraded far from the optimum. The throughput of TCP is limited by its window size and round trip delay time(RTT), and the ideal throughput of TCP with 64Kbyte maximum window size over a trans-Pacific ocean link such as TPC-4 is only about 2.6Mbps for a 45Mbps DS3 line with 200ms RTT. In order to achieve high throughput of inter-LAN communications via a high speed, long delay WAN, such as B-ISDN, we propose a multi-connection TCP, and verify its performance by applying it to FTP. A multi-connection FTP can establish up to 16 connections simultaneously, and transfer the data in parallel. The experiment shows multi connection TCP (16 connections and 24Kbytes of each window size) achieves about 8.5Mbps end-to-end throughput via an emulated OC3 155Mbps WAN with 200ms RTT.

Keywords

TCP, transport protocol, high speed network, long delay network, ATM

1 INTRODUCTION

ATM network services are now available in many countries, and data communication networks, such as enterprise networks, are becoming faster and more world-wide oriented. It is, however, difficult to achieve high throughput in a high speed and long delay environment. The performance of window-based, flow-controlled and connection-oriented transport protocols, such as TCP, is extremely degraded due to the short window size in such an environment. The window size of the ordinal TCP implementation is limited to 64Kbytes from the protocol specification[1], and almost all applications, including FTP, uses a TCP window size of less than 32Kbytes. The other problem, for high performance throughput, is traditional error control mechanism of TCP. The existing TCP uses simple re-transmission of the remaining data on an errored packet which is detected. In order to obtain high performance of error protection in a high-speed and long delay environment, a more efficient method, such as a selective acknowledge re-transmission mechanism, is necessary[2].

In order to resolve these problems, several solutions have been investigated. The first solution is to develop a completely new protocol which works efficiently even in a high-speed, long delay network and this is one of the best methods for the future. XTP[3] is one of those protocols. However at present, these new protocols have not been fully standardized, and there is no compatibility with the existing TCP. They can be implemented only in a few types of OSs, hence will not be used widely in the market. The second solution is to enhance the existing TCP protocol[4]-[7]. It should be noted, however, that the window size will be limited up to several hundreds of Kbytes due to the available resource of workstations, and cannot be easily expanded to an order of Mbytes. In addition, the expanded TCP does not support a selective re-transmission mechanism. The extended TCP method is evaluated by [8] in an error free environment with a moderate delay (up to 68ms).

The third solution, proposed and evaluated in this paper, is to establish multiple TCP connections simultaneously to obtain high throughput as an aggregated performance. The idea of a multiple connection mechanism itself has been commonly used in the lower layers, e.g., a multi-link procedure[9] in the HDLC datalink layer, and a digital channel aggregation[10] in the physical layer. In a transport layer, multi-connection TCP is proposed here to achieve high throughput on an end-to-end basis.

This paper presents a multiple connection TCP mechanism, which differs from the previous approaches as follows : (1) the transport layer contains multiple windows, each of which corresponds to a connection, (2) each connection may follow a standard TCP mechanism, (3) this method can also work like a selective acknowledgment mechanism in an erroneous environment, (4) the TCP slow start mechanism in this method works more rapidly than that in the extended TCP method. Furthermore, this solution could be applied to the extended TCP method to obtain higher performance.

We have implemented a multi connection mechanism to evaluate the performance of a data transfer method in a high-speed, long delay network. Implementation has been firstly done as an FTP application level, i.e., Multi-connectioned FTP(MFTP). MFTP aggregates the multiple TCP connections. In the following Section 2, we clarify the problems of TCP dynamics over a high-speed, long delay network, and examine several solutions of the problems in detail. In Section 3, we present the mechanism of MFTP and discuss features and applicability of the multi-connection method to other applications such as HTTP. In Section 4, experiments on MFTP and results are shown. In Sections 5 and 6, a discussion and conclusion are given.

2 TCP OVER HIGH SPEED, LONG DELAY NETWORKS

2.1 Problems of TCP dynamics

Because the TCP was not designed to be used in recent high-speed, long delay network, such as ATM wide area networks, the maximum window size is insufficient for those network applications. The maximum window size of TCP is specified in [1] and is limited to 64Kbytes. In order to make use of network bandwidth efficiently, the round trip time(D) bandwidth(BW) products(D *BW) of the network must be smaller than the maximum window size of TCP. The ideal throughput(Ti) of TCP, where the bandwidth is not limited, is obtained by the equation(1).

$$Ti = W / D \qquad (1)$$

Ti(bps) : throughput of TCP
W(bit) : maximum window size of TCP connection
D(s) : Round Trip Time (RTT)

For example, using a DS3(45Mbps) link between the United States and Japan(the round trip time is about 200ms via submarine cable), the D*BW product is about 9Mbits(1.12Mbytes), which is much larger than the TCP maximum window size. In this case, a situation called window close occurs. This causes degradation of network utilization, and the throughput with a 64Kbyte TCP window size achieves only about 2.6Mbps. The bottleneck of the throughput is the shortage of TCP window size.

2.2 Features and advantages of multi connection method

In order to reduce the network idle time caused by window close, there are three approaches. The first is a completely new protocol for a high-speed, long delay network such as XTP. The second is the extension of TCP window size using TCP options described in [7], and the third is the aggregation of multiple TCP connections.

The third one, which is called multi connection TCP mechanism, lets an application use the multiple existing TCP connections in the aggregate. This multi connection mechanism could apply to the extended TCP described above as well as the traditional TCP. Furthermore, this mechanism supports a pseudo selective acknowledgment mechanism.

In the case of a single TCP connection, when an error is detected, a go-back-N error control mechanism starts and re-transmits all data after errored data. As mentioned in [5], it may cause the degradation of throughput to expand the window size in an erroneous environment. However, in the case of the multi connection method, if data transmit errors occur in some connections, the go-back-N error control mechanism starts simultaneously only for those connections and re-transmits all data after errored data for each errored connection respectively. This speeds up the re-transmission.

When all the errors have been re-transmitted, TCP starts a slow start. As shown in Figure 1, the time Tr to recover from the slow start condition to the congestion control mode of a multi

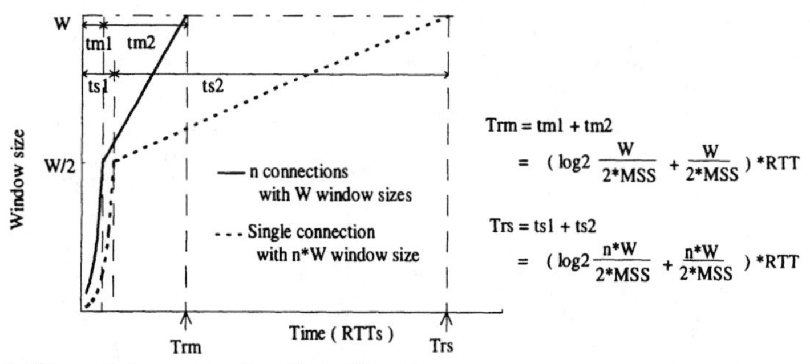

Figure 1 Illustrated recovery time from Slow Start.

TCP connection (n connections, each with window size W and maximum segment size MSS) is much shorter than a single TCP connection with window size n*W and maximum segment size MSS. The former consists of n independent connections with maximum window size W in parallel. Tr for each congested connection is given as (log2(W/(2*MSS))+ W/(2*MSS)) times RTT, which is the Tr for the former. Tr for the latter case is given as (log2(n*W/(2*MSS))+n*W/(2*MSS)) times RTT. This means that Tr for the former is almost 1/n of the latter Tr for large W/MSS.

These advantages described above show the multi-connection method is more durable than single connection method in an erroneous environment.

Considering software portability at present and applicability to extended TCP in the future, we also present the application layer multiple TCP connection method as a near term solution to realize high performance file transfer application over high speed, long delay networks.

3 MULTI CONNECTION TCP MECHANISM

3.1 Two Implementations of Multi connection TCP mechanism

There are two ways to implement the multi TCP connection mechanism as in shown in Figure 2. One is to implement this mechanism inside the application layer, such as Multi-connectioned FTP(MFTP). The other is to implement it inside the transport layer as a Multi TCP connection sub-layer which provides an existing socket-like interface to the applications. The mechanism of multi TCP connection is the same in both implementations, but a difference exists in the interface to applications.

We first developed MFTP to evaluate a multi-connection mechanism due to the following reasons. The traffic of HTTP applications is increasing exponentially in the INTERNET but the FTP[11] is still one of the highest traffic applications[12]. It has become popular in exchanging larger files, such as graphical images or photos between sites via high speed intercontinental WANs. Furthermore, in closed user networks, such as enterprise networks, many kinds of huge amount of data are transferred world wide. In such cases, a high performance file transfer protocol, which still works well over high speed, long delay networks, is required. Therefore, we firstly applied the multi TCP connection mechanism to FTP to evaluate this mechanism. We implemented the core of the mechanism in the FTP application layer, and called it Multi connectioned FTP(MFTP).

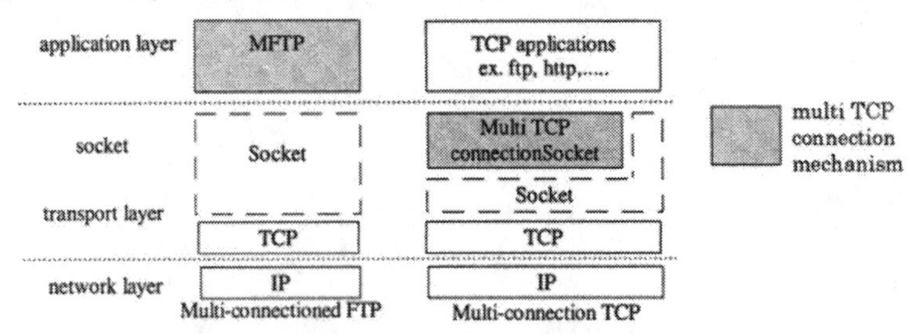

Figure 2 Two typical implementations of multi TCP connection mechanism.

MFTP uses multiple existing TCP connections to utilize the aggregated large window size. The theoretical throughput of MFTP is obtained by the equation (2).

$$Tm = N * W / D \qquad (2)$$

Tm(bps) : throughput of MFTP
N : the number of TCP connections used in MFTP
W(bit) : window size of each TCP connection
D(s) : Round Trip Time

3.2 Functional Blocks of Multi-connectioned FTP

The functional blocks of MFTP are illustrated in Figure 3. The shaded parts of Figure 3 are new additions or modifications to the existing FTP to support the MFTP. We used the source code of FTP programs of FreeBSD as the basis of MFTP. The newly added part of the source code is about 200 steps in each user and server program. The MFTP is upper compatible to the existing FTP. The existing FTP uses the Telnet protocol to send and receive FTP commands. The MFTP adds new commands to establish multiple connections for data transfer. The function of each block is as follows.

● **User Interface** : This module inputs and outputs the MFTP commands between the user and User PI. The user interface is the same as the existing FTP.

● **User/Server Protocol Interpreter(PI)** : User PI initiates the control connection to the Server PI and interprets FTP commands. MFTP PIs are an additional function to interpret the new MFTP command.

● **User/Server Data Transfer Process(DTP)** : User and Server DTP implement the interface to the file system and control TCP connections according to commands from User/Server PI. They contain the following functional blocks.

· Parameter Collection function

This function exists only in User DTP and gathers the parameters which are necessary to determine the appropriate number of TCP connections for a specific destination. It uses a ICMP Echo/Reply function to measure the RTT. Parameters are stored to use on the next occasion to access the same destination.

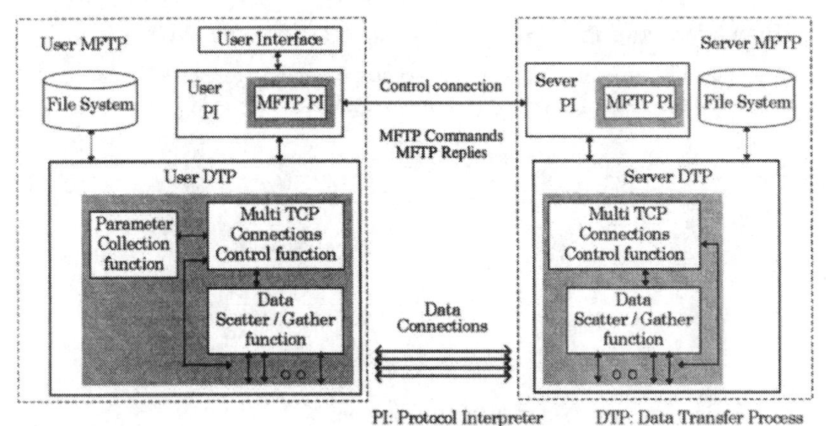

Figure 3 Functional blocks of multi-connectioned FTP (MFTP).

- Multi TCP Connections Control function
  This function calculates the number of TCP connections for data transmission according to equation(2) using parameters which were collected by the Parameter Collection function. The calculated number of TCP connections may be overridden by a MFTP user using a MFTP command. This function negotiates the number of connections between the user and server DTP and establishes, controls or disconnects multiple TCP connections.
- Data Scatter and Gather function
  At the sender's site, this function sequentially scatters the data to be transmitted into N TCP connections block by block. At the receiver's site, this function obtains blocks of data from N TCP connections and reconstructs them into the original data. The receiver's site needs a large size of receive buffer to reconstruct the gathered data in a correct sequence.

3.3 Two methods for Data scattering/Gathering

There are two ways to scatter data as shown in Figure 4. One is the static assignment method and the other is the dynamic assignment method.

In the static assignment method, each data block is assigned to each connection permanently before the transmission. This method is easy to implement and the load of the scattering process is low. However, the total performance of multi TCP connections is degraded by the one with the lowest performance, and this method requires a larger size of receive buffer to deliver received data to an FTP application in sequence. It is difficult to set up a stream lined service for applications using this method.

In the dynamic assignment method, each data block is assigned to the first non-blocked connection found at that time. This method requires a smaller buffer size than the former to reconstruct the data in sequence and the processing time becomes shorter.

In our implementation, the MFTP uses the dynamic assignment method considering the applicability of this mechanism to other applications, such as HTTP.

4 EXPERIMENTS AND RESULTS

4.1 Configurations and Protocol Stacks of the Experimental Network

The configurations and protocol stacks of the experimental network are shown in Figure 5. The performance evaluation of MFTP was made over the interconnected ATM LANs. MFTP user

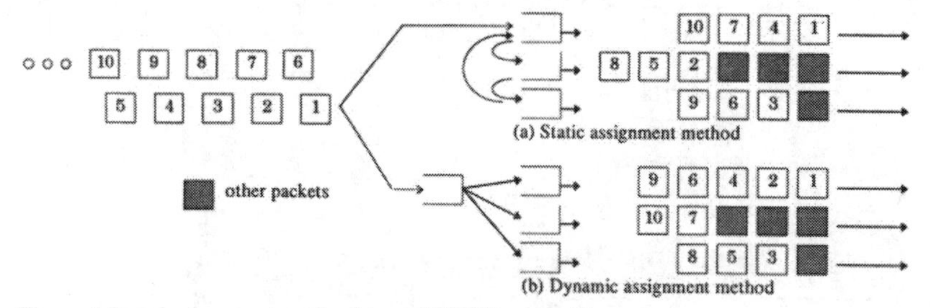

Figure 4 Data assignment method to multi TCP connections.

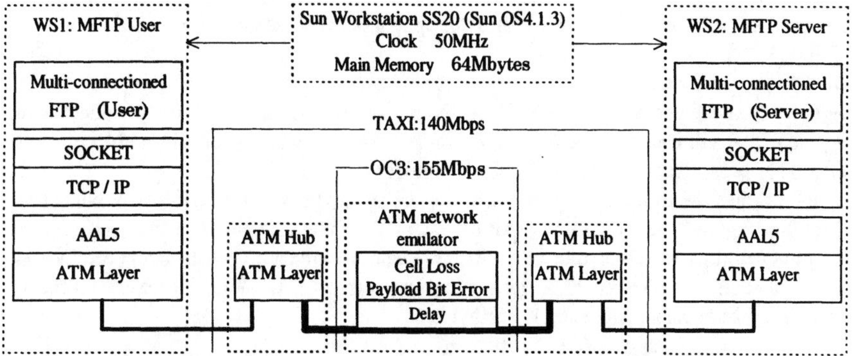

Figure 5 Configurations and protocol stacks of the Experimental Network.

and server programs were installed in WS1 and WS2. These WSs are SUN SparcStation20(Sun OS 4.1.3), and their specifications are depicted in Figure 5. WS1 and WS2 were accommodated in ATM-Hubs(Fore Runner ASX-200) using a 140Mbps TAXI interface respectively. These ATM-Hubs were interconnected via the ATM network emulator[13] using OC3 interfaces to introduce the propagation delay, ATM cell loss and payload bit errors artificially. WS1 and WS2 used AAL5(VBR) over PVC.

4.2 Testing Parameters

The throughput of MFTP was measured by transferring 16Mbyte data files from WS1 to WS2. In order to reduce the effect of disk access speed, the user side put the test data to '/dev/null' on the server side. The MFTP worked in a block transfer mode. The parameters of measurements are listed in Table 1. The existing FTP of SUN OS 4.1.3 adopts the value of 24Kbytes as the maximum window size, so, we used the value of 24Kbytes as the maximum window size in these measurements. To investigate the relation between the maximum window size and the throughput for reference, MFTP throughput is measured in various windows in an error free and RTT 100ms environment.

Each measurement was performed over 10 times and the average of the results was calculated. However, there was little variance in the measured data, and each standard deviation was at most under 10% from the average for each value.

4.3 Results of MFTP throughput

MFTP throughputs measured in an error-free environment are shown in Figure 6. The throughputs for file transfer in different RTTs are plotted for each number of connections. The

Table 1 Measurement parameters

Parameters	Range
Number of connections	1~ 16
Inserted Round Trip Time (ms)	0 ~ 500
Cell Loss Ratio	$2.4*10^{-4} \sim 3.8*10^{-6}$
Payload Bit Error	$4.8*10^{-7} \sim 3.0*10^{-8}$
Window size / connection	24Kbytes,(4K ~ 51Kbytes)

maximum window size of each connection is 24Kbytes for an RTT of 0~500ms. In Figure 7, the relation between maximum window size and throughput of MFTP is shown in an environment of RTT 100ms.

MFTP throughputs measured in an erroneous environment are shown in Figures 8 to 10. In order to evaluate the performance of MFTP in the erroneous environment, two types of artificial errors were injected. One is cell loss error to simulate the error which occurs in ATM switches, and the other is payload bit error to simulate the link error.

Figure 8 shows the throughputs in a cell loss error injected environment for each number of connections. Figure 9 plots the throughput for each RTT. Figure 10 plots the characteristics of the throughput for each error rate in a RTT 200ms environment. The throughput values in Figure 10 are normalized by that of 16 connections in an error free and RTT 200ms environment. The window size of each connection is 24Kbytes.

4.4 Multi connection TCP v.s. single large window TCP

To compare the performance of multiple small windows TCP to single large window TCP in erroneous environment, another experiment was performed using 'SUN OS TCP-LFN' option. It supports the extension TCP window size and dose not support selective acknowledgment. The experiment was performed via bridged interconnected ethernet LAN. The testing parameters and the results are listed in Table 2.

Table 2 Throughput (kbps) of Multi connection TCP and single big window TCP

	total window size (Kbytes)				
	64	128	256	384	512
Multi connection TCP (n*64Kbyte)	762	848	883	1187	1465
Single big window TCP(TCP extension)	762	770	764	751	747

added bit error ratio = $3.0*10^{-7}$ added RTT = 200ms

5 DISCUSSION

5.1 Multi connection TCP mechanism in the error free environment

In error-free environments without any inserted delay, the throughput of MFTP in the experimental network achieved over 32Mbps as shown in Figure 6 using any number of connections. Because an ATM link(OC3:155Mbps) is faster than the ATM LAN (TAXI:140Mbps), there is no bottleneck in the ATM link in the experimental network, so it is seen that 32Mbps is the maximum performance of MFTP with the workstations used in our experiments.

In Figure 7, as the maximum window size is increasing, the throughput is increasing step by step. It is considered that the length of TCP packet affects the characteristics of the throughput. The maximum length of IP datagram(MTU size) and TCP packet(MSS size) are 9188bytes and 9148bytes respectively in the ATM network. In the experimental network, the average length of TCP packets used in MFTP is about 9Kbytes calculated by transferred data size and the number of TCP packets. This means that the TCP used in this MFTP experiment sends data near the MSS size only, therefore TCP does not always utilize the whole window size. In the case where the maximum window size is set to 24Kbytes, the actual window size, which we call segmented window size, works out to about 18Kbytes. The measured throughput therefore increases discretely as shown in Figure 7.

5.2 Effects of multiple connections over long delay network

As shown in Figure 6, when the round trip time is larger than 200ms, the throughputs increase linearly, so the multi TCP connection mechanism works effectively. However, the theoretical

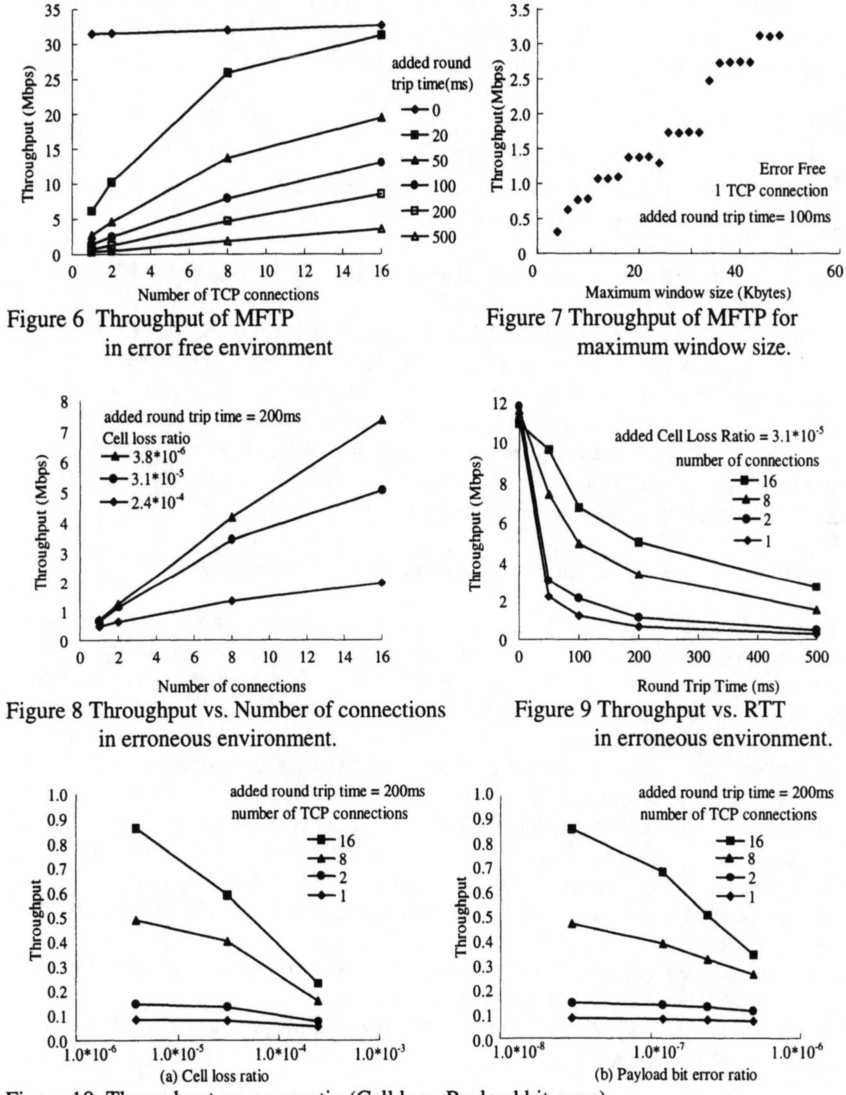

Figure 6 Throughput of MFTP in error free environment

Figure 7 Throughput of MFTP for maximum window size.

Figure 8 Throughput vs. Number of connections in erroneous environment.

Figure 9 Throughput vs. RTT in erroneous environment.

Figure 10 Throughput vs. error ratio (Cell loss, Payload bit error).

throughputs obtained by equation(2) using the maximum window size and the added round trip time are higher than the measured throughput. For example, according to the equation(2), MFTP should achieve 15.7Mbps using 16 connections under the conditions of 200ms of round trip time and 24Kbytes of maximum window size for each connection, while the actual throughput achieved is only 8.5Mbps. It is confirmed that there is no packet loss in the error free environment, hence this degradation is not caused by re-transmission. Equation(2) can obtain only the ideal maximum throughput using the maximum window size and the added round trip time. To find the practical throughput Tp, other factors, for example, the protocol processing time at the work stations, the segmented window size discussed in the last of clause 5.1, should be considered. Equation(2) is modified to (2)' to reflect these factors.

$$Tp = k * N * W' / D \qquad (2)'$$

Tp(bps)	: Real throughput of MFTP.
N	: Number of TCP connections used in MFTP.
W'(bit)	: Segmented window size of each TCP connection.
D(s)	:Round trip time.
k	:System parameter of the experimental workstations and network.

The segmented window size W' is considered to be about 18Kbytes as depicted in Figure 7.

k is a system parameter of the experimental network. *k* is about 0.75 calculated from the measured throughput data for 16 TCP connections, RTT 200ms and the segmented window size of 18Kbytes. Using this value of *k*, equation(2)' also matches the measured data for RTT 500ms. The equation(2)' with the same *k* fits the lower throughput case compared to the maximum throughput of the experimental network(32Mbps). Equation(2)' also obtains the optimal number of connections when the expected throughput is set to Tp.

5.3 Multi connection TCP mechanism in erroneous environment

In the environment RTT 200ms, the throughput increases as the number of connections increases as shown in Figure 8. Except for the no delay environment, the throughput increases as the number of connections increases as shown in Figure 9. Hence, even in the erroneous environment, the multi connection method worked effectively. However, the increasing rate is not linear in high error rate environments.

In the TCP protocol, the error detection is carried out packet by packet. In order to compare the characteristics of MFTP throughput for cell loss error and payload error, the cell loss error rate and pay load error rate should be converted to TCP packet error rate. The average length of TCP packet used in this experiment is about 9Kbytes as described before. A TCP packet consists of 188 cells(48bytes/cell) or consists of 72kbits. Converting the X axis to packet error rate according to the number of cells or bits of a TCP packet, Figure 10 (a) and (b) depict almost the same characteristics.

Furthermore, as listed in Table 2, Multi connection TCP obtain higher throughput than that of single connection TCP which has the same total window size.

From the results of the experiments, it is considered that the MFTP still works effectively in erroneous long delay networks.

6 CONCLUSION

This paper has discussed a multi-connection method at the application layer as well as the transport layer to increase the data transfer throughput in the environment of high-speed, long delay networks. We have developed a multi-connection method for FTP, MFTP, to evaluate the performance of data transfer. Our MFTP has the following features and advantages.

1. MFTP uses multiple existing TCP connections in aggregate, so it is easy to implement it in any other work stations.
2. MFTP can adjust the number of connections flexibly using the parameter collection function.
3. The multi connection method applied to a file transfer protocol works effectively. MFTP has achieved over 8.5Mbps end-to-end throughput using 16 connections of 24Kbyte maximum window size via the emulated ATM WAN of an OC3 155Mbps link with 200ms round trip delay.
4. MFTP works effectively even in an erroneous environment, and it is considered that the multiple TCP connection method is more durable than a single large window mechanism such as the TCP extension.

The multi-connection handling mechanisms could be applied to a TCP socket sub-layer with the above advantages from 2 through 4. In this case any TCP applications, such as HTTP, can increase the throughput in the environment of high-speed, long delay networks.

We conclude that the multiple-connection method is, at present, a suitable and desirable solution for the high performance data transfer application over high speed, long delay networks such as inter national ATM networks. We are now implementing the multi-connection mechanism over extended TCP to gain a much greater increase of throughput.

7 ACKNOWLEDGMENT

The authors wish to thank Dr. Y. Urano and Mr. S. Fukumitsu of KDD R&D Laboratories for their continuous encouragement and support.

8 REFERENCES

[1] J. Postel, "Transmission Control Protocol", RFC793, Internet Engineering Task Force, Sep. 1981.

[2] S. O'Malley, L. Peterson, "TCP Extensions considered harmful", RFC1263, Internet Engineering Task Force, Oct. 1991.

[3] XTP Forum, "Xpress Transport Protocol Specification", XTP Revision 4.0, March 1995

[4] V.Jacobson and R.Braden, "TCP Extensions for Long-Delay Paths", RFC1072, Internet Engineering Task Force, Oct. 1988.

[5] R.Fox 'TCP Big Window and Nak Options", RFC1106, Internet Engineering Task Force, Jun. 1989.

[6] V.Jacobson, R.Braden and L.Zhang, "TCP Extension for High-Speed Paths", RFC1185, Internet Engineering Task Force, Oct. 1990.

[7] V.Jacobson, R.Braden and D.Borman, "TCP Extensions for High Performance", RFC1323, Internet Engineering Task Force, May 1992.
[8] C.Villamizar and C.Song, "High Performance TCP in ANSNET", ACM SIGCOMM Computer Communication Review, pp.45-60, Vol.24 Num.5, Oct. 1995.
[9] ISO7478 Information processing systems - Data communication - Multilink procedures
[10] ISO/IEC13781 Information technology - Telecommunications and information exchange between systems - Private Integrated Services Network - Digital Channel aggregation
[11] J. Postel, J. Reynolds, 'FILE TRANSFER PROTOCOL (FTP)", RFC959, Internet Engineering Task Force, Oct. 1985.
[12] V. Paxon, "Empirically Derived Analytic Models of Wide-Area TCP Connections", IEEE/ACM Transactions on Networking, vol.2, No.4, Aug. 1994.
[13] K.Yamazaki, T.Nakajima and S.Hayakawa, "Considerations on Network Performance of 64kbit/s-Based Services in an ATM Network", IEICE TRANS. COMMUN., VOL. E78-B, NO.3 MARCH 1995.

38

Two Heuristics for Multicasting in ATM Networks

F. Bernabei, A. Coppi, R. Winkler

Fondazione Ugo Bordoni
Via B. Castiglione, 59 - 00142 Roma Italy
Fax: +39 6 54804404; e-mail: {bernabei, wnk}@fub.it

Abstract

This paper deals with multicasting in ATM networks. An overlaid network is proposed, that includes the ATM switches that provide the required signaling and information replication capabilities. These switches are interconnected by means of point-point Virtual Paths and run a hop-limited point-point routing scheme to configure the point-multipoint connections. The hop-limit constraint is embedded in the two considered multicasting algorithms, designed for operation in directed graphs. One algorithm modifies the well known KMB heuristic, originally developed for undirected graphs and now able to operate on directed graphs with the hop limit constraint. The other one modifies the Dijkstra algorithm, by adopting a cost function that considers the link capacity instead of its length. The performance of the two heuristics are quite close under many different operating conditions, but the modified Dijkstra has a lower algorithmic complexity.

Keywords

Multicast algorithms, hop limited routing, ATM, directed graphs

1 INTRODUCTION

The availability of efficient point-multipoint information transport facilities is one of the key issues for multiparty communications in ATM networks. The term "multicasting" has entered in the common vocabulary to refer to the transport of point-multipoint information using a single user to network transaction (in datagram networks) or a single connection (in virtual channel oriented networks).

This paper proposes an architecture and compares two heuristics for multicasting in ATM networks.

The architecture consists of an overlaid logical network, the *Multicast Logical Network* (MLN), composed of those ATM switches that provide the capabilities perform multicasting, i.e. the replication of user information and the in-call control of point-multipoint connections. These *Multicast Capable Switches* (MCS's) are connected by means of Virtual Paths (VP's), that traverse transparently the ATM switches that do not replicate information flows and are unable to control point-multipoint connections. The MCS's cooperate to establish, release and modify the point-multipoint connections, following the users' requests and according to the stream characteristics and the status of the MLN. Each user connects to one MCS, under the control of its Local Exchange. Thus, a point-multipoint connection is implemented by means of a two-tier architecture: one tier is the actual point-multipoint connection inside the MLN, the other tier is the set of point-point connections between the communicating parties and the relevant MCS's.

To route a point-multipoint connection the MCS's run a real time point-point routing scheme and a multicasting heuristic. The point-point routing scheme may be derived from the those currently adopted for evolving POTS and SS7 networks [Ash, 1994]. These mechanisms limit the maximum number of hops between source and destination, an effective way to simplify the processing and to reduce the number of switches to be traversed.

The multicasting heuristic uses the information provided by the routing scheme about the existing and suitable point-point routes, to establish a point-multipoint connection rooted at the MCS relevant to the source of the multicast stream and spanning the MCS's relevant to the stream destinations. This paper considers two heuristics: MaxCap and KMB*. MaxCap selects the widest (in terms of available bandwidth) suitable and hop limited point-point routes between the source MCS and each destination MCS and combines them to form a loopless point-multipoint connection. KMB* modifies the widely considered KMB algorithm [Kou, Markowsky, Berman, 1981], to introduce the hop limit and allow for application to directed graphs whose edge weights represent the available link capacity. The performance of MaxCap and KMB* are compared by means of simulation on random graphs in a static communication scenario, that does not consider the possible addition/dropping of MCS's to/from a connection, resulting from the reconfiguration of the active calls.

While architectures for multicasting in ATM networks have not yet been proposed in the technical literature, several algorithms for multicasting has been discussed, mostly with reference to datagram networks [Deering, 1990; Bharat-Kumar, 1983; Chow, 1991; Ballardie, 1993; Kompella, 1993; Deering, 1994]. Moreover, a vast amount of literature on exact algorithms and heuristics is available to solve the "Steiner problem in graphs" [Winter, 1987], that approaches this problem from the point of view of graph theory. These solutions cannot be applied to ATM networks for two main reasons. First, resource reservation is not performed in the case of connectionless data transfer but is required by ATM; as a consequence, algorithms that work only on undirected graphs, whose link weight do not represent the bandwidth availability, are not suitable for ATM multicasting. The undirected graph assumption is so established that it has been applied also in papers, e.g. [Jiang, 1992], that identify the available bandwidth as the key metric to solve the problem. Second, multicasting in datagram networks is approached by considering point-point routing schemes, e.g. reverse path routing, that are commonly used for data networks but do not suit to ATM networks. These issues are taken into consideration for the definition of MaxCap and KMB*.

The section 2 describes the network architecture. The two heuristics are formalized in the section 3. The models used for the simulation are described in the section 4. The performance results are discussed in the section 5 and the conclusions are given in the section 6.

2 NETWORK ARCHITECTURE

The MLN is a logical network of MCS's and VP's, with topology and coverage area depending on the desired grade of connectivity, on the offered traffic and on the supported communication services. The MCS's receive and process all the requests for multipoint connection establishment, rearrangement and abatement and translate them into the most appropriate route configuration, according to the multipoint stream specification and to the network status.

MCS's interact one another to obtain information about the status of the MLN and the location of other MCS's. This is part of the point-point routing scheme, that adopts a limit on the maximum number of hops on a route, a useful approach to simplify the processing and save bearer resources under the assumption that a low hop count in the logical network implies a low hop count in the physical network.

As shown in Figure 1, one point-multipoint ATM connection is established by using a two-tier architecture, that combines one unidirectional point-multipoint connection with several bidirectional point-point ones. The point-multipoint connection is rooted at the MCS relevant to the stream source and spans the MCS's relevant to its destinations; a point-point connection attaches one communicating party to the relevant MCS's, under the control of the Local Exchange that may consider either static (e.g. location based) or dynamic (e.g. load controlled) association criteria. The main advantage of this architecture is the reduction of signaling traffic with respect to a single-tier architecture: this one implies the rearrangement of a point-multipoint connection each time an user adds to or drops from it, whereas the two-tier architecture requires this each time an MCS adds to or drops from a connection.

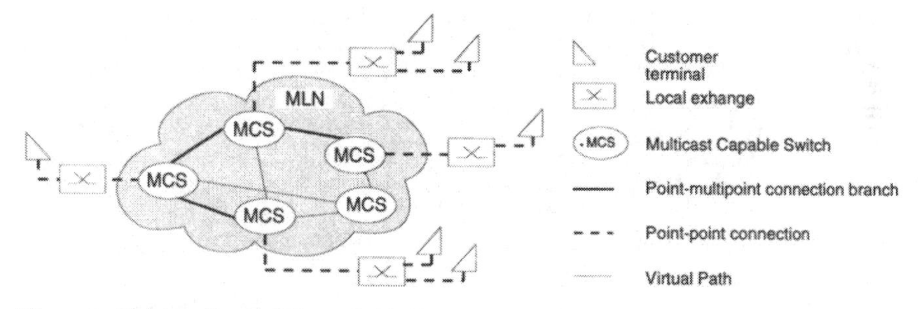

Figure 1 Multicast Logical Network Architecture.

The architecture of a MCS is sketched in Figure 2, for the case a Replication Server enhances a basic ATM Local Exchange with the capabilities required by multicasting. The Replication Server is composed of a number of entities, among which Figure 2 shows: a copy module in the User Plane to replicate the ATM cells, modules in the Control Plane to implement the [ITU-T, 1994; ITU-T, 1995] point-multipoint signaling procedures and the Multipoint Connection Controller, to collect the available arrangements to establish or modify a point-multipoint connection and select the best one.

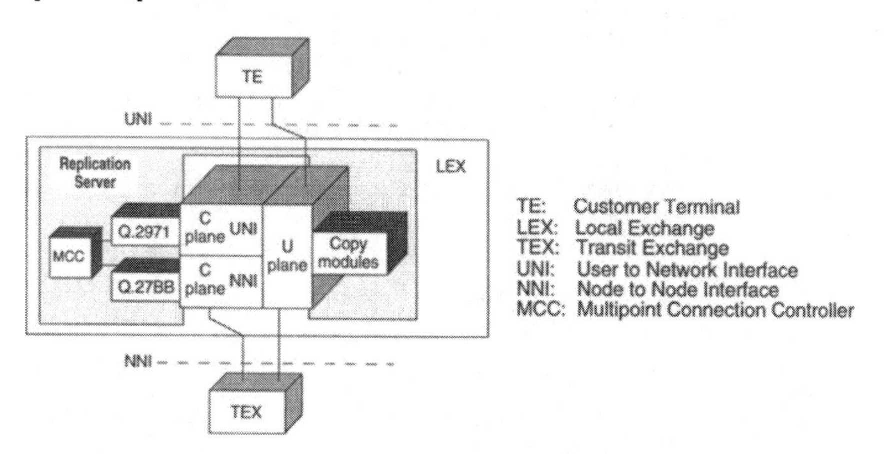

Figure 2 Architecture of a Multicast Capable Switch implemented as a server.

3 HEURISTICS FOR POINT-MULTIPOINT ROUTING

The overlaid logical network of MCS's may be represented by a directed labeled graph $G=(V, E, C)$, wherein V is the set of vertices associated to the MCS's, E is the set of edges associated to the VP's between the MCS's and C is a function that labels the edge (v_i, v_j) with its available capacity C_{ij}. With reference to this graph, we formulate two heuristics for hop-limited, point-multipoint routing in directed graphs: KMB* and MaxCap. These heuristics are parametric to maximum number of hops (H) and aim at determining a loop-free point-multipoint route, rooted at a vertex v_0 (the MCS relevant to the stream source) and spanning a set of vertices $S \subseteq V$ (the MCS's relevant to the stream destinations).

KMB* modifies the KMB algorithm [Kou, Markowsky, Berman; 1981] to take into account the value of H and the link orientation. An important difference between KMB and KMB* is that the former does not make any distinction between the vertices in S, as all of them play the role of both sender and receiver, whereas the latter clearly separates the source of a stream from the set S of its destinations. KMB* operates according to the following steps.

1. A complete directed labeled graph $G'=(V', E', C')$ is constructed from G in such a way that $V'=S \cup \{v_0\}$ and, for every edge $e_{ij} \in E'$, the associated label C'_{ij} is set equal to the available capacity of the widest path from v_i to v_j in G. If more than one widest paths exist, then the one with minimum number of hops is selected (the reader may note that this last aspect is new with respect to the original KMB and allows for a lower cost in terms of consumed bandwidth). G' is constructed as follows:

1.1. let: $V'=S \cup \{v_0\}$; $E'=\{e_{ij}=(v_i, v_j) \mid v_i, v_j \in V'\}$; $C'_{ij}=0$ for all the pairs (v_i, v_j)

1.2. for each vertex $v_k \in V'$

1.2.1. $B = V'-\{v_k\}$ is the set that holds the non visited nodes; each vertex $v_i \in V'$ is associated a couple (H_{ki}, D_{ki}), where H_{ki} and D_{ki} are the number of hops and the available capacity of the widest path from v_k to v_i, respectively. (H_{ki}, D_{ki}) is initialized at

$$(H_{ki}, D_{ki}) = \begin{cases} (0, \infty) & \text{if } i = k \\ (1, C_{ki}) & \text{if } i \neq k \text{ and } e_{ki} \text{ exists} \\ (\infty, 0) & \text{otherwise} \end{cases}$$

1.2.2. while the set $V' \cap B$ is not empty

1.2.2.1. find the index j such that $D_{kj} = \max_{v_i \in B} \{D_{ki}\}$; if more than one vertices v_i have the same maximum value of D_{kj} then select at random one among those with minimum H_{ki}; let $C'_{kj} = D_{kj}$.

1.2.2.2. let $B = B - \{v_j\}$.

1.2.2.3. each couple (H_{ki}, D_{ki}) of $v_i \in B$ is updated if $H_{kj} < H$ and the edge e_{ji} exists; the new value is:

$$(H_{ki}, D_{ki}) = \begin{cases} (H_{kj}+1, \min(D_{kj}, C_{ji})) & \text{if } \min(D_{ki}, C_{ji}) > D_{ki} \\ (H_{kj}+1, D_{kj}) & \text{if } \min(D_{ki}, C_{ji}) = D_{ki} \text{ and } H_{kj}+1 < H_{ki} \\ (H_{kj}, D_{kj}) & \text{otherwise} \end{cases}$$

2. Find the spanning tree $T_2 = (V_2, E_2)$ of G' with widest path. When inserting a vertex in T_2, only the edges that respect the limit on the number of hops are considered among those that have maximum capacity. Formally, each edge e_{ki} of the complete graph G' is associated to the couple (H_{ki}, D_{ki}), associated to the vertex v_i in the step 1.2. when v_k is considered. Moreover, each vertex v_j is given a value h_j, representing the number of hops

necessary to reach the vertex v_j from the vertex v_0 following the path imposed by T_2. The procedure to build T_2 is as follows:

2.1. Let $V_2 = \{v_0\}$, $E_2 = \{\}$, $h_0 = 0$.

2.2. While the set $V' - V_2$ is not empty:

 2.2.1. define the set $E_3 = \{e_{ij} = (v_i, v_j) \mid v_i \in V_2, v_j \in V' - V_2, h_i + H_{ij} <= H\}$

 2.2.2. select the edge $e_{ij} \in E_3$ with maximum D_{ij}; if more than one edges with maximum capacity exist, then select at random one among those with minimum H_{ij}.

 2.2.3. let $V_2 = V_2 + \{v_j\}$; $E_2 = E_2 + \{e_{ij}\}$; $h_j = h_i + H_{ij}$

3. Construct the subgraph G_S of G by replacing each edge in T_2 by its corresponding path in G. For this purpose, one edge each in T_2 is considered at a time and, if the corresponding path leads to a vertex already reached by another path in G, then all the vertices and edges belonging to the longest path are eliminated. This may occur since intermediate vertices on a path in G may belong to the set S and is to be avoided for T_2 to be a tree. In order to remove loops we have chosen to eliminate the longest paths.

MaxCap implements a modified, hop limited, Dijkstra based routing that considers the link capacity to select the branches of a point-multipoint connection. For this purpose, it relies on the routing scheme proposed in [LeBoudec, Przygienda, Sultan, 1994], that incorporates a "widest path metric" into the well known Djikstra shortest path routing scheme. A similar approach is also used in [Shacham, 1992], to compute the maximum bandwidth shortest path for a point-multipoint connection, but it does not consider the hop limit and the directed graph model. The point-multipoint connection is built by assembling, loop-free, a set of point-point routes, each one meeting the constraint on the maximum number of hops.

MaxCap mainly differs from KMB* for a modification of the step 1 and the lack of step 2. The step 1 is modified so that, in the step 1.2, only $v_k = v_0$ is considered instead of taking into account all the vertices $v_k \in V'$. Thus, MaxCap simplifies KMB* by selecting, at the end of step 1., the tree rooted at v_0 and composed of all the edges outgoing from the root and relevant to the vertices in S. It results that MaxCap selects the minimum length widest available path, whereas KMB* selects the widest paths so as to reduce the global cost in terms of consumed capacity, that is obtained by sharing as much as possible the links to the destinations.

The algorithmic complexity is an important index of performance of tree-building heuristics. From the above description it comes out that the complexity of MaxCap is lesser than that of KMB* for a factor $|V|$. In fact, the complexity of KMB* is $|S|\,|V|^2$, because its three steps have complexity of the order of $|S|\,|V|^2$, $|S|^3$ and $|S|\,|V|$ and $|S| \leq |V|$, and the resulting complexity of MaxCap is $|S|\,|V|$, that is the complexity of each of its steps.

4 PERFORMANCE METRICS AND SYSTEM MODEL

The performance of MaxCap and KMB* are evaluated by means of simulation on random graphs. Taking advantage of the adopted two tier multicasting architecture, these graphs need to represent only the MLN topology and the connections between the customers and the MCSs', disregarding the other switches and links. The following indices of performance are considered to evaluate the performance of MaxCap and KMB*:

- P: percentage of the communicating parties that are not attached to one point-multipoint connection;
- S: number of stream copies made by the source and transit MCS's for each point-multipoint connection;
- C_S: number of consecutive connections set-up with $P = 0$.

P measures the party blocking probability in a point-multipoint connection; S gives an idea on the processing and traffic load on the MCS's; C_S measures the number of multicast connections that may be concurrently accommodated by the considered heuristics.

To evaluate the above parameters for a number of different MLN mesh topologies, the random graph model proposed in [Waxman, 1988] has been modified to generate fully meshed networks and to make the edge probability independent of its length. The edge probability function is $P_{u,v} = G \exp \{[-d(u,v) * R] / [D - d(u,v)]\}$, wherein $d(u,v)$ is the Euclidean distance between the MCS's u and v, D is the maximum possible distance between any two MCS's, R and G are defined in [0, 1] and give the ratio of short to long edges and the grade of connectivity of the MLN, respectively. The values of R and G considered for the performance evaluation are shown in Table 1.

Table 1 Random mesh MLN's

	mesh #1	mesh #2	mesh #3	mesh #4	mesh #5	mesh #6
G	1	1	1	0.5	0.5	0.5
R	0	0.5	1	0	0.5	1

The N MCS's in a MLN are randomly located in the (x,y) plane. The edges are not weighted, to represent a MLN made of homogeneous MCS's that present the same level of load and usage cost. The links are weighted by the available capacity of the relevant VP, drawn from the probability distributions shown in Table 2 (the values are expressed in stream units, the maximum VP capacity is assumed equal to 10 stream units).

Table 2 Link capacity probability distribution

Link capacity	D1	D2	D3	D4
1	0.3	0.1	0.3	0
3	0.4	0.6	0.6	1
10	0.3	0.3	0.1	0

Each multicast group has a constant number M of members. Two different multicast groups are considered: one is related to a multiparty conference communication, where each member transmits to and receives from all the other members (M stream sources per group and M-1 destinations per stream); the other group is related to a distributive communication from a single source to M-1 destinations. A multicast group is created by randomly generating M pairs of {MCS number, party number}. The members of a group are attached to different MCS's. Information streams are modeled as constant bit rate flows with unitary capacity.

No connectivity check is carried out after the generation of a MLN and a multicast group and it may occurs that some members of a group are not reachable from the others. That models the possible network inability in accommodating all the parties in a call, as a result of the bandwidth previously allocated to other in progress communications. The percentage of unreachable destinations (P_{DU}), measured over *1500* pairs {random network, random group} for $M = 5$, is shown in Table 3 for the six considered MLN topologies. The values of P_{DU} are independent of the multicasting heuristic and of the link capacity, as they are evaluated for each stream as if it were the first one to be allocated for the group.

Table 3 P_{DU} values for the six considered MLN's

	mesh #1	mesh #2	mesh #3	mesh #4	mesh #5	mesh #6
P_{DU}	0.0	0.0017	0.0442	0.0487	0.2202	0.4013

5 PERFORMANCE RESULTS

To evaluate the performance of the two multicast heuristics, 1500 trials are carried out for each combination of the values of G, R, N, M and H and for two different communication services: videoconferencing and TV distribution.

As for videoconferencing, MaxCap and KMB* are applied at each trial to one pair of random MLN and multicast group. The M streams are sequentially considered according to their order of generation and every member of the multicast group becomes, in turn, the root of a point-multipoint connection; as a consequence, the ith stream in a sequence finds the network status modified by the previous i-1 streams ($2 \leq i \leq M$). The outcome of each stream allocation by each heuristic is stored for post processing, so as to obtain the mean value of P and S for the ith stream ($1 \leq i \leq M$).

Tables 4 to 8 compare the P performance of KMB* and MaxCap for N=10, M = 5 and H=2, varying the link capacity distribution. P is always equal to zero for the mesh number 1 and the relevant valus are not shown in the Tables 5 to 8. The tables show that:

- KMB* and MaxCap have the same performance for the stream number 1 (see Table 4) and this is the value of P_{DU} reported in Table 3; in fact the step 1 of the two heuristics ensures that a solution is found whenever it exists; this solution is related to the mesh connectivity and is independent of the amount of available capacity;
- for the streams 2 to 5 a slight difference is appreciated in favor of KMB* only in a limited number of cases (shown unshaded in the Tables 5 to 8); that means that KMB* is more able than MaxCap to avoid the utilization of links with very low capacity and to aim at an overall equalization of the capacity of the links in an MLN;
- P is higher for D1 and D3, that have highest probability of unitary bandwidth links;
- the MLN connectivity degree affects heavily the P values: the full mesh MLN (mesh 1) accommodates always all the streams in a multicast group, whereas the mean value of P approaches 1E-2 in the mesh 4 and is higher than 4E-1 in the mesh 6;
- the last streams in a group suffer for a high mean value of P, for the reduced bandwidth in the MLN; this is more evident in the highly connected topologies, where the stream 5 has P values that are even 3 - 4 times larger that those of the stream 1; that is less evident in the case of meshes with low connectivity because P is high even for the stream 1;
- there is no difference between KMB* and MaxCap if D4 is applied; consequently, the Table 8 refers only to MaxCap.

Table 4 KMB*, MaxCap: P performance for the stream number 1 (N=10, H = 2, any D)

Mesh # 1	Mesh # 2	Mesh # 3	Mesh # 4	Mesh # 5	Mesh #6
0.0	0.0017	0.0442	0.0487	0.2202	0.4013

Table 5 KMB* vs. MaxCap: P performance (N=10, D1, H = 2)

Mesh #	KMB*				MaxCap			
	stream 2	stream 3	stream 4	stream 5	stream 2	stream 3	stream 4	stream 5
2	0.0027	0.0032	0.0033	0.0065	0.0027	0.0032	0.0033	0.0065
3	0.0517	0.0540	0.0573	0.0672	0.0518	0.0540	0.0573	0.0673
4	0.0580	0.0650	0.0752	0.0832	0.0580	0.0653	0.0755	0.0837
5	0.2388	0.2497	0.2713	0.2872	0.2390	0.2497	0.2718	0.2877
6	0.4228	0.4357	0.4475	0.4782	0.4228	0.4357	0.4477	0.4785

Table 6 KMB* vs. MaxCap: P performance (N=10, D2, H = 2)

Mesh #	KMB*				MaxCap			
	stream 2	stream 3	stream 4	stream 5	stream 2	stream 3	stream 4	stream 5
2	0.0027	0.0030	0.0035	0.0043	0.0027	0.0030	0.0035	0.0043
3	0.0449	0.0460	0.0487	0.0510	0.0449	0.0465	0.0488	0.0512
4	0.0518	0.0530	0.0575	0.0598	0.0518	0.0530	0.0575	0.0598
5	0.2210	0.2215	0.2283	0.2372	0.2210	0.2215	0.2283	0.2372
6	0.4085	0.4095	0.4127	0.4227	0.4085	0.4095	0.4127	0.4230

Table 7 KMB* vs. MaxCap: P performance (N=10, D3, H = 2)

Mesh #	KMB*				MaxCap			
	stream 2	stream 3	stream 4	stream 5	stream 2	stream 3	stream 4	stream 5
2	0.0032	0.0037	0.0043	0.0067	0.0032	0.0037	0.0043	0.0067
3	0.0517	0.0540	0.0580	0.0685	0.0518	0.0540	0.0582	0.0687
4	0.0580	0.0650	0.0752	0.0845	0.0580	0.0653	0.0755	0.0850
5	0.2388	0.2497	0.2722	0.2892	0.2390	0.2497	0.2727	0.2893
6	0.4228	0.4357	0.4485	0.4790	0.4228	0.4357	0.4487	0.4793

Table 8 MaxCap: P performance (N=10, D4, H = 2)

Mesh #	stream 2	stream 3	stream 4	stream5
2	0.0025	0.0027	0.0030	0.0035
3	0.0447	0.0447	0.0449	0.0452
4	0.0503	0.0507	0.0510	0.0545
5	0.2217	0.2221	0.2224	0.2230
6	0.4020	0.4028	0.4032	0.4032

Figures 3 to 6 show the influence of H on the P performance of MaxCap; similar values are also obtained for KMB*. These figures show that.

- there is almost no benefit in increasing the value of H when the MLN is highly connected; this is line with the present trends in the design of point-point routing mechanisms for signaling and multiservice intelligent networks, that take advantage of the network full connectivity to rely on a non-hierarchical two-hop limited scheme;
- increasing H is a proper way to boost the performance when the mesh is poorly connected; the performance gain is around 10% for the mesh number 6 and the stream number 5, when going from H=2 to H=4, and goes up to more than 100% for the mesh number 6 and the stream number 1, from H=2 to H=4;
- medium connected MLN's with a few tens of MCS's do not present any advantage when going from H=3 to H=4; the improvement is remarkable from H=2 to H=3;
- D2 and D4 (that feature a lower probability of links with unitary capacity but do not have overall higher average bandwidth) have slightly lower values of P and a reduced variability of P with H than D1 and D3.

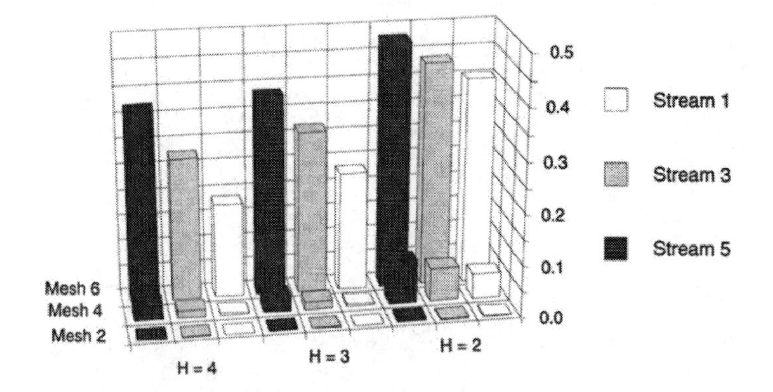

Figure 3 P performance of MaxCap for variable H (N=10, M=5, D1).

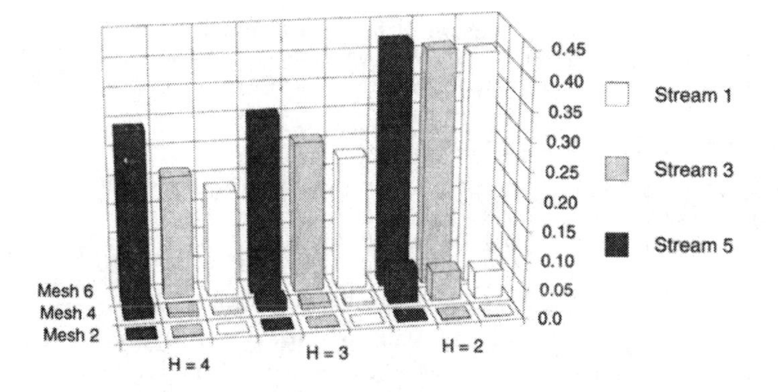

Figure 4 *P* performance of MaxCap for variable *H* (*N*=10, *M*=5, D2).

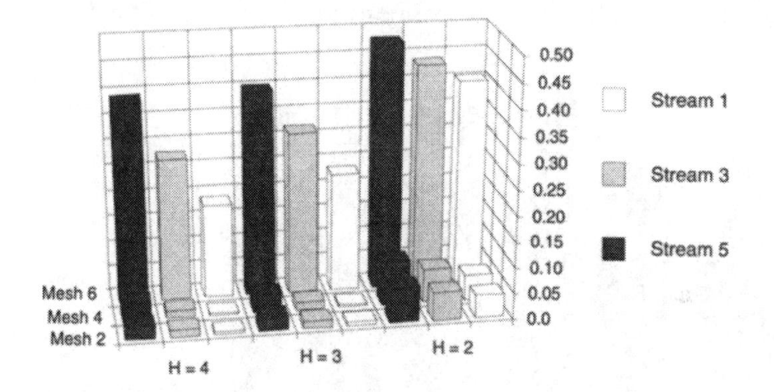

Figure 5 *P* performance of MaxCap for variable *H* (*N*=10, *M*=5, D3).

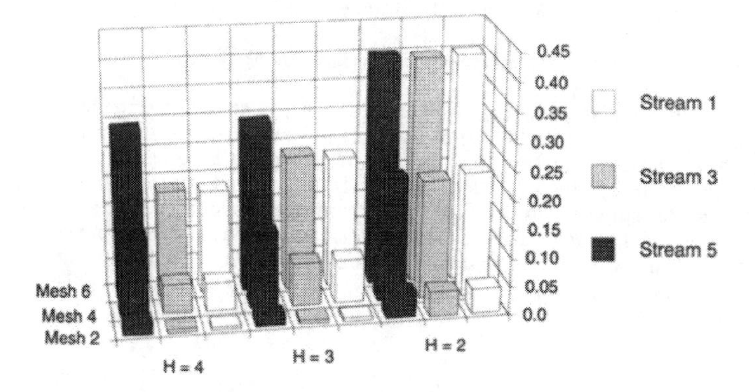

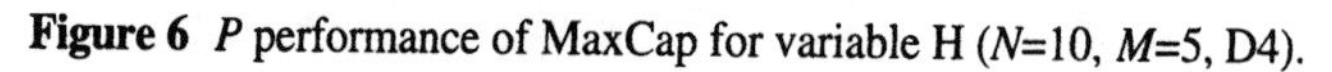
Figure 6 *P* performance of MaxCap for variable H (*N*=10, *M*=5, D4).

Table 9 compares the performance attained with N=20 and M=5 to those relevant to N=10 and M=5. The comparison is shown only for MaxCap, as these values are also representative of KMB* performance under the same conditions. $P = 0$ for every stream in the mesh 1 and 2.

Table 9 *P* performance of MaxCap with variable N (D2, H=2)

Mesh #	N = 10					N = 20				
	stream 1	stream 2	stream 3	stream 4	stream 5	stream 1	stream 2	stream 3	stream 4	stream 5
3	0.005	0.005	0.005	0.006	0.005	0	0	0	0	0
4	0.003	0.003	0.002	0.002	0.003	0	0	0	0	0
5	0.056	0.059	0.063	0.060	0.063	0	0.001	0.002	0.003	0.003
6	0.195	0.198	0.194	0.195	0.204	0.024	0.024	0.027	0.029	0.041

Figure 7 plots the mean value of S for MaxCap and KMB* for N=10, M=5 and D1. Again, KMB* presents a slight advantage over MaxCap and the difference increases with H. Additional information may be obtained from Figure 8, that plots the mean number of copies made by the root of the point-multipoint connection in the scenario of Figure 7: KMB* builds slightly shorter paths with a higher degree of link sharing among the destinations of a group and involves less extra-MCS's in addition to those relevant to the group members.

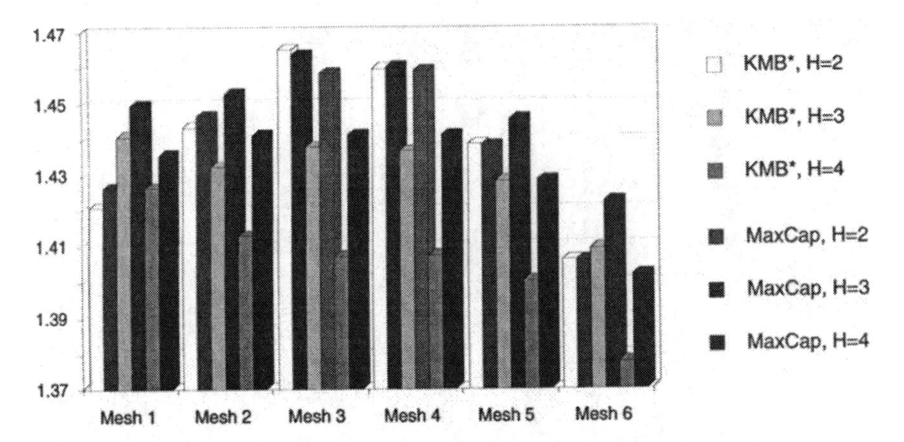

Figure 7 MaxCap vs. KMB*: S performance for variable H (N=10, M=5, D1).

In the case of distributive communication service, just one stream is generated for each multicast group and, consequently, one connection is to be established. A full mesh MLN, whose links have capacity equal to 10, is generated to support a sequence of long-lived connections, each one obtained by randomly generating a multicast stream with M-1 destinations. The ith multicast stream finds the network status modified by the previous k streams ($1 \leq k < i$) and, for each heuristic, the simulation terminates in correspondence of the first blocked stream ($i = C_S$), i.e. the first time the heuristic fails in successfully covering the set of M-1 destinations of a multicast stream. Again, 1500 trials are repeated to obtain the distribution function of C_S.

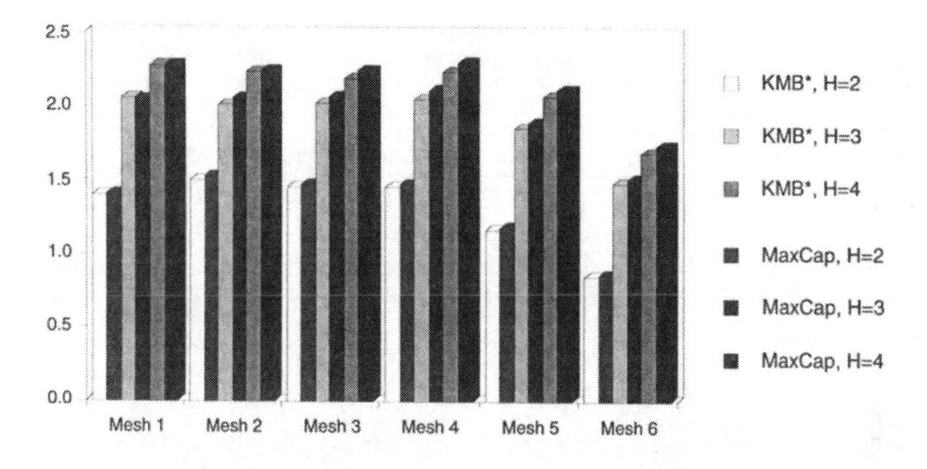

Figure 8 MaxCap vs. KMB*: average number of copies made by the root MCS for variable H (N=10, M=5, D1)

Figure 9 shows plots the distribution function of C_S, for a full mesh with N=10, initial link capacity equal to 10, M=5 and H = 2, 3. As a consequence of its better utilization of the link capacity and shorter path length, KMB* is able to accommodate more concurrent connections than that MaxCap; anyway, for a given probability, the difference between the number of established connections is, in any case, lower than 3%.

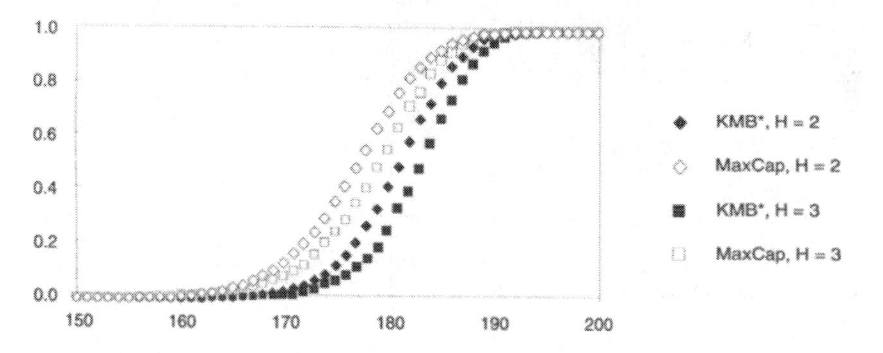

Figure 9 Distribution function of C_S for MaxCap and KMB*, with H = 2, 3(Mesh 1, M = 5)

CONCLUSIONS

This paper proposes an architecture and two heuristics for multicasting in ATM networks, in presence of a constraint on the maximum number of hops on the routes that form the point-multipoint connection and modeling the considered network by means of a directed graph. One heuristic modifies the well known KMB algorithm for solving the Steiner tree problem in networks; the other one relies on a simple modification of the Dijkstra mechanism to take into account the link capacity in the construction of the connection. The performance evaluation shows that the two heuristics behave similarly under many different operating conditions, with just some slight advantage attained by the modified KMB algorithm; this has to be traded off against its higher algorithmic complexity. Further work is required to compare these heuristics in a dynamic environment, more suitable to represent the addition/dropping of users to/from an active connection. Moreover, additional analysis is required to evaluate alternative designs for the Multicast Logical Network topology and architecture (e.g. placement of the Multicast Capable Switches) and to understand the implications of the resulting logical network architecture on the physical network.

ACKNOWLEDGEMENT

Work carried out in the framework of the agreement between the Telecom Italia and the Fondazione Ugo Bordoni

REFERENCES

G.R.Ash, J.S.Chen, A.E.Frey, B.D.Huang: "Real Time Network Routing in a dynamic class-of-service network", Proceedings of ITC 13, pp. 187-194

T.Ballardie, P.Francis, J.Crowcroft: "Core based tree"; ACM SIGCOMM '93; pp. 85-95

K.Bharat-Kumar, J.M.Jaffe: "Routing to multiple destinations in computer networks"; IEEE Transactions on communications; Vol. COM-31, No. 3, March 1983, pp. 343-351;

C.H.Chow: "On multicast path finding algorithms"; IEEE INFOCOM 1991, pp. 1274-1282

S.E.Deering, D.R.Cheriton: "Multicast routing in datagram internetworks and extended LAN's"; ACM Transaction on Computer Systems, Vol. 8, No.2, May 1990, Pages 85-110

S.Deering, D.Estrin, D.Farinacci, V.Jacobson, C.Liu, L.Wei: "An architecture for wide-area multicast routing"; Proceedings of ACM SIGCOMM '94; pp.126-135

ITU-T Draft Recommendation Q.2971: "Point to multipoint call/connection control"; Geneva Sept. 1994

ITU-T Draft Recommendation Q.27BB "Point-to-multipoint B-ISUP"; 1995

J.Y.LeBoudec, B.Przygienda, R.Sultan,: "Routing metric for connections with reserved bandwidth"; 12th EFOC&N conference, June 1994, pp. 84-87

X.Jiang: "Routing broadband multicast streams"; Computer Communications, Vol. 15. No. 1, pp. 45-51

V.P.Kompella, J.Pasquale, G.C.Polyzos: "Multicast routing for multimedia communication"; IEEE/ACM Transaction on networking, Vol. 1. No. 3, June 1993, pp. 286-292

L.Kou, G.Markowsky, L.Berman: "A fast algorithm for Steiner trees"; Acta Informatica, 15, 141-145, 1981

N.Shacham: "Multipoint communications by hierarchically encoded data"; IEEE Infocom '92; paper 9A.4

B.Waxman: "Routing of multipoint connections"; IEEE JSAC, Dec. 1988, pp.1617-1622

P.Winter: "Steiner Problems in Networks"; Networks, Vol. 17 (1987), pp. 129-167

39

Traffic Engineering for VBR Video with Long-Range Dependence

K.R. Krishnan
Bellcore
MCC-1C-353R
445 South Street
Morristown, NJ 07960-6438, USA
Tel: 201-829-2891
Fax: 201-829-2645
e-mail: krk@bellcore.com

Gopalakrishnan Meempat
Bellcore
NVC-3X-311
331 Newman Springs Road
Red Bank, NJ 07701-5699, USA
Tel: 908-758-4034
Fax: 908-758-4371
e-mail: gopal@bellcore.com

ABSTRACT
Recent studies at Bellcore have shown that traffic rates in high-speed data networks exhibit 'long-range dependence' in their correlation, characterized by the so-called Hurst parameter H assuming values in the range $0.5<H<1$. This corresponds to correlation that has a smaller asymptotic rate of decay than in Markovian processes, for which H=0.5. Simulations suggest that link-engineering based on Markovian traffic models could, in some situations, underestimate the bandwidth requirements of the actual traffic. However, another Bellcore study presents results in which a Markovian traffic model has been successful in determining bandwidth requirements for Variable-Bit-Rate video sources, which have $H>0.5$.

We report here the results of an investigation undertaken to bridge the gap in our understanding of the results of these two studies, and to analyze the effect of the Hurst parameter on link-engineering for ATM traffic. Our investigation has shown that a curious 'scaling' property of streams with long-range dependence, derived in another recent Bellcore study, helps to bridge the gap by accounting for the results of both studies. The analysis suggests that a high value of H is not, in itself, a reason to suppose that Markovian traffic models will lead to under-engineering of bandwidth on ATM links.

Keywords
ATM, long-range dependence, multiplexing, effective bandwidth

1. Introduction

The starting point of the investigation reported here was the need to bridge a gap in our understanding of the results reported from two studies, Heyman et al (1992) and Leland et al (1993), on traffic streams that exhibit the property of 'long-range dependence' in the behavior of the auto-correlation of their instantaneous arrival rates.

On the one hand, the study by Leland, Taqqu, Willinger, and Wilson (1993) established, on the basis of extensive traffic measurements, that traffic streams in high-speed networks exhibit 'long-range' dependence in their auto-correlation, characterized by an auto-correlation with a slower asymptotic decay than in standard Markovian traffic models. This property of 'persistent' correlation was described by the *Hurst* parameter H of the stream, with $H = 0.5$ for Markovian streams and $H>0.5$ for streams with long-range dependence. It was further shown, both by analysis [Norros (1994)] and simulation [Erramilli et al (1994a)], that engineering formulae (such as effective bandwidth for connections) based on Markovian models could, in *some cases*, result in under-engineering.

On the other hand, in studies of the bandwidth needed on an ATM link for carrying Variable Bit Rate (VBR) video-traffic, Heyman, Tabatabai, and Lakshman (1992) and Elwalid, Heyman, Lakshman, and Mitra (1995) have shown that, *even though the Hurst parameter of the VBR stream has been determined to be about* 0.7 [Beran et al (1995)] (and thus corresponds to the case of long-range dependence), effective bandwidth formulae derived for Markovian models are, nevertheless, successful in determining the bandwidth required to support the VBR streams.

Clearly, a better understanding of the significance of the Hurst parameter is needed in order to be able to explain and account for the results of both studies. In the course of our investigation of ATM link-engineering for traffic streams with long-range dependence, we we have shown that a curious 'scaling' property of streams with long-range dependence, derived in recent work by Krishnan (1995), helps to bridge the gap between the results of the two studies, and also suggests that a high value of H is not, in itself, a reason to suppose that Markovian traffic models will lead to under-engineering of bandwidth on ATM links.

2. Question to be Investigated

The study by Heyman et al (1992) makes use of an actual trace of the traffic generated by a single source during a video-conference call of 32.3 minutes duration. The original trace consists of 48500 frames, each frame being 40 milliseconds in duration. The traffic data consist of a sequence of 48500 integers, corresponding to the numbers of cells in the frames. The frame-size sequence is shown to correspond to a stationary sequence. The mean frame-size is 130.3 cells, the maximum frame-size is 629 cells, and the variance of frame-size is 5536.9. With a frame interval of 40 milliseconds and with 64 bytes per cell, the VBR stream has thus a mean rate of 1.668 Mb/s and a peak rate of 8.05 Mb/s. The marginal distribution of frame-size is shown to be well-approximated by a gamma distribution with a scale parameter of 0.02353 and a shape parameter of 3.066. In simulations to determine the performance obtained with the VBR video-source, we have made use of the subsequence consisting of the first 45000 frames of data, which corresponds to a 30-minute video-call.

Beran, Sherman, Taqqu, and Willinger (1995) have analyzed this VBR data stream and have concluded that it exhibits long-range dependence, with a Hurst parameter of H=0.7, considerably larger than the value 0.5 applicable for Markovian streams. However, Heyman et al (1992) have constructed a Markovian model, called the Discrete Auto-Regressive (DAR) model, that provides a good match both to the distribution of frame-size and to the auto-correlation between successive frame-sizes, extending over about 150 frame-lags.

Elwalid et al (1995) extend the effective bandwidth theory developed for Markovian traffic streams to the DAR model of the VBR stream, to determine the number of independent, simultaneous VBR sources that can be supported with a cell-loss rate of 1.0e-6 by an ATM link with a bandwidth of 45 Mbits/sec. Event simulations are carried out using time-shifted versions of the actual 30-minute data stream to confirm that the bandwidth of 45 Mbits/sec can, in fact, support the number of sources determined on the basis of the Markovian DAR model of the VBR trace.

Thus, the example considered by Elwalid et al (1995) is at least one instance in which link-engineering based on Markovian models turns out to be adequate for traffic known to possess long-range dependence in its correlations. The question arises, therefore, whether there are situations in which effective bandwidth determined from Markovian models will be adequate for traffic with long-range dependence. The goal of this paper is to investigate this question and present the answers, though incomplete, that have emerged from our investigation.

3. Preliminary Matters

As noted earlier, the actual VBR data available to Heyman et al (1992) consists of the traffic generated by a *single* video-source of 45000 frames. In order to consider the combination of multiple *independent* streams, Heyman et al (1992) use phase-shifted versions of the original 45000-frame stream, (regarded as a cyclic sequence), to represent independent traffic streams. The phase-shifts themselves were chosen in a random manner over the span of 45000 frames; the idea behind this procedure is that the correlation between X_k, the size of frame k, and X_{k+m}, the size of frame $(k+m)$ becomes negligible for large enough values of the lag m, and thus the stream $\{X_k\}$ and the stream $\{X_{k+m}\}$ are effectively uncorrelated over a large span. In fact, a minimum value for the phase-separation could be selected by examining the decay of the correlation in the original data stream with the lag m, and requiring all phase-separations to be larger than a value of m for which the correlation becomes negligible. Generating independent realizations of the DAR *representation* of the original trace, however, is quite straightforward, with the use of different non-overlapping sections taken from a long stream generated by a pseudo-random seed.

3.1 Cell-Loss Rate Criterion for Link-Engineering

It is common in papers on ATM to consider extremely low cell-loss rates, of the order of 1.0e-6 to 1.0e-9, as the performance criterion to be met on an ATM link. It is often difficult to ascertain, by means of simulation over a limited duration, whether such low, long-term loss-rates are being met or not. For example, the VBR traffic data for the single source considered here contains a total of 5,818,500 cells in its 45000 frames over its 30-minute duration. A loss-rate of 1.0e-6 would correspond to the loss of *just 6 cells* over the entire 30-minute duration. Since losses often occur in bursts, and not at a steady rate over the finite duration of interest, the position occupied by a single burst in the arrival stream over the 30-minute duration could make the difference between meeting or violating the grade-of-service. In particular, in such cases, there can be no reasonable confidence levels associated with the results of simulation.

Since the actual VBR traffic stream available to us is limited to the 30-minute trace, we have used a loss-rate of 1.0e-3 as the engineering criterion in some of the results presented below. This loss-rate corresponds to the loss of about 5800 cells per source over the 30-minute duration, which is about 45 times the average frame-size of the stream.

4. Effect of Long-Range Dependence on Performance

4.1 Effective Bandwidth and Multiplexing Gain for the Zero-Buffer Case

We use the following general definition for G, the multiplexing gain achieved when n independent streams are combined for transmission on a single transmission link:

First, consider a single traffic source, and let

R_p = the peak rate of the source,

R_m = the mean rate of the source,

Given a buffer-size B and a desired quality-of-service q (cell-loss rate, cell-delay and jitter, etc), let

C_s = link bandwidth needed to meet desired quality of service for the single source.

We know that $R_m \leq C_s \leq R_p$. The statistical variations in the transmitting rate of the traffic source may enable us to meet the desired quality-of-service with a link bandwidth smaller than the peak rate R_p. The statistical gain achieved for a single source with the use of C_s as the effective bandwidth is defined to be

$$G_s = R_p/C_s \quad (1)$$

We now consider multiplexing the realizations of n such independent sources for transmission on a single link. The independence in the statistical variations of the n individual sources offers the potential for achieving a further reduction in the bandwidth required for the combined stream. Let

$\overline{C}_n$ = link bandwidth needed to meet desired quality of service for the multiplexed stream of n sources

Then, we define the total multiplexing gain G as follows:

$$G = \frac{nR_p}{\overline{C}_n} \qquad (2)$$

$$= [\frac{nC_s}{\overline{C}_n}][\frac{R_p}{C_s}]$$

$$= [\frac{nC_s}{\overline{C}_n}]\, G_s$$

= [multiplexing gain *across* sources][statistical gain in a *single* source]

Thus, we see that the total multiplexing gain due to the independent combination of identical sources can be resolved into two factors, one expressing the efficiency of statistical multiplexing of independent streams (gain *across* sources) and the other expressing the advantage gained (by means of buffering) from the statistical rate variations *within* a source.

One can obtain a conservative estimate of the bandwidth needed for a traffic stream by supposing that delayed cells cannot be buffered (i.e. B=0), and thus insisting that the link-bandwidth $\overline{C}$ be chosen such that the probability of the rate r of the stream ever exceeding $\overline{C}$ is smaller than the probability of loss specified for the quality-of-service. Clearly, this calculation for a stationary process makes use only of the marginal distribution of the rate r, and is unaffected by the properties of the auto-correlation, and thus by the presence or absence of 'long-range dependence'. Heyman et al (1992) have shown that the distribution of frame-sizes in the video-stream (a measure of the average cell-arrival rate over the duration of a frame) is well-represented by a gamma distribution with a scale parameter $\lambda = 0.02353$ and a shape parameter s =3.066. For a single source, the use of the above gamma distribution of frame-size gives a link-bandwidth C = 10.5 Mb/sec for the zero-buffer case, when the probability of cell-loss is not to exceed 1.0e-6. Of course, since the single source has the *known* peak-rate of 8.05 Mb/sec, the assumed marginal distribution is not quite accurate for this case, and we use just the peak-rate. However, when one considers the multiplexing of 15 independent streams, as in Heyman et al (1992), the assumption of the gamma distribution for the single source enables us to conclude that the frame-size distribution for the multiplexed stream is again a gamma distribution, with a scale parameter of λ and a shape parameter of 15s. For a loss-probability of 1.0e-6, the link-bandwidth for the zero-buffer case is easily determined to be 46.6 Mb/sec, which is much smaller than 15 times 8.05 Mb/sec (the bandwidth determined for a single source). In fact, the multiplexing gain G for this case is $15*8.05/46.6 = 2.59$. Thus, even when the effective bandwidth is based entirely on the statistics of the arrival rate, with no consideration of buffers, and *even when it is necessary to allocate the peak rate for a single source* (as in this example), it is still possible to realize significant multiplexing gain when multiple *independent* sources are combined.

The effective bandwidth of 46.6 Mb/sec for the zero-buffer case for 15 sources corresponds quite closely to the result obtained by Elwalid et al (1995) that 15 sources can be supported with a link-bandwidth of 45 Mb/sec with a buffer of 5000 cells (which is about 39 times the mean frame-size). It thus appears that, for the VBR streams studied by Heyman et al (1992), for small and moderate buffer-sizes, *the multiplexing gain is essentially achieved 'across' sources, rather than from the statistical rate variations within a source.* For the zero-buffer case, this component of multiplexing gain is essentially determined by the marginal distribution of the cell-arrival rate, and is uninfluenced by the presence or absence of 'long-range dependence' in the auto-correlation. Thus, the VBR data-stream studied by Heyman et al (1992) is an example in which significant multiplexing gain is achieved for multiple sources in the presence of long-range dependence.

4.2 The DAR Approximation for VBR Traffic

We now come to the central question of our investigation: if, in general, long-range dependence in a traffic stream *is* important for queue-performance, as suggested by the results of Erramilli et al (1994a), how do we explain the successful use of Markovian DAR models for VBR video-traffic, in Heyman et al (1992) and Elwalid et al (1995), for the determination of effective link-bandwidth for a specified cell-loss rate?

We examine below a few of the factors which we believe to be pertinent to this question.

4.2.1 The Correlation-Structure in the VBR Video-Stream Studied

As noted earlier, Beran et al (1992) exhibits long-range dependence, with a Hurst parameter H=0.7. The Hurst parameter, however, is an *asymptotic* property of the auto-correlation. For traffic streams that are only *approximately* self-similar [Leland et al (1993)], (unlike fractional Brownian motion [Norros (1994)], which is *exactly* self-similar), the behavior of auto-correlation for small time-lags could be quite different from its eventual behavior. In fact, for the VBR traffic stream in question, it is shown in Heyman et al (1992) that the auto-correlation of frame-size, *over an initial range of about 150 frame-lags*, is very accurately matched by the exponential decay associated with the DAR model. On the other hand, the fact that the *initial* behavior of the auto-correlation of the VBR stream is quite different from its *asymptotic* behavior is confirmed by the 'variance-time' plot for this stream appearing in Beran et al (1995).

Thus, the video-stream under study is consistent with a Markovian model for auto-correlation over the span of the first 150 frame-lags; the same property has also been shown to hold for other video-streams studied by Heyman et al in (1994). A lag of 150 frames corresponds to a time-lag of 6 seconds. Thus, in situations where 'memory' beyond 6 seconds has no significant effect on the buffer-queue, one expects that it is safe to ignore the asymptotic behavior of the auto-correlation. However, it is hard to define precisely what such 'safe' situations are, except to note that low levels of utilization correspond to small buffer backlogs, and thus to 'safe' conditions.

4.2.2 A Scaling Property of Streams with Long-Range Dependence

In recent work by Krishnan (1995), it has been shown that the fractional Brownian model [Norros (1994)] for traffic with long-range dependence exhibits the following property: given two such streams with the same mean data-arrival rate λ, the same *fixed* (*variance/mean*) ratio α for arrivals per unit time, and two different Hurst parameters H_1 and H_2, with $H_1<H_2$, then, for a given buffer-level b and a given buffer-overflow probability p (in an infinite buffer), there exists a value λ^* of λ such that for $\lambda>\lambda^*$, the stream with the higher Hurst parameter requires *smaller* bandwidth to meet the specified grade-of-service. Thus, there appears to be a 'crossover' in the relative performance of the two streams with increasing load (for a fixed *variance/mean* ratio). This behavior has been confirmed in recent simulations of queue-behavior with traffic generated from the fractional Brownian model [Erramilli (1994b)].

When we consider the multiplexing of *independent* sources of identical statistical characteristics, this result suggests that when enough streams are multiplexed, a high-H source requires less bandwidth than a low-H source. Of course, strictly speaking, the result holds only for fractional Brownian models, which are *exactly* self-similar, whereas the VBR video-source studied in Heyman et al (1992) is only asymptotically self-similar with H=0.7. Nevertheless, we decided to investigate whether the 'crossover' effect can be observed with the VBR source. The DAR model in Heyman et al (1992) for the VBR source, it should be noted, has the same mean load and variance of frame-size as the actual source, but has H=0.5, since it is a Markovian model.

In devising a satisfactory procedure to check for the crossover effect, we had to contend with the fact that the only data available for VBR traffic consisted of a *single*, 30-minute trace for a single source. To simulate the multiplexing of an ensemble of such independent sources, Heyman et al (1992) proposed the use of *phase-shifted copies of the single realization*. Even though the combination of phase-shifted streams may be regarded, for some choices of phase-shifts, as an acceptable approximation to the multiplexing of independent sources, it is nevertheless true that the actual *performance* obtained with such a combined stream can be quite sensitive to the set of phase-shifts used. Therefore, it becomes necessary to consider simulations for various choices of phase-shifts in order to arrive at representative results.

Heyman et al (1992) also propose a more detailed Markov Chain (MC) model for VBR traffic which offers greater accuracy than the DAR model, *especially when the number of multiplexed sources is small*. For this reason, we repeated the above steps for the more elaborate MC model also, but, owing to space limitations, present only the DAR results.

Thus, we proceeded as follows in our experiment:

1. For various values of n in the range $1 \leq n \leq 20$, we multiplexed n DAR streams, generated by multiplexing phase-shifted realizations of the DAR model of the source, for *two different choices of phases*, both of which produced acceptable approximations to independent sources. For each value of n, and for each of the two choices of phase-shifts, we determined the effective bandwidth required for the corresponding multiplexed stream, to realize a buffer-overflow probability of 1.0e-3 for a buffer-level of 5000 cells. In order to obtain a large enough sample for the estimation of buffer overflows, we chose to work with the overflow-probability of 1.0e-3 rather than 1.0e-6. *The effective bandwidth was determined by means of actual simulation* because the algorithm of Elwalid and Mitra (1993) used for the results in Elwalid et al (1995) was unavailable to us at the time this work was performed.
2. Using the effective bandwidth determined for n multiplexed DAR streams, we determined the performance that is realized when n *VBR sources* are multiplexed, with the streams generated, once again, by suitable phase-shifting of the single-source trace that is actually available.

The results (for the infinite-buffer case) for the VBR-DAR comparison are shown in Figures 1a-d and in Figures 2a-d, for the two choices of phases used for multiplexing n sources, for different values of n. Of course, Figures 1a and 2a, for $n = 1$, are the same. In Figures 1 and 2, we observe that the performance obtained is quite sensitive to the choice of phases, but, in both cases, when n (the number of multiplexed streams) is greater than 11, the performance of the actual VBR source (H=0.7) is better than that of the DAR model (H=0.5). For several values of n below 11 (not included in the Figures), the relative performance was found to depend on the choice of phases, and rather erratic.

On the basis of the results in Figures 1 and 2 (and similar results obtained with the MC model), it appears safe to conclude that the results are consistent with the property, proved for fractional Brownian Motion by Krishnan (1995), that for a large enough number of multiplexed streams, the performance of a high-H source becomes better than that of a low-H source.

We also present, in Table 1, a VBR-DAR performance comparison (for the phases corresponding to Figure 1) for the case of a *finite* buffer of 4500 cells, with the effective bandwidth determined for the multiplexed DAR streams for a loss-probability of 1.0e-3. The results show that, for the finite-buffer case also, the performance of the VBR source is better than that of the DAR model for $n > 11$.

Elwalid et al (1995) consider 15 or more sources in their investigations, a number greater than the 'critical' point observed in the above results. (Their bandwidth calculations were made for a cell loss-rate of 1.0e-6, while we used a loss-rate of 1.0e-3, but this has a minor effect on the precise point of the crossover). In other words, the results suggest that *the results of Heyman et al (1992) and of Elwalid et al (1995) correspond to a regime in which the Markovian model (DAR or MC) for determining effective bandwidth provides a conservative estimate for the bandwidth needed for the actual VBR video-sources.* This is in agreement with the simulation results in Elwalid et al (1995), which show that more sources can be supported than predicted by the DAR model, for the numbers of sources considered there.

TABLE 1: Loss Probabilities for Multiple Sources (Finite Buffers)

Number of sources	Bandwidth (Mbps)	Loss probability	
		VBR	DAR
1	4.27	2.195 e-3	3.308 e-4
5	14.50	5.776 e-4	1.141 e-4
8	21.62	3.336 e-4	5.421 e-5
10	24.25	4.017 e-4	3.955 e-5
11	27.72	1.506 e-5	2.557 e-5
15	35.68	0.0	3.199 e-5
20	44.30	0.0	2.819 e-5

4.2.3 Finite-Buffer Effects for Traffic with Long-Range Dependence

The above simulations demonstrating 'crossover' assumed infinite buffers (the case for which the crossover effect was proved by Krishnan (1995)), while Elwalid et al (1995) worked with finite buffers of various sizes in their simulations. With Markovian models, there is reason [Avi-Itzhak and Halfin (1994)] to expect that estimating the loss from a finite buffer by the corresponding overflow probability in an infinite buffer is a conservative approximation; i.e., the loss rate from a finite buffer of size B is smaller than the 'overflow' rate above level B in an infinite buffer. A comparison of the above simulation results with those of Table 1 shows a similar relation for long-range dependent traffic also. Moreover, recent research [Rao et al (1996)] suggests that finite buffers can, in certain cases, mitigate the ill-effects of long-range dependence on buffer queues, rendering Markovian approximations acceptable. These results will be discussed in a future publication.

5. Conclusion

Recent Bellcore studies [Leland et al (1993) and Erramilli et al (1994a)] have demonstrated that traffic streams in high-speed networks exhibit long-range dependence in their auto-correlation, characterized by correlation that decays at a smaller rate than in Markovian streams. It has also been shown [Erramilli et al (1994a)] that there are cases in which link-engineering based on Markovian models would underestimate the actual bandwidth requirements for streams with long-range dependence. However, there is also an example [Elwalid et al] of the successful application of Markovian models to determine the bandwidth requirements of VBR video-sources, which are known to exhibit long-range dependence. Thus, there is a need to bridge the gap in our understanding of the effects of long-range dependence on traffic enginnering in order to account for both results. Spurred by this need, we undertook an investigation to gain a fuller understanding of the effect of the Hurst parameter H on link-engineering.

During the course of our investigation, Krishnan (1995) discovered the following curious 'scaling' property of streams with long-range dependence that helps to bridge the gap between the results of Leland et al (1993) and those of Heyman et al (1992) and Elwalid et al (1995): for *exactly* self-similar traffic streams described by the fractional Brownian model [Norros (1994)], when a sufficiently large number of sources are multiplexed, a high-H source requires less bandwidth than a low-H source [Krishnan (1995)]. From the results presented here, it appears that this effect, derived for the fractional Brownian model, is also at work in the VBR video sources considered by Heyman et al (1992). On the basis of these results, we have offered an explanation that accounts for the results of both Leland et al (1993) and Heyman et al (1992). The analysis suggests that a high value of H is not, in itself, a reason to suppose that Markovian traffic models will lead to under-engineering of bandwidth on ATM links.

6. Acknowledgement

We thank A. Erramilli, D.P. Heyman, T.V. Lakshman, and W. Willinger for their helpful advice at several stages of this investigation.

7. REFERENCES

Avi-Itzhak, B. and Halfin, S. (1994) A method for call admission control in ATM networks with heterogeneous sources, Bellcore Technical Report.
Beran, J., Sherman, R., Taqqu, M.S. and Willinger, W. (1995) Long-range dependence in variable-bit-rate video traffic, IEEE Transactions on Communications, **43**, 1566-1579.
Elwalid, A.I. and Mitra, D (1993) Effective bandwidth of general Markovian traffic sources and admission control of high speed networks, IEEE/ACM Transactions on Networking, **1**, 329-343.
Elwalid, A.I., Heyman, D.P., Lakshman, T.V. and Mitra, D (1995) Fundamental bounds and approximations for ATM multiplexers with applications to video teleconferencing, IEEE Journal on Selected Areas in Communications, **13**, 1004-1016.
Erramilli, A., Narayan, O. and Willinger, W. (1994a) Experimental queueing analysis with long-range dependent packet traffic, Bellcore Technical Report.
Erramilli, A. (1994b), personal communication.
Heyman, D.P., Tabatabai, A. and Lakshman, T.V. (1992) Statistical analysis and simulation study of video teleconference traffic in ATM Networks, IEEE Transactions on Circuits and Systems for Video Technology, **2**, 49-59.
Heyman, D.P., Tabatabai, A., Lakshman, T.V. and Heecke, H. (1994) Source models for VBR broadcast-video traffic, Proceedings of IEEE INFOCOM'94, 664-671,
Krishnan, K.R. (1995) A new class of performance results for fractional Brownian traffic model, Bellcore Technical Report.
Leland, W.E., Taqqu, M.S., Willinger, W. and Wilson, D.V. (1993) On the self-similar nature of ethernet traffic, Proceedings of ACM SIGCOMM'93, 183-193.
Norros, I. (1994) A storage model with self-similar input, Queueing Systems,**16**, 387-396.
Rao, B.V., Krishnan, K.R. and Heyman, D.P. (1996), to be published.

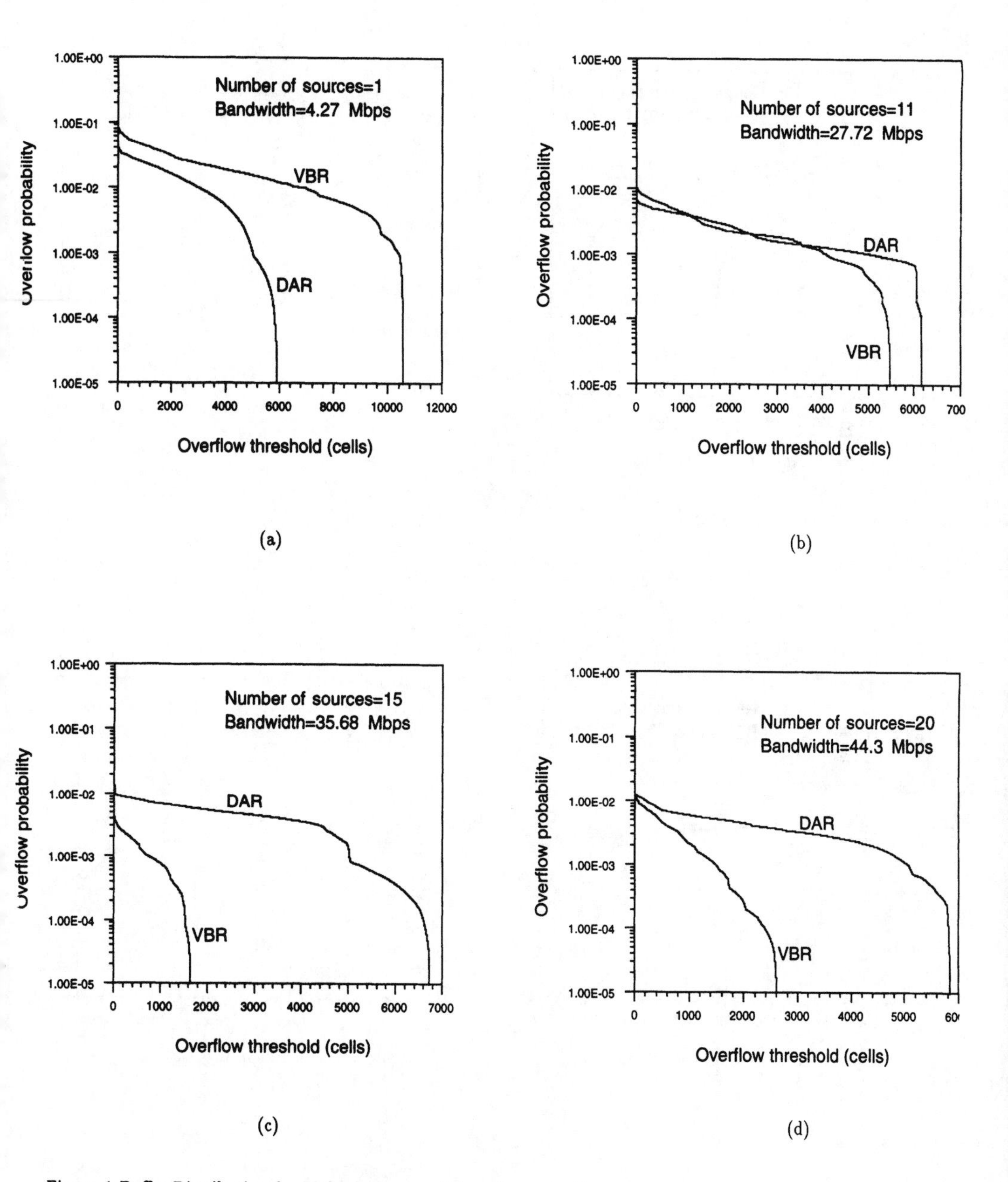

Figure 1 Buffer Distribution for Multiple Sources (Phase Sequence 1).

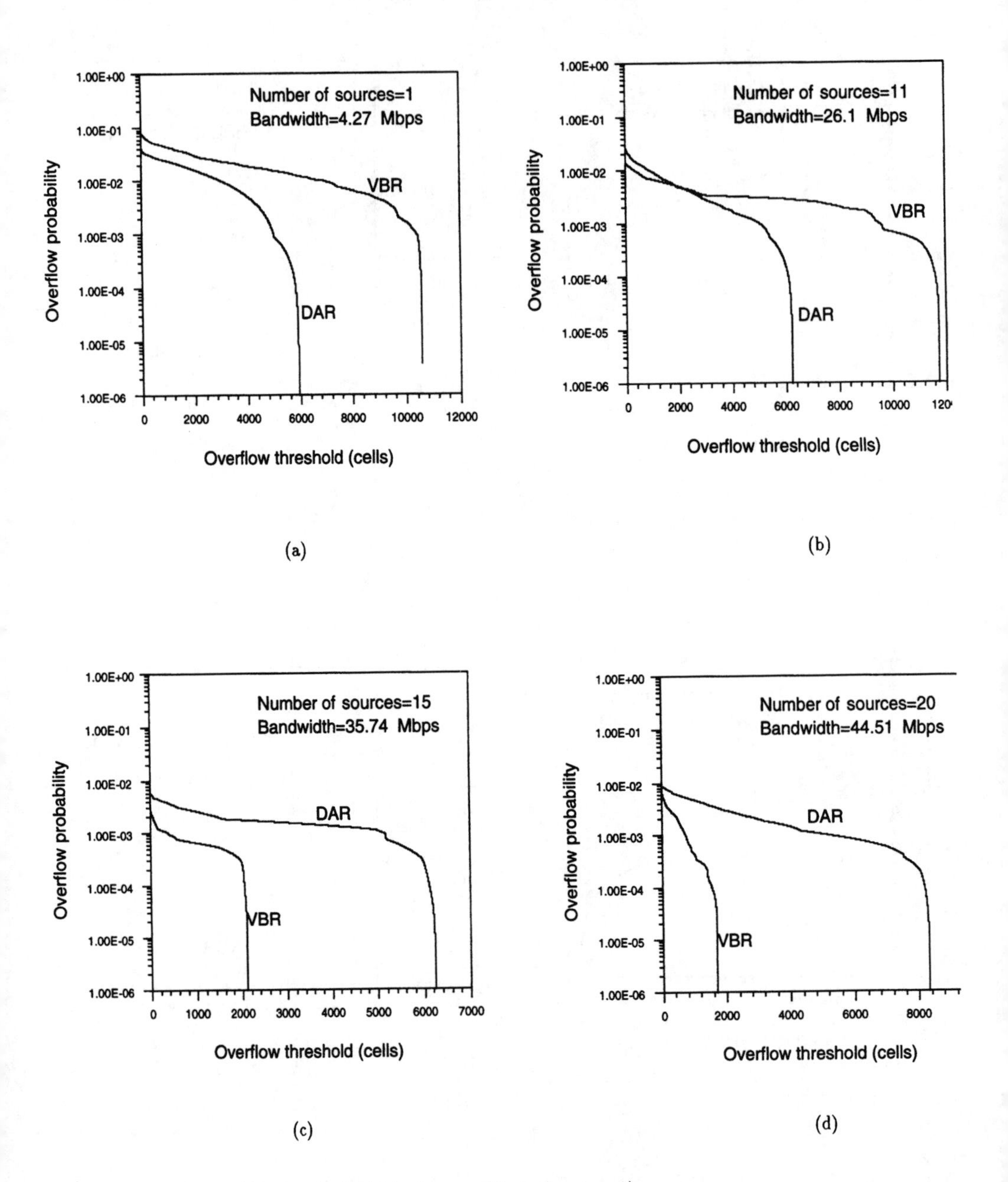

Figure 2 Buffer Distribution for Multiple Sources (Phase Sequence 2).

40
An Efficiency Prediction Method for ATM Multiplexers

Kalyan S. Perumalla[†]*, C. Anthony Cooper*[*] *and Richard M. Fujimoto*[†]
[*]*Bell Communications Research, Red Bank, NJ 07701-5699.*
Phone: +1 908 758 2144.
[†]*College of Computing, Georgia Institute of Technology*
Atlanta, GA 30332-0280. Phone: +1 404 894 5620.
Fax: +1 404 894 9442. email: kalyan@cc.gatech.edu

Abstract
This paper describes a methodology for conservatively predicting the traffic carrying efficiency of an ATM multiplexer that is operating in compliance with realistic performance objectives. Our approach uses a multi-phased treatment of this important problem. In the first phase, a synthetic model of bursty traffic sources is used to construct a set of loading curves that predict worst-case statistical multiplexing efficiencies as a function of certain parameters which characterize both the traffic sources and the multiplexing system under consideration. In the second phase, a set of real traffic sources is matched with these same parameters through appropriate traffic measurements, and a prediction of achievable statistical multiplexing efficiency is obtained by the appropriate use of these loading curves. This technique has been successfully tested on real traffic source data obtained from Local Area Networks, and which has been shown to possess the "self-similar" temporal characteristic that is known to present significant challenges for statistical multiplexing. This methodology was developed with the aid of a high-performance, parallel simulator that is used both for the construction of representative loading curves and for demonstrating the conservative nature of the predicted multiplexing efficiencies that result when this methodology is applied to samples of real traffic. When augmented with a suitable set of routine traffic measurements, it is anticipated that the methods described here can play a significant role in practical ATM network dimensioning processes.

Keywords
ATM multiplexers, multiplexing efficiency, parallel simulation, Markov Chain models, Ethernet LAN traces, self-similar

1 INTRODUCTION

The deployment of telecommunications network equipment based upon Asynchronous Transfer Mode (ATM) technology poses a significant challenge with respect to the network dimensioning methodology needed to achieve economic levels of network equipment usage while simultaneously maintaining satisfactory levels of network performance. The methodology described in this paper addresses a key element of this ATM network dimensioning challenge — the prediction of ATM multiplexing efficiencies achievable under realistic performance objectives while operating with traffic loads imposed by real traffic sources. This methodology is most directly applicable for dimensioning ATM networks and network equipment used in support of Permanent Virtual Connection applications. When used in combination with additional methods for treating the blocking experienced by ATM connections of diverse capacities, one may reasonably anticipate the extension of this methodology for dimensioning ATM networks supporting Switched Virtual Connection applications.

A variety of approaches have been advanced for addressing this ATM network dimensioning challenge, of which Bensaou (1990), Hui (1990) and Elwalid (1995) could be viewed as representative. In addition, significant evidence exists concerning the adverse performance impacts of improperly dimensioned statistical multiplexers operating on traffic that possesses the self-similar temporal characteristic (Leland (1994), Erramilli (1994)). It is therefore prudent for any ATM dimensioning methodology to include verification of its multiplexer performance predictions by testing with real traffic loads that possess this self-similar property. Such verification is provided for the method presented here. We remark that an extension of this measurement-oriented verification process can establish a basis for using routine traffic measurements in support of ATM network dimensioning.

Our study develops in several phases a method that predicts maximum safe load levels for ATM multiplexers. This methodology is developed using several distinct types of traffic source models and a high capacity simulator that is capable of completing simulation runs on the order of 10^{10} cell transfers in less than one hour of real time (Nikolaidis (1993), Nikolaidis (1994), Fujimoto (1995)). The unique benefits obtainable with this high capacity simulator become apparent when examining the clumped nature of cell losses.

In the first phase of this study, a traffic source model based upon a low order Markov Chain is used in a simulation-based process to develop a set of loading curves that are intended to provide conservative, or worst-case, predictions of multiplexer efficiency. These loading curves are parameterized by certain dimensionless ratios relating to both multiplexer and traffic source characteristics. In the second study phase, predictions of multiplexing efficiency based on these loading curves and suitable parameter matching techniques are validated using results from simulations driven by samples of Local Area Network (LAN) traffic.

The remainder of this paper is organized as follows. First, our goals and approach are described and the rather general statistical multiplexing model used for this study is identified. The Markov Chain-based synthetic traffic source model used in the initial phase of this work is then cited, together with the preliminary simulation-based results used to select "worst case" parameters for this traffic source model. Next considered are simulation techniques based on this initial traffic source model and the generation of loading curves. We then describe the simulation technique used in the second phase of this study, which uses a traffic source model based upon Ethernet LAN traffic samples. Comparison between the multiplexing efficiencies predicted from the loading curves and the multiplexing efficiencies achievable with the LAN traffic samples demonstrates the conservative character of this methodology.

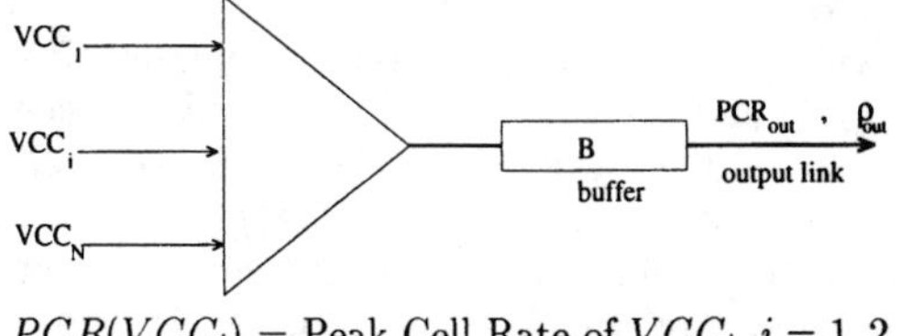

$PCR(VCC_i)$ = Peak Cell Rate of VCC_i, $i = 1, 2, \cdots, N$
PCR_{out} = Peak Cell Rate supportable on output link
$G = \frac{\sum_{i=1}^{N} PCR(VCC_i)}{PCR_{out}}$

Figure 1 Statistical Multiplexer Model used in this Study.

2 GOALS AND APPROACH

This methodology's primary goal is to conservatively predict the statistical multiplexing gain and output link utilization of an ATM statistical multiplexer that can be achieved with real traffic while complying with established performance objectives, particularly for the recognized Cell Loss Ratio (CLR) performance parameter (ANSI Standard (1994)). The traffic of interest here has an intensity that varies randomly in time between zero and a maximum value that is described by a Peak Cell Rate (PCR). Such traffic can include the output of various data or video traffic sources; it has been referred to as Statistical Bit Rate traffic; and certain additional quantifying parameters* for this traffic have been introduced in international standards (ITU (1995)). The CLR performance objective of 10^{-7} is used here as a realistic and previously documented value for the cell loss impairment experienced by a data-oriented application when carried by an ATM virtual connection across one ATM multiplexer (Bellcore (1995)).

A secondary goal of this research is to better characterize the cell loss events occurring at an ATM statistical multiplexer. Such characterization should have sufficient accuracy to support the satisfactory interpretation of simulation results with respect to the observed cell losses on a given simulation run, and the resulting inference about the likely range of CLR values.

Our approach is based upon the use of two distinct types of traffic source models. Each type of traffic source model is used in a separate study phase to represent inputs to an ATM statistical multiplexer. The rather general multiplexer model shown in Figure 1 is used throughout this study. This statistical multiplexer can be viewed as having N inputs, which are interpreted as ATM connections, or more specifically, Virtual Channel Connections (VCCs), and a single output of known capacity that is characterized by its Peak Cell Rate, PCR_{out}, and that is assumed to be supplemented with an output buffer of known size, identified as B cells. The details of internal data transfer between the inputs and the output of this multiplexer are submerged for purposes of this study, and so it is only necessary to assume that any queuing at input buffers is either negligible or else

*Relevant parameters are the Sustainable Cell Rate (SCR), in cells/second, and the Maximum Burst Size (MBS), in cells.

approximated by an equivalent amount of output buffering. It is generally desirable to achieve large values for N and for two related measures of multiplexer efficiency, which are the statistical multiplexing gain and the output link utilization. It is taken as mandatory that the CLR for this system remain less than its objective value, so that all such efficiency measures are maximized subject to this constraint.

Three dependent variables are useful for characterizing the efficiency of an ATM multiplexer. If we let N be the number of inputs, or traffic sources, driving the multiplexer, these dependent variables are:

N^* = maximum value of N that meets the CLR performance objective for a system configuration

G = multiplexing gain, which is relatable to N^* as $G = \frac{N^* \cdot PCR(VCC)}{PCR_{out}} = \frac{N^*}{K}$ (1)

ρ_{out} = output link utilization, which is relatable to G as $\rho_{out} = \frac{G \cdot SCR(VCC)}{PCR(VCC)} = \frac{G}{b}$

where it is convenient to assume that $PCR(VCC_i) = PCR(VCC)$ and $SCR(VCC_i) = SCR(VCC)$ for $i = 1, \ldots, N$, and where K and b are dimensionless ratios whose definitions follow from the relations provided in (1).

The first type of traffic source model used here is based upon a low order Markov Chain that includes sufficient detail to permit independent adjustment of the first order and second order burst length statistics for traffic sources having ON/OFF emission characteristics. A four-state Markov Chain is selected for this purpose. Most of the parameters associated with this source model and the associated multiplexer are next set to "worst-case" values, which are intended to yield the most pessimistic simulation-based results possible for the largest achievable N^*, G and ρ_{out} when the multiplexing system is operating in compliance with its CLR performance objective. Then the remaining parameters of this worst-case configuration (most of which are expressed as dimensionless ratios) are varied to produce a set of loading curves. It is hypothesized that, when the statistics of real traffic sources are appropriately matched with the parameters that index these loading curves, safe predictions can be made concerning the largest values of N^*, G and ρ_{out} that will be achievable.

The second type of traffic source model used in this study incorporates time-based records, or traces, of Ethernet LAN traffic (Leland (1994)). We select this type of traffic for testing the loading curve predictions because such traffic has been shown to possess the self-similar temporal characteristic that makes its statistical multiplexing difficult (Leland (1994), Erramilli (1994)).

Simulation of high speed ATM multiplexers typically requires long execution times, owing to the large number of cell transfer events to be simulated in support of a practical CLR performance objective. While conventional sequential simulators are capable of simulating on the order of 10^7 cell transfers in a reasonable amount of processing time, our results indicate that significantly higher simulation capacity is desired to study cell losses commensurate with a CLR objective of 10^{-7} and to provide adequate consideration for the demonstrably nonindependent nature of cell losses.

Our research makes use of appropriately structured parallel simulation techniques that extend the methods of Fujimoto (1995) and allow about 10^{10} cell transfers to be simulated in about one hour of wall clock time. Our implementation is based on a 32 processor Kendall Square Research shared-memory multiprocessor. This simulator allows the ATM multiplexer's input links to be driven by a variety of source models. The two source models used in the current study are a four-state Markov Chain model, and a LAN trace-driven model.

3 ROLE OF A FOUR-STATE MARKOV CHAIN SOURCE MODEL

A four-state Markov Chain model having an ON/OFF operating mode is selected as our initial traffic source model because it provides several desirable features. First, it permits the independent adjustment of both first order and second order burst-oriented statistics by suitable treatment of the sojourn times in each of its four states. Its characteristics and application to this type of study are reasonably well understood (Eliazov (1990)). Furthermore, it may support analytically tractable analysis of the queuing aspects associated with the multiplexing system under consideration. Since our goal is to relate intermediate results in the form of loading curves obtained with this source model to the statistical multiplexing efficiencies achieved with real traffic sources (and for which analytically tractable models are not generally available), the focus of this work is oriented towards simulation-based analysis.

In general, each of the dependent variables N^*, G and ρ_{out} will be a function of a number of independent variables that characterize various aspects of the traffic sources and multiplexing system under consideration. The more significant of these independent variables are:

- B = size of output buffer, in cells
- K = capacity factor for a traffic source and this output link = $\frac{PCR_{out}}{PCR(VCC)}$
- CLR_{obj} = Cell Loss Ratio objective = 10^{-7} for this study
- MBS = Maximum Burst Size = bound on a traffic source's active period or burst length
- $c^2(A)$ = squared coefficient of variance of the active (ON) period of a traffic source
- $c^2(S)$ = squared coefficient of variance of the silent (OFF) period of a traffic source
- $m(A)$ = mean active (ON) period of a traffic source, in cell emission times referenced to $PCR(VCC)$
- $m(S)$ = mean silent (OFF) period of a traffic source, in cell emission times referenced to $PCR(VCC)$
- b = burstiness factor of a traffic source = $\frac{PCR(VCC)}{SCR(VCC)} = \frac{m(A)+m(S)}{m(A)}$

Observe that any two of the last three parameters listed are sufficient. We elect to use $m(A)$ and b.

Consider now the functional dependencies of N^*, G and ρ_{out}. The value of N^* can be expressed as

$$N^* = N^*(\, B,\, K,\, CLR_{obj},\, MBS,\, c^2(A),\, c^2(S),\, m(A),\, b\,) \tag{2}$$

and analogous dependencies exist for G and ρ_{out}. We wish to reduce the number of independent variables in (2). This reduction can be accomplished by fixing certain parameters as system constants, and by setting other parameters to values that yield the lowest, or most pessimistic value, of N^*. The selection of model parameters to yield such a worst-case estimate is consistent with our objective of establishing a set of loading curves that can be used to provide conservative estimates of statistical multiplexing efficiency. It can be deduced from the relations provided in (1) that any selection of independent variable values to maximize N^* will also maximize G and ρ_{out}.

The independent variables B, CLR_{obj} and $m(A)$ can be fixed as constants. The value of B is set by the available multiplexing equipment. CLR_{obj} is a limiting value fixed by the operator of that multiplexing equipment, and is assumed here to be 10^{-7}. The value of $m(A)$ is set to approximate the mean Protocol Data Unit length associated with the type of traffic sources under consideration. Since these results are intended for validation against samples of Ethernet LAN traffic, $m(A)$ is taken here to be 28 cells†.

The independent variables MBS, $c^2(A)$ and $c^2(S)$ are selected to yield worst case values of N^*, as follows.

A number of considerations, including a preliminary set of simulation runs, indicate that N^* generally increases as MBS is made smaller. Therefore setting MBS to infinity results in the worst-case treatment of this independent variable. Since the source model based on a four-state traffic Markov Chain yields unbounded burst lengths, such worst case treatment is provided.

Preliminary simulation runs, as well as prior investigation (Eliazov (1990)), indicate that setting $c^2(A) = c^2(S) \equiv c^2$ has little affect on the resulting value of N^*. Hence this is done here. It remains to select appropriate values for c^2.

From simulation runs using representative parameter settings, the variation of N^* with respect to c^2 was observed (the details of these sample runs are omitted here due to space limitations). From such experiments, it was observed that after an initial decrease in N^*, a point exists beyond which further increases in c^2 do not yield a significant further decrease in N^*. The value of c^2 at this point is selected for establishing simulation-based estimates of N^*, G and ρ_{out}. Once this is accomplished, the functional dependency of the dependent variables upon c^2 is eliminated.

Simplified Functional Dependencies

After either fixing or selecting worst-case values for many of the independent variables in the manner described, N^* and the other dependent variables can be represented as functions of a reduced set of parameters. Hence functional dependencies of the type reflected by (2) can be replaced with

$$G = G(b, K); \quad N^* = N^*(b, K); \quad \rho_{out} = \rho_{out}(b, K) \tag{3}$$

With (3), these three measures of multiplexing efficiency are described in terms of the burstiness factor of the traffic sources and the capacity factor, which depends upon characteristics of both the traffic sources and the multiplexing equipment.

†The useful payload carried in 28 cells depends to some extent on the protocols used, but it is in the neighborhood of 1,300 octets. which is less than the 1,518 octet maximum length of an Ethernet packet.

4 SIMULATION ASPECTS AND LOADING CURVES

We now examine some further aspects of this simulation including characterization and treatment of cell loss events, and exhibit some representative loading curves for the initial phase of this methodology. Our simulator software is structured so that the simulated cell transfers (and cell loss events) associated with a single run are captured in a number of separately treatable "segments".

For one simulation run of about 10^{10} cell transfers, our simulator uses roughly 1,500 segments, each of which contains approximately 8×10^6 cell transfers. The observed CLR for one simulation run is calculated as the ratio of the number of cells lost to the total number of cells offered during that run.

On runs for which this observed CLR is in the neighborhood of 10^{-7} , it is found that all cell losses occur in a very few (typically less than 10) of these roughly 1,500 available segments. This is evidence of the tendency of cell losses to be clumped.

In our study, we make use of a "stopping rule" that permits the conclusion with reasonable confidence that $CLR_{obj} \equiv 10^{-7}$ is not exceeded on a particular simulation run, or on its indefinite extension. For a given set of independent variables, an efficient binary search procedure is used to select N for each simulation run, and to control convergence to N^* under the reasonable assumption that CLR is monotonically nondecreasing with N.

Loading Curves

We refer to a plot of N^* as a function of the capacity K (with all other independent variables being held constant) as a *loading curve*. Loading curves were generated for each of the three dependent variables over a range of burstiness factors and output buffer sizes. Other independent variables are either fixed or extremized as previously described. Figures 2, 3 and 4 show the set of loading curves obtained for N^*, G, and ρ_{out}, respectively, when $B = 3000$, and b takes selected values within a representative range.

5 ROLE OF A LAN TRACE-BASED SOURCE MODEL

Some measurement-based data for real traffic sources that might be statistically multiplexed on ATM networks are available in the form of suitably collected traces of Ethernet LAN traffic (Leland (1994), Erramilli (1994)). Such data are particularly useful for testing the applicability of our loading curves because such data have been shown to exhibit the self-similar characteristic that makes statistical multiplexing difficult.

Six such traces were examined in the second phase of this study, with each trace consisting of at least 10^6 Ethernet packets, and being equivalent to between 10^7 and 10^8 cells. The relevant information abstracted from such a trace are the time of a particular packet's occurence and the length of that packet. While the following results are framed for conciseness in terms of a single such trace, they generally apply to each of the Ethernet LAN traces that we have examined to date.

For the purpose of simulation using trace-based traffic sources, each source (or VCC as illustrated in Figure 1) effectively uses its own copy of the trace. To minimize time correlation across multiple sources, the starting poing within this trace is randomly and independently selected for each source. If and when the trace data is exhausted by any particular source, a second random starting point is selected. Because the random selection

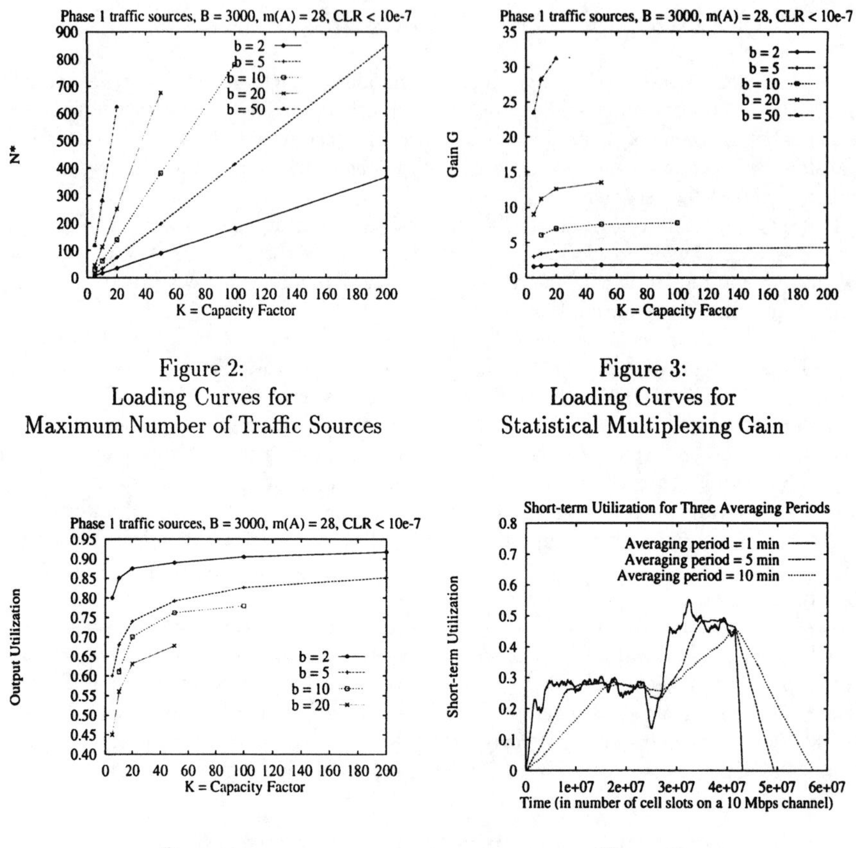

Figure 2:
Loading Curves for
Maximum Number of Traffic Sources

Figure 3:
Loading Curves for
Statistical Multiplexing Gain

Figure 4:
Loading Curves for
Output Link Utilization

Figure 5:
Short-term Utilization of a Representative
Trace of LAN Traffic

of starting points is confined by our procedures to the initial portions of this trace, no source uses the trace more than twice when generating results reported here.

Some Statistical Properties of an Examined Ethernet Trace

As a preliminary step, the autocorrelation function for the basic trace was determined. This autocorrelation function indicates that some correlation does exist at the mean separation between the random starting times for the LAN trace copies used by different sources. We conclude from such correlation that the results of multiplexing simulations driven by these sources will understate to some extent the maximum number of such sources, N^*, achievable when operating in conformance with the loss objective, CLR_{obj}. This will result in somewhat conservative values for the three dependent variables, N^*, G and ρ_{out}, used here to characterize multiplexing efficiency. It also highlights, however, the need to further pursue such investigations with considerably larger bases of real traffic data.

A number of statistics were obtained for the Ethernet trace under consideration after a preliminary conversion from packet-oriented data to cell-oriented data. This conversion is accomplished by treating each packet as a cell burst having suitable length and occurring at a $PCR(VCC)$ that is equivalent to 10 Mbits/second (for the Ethernet LAN). The average utilization, ρ_{source}^{long}, of this converted trace over its 30 minute duration is 0.33. The following measured values for this converted trace are noted:

- $m(A) = 24.8$ cells • $m(S) = 49.7$ cells • $c^2(A) = 4.63$ • $c^2(S) = 3.80$

Figure 5 displays three plots of the short term utilization for this converted trace that were obtained by averaging over sliding windows of 1 minute, 5 minutes and 10 minutes, respectively. A point on such a utilization plot is found through dividing the number of trace-originated cell slots occurring in a particular window (whose right end point is plotted on the abscissa) by the number of cell slots available in that window at the selected $PCR(VCC)$. The primary interest here concerns the central portions of these plots, with the end ramps being artifacts of data exhaustion in the sliding window averaging procedure. These plots of Figure 5 will be relevant when matching the multiplexing efficiencies achieved with trace-based sources to the efficiencies predicted by the loading curves given in Figures 2, 3 and 4.

Determining Safe Load Levels for Traffic Sources Based on Ethernet Traces

The next task is to determine safe traffic load levels, or equivalently, the maximum multiplexing efficiency that is safely achievable with a particular set of trace-based traffic sources. Except for the use of a different traffic source model, the maximum achievable multiplexing efficiencies for traffic sources based on one or more Ethernet traces are determined using methods similar to those presented in Section 4. For fixed values of all independent variables (e.g., $B = 3000$), the number of trace-based sources, N, is varied for each of a number of simulation runs. The largest admissible N, call it N^*, is next identified with a binary search procedure. Then knowing N^*, (1) is used to evaluate the remaining dependent variables, G and ρ_{out}.

Performing the above procedure for selected values of $K = 20$, 50 and 100 and plotting the results gives the desired safe load levels. These are shown for N^*, G and ρ_{out} as the solid curves in, respectively, Figures 6, 7 and 8.

Relation of Loading Curve Predictions to Ethernet Trace Results

Consider now the use of loading curves as provided in Figures 2, 3 and 4 for predicting the multiplexing efficiencies achievable with Ethernet-based traffic sources. The solid curves for N^*, G and ρ_{out} shown respectively in Figures 6, 7 and 8 are the results obtained with traffic sources modeled from an Ethernet trace. They are the results that one would wish to predict from these loading curves.

To apply these loading curves from the first study phase, it is necessary to identify the appropriate burstiness factor, b, to associate with the Ethernet-based traffic sources. The relation

$$b = \frac{1}{\rho_{source}} \tag{4}$$

implies that an appropriate value of b can be found from a properly selected value of ρ_{source}. But for most real traffic sources, ρ_{source} depends upon both the length of the

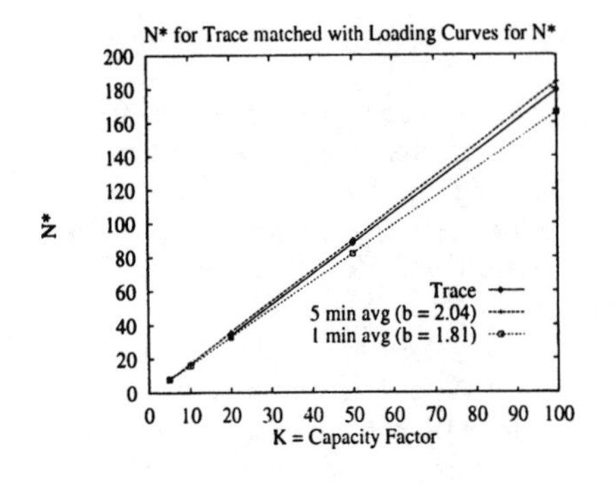

Figure 6:
Matching Trace and Loading Curve Results for Maximum Number of Traffic Sources

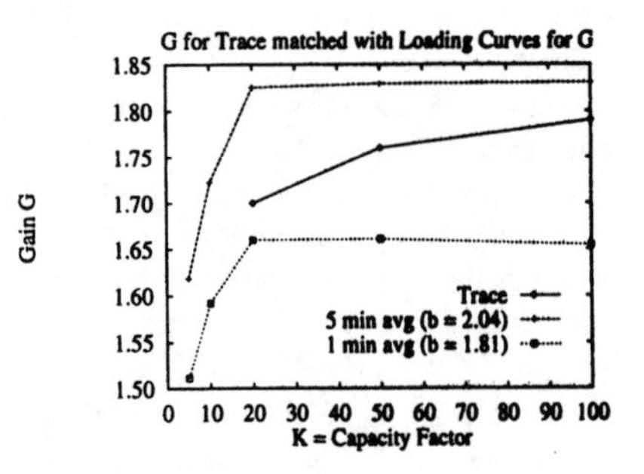

Figure 7:
Matching Trace and Loading Curve Results for Statistical Multiplexing Gain

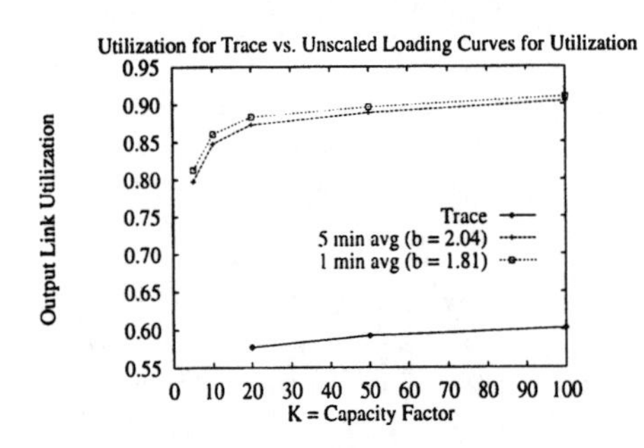

Figure 8:
Trace and *Unscaled* Loading Curve Results for Output Link Utilization

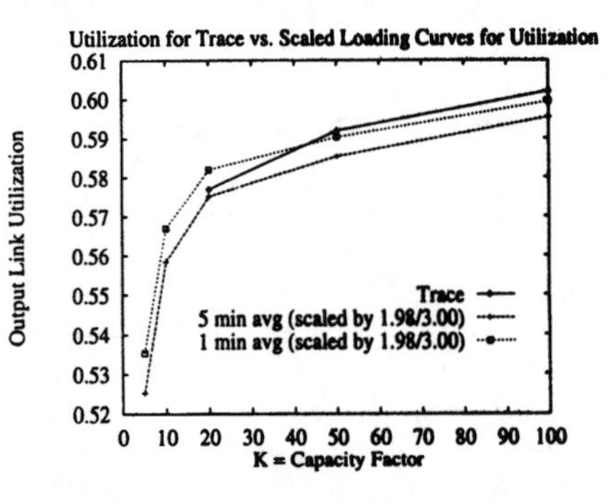

Figure 9:
Trace and *Scaled* Loading Curve Results for Output Link Utilization

averaging interval used to measure this quantity and the placement (or starting time) of that interval.

Several considerations are relevant here. First, for a given averaging interval length, a reasonable and conservative approach is to select the maximum value of ρ_{source} over an available period of observation. Second, the length of this averaging interval can be selected to yield a reasonable match between the results derived from a given set of data and the prediction methodology under test. The effectiveness of this prediction methodology can then be gauged by the degree to which this process can be reproduced with other sets of data and the same length of averaging interval.

By using plots of the type shown in Figure 5, maximum values of ρ_{source} can be identified for various averaging interval lengths. Then, based on the maximum value of ρ_{source} for each plot, the corresponding value of b is determined with the help of (4). This value of b and a specified value of K can then be used to identify a corresponding point on a loading curve for N^*, G or ρ_{out}. This corresponding point is, of course, dependent upon the length of averaging interval used to generate the plot of ρ_{source}.

The execution of this process is demonstrated using the ρ_{source} plots from Figure 5 that correspond to the 1 minute and 5 minute averaging intervals. The resulting values of b are 2.04 for the 5 minute averaging interval and 1.81 for the 1 minute averaging interval. Then the initial loading curves in Figures 2, 3 and 4 are interpolated (or extrapolated) to yield the pairs of dashed curves shown in Figures 6, 7 and 8.

Comparing the 1 minute-based extrapolated loading curve with the trace-generated curve in Figure 6 shows a conservative prediction for N^*. A similar comparison for Figure 7 shows that a conservative prediction for G is achieved with the 1 minute-based extrapolated loading curve. Furthermore, we note that conservative predictions are found for N^* and G obtained from similar analyses using 1 minute averaging intervals for the remaining five Ethernet traces examined in this study.

In distinct contrast, the trace-generated curve for ρ_{out} shown in Figure 8 does not match either the 1 minute or 5 minute-based interpolated loading curves. This mismatch is not unexpected because related phenomena have been observed, whereby low order Markov Chain models significantly overstate the utilization levels of self-similar traffic that can be safely handled. The proper method for applying the initial loading curves for ρ_{out} is through an appropriate scaling procedure.

Scaling of the Loading Curve for Output Utilization

For steady state operation with negligible cell loss, it follows from continuity considerations that $N \cdot \rho_{source} \cdot PCR(VCC) = \rho_{out} \cdot PCR_{out}$. Setting N and ρ_{out} at their maximum permissible values yields

$$\rho_{source} = \frac{K}{N^*} \cdot \rho_{out} \tag{5}$$

Since ρ_{source} differs between the four-state Markov Chain model (and hence the loading curves) and the trace-based model, this relation indicates that ρ_{out} must differ proportionally as the source model changes. Interpolation of the b values shown in Figure 6 for the (dashed) 1 minute and 5 minute curves yields the corresponding value for this figure's solid curve as $b_{source} = 1.98$. As b_{source} has been matched to the loading curves – whose long term values of b are determinable from relations given in Section 3 – it is clear that ρ_{out} must be scaled by the ratio of the b_{source} to the long term value of b for the trace-based sources. Using the previously cited value $\rho_{source}^{long} = 0.33$ and (4) yields this quantity, $b_{source}^{long} = 3.00$.

Figure 9 replots ρ_{out} after this rescaling. The reasonable agreement shown here is also obtained for the other five Ethernet traces processed to date. We remark that for measurement-based applications, it appears possible to avoid the significant additional effort needed to evaluate b_{source} by approximating it with the reciprocal of the more easily measurable peak value of the appropriate short term utilization.

6 CONCLUSIONS AND EXTENSIONS

This paper describes a method for conservatively predicting the traffic carrying efficiency of ATM statistical multiplexers that are operating in compliance with a given cell loss objective. A suitable set of loading curves are first constructed using a traffic source model based upon a specific, low order Markov Chain. These loading curves are intended to support estimates of the highest multiplexing efficiency achievable when operating with real traffic sources. These loading curves are properly applied by using our identified scaling method, together with appropriate traffic measurements. The resulting predictions of multiplexing efficiency provided by this methodology have been shown, based on a limited amount of self-similar traffic data, to be conservative. This methodology's practical application is based upon a set of traffic measurements whose general availability may be reasonably anticipated, but which is not assured at present time. Additional refinement of this method is expected to further test its effectiveness on a larger base of traffic data and to better incorporate available traffic measurements.

REFERENCES

ANSI Std. T1.511-1994, *B-ISDN ATM Layer Cell Transfer - Performance Parameters.*

GR-1110-CORE (1995) *Broadband ISDN Switching System Generic Requirements*, Issue 1/Revision 2, (Bellcore).

Bensaou, B., Guibert, J., Roberts, J. W. (1990) Fluid Queuing Models for a Superposition of On/Off Sources. *7th ITC Specialist Seminar*, Paper 9.3.

Eliazov, T. E., Ramaswami, V., Willinger, W., Latouche, G. (1990) Performance of an ATM Switch: Simulation Study. *Proceedings of the IEEE Infocom '90*, San Francisco.

Elwalid, A., et al (1995) Fundamental Bounds and Approximations for ATM Multiplexers with Applications to Video Teleconferencing. *IEEE Journal on Selected Areas in Communications*, Vol. 13, No. 6, 1004–1016.

Erramilli, A., Gordon, J., Willinger, W., (1994) Applications of Fractals in Engineering for Realistic Traffic Processes. *Proceedings of the 14th International Teletraffic Congress*, 35–44, Amsterdam: Elsevier.

Fujimoto, R. M., Nikolaidis, I., Cooper, C. A. (1995) Parallel Simulation of Statistical Multiplexers. *Journal of Discrete Event Dynamic Systems — Theory and Applications*, Vol. 5. 115–140.

Hui, J. Y. (1990) Switching and Traffic Theory for Integrated Broadband Networks. Boston: Kluwer.

Draft Revised ITU Recommendation I.371 (1995) Traffic Control and Congestion Control in B-ISDN. ITU Study Group 13 Temporary Document Number 71(P), Geneva.

Leland, W. E., Willinger, W., Taqqu, M. S., Wilson, D. V. (1994) Statistical Analysis and Stochastic Modeling of Self-Similar Datatraffic. *Proceedings of the 14th International Teletraffic Congress - ITC14*, 319–328, Amsterdam: Elsevier.

Nikolaidis, I., Fujimoto, R. M., Cooper, C. A. (1993) Parallel Simulation of High-Speed Network Multiplexers. *32nd IEEE Conference on Decision and Control*, 2224–2229.

Nikolaidis, I., Fujimoto, R. M., Cooper, C. A. (1994) Time-Parallel Simulation of Cascaded Statistical Multiplexers. *ACM Sigmetrics Conference on Measurement & Modeling*, 231–240.

41

Buffer dimensioning for rate adaption modules

B. Steyaert and H. Bruneel
SMACS Research Group, Lab. for Communications Engineering, University of Ghent, Sint-Pietersnieuwstraat 41, B-9000 Gent, Belgium. email : {bart.steyaert,herwig.bruneel}@lci.rug.ac.be

Abstract

In a telecommunication network, information (represented by fixed–length *packets*) sent by a source typically passes through a number of nodes before reaching its final destination. In a variety of networks, the communication links that interconnect multiple nodes do not necessarily transmit the same amount of information per time unit. In particular, when the speed of an incoming link(s) in a node exceeds the speed of the outgoing link(s), buffering of packets must be provided in order to avoid excessive packet loss. In this report, we examine the problem of dimensioning such a *rate–adaptation* buffer. The packet arrival stream is described by characterizing the length of consecutive *active* and *passive* periods (i.e., a series of consecutive slots during which packets, respectively no packets, are generated); the former quantities can have any distribution, while the latter are assumed to be geometrically distributed. Using a generating–functions approach, an expression for the steady–state probability generating function of the buffer occupancy is derived. From this result, a closed–form expression for the tail distribution of the buffer occupancy is derived, that is practical and easy to evaluate; this latter quantity is especially useful for buffer dimensioning purposes. In addition, an accurate approximations for this quantity, that reduces all numerical calculations to an absolute minimum, is established as well.

Keywords

Rate adapter, generating functions, tail approximation

1 SYSTEM DESCRIPTION

Let us consider a communication network, where information is segmented into fixed–length packets, and transmissions, as well as arrivals of packets are synchronized with respect to an infinite set of periodic (i.e., equidistant) time instants. The time period elapsed by two consecutive time instants is referred to as one *slot*, and one slot

---

The authors wish to acknowledge the support of Alcatel Bell Telephone Mfg. Co. (Antwerp, Belgium), and the National Fund for Scientific Research (NFWO).

suffices for the transmission of exactly one packet, a situation which, for instance, occurs in an ATM–based B–ISDN network, where a packet then represents one ATM cell of 53 bytes (De Prycker, 1991).

The packets generated by a source typically pass through a number of nodes before reaching their final destination, and we consider the situation where at some point in the network, the amount of information per time unit that can arrive on the incoming link (*input link*) exceeds the amount of information that can be transmitted per time unit on the outgoing link (*output link*). This could be caused, for instance, by a difference in transmission speed between input and output link due to different clock rates, or because, due to some internal representation mode in a switching network, bits are added to the original packets before sending them through the network. In any case, if packet loss is to be avoided, incoming packets must be buffered in a so–called rate–adaptation buffer, and the difference in transmission rate can be incorporated in the corresponding discrete–time queueing model by choosing a different time scale, i.e., slot length, on the input and output link (meaning that we obtain longer slots on the output link compared to the input link), a situation which is depicted in Figure 1.a,b together with some additional quantities still to be defined in this and the following section. In the remainder of the paper, a slot corresponding to a time unit on the Input (Output) Link will be referred to as an IL (OL) slot, and, similarly, the associated Time Scale on the Input (Output) Link will be denoted by ILTS (OLTS).

The packet arrival process on the input link will be characterized by specifying the lengths of successive *passive* and *active periods*, being defined as a number of consecutive slots during which there is no packet arrival, respectively a number of consecutive slots each carrying exactly one packet (as we already mentioned, packet arrivals are synchronized with respect to the IL slot boundaries). In particular, we let the random variables b_n and i_n, $n \geq 1$, represent the lengths of successive active and passive periods, where at some initial time instant t=0, the first active period is initiated, each active period being followed by a passive period. It will now be assumed that $\{b_n \mid n \geq 1\}$ and $\{i_n \mid n \geq 1\}$ are two sets of i.i.d. random variables; in addition, the elements of these two sets of random variables are assumed to be mutually independent. This implies that the probability mass function of any random variable b_n (i_n) can be represented by one common probability generating function B(z) (I(z))

$$B(z) \triangleq \mathbf{E}[z^{b_n}] \quad , \quad I(z) \triangleq \mathbf{E}[z^{i_n}] \tag{1}$$

where **E**[.] denotes the expected value of the tagged quantity. In the analysis throughout the following sections, B(z) can take any form, whereas this is not the case as far as I(z) is concerned. For our purposes, it is sufficient to assume that I(z) has a geometric form with parameter α and mean $1/(1-\alpha)$, and thus can be written as

$$I(z) = \frac{(1-\alpha)z}{1-\alpha z} \quad , \tag{2}$$

although a more general, rational form for I(z) could also be taken into consideration.

The analysis presented in this paper, is a first step in the study of the rate–adaptation buffer–dimensioning problem. To that extent, we assume that the ratio of the transmission rate versus the arrival rate can be written as the fraction of two integers

$$\frac{Transmission\ Rate}{Arrival\ Rate} = \frac{k-1}{k} \quad , \quad k \geq 2 \quad , \tag{3}$$

meaning that during the time period required to transmit k–1 packets, exactly k packets could arrive. This assumption thus covers a considerable range of possible values for the above mentioned ratio, while keeping the analysis tractable. Now, assuming that the initial time instant t=0 coincides with both an IL and an OL slot boundary, then when the packet arrival process, originally generated on an ILTS–basis, is converted to the OLTS, the beginning (and ending) of a packet arrival will coincide with any of the k time instants within an OL slot that arise from devising the OL slot into k time periods of equal length; such a time instant will be referred to as a *microslot boundary*, while the time period elapsed between two successive microslot boundaries, is called a *microslot*. This is illustrated in Figure 1.a,b for k=6.

To the best of our knowledge, the buffer dimensioning of a rate adapter module has received only little attention in the literature. In Rothermel (1992), for a Bernoulli arrival process, (i.e., geometrically distributed active and passive periods, with parameters σ and $1-\sigma$ respectively, where σ is the load of the arrival stream on the ILTS), a simple approximate procedure was developed, which neglects the equidistant nature of slots during which multiple cell arrivals can occur on the OLTS. In Michiel (1990), also for a Bernoulli arrival stream, approximate results were derived that are sufficiently accurate, as long as the difference between input and output rates remains sufficiently small (less than 2%).

2 SYSTEM EQUATIONS

Since the transmission of packets is synchronized to the OL slot boundaries, and as such is essentially based on the OLTS, the packet arrival process, described by the random variables b_n and i_n, must be translated into corresponding quantities describing the arrival process on the OLTS. Let us therefore define $b_{n,o}$ and $i_{n,o}$ as the random variables describing the lengths (i.e., the numbers of OL slots) of successive active and passive periods on the OLTS, where an active (passive) period on the OLTS is defined as a number of consecutive OL slots during which at least one (no) packet arrives. Since it is quite possible that a packet arrival crosses an OL slot boundary (see Figure 1.b), it should be indicated that a packet is considered to be in the buffer only when its arrival is completed; therefore, the OL slot of arrival of a packet is the slot during which its arrival has ended. Also, note that during an OL slot, there are either 0, 1 or 2 packet arrivals. It is now possible to express $b_{n,o}$ and $i_{n,o}$ in terms of b_n and i_n respectively. For that purpose, let us also define R_n and P_n, $0 \leq R_n, P_n \leq k-1$, as the discrete random variables that represent the position of the microslot boundary within an OL slot that coincides with the beginning (on the ILTS) of the n–th active, respectively the n–th

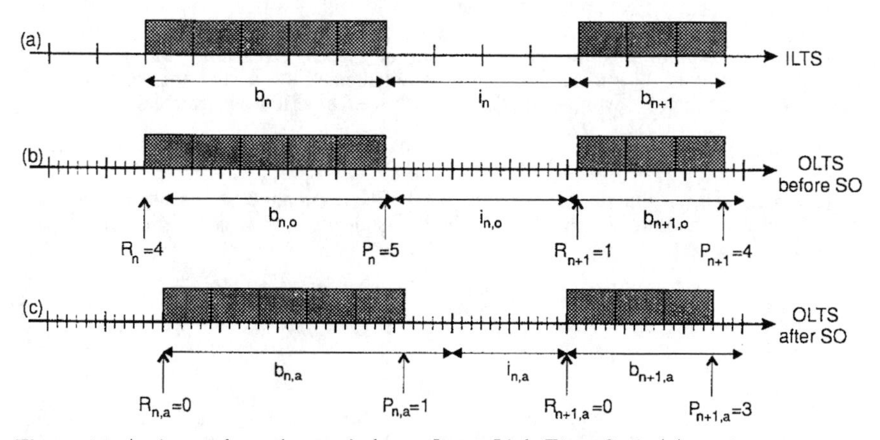

Figure 1 Active and passive periods on *Input Link Time Scale* (a), and on the *Output Link Time Scale* before (b) and after (c) the *Shift Operation.*

passive period. $R_n=0$ ($P_n=0$) would correspond with the beginning of an OL slot, and successive microslot boundaries within an OL slot are numbered sequentially, up to k–1, as shown in Figure 1.b. Obviously, the value of the steady–state joint probabilities

$$p(i,j) \triangleq \lim_{k\to\infty} \text{Prob}[R_n=i, P_n=j] \quad , \tag{4.a}$$

only depends on the number of IL slots enclosed by these two time instants (i.e., the length of b_n), and therefore only depends on the difference i–j. Consequently, defining $p_i = p(i,0)$, it can be verified that

$$\begin{aligned} p_i &= \frac{1}{k}\text{Prob}[(b_n \bmod k)=i] = \frac{1}{k}\sum_{j=0}^{\infty} b(jk+i) \quad , \quad 0 \le i \le k-1 \quad , \\ p(i,j) &= p_{i+k-j} \quad , \quad 0 \le i \le k-1 \text{ and } i < j \le k-1 \\ p(i,j) &= p_{i-j} \quad , \quad 0 \le i \le k-1 \text{ and } 0 \le j \le i \quad , \end{aligned} \tag{4.b}$$

where $b(j)$, $j \geq 1$, denotes the probability mass function of the random variable b_n, and (x mod y) represents the remainder of the fraction x/y of the integers x and y. From (4.b), it is also found that

$$\text{Prob}[R_n=i] = \text{Prob}[P_n=j] = 1/k \quad , \quad 0 \le i,j \le k-1 \quad , \tag{4.c}$$

as could be expected. After carefully examining the problem, the following relations between $\{b_{n,o}, i_{n,o}\}$ and $\{b_n, i_n\}$ were established :

$$b_{n,o} = b_n + \theta(R_n=0) + \theta(R_n=1) - ((b_n+k-R_n) \text{ div } k) \tag{5.a}$$

$$i_{n,o} = i_n + \theta(P_n=0) - ((i_n+k+1-P_n) \text{ div } k) \tag{5.b}$$

where (x div y) denotes the integer part of the fraction x/y, and where the indicator function $\theta(.)$ equals one if the Boolean argument is true, and zero otherwise.

Since we are interested in dimensioning the rate–adaptation buffer, let us define the random variable v_n as the buffer occupancy (i.e., the number of packets in the buffer, including the one that is currently being transmitted, if any) at the beginning of the first OL slot of the n–th active period (on the OLTS), and, similarly, u_n as the buffer occupancy at the beginning of the first OL slot of the n–th passive period. With the previous definitions and assumptions, a number of system equations that relate the sequence of random variables u_n and v_n can be established. First of all, since exactly one packet can be transmitted during each OL slot of a passive period, we find that

$$v_{n+1} = (u_n - i_{n,o})^+ \quad , \tag{6.a}$$

where $(.)^+ \triangleq \max\{0,.\}$. On the other hand, the buffer occupancy at the beginning of a passive period can be expressed in terms of the buffer occupancy at the beginning of the preceding active period as

$$u_n = (v_n - 1)^+ + a_n + 1 \quad , \tag{6.b}$$

where

$$a_n \triangleq b_n - b_{n,o} \quad , \tag{6.c}$$

is the number of packets that must be buffered during the active period due to the difference in arrival and transmission rate. Under the assumptions described in the previous section, it is not difficult to see from (5.a,b) and (6.a–c) that consecutive pairs of random variables $\{v_n, R_n\}$, $\{u_n, P_n\}$, for increasing values of n, form a two–

dimensional Markov chain. However, we will not try to solve the system equations for the exact problem. Instead, we first execute a so–called *Shift Operation* (SO), which means that we let the the first packet of a new sequence coincide with the beginning of the first OL slot of the corresponding active period, as shown in Figure 1.c. In the remainder, the random variables (b_n, i_n, R_n, P_n) will be tagged with the subscript $_a$ if necessary, to indicate that we consider these random variables *after* the SO has been executed. This SO implies that, since $R_{n,a}=0$, consecutive random variables v_n and u_n, for increasing values of n, form a Markov chain, thereby considerably reducing the complexity of the solution.

Note that, compared to the exact problem, this SO has almost no influence on the length of passive and active periods (i.e., a difference in length of at most one slot), and therefore, we expect to find a very accurate approximation while keeping the analysis tractable. First of all, when combining (6.c) and (5.a) while setting $R_n=0$, we obtain

$$a_n = b_n - b_{n,a} = (b_n \text{ div } k) \quad . \tag{7}$$

Using some standard techniques related to the calculation of z–transforms, it can be shown that the associated probability generating function A(z) is then given by

$$A(z^k) = \frac{1}{k}\sum_{s=0}^{k-1} \frac{1 - z^{-k}}{1-(\mu^s z)^{-1}} B(\mu^s z) \quad , \quad \mu \triangleq \exp\{\frac{2\pi\iota}{k}\} \quad , \tag{8}$$

where ι is the imaginary unit

In addition, the SO can also alter the length of the passive periods, as becomes clear from Figure 1.b,c. The random variable $i_{n,a}$ that represents the length of the n–th passive period *after* the SO, is related to $i_{n,a}$ by

$$i_{n,a} = i_{n,o} + \theta(R_n=1,P_n=1) - \theta(R_n \geq 2,P_n=0) - \theta(R_n \geq 2,P_n>R_n) \quad , \tag{9}$$

and system equation (6.a) now becomes

$$v_{n+1} = (u_n - i_{n,a})^+ \quad . \tag{10}$$

The combination of (9) and (5.b) leads to an expression for $i_{n,a}$ in terms of i_n, the length of the original passive period on the ILTS. This relation can be transformed into a relationship between z–transforms, and it can be shown that the corresponding probability generating function $I_a(z)$ satisfies

$$I_a(z^k) = \frac{1}{k}\sum_{s=0}^{k-1} (\mu^s z^{k-1})^{-1} \left[\frac{1 - z^k}{1 - \mu^{-s} z}\right]^2 I(\mu^s z^{k-1})\, P(\mu^s z^{-1}) \quad , \quad P(z) \triangleq \sum_{i=0}^{k-1} z^i p_i \quad . \tag{11}$$

where the p_i's and I(z) were defined in (4.b) and (1) respectively. Expression (11) for $I_a(z)$ is convenient as far as the calculation of the mean value and higher order moments is concerned. However, as will become clear in the following analysis, we especially require the probability mass function that corresponds to $I_a(z)$. Assuming that I(z) indeed satisfies (2), $I_a(z)$ can be transformed into

$$I_a(z) = \frac{1-\alpha}{k-1}\sum_{t=0}^{k-2} \frac{s(t)}{1 - \nu^{tk}\alpha^{k/(k-1)} z} - s_0 \quad , \quad \nu \triangleq \exp\{\frac{2\pi\iota}{k-1}\} \quad , \tag{12.a}$$

$$s_0 \triangleq \frac{1-\alpha}{\alpha}\left\{\sum_{j=k-1}^{2(k-1)} (2k-j-1)p_{j-k+1} + p_0/\alpha\right\}$$

$$s(t) \triangleq \left[\frac{1 - \alpha^{-k/(k-1)}\nu^{-kt}}{1 - \alpha^{-1/(k-1)}\nu^{-t}}\right]^2 P(\alpha^{1/(k-1)}\nu^t) \quad . \tag{12.b}$$

3 THE GENERATING FUNCTION OF THE BUFFER OCCUPANCY

Now we have everything at hand to be able to calculate expressions for $U_n(z)$ and $V_n(z)$, the generating functions corresponding to u_n and v_n respectively. First of all, system equation (6.b) can be transformed into a relation between z–transforms, which, in view of the statistical independence of a_n and v_n, leads to

$$U_n(z) = A(z)\{V_n(z) + (z-1)V_n(0)\} \quad . \tag{13}$$

On the other hand, from (10) and the statistical independence of u_n and $i_{n,a}$, we obtain

$$V_{n+1}(z) = U_n(z)\, I_a(\tfrac{1}{z}) + \sum_{i=1}^{\infty} \{I_i(1) - z^i I_i(\tfrac{1}{z})\} \,\mathrm{Prob}[u_n=i]$$

$$I_i(z) \triangleq E[z^{i_{n,a}} \mid i_{n,a} \geq i]\, \mathrm{Prob}[i_{n,a} \geq i] \quad , \quad 1 \leq i \ ,$$

which, due to expression (12.a) for $I_a(z)$ equals

$$I_i(z) = \frac{1-\alpha}{k-1} \sum_{t=0}^{k-2} s(t) \left\{ \frac{(xz)^i}{1-xz} \right\}_{x=\nu^{tk}\alpha^{k/(k-1)}} \ .$$

Combining the previous expressions for $I_i(z)$ and $V_{n+1}(z)$, we find

$$V_{n+1}(z) = U_n(z) I_a(\tfrac{1}{z}) + \frac{1-\alpha}{k-1} \sum_{t=0}^{k-2} s(t) \left\{ \frac{U_n(x)}{1-x} - \frac{U_n(x)}{1-xz^{-1}} \right\}_{x=\nu^{tk}\alpha^{k/(k-1)}} \ . \tag{14}$$

The buffer occupancy will typically reach its highest values just after an active period, i.e., at the beginning of a passive period. It is appropriate to use the distribution of the buffer occupancy at these worst–case time instants for buffer–dimensioning purposes. We will therefore establish an expression for U(z), the steady–state probability generating function describing the buffer occupancy at the beginning of a passive period. The system will reach a steady–state only if the equilibrium condition is satisfied, meaning that ρ, the mean number of packet arrivals per slot (on the OLTS) in the rate adapter buffer must be less than 1, which is equivalent to requiring that

$$I_a'(1) > A'(1) \quad , \tag{15}$$

where primes denote derivatives with respect to the argument. In other words, the mean length of a passive period on the OLTS must exceed the mean number of packets that are accumulated during an active period. Equations (13) and (14) now lead to the following expression for U(z) :

$$U(z) = \frac{z(z-1)A(z)\sum_{j=0}^{k-2} r_j z^j}{z^{k-1} - (\alpha^k + A(z)\, I_a(\frac{1}{z}) Q(z))} \quad , \quad Q(z) \triangleq z^{k-1} - \alpha^k \ . \tag{16}$$

While deriving (16), we have used the property that, just after an active period, there is always at least one packet in the buffer, implying that U(0)=0. The constants r_j, $0 \leq j \leq k-2$, that occur in (16) are linear combinations of the unknowns $U(\nu^{tk}\alpha^{k/(k-1)})$, $0 \leq t \leq k-2$, and V(0), where V(z) is the steady–state limit of $V_n(z)$. From Rouché's theorem, one can show that the denominator

$$D(z) \triangleq z^{k-1} - (\alpha^k + A(z) I_a(\tfrac{1}{z}) Q(z)) \tag{17}$$

of expression (16) for U(z) has k–1 zeros inside the unit disk (including z=1). Without giving a full prove, note that the zeros of Q(z) cancel the poles of $I_a(1/z)$, which leads to the observation that D(z) is analytic inside the complex unit disk, a necessary condition for applying Rouché's theorem. In the remainder, we we will denote by z_j, $1 \leq j \leq k-2$, the zeros of D(z) inside the unit disk and different from 1. Since U(z) is analytic inside the complex unit circle, these zeros must also make the numerator of (16) zero, and combined with the normalization condition, this completely determines the (k–2)–th polynomial in the numerator. We thus obtain

$$U(z) = D'(1)\frac{z(z-1)A(z)}{D(z)}\prod_{j=1}^{k-2}\frac{z - z_j}{1 - z_j} . \tag{18}$$

where the coefficient $D'(1) = Q(1)(I_a'(1)-A'(1))$ guarantees the normalization of U(z). From this expression for the probability generating function, we readily obtain the moments of the buffer contents by taking the appropriate derivatives with respect to z for z=1 of U(z). In this paper, we focus attention on the tail distribution of the buffer contents, a quantity which is very useful for buffer dimensioning purposes. Whatever performance characteristic we are interested in, the time consuming part of all numerical calculations remains finding the zeros z_j of D(z) inside the complex unit disk. Therefore, we now show that these quantities can be accurately approximated by

$$z_j \cong \alpha^{k/(k-1)}\nu^j \quad , \quad 1 \leq j \leq k-2 \quad , \tag{19}$$

Indeed, from expressions (12.a) and (16) for $I_a(z)$ and Q(z), it readily follows that

$$\left|D(\alpha^{k/(k-1)}\nu^t)\right| = (1-\alpha)\alpha^k\left|A(\alpha^{k/(k-1)}\nu^t)\right|\left|s(i)\right| \quad , \quad 1 \leq t \leq k-2 \quad , \tag{20}$$

where the integer i is such that $\nu^{ik} \equiv \nu^t$. Now, since

$$|s(i)| \leq \sum_{i=0}^{k-1} p_i \sum_{j=0}^{2(k-1)} \min(j+1,2k-j-1)\alpha^{-j/(k-1)} = \frac{1}{k}\left[\frac{1 - \alpha^{-k/(k-1)}}{1 - \alpha^{-1/(k-1)}}\right]^2 ,$$

and $|A(z)|<1$ if $|z|<1$, the above relations explain why (19) forms a very good approximation for the z_j's, which, due to the presence of the factor α^k/k, becomes better as k increases. The accuracy of this approximation is illustrated in Section 5.

4 TAIL DISTRIBUTION OF THE BUFFER OCCUPANCY

As has already been indicated in various papers (Woodside (1987), Desmet (1992), Sohraby (1992), Bruneel (1993)), if the probability generating function of the buffer occupancy has nothing but non–essential singularities of order 1 (i.e., simple poles), then the tail of the buffer–occupancy distribution can be approximated very accurately by a geometric form, implying in our specific case that the probability that the buffer occupancy just after an active period exceeds an integer threshold U, is given by

$$\text{Prob}[u>U] \cong \frac{-Cz_0^{-U-1}}{z_0 - 1} , \tag{21}$$

where z_0 is the pole of U(z) with the smallest modulus (i.e., the solution of D(z)=0 outside the unit disk with the smallest modulus), which is a real and positive quantity larger than 1. Due to the residue theorem, the constant C in the above expression is equal to

$$C \triangleq \lim_{z\to z_0} (z-z_0)U(z) = z_0(z_0-1)A(z_0)\frac{D'(1)}{D'(z_0)}\prod_{j=1}^{k-2}\frac{z_0 - z_j}{1 - z_j} . \tag{22}$$

Using approximation (19) for the z_j's, a close approximation for C avoiding the calculation of these quantities can be derived. We obtain

$$C \cong z_0(z_0-1)A(z_0)\frac{D'(1)}{D'(z_0)}\frac{z_0^{k-1}-\alpha^k}{z_0-\alpha^{k/(k-1)}}\frac{1-\alpha^{k/(k-1)}}{1-\alpha^k} \quad . \tag{23}$$

From equations (21) and (23), an approximation for the geometric–tail limit of the buffer occupancy distribution is obtained, which is easily evaluated, since it merely requires the calculation of z_0. As will be shown in the next section, this approximation is extremely close to the actual values of the geometric–tail approximation, calculated by combining (22) and (23).

5 SOME NUMERICAL EXAMPLES

From now on, we let σ denote the packet arrival rate on the input link. The rate adapter model previously described is completely specified, once the value of the parameter k (which determines the difference between input and output rate), and the active and passive period distributions (or, equivalently, B(z) and I(z)) are given. Up to now, B(z) could have any form, and in this section, we consider two cases, where either the input process is the output of a discrete–time M/D/1 queue with load σ (in the remainder referred to as *M/D/1–like arrivals*), or a model where passive and active periods are statistically independent and both geometrically distributed (in the remainder referred to as *geo–like arrivals*), with parameters α and β respectively, satisfying $\sigma = (1-\alpha)/(2-\alpha-\beta)$. Note that in the case of M/D/1–like arrivals, the lengths of active and passive periods indeed are statistically independent.

First of all, for M/D/1–like arrivals, the parameter α characterizing I(z) in (2) equals $\exp\{-\sigma\}$, the probability of having no packet arrivals during a slot in the M/D/1 queue, and on the other hand, it has been derived in Bruneel (1993) that in this case B(z) is given by

$$B(z) = \frac{R(z)/z - \alpha}{1-\alpha} \quad , \tag{24.a}$$

where R(z) is implicitly defined by the equation

$$R(z) = z\exp\{\sigma(R(z)-1)\} \quad . \tag{24.b}$$

Furthermore, in Steyaert (1993) it was shown that for a generating function satisfying (24.b), the corresponding probability mass function, here denoted by $r(j)$, is given by

$$r(j) = \sigma\frac{(j\sigma)^{j-2}}{(j-1)!}\exp\{-j\sigma\} \quad , \quad j \geq 1 \quad . \tag{24.c}$$

From (24.a,c), the probability mass function corresponding to B(z) is readily obtained.

In the case of geo–like arrivals, I(z) is still given by (2), while B(z) now equals

$$B(z) = \frac{(1-\beta)z}{1-\beta z} \quad . \tag{25}$$

An interesting quantity is the parameter L, which is defined as the ratio of the mean length of a passive (or, equivalently, active) period versus the mean length of a passive (active) period in the case of M/D/1–like arrivals and equal values of the load σ of the input process. Consequently, L=1 means that the geometrically distributed active and passive periods have the same average length as for the output process of an M/D/1 queue with equal load, and considering increasing values of L while keeping the ratio $\sigma = B'(1)/(I'(1)+B'(1))$ constant implies that the arrival rate in the rate adapter module remains constant, while the 'variability' in the arrival process increases. From the previous, we find that the parameter L can be calculated from

$$L = \frac{1 - \exp\{-\sigma\}}{1 - \alpha} \quad . \tag{26}$$

The packet arrival process in the geo–like arrivals case can now be characterized by the pair (σ,L) instead of (α,β), for given values of k.

The results obtained throughout this paper have been illustrated in Figs. 2–8. In Figs. 2–3, for various values of k and $\rho=\sigma.k/(k-1)$, we have plotted the 'exact' geometric–tail approximation for Prob[u>U], obtained from (21) and (22) (full line) together with the simplified result obtained from (21) and (23) (marks), in the case where the arrivals process is either M/D/1–like (Fig. 2), or geo–like (Fig. 3) with L=1. First of all, it is found that, whatever the type of arrival process and whatever the values of the parameters characterizing the arrival process, no difference between the 'exact' and approximate results can be observed. These curves show that approximation (19) for the z_j's is extremely accurate, and can be used without any restriction. Furthermore comparing the curves of both figures, we may conclude that, although the active period distribution has the same mean value in both cases when considering equal values of ρ and k (due to L=1; note that this also implies the passive periods have identical geometric distributions), a substantial difference between both cases exists, i.e., second–order effects (such as higher–order moments of the active periods) still have a considerable impact on the buffer behavior of the rate–adapter, and, therefore, cannot be neglected in the buffer dimensioning process.

In Figs. 4–6, we have plotted the geometric–tail approximation for Prob[u>U] in the M/D/1–like arrivals case, for constant values of ρ and various values of k. It is observed that when, while large differences occur for low values of ρ, these diminish as ρ

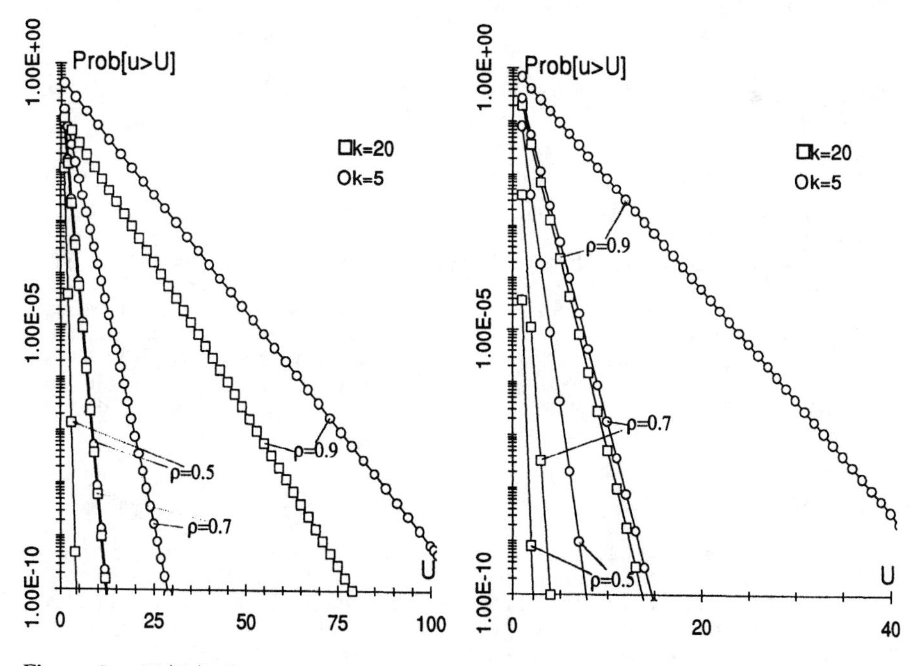

Figure 2 M/D/1–like arrivals, k=5,20, ρ=0.5,0.7,0.9.

Figure 3 Geo–like arrivals, k=5,20, ρ=0.5,0.7,0.9, L=1.

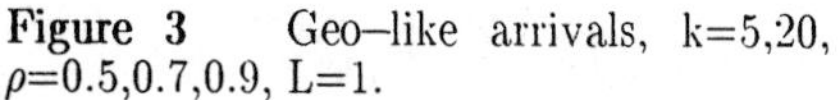

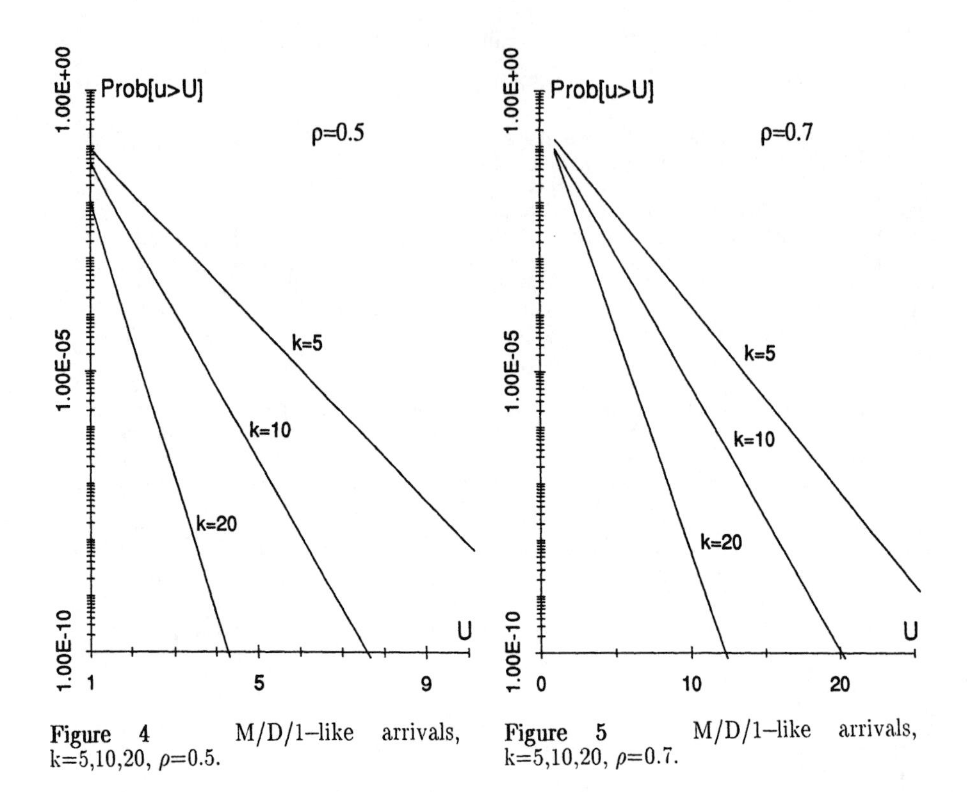

Figure 4 M/D/1–like arrivals, k=5,10,20, ρ=0.5.

Figure 5 M/D/1–like arrivals, k=5,10,20, ρ=0.7.

increases, i.e., the results for the tail distribution become less sensitive to the exact value of the output rate/input rate ratio. Finally, in Figs. 7–8, we examine the impact of increasing values of L on the buffer requirements of the rate adapter, in the geo–like arrivals case. The case L=1 was plotted in Fig. 3; in Figs. 7–8, for various values of k and ρ, we considered values of L equal to 5 and 10. It is observed that, when comparing the respective curves for equal values of k and ρ, the required buffer space increases as L increases. Keeping in mind the conclusions in the discussion concerning Figs. 2–3, this is hardly surprising, since, again, increasing values of L implies increasing variability in the of the arrival process, and we observe that the impact of increasing values of L on the buffer beahvior is quite severe.

6 SUMMARY

In this paper, we have tackled the problem of dimensioning a rate adaption module, which arises in a node of a telecommunication networks if the arrival rate on the input link exceeds the transmission rate of the output link. Based on a generating functions approach, we obtained expressions for the mean and tail distribution of the buffer occupancy, which are easy to evaluate, since, due to the approximation for the z_j's, numerical calculations are reduced to a minimum. From the numerical results we may conclude that (1) even for small differences between input and output rate (less than 5%), the involved buffers can become quite large, and (2) second order effects, such as higher order moments of the lengths of active and passive periods are not negligible.

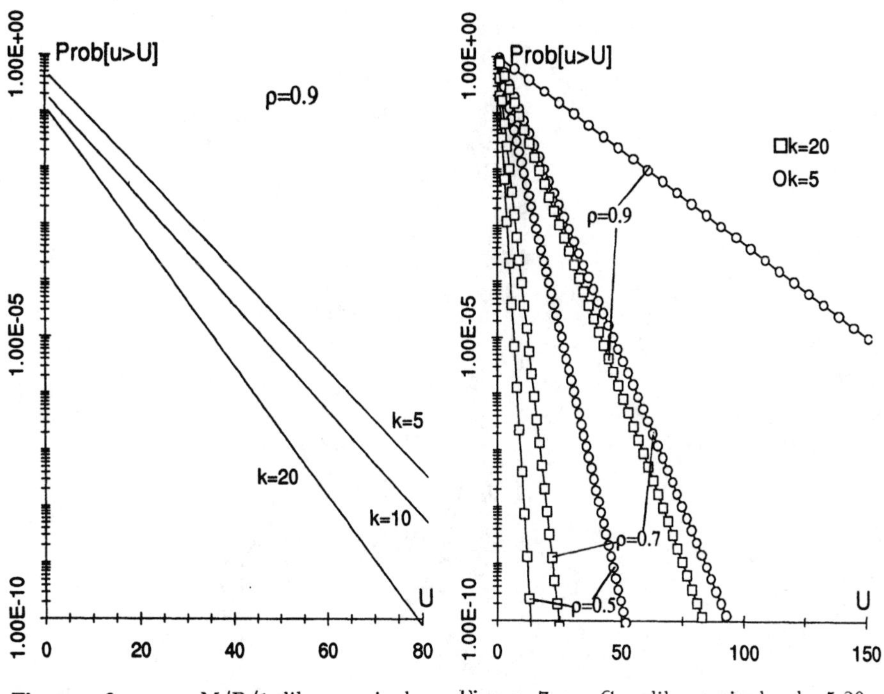

Figure 6 M/D/1–like arrivals, k=5,10,20, ρ=0.9.

Figure 7 Geo–like arrivals, k=5.20, ρ=0.5,0.7,0.9, L=5.

7 REFERENCES

Bruneel, H., and Kim, B.G. "Discrete–time models for communication systems including ATM" (Kluwer Academic Publishers, Boston, 1993).

De Prycker, M. "Asynchronous Transfer Mode – Solution for Broadband ISDN", Ellis Horwood Ltd. (England, 1991).

Desmet, E.; Steyaert, B.; Bruneel, H. and Petit, G.H. "Tail distributions of queue length and delay in discrete–time multiserver queueing models, applicable ATM networks", *Proceedings of the 13th ITC* (Copenhagen, 1992), Vol. *Queueing, Performance and Control in ATM*, pp. 1–6.

Michiel, H. and Petit, G.H. "Dimensioning of a rate adapter", Alcatel–Bell Research Report ATG_029/HM_901019 (Antwerp, 1990).

Rothermel, K. "Traffic studies of transmission bit rate conversion in ATM networks", *Proceedings of the 13th ITC* (Copenhagen, 1992), Vol. *Queueing, Performance and Control in ATM*, pp. 65–69.

Sohraby, K. "On the asymptotic behavior of heterogeneous statistical multiplexer with applications, *Proceedings of Infocom'92*, pp. 839–847.

Steyaert, B. and Bruneel, H. "Delay characteristics of a CBR source with priority background traffic", LCI Research Report 18, June 1, 1993.

Woodside, C. and Ho, E. "Engineering calculation of overflow probabilities in buffers with Markov–interrupted service", *IEEE Trans. on Commun.*, Vol. COM–35 (1987), pp. 1272–1277.

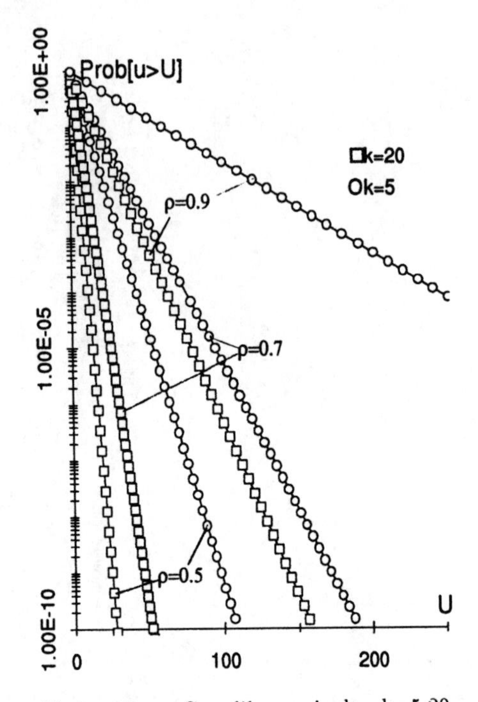

Figure 8 Geo–like arrivals, k=5,20, ρ=0.5,0.7,0.9, L=10.

42

A Diameter Based Method for Virtual Path Layout in ATM Networks

Jimmy P. Wong, Rajesh K. Pankaj
University of Toronto
10 King's College Road, Toronto, Canada M5S 1A4. Phone: +1 416 978 0806.
Fax: +1 416 978 4425. email: wongja,pankaj@comm.toronto.edu

Abstract

ATM networks use virtual paths (VPs) to route information from source to destination. By using VPs, the call set-up and switching costs can be reduced. In this paper, we consider the problem of selecting where VPs should be placed. We propose an algorithm that is based on reducing the network diameter. The performance of this *diameter* method is compared with another heuristic that uses a *clustering* algorithm. Using simulation, it is shown that the Diameter method performs better than the Clustering method in reducing the average connection cost.

Keywords

ATM, Virtual Paths, Network Management

1 INTRODUCTION

The use of virtual paths is an important concept associated with Asynchronous Transfer Mode (ATM) (Ohta, 1988), (Burgin, 1991), (Sato, 1990). A virtual path (VP) is a logical connection between two nodes in a network. Each VP consists of a sequence of one or more physical links. A network of VPs forms a higher layer that is logically separate from the underlying physical network. From this VP network, individual connections, or virtual channels (VCs), can be routed. Each VP can be used by many virtual channels. That is, VCs are multiplexed together on to a VP and transported with a common identifier, called a Virtual Path Identifier (VPI). Individual VCs within a virtual path are distinguished by their Virtual Channel Identifier (VCI).

The advantages of using virtual paths include the following. At call set-up, the routing tables at the intermediate nodes of an existing VP do not need to be updated. This reduces the call set-up delays. By grouping many connections (VCs) into single units (VPs), the switching costs are reduced. In particular, at the intermediate nodes of a VP, only the VPI label of a VC needs to be processed. As shown in (Burgin, 1991), more than 90% of the processing time can be saved if virtual paths are used. However, this improvement can only be obtained if the VPs are assigned efficiently. That is, given a network topology and traffic distribution, establish a system of virtual paths so that the network performance is optimized. This involves finding, for each VP, the VP end-nodes (or terminators), the actual route between the terminator nodes, and the path capacity.

In what sense is a VP network optimal? In papers by Cheng and Lin (Cheng, 1994) and Ahn *et al.* (Ahn, 1994),

VPs are assigned to minimize the call blocking probability. Chlamtac *et al.* (Chlamtac, 1993) use a *Clustering* algorithm to establish VPs. Another issue in assigning VPs is the effect of failures in the network. Murakami and Kim (Murakami, 1994) propose a VP routing scheme which minimizes the expected amount of lost flow due to a network failure.

In this paper, we propose a solution to a simplified form of the above VP assignment problem. Given a network topology and traffic demands, a VP network is established to minimize the *average connection cost* under the constraint that the number of VPs assigned in the network is limited. We then also consider this problem under an additional constraint that the number of VPs on each link is less than a prespecified bound. The bounds on the number of VPs in the network and the number of VPs over a link are important consideration because of the limited number of available VPI addresses. If separate VP layouts are needed for different classes of service supported by the network, the limit on the number of VPI addresses can become a significant constraint. In our algorithm, we consider a single set of VPs. If different VP layouts are required for different classes of service, the algorithm should be run separately to determine the VP layout for each class of service. A bound on the number of VPs traversing a given link also limits the effect of a single link failure on the VP layout.

We consider only the problem of finding the VP terminators and the actual path between the end-nodes of each VP. We do not determine the bandwidth allocated to each VP. The work on bandwidth assignment in conjunction with the algorithms proposed in this paper is under progress.

Note that the *cost* associated in routing a connection can be a general cost function. However, one advantage of the use of virtual paths is a reduction in call set-up and switching costs. Therefore, in this paper, we consider a cost function that is representative of the call set-up and switching costs.

This paper is organized as follows. In Section 2, the VP assignment problem is defined. Section 3 describes the proposed VP layout algorithm and analyzes its computational complexity. In Section 4, some simulation results are presented to compare the proposed method with the Clustering algorithm proposed in (Chlamtac, 1993). Section 5 concludes the paper.

2 PROBLEM FORMULATION

The VP assignment problem is formulated as follows.

1. Let an undirected graph $G = (V, E)$ represent the physical network, where V denotes ATM switches and E represents physical links connecting the nodes. Let graph G have $|V| = N$ nodes and an arbitrary number of edges. We assume the network is connected.
2. The amount of traffic from node i to node j is given by ρ_{ij}. Let the total amount of traffic in the network be

$$\rho_T = \sum_{i,j \in V} \rho_{ij}. \tag{1}$$

3. A cost function $C(i,j)$ represents the cost of routing one unit of traffic from node i to node j. Assuming a connection transmits one unit of traffic, $C(i,j)$ represents the cost of routing the connection from i to j. As discussed previously, we assume $C(i,j)$ is associated with the call set-up and switching costs. Note that $C(i,j)$ is a function of the VPs assigned to the network.
4. Let M be the number of VPs assigned to the network. Each VP is defined by a terminator pair, (s,t), and the actual route from s to t. The cost of using VP (s,t) is given by

$$C_{VP}(s,t) = \beta + \alpha\, d(s,t) \tag{2}$$

where $d(s,t)$ is the number of physicals links traversed by the VP. The cost of using a physical link by itself is 1. Parameter β is a fixed cost in routing over the VP. This includes the cost associated with the selection of an available VCI label for the connection during call set-up and also the cost of processing of the VCI label at the end-nodes of the VP. $\alpha\, d(s,t)$ is a cost that is proportional to the length of the VP. It represents the cost of translating the VPI field at the intermediate nodes of the VP.
The variables α and β are chosen so that $C_{VP}(s,t) < d(s,t)$. That is, routing a connection over the VP is less costly than routing over the corresponding physical links. Finally, $C_{VP}(s,t)$ should not be less than $d(s,t)$ if node s is adjacent to node t. The cost of routing over a one-link VP should be greater than or equal to the cost of routing over the physical link itself.
5. Each physical link, e_k, in the network is restricted to having a maximum of η_k VPs routed over it. This constraint is motivated by two possible reasons. First, the number of available VPI addresses may be limited which puts a limit on the number of VPs that can traverse that link. Second, by limiting the number of VPs that can be on a physical link, we limit the sensitivity of the VP layout network in the event of a link failure.
The first of the two reasons can be important if we see the VP layout algorithm running separately for each class of service. If a larger number of classes of service are supported, each class will have only a limited number of VPI addresses. We define the *maximum link load* to be the largest number of VPs going through the same physical link.

Given the above constraints, assign a set of VPs to minimize

$$\bar{C} = \frac{\sum_{i,j \in V} \rho_{ij} C(i,j)}{\rho_T}. \tag{3}$$

$\bar{C}$ represents the average connection cost of the network given a VP layout. Note that if we let $D(i,j) = \rho_{ij}C(i,j)$, minimizing

$$\sum_{i,j \in V} D(i,j) \tag{4}$$

is equivalent to minimizing Eqn. 3.

In the next section, we propose an algorithm which attempts to minimize $\bar{C}$ while maintaining a small maximum link load.

3 VP ASSIGNMENT ALGORITHMS

In this section, we present a heuristic for assigning virtual paths that is based on reducing the network diameter. We will refer to it as the *Diameter method*. As a comparison, we also briefly discuss a method proposed in (Chlamtac, 1993). The computational complexity of both methods are then analyzed.

3.1 Diameter Method

In the Diameter method, a VP is assigned if it reduces the *diameter* of the network. The diameter of a network is the *distance* between the node-pair that is farthest apart. Distance can be defined as the minimum number of hops between the two nodes, the propagation delay, or some other measure. Therefore, depending on what *cost* the VP assignment is set to minimize, an appropriate distance measure is chosen.

The Diameter algorithm is iterative. In each iteration, a new VP is selected. Therefore, to assign M VPs, M iterations are performed. For each virtual path, the VP terminator-pair is first chosen, and then the actual route is selected.

The selection of a VP proceeds as follows. Let $G = (V, E)$ represent the physical network. Let S_0 be the set of VPs selected in the previous iterations. We define $E' = E \cup S_0$ and a weighted digraph $G' = (V, E')$. G' represents the VP network overlayed on top of the physical network. Let g_k and g'_k be the weights associated with edges in G and G', respectively. The edges in G' associated with physical links are assigned a weight $g'_k = 1$. An edge representing virtual path (s,t) is assigned a weight $g'_k = C_{VP}(s,t)$ (refer to Eqn. 2).

Define another graph H and let the edges have weight h_k. H keeps track of the physical links on which new VPs can be routed. Note that VPs may not be routed over links that already have the maximum number of VPs routed over them. Let η_k be the maximum of VPs that can use physical link e_k. Let n_k be the number of VPs using link e_k. Let $Q = \{(i,j) \mid i,j \in V\}$ be the set of node-pairs in graph H where there exists no path from i to j. Initially, graph $H = G$, and hence is connected. However, when a link, e_k, cannot support any more VPs ($n_k = \eta_k$), it is removed from H. Therefore, H may become disconnected as VPs are added to the network. Q is the set of node-pairs where a path no longer exists due to the maximum allowable VP per link constraint.

In the Diameter algorithm, graph G' is used to find the VP terminator-pair and graph H is used to find the actual route. Define

$$D(i,j) = \rho_{ij} C(i,j) \tag{5}$$

as the *distance* from node i to node j in graph G'. Recall ρ_{ij} is the amount of traffic from i to j. Let $C(i,j)$, the cost of routing a unit of traffic from i to j, be the weight of the shortest path from i to j in graph G'.

The VP terminator-pair is selected as follows. A node-pair (s,t) that is maximally separated, according to the distance measure $D(s,t)$, is selected. By assigning a VP at (s,t), we can reduce the distance between the node-pair, and hence the diameter of the network can be reduced.

Once a VP terminator (s,t) is chosen, the actual route can be selected. Using graph H, a minimum weight path is found from s to t. Recall that each physical link e_k in the network is restricted to supporting only η_k VPs. Initially, edges in graph H have a weight $h_k = 1$. However, when a link is used by the maximum allowable number of VPs, the edge weight is changed to $h_k = \infty$. Therefore, these edges are removed from consideration when the paths for subsequent VPs are computed. Note that if a path in graph H cannot be found for the VP (s,t), then that node-pair is ignored and we proceed to find another terminator-pair.

The above process is repeated using the updated graphs G' and H to find the next VP terminator-pair and VP route, respectively. The Diameter method is summarized below.

Diameter Algorithm

Let graph $H = G(V,E)$. Edges in G and H have weight $g_k = 1$ and $h_k = 1$, respectively. Let $S_0 = \{\}$; initially no VPs are selected. Let $Q = \{\}$; initially graph H is connected. The number of VPs using link e_k is initially zero ($n_k = 0$ for all links). For each VP do:

1. Let $G' = (V, E \cup S_0)$. VP terminator-pairs that have been previously selected are added to G to form a new graph, G'. Note that physical links have weight $g'_k = 1$, while VP edges have weight $g'_k = C_{VP}(s,t)$.
2. Using graph G', find

$$D_{\max} = \max_{(i,j) \notin S_0 \text{ or } Q} D(i,j)$$

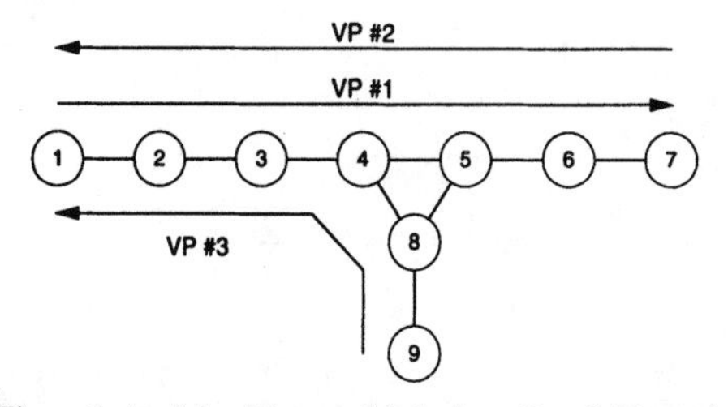

Figure 1 Applying Diameter Method to a 9-node Network

and the corresponding the node-pair, $x = (i, j)$, with $D(i, j) = D_{\max}$.

3. A VP terminator-pair is selected at $y = (s, t) = (i, j)$.
4. Find the shortest path and the corresponding shortest distance, $h(s, t)$, from node s to node t in graph H. If $h(s, t) < \infty$, then a feasible path was found. For each edge e_k used by the path do:

- $n_k = n_k + 1$
- If $n_k = \eta_k$, then set $h_k = \infty$

If $h(s, t) = \infty$, then no feasible path was found. Therefore, a VP cannot be placed at (s, t). Let $Q = Q \cup \{(s, t)\}$ and go to Step 2.

5. Let $S_0 = S_0 \cup \{(s, t)\}$. VP (s, t) is added to the set of selected VPs, S_0.

As an example, we apply the Diameter algorithm to a 9-node 9-link network (Figure 1). We assume the traffic distribution is uniform ($\rho_{ij} = 1$ for all node-pairs) and the VP cost parameters are $\beta = 0.9$ and $\alpha = 0.1$. There is no restriction on the maximum number of VPs per link ($\eta_k = \infty$ for all links). Since node-pairs $(1, 7)$ and $(7, 1)$ are equally far apart, we arbitrarily place the first VP at $(1, 7)$. The second and third VPs are placed at $(7, 1)$ and $(9, 1)$, respectively.

The Diameter method reduces the diameter of the network in an attempt to minimize the average connection cost. However, assigning a VP between a node-pair that is farthest apart in the network is, in general, not optimal. For example, consider the 7-node 1-D network* in Figure 2. Assume the traffic distribution is uniform, $\beta = 0.9$ and $\alpha = 0.1$. For simplicity, we assign a single bidirectional VP. Using the Diameter method, a VP is placed at $(1, 7)$. However, the optimal VP location to minimize the average connection cost is at $(2, 6)$. Therefore, assigning a VP end-to-end in the 1-D network is not optimal.

By placing the VP terminators away from the end-nodes of the 1-D network, we can reduce the average connection cost further. But where is the optimal location?

Given an N node 1-D network with an end-to-end length equal to $\mathcal{L}$, the optimal location of a single bidirectional VP can be found. If we assume the traffic distribution is uniform, as $N \to \infty$, the optimal VP placement occurs

*In a 1-D network with nodes $\{1, 2, \ldots, N\}$, node i is connected to nodes $(i - 1)$ and $(i + 1)$.

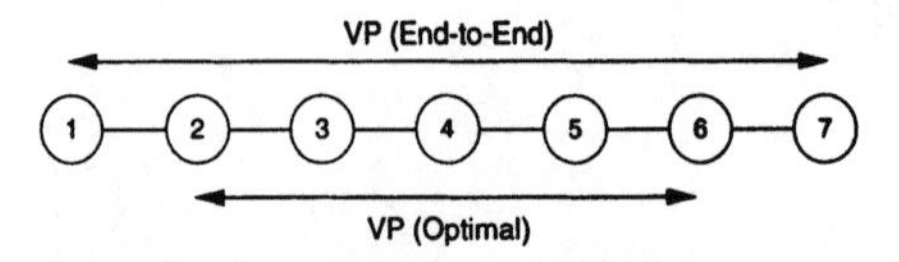

Figure 2 7-node 1-D Network

at (s, t), where s and t are the two nodes which are at a distance $a\mathcal{L}$ away from the end-nodes of the 1-D network. Parameter a is found using the following equation (see (Wong, 1995) for details).

$$a = \frac{(\alpha' + \beta' - 1)\left(-1 + 2\alpha' + \alpha'^2 + \sqrt{2\,(1+\alpha'^2)}\right)}{2\,(\alpha' - 1)\,(1 + 4\alpha' + \alpha'^2)} \tag{6}$$

where $\alpha' = \alpha/\mathcal{L}$ and $\beta' = \beta/\mathcal{L}$. By using Eqn. 6, the optimal (or near-optimal) location of a VP can also be found for a 1-D network with $N < \infty$. Note that parameter a was computed under the assumption that $N = \infty$ and therefore it may not be optimal for finite N.

3.2 Modified Diameter Method

In Section 3.1, it was shown that assigning a VP end-to-end in a 1-D network is not optimal. In an attempt to improve the performance of the algorithm, we extend the results from Eqn. 6 to the Diameter method for general networks with uniformly distributed traffic.

Instead of assigning a VP between node-pairs that are furthest apart in the network, a VP is established between node-pairs that are *slightly* less separated than the maximum amount. The algorithm is implemented as follows. A node-pair (i, j) that is maximally separated, according to the distance measure $D(i, j)$, is chosen. Let $D_{max} = D(i, j)$ and let $\mathcal{P}$ represent the shortest path from i to j with length D_{max}. Consider path $\mathcal{P}$ as a 1-D network with length D_{max}. A VP is placed at (s, t) according to the results of Eqn. 6. That is, s and t are at a distance aD_{max} from the end-nodes of $\mathcal{P}$.

The modified Diameter method is summarized below. Note that only Step 3 in the algorithm is changed.

Diameter Algorithm (Modified)

Let $S_0 = \{\}$ and let $Q = \{\}$. Let $n_k = 0$ for all links. For each VP do:

1. Let $G' = (V, E \cup S_0)$.
2. Using graph G', find

$$D_{\max} = \max_{(i,j) \notin S_0 \text{ or } Q} D(i, j)$$

   and the corresponding the node-pair, $x = (i, j)$, with $D(i, j) = D_{\max}$.
3. Let $\mathcal{P}$ be the path from i to j in G' with length $D_{\max}$. Model path $\mathcal{P}$ as a 1-D network with uniform traffic and assign the VP terminator-pair (s, t) according to Eqn. 6.
4. Find the shortest path from node s to node t in graph H. If no feasible path is found, a VP cannot be placed at (s, t). Let $Q = Q \cup \{(s, t)\}$ and go to Step 2.

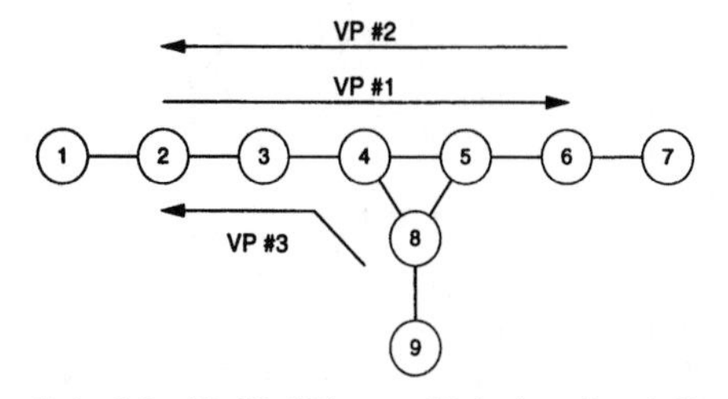

Figure 3 Applying Modified Diameter Method to a 9-node Network

5. Let $S_0 = S_0 \cup \{(s,t)\}$. VP (s,t) is added to the set of selected VPs, S_0.

As an example, consider the same network from Figure 1. Using the Modified Diameter algorithm, VPs are placed at $(2,6)$, $(6,2)$, and $(8,2)$ (Figure 3). Note that in most cases, there is not a node which is exactly a distance aD_{max} from the end-node of $\mathcal{P}$. Therefore, we choose the node that is nearest to that position.

In summary, for a general network with a uniform traffic distribution, the above algorithm can be used to assign VPs. However, when the traffic distribution is non-uniform, it is difficult to determine the optimal placement of a VP in a 1-D network. For simplicity, under non-uniform traffic, the VP is placed at the end-nodes of $\mathcal{P}$, i.e., the Diameter algorithm in Section 3.1 is used.

Both the Diameter and Modified Diameter methods were simulated over a variety of networks. Simulation results are presented in Section 4. We also simulated the Clustering method (Chlamtac, 1993) to compare its performance with the Diameter method.

The Clustering algorithm was proposed by Chlamtac, Faragó, and Zhang (Chlamtac, 1993) (Chlamtac, 1994). In the Clustering method, VPs are chosen which are far away from each other. In each iteration, a new VP is selected which is farthest away from existing VPs. In what sense is a VP *far away* from another VP? Define $D(x,y)$ as the *distance* between node-pairs $x = (i,j)$ and $y = (s,t)$. Let

$$D(x,y) = \rho_{ij}\,[d(i,s) + C_{VP}(s,t) + d(t,j)] \tag{7}$$

where $d(a,b)$ is the minimum physical hop distance from node a to node b and $C_{VP}(s,t)$ is the cost of a connection using VP (s,t). VPs can be selected that are far apart according to Eqn. 7 (see (Chlamtac, 1993) for details).

3.3 Computational Complexity Analysis

In this section, we analyze the computational complexity of the Diameter method and the Clustering algorithm.

In Step 2 of the Diameter algorithm, finding the node-pair that is farthest apart consists of first finding the distance between every node-pair. The time taken by Warshall-Floyd's shortest path algorithm to compute these distances is $O(N^3)$,† where N is the number of nodes in the network. After finding the VP terminator-pair, computing the actual route using Dijkstra's method requires $O(N^2)$. Therefore, each VP requires $O(N^3)$ computations. Since we are assigning M VPs, the overall complexity of the Diameter method is $O(N^3M)$.

†A function $f(n)$ is $O(g(n))$ if $\exists c > 0$ such that $f(n) < cg(n)$ for all n sufficiently large.

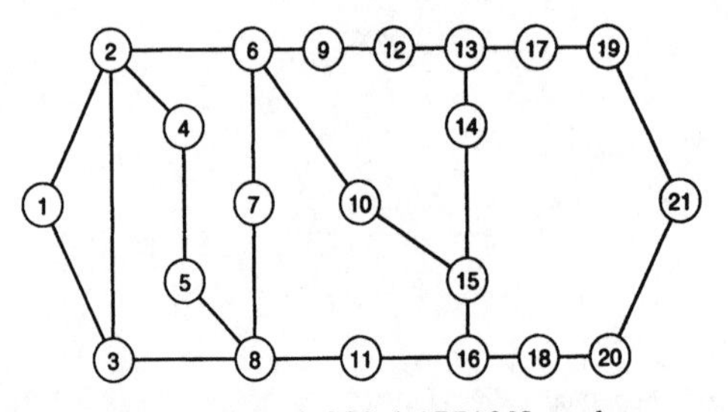

Figure 4 21-node 26-link ARPA2 Network

The computation complexity of the Clustering method is as follows. For the distance measure given in Eqn. 7, it can be shown that the complexity of the Clustering algorithm is $O(N^3)$.

From the above analysis, the Clustering algorithm has a lower computational complexity. However, in the following simulations, we show that the Diameter method performs better than the Clustering algorithm in reducing the average connection cost. If we assume that the time a VP network remains fixed is in the order of hours, the better performance the Diameter method outweighs the lower computational cost of the Clustering algorithm.

4 SIMULATION RESULTS

In this section, simulation results are presented to compare the performance of the Diameter method with the Clustering algorithm. The VP assignment algorithms are applied to some general networks under uniform and non-uniform traffic. The average connection cost, $\bar{C}$, is found as a function of the number of VPs, M, assigned to the network. In addition, we compare the algorithms' sensitivity to link failures.

4.1 Uniform Traffic

A uniform traffic distribution is defined as follows. The amount of traffic from node i to j is $\rho_{ij} = 1$ for all node-pairs. In the following simulations, the VP cost parameters are $\beta = 0.9$ and $\alpha = 0.1$, and the number of VPs per link is unrestricted.

Results for a 21-node 26-link ARPA2 network (Figure 4) is given in Figure 5. We see that the Diameter method performs as well or better than the Clustering method for all M. $\bar{C}$ becomes smaller as more VPs are assigned to the network. However, successive VPs decrease $\bar{C}$ by a smaller and smaller amount.

Figure 5 also shows the performance when VPs are assigned using the Modified Diameter method as opposed to the Diameter method. For small M, the Modified Diameter method gives better performance over the Diameter method. However, for large M, there is very little difference between the two methods. In some cases (e.g.,. $M =$ 10), the Diameter method actually performs better than the modified method. This occurs because the Modified Diameter method is optimal only in the sense of establishing a single VP in a 1-D network.

From these results the Diameter method performs better than the Clustering algorithm under a uniform traffic distribution (refer to (Wong, 1995) for more results). Note that we do not guarantee that the Diameter method will

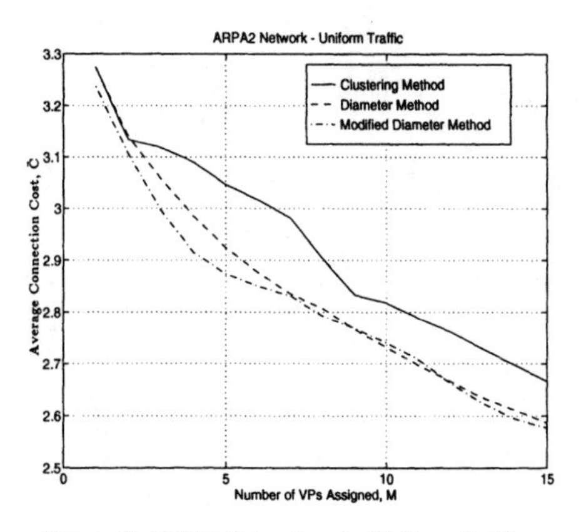

Figure 5 ARPA2 Network under Uniform Traffic.

always perform better than the Clustering method. For some networks, the Clustering algorithm gave a lower average connection cost.

The benefits of using the Modified Diameter method over the Diameter method is more pronounced in large networks. In a small 1-D network, there is little difference between assigning a VP optimally as opposed to end-to-end. For example, the benefit of an optimal VP assignment is much larger in a 20-node network than in a 3-node network. Therefore, in the ARPA2 network, where the average distance between two nodes is small, the performance is approximately the same for the two methods. However, for a large network, where the typical distance between two nodes is larger, the benefits of the Modified Diameter algorithm is emphasized (see (Wong, 1995)).

4.2 Non-Uniform Traffic

In this section, we compare the average connection cost of the two VP assignment algorithms given a non-uniform traffic distribution. We define a non-uniform traffic distribution as follows. The amount of traffic from node i to node j is:

$$\rho_{ij} = U_{ij} \text{ for all } i, j \in V$$

where U_{ij} is a random variable uniformly distributed over $[0, 1]$. Note that the amount of traffic of each node-pair is chosen independently of the other node-pairs. Also, there is no restriction on the number of VPs that can use a particular link.

A number of non-uniform traffic distributions were generated. For a given network, the average connection cost was obtained for each of these traffic distributions. The mean and 90% confidence interval of $\bar{C}$ are shown in the following figures.

Figure 6 show results for the 21-node 26-link ARPA2 network. As in the uniform traffic case, the Diameter method performs better than the Clustering method.

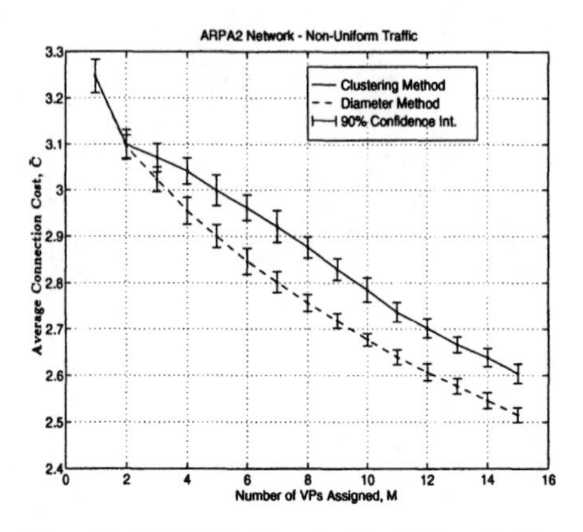

Figure 6 ARPA2 Network under Non-Uniform Traffic.

4.3 Limit on Number of VPs per Link

In the previous simulations, we have not limited the number of virtual paths that can use a physical link. In this section, we assume that the number of virtual paths that can use a physical link must be less than a prespecified limit. This limit is likely to be imposed by the limited number of VP addresses available. The limit may also be imposed to reduce the sensitivity of the VP layout to a link failure. We compare the performance only the Diameter algorithm has this restriction. The Clustering method can assign any number of VPs on a link.

In the following, we apply both algorithms to a number of 40-node 80-link networks. The maximum link load, L, is found from the resulting VP networks. These simulations are performed under uniform traffic and without restriction on the number of VPs per link. In Figure 7, the mean and 90% confidence interval of L is found as a function of M, the number of VPs assigned to the network.

As the number of VPs (M) increases, L also increases. In general, using the Clustering method results in a VP layout that has a lower maximum link load compared to the unrestricted Diameter method, and hence is less sensitive to link failures. This occurs because the Clustering algorithm assigns VPs which are far apart from each other. Therefore, the VP routes are less likely to have common links which contribute to a higher link load.

In the following simulations, we limit the maximum number of VPs that can use a link. Let $\eta_k = n$ for all links. That is, each physical link can support a maximum of n VPs. Note that this restriction applies only to the Diameter algorithm. We can extend this restriction to the Clustering method by modifying its algorithm. However, in the simulations, this was not done.

Figure 8 is a graph of $\bar{C}$ vs. n for a 21-node 26-link ARPA2 network under a non-uniform traffic distribution using the Diameter method. The number of VPs assigned is $M = 5, 10, 15$. As n decreases, the average connection cost increases.

When we restrict the number of VPs per link, VPs that previously used a certain path must be routed to another path that is possibly longer. This increases the VP cost and hence the average connection cost may also increase. More importantly, the VP restriction may actually prevent a VP from being established if no feasible route is found. The Diameter algorithm is forced to choose a less beneficial VP, resulting in a higher average cost.

For comparison, the graph also shows $\bar{C}$ and L using the Clustering method (shown by the 'o' on the graph). When

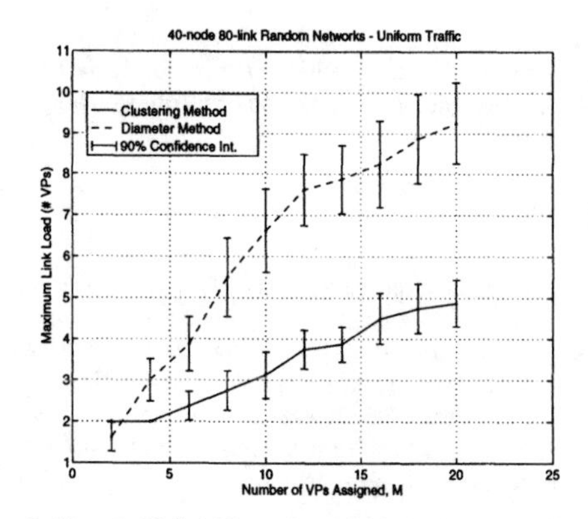

Figure 7 40-node 80-link Networks with Limited # VPs per link.

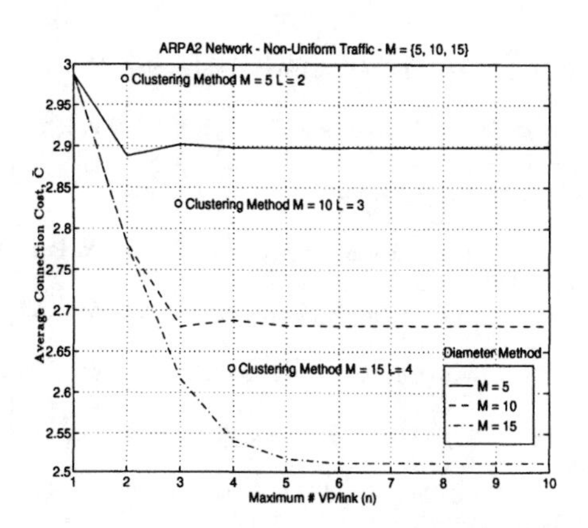

Figure 8 ARPA2 Network under Non-Uniform Traffic.

the VP assignment is unconstrained (n large), the Diameter method performs better than the Clustering method. However, even in some instances when the Diameter algorithm is restricted (n small), it still performs better than the Clustering method. For example, consider the $M = 15$ results. The VP assignment using the Clustering algorithm gives an average connection cost of $\bar{C} = 2.63$ and a maximum link load of $L = 4$. With $n = 3$, the average cost, using the Diameter method, is $\bar{C} = 2.62$. In this case, the Diameter method provides better performance in both average connection cost and link failure sensitivity.

In selecting the location of a VP, the Diameter algorithm uses the information from the currently assigned VPs. It does not consider the possibility of subsequent VPs. For example, given that we wish to establish two virtual paths in a network, the first VP is assigned under the assumption that only one VP is to be selected. A more efficient method is to consider assigning both VPs simultaneously in an attempt to minimize the average cost. However, this becomes more complex as M increases. The Diameter method trades off optimality for simplicity.

5 CONCLUSION

In this paper, we presented an algorithm for finding a system of VPs. Given a network with traffic distribution ρ_{ij}, the location of the VP terminators and the actual path between the end-nodes for each VP was found. Conceptually, the algorithm is simple and easy to implement. It has good performance in reducing the average connection cost while also limiting the number of VPs that can be routed on a physical link. The algorithm was compared with the Clustering algorithm proposed in (Chlamtac, 1993). It was shown that the our method has a slightly higher computational complexity than the Clustering method but performed better in terms of reducing the connection cost. Therefore, it is a good candidate for practical implementation.

REFERENCES

Ahn, S., Tsang, R. P. and Du, D. H. C. (1994) Virtual Path Layout Design on ATM Networks. *INFOCOM'94*. 192-200.

Burgin, J. and Dorman D. (1991) Broadband ISDN Resource Management: The Role of Virtual Paths. *IEEE Communications Magazine*, 29(1991).

Cheng, K.-T. and Lin, Y.-S. (1994) On the Joint Virtual Path Assignment and Virtual Circuit Routing Problem in ATM Networks. *GLOBECOM'94*. 777-782.

Chlamtac, I., Faragó, A. and Zhang, T. (1993) How to Establish and Utilize Virtual Paths in ATM Networks. *Proceedings of IEEE ICC'93*. 1368-1372.

Chlamtac, I., Faragó, A. and Zhang, T. (1994) Optimizing the System of Virtual Paths. *IEEE/ACM Transactions on Networking*. 581-586.

Murakami, K. and Kim, H. S. (1994) Near-Optimal Virtual Path Routing for Survivable ATM Networks. *INFOCOM'94*. 208-215.

Ohta, S., Sato, K. J. and Tokizawa, I. (1988) A Dynamically Controllable ATM Transport Network Based on Virtual Path Concept. *GLOBECOM '88*. 1272–1276.

Sato, K. I., Ohta, S. and Tokisawa, I. (1990) Broad-Band ATM Network Architecture Based on Virtual Paths. *IEEE Transactions on Communications*, Vol. 38, No. 8.

Wong, J. P. (1995) *A Diameter Based Method to Establish Virtual Paths in ATM Networks*. Master's Thesis, University of Toronto.

43
Principles of ATM Topology Design and Cost Modeling

G. Ruth and J. Etkin
GTE Laboratories
40 Sylvan Rd., Waltham, MA,02254, USA grr1@gte.com (617) 466 2446 ; je00@gte.com (617) 466 2039 ; fax (617) 466 2650

Abstract
It is highly important to design a cost effective topology of the ATM network that will support the long term forecast of broadband services. A conceptual approach towards such a design is presented in the paper, which includes the basic topological model and its major parameters. Emphasis is given to the development of a cost model of an ATM network. The costing methodology and assumptions are covered, followed by a discussion of findings and sensitivity analysis. The conclusions are that the cost of the switch access transport and cloud is relatively minor, as compared to the cost of the switching equipment and the special access lines. Another conclusion emphasizes the need for usage based tariffs.

Keywords
ATM network, topology design, cost modeling

1 INTRODUCTION

It is expected that the ATM network will be a major transport mechanism in the future for broadband services, carrying data, video and voice. However, at the current stage the actual traffic requirement is uncertain. Moreover, the design of the ATM functionality is driven by two competing and sometime contradicting players: the telecommunication industry and the data networking/ computer industry [Decina, 1995], where a third player, the entertainment industry, demands its share. To make things even more complicated, there are certain legal restrictions, at least in the U.S, that divide responsibility for local telecommunications service and long distance carriers among different players. The result of all of these constraints is a limited effort in network planning, concentrating mainly on local optimization and ignoring the big picture. Moreover, the designers emphasize maximizing the switch performance by incorporating sophisticated screening techniques and introduction of access control. Thus, conflicting interests among the parties involved in implementing ATM networks preclude the focused cooperation necessary to develop the most cost efficient network designs.

A topological design of the ATM network should consider the traffic requirements for integrated data, video and voice. In this respect the design is not merely an "ARPANET revisited" but rather should consider the diverse requirements in terms of high bandwidth and real time requirements of the network and its applications. Moreover, a correct design of an ATM network topology should consider the new degree of freedom in the dimensioning of the ATM network: a number of logical subnetworks can exist on top of the physical architecture, sharing the same physical transmission and switching capacities [Fargo, 1995]. The design of the topology should also consider the requirement for different quality of service levels (QOS) for various services . Most network architects consider the topology design problem so complex and in such a state of flux, that they simply ignore the issue.

The paper is organized as follows. First the conceptual topological model and its major parameters are presented. The rest of the paper makes use of a cost modeling example, based on a typical carrier network. We present the methodology of building a cost model of an ATM network and its assumptions. There follows the design of the cost model and a discussion of our findings. We conclude with a discussion about our future work on this subject.

2 THE CONCEPTUAL MODEL

Conceptually, the network is divided into backbone and local access parts. The first step in the design of an ATM network is the design of the backbone topology. This takes into account current and future traffic demands, availability of fiber connectivity, and the cost/tariff. This design problem is very complex, since some of the factors are not known at the time of the design (e.g. traffic demands for new network.) [Gerla, 1989; Minoux, 1989].

Generally, the design problem is characterized as:

Given:	Location of the end users computing systems
	Traffic matrix
	Cost matrix
Performance constraints:	Reliability/QOS
	Delay/Throughput
Variables:	Topology
	Line capacities
	Flow assignment
Goal:	Minimize cost
	Allow Growth

The approach used to design the ARPANET [Tanenbaum, 1981] was to generate a starting network topology and let it obey the connectivity and the delay constraints, by assigning to it flow and capacity. The cost of this network is then minimized by a set of perturbation heuristics. The backbone network topology was modeled by a directed graph G= (N,L), where N is the set of nodes and L is a set of links [Alavi, 1985]. The network graph is assumed to be strongly connected, i.e. each node is connected by a directed path to any other node [Chlamtac, 1993]. However, the connectivity problem to be solved in the design of ATM topology is different from the generic model. One difference is that we do not have directed links in the simplest sense, but rather SONET rings, where the direction of flow can be changed if the fiber link is cut. Another change relates to the hierarchical solution of the local access problem. The trend in ATM network planning can lead to three tiers of WAN switches (core, backbone, access) and four tiers of local switches (campus backbone, power workgroup, mainstream workgroup, remote site) [McQuillan, 1995]. The mix among these tiers should be designed appropriately for minimized cost at certain level of reliability and efficiency. The traffic profiles for the various services should also be evaluated to establish an end to end traffic model. Topology design should also consider the growth requirements and the deployment strategy.

3 ATM NETWORK ARCHITECTURE FOR THE COST MODEL

The architecture of the ATM service network is shown in Figure 1. Subscribers connect through an access network to a backbone of interconnected (but not necessarily fully meshed) ATM switches. The access network consists of dedicated circuits between the subscriber and the backbone switches through a fiber network. The customer premises equipment (CPE) has an optical interface for a fiber pair. The special access line between the CDL (Customer Designated Location) and the SWC (Serving Wire Center) is a dedicated fiber pair, regardless of the access bandwidth (DS3 or OC3). From the SWC the subscriber circuit is backhauled over interoffice transport to the nearest CO containing an ATM switch. ATM switches are trunked together with OC3 circuits (OC12 or OC48 in the future), making use of the fiber distribution system. Interoffice transport is typically provided by multiplexed OC48 or OC12 SONET links.

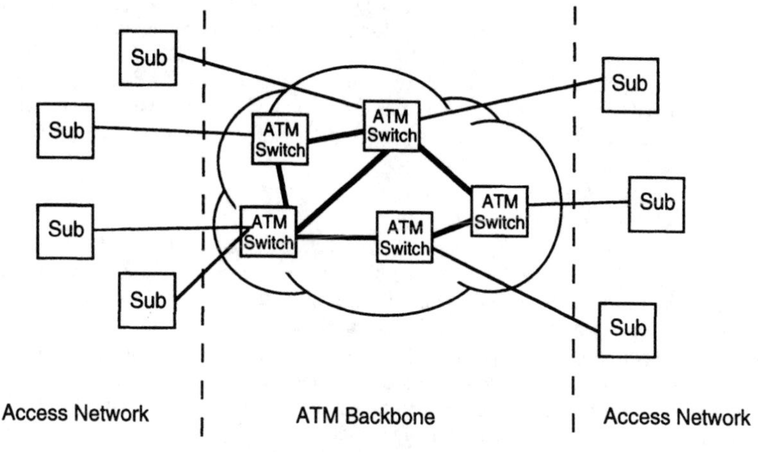

Figure 1 ATM Network Architecture.

4 ASSUMPTIONS

The following assumptions are made in the development of the cost model .

- Only DS3 and OC3 access rates are treated.
- The ratio of DS3 to OC3 subscribers is assumed to be 3:2.
- Traffic assumptions
  - 50% of all ATM traffic is local (in and out of the same switch). 50% is nonlocal (travels between ATM switches, evenly distributed).
  - The peak period average line utilization for an ATM port is 25%.
- The only CPE included in cost calculations is a customer premises fiber panel.
- Amortization
  - The assumed lifetime for transport related assets (e.g. fiber, SONET) is 10 years.
  - The assumed lifetime for switching assets (e.g. ATM) is 5 years.
- Transport
  - No extra equipment is necessary for transport to and between switches.
  - All fiber optic cable (FOC) is assumed to be buried.
  - The percentage of rocksaw FOC installation is 50%
  - All interoffice trunks are OC48s over SONET rings.
  - Interswitch trunking (between ATM switches) is *costed* as multiples of OC3, even though it is provisioned by higher bandwidth circuits.

Figures for average utilization, average local traffic and average nonlocal traffic are suggested by experience with similar packet technologies such as X.25, IP and SNA. Experience to date with ATM is insufficient to confirm their plausibility.

5 COSTING METHODOLOGY

The cost of providing a service is divided into two categories: a nonrecurring cost (NRC) and monthly recurring costs (MRCs). A NRC is viewed as the cost of setting up the service and is recovered by a one time charge at the time of subscription. A MRC is seen as the monthly cost of ongoing provision of the service and is recovered by corresponding monthly charges. The present model is concerned only with the calculation of MRCs.

5.1 Monthly Recurring Costs

The total cost of an asset may be viewed as the sum of all of the costs associated with that asset over time. These costs include the actual purchase price, engineering, installation, maintenance, administration, etc. In calculating the true cost of an asset, one must take into account the fact that a dollar spent in the future is "cheaper" than a dollar spent today, because the former can be invested until it has to be spent. Therefore, in calculating the present value (PV) of a dollar spent in the future, that dollar is discounted by the rate of return, r, that it may be expected to earn before the future arrives. Thus, the *PV* of an asset, *A*, with an associated flow of future costs $C_0, C_1, C_2, \ldots C_n$ in years 0 (now), 1, 2, 3 ... n, respectively, is

$$PV(A) = \sum_{t=0}^{n} \frac{C_t}{(1+r)^t}$$

A service provider can recover the value of a capital asset through a stream of equal monthly charges over the useful life of the asset. The monthly recurring cost (MRC) of that asset is defined as the monthly charge that will make the present value of this stream *exactly* equal to the present value of the future stream of costs associated with the asset.

In this study the MRC has been calculated by straightline amortization of all direct cost outlays (capital cost, installation and engineering) loaded by a factor of 100% for overhead (administrative cost, taxes, etc.) The portion of the base cost that must be recovered every month (the Monthly Amortization Factor) for transport and switching assets is given by:

$$MAF_{transport} = \frac{2}{10 \times 12} = 1.667\%$$

$$MAF_{switching} = \frac{2}{5 \times 12} = 3.333\%$$

5.2 Cost Elements

The cost of providing ATM service for one subscriber may be decomposed into the following:

- Special Access Line (SAL) cost – a dedicated DS3 line or an OC3c line between the subscriber's premises and the serving wire center (SWC)
- CO cost, comprised of two components:
  - switch cost – the per user cost of the switch exclusive of ports
  - switch port – the per user cost of the interface modules and equipment.
- Transport cost
  - switch access facilities – the cost of transport from the SWC to the ATM switch
  - cloud transport facilities – the cost of transport among ATM switches

5.2.1 SAL Costs

The cost of providing DS3 and OC3c access lines is taken as that of providing these services to the SWC. DS3 and OC3c access are not generally tariffed services. Therefore, the cost of these SALs has been calculated using fiber optic cost models.

The cost of a special access line is decomposed into three components:

- CPE – the fiber patch panel at the CDL
- Fiber – the prorated cost of installing a cable (FOC) from the CDL to the SWC.
- CO costs – the cost of a fiber patch panel and fiber termination at the SWC.

Assuming that a 12 fiber FOC is used to connect the CDL to the SWC, the portion of the fiber cost due to the SAL consisting of one fiber pair plus two spares is (2+2)/12 = 1/3. Thus, the cost of a SAL is given by

$$C_{SAL} = 2 \bullet C_{FOP} + C_{FT} + \frac{C_{fiber}}{3}$$

where C_{FOP} is the cost of fiber optic panel

C_{FT} is the cost of the fiber terminal for the service (DS3 or OC3)
C_{fiber} is the cost of the installed FOC, given by:

$$C_{fiber}(d) = d \bullet FOC_{12}$$

where d is the length in miles and
FOC_{12} is the installed cost per mile of a FOC with 12 fibers.

FOC_{12} includes the cost of material, installation, and engineering, all loaded by an OSP (outside plant) support factor. The installed cost per mile depends not only on the number of fibers in the cable, but also on: whether the cable is aerial, buried, or underground; the portion of the cable for which rocksaw installation is necessary; and the labor rates for the market in which the FOC is installed.

5.2.2 CO Costs

The switch port costs are decomposed into the following components.

- The prorated cost of the switch equipment itself (i.e. the cost of the switch divided by the number of subscribers). This includes the switch (with backplane, power supply), redundant parts, spare parts, software, engineering, installation maintenance and network management costs.
- The prorated cost of the ATM interface modules themselves. This is calculated by dividing the cost of a module by the number of user ports it provides.

Per user access line termination costs are included in the SAL costs.

5.2.3 Interoffice Circuit Costs

The ATM transport costs (switch access and cloud transport) are based on interoffice circuits. Interoffice circuit costs are calculated based on multiplexed OC48 service on 12 fiber FOC over SONET rings. The cost of an interoffice DS3 or OC3 circuit is decomposed into three parts:

- Fiber – the prorated cost of installed cable (FOC) from one CO to another CO.
- CO costs – the prorated cost of CO equipment for a SONET interoffice OC48 circuit
  - two fiber patch panels plus one fiber patch panel for each intermediate CO (one through which the fiber passes passively)
  - fiber termination – a SONET OC48 add/drop multiplexer (ADM)
- regeneration – if the distance between SONET ADMs exceeds 25 miles.

As above, the OC48 circuit uses a fiber pair plus two spares, so the portion of the OSP fiber cost due to the OC48 is (2+2)/12 = 1/3. The portions of the OC48 cost due to a DS3 or OC3 circuit multiplexed over it are

$$f_{mux}(DS3, OC48) = \frac{1}{48}$$

$$f_{mux}(OC3, OC48) = \frac{1}{16}$$

Thus, the cost $C_{ckt}(d)$ of an interoffice circuit of length d is given by:

$$C_{ckt}(d) = (2 + p) \bullet C_{FOP} + f_{mux} \bullet \left(C_{ADM} + \frac{C_{fiber}(d)}{3} + C_{regen}(d) \right) \qquad (1)$$

where p is the number of passthrough COs
C_{FOP} is the cost of fiber optic panel
C_{ADM} is the cost of the SONET ADM
$C_{fiber}(d)$ is the cost of an installed FOC of length d
$C_{regen}(d)$ is the cost of regenerators over distance d

As in section 5.2.1, the cost of the installed FOC is given by

$$C_{fiber}(d) = d \bullet FOC_{12}$$

5.2.4 ATM Switch Access Transport Costs

A switch access (SWC to ATM switch) circuit is dedicated to a single user. That is, if there are n DS3 ATM subscribers connected to an SWC, n DS3 circuits to the switch site are needed. Therefore, the per user cost of switch access transport cost is given by applying the formula in equation (1) of the previous section to the weighted average access distance $\bar{d}_A$. $\bar{d}_A$ was assumed to be 5 miles.

If per CO forecasts were available, the weighted average access distance could be calculated as follows. The distance (in "air miles") is calculated between every SWC and the ATM office to which it is homed (the CO where the corresponding ATM switch resides). Then each distance is weighted by the fraction of all ATM demand for the associated CO. That is, the weighted distance $\hat{d}_i$ for the *i*th office is calculated as:

$$\hat{d}_i = \frac{d_i \bullet F_i}{D_{total}}$$

where d_i is the distance in miles
F_i is the traffic forecast for the *i*th office
D_{total} is the total network demand (the sum of all subscribers' access rates)

The average switch access transport facility length, $\bar{d}_A$, is calculated as the sum of the weighted distances.

$$\bar{d}_A = \sum_i^{\text{all COs}} \hat{d}_i$$

5.2.5 ATM Cloud Transport Costs

Unlike switch access transport, cloud transport (interswitch trunking) is shared by all users and can thus be sized to take advantage of the efficiencies of statistical multiplexing all traffic flows. In particular the requirement for interswitch bandwidth is mitigated by the average peak utilization of subscriber access lines (25%) and further by the fraction of traffic that is nonlocal (50%.) Furthermore, the cost of cloud transport should be multiplied by the average number of hops a cell traverses from the switch it enters to the switch it leaves. In the market area studied the network diameter (the maximum number of hops in the shortest path between any two nodes) is small, typically one or two. For purposes of this model the average path length for nonlocal traffic in hops, $\bar{h}$, is approximated as follows:

$$\bar{h}(n) = \left\{ \begin{array}{c} 0 \\ 1 \\ \frac{1}{n}\sum_{i=1}^{n}\frac{1}{n-1}\sum_{j \neq i} h_{\min}(s_i, s_j) \end{array} \right\} \begin{array}{l} \text{for n} = 1 \\ \text{for n} = 2 \\ \text{for n} \geq 3 \end{array}$$

where $h_{\min}(a,b)$ is the minimum number of hops between switch *a* and switch *b*.
n is the number of nodes (ATM switches) in the network.

For networks of more than 2 nodes the nonlocal traffic is assumed to be evenly distributed.

Finally, since every traffic flow is counted twice (once on the sending end and once on the receiving end), a factor of 1/2 must be introduced to compensate for double counting. The total required volume *V* for interswitch trunking is therefore:

$$V = \frac{f_u \bullet f_{nl} \bullet \bar{h} \bullet D_{total}}{2} \quad (2)$$

where f_u is the subscriber utilization factor (average peak utilization)
f_{nl} is the cloud utilization factor (average percentage of nonlocal traffic).

The number of OC3 interswitch trunks required for a network of n switches is :

$$T = Max\left(\left\lceil \frac{V}{155Mbps} \right\rceil, n\right) \tag{3}$$

assuming that at a minimum the switches are connected in a ring. The total cost C_C of cloud transport is then:

$$C_C = \sum_{i=1}^{T} C_{ckt}(d_i)$$

where d_i is the length of the ith trunk

If D_{total} is expressed as:

$$D_{total} = 45 \bullet D_{DS3} + 155 \bullet D_{OC3}$$

where D_{DS3} is the total number of DS3 subscribers, and

D_{OC3} is the total number of OC3 subscribers

then the average portions, t_{DS3} and t_{OC3}, of the cloud transport cost per DS3 and OC3 subscriber, respectively, are:

$$t_{DS3} = \frac{45 \bullet D_{DS3}}{D_{total}} \bullet \frac{1}{D_{DS3}} = \frac{45}{D_{total}}$$

$$t_{OC3} = \frac{155}{D_{total}}$$

The average cloud cost per DS3 subscriber is thus $t_{DS3} \bullet C_C$ and the average per OC3 subscriber is $t_{OC3} \bullet C_C$.

6 FINDINGS

This section describes how the per subscriber cost of providing ATM service varies with the following key factors: number of users, network configuration and vendor equipment.

6.1 Cost Components

The cost breakdown for ATM service is shown in Table 1 based on a typical configuration (see Fig. 2.) It was assumed that all switch ports were utilized.

Special Access Lines

For a given subscriber and rate, the only variation possible in the SAL is the number of fibers in the fiber optical cable (FOC) between the subscriber site and the SWC. The SAL requires 4 fibers (a pair plus two spares). The cost of these 4 fibers is prorated according to the cost of the FOC and the number of fibers it contains. Because the only factor in this cost is the fiber itself, and 4 fibers are required regardless of the SAL's bandwidth, the absolute cost of DS3 and OC3c SALs is virtually identical.

Switch Port

The cost of a switch port for a given service rate depends on the cost of the switch product and the number of subscriber ports per switch. Interswitch trunk ports are subtracted from the total number of switch ports to arrive at the number of switch ports available for subscribers.

In general, the number of subscribers per switch depends in turn on the total number of users *vs*. the number of switches deployed.

Table 1 Relative Sizes of ATM Service Cost Components.

Cost Component	*varies with*	*% of Cost*	
		DS3	OC3

SAL	FOC size	44%	29%
Switch port	number of users topology vendor	47%	52%
Switch access transport	topology interswitch service	7%	15%
Cloud transport	topology technology	2%	4%

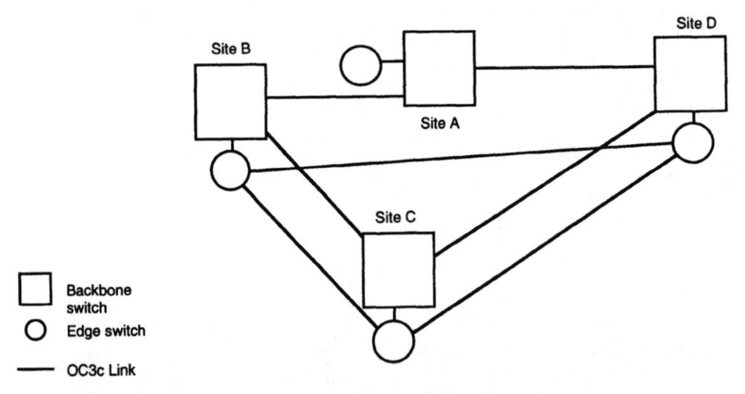

Figure 2 Typical ATM Network Configuration.

Switch Access Transport
This is the cost of backhaul, that is, of data transport for a subscriber from his SWC to the ATM switch to which that CO is homed. To provide DS3 access the ATM cloud, he must have, in addition to a DS3 SAL to his SWC, a dedicated DS3 circuit from the his SWC to the CO where the ATM switch is. This uses a fraction of the interswitch service (e.g. $^1/_{48}$ of an OC48 interoffice circuit) and its cost is prorated accordingly.

Cloud Transport
This is the cost of data transport between ATM switches. Because ATM statistically multiplexes user traffic, this cost is mitigated by the subscriber utilization factor and the cloud utilization factor as explained above. Assuming that the subscriber utilization factor is 25% and the cloud utilization factor is 50%, then on the average the interswitch transport requirement is only 6.25% (one half of 0.25 • 0.5) of the total subscriber access capacity (see equation 2).

Reducing Transport Cost
If the cost of bandwidth/mile from the SWC to the ATM switch is the same as (or greater than) the cost of bandwidth/mile between ATM switches, then locating the ATM switches closer to the subscribers can reduce the transport cost by:

- shortening the path that local traffic has to take and
- shortening the dedicated part of intersubscriber paths and sending more traffic over interswitch links (which take advantage of fractional utilization)

However, if moving the ATM switches closer to the subscribers leads to more switch sites, the following factors will contribute to increased transport cost:

- greater total interswitch transport miles
- an increase in the percentage of traffic that is nonlocal

Additional Factors
The exact values of the following parameters have been approximated:

- average peak percentage of traffic that is nonlocal traffic
- average peak percentage utilization of a subscriber port
- average path length (in hops)

These factors are beyond the control of the network operator.

6.2 Cost Variation with Equipment Vendor

Equipment of two ATM switch vendors is used in the model. The relative capacities for the basic ATM switch configurations are shown in Table 2.

Table 2 Relative switch capacities.

	Switch A	*Switch B*
bandwidth	4.95 Gbps	2.48 Gbps
I/O slots	16	16
maximum DS3 ports	64	16
maximum OC3c ports	32	16

Each switch has enough bandwidth to support a full population of OC3 interface. Switch B, supports one interface per module and one module per slot, regardless of interface bandwidth. Every pair of slots in Switch A may be used to provide either 4 OC3c or 8 DS3 interfaces.

The relative installed first costs of the two vendors' switching equipment was compared in two ways. First, the cost of comparably equipped switch configurations was calculated for each vendor. Second, the cost of a equivalent 3 city networks was calculated. The ratio between the costs of switch A and switch B were: 1.18 per DS3 port, 1.53 per OC3 port and 1.21 per city. (Note: these figures do not include the cost of transport.)

The cost of Switch A OC3 ports does not compare as favorably as that of DS3 ports because the former cannot be populated as densely as the latter on Switch A.

6.3 Cost Variation with Number of Subscribers

The actual cost of proving ATM service in a given area will depend on subscriber demand. Figure 3 illustrates how the per subscriber monthly cost of DS3 ATM services varies with the number of subscribers for the 4 node network configuration. The cost figures shown do not include transport cost components. (The transport costs are basically constant regardless of the number of subscribers.)

Notice that as it becomes necessary to install multiple Switch B's at each site, the average per subscriber cost rises slightly because some subscriber ports must be sacrificed to provide intercommunication among switches in a site cluster. The Switch A cost per subscriber continues to improve when a cabinet must be added to a switch because an extra cabinets is less expensive than a new switch.

Figure 4 shows cost as a function of the number of subscribers for OC3 ATM service in the same 4 node network configuration.

6.4 Cost Variation with Peak Utilization

An ATM network is built to accommodate the traffic load of the peak hour. By definition the total traffic load T_{max} during the peak hour is expected to be

$$T_{\max} = f_u \bullet D_{total}$$

where f_u is the subscriber utilization factor (average peak utilization).

D_{total} is the total network demand (the sum of all subscribers' access rates).

For the calculations of this example the subscriber utilization is assumed to be 25%. As this factor increases, extra interswitch trunking must be added to handle the peak load and extra trunking ports must be added to the switches. The result will be increased cost per subscriber. Likewise, if the subscriber utilization factor is smaller than 25%, the per subscriber costs will go down. (Note that the cost variation derives entirely from increased nonlocal traffic; the ATM switches are all designed to handle 100% utilization.)

Figure 5 shows the variation of the per subscriber cost of OC3 service with the subscriber utilization factor. This analysis uses the 4 node network configuration with two cabinets of Switch A at each site. Unlike Figures 3 and 4 all transport costs are included.

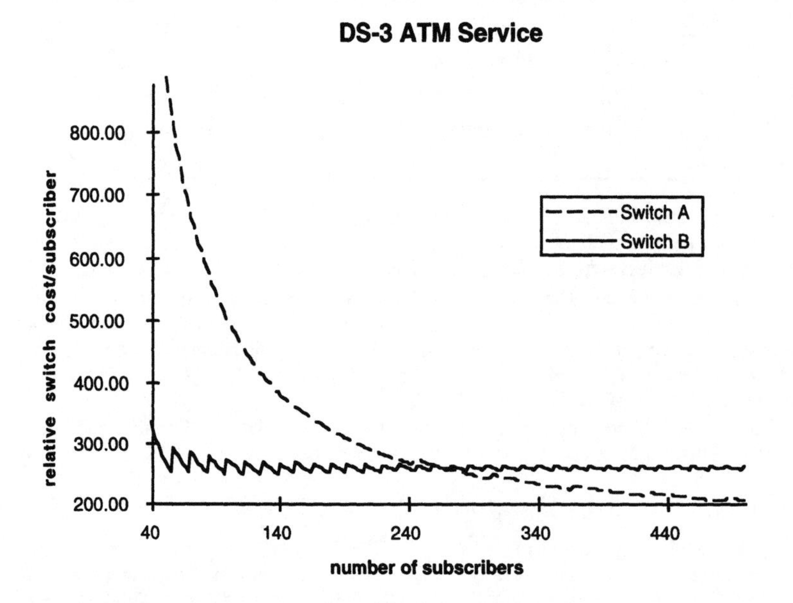

Figure 3 Switch Cost of DS3 ATM Service vs. Number of Subscribers.

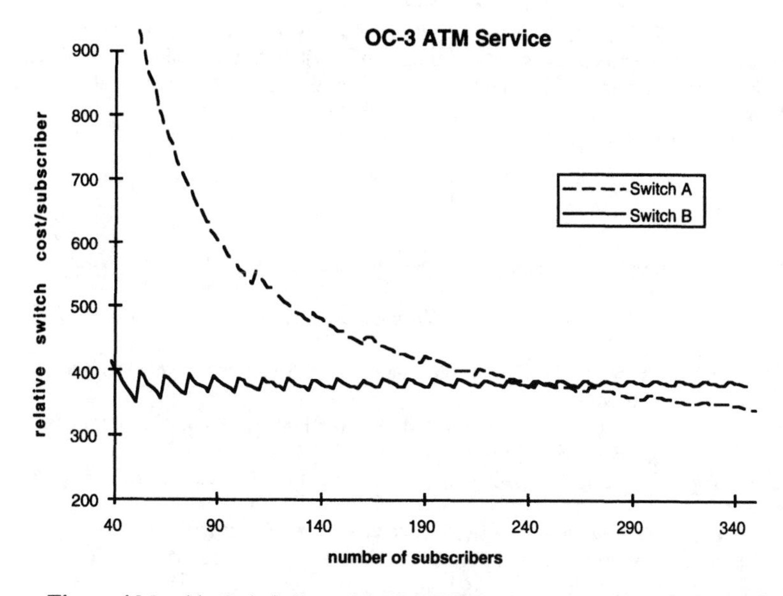

Figure 4 Monthly Switch Cost of OC3 ATM Service vs. Number of Subscribers.

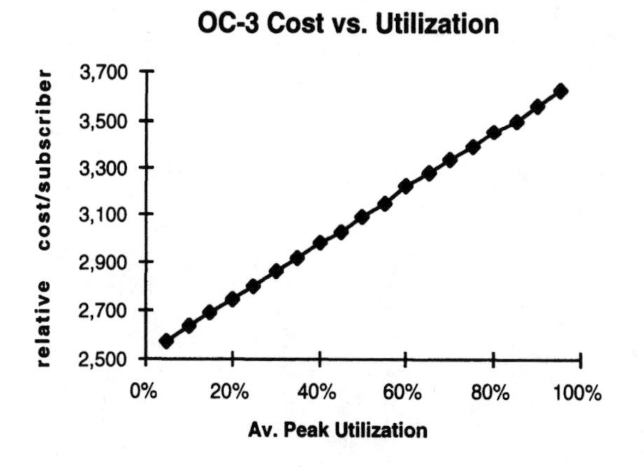

Figure 5 Monthly Cost of OC3 ATM Service vs. Average Peak Utilization.

7 CONCLUSIONS

The significant findings of the cost modeling example are as follows.

- The dominant factors in the cost of providing ATM service are the cost of the SAL and the cost of the switching equipment.
  - The per subscriber cost of the switch itself varies significantly with the number of

subscribers. Underutilized switches are very expensive.

- The costs of the SAL and the switch ports are invariant with the number of subscribers.
- The costs for switch access transport and cloud (interswitch) transport are relatively minor.
- Per subscriber costs vary considerably with average peak utilization, indicating that flat rate ATM tariffs are not equitable to all subscribers. Cost recovery schemes should take into account a subscriber's offered traffic load and charge accordingly.

REFERENCES

Y. Alavi, "Graph Theory With Applications to Algorithms and Computer Science," Wiley Interscience, 1985.

I. Chlamtac et al., "How to Establish and Utilize Virtual Path in ATM Networks," ICC93, Geneva, Switzerland, pp. 1368-1372.

M. Decina, "Which Way Towards the Global Information Infrastructure?" Keynote speaker, EUROMICRO95, Como, Italy.

A. Fargo et al., "A New Degree of Freedom in ATM Network Dimensioning: Optimizing the Logical Configuration," IEEE JSAC, Sept. 1995, pp 1199-1206.

M. Gerla et. al., "Topology Design and Bandwidth Allocation in ATM Nets," IEEE JSTC, Oct. 1989, pp 1253-1262.

J. McQuillan, "Strategic Perspectives on ATM," Keynote speaker, Cascade Forum 1995.

M. Minoux, "Network Synthesis and Optimum Network Design Problems: Models, Solution Methods and Applications," The Networks Journal, John Wiley & Sons, May 1989, pp. 313-360.

A. Tanenbaum, "Computer Networks," Prentice Hall, 1981.

44

On the Joint Topological, Dimensioning and Location Problem for Broadband Networks

Steven Chamberland [a], Odile Marcotte [b], Brunilde Sansó [a]

[a]GERAD, CRT and Department of Mathematics and Industrial Engineering, École Polytechnique de Montréal, C.P. 6079, Succ. Centre-Ville, Montréal (Québec), Canada H3C 3A7

[b]GERAD and Department of Computer Science, Université du Québec à Montréal, C.P. 8888, Succ. Centre-Ville, Montréal (Québec), Canada H3C 3P8

Abstract
In this paper we tackle the problem of jointly finding the optimal location of the ATM switches, the topology of the backbone and local access networks, the dimension of switches and links between switches. We propose a mixed 0–1 integer programming formulation for the problem. Computational results for two examples will be presented and discussed.

Keywords
Topological design, dimensioning, location problem, backbone and local access networks, network devices

1 INTRODUCTION

The development of traditional communication networks was achieved over a long period of time. Network design techniques were progressively used as the networks increased in size to address some of the issues that were involved in the network development. This does not seem to be the case for broadband networks: the concepts and technicalities underlying those networks are being defined beforehand. The system planners have therefore a unique opportunity to design the system from scratch and to implement design tools, so that design decisions for those networks can be carefully evaluated.

Traditionally, the communication network design process has been partitioned into several subproblems that we schematically present in Figure 1. Many exact and heuristic algorithms have been proposed for all subproblems but, in general, optimal solutions to all subproblems do not provide an optimal solution to the global problem.

The objective of this paper is to present a model that addresses the broadband network design problem as a whole. The paper is divided as follows. Section 2 is dedicated to a literature review. The problem is formulated in Section 3. A simplified formulation is proposed in Section 4. Some computational results are presented in Section 5, followed by conclusions and suggestions for further research in Section 6.

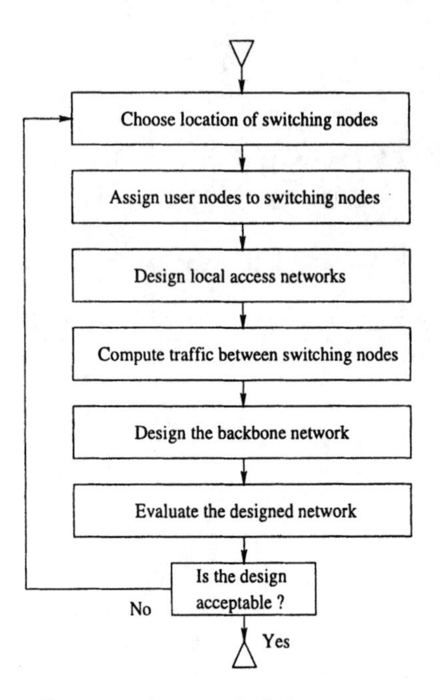

Figure 1 A network design process.

2 LITERATURE REVIEW

As we mentioned in Section 1, network design approaches have rarely been addressed in a global manner. A good overview of traditional telecommunications network design techniques and algorithms may be found in Sharma (1990) and Kershenbaum (1993).

The global approach has been considered only by a few authors. Gersht and Weihmayer (1990) proposed a mixed 0–1 integer programming model for the overall design of a data network that integrates routing, capacity assignment, transmission facility selection and the topological design subject to performance and reliability constraints. The authors showed that the model leads to a natural decomposition of the problem into two subproblems that can be solve sequentially. An algorithm based on a greedy heuristic is proposed to solve the model. Gavish (1992) proposed a nonlinear combinatorial optimization model for the overall problem of designing a computer communication network. The model involves decisions on where to place the switches and how to select the set of backbone links, how to link user nodes to the switches and how to route the messages. A Lagrangean relaxation is proposed to solve the model. Chung, Myung and Tcha (1992) presented a quadratic 0–1 programming model for the overall design of a distributed network with a two-level hierarchical structure, where the embedded backbone network is full-meshed and the local access networks are stars. The authors proposed an equivalent linearized version of the model. A dual-based lower bounding procedure incorporated in a branch-and-bound solution method is proposed. Kim and Tcha (1992) investigated the topological design problem of a two-level hierarchical network where the backbone network is a tree and the local access networks are stars. The problem is modeled as a mixed 0–1 integer program and solved by a branch-and-bound algorithm. This work was extended by Lee, Ro and Tcha (1993), who considered ring-shaped backbone networks. Lee, Qiu and Ryan (1994) modeled the two-level hierarchical network design problem as a degree constrained node-weighted Steiner tree problem. The authors proposed a heuristic procedure to find feasible solutions. The procedure is incorporated into a branch-and-cut solution method.

Although all the aforementioned articles have taken a global approach to the network design problem, none of them has incorporated the special structure of network devices such as switches, multiplexers and links. For instance, no degree constraints on the switching nodes are considered. In fact, to our knowledge, the overall communications network problem, including the structure of the devices, has not been treated so far because it is enormously complex and hard to solve.

In the next section, we propose a model for the overall broadband network design problem.

3 PROBLEM FORMULATION

The general philosophy behind the model is to consider the structure of the devices in the overall design procedure for broadband networks. These devices are switches, multiplexers and communication links. Before formally presenting the model, we underline the functionality and the structure of these devices.

Switching nodes

The main tasks for an ATM switching node are VPI/VCI translations and the cell transport from its input port to its dedicated output port. The general model of a switching node is presented in Figure 2. We make some assumptions concerning the switches. First, we assume that a connection involves exactly one input port and one output port. In the following, a pair of input/output ports is considered to be an entity and is simply called a port. Second, we make the assumption that the ports of a switching node may have different speeds. Typically, two different speeds at the physical layer are mentioned in the literature: STM-1 (155.520 Mbit/s) and STM-4 (622.080 Mbit/s). Third, we assume that a switching node has a global capacity. 10 Gbit/s was mentioned by our colleagues in the industries. Finally, we assume that a switching node cannot have more than n ports. This assumption is translated by a degree constraint for the switching nodes.

The cost of a switch consists of two terms: a fixed cost and the cost of the ports. The fixed cost involves the cost for establishing the switch and the cost of the switch without the ports. The cost of the ports is simply the sum of the costs of all ports, where the cost of a port is a function of its speed.

For more details about broadband switching nodes the reader is referred to Händel, Hubert and Schröder (1994).

Multiplexers

Formally, this device multiplexes several signals originating from different users into a single access link. Because we are working at the physical layer, we assume the use of STM multiplexers. For STM multiplexers, no concentration takes place. The general model of a STM multiplexer is presented in Figure 3. We make the assumption that at most one level of multiplexing is used.

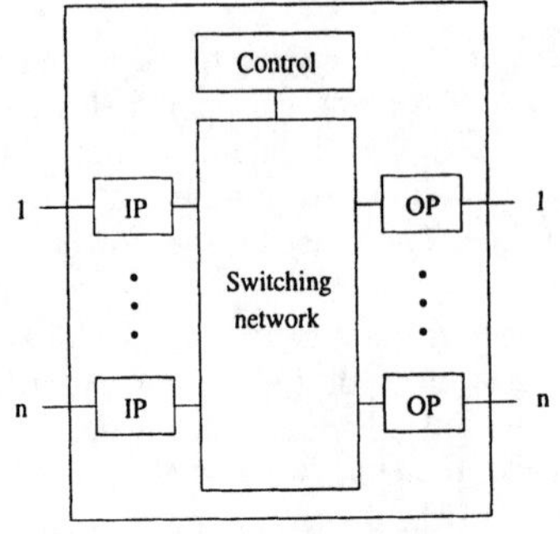

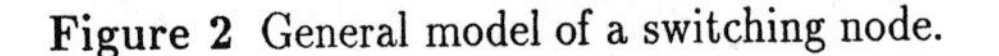

Figure 2 General model of a switching node.

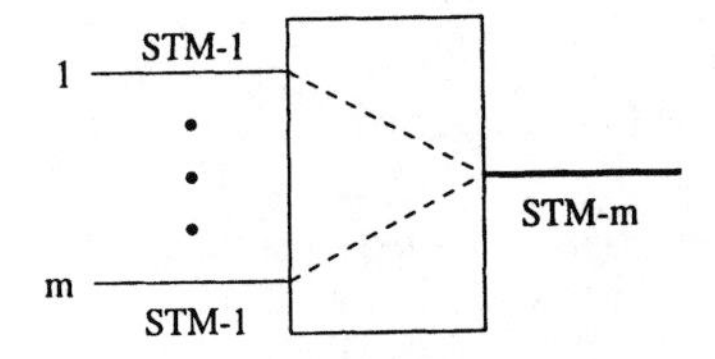

Figure 3 General model of a STM multiplexer.

Communication links

A communication link carries signals in a full-duplex fashion between two nodes of the network. We make the assumption that the communication links may have different speeds. Typically, two different speeds are mentioned in the literature: STM-1 and STM-4. Moreover, we assume that each user node is connected to the backbone network through a STM-1 link.

3.1 The model

The following notation is used throughout the paper.

- M the set of user nodes;
- N the set of candidate locations for the switching nodes;
- R the set of port types;
- S the set of ports for a single switch;
- K the total capacity [Mbit/s] of a switching node;
- c^r the speed [Mbit/s] of a port of type $r \in R$;
- c the speed [Mbit/s] of the links used to connect the user nodes to the backbone network;
- m^r the maximum number of input links for a multiplexer when the speed of its output link is c^r.

Let v_{jb}^{rs} be a 0–1 variable such that $v_{jb}^{rs} = 1$ if and only if the port $s \in S$ of the switching node at location $j \in N$ is of type $r \in R$. If $b = 0$, the port is used by users whereas if $b = 1$, the port is used for a connection between switching nodes. Also, let u_j be a 0–1 variable such that $u_j = 1$ if and only if a switching node is established at location $j \in N$.

Next, we present the model in three parts: the physical constraints, the flow constraints and the objective function.

Physical constraints

We include under this title the set of constraints that are related to topology, location and the choice of the devices in the system. Let x_{ij}^{rs} be a 0–1 variable such that $x_{ij}^{rs} = 1$ if and only if the user node $i \in M$ is connected, via a multiplexer, to port $s \in S$ of type $r \in R$ of the switching node at location $j \in N$. Also, let y_{jk}^{rst} be a 0–1 variable such that $y_{jk}^{rst} = 1$ if and only if the port $s \in S$ of the switching node at location $j \in N$ is connected to the port $t \in S$ of the switching node at location $k \in N$, $j < k$, where both ports are of type $r \in R$. Now we present the physical constraints.

- Each user node is connected to exactly one switching node.

$$\sum_{j\in N}\sum_{s\in S}\sum_{r\in R} x_{ij}^{rs} = 1, \quad i \in M. \tag{1}$$

- Each port used is of exactly one type and a port can be used either to connect to user nodes or to a switching node but not by both.

$$\sum_{r\in R}(v_{j0}^{rs} + v_{j1}^{rs}) \le u_j, \quad j \in N, \ \ s \in S. \tag{2}$$

- The sum of the speeds of the ports is at most the total capacity of a switching node.

$$\sum_{s\in S}\sum_{r\in R} c^r(v_{j0}^{rs} + v_{j1}^{rs}) \le K u_j, \ \ j \in N. \tag{3}$$

- The maximum number of users connected to a switching node via the same multiplexer equals the number of inputs of the multiplexer.

$$\sum_{i\in M} x_{ij}^{rs} \le m^r v_{j0}^{rs}, \quad j \in N, \ \ r \in R, \ \ s \in S. \tag{4}$$

- A port is used for a user connection if at least one user is connected to it.

$$x_{ij}^{rs} \le v_{j0}^{rs}, \quad i \in M, \ \ j \in N, \ \ r \in R, \ \ s \in S. \tag{5}$$

- A port is used for a connection between switching nodes if another port is connected to it.

$$y_{jk}^{rst} \ \le \ v_{j1}^{rs}, \quad j < k, \ \ j,k \in N, \ \ s,t \in S, \ \ r \in R, \tag{6}$$

$$y_{jk}^{rst} \ \le \ v_{k1}^{rt}, \quad j < k, \ \ j,k \in N, \ \ s,t \in S, \ \ r \in R. \tag{7}$$

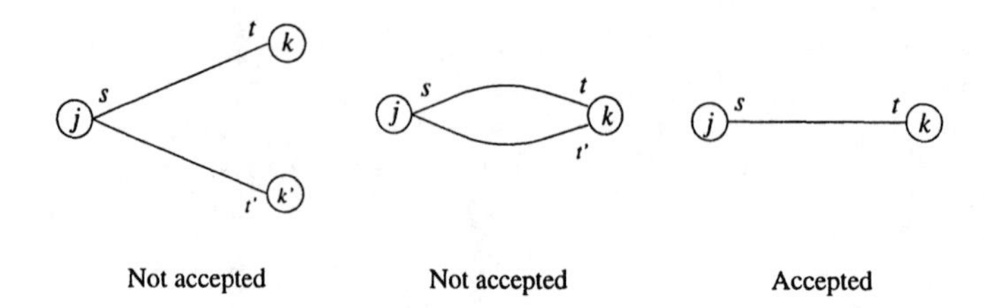

Figure 4 Illustration of a connection between switching nodes.

- A connection between switching nodes is carried out by exactly one port to another port (see Figure 4.)

$$\sum_{\substack{k\in N\\k>j}}\sum_{t\in S} y_{jk}^{rst} + \sum_{\substack{k\in N\\k<j}}\sum_{t\in S} y_{kj}^{rts} \le v_{j1}^{rs}, \quad j\in N,\ r\in R,\ s\in S. \tag{8}$$

The flow constraints

Let $O \subseteq M$ be the set of origin nodes, $D \subseteq M$ the set of destination nodes and d^{pq} the demand in [Mbit/s] from the node $p \in O$ to $q \in D$. Also, let f_{ab}^{pq} be the flow in [Mbit/s] from the node $p \in O$ to $q \in D$ that uses link (a,b) in the network, where node a (respectively b) may be a user node or a switching node. Finally, we denote C_{ab} the capacity of link (a,b) in [Mbit/s]. We now present the set of flow constraints.

- Flow conservation constraints.

$$\sum_{j\in N} f_{ij}^{pq} - \sum_{j\in N} f_{ji}^{pq} = \begin{cases} d^{pq} & \text{if } i=p \\ -d^{pq} & \text{if } i=q \\ 0 & \text{otherwise} \end{cases}, \quad i\in M,\ p\in O,\ q\in D, \tag{9}$$

$$\sum_{i\in M} f_{ji}^{pq} - \sum_{i\in M} f_{ij}^{pq} + \sum_{k\in N} f_{jk}^{pq} - \sum_{k\in N} f_{kj}^{pq} = 0, \quad j\in N,\ p\in O,\ q\in D. \tag{10}$$

- Capacity constraints.

$$C_{ij} = \sum_{s\in S}\sum_{r\in R} c x_{ij}^{rs}, \quad i\in M,\ j\in N, \tag{11}$$

$$C_{jk} = \sum_{s\in S}\sum_{t\in S}\sum_{r\in R} c^r y_{jk}^{rst}, \quad j<k,\ j,k\in N. \tag{12}$$

Hence,

$$\sum_{p\in O}\sum_{q\in D} f_{ij}^{pq} \le C_{ij}, \quad i\in M,\ j\in N, \tag{13}$$

$$\sum_{p\in O}\sum_{q\in D} f_{ji}^{pq} \le C_{ij}, \quad i\in M,\ j\in N, \tag{14}$$

$$\sum_{p\in O}\sum_{q\in D} f_{jk}^{pq} \le C_{jk}, \quad j<k,\ j,k\in N, \tag{15}$$

$$\sum_{p\in O}\sum_{q\in D} f_{kj}^{pq} \leq C_{jk}, \quad j<k, \ j,k \in N. \tag{16}$$

- Nonnegativity constraints.

$$f_{ij}^{pq}, f_{ji}^{pq} \geq 0, \quad i \in M, \ j \in N, \ p \in O, \ q \in D, \tag{17}$$

$$f_{jk}^{pq}, f_{kj}^{pq} \geq 0, \quad j<k, \ j,k \in N, \ p \in O, \ q \in D. \tag{18}$$

The objective function

Let α_{ij} be the cost of the access link between node $i \in M$ and $j \in N$, β_{jk}^{r} the cost of the backbone link between node $j \in N$ and $k \in N$, $j < k$, for a speed c^r. Also, let δ^r be the cost of a port of type $r \in R$ and γ_j the cost for establishing a switching node at location $j \in N$. Finally, we denote ϵ^r the cost of a multiplexer plus the cost of its output link to the port of type $r \in R$.

- The total transmission link cost (Z_L) equals the cost of the links of the local access networks plus the cost of the links of the backbone network.

$$Z_L(\mathbf{x},\mathbf{y}) = \sum_{i\in M}\sum_{j\in N}\sum_{s\in S}\sum_{r\in R} \alpha_{ij} x_{ij}^{rs} + \sum_{j\in N}\sum_{\substack{k\in N\\k>j}}\sum_{s\in S}\sum_{t\in S}\sum_{r\in R} \beta_{jk}^{r} y_{jk}^{rst}. \tag{19}$$

- The total cost for the switching nodes (Z_N) equals the cost for establishing the nodes plus the cost of the ports.

$$Z_N(\mathbf{u},\mathbf{v}) = \sum_{j\in N} \gamma_j u_j + \sum_{j\in N}\sum_{s\in S}\sum_{r\in R} \delta^r (v_{j0}^{rs} + v_{j1}^{rs}). \tag{20}$$

- The total cost for multiplexing (Z_M) equals the cost of the multiplexers plus the cost of the output lines to the switching nodes.

$$Z_M(\mathbf{v}) = \sum_{j\in N}\sum_{s\in S}\sum_{r\in R} \epsilon^r v_{j0}^{rs}. \tag{21}$$

Then the total cost of the network, i.e., the objective function of the problem is $Z_1(\mathbf{x},\mathbf{y},\mathbf{u},\mathbf{v}) = Z_L(\mathbf{x},\mathbf{y}) + Z_N(\mathbf{u},\mathbf{v}) + Z_M(\mathbf{v})$.

It should be noted that the cost of a multiplexer with one input link is 0, and this corresponds to a direct connection between the user and the switch.

The problem (P1)

The problem **(P1)** is to minimize $Z_1(\mathbf{x},\mathbf{y},\mathbf{u},\mathbf{v})$ subject to (1-16) and $\mathbf{v} \in B^{2|N||R||S|}$, $\mathbf{u} \in B^{|N|}$, $\mathbf{x} \in B^{|M||N||R||S|}$, $\mathbf{y} \in B^{\frac{|N|}{2}(|N|-1)|R||S|^2}$, $\mathbf{f} \in R_{+}^{|N|(|N|+|M|-1)|O||D|}$, where B^n is the set of n-dimensional binary vectors and R_{+}^{n} the set of nonnegative real n-dimensional vectors.

4 SIMPLIFIED FORMULATION

Problem **(P1)** has the following drawback: the number of constraints and binary variables is extremely large. As a result, the problem is very hard to solve. In this section we propose a simplified formulation of the problem for which the integrality constraints apply to the **u** and **v** variables only. The idea is to drop the flow constraints and add tree constraints to the backbone network. Then no dimensioning is needed at this level.

Physical constraints

Let r' be the port type used in the backbone network and y_{jk}^{st} a 0–1 variable such that $y_{jk}^{st} = 1$ if and only if the port $s \in S$ of the switching node at location $j \in N$ is connected to the port $t \in S$ of the switching node at location $k \in N$.

In the simplified formulation we keep the physical constraints (1-5) and add the following ones.

- A port is used for a connection between switching nodes if another port is connected to it.

$$y_{jk}^{st} \le v_{j1}^{r's}, \quad j < k, \ j,k \in N, \ s,t \in S, \tag{22}$$

$$y_{jk}^{st} \le v_{k1}^{r't}, \quad j < k, \ j,k \in N, \ s,t \in S. \tag{23}$$

- A connection between switching nodes is carried out by exactly two ports, one at each node.

$$\sum_{\substack{k \in N \\ k > j}} \sum_{t \in S} y_{jk}^{st} + \sum_{\substack{k \in N \\ k < j}} \sum_{t \in S} y_{kj}^{ts} \le v_{j1}^{r's}, \quad j \in N, \ s \in S. \tag{24}$$

- The backbone network is a tree. Then the number of backbone links equals the number of switching nodes minus 1.

$$\sum_{j \in N} \sum_{\substack{k \in N \\ k > j}} \sum_{s \in S} \sum_{t \in S} y_{jk}^{st} = \sum_{j \in N} u_j - 1. \tag{25}$$

- Since the backbone network is a tree, no cycle must exist in the backbone network. Therefore, anti-cycle constraints are necessary. Constraints of this type may be generated adaptively, as needed (see Lee, Qiu and Ryan (1994) for more details).

$$\sum_{j \in H} \sum_{\substack{k \in H \\ k > j}} \sum_{s \in S} \sum_{t \in S} y_{jk}^{st} \le \sum_{j \in H \setminus \{l\}} u_j, \quad l \in H, \ H \subset N, \ |H| \ge 2. \tag{26}$$

The objective function

The objective function of the simplified formulation, $Z_2(\mathbf{x}, \mathbf{y}, \mathbf{u}, \mathbf{v})$, is the same as that for problem **(P1)**, except that we remove the summation over r in the computation of the cost of the backbone links and replace β_{jk}^{r} by β_{jk}, the backbone link cost between node $j \in N$ and $k \in N, j < k$.

The problem (P2)

The simplified formulation, called **(P2)**, is to minimize $Z_2(\mathbf{x}, \mathbf{y}, \mathbf{u}, \mathbf{v})$ subject to (1-5), (22-26) and $\mathbf{v} \in B^{2|N||R||S|}, \mathbf{u} \in B^{|N|}, \mathbf{x} \in R_{+}^{|M||N||R||S|}, \mathbf{y} \in R_{+}^{\frac{|N|}{2}(|N|-1)|S|^2}$

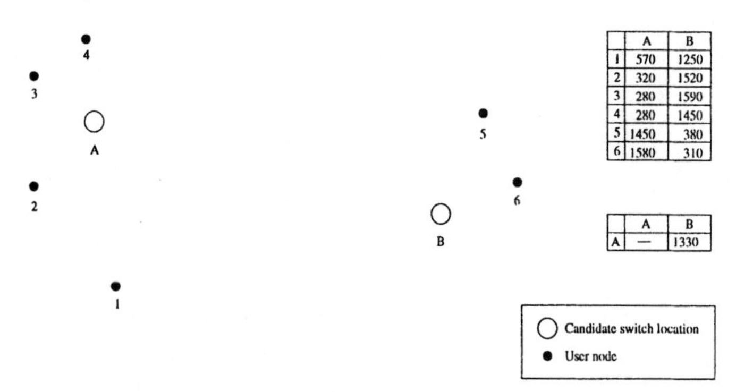

	A	B
1	570	1250
2	320	1520
3	280	1590
4	280	1450
5	1450	380
6	1580	310

	A	B
A	—	1330

Figure 5 Input for example A.

5 COMPUTATIONAL RESULTS

The two models were solved for two different examples, called A and B, using the CPLEX Mixed Integer Optimizer (see the CPLEX user's manual (1993) for more information about CPLEX). Note that the algorithm used by the CPLEX Mixed Integer Optimizer is the branch-and-bound algorithm (see Nemhauser and Wolsey (1988) for more details about the branch-and-bound algorithm). For the computing platform, we used a Sun Server1000.

In the examples, the cost of any link is proportional to the distance between the end-points. The costs of STM-1 and STM-4 links are 10 $/mile and 20 $/mile respectively. The base cost of a switch is 3000 $ whereas the costs of STM-1 and STM-4 ports are 500 $ and 1000 $ respectively. Also, the cost of a multiplexer with four STM-1 input links and a STM-4 output link is 500 $. The maximum number of ports for a switching node is 8.

Example A

Example A is used to test model **(P1)**. The location of user nodes, the candidate locations for the switching nodes and distances in miles are illustrated in Figure 5. For example, the distance between user node 1 and the candidate switch location B is 1250 miles and that between candidate switch location A and B is 1330 miles. Moreover, the user nodes are attached to the backbone network through STM-1 links. The demand between each pair of users is 1 Mbit/s and the total capacity of a switching node is 10 Gbit/s.

The optimal solution is illustrated in Figure 6 and the cost of this solution is 44 200 $. The solution time is 1985.56 sec. The reader can verify that this is the optimal solution.

Example B

Example B is used to test model **(P2)**. The location of user nodes, the candidate locations for the switching nodes and distances in miles are illustrated in Figure 7. The user nodes are attached to the backbone network through STM-1 links, and the switching nodes are connected through STM-4 links.

The optimal solution is illustrated in Figure 8 and the cost of this solution is 90 000 $. The number of nodes in the branch-and-bound tree is 86 and the solution time is 4.60 sec.

It should be noted that even though problem **(P2)** is more tractable than problem **(P1)**, **(P2)** becomes hard to solve for large networks. In fact, this problem is harder than the

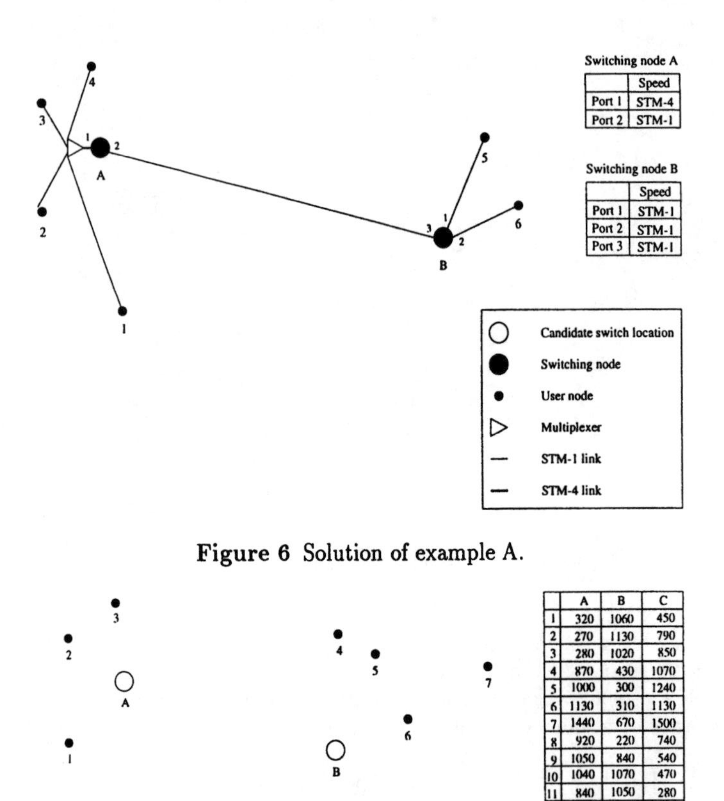

Figure 6 Solution of example A.

	A	B	C
1	320	1060	450
2	270	1130	790
3	280	1020	850
4	870	430	1070
5	1000	300	1240
6	1130	310	1130
7	1440	670	1500
8	920	220	740
9	1050	840	540
10	1040	1070	470
11	840	1050	280
12	860	1140	350
13	680	1190	380

	A	B	C
A	—	860	560
B	—	—	830

Candidate switch location
User node

Figure 7 Input for example B.

degree constrained node-weighted Steiner tree problem presented by Lee, Qiu and Ryan (1994), which has been proved to be NP-Hard. Because of this complexity, we may not expect to find an efficient algorithm to solve these problems to optimality.

6 CONCLUSIONS AND FURTHER WORK

In this paper we have defined the design of broadband networks as a global problem that includes the location problem, the topological design of the backbone and access networks

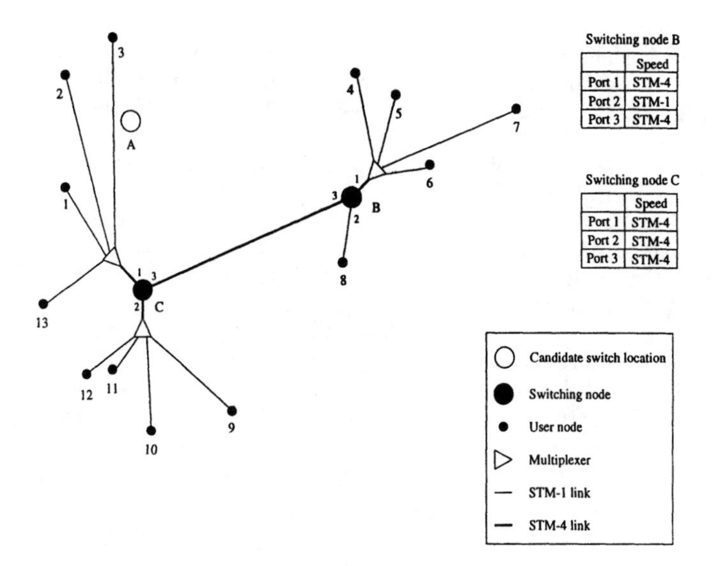

Figure 8 Solution of example B.

as well as the dimensioning of switches and the links between switches. Technical details of the devices have been taken into account. Two models were proposed and one example solved for each model.

Since the problem is NP-hard, it is clear that exact algorithms will be very time consuming. Nevertheless, we plan to tackle several solution approaches based on the constraint generation technique, in order to find the optimal solution of the models for networks of fair size (in the tens of switching nodes). This approach will help planners find good bounds for custom-made heuristics used by the industry.

7 ACKNOWLEDGMENTS

The authors would like to express their gratitude to M. Beshai of BNR for suggesting the problem and for the stimulating discussions on this topic. This work was supported by NSERC and by the Quebec granting agency FCAR.

8 REFERENCES

Chung, S., Myung Y. and Tcha D. (1992) Optimal Design of a Distributed Network with a Two-Level Hierarchical Structure. *European Journal of Operational Research*, **62**, 105–115.

CPLEX Optimization, Inc. (1993) Using the CPLEX Callable Library and CPLEX Mixed Integer Library.

Gavish, B. (1992) Topological Design of Computer Communication Networks — The Overall Design Problem. *European Journal of Operational Research*, **58**, 149–172.

Gersht, A. and Weihmayer, R. (1990) Joint Optimization of Data Network Design and Facility Selection. *Journal on Selected Areas in Communications*, **8**, 1667–1681.
Händel, R., Hubert, M.N. and Schröder, S. (1994) ATM Networks: Concepts, Protocols, Applications. Addison-Wesley.
Kershenbaum, A. (1993) Telecommunication Network Design Algorithms. McGraw-Hill Series in Computer Science.
Kim, J. and Tcha, D. (1992) Optimal Design of a Two-Level Hierarchical Network with Tree-Star Configuration. *Computers and Industrial Engineering*, **22**, 273–281.
Lee, Y., Qiu, Y. and Ryan, J. (1994) Branch and Cut Algorithms for a Steiner Tree-Star Problem. *US West Technologies*, Working Paper.
Lee, C., Ro, H. and Tcha, D. (1993) Topological Design of a Two-Level Network with Ring-Star Configuration. *Computers and Operational Research*, **20**, 625–637.
Nemhauser, G.L. and Wolsey, L.A. (1988) Integer and Combinatorial Optimization. Wiley.
Sharma, R.L. (1990) Network Topology Optimization. Van Nostrand Reinhold.

45

Design and Performance Issues of Protection Virtual Path Networks

P.A. Veitch†, D.G. Smith†, and I Hawker‡

† Communications Division,
Department of Electronic and Electrical Engineering,
University of Strathclyde,
Glasgow G1 1XW,
Scotland, U.K.
Tel: (0141) 552 4400
Fax: (0141) 552 4968
Email: {pveitch,g.smith}@comms.eee.strathclyde.ac.uk

‡ BT Laboratories,
Martlesham Heath,
Ipswich IP5 7RE,
England, U.K.

Abstract

This paper addresses the problem of allocating protection Virtual Path (VP) routes in an ATM network to enable rapid restoration from span or node failures. It is shown how the minimisation of one resource, spare capacity, comes at the expense of another, the number of redundant Virtual Path Identifiers (VPIs). Furthermore, because spare capacity reduction is achieved by allowing longer backup path routes in terms of the number of links used, performance issues related to delay and reliability become increasingly important. After formulating appropriate design and performance metrics pertaining to a protection VP network design, it is illustrated how they are traded off in a design space where spare capacity minimisation is performed.

Keywords: Virtual Paths; Survivability; Network Design

1 Introduction

Broadband transport networks based on Asynchronous Transfer Mode Virtual Paths (ATM VPs) are expected to support both public and private service connections comprising a variety of traffic classes (Aoyama et al, 1993). It has been suggested that the construction of backbone architectures to enable the wide area support of such traffic, will involve the phasing in of Virtual Path crossconnect nodes to replace synchronous path crossconnecting equipment, thus reducing the costs of crossconnect nodes and easing network management tasks like routing and capacity reconfiguration (Miki, 1993). Another attraction of the ATM VP-based architecture is the potential simplicity and speed with which restoration may be executed (Kawamura et al, 1994). Trunk network restoration has been deemed of paramount importance in light of a series of crippling network disasters in recent years (McDonald, 1994). The threat of sizeable revenue losses to operators and service providers, coupled with subscriber dissatisfaction as a result of cable break or node failure, spurred the development of operational centralised restoration systems in the U.S. (Chao et al, 1994) and Japan (Yamagishi et al, 1995). Ongoing research into distributed control mechanisms has aimed at achieving rapid restoration, typically of the order of 1 or 2 seconds, with techniques proposed for Synchronous Digital Hierarchy (SDH) networks and ATM networks alike (Wu, 1992).

A promising approach to VP restoration in a meshed ATM backbone network which combines features of simplicity, resource efficiency and rapid response times, is that of assigning a disjoint protection (or backup) VP to a working VP route (Kawamura et al, 1994, Veitch et al, 1995(b)). A centralised controller is responsible for the choice of protection route, and the relevant information is subsequently downloaded to the appropriate crossconnects and path terminators. In the event of failure, the VP network can autonomously recover by simple Operations, Administration and Maintenance (OAM) protocols to enable switching from working to protection paths (Veitch et al, 1995(a)).

These matters are outwith the scope of this paper; rather, the concern here is the design of protection Virtual Path networks. The design of protection paths which share network capacity between different possible failures has been examined in the context of synchronous crossconnect network structures (Coan et al, 1991, Lubacz and Tomaszewski, 1994). Since these methods apply to paths which are switched according to timeslot allocation within a hierarchical framing arrangement, the principal overhead is spare transmission capacity. The minimisation of this resource is therefore fundamental to survivability planning for SDH architectures. An ATM Virtual Path is a logical entity allowing link-by-link cell switching according to a Virtual Path Identifier (VPI) in the cell header. In establishing protection VPs in a network, spare transmission capacity may be shared between possible failures, as in the SDH case. However, VPIs associated with protection routes must also be reserved, meaning two distinct resources have to be managed in a survivable ATM network.

The design of protection routes subject to spare capacity minimisation will invariably yield paths which do not use the minimum possible number of spans, also called non shortest hop paths. This use of longer paths spreads or *balances* the demand for spare capacity around diverse locations in the mesh topology. This mode of protection design could have serious implications in an ATM network. The more links used in a protection path, the greater the VPI redundancy, meaning the spare capacity resource is traded off against the quantity of reserved VPIs. Another consideration is that of user-perceived quality of service following restoration by switching to a protection VP. A call admission policy which admits Virtual Channels (VCs) to a system of VPs, will involve negotiation of traffic parameters which specify a tolerable level of delay and cell loss (Gupta et al, 1992). The VC will be accepted/rejected given the delay characteristics of VPs in the proposed route. Part of this delay will be fixed due to the number of hops used in each VP of the VC route. Hence, although a VP may be restored by switching to a backup route, the quality of service provided to certain VCs multiplexed on the VP may be degraded due to violation of delay bounds negotiated at call set-up. A further implication of long protection paths is that of inferior reliability. The more hops employed in a backup route, the greater the probability that if a double span failure occurs, it will affect both working and protection routes, thus isolating the associated end-to-end connections. Hence, there are performance issues as well as design concerns related to protection VP network design.

The remainder of the paper layout is now presented. In section 2, metrics are formulated to enumerate the various tradeoffs in a given network topology comprising a layout of working and protection VPs. Section 3 will describe a heuristic design procedure which aims to minimise the cost of a single resource, spare capacity, and the impact that this has on other metrics will be demonstrated for a 20 node mesh network model. Section 4 will discuss the ramifications of the analysis in the context of Virtual Path protection network design, thus concluding the paper.

2 Design and performance metrics

The physical network is described as a graph $G(V, E)$, whereby V is the set of m vertices representative of ATM nodes, and E is the set of n edges representing inter-nodal spans. A single vertex is denoted v ($v \in V$) whilst a single edge is symbolised as e ($e \in E$). The working capacity of an edge e is denoted W_e, whilst the spare capacity is S_e. The logical network is described by the set of k paths, P, whereby a single path π ($\pi \in P$) is the collection of edges traversed. The capacity of a path π is C_π. The set of protection routes is defined as $\hat{P}$, with a protection path pertaining to a working path π, denoted $\hat{\pi}$. It is assumed that the resource provisioning pertains to spans between ATM VP crossconnects, each of which comprises a single bidirectional fibre link.

Spare Capacity Ratio (SCR)

We define the SCR as the ratio of the aggregate spare capacity in the network, to the aggregate working capacity. The working capacity of an edge is found by summing the capacities of constituent paths:

$$W_e = \sum_{\pi \in P, e \in \pi} C_\pi. \tag{1}$$

Hence, the total working capacity is found by summing (1) over the set of network edges. Depending on the edge which has failed in the network, the required spare capacity on the remaining edges varies. This is because a different set of working paths will be affected by each possible failure, hence a different reconfiguration is performed in each case. Letting S_e^f symbolise the spare capacity required on edge e due to failure of edge e_f, we have:

$$S_e^f = \sum_{\pi \in P, e_f \in \pi, e \in \hat{\pi}} C_\pi. \tag{2}$$

Now, the provisioning of spare capacity on each edge must account for the edge failure which will yield the greatest demand for rerouted traffic, giving:

$$S_e = max\{S_e^1, S_e^2, ..., S_e^n\}. \tag{3}$$

It should be stressed that no attempt is made at capacity modularisation so as to conform to specific transmission systems. The value of SCR is subsequently found by dividing the total spare capacity by the total working capacity:

$$SCR = \frac{\sum_{e \in E} S_e}{\sum_{e \in E} W_e}. \tag{4}$$

Mean VPI Redundancy (MVR)

As already stated, when protection routes are designed, Virtual Path Identifiers (VPIs) must be reserved for the appropriate links. The total number of idle VPIs is a function of the number of edges used in each protection route. Letting $L(\hat{\pi})$ be the length (number of edges used) of a specific protection path, $\hat{\pi}$, the total number of idle VPIs, denoted N^v, is found from:

$$N^v = \sum_{\hat{\pi} \in \hat{P}} L(\hat{\pi}). \tag{5}$$

Now, given the total number of VPIs, the MVR may be found by dividing N^v by n (the number of edges) giving the mean VPI redundancy per edge; this quantity can then be normalised to the maximum number of VPIs per link (4096), yielding:

$$MVR = \frac{N^v/n}{4096}. \tag{6}$$

Path Elongation Factor (PEF)

This is simply the ratio of the mean length of a VP rerouted during failure restoration to the mean working path length:

$$PEF = \frac{\sum_{\hat{\pi} \in \hat{P}} L(\hat{\pi})}{\sum_{\pi \in P} L(\pi)}. \tag{7}$$

Mean Switch-pair Availability (MSA)

We assume that each switch pair is connected by a VP, with a span disjoint protection VP. The availability of any switch pair is defined here as the proportion of time that a direct connection (i.e. a VP) exists between the end nodes, hence is dependent on the availabilities of both the working and protection VP. Assuming span failures only, the availability of a path π is the probability that at any moment, the $L(\pi)$ constituent spans are simultaneously *up*. If ρ denotes the probability that a span is *down* therefore, and the value is uniform for each span, the availability of a path π is written:

$$A(\pi) = (1 - \rho)^{L(\pi)}. \tag{8}$$

If the unavailability of path π, written $U(\pi)$, is found from $1 - A(\pi)$, and given that an origin-destination pair has a working path π, and protection path $\hat{\pi}$, the availability of an o-d pair may be expressed in terms of joint unavailability as:

$$A(o, d : \pi, \hat{\pi}) = 1 - U(\pi).U(\hat{\pi}). \tag{9}$$

Whereby $U(\hat{\pi}) = 1 - A(\hat{\pi})$ and $A(\hat{\pi})$ is found by substituting $L(\hat{\pi})$ for $L(\pi)$ in equation (8). Now, averaging the k switch pair availabilities, we obtain the MSA metric:

$$MSA = \frac{\sum_{\pi \in P, \hat{\pi} \in \hat{P}} (1 - (U(\pi).U(\hat{\pi})))}{k}. \tag{10}$$

3 Effects of spare capacity minimisation

The aim of this section is to study the effects of spare capacity minimisation (indicated by the SCR metric) on the MVR, PEF and MSA metrics. Since the design problem is NP complete (Lubacz and Tomaszewski, 1994), heuristic procedures or stochastic techniques like simulated annealing may be applied to networks of reasonable size. We have developed a simple heuristic based on iterative edge weighting which produces a suboptimal solution. A shortest span disjoint path algorithm is applied to each working path in sequence, however weights of edges are adjusted to reflect the current load of protection paths which would become active along with the current path being considered. This design heuristic produces some protection routes which are non shortest hop paths, thus in terms of spare capacity, the layout is *balanced*. On the other hand, if shortest hop paths were found by maintaining unity edge weights, the protection network design could be deemed *unbalanced* where spare capacity is concerned. To view the gradual progression from an unbalanced to a balanced network design which minimises spare capacity, the design and performance metrics at each point in the design space may be plotted. We do this by considering the unbalanced design to be the most constrained in terms of a parameter called hop count, symbolised H. As we tend towards a balanced design with longer protection paths, the value of H is increased, so relaxing the design constraint. We can incorporate this into our design procedure by comparing the length of a protection path, $L(\hat{\pi})$, found by the heuristic, with H. If $L(\hat{\pi}) \geq H$, then the path is replaced with the shortest hop path, if not already of that type. By repeating the design procedure for different values of H, we effectively traverse the design space from an unbalanced to a balanced solution. The effect that this has on the SCR metric for a 20 node mesh network, with 190 VPs and a protection path per working route, can be seen in Figure 1. The choice of a mesh network reflects the nature of actual trunk network topologies.

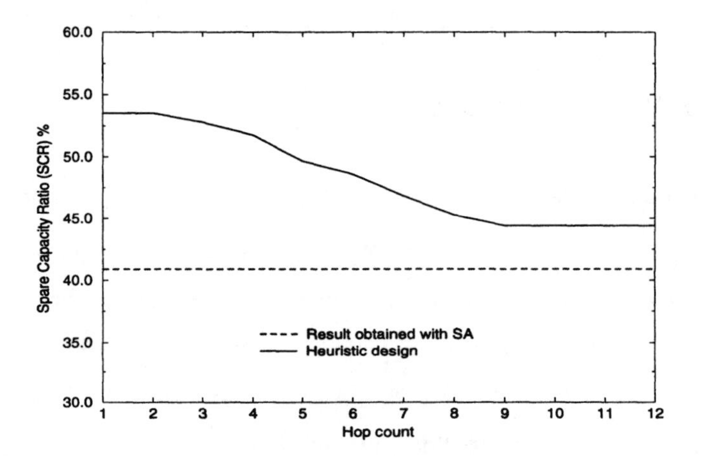

Figure 1: SCR for increasing relaxation of hop count constraint.

Also plotted is the end result of a Simulated Annealing (SA) program which operates on the same network configuration to minimise the required spare capacity. The result obtained with SA is assumed to be very close to the optimum. The beginning of the solid line, where H is low, represents the shortest hop disjoint protection path design which yields the highest value of SCR. As H increases, the SCR is reduced, and eventually the result is that yielded by the heuristic design procedure; since H is so high, no path lengths exceed H, so the choice of protection route is never overridden with a shortest hop backup. The heuristic design produces a suboptimal result in terms of SCR, yet improves the unbalanced design by 17%. Figures 2 and 3 show the corresponding results for the MVR and PEF metrics. As shown, the optimal result where spare capacity is concerned, is the least favourable result where VPI redundancy and path elongation are of importance, since these values increase as the hop count constraint is relaxed.

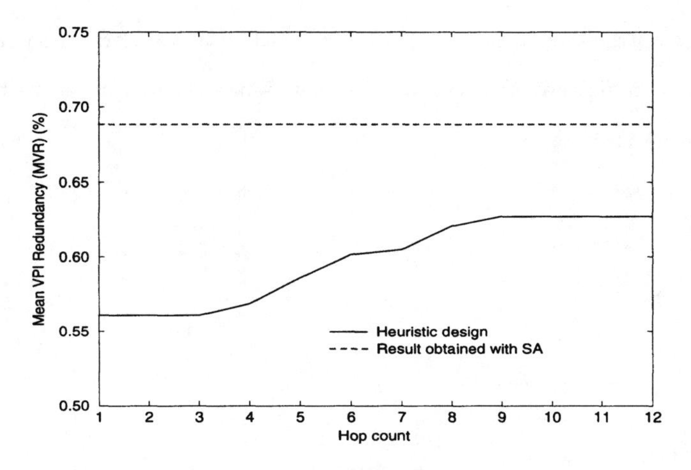

Figure 2: MVR for increasing relaxation of hop count constraint.

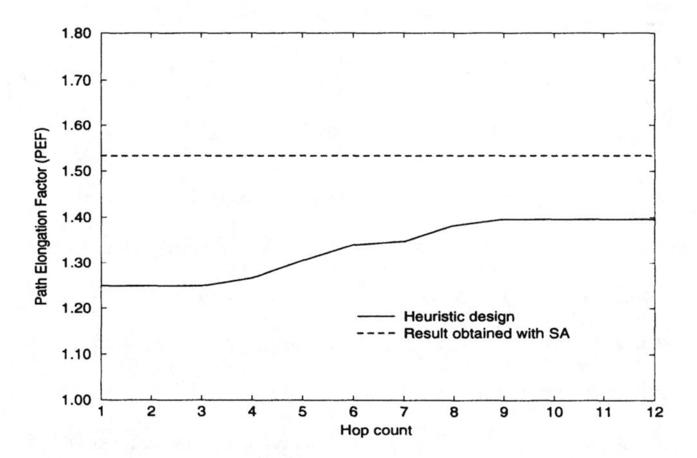

Figure 3: PEF for increasing relaxation of hop count constraint.

As previously discussed, path elongation following restoration may alter the quality of service provided to some subscribers, and indeed the negotiated levels may be breached due to the extra delay incurred by the added hops in the VC route. Obviously, the

PEF metric does not explicitly signify what the extra delay is, or which subscribers may be affected the most. Rather, it indicates the expected level of path elongation following restoration, hence certain fixed delay parameters can be employed to determine a minimum expected degradation under these conditions.

The mean switch-pair availability (MSA) metric plotted for each protection network design is shown in Figure 4. As shown, three separate values of ρ, the probability that at any moment in time a span is down, are used.

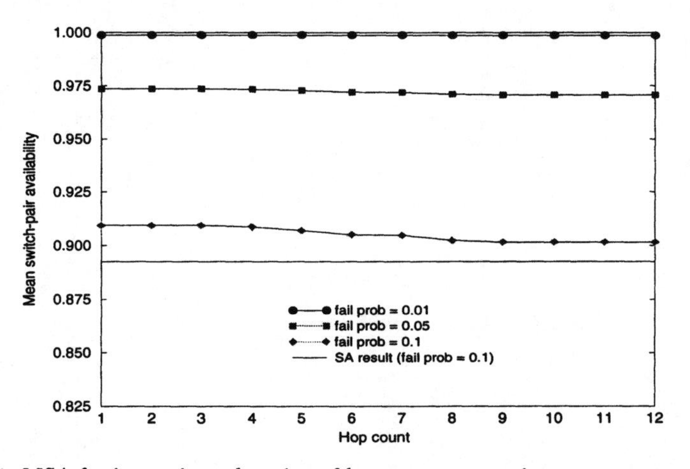

Figure 4: MSA for increasing relaxation of hop count constraint.

For a very low span failure probability like 0.01, the impact of removing the shortest hop path constraint is negligible due to the very small probability that two spans, one in the working path, and one in the protection path, are simultaneously *down*. As the value of ρ increases meanwhile, the impact that the protection network design has on average availability becomes notable. For $\rho = 0.05$, the shortest hop path design yields a mean switch pair availability of 0.973502, whilst that for the balanced design is 0.970664. Although seemingly small, this translates to an extra 24 hours of availability in a year for the switch pairs protected by shortest hop disjoint paths. For the relatively high value of $\rho = 0.1$, this difference becomes approximately 3 days in the year, and between the shortest hop design and the SA result there is approximately 6 days difference.

4 Discussion and concluding remarks

This paper has discussed the importance of network restoration techniques to improve survivability in the context of broadband transport networks based on Virtual Paths. The method of assigning disjoint protection VPs in advance of span or node failure was advocated, and various design and performance metrics influenced by the protection network layout were formulated. By virtue of a heuristic design procedure which aims to minimise spare capacity, it was shown how the reduction in this resource came at the expense of another, the level of VPI redundancy. Furthermore, the protection network designs, albeit *balanced* in terms of spare capacity requirements, exhibited longer paths than would be produced by a shortest hop disjoint path algorithm. This implies that during post-restoration conditions, extra delays may be incurred by rerouted connections, with the possibility of traffic contract violations where path elongation is excessive. Another adverse effect of longer-than-necessary protection paths is the inferior reliability of such routes, since more network elements are traversed. The design of protection VP layouts for ATM networks is thus complicated by conflicting design and performance requirements. This could be partially alleviated by splitting VP traffic into two categories. The first would comprise traffic which is either delay sensitive, requires ultra-high reliability, or both. Such VPs could invariably be protected by a shortest hop disjoint path, so that the effects of extra delay are minimised and the availability maximised. The second category could constitute all other traffic which is neither delay sensitive, nor requires special levels of reliability. The protection paths for such traffic need not be constrained to be shortest hop routes. To conclude, it is important that operators, service providers, and subscribers of broadband communications systems reach agreement on, firstly, the extent to which service restoration is provided, and secondly, the quality of service provided during post-failure/pre-repair conditions. This paper highlighted the intrinsic link between these matters where pre-design of a protection network is concerned.

References

Aoyama, T., Tokizawa, I. and Sato, K-I. (1993). ATM VP-Based Broadband Networks for Multimedia Services. IEEE Communications Magazine, April, 30-39.

Chao, C-W., Fuoco, G. and Kropfl, D. (1994). FASTAR Platform Gives the Network a Competitive Edge. AT&T Technical Journal, July/August, 69-81.

Coan, B.A., Leland, W.E., Vecchi, M.P., Weinrib, A. and Wu, L.T. (1991). Using Distributed Topology Update and Preplanned Configurations to Achieve Trunk Network Survivability. IEEE Transactions on Reliability, October, 404-416.

Gupta, S., Ross, K. and El Zarki, M. (1992). Routing in ATM Virtual Path Based Networks. Proc. IEEE Globecom '92, Orlando, Florida, 571-575.

Kawamura, R., Sato, K-I. and Tokizawa, I. (1994). Self-Healing ATM Networks Based on Virtual Path Concept. IEEE Journal on Selected Areas in Communications, January, 120-127.

Lubacz, J. and Tomaszewski, A. (1994). A Lower Bound Based Approach to the Design of Meshed Networks with Protected Transmission Paths. Proc. ITC 14, Antibes, France, 1445-1453.

McDonald, J.C. (1994). Public Network Integrity- Avoiding a Crisis in Trust. IEEE Journal on Selected Areas in Communications, January, 5-12.

Miki, T. (1993). Optical Transport Networks. Proceedings of the IEEE, November, 1594-1609.

Veitch, P.A., Smith, D.G. and Hawker, I. (1995(a)). A Distributed Protocol for Fast and Robust Virtual Path Restoration. Proc. IEE 12th UK Teletraffic Symposium, Windsor, England, 21/1-21/10.

Veitch, P.A., Smith, D.G. and Hawker, I. (1995(b)). The Design of Survivable ATM Networks, in Performance Modelling and Evaluation of ATM Networks, Volume 1 (Ed. D.D. Kouvatsos), 517-534. Chapman & Hall, London.

Wu, T-H. (1992). Fiber Network Service Survivability. Artech House.

Yamagishi, K., Sasaki, N. and Morino, K. (1995). An Implementation of a TMN-Based SDH Management System in Japan. IEEE Communications Magazine, March, 80-85.

46

Integration of Distributed Restoration Procedures in the Control Architecture of ATM Cross-Connects

Hans Vanderstraeten, Dominique Chantrain, Gzim Ocakoglu, Leo Nederlof, Luc Van Hauwermeiren
Alcatel Corporate Research Centre, Location Antwerp
Francis Wellesplein 1, B-2018 Antwerpen, Belgium, tel. +32/3-240.79.05, fax +32/3-240.99.32 E-mail : hvds@rc.bel.alcatel.be

Abstract

It has been claimed that Distributed Restoration Algorithms (DRAs) have considerable advantages in terms of speed compared to centralised restoration schemes. However, these statements have always been based on simulations of the algorithms only. In order to study DRAs in a real environment, a testbed for experimenting with restoration in ATM networks has been realised. The testbed consists of five 8x8 STM1/ATM cross-connects, and is supervised by a central network management system. Hereby, provisions have been made to incorporate a maximum of flexibility, in order to test various restoration mechanisms, while striving for optimal performance.

The DRA has been integrated into the testbed environment and its robustness has been demonstrated for various network configurations of the testbed. Early results show that it is indeed possible to find the alternative routes within the call dropping time limit (± 2 seconds).

Keywords

Network survivability, distributed restoration algorithms, ATM cross-connects

1 INTRODUCTION

In broadband networks, more and more traffic is concentrated on fewer network elements and fibre routes. This implies that network outages (cable cuts, switching system failures, ...) will cause huge losses of revenue for operators, service providers and commercial users. Therefore, network survivability is essential for broadband systems.

Various approaches for survivability exist (Wu, 1995; Nederlof, 1995). Automatic protection switching is the most simple approach, and is commercially available today. This approach is also used in line switched ring systems. Path protection switched SONET/SDH rings switch to the backup ring on the SONET/SDH path level.

Recently, distributed restoration techniques received considerable attention. Although potentially slower than protection switching to preplanned dedicated spare resources, they have the advantage to operate on meshed and therefore also flexible topologies. They start from the assumption that each node has only limited knowledge of the network topology. Storing too much knowledge at each node might cause problems of consistency among distributed databases when changes occur in the network. Therefore, the distributed restoration algorithms build up network information (topology and location of spare resources) required to restore the failed paths *after* the failure has occurred, guaranteeing an up to date view of the network. Although the gathering of information before restoration consumes time, all proposed distributed restoration algorithms have the explicit ambition to prevent call dropping.

Grover et al. presented the first distributed restoration algorithms (Grover, 1987; Grover, 1990), later followed by several other research groups (Han Yang, 1988; Komine, 1990; Chow, 1993a; Chow, 1993b). All these algorithms are targeted at networks based on the SONET/SDH transmission standards. Kawamura et al. (1992) developed an algorithm for ATM VP networks, which they proved were more efficient than their STM counterparts due to the properties of ATM.

This paper reports on the development of a Distributed Restoration Algorithm (DRA) for ATM networks, and the integration of the algorithm in a network of real ATM cross-connects. Section 2 summarises the objectives of the research on survivability. As will be shown, these objectives will influence some decisions taken for the integration. Section 3 introduces briefly the DRA to be integrated in the ATM cross-connects. Its performance characteristics as derived by the simulation are shortly discussed. The ATM cross-connect and the testbed are presented in Section 4, while Section 5 discusses the integration of the DRA in the cross-connect architecture. Network management issues are shortly addressed in Section 6. The conclusion summarises the current status of the research, including early results on the testbed, and gives an overview of the issues still to be investigated.

2 OBJECTIVES OF RESEARCH ON NETWORK SURVIVABILITY

Up to now, the performance of distributed restoration algorithms has only been analysed using simulation tools (Bicknell, 1993). The principal parameters in which performance is expressed are the spare capacity efficiency and the speed of restoration. All published studies focused on the performance characterisation on network level. Nodes were considered as black boxes, and the contribution of the internal delays caused by the message processing and cross-connecting were considered as being well defined and fixed. However, it is clear that, given the inherent complexity of STM and ATM cross-connects, both from a hardware and software point of view, this reasoning has to be refined if reliable statements are to be made on the restoration speed of any algorithm in a real network. Recently, Wu and Kobrinski (1993) first expressed concerns on the network simulation-only approach. These authors studied internal STM cross-connect architectures, and showed that optimising overall restoration speed on network level requires an optimum performance of all relevant functionalities inside the cross-connect. This conclusion was confirmed by a study of Kobrinski and Azuma (1993).

An additional shortcoming in current research is that the interaction of any DRA with network management has not been studied in depth. Yet it is of paramount importance that DRAs, although operating autonomously, i.e. independently from the network manager, can be *controlled* by the network management. For instance, the network management has to be able to validate the DRA before it is authorised to take decisions autonomously, and to select a specific DRA out of a set of different protection and restoration techniques/mechanisms, depending on the desired trade-off between the advantages/disadvantages of any DRA for the specific network configuration that is managed.
This paper reports on the integration of a DRA in an ATM cross-connect system. This work is part of a research project on network survivability which has as major objectives :

1. to realise an ATM network platform for experimenting with restoration,
2. to prove that DRAs actually do work in *real* ATM networks, consisting of *real* ATM cross-connects,
3. to allow realistic simulations on existing and future networks of operators, through validation of the node and network models,
4. to determine the characteristics of different proposed restoration schemes (backup VP, span-based DRA, centralised restoration, ...) in real ATM networks,
5. to develop and validate the integration of the autonomous DRA with centralised network management.

These objectives have implications on the specific choices made for the integration. The first objective implies that the integration should for instance allow flexible inclusion of different algorithms in the node, that a general, flexible messaging scheme is preferred over an optimised one (single cell ATM OAM messages), and that the means have to be included to measure the behaviour of the restoration at any time (through inserting/extracting timestamps). The second and third objectives are realised by building the testbed on *commercially available* ATM cross-connects. The specification of the DRA is first validated using the simulation environment. In a next stage, the DRA is integrated into the ATM testbed. Using the network manager, different test scenarios are set up and executed on the network. Additionally, the node model used in the simulation is brought in line with the ATM cross-connect of the testbed (the DRA has to show the same behaviour in terms of several different variables like speed of restoration, sequence of alternative routes found, number of messages generated, ..., on both the testbed and the testbed representation in the simulation tool). In this way, extrapolations to large ATM networks become reliable.
The fourth objective requires that the integration should allow each of the restoration procedures to perform optimally. The first objective will preclude optimal performance, so that measures have to be taken to allow relevant and relatively comparable benchmarking.
The last objective requires that a communication mechanism is foreseen between the node, including the DRA, and the Network Management.

3 DISTRIBUTED RESTORATION ALGORITHM

The Distributed Restoration Algorithm (DRA) is triggered by alarms due to failures of links, and searches for new routes around the failure in a joined effort with the other nodes in the network. Here, we will only explain the general principles of the DRA. For more details, we refer to Nederlof (1995). Extensive simulation results can be found in Vanderstraeten (1996). The proposed algorithm is designed to cope with single link, multiple link, and node failures.

3.1 Working principle

The route search process is based on a flooding of *Request messages*, originating from the *Request Source* (RS) nodes adjacent to the failure as in Figure 1. Intermediate nodes, referred to as *Tandem* (T) nodes, store the request, and forward updated copies of the Request message. As these messages propagate between the nodes, spare capacity that is requested will be explicitly reserved for each message until this capacity is captured as part of an actual alternative route, or released to make place for a different alternative route. A limit is set to the flooding procedure by specifying the maximum number of tandem nodes by which a request message can be remote from the original RS node. This parameter is further referred to as the *hop count*.

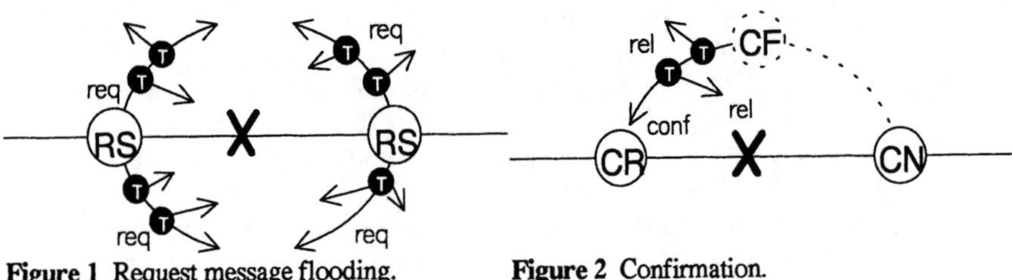

Figure 1 Request message flooding. **Figure 2** Confirmation.

When a route is found, i.e. when request messages meet somewhere between the two RS nodes, the Tandem node where the branches meet, now labelled *Confirm* (CF) node, designates the RS node with lowest ID as *Chooser* (CR) and the other as *Chosen* (CN). The Confirm node sends *Confirm messages* down the candidate alternative route towards the Chooser as illustrated by Figure 2. Any surplus bandwidth reservations are cancelled implicitly, or by sending *Release messages* on obsolete request branches.

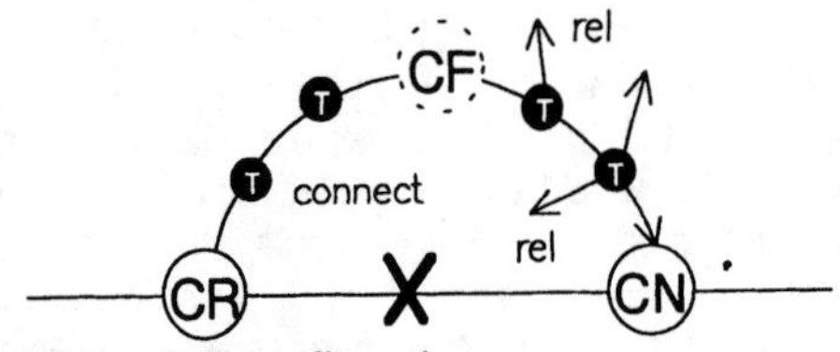

Figure 3 Reconfiguration.

Figure 3 shows the Chooser node that makes a selection of the VPs that can be allocated to candidate alternative routes. Information about the re-routed connections is sent along the route to the Tandem nodes and the Chosen node in *Connect messages*. Each Tandem node between the CF node and the CN node now releases obsolete request branches.

3.2 Simulation results

The proposed DRA has been developed and validated on a discrete event simulator, in which a network model and a node model have been integrated.

The network model for which the results are reported here, is shown in Figure 4. The model has been used by several other authors (Bicknell, 1993). The figures on the links indicate the capacity working(spare). The connections are assumed to have unit bandwidth. In our simulations we assume that each connection corresponds to a single VP.

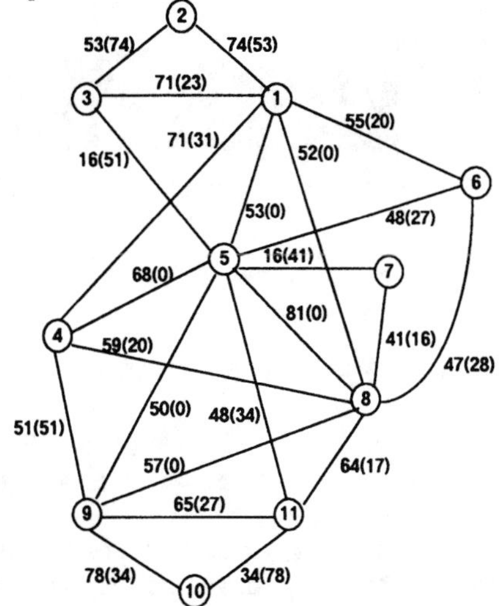

Figure 4 The reference network model used for simulating the distributed restoration algorithm.

By means of a node-level simulation, a suitable node model has been derived in order to describe the aggregate delay caused by the processing and queuing of restoration messages inside the nodes. In the presented simulation results, the message delay on the input and output of each node in the network is taken to be statistically distributed with a mean value between 10 and 20 msec each.

Table 1 illustrates as an example the restoration of the link failure between nodes 4 and 8 of Figure 4. The link originally carried 59 (bi-directional) active connections, which are restored by using the spare capacity in the network and divided among 5 alternative routes. The table also provides the restoration time for each alternative route found in the network (i.e. the time between the detection of the link failure and the arrival of the connect message in the Chosen node). The restoration process is completed for this link failure after 430 msec, which means that no more restoration messages are passing in the network and every extra bandwidth reserved during the restoration process, which does not result in an alternative path, is released after this time. For this particular link failure, the original route between the nodes 4 and 8, is replaced by a set of alternative routes, which carry connections via multiple hops instead of a direct link between two endpoints. The restoration results in an average of 4.71 hops in the connections.

Figure 5 shows the restoration times (complete restoration) for all single link failures in the reference network. The complete duration of the restoration process varies from 180 msec (link failure on 5-7) to 490 msec (link failure on 5-11). The time until the last alternative route is connected varies from 123 msec (link failure on 1-3) to 445 msec (link failure on 5-11). The number of hops has been set in all cases large enough to allow 100 % restoration. The average number of hops on the alternative routes in this analysis is 3.61, while the average restoration time was 303 msec.

It should be emphasised that in these simulations the node model is not comparable yet with the node representation of the real ATM cross-connect, but is more compliant to the model used in other simulations in literature, allowing a meaningful comparison between algorithms (Vanderstraeten, 1996).

Table 1 Link failure restoration between nodes 4 and 8 of Figure 4.

alternative route (indicated by node id)	number of VP connections	restoration time (msec)
4-9-11-8	17	118
4-1-6-8	20	141
4-9-11-5-7-8	10	222
4-9-10-11-5-7-8	6	276
4-9-10-11-5-6-8	6	386
	total = 59	completed after 430 msec

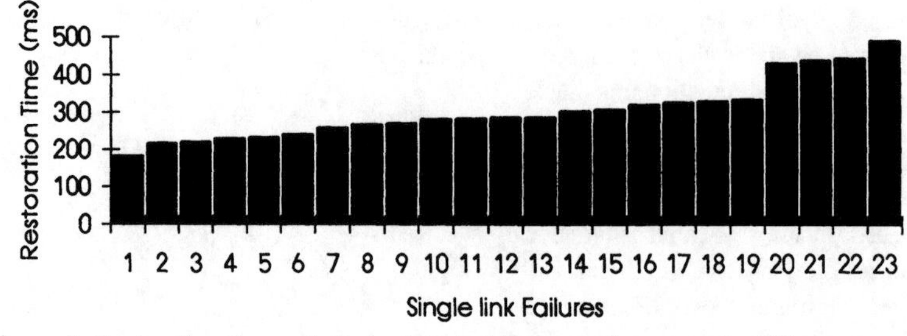

Figure 5 Restoration times of all single link failures on the network of Figure 4.

4 THE TESTBED NETWORK

The ATM cross-connect node used to configure the experimental testbed is an adaptation of the original commercially available cross-connect. A key aspect of this cross-connect is its modularity. This aspect has been fully applied to configure the experimental testbed. Therefore, we first explain the switching and control architecture of the original cross-connect. Afterwards the adaptations and their implications for our testbed application are described.

4.1 The ATM cross-connect

The ATM cross-connect used in the testbed is based on a multipath self-routing switching principle and an internal transfer mode using multislot cells. This switching concept is shown to have a number of advantages (Boettle, 1990), among which the modular extendibility from small to very large systems.
The Termination Link Board (TLK) is the part of the switch that is dependent on the external environment. It performs termination of the external transmission system, the processing required for the external transfer mode (ATM in our case), the conversion between ATM cells and multislot cells, the traffic distribution to and routing from the switching fabric, and resequencing (Banniza, 1991). The board terminates eight STM1/ATM links, and contains, next to the termination logic, also an On-Board Controller (OBC). An on-board mechanism is foreseen to insert/extract cells to/from an external ATM cell stream to/from the OBC. Distribution to and routing from the switching fabric is done using two integrated switching elements. As such, the TLK contains the first stage of the switching fabric.
Subsequent stages of the switching fabric are realised using Switching Modules (SMs), consisting of eight integrated switching elements, structured in a two-stage arrangement. Again, each SM is equipped with an OBC.
Control communications are needed between individual OBCs and the control station (CS) handling call control for services as well as OAM functions at system level. For proper structuring of control communications and loose coupling between transport network and control applications, an intermediate level of control resource has been introduced, the Rack Configuration Controller (RCC), distributed per rack and dedicated to all front-end control functions relating to the OBCs equipped in that rack, but also the power supply and cooling functions.

4.2 The control architecture

Figure 6 shows the control architecture of the ATM cross-connect.
The control functions running on the OBCs of the SMs and TLKs include the initialisation, maintenance and traffic load management. Furthermore, the TLK OBCs fill in the VPI/VCI translation tables, and perform connection control functions (among which is policing).

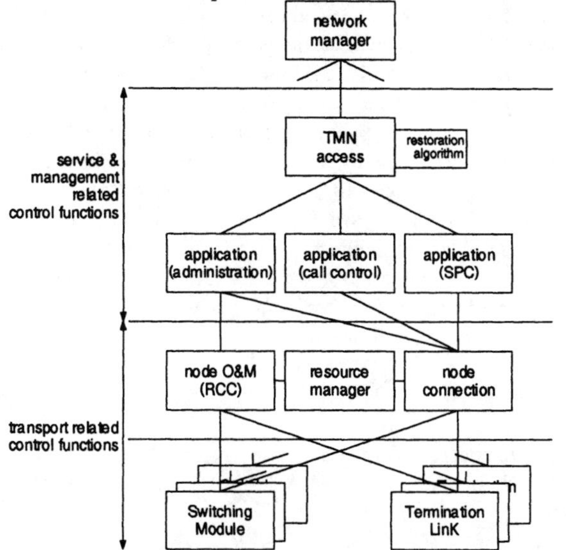

Figure 6 A generic representation of the control architecture of an ATM cross-connect.

The *node O&M* module runs on the RCC, and provides all functionality to allow the RCC to autonomously initialise and monitor the actual hardware equipment of its rack. Maintenance functions in the RCC manage the operational configuration available for traffic. The latter depends on various alarm conditions at rack level and on operational status conditions reported by each OBC about its own board functions and input/output links.
The *node connection* module runs on the CS, and performs resource allocation (VPI/VCI values, and bandwidth) and connection control functions. This module gets its information on free resources from the *resource manager* module (running on CS), which maintains an up to date view on all operationally available resources through the node O&M module.
The *application* modules implement common services offered by the switch. The *Semi-Permanent Connection (SPC)* module appends a time scheduler to a cross-connect point, while the *Administration* module defines external access ports, and maintains user profiles.
The *TMN access* module implements a CMISE interface, and translates the messages and relays them towards the appropriate service modules.
In this way, the control architecture of the cross-connect can be organised as a chain of interacting blocks representing the flow of O&M-related information from the lowest level up to the highest level (being the access for TMN), and a chain of blocks passing connection-related information up to the highest level.

4.3 The ATM cross-connect node in the testbed

The ATM cross-connect node used in the testbed is an adaptation of the original commercially available cross-connect, the latter being described in detail above. The only adaptation consists of reducing the hardware of the cross-connect to the bare minimum, while retaining the complete control architecture, thus exploiting the modularity property of the MPSR switching concept. Since the TLK includes *both* termination and switching functionalities, this board in principle suffices to switch ATM. However, since the control architecture should remain unchanged, the software module running on the RCC is moved to CS. In this way, all maintenance functions to be performed by the RCC still are performed, be it by the CS.
By only reducing the hardware, and preserving the original control architecture, the internal *and* external behaviour of the cross-connect are identical to the commercial version, except for the fact that the size of the switching fabric is reduced to the minimum, and the hardware support functions that are not strictly needed for the switching functionality (such as the control of periphery like fan units and the like) are removed. In this way, the objectives outlined in Section 2 are satisfied, i.e., the testbed allows to study the behaviour of the DRA in terms of speed of restoration and spare capacity usage in circumstances that are directly comparable to the commercially run ATM networks.

4.4 The testbed

The testbed set up for experimenting with restoration consists of five single TLK ATM cross-connects (Figure 7), each terminating eight input/output STM1/ATM links. This arrangement allows the topology to be fully meshed.

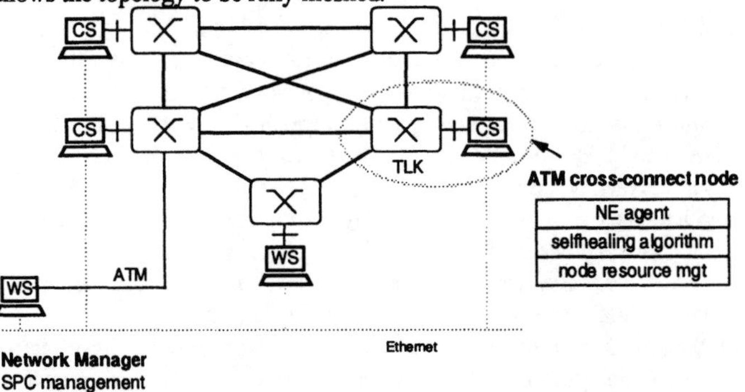

Figure 7 The testbed for experimenting with restoration, consisting of five 8x8 ATM switches, managed by a central network management application.

Communications between the TLK and CS is based on a (point-to-point) proprietary internal control protocol, while communication between the network manager and the agents (physically the CSs) can be either through Ethernet, or through ATM.

Of course, the integration of the DRA in the control architecture, and its interaction with the centralised network management, will require further adaptations. These are extensively discussed in the subsequent sections.

5 INTEGRATION OF THE DRA IN THE CONTROL ARCHITECTURE OF AN ATM CROSS-CONNECT

The choice between possible alternatives for the integration will be determined by what information the DRA needs from its external environment, and by the objectives pointed out in section 2, i.e., as close as possible to real configurations, but allowing experiments.

5.1 DRA's requirements from its external environment

As far as the integration is concerned, only the external information that the DRA needs for proper operation is important, and the mechanism to transfer messages to the neighbouring nodes and the network management have to be considered.
The DRA needs the following information from the node (Nederlof, 1995) :

- the number of external links,
- the nominal bandwidth of each of the links,
- the connections (VP/VC/bandwidth) on each of the links,
- the used and spare bandwidth on each of the links.

From the network manager, the DRA has to obtain the following items :

- its network wide node-id,
- the node-ids of the nearest neighbour nodes,
- a network wide unique id for each of the connections that go through the node,
- the node-ids of the one but nearest neighbour node in both directions for each connection individually.

For the communication between the DRAs a messaging system has to be integrated in the testbed. Obviously, the communication should be optimal in terms of speed. Ideally, the DRA messages should be short enough to fit in a single cell. The intrinsic OAM cell flow foreseen in the ATM standard is the first candidate, since the ATM cross-connects are designed to trap and transfer these cells to the control software efficiently. However, the limited cell size would preclude the insertion of information into the DRA messages that is not strictly needed for the DRA operation as such, but definitely is needed for *monitoring* the behaviour of the DRA for experimental purposes. For instance, several timestamps have to be inserted in the messages to estimate accurately the node model. For this reason, it has been decided to use AAL5 for the DRA messaging. Although this messaging system is less efficient due to the SAR functionality, it does not impose any limitations on the message size. The AAL5 layer has been integrated on the TLK, using the on-board cell drop functionality.
For the communication with the network manager, the AAL5 functionality foreseen for the DRA messaging is reused, and additionally the UDP/IP termination has been implemented. Alternatively, Ethernet/IP/UDP can be used.
Since the DRA has to be able to autonomously delete or create cross-connect points, and to detect alarms, the integration has to be such that at least these operations can be performed.

5.2 Location of the DRA in the control architecture

The location of the DRA should be such that speed of restoration is optimal. The speed of restoration is determined by the execution time of the core algorithm, and by the communication with the environment. The latter is determined by each of the components discussed above : communication between DRAs, and between DRA and the network manager, detecting failures, and setting up and tearing down connections. In order to minimise the total communication time, proper choices should be made for the integration of the DRA, by minimising the individual contribution of each of the components, and by paralleling the components as much as possible.
Some elements of the communication have to be considered to be fixed, e.g., the time needed for the detection and propagation of alarm through the control architecture is determined by the original architecture. Optimisation can only be done by putting the DRA as close as possible to the source of the alarm. The same holds for the creation/deletion of cross-connect points, and receiving/transmitting DRA messages. Thus, the DRA should be located as close to the links as possible, to minimise to message transfer delay within the control architecture.
However, the DRA has to modify routing tables and assign bandwidth to routes as an outcome of the algorithm. Since the DRA only has access to these functionalities if it is implemented on node level or higher, it has to be located at least at node level. Furthermore, the DRA should restore the transport service transparently for the TMN. Combining both requirements leads to the conclusion that the DRA should be located somewhere between the node connection module (refer to Figure 6) and the access to TMN.
The ideal location would thus be the node connection module. However, for the given ATM cross-connect, it is difficult to bring this choice in line with the objective to develop and validate the integration on the autonomous DRA with centralised network management. Indeed, the communication with the network manager, considered to be mandatory in this project, would be more easy if the DRA is integrated higher up, before the dispatching of the network management commands to the different applications. Therefore, it has been decided to integrate the DRA in the TMN-access module. Although this choice is, again, only sub-optimal in terms of core speed of restoration, it does allow to integrate network management in the testbed and the DRA.
Currently, the DRA has been integrated into the testbed environment and its robustness has been demonstrated for various network configurations of the testbed. The very first, and therefore non-optimised, results show that it is indeed possible to find the alternative routes within the call dropping time limit (± 2 seconds).

6 INTEGRATION OF NETWORK MANAGEMENT

Although the distributed restoration mechanism is operating fully autonomously, the network management station still has an important role to play in the whole process. Generally spoken, network restoration mechanisms can be described in 3 phases: pre-restoration, restoration and post-restoration.
During the pre-restoration phase the DRA has to be installed and initialised. Therefore, the network management station passes the relevant network-level information to the nodes (e.g.

the node-ids of the nearest neighbour nodes). Also the parameters of the DRA itself (e.g. the hop count) are initialised. In case more than one restoration mechanism is implemented in the network, the management station initialises the parameters that determine which mechanism will be activated under which circumstances.
During the restoration phase the DRA runs autonomously, and the network management station doesn't play any active role. However, a number of event reports can be generated by the DRA, which allows to monitor the restoration process on-line on the network management station. Also a logging mechanism has been implemented: the DRA can store the event information locally, and this data can be retrieved afterwards by the management station to obtain a complete restoration report, including timing information.
This brings us to the post-restoration phase, where the management station is responsible for collecting the event reports or log records and correlate this data into an overall restoration report, indicating the affected and restored connections, the restoration time, etc.
The implementation of these mechanisms requires adaptations both at the side of the management station (network management applications), and at the side of the DRA.
To allow DRA management data to be exchanged between the management station and the nodes, the MIB of the original ATM cross-connect is extended with a new DRA fragment. The mechanisms of on-line and off-line monitoring (respectively by event reports and log records) are mostly generic and can thus easily be adapted to incorporate also the management of other restoration algorithms.
Next to the management of the restoration system, of course, the necessary management applications are provided to set up SPCs over the network, and to monitor the state of the managed elements.

7 CONCLUSIONS : CURRENT STATUS AND FUTURE WORK

A testbed for experimenting with restoration in ATM networks has been realised. The testbed consists of five 8x8 STM1/ATM cross-connects, and is supervised by a central network management platform. Hereby, provisions have been made to incorporate a maximum of flexibility, in order to test various restoration mechanisms, while striving for optimal performance. Furthermore, the testbed results contain sufficient information to feedback the results of the tests into a simulation environment.
The distributed restoration algorithm, which has previously been reported in literature, has been integrated into the testbed environment, and its robustness has been demonstrated for various network configurations. Early results show that it is indeed possible to find the alternative routes within the call dropping time limit (2 seconds).
In a next stage, the project will extensively measure the contribution to the overall performance of the different time components of the processes the restoration consists of. These results will be used to define a validated representation of the node model behaviour, which can be fed back into the simulation environment. Then, the performance of the distributed restoration algorithm can be assessed for larger ATM networks. The testbed results will also result in requirements how to optimise the algorithm and its implementation into the ATM cross-connect.

The feasibility demonstration of other restoration mechanisms is also envisaged. For this, the necessary measures have been taken to provide the testbed with sufficient flexibility, such that other restoration schemes can be added in both the ATM cross-connect and the network management easily.

REFERENCES

Banniza, T.R., Eilenberger, G.J., Pauwels, B., Therasse, Y. (1991) Design and technology aspects of VLSI's for ATM switches. *IEEE Journal on Selected Areas in Communications*, **9**(8), 1255-1264

Bicknell, J., Chow, C.E., Syed, S. (1993) Performance analysis of fast distributed network restoration algorithms. *IEEE Globecom 93*, 1596-1600

Boettle, D., Henrion, M.A. (1990) Alcatel ATM switch fabric and its properties. *Electrical Communication*, **64** (2/3), 156-165

Chow, C.E., Bicknell, J., McCaughey, S. (1993a) A Fast Distributed Network Restoration Algorithm. *International Phoenix Conference on Computer and Communications*

Chow, C.E., McCaughey, S., Syed, S. (1993b) RREACT: A Distributed Network Restoration Protocol for Rapid Restoration of Active Communication Trunks. *Proc. of 2nd IEEE Network Management and Control Workshop*

Grover, W.D. (1987) The selfhealing network: a fast distributed restoration technique for networks using digital crossconnect machines. *IEEE Globecom 87*, 1090-5

Grover, W.D., Venables B.D., MacGregor M.H. and Sandham J.H. (1990) Development and performance assessment of a distributed asynchronous protocol for real-time network restoration. *IEEE Journal on Selected Areas in Communications*, **9**(1), 112-125

Han Yang, C., Hasegawa, S. (1988) FITNESS: Failure immunisation technology for network service survivability. *IEEE Globecom 88*, 1549-54

Kawamura, R., Sato, K., Tokizawa, I. (1992) Self-healing ATM network techniques utilizing Virtual Paths. *Networks 92*, 129-134

Kobrinski, H., Azuma, M. (1993) Distributed control algorithms for dynamic restoration in DCS mesh networks: performance evaluation. *IEEE Globecom 93*, 1584-8

Komine, H., Chujo, T., Ogura, T., Miyazaki, K., Soejima, T. (1990) A distributed restoration algorithm for multiple-link and node failures of transport networks. *IEEE Globecom 90*, 459-463

Nederlof, L., Struyve, K., O'Shea, C., Misser, H., Du, Y. and Tamayo, B. (1995a) End-to-end survivable broadband networks. *IEEE Communications Magazine*, **33**(9), 63-70

Nederlof, L., Vanderstraeten, H., Vankwikelberge, P. (1995b) A new distributed restoration algorithm to protect ATM meshed networks against link and node failures. *ISS 95*, A8.1

Vanderstraeten, H., Ocakoglu, G., Nederlof, L., Chantrain, D., Van Hauwermeiren, L. (1996) Simulation and implementation of a distributed algorithm for restoration in ATM meshed networks. *Computer Communications*, submitted

Wu, T.H. (1995) Emerging Technologies for Fiber Network Survivability. *IEEE Communications Magazine*, **33**(3), 58-73

Wu, T.H., Kobrinski, H. (1993) A service restoration time study for distributed control SONET digital cross-connect system self-healing networks. *ICC 93*, 893-9

47

Reliability and Performance Analyses of Balanced Gamma Network for use in Broadband Communication Switch Fabrics

H. Sivakumar and R. Venkatesan
Faculty of Engineering and Applied Science
Memorial University of Newfoundland, St. John's, Nfld., Canada A1B 3X5.
Telephone : (709) 737 - 7962, Fax : (709) 737 - 4042
{harinath, venky}@engr.mun.ca

Abstract

Balanced gamma network, a multipath, multistage interconnection network, which features 4×4 switching elements may be considered as a potential candidate for use in broadband communication switch fabrics. This paper studies the fault tolerance and reliability of the balanced gamma network. The three main reliability measures -- terminal reliability, broadcast reliability, and network reliability -- are computed for the balanced gamma network in this paper. The input-output and network mean times to failure of the balanced gamma network are also calculated.

The performance of a multistage interconnection network in the presence of faults gives a better understanding of the dependability of the network than the reliability measures normally used. The performance analysis of the balanced gamma network with faults, under the uniform traffic type is presented. It is shown that the balanced gamma network is single fault-tolerant and robust under multiple faults. Simulation results of the throughput performance of the balanced gamma network with and without faults, under uniform traffic patterns, are also compared in this paper.

Keywords

Balanced gamma network, dependability, faults, fault tolerance, multistage interconnection networks, performance, reliability, switching elements, throughput.

1. INTRODUCTION

ATM (Asynchronous transfer mode) has been identified by ITU-T as the switching system capable of meeting the requirements of B-ISDN (broadband integrated services digital network) such as very high throughput, a low switching delay, a low probability of packet loss, expandability, testability and fault-tolerance, low cost and ability to achieve broadcasting as well as multicasting. ATM is a packet-oriented transfer mode using statistical time division multiplexing techniques [3,13,22]. An ATM switch requires very high speed switch fabrics in order to meet the needs of high throughput. Several switch fabrics have been investigated by researchers and organizations [3].

A multipath, single fault-tolerant MIN called the BG network [24] is discussed in this paper. The BG network has improved throughput performance under several different traffic patterns [19,20,24,25]. In this paper it is shown that the BG network exhibits excellent fault tolerance properties, high reliability and increased throughput performance even in the presence of SE faults.

In Section 2, the BG network is discussed in detail. In Section 3, the fault tolerance properties of the BG network have been highlighted and compared with other fault-tolerant MINs in [21]. The reliability analysis of the BG network is reported in Section 4. In Section 5, the performance of BG network under SE faults has been discussed, followed by the conclusion in Section 6.

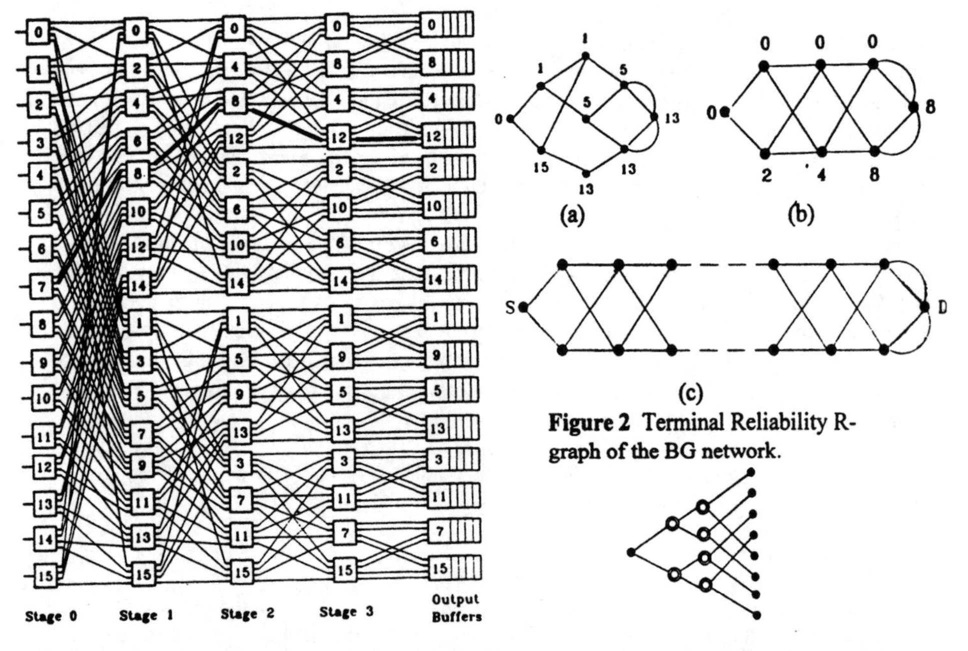

Figure 2 Terminal Reliability R-graph of the BG network.

Figure 1 A 16×16 Balanced Gamma network showing routing of a packet from input port 7 to output port 12.

Figure 3 Reduced Broadcast Reliability R-graph of an 8×8 BG network.

2. THE BALANCED GAMMA NETWORK

The BG network [24] features 4×4 SEs and is derived by enhancing the gamma network. An $N \times N$ gamma network [15] consists of one stage of 1×3 SEs, followed by log_2N-1 stages of 3×3 SEs and finally one stage of 3×1 SEs. Each stage of the gamma network consists of N SEs. The ith SE in the jth stage is connected to SEs i, $(i+2^j)$ mod N, and $(i-2^j)$ mod N, in the $(j+1)$th stage.

An additional link is added to an *i*th SE of *j*th stage of an $N \times N$ gamma network and is connected to SE $(i+2^{j+1}) \bmod N$ in the $(j+1)^{th}$ stage to form the BG network.

Thus an $N \times N$ BG network uses an input stage of 1×4 SEs, $log_2N\text{-}1$ stages of 4×4 SEs and an output stage of buffers, each of which is designed to accept up to 4 packets in each cycle. The buffers are used to store the packets from the network and to feed them to their respective destinations. Each stage comprises of N SEs, numbered 0 to $N\text{-}1$. The stages are numbered 0 to $n\text{-}1$, where $n = log_2N$, as shown in Figure 1. When the output buffers in the BG network are replaced by concentrators, then this network is identical to the kappa network [12]. Each SE in the BG network is addressed as $SE_{i,j}$ $(0 \leq i \leq N\text{-}1,\ 0 \leq j \leq n\text{-}1)$ where i indicates the SE number within a stage and j indicates the stage number.

The BG network uses a reverse destination tag routing scheme [20,21]. The routing tag bits for a cell to be routed from source S to destination D is D in binary. The SEs interpret the tag bits in the reverse order, i.e., the SE in the first stage (stage 0) switches based on the least significant bit (bit 0). In any SE, if the corresponding tag bit is 0 (1), then the packet is routed through either of the two top (bottom) output links. The general routing algorithm employed in the BG network is explained in [21]. An example of a *packet* (referred to as *cell* in ATM nomenclature) being routed from input port 12 to output port 7 in a 16×16 BG network is shown in Figure 1. There are 2^n paths available between any input-output pair in an $N \times N$ BG network [21].

3. FAULT TOLERANCE IN THE BALANCED GAMMA NETWORK

A *fault-tolerant* MIN is one that is able to route packets from input ports to the requested output ports, in at least some cases, even when some of its network components (SEs or links or buffers) are faulty. The *fault model* characterizes all faults assumed to occur, stating the failure modes (if any) for each network component. We choose the following *fault model* to study the *fault tolerance* properties of the BG network :

1. Faults can occur in any network component except the first stage SEs, links connecting the last stage and the output buffers.
2. All faults are random and independent.
3. All faults are permanent.
4. Each faulty component is totally unusable.
5. All faults are detected with success, and the faulty component identified correctly.

The *fault tolerance criterion* is the condition that must be met for the network to be said to have tolerated a given fault or faults. The *fault tolerance criterion* chosen here for the BG network is fulfillment of the condition that there exists at least one path between any input-output pair of the network, which is called as the full access property [2]. The BG network is considered failed when the number and locations of the faulty components prevent the connection between at least one input-output pair of the network. The feature of the BG network that enables *fault tolerance* is the existence of alternate route(s) between every input-output pair in case of network component failures. In order to determine SE failures, each SE continually tests itself, also tests the SEs which are connected to its output links. Using this scheme, failure of an SE and of all the output links connected to it can be combined, which means that we need to consider only SE failures.

The failure of certain pairs of SEs in the BG network will result in loss of full access property for the BG network and so these SEs are called *critical pairs*. The different critical pairs of SEs of a *16×16* BG network are shown enclosed within boxes in Figure 2. The failure of a critical pair in ith stage of a BG network results in packet not being routed to the *(j+1)*th stage.

In the BG network each SE, $SE_{i,j}$, forms critical pair with exactly two other SEs, $SE_{i+2^j,j}$ and $SE_{i-2^j,j}$ (i ∈ {*0,..N-1*} ,j∈{*0,..,n*-1, $(i+2^j)$ and $(i-2^j)$ are mod *N* operations), which are within the same stage of the BG network. There are exactly *N* critical pairs of SEs in each stage of the BG network except in the last stage where the total number of critical pair of SEs is equal to *N/2*. Therefore the total number of critical pairs in an *N×N* BG network is equal to *N×(n-3/2)*. It can be shown that the BG network can tolerate up to *N/2×(n-2)* SE failures without losing full access so long as at least one of the SEs in each critical pair is not faulty [21].

Theorem 1 : *The BG network will lose full access property if and only if any of the critical pairs of SEs fails.*

Proof : Given in [21].

The fault model, fault tolerance criterion and the fault tolerance method used for analyzing the fault tolerance properties of the BG network are the same as those given in [1]. To sum up, the BG network is a *single fault-tolerant* MIN and is robust under multiple faults.

4. RELIABILITY ANALYSIS FOR THE BALANCED GAMMA NETWORK

MINs being considered for use in broadband communication switch fabrics, apart from having high throughput performance, must also be highly reliable as they would be used as real-time communication networks. For real-time systems the principal concerns are time-dependant reliability and the mean time to failure (MTTF) [10].

The reliability of a component (system) is the conditional probability that the component (system) operates correctly throughout the interval $[t_0, t]$ given that it was operating correctly at time t_0 [18]. Exact reliability analysis of a complex system is complicated even under simplifying assumptions of perfect coverage and no repair [5]. The reliability analyses of several MINs have been reported in [5,6,7,14,16,17,23]. There are three basic forms of connections, and consequently three main reliability measures are computed for MINs [18]. These are the terminal reliability, the broadcast reliability and the network reliability. The MTTF which specifies the quality of a system is the expected time that a system will operate before the first failure occurs. The two figures which are usually calculated for MINs are the input-output MTTF and network MTTF. For the reliability analysis the following assumptions are made.

1. The SEs in the first stage and the output buffers are highly reliable.
2. All SEs in the subsequent stages have an identical and constant failure rate of λ, and SE failures are statistically independent.
3. SE faults are permanent and each faulty SE is totally unusable.

4.1 Terminal Reliability

Terminal reliability (TR) is the probability that at least one path exists from a particular input port to a particular output port. TR is always associated with a terminal path (TP) which is a one-to-one connection between an input port (the source) and an output port (the destination). A network is considered failed if it is not able to establish a connection from a given source to a given destination. TR is normally used as a measure of the robustness of a communication network [23].

In the BG network, a packet has an option of being routed through either the regular or the alternate link at each stage. For the evaluation of the terminal reliability of the BG network, the set of paths in the network between the given input-output pair is represented as a directed graph, sometimes referred to as the redundancy graph (or R-graph) [2], with its vertices representing the SEs and the edges representing the connecting links. The R-graph of the BG network is not same for each input-output pair and is dependent on the destination. Certain input-output pairs use only a pair of SEs in each stage in their TP and hence have a lower TR than the other pairs which use more than two SEs at certain stages. The TR of an input-output pair is the lowest when the least significant $\lceil \frac{n}{2} \rceil$ bits are identical. While Figure 2(a) shows the R-graph corresponding to a destination in a *16x16* BG network having the best TR, Figure 2(b) shows an example of the worst case. Figure 2(a) shows the best case TR because it can tolerate upto 2 SE failures in stage 2, while Figure 2(b) shows the worst case TR because it can tolerate only 1 SE failure in stage 2. Figure 2(c) shows the R-graph of an $N \times N$ BG network having the worst case TR. The worst case TR of the BG network is

$$TR(t) = (1-(1-p)^2)^{(n-1)}, \tag{1}$$

where p = reliability of each SE in stages 1 to $n\text{-}1 = e^{-\lambda t}$.

Thanks to rapid advancements in very large scale integrated circuits, component (SE) failure rates of 10^{-6} failures/hour (or better) are quite realistic. The terminal reliability of the BG network for 10 years is given in Table 2. For this and subsequent calculations in this paper, a failure rate of $\lambda = 10^{-6}$ failures/hour is assumed.

One has to remember that the exact terminal reliability is obtained by multiplying the figure in the table by the reliabilities of the components which are assumed to be fault-free, viz., the SE in the first stage, the destination buffers and the links connecting the last stage SEs to the buffers.

Table 1 Terminal reliability of the BG networks.

Network Size(N)	Terminal Reliability
4	0.992956
8	0.985962
16	0.979017
32	0.972121
64	0.965273
128	0.958474
256	0.951723
512	0.945019
1024	0.938362

Table 2 Broadcast reliability of the BG networks.

Network Size(N)	Broadcast Reliability
4	0.985962
8	0.958474
16	0.905776
32	0.808913
64	0.645155
128	0.410382
256	0.166049
512	0.027185
1024	0.000729

4.2 Broadcast Reliability

Broadcast reliability (BR) is the probability that at least one path exists from a particular input port to all the output ports. BR is always associated with a broadcast path (BP) which is a connection from one source to all destinations in the network. BR is usually referred to as the source-to-multiple terminal (SMT) reliability [23. Under this criterion, the network is considered failed when a connection cannot be made from a given input port to at least one of the output ports.

The BR for each input port is identical in the BG network and so the BG network has a uniform broadcast reliability. The equivalent reduced broadcast reliability R-graph for an 8×8 BG network is given in Figure 3 [21]. Each of the composite nodes in Figure 3, shown by a double circle, represents a pair of SEs in the network; the edge between two composite nodes indicates the set of edges between any one of the SEs in the first composite node and any one of the SEs in the second. From the broadcast reliability redundancy graph shown in Figure 3, the broadcast reliability of an $N \times N$ BG network is

$$BR(t) = \prod_{i=1}^{n-1} (1-(1-p)^2)^{2^i} . \quad (2)$$

Table 3 gives the broadcast reliability of the BG network for 10 years.

4.3 Network Reliability

Network reliability (NR) is the probability of maintaining full access capability throughout the network. This measure considers the tolerable average number of switch failures [7]. NR is associated with network path (NP) which is a many-to-many connection, linking sources to many destinations.

The BG network loses full access property only when one or more critical pairs fail. Considering all the possibile combinations of critical pair failures, we arrive at the NR of the BG network as

$$NR(t) = 1 - \sum_{i=2}^{TN} F_i \times (1-p)^i \times p^{TN-i} , \quad (3)$$

where
$TN = N \times (n-1)$ - Total number of simultaneous SE faults.
F_i - Total possible combinations of failure of i SEs causing the BG network to lose full access property where $2 \le i \le TN$.

As N increases the calculation of F_i becomes a very tedious process. The evaluation of NR of a network is known to be an NP-hard problem. Calculation of NR for the BG network for $N=8$ using (3) is done in [21] and this is equal to 0.922639. One alternative approach is to develop approximate methods as the one given in [16]. The lower and upper bounds for NR of the BG network using approximations is given in this paper.

For the BG network to retain full access none of the critical pairs can be faulty. Both SEs in any critical pair are always located within a stage of the network. Therefore, each stage of the BG network has to be reliable for the BG network to be reliable. Stages 1 to $n-2$ are identical as they have N critical pairs and so each of them has the same reliability; stage $n-1$ has only $N/2$ critical pairs and so has a different reliability. Since the reliability of each stage of the BG network is independent of other stages, the NR of the BG network is the product of reliabilities of the each stage (SR) in the BG network. The relability of any stage i $(i \in \{1,2,\dots,n-2\})$ is given by

$$SR(t) = \sum_{i=0}^{N} INF_i p^{N-i} (1-p)^i \quad (4)$$

and the reliability of the last stage (LSR) is given by

$$LSR(t) = \sum_{i=0}^{N} LNF_i p^{N-i} (1-p)^i \quad (5)$$

where INF_i and LNF_i indicate all possible combinations of i SE failures in the intermediate stages and in the last stage respectively, which do not make the BG network lose full access. The NR of the BG network is given by

$$NR_l(t) = SR(t)^{(n-2)} \times LSR(t). \quad (6)$$

The NR of the BG network given by $NR_1(t)$ is slightly less than the actual NR of the BG network because (6) does not take into account the different possible combinations of SE failures in the overall network. The difference in NR given by (3) and (6) is 8.13574×10^{-4} for an *8×8* BG network.

The computation of INF_i and LNF_i is cumbersome for larger values of *N*. Moreover, the values of INF_i and LNF_i for i=0 and 1 have a major contribution to the values of SR and LSR respectively. Therefore we truncate the values of LR and SR at $i = 4$ and get the NR lower bound of the BG network as

$$NR_{LOW}(t) = \left(\sum_{i=0}^{3} INF_i p^{N-i}(1-p)^i\right)^{n-2}\left(\sum_{i=0}^{3} LNF_i p^{N-i}(1-p)^i\right) \tag{7}$$

The computation of F_i is even more complex than INF_i and LNF_i. The values of F_2 and F_3 constitute the major factor in (3). Therefore, the upper bound of NR obtained by approximating NR given in (3) is

$$NR_{UP}(t) = 1 - \sum_{i=2}^{3} F_i \times (1-p)^i \times p^{TN-i} \tag{8}$$

General expressions for INF_i, LNF_i and F_i used in $NR_{LOW}(t)$ and $NR_{LOW}(t)$ are given in [21]. For an 8×8 BG network the values of NR(t), $NR_{LOW}(t)$, and $NR_{UP}(t)$ are 0.92263944, 0.92182587 and 0.94496154 respectively. It can be clearly seen from these values that the NR of an 8×8 BG network is much closer to the lower bound than to the upper bound.

The lower and upper bounds of NR for BG networks of different sizes for 10 years are given in Table 4. It can be seen from the table the lower bound $NR_{LOW}(t)$ of the BG network drops down as the size of the BG network increases. Since NR has been shown to be closer to the lower bound, it can be concluded that the NR of the BG network reduces to quite low values with increase in network size. Since the values of lower bound are nearly zero and those of the higher bound are one for *N*>128, these values have not been included in the above table. It should also be noted that the actual value of NR would be much lower if we take into account the reliabilities of the first stage SEs and of the last stage buffers. However, NR gives the probability that full access is never lost over a period of ten years assuming no repair is possible.

Table 3 Network reliability bounds of the BG networks.

Network Size(N)	$NR_{LOW}(t)$	$NR_{UP}(t)$
4	0.985962	0.986012
8	0.921826	0.944962
16	0.727123	0.971385
32	0.199360	0.999811
64	1.637E-6	1.000000
128	7.27E-19	1.000000

4.4 Input-Output MTTF

The input-output MTTF is the expected time a network will be functional before the failure of at least one of its terminal paths. The input-output MTTF for the BG network is given as

$$MTTF_T = \int_0^{\infty} TR(t)dt\,. \tag{9}$$

Using the TR expression in (1) we have

$$MTTF_T = \int_0^\infty (1-(1-e^{-\lambda t})^2)^{n-1} dt. \quad (10)$$

Table 4 shows the $MTTF_T$ of the BG Network for different values of N.

4.5 Network MTTF

The network MTTF is the expected time a network will be functional before it loses the full access property. The network MTTF for the BG network is given as

$$MTTF_N = \int_0^\infty NR(t)dt \quad (11)$$

Since the exact NR of the BG network is difficult to be computed we use the lower bound given in (7) as this closer to the exact NR than the upper bound given in (8). Subsituting the lower bound experession (7) in (11) we have

$$MTTF_{N-LOW} = \int_0^\infty \left(\sum_{i=0}^{3} INF_i p^{N-i}(1-p)^i\right)^{n-2} \left(\sum_{i=0}^{3} LNF_i p^{N-i}(1-p)^i dt\right) \quad (12)$$

Table 5 shows the MTTF lower bound of the BG Network for different values of N. It can be seen from Table 5, that for $N=1024$, the network loses full access in around 2 months. Therefore, if repair is not possible, each SE of the large sized BG networks should have very low failure rates.

Table 4 Input-Output MTTF (in 10^6 hours) for different sizes of the BG network.

Network Size(N)	Input-Output MTTF
4	1.500000
8	0.916667
16	0.700000
32	0.582143
64	0.506349
128	0.452742
256	0.412421
512	0.380760
1024	0.355094

Table 5 Lower bound Network MTTF (in 10^6 hours) for different sizes of the BG etwork.

Network Size(N)	$MTTF_{N-LOW}$
4	0.916667
8	0.315664
16	0.136992
32	0.063382
64	0.029908
128	0.014223
256	0.006795
512	0.003258
1024	0.001568

5. PERFORMANCE OF THE BG NETWORK IN THE PRESENCE OF SE FAULTS

The various reliability measures explained in the previous section do not provide a clear indication of the dependability of the network. On the other hand, the performance of a network under faulty network components gives us a better insight in this regard.

Performance analysis and simulation results of the BG network under uniform random traffic pattern have been reported in [24] and [20] respectively. Here we present the performance analysis of the BG network in the presence of an SE fault under the uniform random traffic pattern, where one packet is assumed to arrive at each input line during each cycle. The random traffic implies that each output line is equally probable to be requested by the arriving packets.

Let T(N) be the throughput performance of an $N \times N$ BG network without faults under uniform random traffic. Let the failed SE be $SE_{i,j}$ $(j \neq 0)$. The performance of the BG network due to the failure of $SE_{i,j}$ is lower than that of the BG network without the failure of $SE_{i,j}$ because the traffic which normally goes through $SE_{i,j}$ would be routed through the SEs which form critical pairs with $SE_{i,j}$. Due to increased contention at the SEs which form critical pairs with $SE_{i,j}$ the throughput of the BG network drops down. The degradation of performance depends upon the stage in which the SE fault occurs.

5.1 SE fault at stage 1

There is no cell loss at stage 1 of an $N \times N$ BG network if there are no SE faults at stage 1 because each SE in stage 1 receives no more than two cells during each cycle. Due to the presence of an SE fault at stage 1 some cells could be lost at stage 1. Without loss of generality let us consider an SE fault at $SE_{4,1}$ in an $N \times N$ BG network where $N>4$. Due to the SE fault at $SE_{4,1}$, cells which would have normally been routed through it (under no fault condition) from $SE_{4,0}$ and $SE_{3,0}$ would now be routed through $SE_{6,1}$ and $SE_{2,1}$.. $SE_{6,1}$ and $SE_{2,1}$ form critical pairs with the faulty $SE_{4,1}$. Now $SE_{6,1}$ and $SE_{2,1}$ have a possibility of receiving three cells each from stage 0. If all these three cells have the same switching bit for stage 1, then one packet would be dropped. There are four possible combinations in which packets can be lost -- each of $SE_{6,1}$ and $SE_{2,1}$ receiving three packets, where all these three packets have the same switching bit (0 or 1) in each of the SEs. Under full load, 4 out of N incoming packets are lost. The probability $P_{3\text{-stage}1}$ that each of the SEs $SE_{6,1}$ and $SE_{2,1}$ receive three packets and all these three packets having switching bit for stage 1 is given by the expression

$$P_{3-stage1} = \left(\frac{1}{2}\right)^3 \times \left(\frac{1}{2}\right)^3 . \tag{13}$$

Therefore the probability of cell loss at stage 1, $P_{cell\ loss\text{-}stage\ 1}$, is

$$P_{cellloss-stage1} = \frac{4}{N} \times \left(\frac{1}{2}\right)^3 \times \left(\frac{1}{2}\right)^3 . \tag{14}$$

The throughput of an $N \times N$ BG network in the presence of a fault can be approximated as

$$Throughput = T(N) - P_{cellloss-stage1} . \tag{15}$$

The 4×4 BG network constitutes a special case where the stage 1 is the last stage in the network. Due to this fact, throughput of the 4×4 BG network is further degraded by a factor, which can be shown to be 2^{-7} [21]. The throughput of an $N \times N$ BG network in the presence of an SE fault, computed using (15) is given in Table 7. The throughput of the BG network without fault, $T(N)$, under uniform random traffic is taken from [20]. Simulation results of the throughput of the BG network under an SE fault are also provided in Table 6.

It can be clearly seen that the throughput of the BG network in the presence of an SE fault at stage 1 is comparable to that obtained by simulation. It is to be noted that the presence of a single SE fault at stage 1 does not affect the throughput performance of the BG network under uniform random traffic in large networks. This is due to the fact that the probability of cell loss is dependent on the size of the network, as given by (15).

5.2 SE fault at stage j ($j \in \{2,3, ..., n-1\}$)

The throughput analysis of an $N \times N$ BG network ($N > 4$) in the presence of single SE fault in stage j ($j > 1$) is presented in this section. When $SE_{i,j}$ is failed, the packets which are normally routed through $SE_{i,j}$ would now be routed through SEs which form critical pairs with $SE_{i,j}$, which receives packets from four SEs in stage j-1. Of these four SEs, regular links of two SEs are connected to $SE_{i,j}$ and alternate links of the other two SEs are connected to $SE_{i,j}$. Let these SEs be SE1, SE2, SE3 and SE4 respectively. SE1 and SE2 can reroute the packets to be switched to $SE_{i,j}$ through their alternate links. Out of the k packets which arrive at SE3 or SE4 having the same switching bit as that of a packet to be switched to $SE_{i,j}$, k-1 packets are dropped at these respective SEs causing packet loss. This is due to the failure of $SE_{i,j}$. Similarly if SE3 and SE4 have packets to be switched to $SE_{i,j}$ then these packets are dropped at SE3 and SE4 respectively. Therefore, the probability of packet loss at stage j-1 due to failure of $SE_{i,j}$ is

$$P_{CL-STG:J-1} = \frac{4}{N} \times al_{j-1} \tag{16}$$

where al_{j-1} is the probability that the alternate links of stage j-1 carry a packet.
Due to the failure of $SE_{i,j}$ packets are lost even at stage j. If SE1 or SE2 has a packet p_k, which is to be routed to $SE_{i,j}$ then this packet would be routed through the alternate links of SE1 or SE2 causing p_k to go to SEs which form critical pair with $SE_{i,j}$. Let these SEs which form critical pairs with $SE_{i,j}$ be called SEc1, SEc2. If SEc1 and SEc2 already have a minimum of two packets, having the same switching bit as that of p_k for the stage j, then the packet p_k is dropped at stage j. The probability that packet loss occurs at stage j due to the failure of a $SE_{i,j}$ is

$$P_{CL-STG-J} = \frac{4}{N} \times al_{j-1} \times al_j \tag{17}$$

Table 6 Throughput of BG network in the presence of a single SE fault at stage 1.

Networ Size(N)	Throughput without faults T(N)	Throughput with faults by analysis	by simulation
4	1.000000	0.976563	0.978250
8	0.993250	0.985438	0.987125
16	0.985625	0.981718	0.983000
32	0.976250	0.974297	0.974875
64	0.968750	0.967773	0.966391
128	0.959656	0.959168	0.959031
256	0.951191	0.950946	0.950348
512	0.942209	0.942087	0.942387
1024	0.934344	0.934283	0.934273

Table 7 Throughput Analysis of BG network in the presence of a single SE fault at different stages under uniform random traffic.

Networ Size	Analysis results of the performance of the BG network in which SE has failed in stage							
N	2	3	4	5	6	7	8	9
8	0.959	NA	NA	NA	NA	NA	NA	NA
16	0.968	0.967	NA	NA	NA	NA	NA	NA
32	0.967	0.966	0.966	NA	NA	NA	NA	NA
64	0.964	0.964	0.964	0.964	NA	NA	NA	NA
128	0.957	0.957	0.957	0.957	0.957	NA	NA	NA
256	0.950	0.950	0.950	0.950	0.950	0.950	NA	NA
512	0.941	0.941	0.941	0.941	0.941	0.941	09417	NA
1024	0.934	0.934	0.934	0.934	0.934	0.933	0.934	09341

Therefore the overall probability of packet loss due to a failure of $SE_{i,j}$ is the sum of the right hand sides of (16) and (17). Thus the throughput of the BG network under single SE fault at a stage other than it first two stages is

$$Throughput = T(N) - P_{CL-STG-J-1} - P_{CL-STG-J} \tag{18}$$

The throughput performance of the BG network without SE failure, under uniform random traffic *T(N)* is taken from [20]. The values of al_j and al_{j-1} are calculated by the recursive expression given in [24]. The throughput performance of an $N \times N$ BG network using (18) is presented in Table 7. The simulation results closely agree (within 0.2%) with those in Table 7. It can be noted that the throughput of large size BG network is very minimally affected by the presence of single SE fault.

6. CONCLUSION

The fault tolerance properties of the BG network under the chosen fault model and fault tolerance criterion have been studied in this paper. The BG network has been shown to be single fault-tolerant and robust under multiple faults. It has been shown that the BG network fails to satisfy the full access property only due to the failure of certain critical pairs SEs. The reliability analysis of the BG network has been carried out and the different reliability metrics have been computed. From the reliability metrics it can be concluded that either repair should be possible or components used in the BG network should have very low failure rates in order to obtain high network reliability.

It has been shown that the performance of the BG network in the presence of faults gives a better understanding of the dependability of the network than the reliability metrics which have been reported for most of the MINs studied so far. Performance analysis of the BG network in the presence of single SE fault at any stage, under uniform random traffic has been presented in this paper. Results from analysis and simulation exhibit a close match. Performance of the BG network is not significantly degraded when a single SE fault occurs. The throughput performance of the basic BG network is not high enough to meet the specifications of broadband communications. Buffered and enhanced forms of the BG network -- enlargement, dilation and replication. [19,25] -- provide increased throughput and high reliability, thus making them worthy of consideration as potential candidates for use in broadband communication switch fabrics.

7. REFERENCES

[1] Adams, G.B. Agrawal, D.P. and Siegel, H.J. (1987) "Fault-tolerant multistage interconnection networks", IEEE Computer, pp. 14-27.

[2] Agrawal, D.P. and Leu, J.S. (1985) "Dynamic accessibility testing and path length optimization of multistage interconnection networks", IEEE Trans. on Comp., vol. C-34.

[3] Ahmadi, H. et al, (1989) "Survey of modern high-performance switching techniques", IEEE J on Selected Areas in Comm., vol. 7, no. 7, pp. 1091-1103.

[4] Badran, H.F. and Mouftah, H.T. (1990) "Performance of output-buffered broadband switch architectures", Proc. of Canadian Conf. on Elec. and Comp. Engg., no.39.2, Ottawa.

[5] Blake, J.T. and Trivedi, K.S. (1989) "Reliability analysis of interconnections networks using hierarchical composition", IEEE Trans.on Reliability, vol. 38, no. 1, pp. 111-119.

[6] Blake, J.T. and Trivedi, K.S. (1989) "Multistage interconnection reliability", IEEE Trans. on Comp., vol. 38, no. 11.

[7] Booting, C., Rai, S. and Agrawal, D.P. "Reliability computation of multistage interconnection networks", IEEE Trans. on Reliability, vol. 38, no.1, pp. 138-145.

[8] Dias, D.M. and Jump, J.R. (1981) "Analysis and simulation of buffered delta networks", IEEE Trans. on Comp., vol. C-30, no. 4, pp. 273-282.

[9] Goke, L.R. and Lipovski, G.P. (1973) "Banyan networks for partitioning multiprocessor systems", in Proceedings of 1st Annual Symposium on Computer Architecture, pp. 21-28.

[10] Goyal, A. Lavenberg, S. and Trivedi, K. (1987) "Probabilistic modeling of computer system availability", Annals of Operations Research, vol. 8, pp. 285-306.

[11] Jenq, Y.C. (1983) "Performance analysis of a packet switch based on a single-buffered banyan network", IEEE Jof Selected Areas in Comm., vol. SAC-3, no. 6, pp. 1014-1021.

[12] Kothari, S.C. Prabhu, G.M. and Roberts, R. (1988) "The kappa network with fault-tolerant destination tag algorithm", IEEE Trans. on Comp., vol. C-37, pp. 612-617.

[13] Listani, M. and Roveri, A. (1989) "Switching structures for ATM", Comp. Comm., vol. 12, no. 6, pp. 349-358.

[14] Padmanabhan, K. and Lawrie, D.H. (1983) "Fault-tolerance schemes in shuffle/exchange type interconnection networks", Proc. of Int'l Conf. on Parallel Proc., pp. 71-75.

[15] Parker, D.S. and Ragavendra, C.S. (1982) "The gamma network : a multiprocessor interconnection network with redundant paths", Proc. of 9th Annual Symp. on Comp. Arch

[16] Provan, J.S. (1986) "Bounds on the reliability of networks", IEEE Trans. on Reliability, vol.R-35, no.3, pp. 260-268.

[17] Ragavendra, C.S. and Parker, D.S. (1984) "Reliability analysis of an interconnection network", Proc. of the 4th Int'l Conf. on Distributed Computing Systems, pp. 461-471.

[18] Rai, S and Agrawal, D.P. (1987) "Reliability of program execution in a distributed environment", IEEE Trans. on Reliability.

[19] Singh, K.P. and Venkatesan, R. (1993) "Performance evaluation of different buffering schemes for balanced gamma networks", Proceeding of CCECE '93, Vancouver.

[20] Sivakumar. H. and Venkatesan, R. (1995) "Blocking in multistage interconnection networks for broadband packet switch architectures", NECEC '95, St. John's.

[21] Sivakumar, H. (1995) "Performance, fault tolerance and reliability of MINs for broadband packet switch architectures", M.Eng. Thesis, Memorial University of Newfoundland.

[22] Stallings, W. (1992) ISDN and Broadband ISDN, 2nd edition, Maxwell Macmillan Canada.

[23] Varma, A. and Ragavendra, C.S. (1989) "Reliability analysis of redundant-path interconnection networks", IEEE Transactions on Reliability, vol. 38, no.1, pp. 130-137.

[24] Venkatesan, R. and Mouftah, H.T. (1992) "Balanced gamma network - a new candidate for broadband packet switch architectures", IEEE INFOCOM '92, Florence.

[25] Venkatesan, R. (1994) "Performance analysis of kappa networks and enlarged kappa networks under uniform and non-uniform traffic", Proceedings of CCECE '94, Halifax.

[26] Wu, C-L. and Feng, T-Y. (1980) "On a class of multistage interconnection networks", IEEE Trans. on Comp., vol. C-29, no. 8, pp. 694-702.

ATM-based optical access network toward multimedia era

Yasuyuki OKUMURA,
Kazuhito OHNISHI,
Kenji NAKANISHI,
and Tetsuya KANADA
NTT Optical Networks Labs.
1-2356, Take, Yokosuka-shi, Kanagawa, 238-03 Japan
Phone: +81-468-59-5014
Fax: +81-468-55-1283
e-mail: okumura@exa.onlab.ntt.jp

Abstract

This paper proposes an ATM-PON based optical access network architecture and relevant transmission technologies for providing multiple services. The motivation of this network is to realize a service platform and to provide multimedia services as well as POTS and ISDN to both businesses and homes. This paper also describes how the PON architecture will contribute to greater network economy, a smooth evolution from the narrow-band optical access network, and downsizing of the optical network unit (ONU). It also shows the configuration, system parameters, features of the system, hardware, and the experimental results. This ATM-PON system is actually working as a service platform for VOD experiments in our laboratory; more than ten types of servers and settop devices are currently connected. This system is to be used for multimedia experiments in a field environment.

Keywords

ATM, optical access network, multimedia, video on demand, digital CATV, IP routing

1 INTRODUCTION

Dramatic changes in the global marketplace suggest extraordinary telecommunications changes in the not-too-distant future. While basic telephony may remain a centerpiece of telecommunications for many years to come, several new information-networking services promise to have a profound effect on the next-generation telecommunications networks. These new services are being driven by several converging factors. Among them are:changes in the business climate, technology-driven change, and new consumer demands.

These changes validate several new information-networking trends: increased data networking, growth in public-carrier information networking, the use of a variety of diverse network services-on-demand through a common network interface, new applications requiring high-resolution graphics and images, visual telecommunications, and multimedia. In addition, the deployment of new network-based applications requires that networks provide increased service flexibility, a greater variety of services, higher service reliability, and high-speed performance.

At the same time, rapid advances in technology can be seen: the decreasing cost of processing power, decreasing memory costs, the decreasing cost of bandwidth, the increasing use of computing and "user-friendly" graphical interfaces, and the convergence of telecommunications and computing. These changes have not only intensified computing capabilities, but they have also spawned the development of many new applications involving graphical and visual communications. These new applications will greatly intensify interface-bandwidth requirements, a trend that is expected to continue for the foreseeable future.

Given the above situation, we urgently need to construct new computer communication oriented networks that can support various multimedia services. Thus the next generation access networks must provide customers with economical broadband access to any multimedia server and, at the same time, to support the conventional services. As the next generation access network in Japan, Fiber-to-the-home (FTTH) has been and will continue to be intensively pursued with the highest priority. Among FTTH access systems, Asynchronous Transfer Mode-Passive Optical Network (ATM-PON) systems seem to be the most appropriate to provide multimedia services economically and flexibly because ATM technologies are efficient at transporting data for computer communication and because the PON topology is very promising for reducing system cost, e.g. Ballance (1990), Tsuboi (1992), and Takigawa (1993).

In this paper, multimedia service requirements for access networks are first clarified. An ATM-PON system that meets these requirements is then presented together with basic specifications and features. This system will be utilized in multimedia service experiments and the Digital Audio-visual Council (DAVIC) interoperability test.

2 SERVICE CLASSIFICATION

A variety of services will be provided to customers over multimedia networks. Entertainment services such as CATV and VOD are preferred by residential users, while services such as IP routing are preferred by business users. The access network, the infrastructure of the next century, should serve all these users. Services can be categorized into four groups; each places different requirements on the telecommunication network as described in Table 1.

Group A is the contents service group, which includes Video-on-demand (VOD) and other data retrieval services. This group requires downstream access capabilities of up to 6 Mbps which is appropriate for transporting a single MPEG 2 video program. However, the upstream capability is assumed to be 1 Mbps, which requires asymmetric service bandwidth. Residential users will prefer this service group, and business users will also access this service group if useful business information is provided.

Group B is the broadcasting service group which includes CATV services. This service group basically requires the broadcasting bandwidth of about 600 Mbps which is sufficient to broadcast 80 channels x 6 Mbps or 160 channels x 3 Mbps MPEG2 video programs.

Table 1 Service Requirement

Service Type	Group A	Group B	Group C	Group D
Typical service	VOD (Data retrieval)	CATV (broadcasting)	IP routing CSCW Video phone (Data comm.)	POTS / ISDN (Conventiona)
Bandwidth (down)	6 Mbps	622 Mbps (broadcast) 6 Mbps (ch. selection)	1 - 10 Mbps	64 kbps × n
(up)	< 1 Mbps	< 1 Mbps	1 - 10 Mbps	64 kbps × n
Delay	-	-	-	< 1.5 msec (for access sys.)
AAL	type 5	type 5	type 5	other
Characteristics	MPEG 2	MPEG 2	High speed I/F MPEG 1/2	Cell assembly / disassembly

However, to reduce the access bandwidth and thus the cost, only a few channels should be distributed to each customer as requested.

Group C is the two way data communication service group. Examples include broadband IP routing services, Computer-Supported Collaborative Work (CSCW), and Video phones. For example, the broadband IP routing service provides an Ethernet-based UNI to each customer with the 10-Mbps highway being shared by many customers. A customer can access up to 10 Mbps for the tariff of, for example, the 100 kbps bandwidth. The average required access capability for this service is, then, a few Mbps per customer due to bandwidth sharing. The CSCW service, including video conferencing, text and graphics editing, will require around 3 Mbps to support video and conventional data transfer. This service can be achieved by the same access capability as IP routing. Group C will be used for computer-to-computer communication between ordinary customers. The ATM Adaptation Layer (AAL) type 5 will be used for the service groups mentioned above.

Group D is the conventional service group which includes, for instance, telephone and narrowband ISDN services. The bandwidth will be around 64 x N kbps, and will be symmetric for upstream and downstream. This service group usually requires shorter transport delays than the other service groups. For instance, the maximum delay for the access system should be 1.5 ms. The users of this service group are both residential and business users.

3 REQUIREMENTS FOR AN ACCESS NETWORK

3.1 Access network platform

Multimedia networks are based on the architecture which, as shown in Figure 1, consists of information and communication service layers. The latter further consists of the ATM/STM bearer service layer, basic communication service layer such as IP routing and telephony, and the Value Added Network (VAN) service layer. The access network has only the communication service layer function, but it is required to operate as a platform for all types of services including information services. This allows us to define some requirements for the access network.

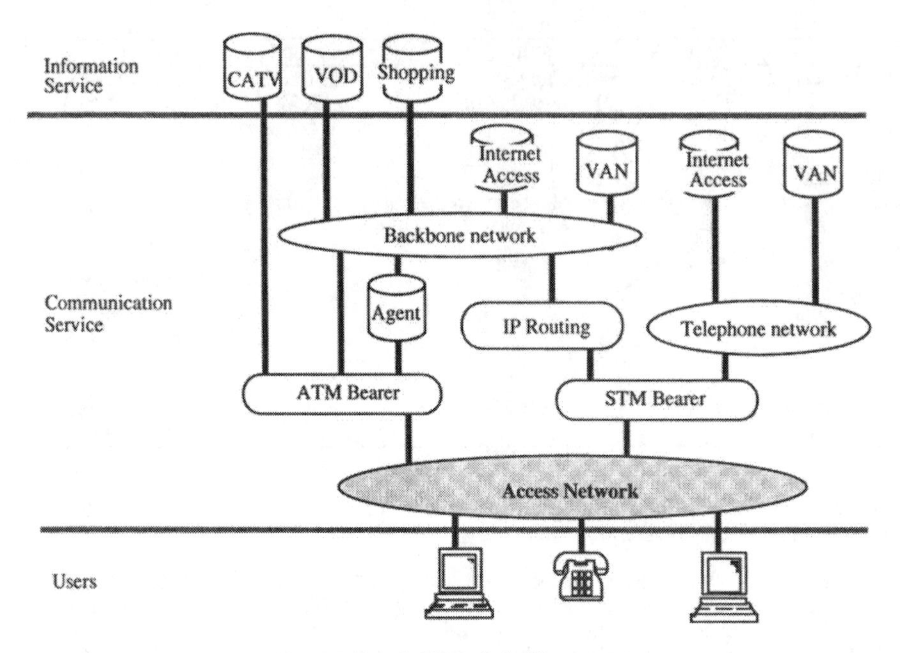

Fig.1 Multimedia Network Architecture

3.2 User Network Interface (UNI)

Group A, B, and C services can be provided in a combined way through a single ATM based UNI as defined by ITU-T Recommendation I. 432 or ATM Forum; different VPs are assigned to different services. In practice, LAN interfaces like 10 Base-T, which correspond to the S Reference Point in the ITU-T standard, may be provided directly.

3.3 Access capabilities

It is assumed that one of the service groups, A, B, and C, plus one conventional service (Group D) must be provided. Thus the access capabilities for one customer is about 7 Mbps including additional capabilities for network use, if only the selected program is transported in the CATV use.

3.4 VP/VC usage

At the UNI, different VPs are assigned to different services. These VPs are connected to corresponding servers through the ATM networks. Therefore, the access networks should basically provide VP connections between UNIs and Service Node Interfaces (SNIs). Between servers and customer terminals, various information streams such as the main contents, signaling, and OAM signals are exchanged by using VCs within the VP. In some services, VC

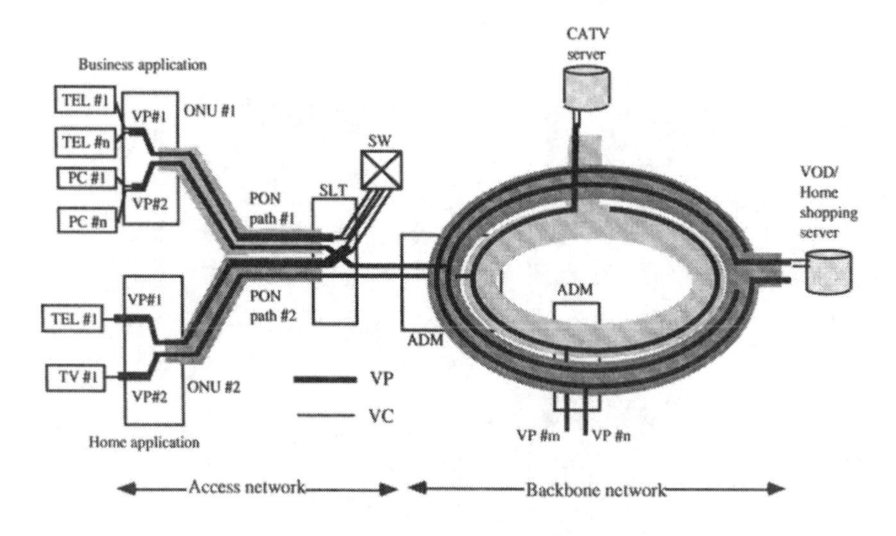

Figure 2 VP/VC allocation example

layer processes such as VC selection and VC multiplexing may be required within the access networks. A possible VP/VC configuration is illustrated in Figure 2.

4 ATM-PON BASED ACCESS SYSTEM

The requirements mentioned in section three have to be satisfied by the access system. Here, an ATM-PON based access system is presented.

4.1 ATM-PON system architecture

The system adopts the PON architecture, which helps reduce network cost, since the SLT and optical fiber are shared by many customers. This also enables a smooth evolution from the narrow-band optical access network using the same configuration. A schematic diagram of the ATM-PON system which meets the requirements mentioned above is shown in Figure 3. The SLT installed in the telephone office contains such functions as an ATM switch for path assignment, PON line terminator (LT), Cell assembler / disassembler (CLAD) to provide conventional services, CATV channel selector, and interface circuit to other intra-office equipment. The Optical Network Unit (ONU) in the customers premises contain such functions as ATM multiplexer (MUX), PON-LT, CLAD, and interfaces to the terminal. The system parameters are shown in Table 2.

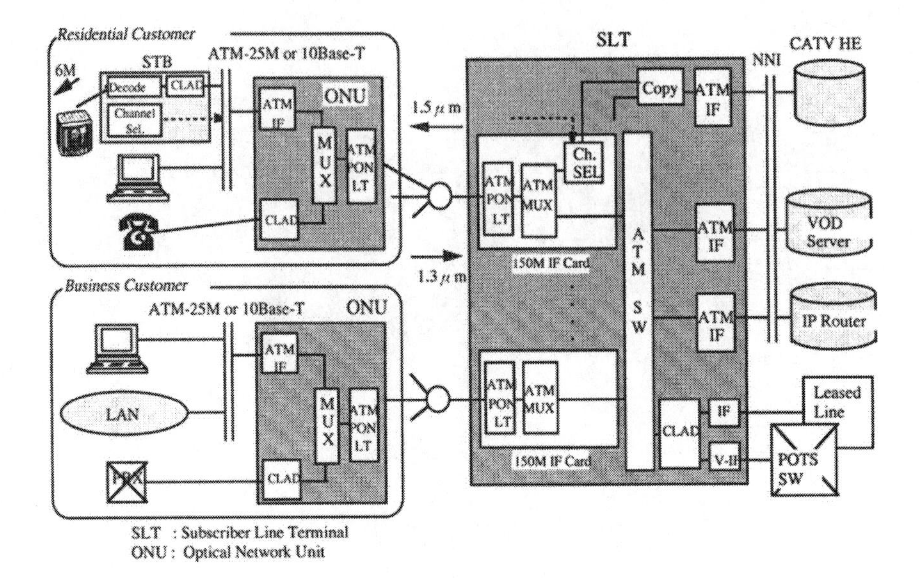

Fig.3 Schematic Diagram of ATM-PON System

Table 2 Outline of ATM-PON system

Items	Contents
Services	ATM, VOD, Digital CATV, IP routing, ISDN/POTS, Leased Line, etc
Fiber topology	PON and Single Star
Line rate	155.52 Mbps
PON payload capacity	SLT: 106 Mbps (Upstream), 118 Mbps (Downstream) ONU (average): 6.6 Mbps (Upstream), 7.4 Mbps (Downstream)
Line code	Scrambled NRZ
Up & Down MUX	Wavelength Division Multiplexing
Wavelength	1.3 μm (Upstream), 1.55 μm (Downstream)
Maximum cable loss	25 dB
Maximum cable length	20 km
SLT capacity	about 500 ONUs / cabinet (1995)

4.2 Frame format

The ATM-PON transmission features are that all downstream signals are broadcasted continuously and an encryption mechanism is adopted to avoid tapping, and that all upstream signals are transmitted as controlled burst signals to avoid collision at the input point of SLT.

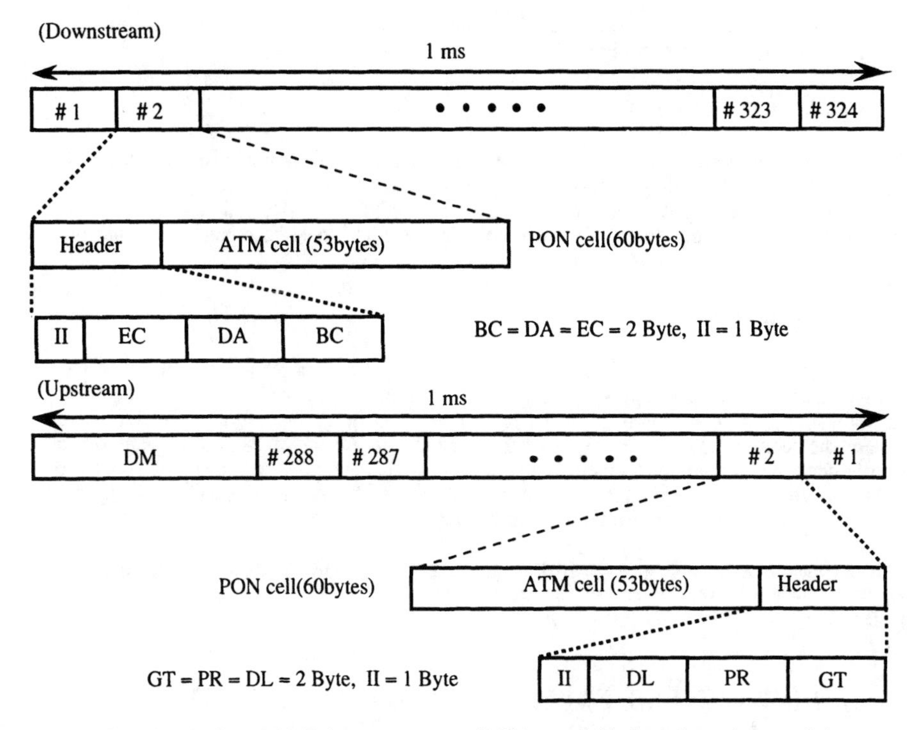

Figure 4 ATM-PON frame format

Regarding upstream flows, the ATM-PON cells from each ONU are received by the SLT as a serial stream of signals, where each signal has its own bit phase and byte phase. Therefore, some bits for bit synchronization (Preamble) and some bits for byte synchronization (Delimiter) are necessary to ensure SLT synchronization with each burst. Since each ONU generates its own burst transmission timing, guard time between the adjacent bursts is also needed to avoid collision. Burst payload contains two types of information: user information and PON-OAM information. The information identification (II) field distinguishes these two types. Consequently, we find that the upstream ATM-PON cell header consists of a preamble (PR), a delimiter (DL), guard time (GT), and information identification (II), as shown in Figure 4.

Concerning downstream flows, a continuous signal is transmitted from the SLT, so that the upstream burst signal overhead bytes such as PR, DL, GT can be omitted. Due to fact that the downstream information is broadcasted to all ONUs, the destination address (DA) must be added to each cell to indicate the destination ONU. Since the information for each ONU should be encrypted, as discussed above, an encryption control (EC) field is added to show the encryption key. In addition, the SLT must indicate the upstream burst transmission timing to each ONU. This requires a burst signal control (BC) field, which contains the ONU identifier. Consequently, we find that the downstream ATM-PON cell header consists of a destination address (DA), encryption control (EC), burst signal control (BC), and information identification (II), as shown in Figure 4.

4.3 CLAD design

One of the most important issues in CLAD design is delay characteristics. To fulfill the delay requirement, the partial filling method is adopted for cell assembly from a regular telephone signal. In this case, one ATM cell conveys only one eighth of its payload to reduce the cell assembly and disassembly delay. The total delay of the ATM-PON becomes approximately 1 ms, comparable to conventional telephone services.

4.4 System features

One of the features of the ATM-PON system is that it provides a simple permanent virtual path (VP) connection that can support all types of services without any complicated signaling software for on-call connection. For the time being, and in the near future, the number of multimedia service customers is expected to be quite limited and most services are of the distribution type. In this environment, the permanent connection of VPs between the customer and the server is simple and realistic. In fact, this ATM-PON system is actually working as a service platform for VOD experiments in our laboratory.

The second feature is that it provides the simple ATM interface specified in ITU-T Recommendation I.432. It requires no carrier modulator/demodulator nor tuner in the STB unlike the Hybrid Fiber Coax (HFC) system.

5 HARDWARE AND EXPERIMENTAL CHARACTERISTICS

The ONU provides ATM and basic ISDN interfaces; it is about the size of a VHS video player, as shown in Figure 5(a). The SLT occupies one shelf, and contains eight PON LT circuits. Four shelves compose one SLT cabinet, which then, contains 32 PON LT circuits, as shown in Figure 5(b). If a 16 branch splitter is adopted in each PON, one SLT set can support up to 512 ONUs.

One of the most important characteristics of the access system is the cell delay and its variation (CDV). The cell delay ranges from 62 to 204 μs, and the CDV is about 140 μs for the upstream cells in the 8 branch case. Regarding downstream flows, the cell delay ranges from 128 to 171 μs, and the CDV is about 50 μs, as shown in Table 3. These values were measured for a 155.52-Mbps interface, and can be regarded as harmless for ATM based multimedia services. The reason for the larger CDV of the upstream cells is that the upstream frame contains the delay measurement field which does not carry any user information cells. This fields duration is 100 ms, and it is used only for measuring the round-trip delay from the SLT to ONU

The other characteristic is the CATV program selection delay. The maximum value measured in the experiments was 300 ms. This value is negligible compared to the software processing delay in the STB.

The bit error rate (BER) was measured, and the results are shown in Figure 6. As shown in this figure, this system is error-free in the 29-dB transmission loss range, which has a 4-dB margin to the objective and covers almost all of the customers from the telephone offices.

(a) ONU (b) SLT

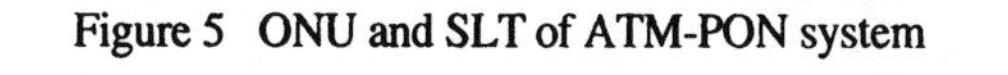

Figure 5 ONU and SLT of ATM-PON system

Table 3 Delay characteristics

Direction	No. of branches per PON	VOD bandwidth [Mbps]	CATV bandwidth [Mbps]	ISDN bandwidth [Mbps]	VOD Delay [μs]			
					Max	Mean	Min	CDV
Down	8	78	48	10	128	146	171	43
	4	106	24	5	128	145	168	40
Up	8	97	--	10	63	87	204	141
	4	88	--	5 Mbps	63	79	187	124

Finally, more than ten types of server-settop pairs are being operationally tested using the ATM-PON system as a platform. In these experiments, MPEG 1 and MPEG 2 transport stream formats are adopted for video coding, and DSMCC-UU for the high layer protocol for video control. Moreover, the ATM-PON will be used in the field trial between two locations; six hundred customers will participate in total. The bandwidth can be flexibly assigned to a customer within the limit of 120 Mbps. Each customer can receive any combination of VOD, CATV, and ISDN services simultaneously provided the total bandwidth does not exceed the limit.

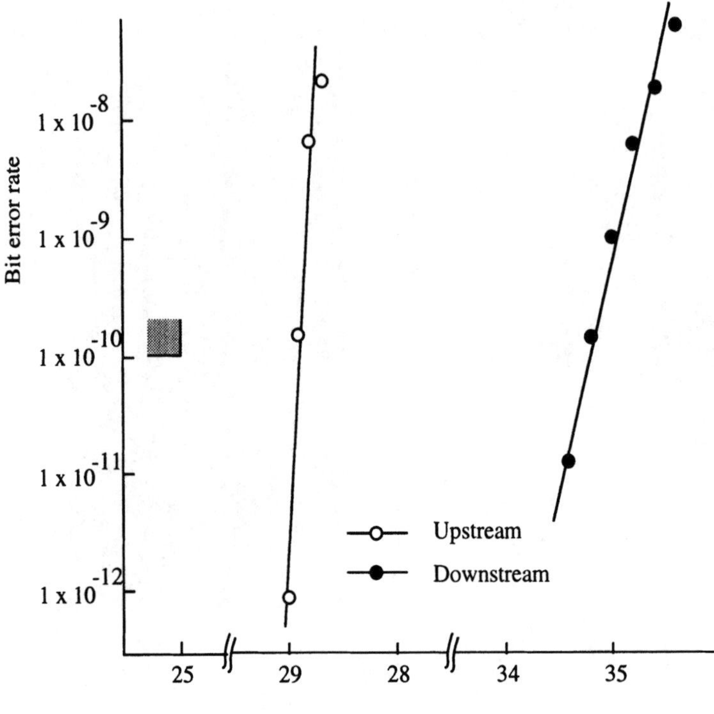

Figure 6 Bit error rate characteristic example

6 CONCLUSION

An ATM-PON based optical access network architecture and relevant transmission technologies were proposed to provide multiple services in a cost effective manner. This system fulfills the service requirements discussed above and is an attractive service platform. This paper also showed the configuration, system parameters, system features , hardware, and experimental results. The ATM-PON system is actually working as a service platform for different pairs ofservers and settops. This paper has provided clear evidence of the architectures advantages and feasibility.

7 ACKNOWLEDGMENT

The authors would like to thank their colleagues especially Dr. Eiji Maekawa, Mr. Yoshihiro Takigawa for their fruitful discussion, and Mr. Shoji Yamamoto and Mr. Eishi Oikawa for their contribution to the experiments.

8 REFERENCES

Ballance, J. W. et al. (1990) A B-ISDN local distribution system based on a Passive Optical Network. *GLOBECOM '90,* **305.5**, 206_210.

Tsuboi, T., Maeda, Y., Hayashi, K. and Kikuchi, K. (1992) Deployment of ATM subscriber line systems. *IEEE JSAC,* **10**, 1448-1458.

Takigawa, Y., Aoyagi, S. and Maekawa, E. (1993) ATM based Passive Double Star system offering B-ISDN, N-ISDN, and POTS. *GLOBECOM ï93,* **1.3**, 14-18.

49

BADLAB^{TM} - A Canadian Broadband Applications and Demonstration Testbed at the Communications Research Centre

J. Michel Savoie, Les Chan, and Robert M. Kuley
Communications Research Centre
3701 Carling Ave., Ottawa
Ontario, Canada K2H 8S2
Tel: (613)998-2489, FAX: (613)998-2753, Email: michel.savoie@crc.doc.ca

James Yuan
OCRInet, Kanata, Ontario

Abdul Lakhani
Telesat Canada, Gloucester, Ontario

ABSTRACT

The Broadband Applications and Demonstration Laboratory is introduced as an R&D facility to demonstrate and test broadband applications and services based on ATM technology. The equipment associated with the facility and its mission are presented along with its connectivity to various network infrastructures, including Satcom. What has been accomplished, the lessons learned, and what lies ahead are described.

1 OVERVIEW

The Broadband Applications and Demonstration Laboratory (**BADLAB**) is an R&D facility, recently established at the Communications Research Centre in Ottawa, which is designed to demonstrate and test Information Highway applications using high-speed Asynchronous Transfer Mode (ATM) technology. The **BADLAB**, together with the various regional and national ATM research networks, plays a significant role in maintaining a Canadian lead in the area of broadband networking.

2 BACKGROUND

The Communications Research Centre (CRC) is Industry Canada's major communications research facility located in Ottawa, Canada. With an annual budget of more than $40 million and over 200 engineers and scientists, CRC conducts leading edge research in a number of key areas including wireless, mobile, satellite, and fibre-optic communications. Early in 1994, a decision was made to establish a central facility or testbed which would develop and test integrated communications systems comprising elements of the above key areas, as well as allow the demonstration of new broadband applications through a variety of telecommunications networks across Canada. With this in mind, the **BADLAB** was created in a 3 000 square foot facility located at the heart of the CRC research campus, and has as its mission the following objectives:

- To test and demonstrate broadband applications over regional, national, and international ATM networks.
- To conduct interoperability and interconnectivity testing between various ATM networks and networking elements.
- To make the facility available to research and educational institutes for trials, demonstrations, lectures, and other collaborative experiments using interactive broadband multimedia.
- To make the facility available to high-technology companies across Canada to assist in the development of applications, equipment, or services which may be of potential commercial value.
- To act as a focal point for the demonstration of CRC-developed broadband technologies.

By all accounts, **BADLAB** has been a success story since it first started operation in March of 1994, conducting or hosting a large variety of groundbreaking broadband network demonstrations within Canada and with other countries in areas such as medicine, education, telerobotics, and engineering. The laboratory has been particularly successful in developing valuable working partnerships with many national and international organizations interested in broadband networking, and this is expected to grow in the coming years as the world's communications infrastructure continues to evolve at a high rate.

3 FACILITY DESCRIPTION

The heart of **BADLAB**'s ATM network is a *Newbridge 36150 MainStreet ATM Access* switch which provides internal and external connectivity as shown in Figure 1. Additional switching capability is provided by a *Newbridge VIVID Workgroup* switch which is attached to the *36150* by means of an OC-3 (155 Mbps) fibre link. The main human interface consists of multimedia workstations connected to the switches via TAXI (100 Mbps) or OC-3. Video and audio streams can also be adapted to ATM cells by an NTSC/JPEG card on the 36150 switch.

A *Cisco 7000* router with an ATM Interface Processor (AIP) is used to route traffic between the ATM network and Ethernet legacy LANs. Virtual ATM ports can be configured on the AIP to support routing of IP traffic over ATM between workstations on different networks. The network management effort can thereby be reduced by setting one PVC from the workstation to the router, with the default route pointing to it at the expense of increased latency due to switching being performed at layer 3.

High quality multimedia places huge demands on a computer system so *Parallax PowerVideo* codecs are used on two of the workstations to off-load the CPU by performing motion JPEG compression and decompression in hardware. A third workstation is equipped with a *SunVideo* card which performs compression in hardware of Cell-B, MPEG-1, and motion JPEG formats. These hardware capabilities are used in

4 NETWORK INFRASTRUCTURES

The **BADLAB** has connectivity to both internal and external broadband networks as well as satellite off-net extension made possible due to a close working relationship with Telesat Canada.

Internal

To facilitate CRC researchers to conduct experiments using ATM broadband networks without the requirement to transfer equipment, fibre was connected to those laboratories which had projects related to the electronic highway. This includes areas such as networking R&D, microelectronics and optical communications, video compression, and satellite communications (Satcom). In addition, the general use facilities such as the auditorium, the conference room, and the CRC Innovation Centre were connected by fibre as illustrated in Figure 2. **BADLAB** serves as the hub of this internal network, and some of these laboratories and facilities have also been equipped with multimedia video conferencing.

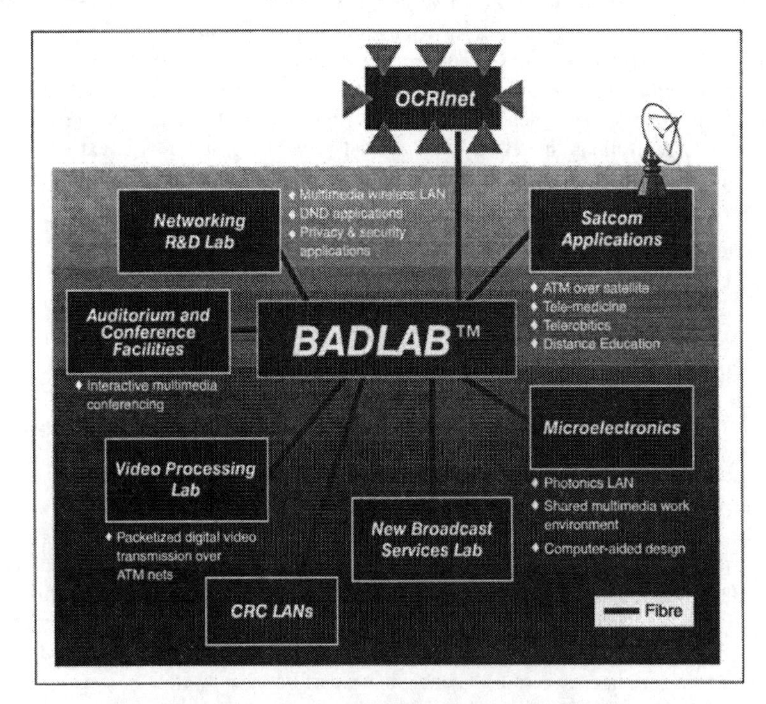

Figure 2. CRC Internal Broadband Network

Regional (Ottawa-Carleton)

CRC is a node of the Ottawa-Carleton Research Institute Network (OCRInet) which is a multivendor ATM metropolitan network linking various major industrial, academic, and government R&D organizations in the region as shown in Figure 3. **BADLAB** access to this network is via two DS-3 (45 Mbps) links. OCRInet has ATM equipment from companies such as Newbridge, Fore, Nortel, Cisco, and General Datacom permitting the investigation of standards-based applications.

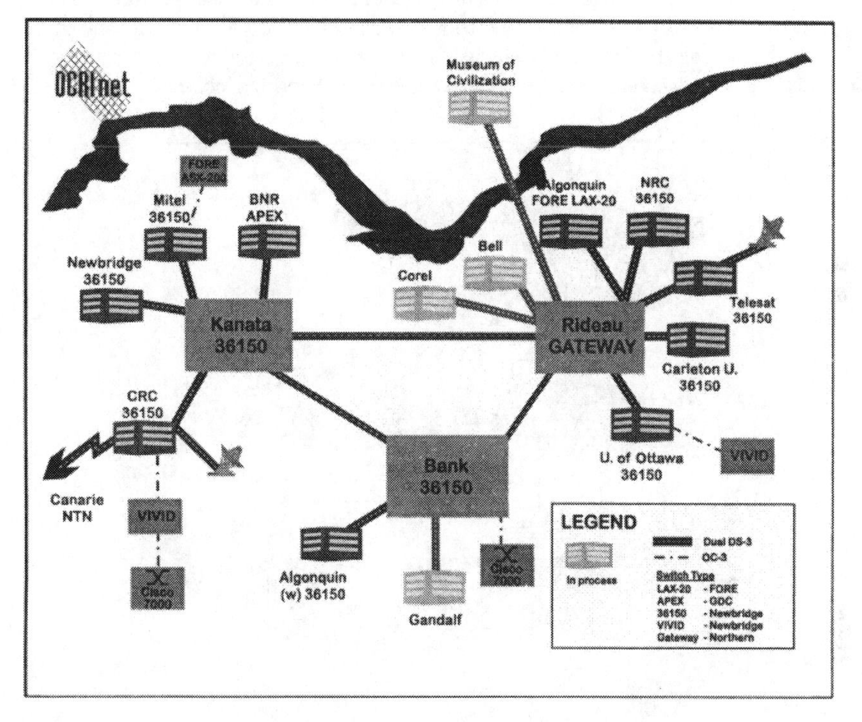

Figure 3. OCRInet Topology

National

OCRInet itself is attached to other Canadian ATM test networks via the CANARIE National Test Network (NTN). Figure 4 shows the present topology of the NTN which offers either DS-3 or OC-3 linkages to the regional test networks. All of the regional networks are in operation with the exception of the Atlantic portion which at the time of writing was still in its formation stages. The NTN is operated by two domestic service providers: Bell Advanced Communications and Unitel.

International

There also exists an OC-3 link from the NTN to the European community via the Teleglobe CANTAT-3 undersea fibre-optic cable. This link facilitates close collaboration between Canadian and European ATM initiatives, and several international demonstrations have already taken place. The landing sites available through the conjunction with desktop conferencing software, which provide an interactive multimedia environment to share the workspace in an effective manner. Software packages used in the **BADLAB** include: *InSoft Communique!* with the *SHARE* option to support collaborative work (e.g. collaborative editing, engineering, etc.); *Sun ShowMe* and *SharedApp* with similar features to *Communique!*, but limited to the *Sun* platform with the *SunVideo* card; *IMA* developed in Germany for telemedicine applications with shared access to medical databases; *Isabel* developed under the RACE program as a computer based distributed collaboration platform; and the *Mbone* conferencing toolset.

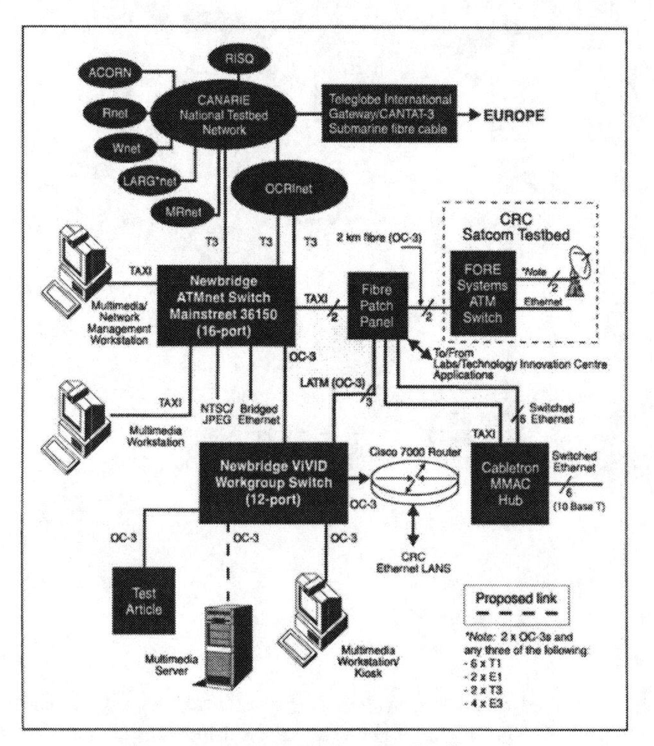

Figure 1. BADLAB Network Configuration

BADLAB also has access to ATM test equipment including: an *Adtech AX-4000 ATM Analyzer* configured with OC-3 and T3 interfaces; and a *GN Nettest InterWATCH 95000 ATM/LAN/WAN Analyzer* configured with Ethernet, Token Ring, T1, E1, T3, E3, and OC-3 interfaces. A *Desknet DS3port hand-held Analyzer* is available to OCRInet members on an 'as-required' basis.

5 ACCOMPLISHMENTS

To date **BADLAB** has been used to demonstrate broadband applications in the following fields:

- **Education:** Distance education has been demonstrated on several occasions. Professor David J. Wright from the University of Ottawa has given lectures on ATM technology to students from the Simon Fraser University and the University of Calgary using the Newbridge NTSC/JPEG service adaptation card. The bandwidth of the virtual circuits associated with the video and audio streams were set at 13 and 1 Mbps respectively. Professor Wright also gave a similar lecture to NorthwesTel engineers located in Whitehorse using the *InSoft Communique!* desktop conferencing application over a T1 ATM Satcom link. Using the NTSC/JPEG card offered good video and audio quality at the expense of higher bandwidth and the need for an external echo cancellation device (e.g. *Coherent Voicecrafter 3000*). The *Communique!* environment added the interactive shared workspace dimension and built in echo cancellation with marginal video and audio quality, albeit at a lesser bandwidth. About 16 fps at 640 X 480 pixels resolution was achieved within 0.8 to 1.2 Mbps depending on the quality factor used.

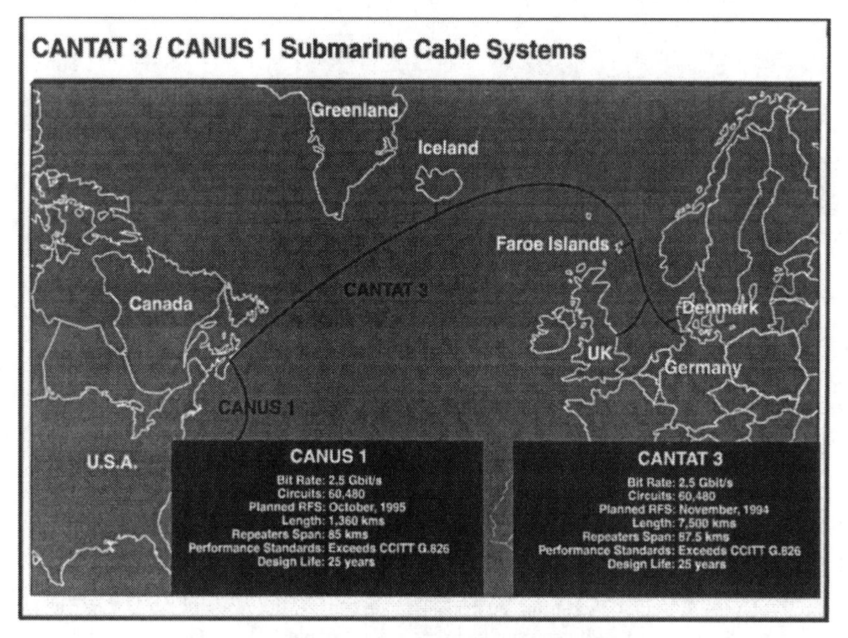

Figure 5. Teleglobe CANTAT-3 Landing Sites

CANTAT-3 facility are depicted in Figure 5 and efforts are underway to forge similar links to the United States, Japan, and other Pacific Rim countries.

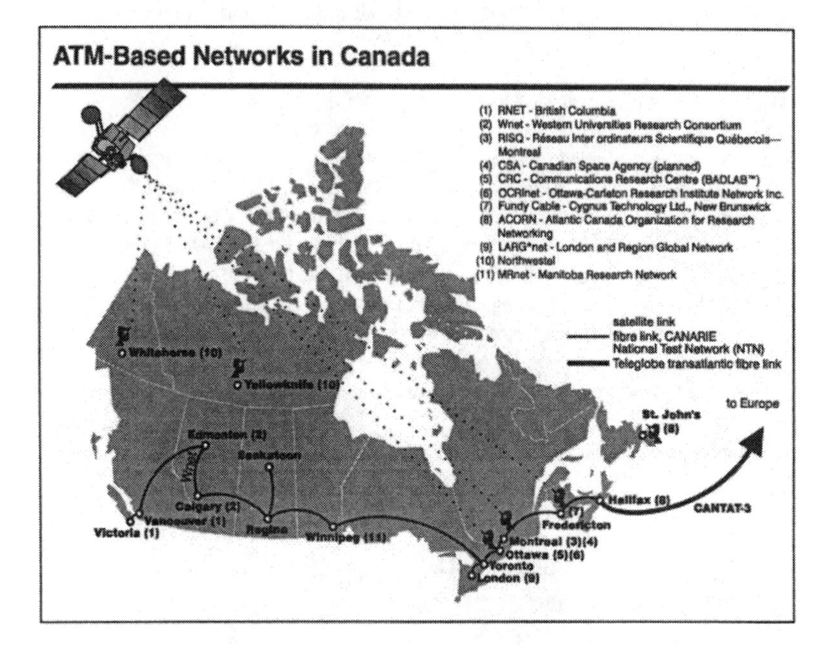

Figure 4. CANARIE National Test Network

Satcom

Satellite communication is allowing to extend the reach of the Canadian ATM initiatives beyond the metropolitan centres. Both the CRC and Telesat have ATM satellite capability, with CRC's Satcom Testbed facility equipped to transmit and receive broadband and narrowband signals in both commercial bands (Ku/12-14 GHz and C/4-6 GHz) as well as the experimental Ka/20-30 GHz band. This facility is equipped with a *Fore ASX-200* ATM switch which is connected to the **BADLAB** via OC-3 links. The Fore switch can be configured to support T1 (1.5 Mbps), E1 (2 Mbps), E3 (34 Mbps), T3 (45 Mbps), and OC-3 (155 Mbps) interfaces to transmit and receive data over a satellite link using appropriate satellite modems. *ATM Access Concentrators* (AACs) from ADC Kentrox have also been used to convert Ethernet packets and streams from H.261 codecs into ATM cells for transmission over satellite. Typical equipment configurations to provide ATM network extensions over satellite are shown in Figure 6.

Through these and planned external network infrastructures, the researchers at CRC can collaborate with similar facilities virtually anywhere in the world.

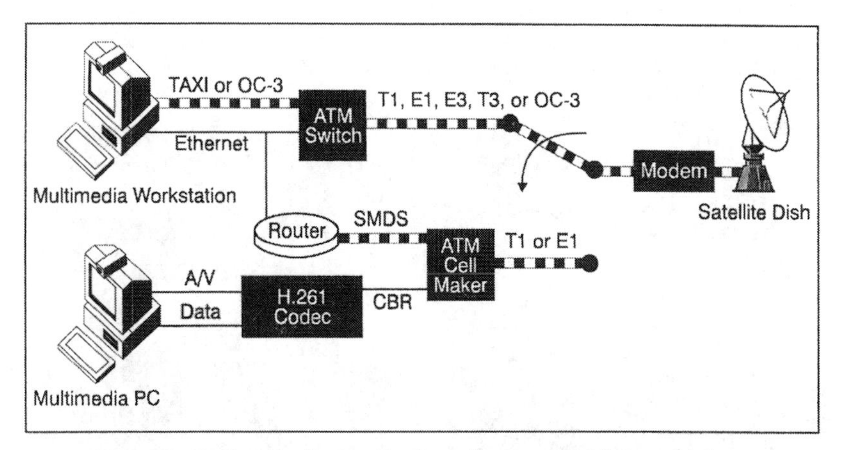

Figure 6. Typical Equipment Configuration For ATM Satcom Links

- **Medical:** Telemedicine using interactive multimedia conferencing tools supporting patient diagnosis and physician consultations between different parts of the country and European medical institutes have been demonstrated. Doctors from the Ottawa Heart Institute consulted with their colleagues from the Berlin Heart Institute at the G7 Summit held in Brussels in February, 1995 and at TELECOM '95 held in Geneva using Teleglobe's Cantat-3 facility. The *IMA* software was used to exchange biotelemetry information from the EVAD artificial heart and high-quality medical images as depicted in Figure 7. Although this environment provides good quality video, it is at the expense of lower frame rates and does not support echo cancellation. The shared pointer tool is somewhat limited in functionality and exhibits a slow response when used with the shared medical images.
- **Telerobotics:** Aastra Aerospace demonstrated the remote control of a robotic vehicle mounted with a camera to achieve telepresence. The robotic vehicle was located in the **BADLAB** while being controlled from the Canadian Space Agency's headquarters in St. Hubert, Quebec over a T1 ATM Satcom link. Personal computers were used to implement the control scheme while workstations running the *InSoft Communique!* software were used to provide the video feedback. The Satcom link, using Telesat's Anik E satellite, provided the desired transport delay needed to evaluate and demonstrate algorithms suitable for controlling a robotic arm on the planned international Space Station from the ground. The experiment proved the feasibility of ground-based control of a robotic device in space and the use of multimedia applications to enhance the human interface. A similar demonstration was held with McGill University as one of the control sites.
- **Tourism:** MPR Teltech demonstrated an ATM Client/Server application developed using their *ATM Vision* multimedia authoring tool. A tourist kiosk located in the **BADLAB** was used to access a multimedia server located at Newbridge by means of native ATM calls to set up the necessary switched virtual circuits. The application allowed a prospective traveller to navigate through video clips of possible vacation and

resort destinations and to engage in a live video conference with a travel agent to finalize a business transaction. Further work is underway to adapt this technology to support interactive multimedia retrieval systems for government services.

Figure 7. International Telemedicine Demonstration

- **News:** An electronic version of the Ottawa Citizen newspaper developed at the University of Ottawa was accessed from the **BADLAB** using a modified MOSAIC environment which supports on demand video and audio clips that are played as the data is being received in real time. This demonstrated how bandwidth intensive applications could be readily ported to run over an ATM network with improved performance and increased user satisfaction.
- **Public Relations:** Virtual presence to the **BADLAB** has been demonstrated using the *InSoft Communique!* desktop conferencing application over terrestrial or satellite links to promote the CRC and the **BADLAB** at trade shows and major events. This has been very successful as a marketing tool for CRC technologies and R&D services. Also the *Isabel* application was used to achieve virtual presence to the EXPLOIT Open House held in Basel in November, 1995. *Isabel* provides good video at low frame rates, doesn't support echo cancellation, but supports slide presentations with an effective shared pointer.

Figure 8 depicts multiple applications (distance education, telemedicine, telerobotics, and Internet access) which have been demonstrated in collaboration with Telesat over an E1 ATM Satcom link.

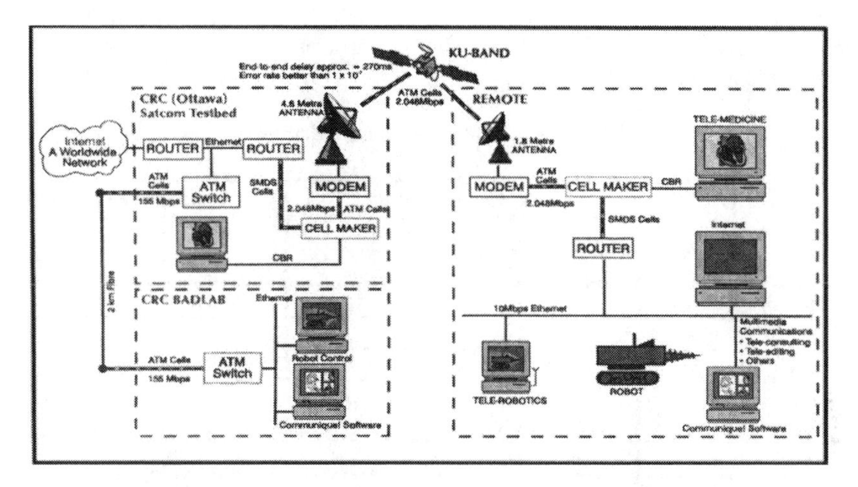

Figure 8. Multiple Applications Over E1 ATM Satcom Link

6 TESTS CONDUCTED AND CONCLUSIONS

Satcom Characterization

A high bit error rate is prevalent in a satellite environment which consequently can be quite detrimental to ATM cell transmission as higher data rates are being contemplated for ATM Satcom links. Telesat has experimented with T1 and E1 ATM Satcom links to assess the impact of burst errors and how to overcome their effect (Lakhani, 1995). One solution has been to optimize the design of the Satcom links to support multimedia applications. The other approach is to distribute the effect of the burst errors over longer intervals of time so that the cell error correcting mechanism can correct one bit errors in the header cell and the higher layer protocols correct the errors in the payload. The former approach has been very successful for multimedia applications at these lower data rates. The later approach may be necessary at higher data rates. CRC engineers are investigating this issue and have conducted ATM Quality of Services (QoS) measurement tests to characterize a T3 ATM link over Telesat's Anik E satellite between CRC and the USAF Rome Laboratory in New York (Nourry, 1995). Cell Loss Ratio (CLR), Cell Miss-insertion Ratio (CMR), and Cell Error Ratio (CER) measurements were gathered for QPSK ¾ and 8PSK 2/3 modulations using the EF Data Modem. The *Adtech AX-4000* was used to generate and analyze the data. More work is in progress using a T3 ATM link over the U.S. Advanced Communications Technology Satellite (ACTS) operating in the experimental Ka/20-30 GHz band.

Interoperability

Tests were conducted to identify the specific RFC 1483 encapsulation schemes supported by the variety of ATM equipment in the **BADLAB**. The results have been tabulated in an interoperability matrix which is posted on OCRInet's Web site (http://www.ocrinet.ca/ocrinet/r1483.html) and included as Table 1. This matrix is continually being updated as new results become available. The *Adtech AX-4000* was used to analyze the data obtained through the course of the interoperability tests conducted in the lab.

Table 1. RFC 1483 Compliance Matrix

Device Type	LLC Encapsulation			VC Mux Encapsulation	
	Routed	Bridged		Routed	Bridged
		FCS	no FCS		
Fore SBA-200 NIC	yes			yes	
Cisco 7000 AIP	yes			yes	
VIVID S-Bus NIC			yes		
Newbridge 36150 Ethernet Card			yes		
GDC APEX			yes		

Lessons Learned

In trying to interconnect a variety of equipment and running applications over an ATM network we have learned the following:

- **Traffic Shaping:** When using AAL-5 to carry TCP/IP transactions, the variable bit rate traffic is quite bursty and can result in discarded cells if adequate precaution is not taken. A peak rate must be set at the end stations when: Network Interface Cards (NIC's) are of different transmission rates (e.g. OC-3 and TAXI), allocated bandwidth within the network is less than the transmission rate of the NIC and policing is enabled, or the network trunk cards have a lower transmission rate than the NIC's.
- **Traffic Policing:** The averaging algorithm used to implement traffic policing is not standardized. When connecting equipment from different vendors with the same peak rate being set throughout the connection, discarded cells have been observed. To get around this problem, the peak rate within the network was doubled w.r.t. the peak rate set at the end nodes.
- **RFC 1483/1577:** These recommendations are being loosely interpreted by the manufacturers and the associated terminology being used is inconsistent amongst the vendors. With RFC 1483, there is a multiplicity of choices within the recommendation and usually a manufacturer will choose to implement only one of the choices which may not be the one opted for by another manufacturer resulting in lack of interoperability.

7 FUTURE WORK

As the existing standards evolve and new ones emerge there will be some growing pains and interoperability issues for years to come especially in the areas of signalling, LAN Emulation (LANE), Next Hop Resolution Protocol (NHRP), Multi Protocol Over ATM

(MPOA), traffic management, etc. Interoperability testing will play a key role and the **BADLAB** is well positioned to continue and extend this aspect of its mandate to conduct distributed interoperability testing over ATM networks with its collaborating partners such as MPR Teltech, NorthwesTel, Teleglobe, Telesat, and TRLabs, as well as with international organizations. New applications that are being developed which can utilize the full features of ATM by specifying the required QoS through an appropriate Application Programming Interface (API) to the NIC's will require more testing to select the correct QoS to achieve optimal performance for a given application.

Other related R&D projects which are ongoing at CRC and which will be demonstrated in **BADLAB** during the coming months in the context of broadband ATM communications include the development of wavelength-division multiplexing optical networking technology to increase the information-carrying capacities of existing LANs; experimental trials to transport MPEG-2 encoded data over the networks in order to evaluate the QoS offered by ATM; the use of infrared optical and EHF radio techniques for wireless broadband terminal connectivity; and EHF Satcom for the transport of broadband multimedia to remote locations.

8 REFERENCES

Lakhani, A. and Poulos B. (1995) Can ATM Technology Work on Satellites? Yes! It Can! In Proceedings of the Telecon '95 Conference sponsored by the Canadian Business Telecommunications Alliance (CBTA), Vancouver, Sept. 10-14, 1995.

Nourry, G. (1995) An ATM-based C2 Demonstrator. In Proceedings of the Command and Control Conference sponsored by the Canadian Defence Preparedness Association (CPDA), Ottawa, Sept. 26, 1995.

50
Traffic Management and Priority Control in an ATM Test-Bed

Antonio DeSimone, Bharat T. Doshi, Subra Dravida, Hongbin Ji
AT&T Bell Laboratories
Holmdel, NJ 07733-3030

Abstract

A field grade ATM test bed has been built to understand the role of ATM technology in providing network infrastructure for transporting current and emerging services. Physical, ATM, AAL, and application level testing and measurements are being used to obtain this understanding. One aspect undergoing investigation is the role of cell level traffic controls. In this, we discuss the testing, measurements, and analysis to investigate how ATM traffic controls deployed in switches can be used to manage the resources in a broadband ATM network. Traffic contracts based on UPC and traffic policing at the network edge have been tested to understand how cell-level controls on a per-virtual-circuit basis affect CBR and data applications. The effects of priority controls as implemented in the test-bed have also been studied to understand how diverse applications (voice over CES, bursty data) can share network facilities.

Keywords

Asynchronous transfer mode (ATM), Broadband ISDN, Traffic management, TCP/IP, Priority control.

1 INTRODUCTION

Broadband ISDN networks are being developed to carry multimedia traffic, including data, audio and video (McDysan and Spohn, 1995). Providing multiple services over Broadband ISDN networks requires characterization of the traffic demands from the applications, both to understand the effects of impairments on the applications and to understand how ATM-based traffic control mechanisms can be used to manage resources in the network and to provide quality of service (QoS) required by individual applications.

Generally speaking, the traffic may be categorized into non-real-time traffic (e.g. bursty data transfer) and real-time traffic (e.g. circuit emulation and video services). As indicated by their names, non-real-time can tolerate longer delay while real-time applications usually have strict requirements on cell delay variation (CDV). Furthermore, non-real-time applications may be able to tolerate larger cell loss probability since retransmission procedures such as go-back-N and selective repeat (SR) protocol may be used to provide recovery from cell losses. However, for some real-time services, strict requirement on cell loss probability may be necessary since error recovery is impossible.

Specifically, the traffic may be classified into five categories: constant bit rate (CBR), real-time variable bit rate (RT-VBR), non-real-time variable bit rate (NRT-VBR), unspecified bit rate (UBR) and available bit rate (ABR). Correspondingly, the architecture for services provided at the ATM layer consists of these five service categories (ATM Forum, 1995).

The CBR service category is intended for real-time applications which require tightly constrained delay and delay variation. Cells which are delayed beyond the value specified by cell transfer delay (CTD) will be of significantly reduced value to the application. Typical applications for CBR are interactive video and audio, and private line service (Circuit Emulation) at various bandwidths.

The real time VBR service category is intended for real-time applications which require tightly constrained delay and delay variation and sources are expected to transmit at a rate which varies with time. In other words, the source can be described as bursty. Typical applications for RT-VBR are interactive audio and video for which the end-system can benefit from statistical multiplexing by sending at a variable rate and can tolerate some cell losses (e.g. MEPG2).

The non-real-time VBR service category is intended for non-real time applications which have bursty traffic characteristics and can be characterized in terms of peak cell rate (PCR), sustainable cell rate (SCR), and maximum burst size (MBS). Typical applications for NRT-VBR are response time critical transaction processing (e.g banking transactions) and Frame Relay interworking over ATM.

The unspecified bit rate (UBR) service category is intended for non-real-time applications which do not require tightly constrained delay and delay variation. Examples of such applications are traditional computer communications applications such as file transfer and email.

The available bit rate (ABR) service category is intended for the applications for which the end-system will control the source traffic rate in response to changing ATM layer transfer characteristics. It is expected that an end-system which adapts its traffic in accordance with the feedback will experience a low cell loss ratio and obtain a fair share of the available bandwidth according to the network specific allocation policy. Typical ABR applications are data communications applications which can tolerate significant delay but can suffer adversely from cell losses (e.g. file transfers in which each cell loss may result in retransmission of one or more packets which are much larger than a cell).

Our ATM test-bed includes large central-office ATM switches, small ATM-LAN switches, traditional telephony equipment (PBXs), data networking equipment and a variety of end systems running applications. We have deployed applications based on Circuit Emulation, several VBR video and data applications, and LAN interconnection.

For this set of applications, one of the objectives is to relate traffic characterization to call admission control rules. Given the traffic characterization for various sources, what combination is simultaneously allowable into the network that can meet QoS requirements of all the admitted sources.

The main challenge arises from the need to handle very bursty and highly unpredictable data sources along with highly predictable and smooth Constant Bit Rate (CBR) sources and with other relatively smooth data sources. Two simple mechanisms used to manage diverse traffic sources are call acceptance/denial controls and traffic policing based on a leaky-bucket traffic characterization, and a system of loss/delay priorities in the network switches. These controls are appropriate to CBR and VBR services while ABR services rely largely on reactive controls. We focus here on experimental tests of two controls above for circuit-emulation applications using CBR services and data applications using VBR services, although at a higher level reactive controls may be implemented, for example in the end-to-end transport protocol, with TCP as the dominant example.

2 TEST-BED CONFIGURATION

We have deployed test-beds with ATM network equipment and end systems to exercise the technology both from a functionality and performance point of view, and to understand the operational and management aspects of ATM in a multi-carrier environment. Figure 1 shows the configuration of our ATM Test-bed. The networking equipment is focused around an AT&T Globeview 2000 ATM switch, with an aggregate switching capacity of 20 Gb/s. Smaller ATM switches from Fore Systems and GDC and a variety of terminal adapters have been deployed to provide interfaces for workstations and for circuit-based equipment such as PBXs. Figure 1 shows the configuration inside one laboratory. Similar configuration at other laboratories connected to this one via DS3 and/or OC-3 links between GV-2000 provides a field grade ATM test bed. We can also loop back to the laboratory in Figure 1.

3 TRAFFIC CHARACTERIZATION

Given the dominance of TCP in the end-system protocols for data applications, one of our objectives here is to study TCP behavior over the ATM network with UPC functions. First, we need to understand the traffic generated by TCP software for workstations directly attached to an ATM network to determine what leaky-bucket parameters characterize the traffic stream. In particular, it is important to understand the PCR, CDVT, SCR and BT (Burst Tolerance) parameter settings (defined below) that do not degrade TCP throughput, as a function of the TCP protocol parameters, access line speed and the channel conditions in the network.

As defined by ATM Forum UNI specification version 3.0, the traffic is described by the following parameters: a mandatory peak cell rate (PCR) in cells/second in conjunction with a cell delay variance tolerance (CDVT) in seconds, and an optional sustainable cell rate (SCR) in cells/second in conjunction with a maximum burst size

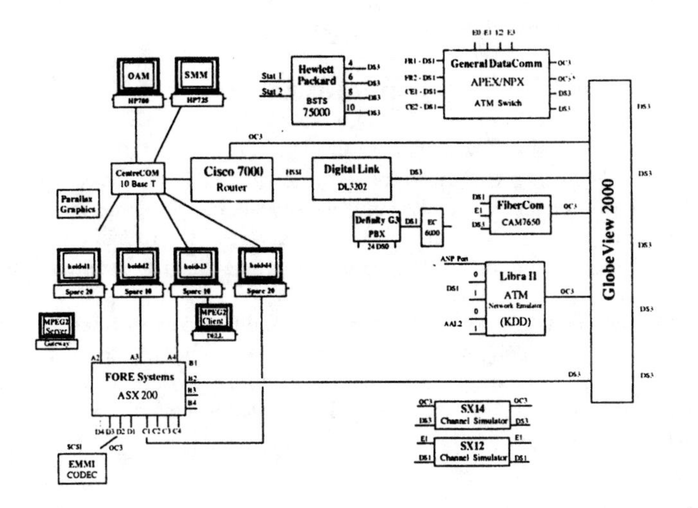

Figure 1 Configuration of ATM Test-Bed

(MBS) in cells. The above parameters are given by PCR $= \frac{1}{T}$ with T the minimum inter-cell spacing in seconds (i.e. the time interval from the first bit of one cell to the first bit of the next cell). The CDVT is defined as the maximum variation in delay over the access network. CDVT $= \tau$ is the maximum cell delay variation expected, and is used to size the bucket in the UPC using the generic cell rate algorithm (GCRA). Thus, the number of cells that can be sent back-to-back at the access line rate is $\frac{\tau}{T} + 1$.

For a periodic on-off source transmitting at the PCR during the on period, the SCR is the maximum average rate that can be sustained. MBS is the maximum number of cells that the on-off source can send at the peak rate.

3.1 Configuration for Traffic Characterization Measurements

We collect the data for traffic characterization of TCP using a pair of Sun SPARC-station 10 workstations which host Fore adapter cards, and are connected to a Fore ASX-200 ATM switch over 140 Mb/s TAXI links. The AT&T Globeview 2000 is a central-office ATM switch, connected to the Fore switch over a 45 Mb/s DS-3 link. We use a DS-3 channel simulator to control the channel delay and error characteristics, thus mimicking long propagation delays as well as transmission impairments. We specify the MSS of TCP as 9148 bytes. Both of the sending and receiving buffer size of TCP is equally set as 16K, 24K, and 48K bytes. The UPC parameters are studied at the input to the Globeview 2000 switch.

3.2 UPC-based Traffic Characterization for TCP

The PCR and SCR traffic parameters are formally defined in terms of a virtual scheduling algorithm, called GCRA, which is equivalent to a leaky bucket algorithm, in ITU/CCITT Recommendation I.371 and the ATM Forum UNI specification.

The PCR is modeled as a leaky bucket drain rate and the CDVT defines the bucket depth for peak rate conformance checking on either the CLP=0 or the combined CLP=0+1 flows.

The SCR is modeled as a leaky bucket drain rate, and the burst tolerance, which is proportional to the MBS, defines the bucket depth for sustainable rate conformance checking on either the CLP=0, CLP=1, or CLP=0+1 flows. The burst tolerance, which defines the SCR bucket depth, is defined by the following formula in the ATM Forum UNI version 3.0 specification:

$$BT = (MBS - 1)\left(\frac{1}{SCR} - \frac{1}{PCR}\right) \tag{1}$$

The burst tolerance, or bucket depth, for the SCR is not simply the MBS because the sustainable rate bucket is draining at the rate given by SCR. The formula above is simply calculating the bucket depth required for MBS cells arriving at a rate of PCR while draining out at the rate of SCR. The impact of CDVT on the SCR is to increase the MBS.

Given PCR, SCR, and bucket size, we use the virtual scheduling algorithm to calculate the number of non-conforming cells in the traffic stream. Thus, we may compute cell loss probability for the given traffic stream.* We can then ask the question, what is the minimum bucket size to guarantee a given cell loss tolerance?

As an example, Figure 2 illustrates minimum bucket size versus SCR for cell loss rates of $10^{-1}, 10^{-2}$ and 10^{-3} for the case there is no channel delay and the TCP/IP send/receive buffers are 16 Kbytes. The figure confirms that the minimum bucket size which is required to meet a given cell loss requirement decreases as the drain rate increases. If the cell loss tolerance is relaxed, the minimum bucket size will also be reduced. What is striking is that for modest loss rates† a well defined breakpoint exists in the BT-SCR curve.

In Table 1, we summarize leaky-bucket traffic descriptors under different TCP/IP sender/receiver buffer and channel delay. SCR is chosen so that BT will not decrease significantly when the drain rate is increasing. In other words, for each cell loss value

*It is important to note that this procedure gives us only an estimate of the cell loss probability for the *given* traffic stream. In a real system, the TCP would change its behavior in response to loss. This effect is what we measure when we do UPC tests with a real TCP source.

†While a loss rate of 10^{-2} is not normally considered "modest" in the context of Broadband ISDN, here we are talking about TCP, and on the Internet, loss rates of 5% are common under normal operating conditions.

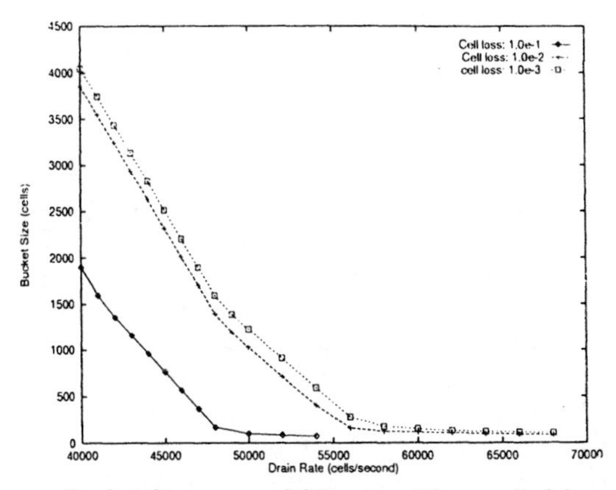

Figure 2 Minimum Bucket Size versus SCR rate with no cell delay and bit error and buffer size of 16K bytes

TCP s/r buffer	Delay (ms)	TCP (Kbytes/s)	SCR (cells/s)	BT (cells) for CLR 0	10^{-3}	10^{-2}	10^{-1}
16 Kbytes	0	2297.60	58000	180	177	125	70
	125	62.76	1400	340	336	330	298
24 Kbytes	0	2675.18	61000	520	510	168	85
	125	70.94	1500	450	445	360	320
48 Kbytes	0	3707.07	84000	845	842	637	150
	125	162.83	3500	910	908	843	678

Table 1 SCR and BT values for TCP with 9188-byte MTU

of interest, the SCR is set at the breakpoints in the BT-SCR curve. The table also shows the TCP throughput achieved for each measurement. Note that the SCR chosen in this way is larger than the measured TCP throughput.

The measured TCP throughput is an average over a relatively large (compared to the MBS) data transfer: approximately 6 MBytes for the data in Table 1. The burstiness that can be accommodated by a leaky-bucket traffic descriptor is over relatively short times, and as the TCP traffic is changing over time, the leaky-bucket descriptor is dominated by some "worst-case" burstiness seen during the data transfer. The value depends on the cell loss rate that is acceptable, but generally, the SCR needs to be be significantly larger than the TCP throughput to accommodate the extreme traffic behavior in the TCP stream. Also, in general, the SCR to throughput ratio at the break point is smaller for larger window sizes: the traffic for larger window size is less bursty. The implication for call admission control is discussed next.

3.3 Role of Traffic Characterization in Call Admission Control

We will now discuss the role of traffic characterization in designing rules for call admission control (Doshi, 1994). In the previous section, we have shown TCP data traffic characterized by leaky bucket parameters. It can be observed that the leaky bucket parameters depend upon the application traffic pattern and tolerance for cell loss. For example, in Figure2, for a RTT of 125 ms and a bucket size of 2000 cells, the drain rates are 1100, 1230 and 1250 for cell loss rates of 10^{-1}, 10^{-2} and 10^{-3} respectively. The number of simultaneous calls that can be set up would depend upon the cell loss objective that the application can tolerate. In order to illustrate these tradeoffs, we have assumed the drain rates stated earlier to determine number of calls admitted. Given a specific drain rate (say 1100 cells per second), a bucket size of 2000 cells and peak rate equal to 40.7 Mb/s, we intend to compute the number of simultaneously allowable calls on to a trunk of speed 150 Mb/s (OC-3). Starting with the leaky bucket parameters, we assume a worst-case on-off source and compute the number of allowable calls into an unbuffered OC-3 trunk. When multiple calls are simultaneously admitted on to a trunk, cell loss can occur. Therefore losses in the network contribute to the total loss rate, in addition to loss from UPC violations. We have computed number of allowable calls assuming two different loss objectives at the OC-3. These results are shown in Figure 3.

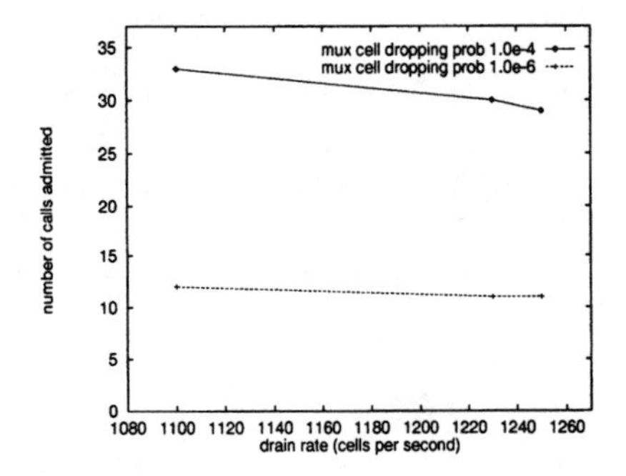

Figure 3 Number of Calls that can be supported on an OC-3 trunk

For a given multiplexer loss objective, the number of allowable calls goes down as the drain rate increases (or the UPC violation probability decreases). For a given drain rate, number of allowable calls increases as the multiplexer loss rate increases. While the above calculations have been performed on the assumption of worst case on-off sources, our objective during the course of the trial is to test, model and validate such results in the actual test bed. The ratio of SCR to useful throughput indicates the inefficiency of using VBR for these bursty traffic sources. ABR service provides an alternative approach to bursty sources aimed at improving network efficiency. Understanding the tradeoffs between ABR and VBR is an objective of further work.

4 MULTIPLEXING OF CBR TRAFFIC WITH VBR DATA

Mixing circuit emulation traffic with strict real-time performance and cell loss requirements with bursty data traffic requires differentiating the traffic with respect to scheduling (to manage delay) and buffering (to manage loss). The Globeview 2000 switch has multiple levels of priority that allow differential treatment of CBR and VBR virtual circuits. Cells from CBR VCs are given priority in scheduling, and buffering is reserved for such cells.

In this section, we consider multiplexing of CBR traffic with VBR data. The CBR traffic is generated by HP75000 test set while the VBR data traffic is generated the workstations using TCP/IP stack.

In general, priority control can help achieve the full range of QoS such as loss and delay required by multimedia applications. This may be accomplished by what is also called priority queueing, service scheduling, or fair queueing. Basically, multiple queues are implemented in the switch, such that traffic on certain VPC/VCCs that are not tolerant of delay can "jump ahead" of those that are more tolerant of delay. For one experiment, we us the loss/delay priorities offered by GV2000 switch. That is, CBR is given delay priority over VBR and buffer space reservation is used to give some loss priority to CBR traffic.

4.1 System Configuration

We use HP75000 as CBR traffic generator as well as the measurement test set. VBR traffic is TCP/IP data traffic, which is generated by `ttcp`. The test configuration is shown in Figure 4.

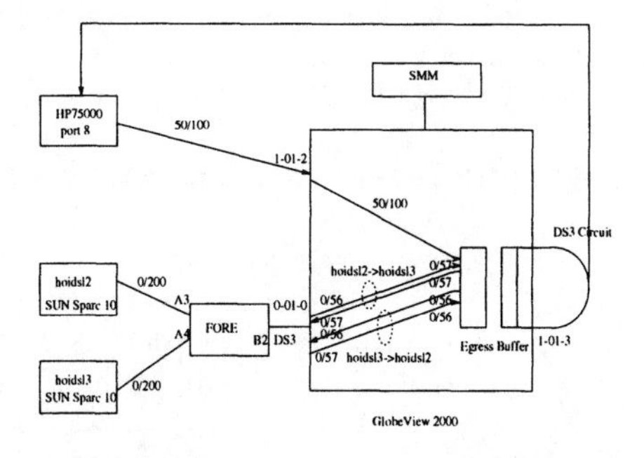

Figure 4 Configuration for Multiplexing CBR with VBR

It illustrates multiplexing of two traffic streams (CBR and VBR) into the ATM switch. The HP75000 equipment generates CBR traffic at the rate of 24.48 Mbps (equivalent to 12 E1). The CBR traffic goes through VPI/VCI at 50/100 coming into GlobeView port 1-01-2. Then it is taking the same VPI/VCI and arriving at the egress buffer. At last, it is drained by port 1-01-3 at DS3 speed.

The VBR data traffic is generated at `hoidsl2` by running `ttcp`, a program producing TCP traffic. It goes through 0/200 at port A3 of Fore Switch and comes out at port B2 to DS3 line to GlobeView at port of 0-01-0. The traffic takes 0/56 mapped to 0/57 and then arrives at the the same Egress buffer as CBR traffic (at this critical point, CBR and VBR traffic will compete for buffer resources). At outgoing port 1-01-3, CBR and VBR traffic will compete for DS3 bandwidth. Then the VBR traffic comes back at 0/57 to port B2 of Fore Switch. Finally, it arrives at `hoidsl3` at port A4, 0/200. The reverse pair of channels is used for TCP acknowledgement from `hoidsl3` to `hoidsl2`.

4.2 CBR Traffic Performance Improvement under Priority Control

We use HP75000 to collect the traffic at the receiver side. The overhead gives the timestamp at which the ATM cell arrives. And the payload gives the time when the cell was sent out. The maximum transmission unit (MTU) of TCP/IP is set at 9188 bytes while the sender/receiver window size is set at 48 Kbytes.[‡]

In this section, we only analyze the statistics of CBR traffic. Moreover, our primary concern is on cell delay and cell delay variation (CDV).[§]

In Table 2, we summarize the statistics of CBR traffic ATM cell interarrival time with/without priority control. As expected, we see that mean of cell interarrival time is almost the same, which is mainly determined by the CBR transmission speed. But the variance of cell interarrival time reduces dramatically when higher priority is assigned to CBR traffic.

In Table 3, we summarize the statistics of CBR trarffic ATM cell delay time with/without priority control. Once again, we see that mean and variance of cell delay time is remarkably reduced when higher priority is assigned to CBR traffic. Also, at least for this mix, the priority allows the CBR cell delay per switch to be well within the recommended bounds (a few hundred microseconds).

4.3 TCP/IP Throughput Degradation under Priority Control

We measure TCP data throughput as shown in Table 4. When TCP data and CBR are assigned the same priorities, (i.e. VBR data fairly compete for resources with CBR traffic), the average throughput of TCP is 14.188192 Mbps. When higher pri-

[‡]Since in the TCP header the window size is represented by 16 bits, the maximum window size is theoretically 64 Kbytes. Due to the limitation in SUN OS, the actual maximum window size is about 50 Kbytes.

[§]Since the default buffer size of 4095 cells in Globeview 2000 is used in the test, by simple computation, we see that there is no cell loss for multiplexing CBR with TCP VBR traffic.

	Priority Control	min	max	mean	var
1	without priority	10.2	42.6	16.9596	97.0047
	with priority	10.2	32.3	17.3209	51.4392
2	without priority	10.2	52.8	17.1553	99.7431
	with priority	10.2	32.3	17.318	51.1028
3	without priority	10.2	52.8	16.8476	95.5805
	with priority	10.2	32.3	17.3259	32.7453
4	without priority	10.2	51.6	18.1882	105.304
	with priority	10.2	32.3	17.3181	54.1043
5	without priority	10.2	42.6	17.5242	99.941
	with priority	10.2	32.3	17.3234	53.5533

Table 2 Statistics of Cell Interarrival Time (microseconds)

	Priority Control	min	max	mean	var
1	without priority	4455.5	9746.4	7195.78	1963.07
	with priority	69.3	83.6	78.925	19.8487
2	without priority	4516.1	8975.1	6859.25	1438.32
	with priority	53.3	86.8	78.8609	53.0899
3	without priority	4670.5	10012.8	7212.77	1676.96
	with priority	47.5	90.3	79.266	12.7276
4	without priority	4471.7	9026.4	6769.13	1552.93
	with priority	53.0	86.8	78.1258	55.1189
5	without priority	4563.2	9073.7	5928.91	1414.59
	with priority	53.9	88.8	80.2574	59.5334

Table 3 Statistics of Cell Delay Time (microseconds)

ority is assigned to CBR traffic, TCP throughput is degraded to 13.631808 Mbps. The amount of 0.566384 Mbps TCP performance degradation is sacrificed for performance improvement of CBR service in terms of cell delay and variance.

The TCP throughput is drawn in Figure 5, which shows that TCP throughput is reduced by a small amount when higher priority is given to CBR traffic.

It should be noted that the whole bandwidth of DS3 physical layer can be fully utilized when the multiplexed CBR and TCP/IP VBR traffic have the same priority.

priority control	TCP Throughput (Mbps)					Average
same priorities	14.1888	14.18776	14.1884	14.18768	14.18832	14.188192
CBR higher	13.62984	13.62848	13.63128	13.63312	13.63632	13.631808

Table 4 End-to-End TCP Throughput Measurement and Average under priority control

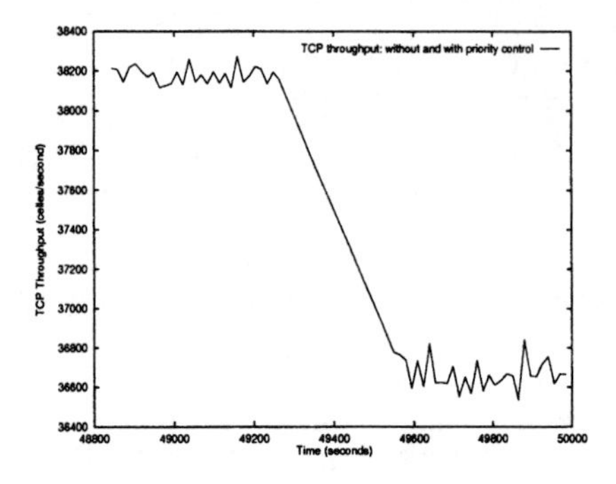

Figure 5 TCP throughput measurement

The performance of CBR traffic can be significantly improved by assigning higher priority to CBR traffic. In this case, the VBR traffic will degrade its throughput by a small amount. An important aspect of future studies will involve experimentation with TCP window sizes and/or other retransmission protocols (e.g. SSCOP) to see if the VBR throughput loss can be eliminated.

5 CONCLUSION AND FUTURE WORK

In this paper, we have presented results from measurements taken in a field grade ATM test bed to understand traffic management issues associated with efficient provisioning of resources while protecting the network from traffic overloads, as well as providing good performance to the applications. TCP/IP traffic over ATM network is characterized based on leaky-bucket parameters. We also show that quality-of-service of different applications can be guaranteed by appropriate priority control.

Our future work is to study multiplexing of audio, video and data applications. For these multimedia applications, we need to address traffic management related issues such as traffic shaping and usage parameter control. Further studies are needed on QoS guarantees for every admitted application through appropriately allocating network resources in terms of bandwidth and buffer.

REFERENCES

The ATM Forum Technical Committee. Traffic management specification version 4.0. Technical Report 95-0013R8, ATM Forum, October 1995.

B. T. Doshi. Deterministic rule based traffic descriptors for ATM/B-ISDN: worst case behavior and connection acceptance control. In *IJDACS*, 1994 and *Proceedings of MILCOM 94*.

Cathy Fulton and San qi Li. Delay jitter correlation analysis for traffic transmission on high speed networks. In *Proceedings of INFOCOM '95*, 1995.

David E. McDysan and Darren L. Spohn. *ATM Theory and Application.* McGraw-Hill, New York, 1994.

51

Service Clock Recovering in CBR Services: Adaptive vs SRTS

Luis A. Merayo, Javier Alonso and Jesús Mariño

Telefónica Investigación y Desarrollo
Emilio Vargas, 6. 28043 Madrid (Spain)

Tfn: +34 1 337 46 33
Fax: +34 1 337 45 02
e-mail: lmerayo@tid.es

Abstract

One requirement for ATM networks is the support for the provision of any constant and continuous bit rate service. This paper examines the evolution of the different alternatives considered within the standardization bodies for the service provision (AAL types 1 and 5) and it will focus on the features of the two service clock recovery methods (Adaptive and SRTS). An implementation of the Adaptive method is presented and evaluated in both ideal and stressed network conditions. Results show that, under the related test conditions, the adaptive method provides a high quality recovered clock that fulfills the ITU-T G.823 Recommendation for frequencies above 5 mHz.

1.-Introduction

The ATM bandwidth on demand philosophy, that gives ATM its flexibility, is based on two main characteristics: bit rate decoupling and fix length transfer data unit. The use of ATM as the single transfer mode for all the (present and future) services requires to enhance the features provided by the ATM layer in order to support the functions required by service dependent upper layers, mainly synchronization capabilities for real time applications, preservation of data integrity and mapping between the different data units. The best place to incorporate these additional functions is the ATM Adaptation Layer (AAL).

The specification of the different types of AALs has suffered a considerable transformation during the ITU-T standardization process. Initially, a vertical mapping was defined in Recommendation I.362 among the upper layer service, a specific Service Class, the associated Quality of Service Class and the Adaptation Layer type. In order to minimize the number of AAL protocols (or types), a service classification based on three basic parameters (timing relation between source and destination required or not, service bit rate constant or variable and connection or connectionless oriented mode).

The resulting mapping between service classes, service parameters and AAL types was the following well know table:

AAL Type	**1**	**2**	**3**	**4**
Service Class	**Class A**	**Class B**	**Class C**	**Class D**
Timing relation between source and destination	Required	Required	Not required	Not required
Bit rate	Constant and continuos	Variable	Variable	Variable
Connection mode	Connection-oriented	Connection-oriented	Connection-oriented	Connectionless

As a result of the mapping, the Recommendation I.363 started the specification of four different AAL (but as a matter of fact only two ones survived): the AAL type 1 devoted to support any constant and continuos bit rate service, the AAL type 2 which specification process never concluded due to the lack of VBR service requirements, the AAL type 3 and AAL type 4 were combined in a single AAL type 3/4. Some strong players in standardization bodies, driven by market suitability considerations, proposed a new simple and efficient AAL (in terms of process complexity and minimum overhead): the AAL type 5.

Once the AAL type 5 got in competition, the old AAL 3/4 was progressively replaced (only CBDS/SMDS services require it). End system manufacturers, coordinated by ATM Forum, pushed the use of such AAL type 5 as far as possible following the approach of a very simple Common Part Convergence Sublayer and Service Specific Convergence Sublayers devoted for each specific service. Therefore, the vertical identification between Service and QoS classes and AAL types were transformed into an horizontal approach with an unique AAL type 5 but as many service specific sublayers as final services.

The result of some experiences dealing with the QoS provided by the AAL type 5 horizontal approach (mainly the impact of CDV and cell losses on the recovered service clock) made to some major players to reconsider the hypothesis of a single common AAL 5 for all kind of services, coming back to the AAL1 approach. That is accepted today is a mixed solution: the major part of CBR services can be supported by both AAL type 1 and AAL type 5 (ATM Forum AMSIA, ITU-T J.82) with different performances depending on the AAL used. Only the services with strong timing requirements (e.g. circuit transport) remain out of this evolution and have to be only provided over AAL type 1.

The functions performed by the AAL type 1 in order to support the CBR service requirements are:

- segmentation and reassembling of user information into 47 bytes AAL-1 SAR-PDU's

- blocking and debloking of user information in order to be transparent to any upper layer structure (e.g. octet alignment)
- compensation of cell delay variation (e.g. output buffering)
- handling of cell payload assembly delay (e.g. partially filled cells)
- detection, and correction if possible, of lost and misinserted cells (e.g. FEC and interleaving)
- source clock recovery at the receiver (Adaptive or SRTS methods)
- recovery of the source data structure at the receiver (e.g. SDT method for frame recovery)
- monitoring and handling of AAL-Protocol Control Information for bit errors
- monitoring of user information field for bit errors and possible corrective action

This paper will focus on the features of the two source clock recovery methods recommended by the standardization bodies: Synchronous Residual Time Stamp (SRTS) and Adaptive. An implementation of the last one (Adaptive) is described and evaluated under both ideal and stressed (CDVs, losses) network conditions, concluding that, with an adequate implementation, the performances attained with the Adaptive method accomplish the majority of the CBR service requirements.

The paper is organized as follows: section 2 deals with the requirements for a reliable clock recovery mechanism, in section 3 the features of both SRTS and Adaptive methods are compared, an implementation of the Adaptive is presented and evaluated in section 4 and the main conclusions are summarized in section 5.

2.-Requirements for a reliable clock recovery mechanism

Real time applications, most of them still coming from the 64 Kbit/s synchronous world, lie on terminals requiring not only data but also the clock reference at which data has been produced (e.g. H.320, MPEG-1 and MPEG-2, ...). Other services requiring timing references between transmitter and receiver are the circuit emulation in both synchronous (G.702 and G.709) and plesiochronous (G.703) scenarios. The need for the provision of this type of services over ATM is not too urgent today (in public scenarios the current situation is based on small ATM islands into a synchronous/plesiochronous sea), but the progressive integration of both worlds requires these services. In fact, only the plesiochronous option (in which there are not relationship between service and network clocks) needs the clock recovering because in the synchronous case the receiver obtains the service clock directly from the network. Taking into account the evolution of the network infrastructure toward the synchronous scenario, it seems not risky to assume that the users of the service clock recovering function are terminals more than network nodes (it is assumed that the circuit emulation service will be provided at network nodes, not at final user terminals). This is a major issue due to the fact that the requirements on the recovered clock quality are difference in both cases.

The target features required to a clock recovery method should be:

- To be resilient and simple in order to support different QoS of ATM layer (cell insertions, loses and CDV) simplifying and making cheap its implementation.

- The system must offer fault tolerant characteristics: even in the case of data losses (caused by starvation or overflows in the CDV compensation buffer) the clock still has to be provided.

- The quality of the recovered clock in terms of jitter and wander must be the adequate for the service users. Recommendation G.823 is usually taken as a reference for the recovered clock quality although, strictly speaking, the accomplishment of this recommendation is only required for circuit emulation services.

- The system must offer an adequate and flexible (programmable) trade-off between hold-in range (range of frequencies of the input clock the system will stay in lock once it is locked), capture capability (range of frequencies of the input clock over which the system locks), capture time and capability to attenuate and filter small variations on the incoming clock.

3.-SRTS and Adaptive methods evaluation

The SRTS method [1,2] is based on the availability of a common reference clock in both source and destination sides. The difference between service and reference clocks (i.e. RTS value) is conveyed by the sender and transported in the ATM cells (CSI bit of the AAL-1 header containing). The receiver reconstructs the service clock from a PLL system driven by a reference timing signal obtained from the RTS value and the common clock reference.

The Adaptive method does not require the existence of a common reference clock. The receiver clock runs freely at a frequency $F \pm \Delta F$ obtained from a VCXO module with a nominal frequency equal to the transmitter one (service clock).The VCXO control signal is obtained from the long term variations in the filling level of the CDV compensation FIFO. If the FIFO tends to fill up, the VCXO output increases slowly (upper value F+DF); if the FIFO tends to empty , the VCXO decreases (lower value F-DF).

Both methods are able to support different QoS of the ATM layer (missinsertions, cell losses and CDV) and offer fault tolerance characteristics (loss of data does not means loss of clock). Its main difference lies on the better performances in terms of low frequency clock stability (wander) that the SRTS method offers. In fact, the SRTS method is able to afford a high quality recovered clock (e.g. compliant with G.823 Recommendation.) if the network provides a reference clock with similar characteristics. The need of a synchronous overlayed network makes this option much more expensive (than the Adaptive method) and creates implementation problems in multi-PNO environments where all the reference clocks must be G.823 complaint and the long term variations between them must be compliant to G.811. As an alternative, GPS-based mechanisms able to provide a common pan-synchronization scenario are being studied and evaluated (e.g. EURESCOM P.513 project).

Just to conclude this first evaluation of the two methods, it should be noticed that the SRTS is protected by a patent, strange situation because it seems not reasonable that one algorithm that is recommended by the standardization bodies requires simultaneously a royalty based payment.

4.-Adaptive method implementation and evaluation: a study case

As a part of a Service Multiplexer including both LAN interconnection and circuit emulation services, an
type 1 module with service clock recovering based on Adaptive and SRTS methods was developed in Telefónica I+D.

The module that implements the Adaptive method, which block diagram is shown in figure 1, has a structure similar to a classical PLL.

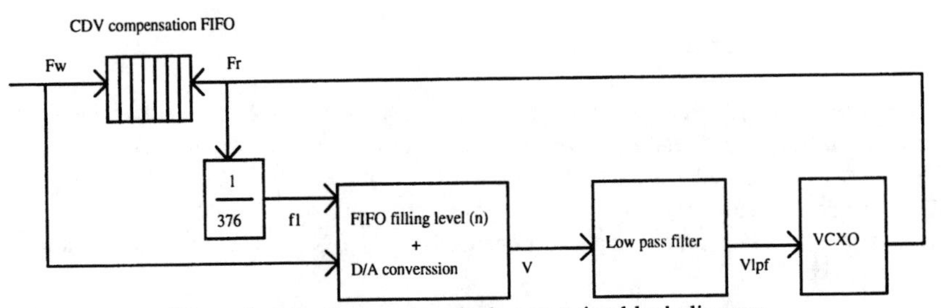

Figure 1. Adaptive service clock recovering block diagram

The AAL type 1 SAR-PDU's (47 bytes of the cell payloads once the cell sequence number correctness has been verified) are written (PDU clock Fw affected by CDV and cell losses) in the CDV compensation FIFO, FIFO that is read with the recovered service clock (Fr). The FIFO filling level is calculated in real time and this value is converted from digital to analog. This block is equivalent to the phase detector in classical PLLs: its output V is a voltage proportional to the FIFO filling level n, value that is obtained from the Fw and Fr frequencies. Given that Fw is the SAR-PDU writing frequency and Fr the recovered (bit) service clock, a factor of $\frac{1}{376}$ (number of bits in the 47 bytes SAR-PDU) has to be applied. The transfer function of this module is $V(s) = K_1 \, (Fw\text{-}f_1)$.

The low pass filter, based on an active filter which transfer function is $F(s)=\frac{1+s\cdot K_2}{s\cdot K_3}$, has as a function to pass the average voltage of the phase-detector (FIFO filling level) to the voltage-controlled oscillator VCXO, which transfer function is $\frac{K_4}{s}$.

Consequently, the transfer function of the complete system is:

$$H(s)=\frac{K_1\cdot K_4}{376\cdot K_3}\cdot\frac{1+K_2\cdot s}{s^2+\frac{K_1\cdot K_2\cdot K_4}{376\cdot K_3}\cdot s+\frac{K_1\cdot K_4}{376\cdot K_3}}$$

that corresponds to a second order system with a zero in $s=-\frac{1}{K_2}$. Because of stability requirements (system unstable when the zero is close to the imaginary axis), $K_2 \langle\langle 1$. The natural frequency $w_n=\sqrt{\frac{K_1\cdot K_4}{376\cdot K_3}}$ and dumping factor $d=\frac{1}{2}\cdot K_2\cdot\sqrt{\frac{K_1\cdot K_4}{376\cdot K_3}}$ values were selected (31.4 rad/s and 1.2 respectively) in order to have a narrow hold-in range and a capability to attenuate small variations on the incoming clock, providing an output clock that follows slowly the variations around the central frequency of the input one.

In order to measure the quality of the recovered clock (jitter and wander) in a 2.048 Mb/s circuit emulation service, and to investigate the influence of the QoS offered by the ATM layer (variable delay and cell losses) in the recovered clock, the following configuration (figure 2) was set up:

- A 2.048 Mb/s G.703 circuit emulation equipment developed in Telefónica I+D (AAL-1 transmitter and receiver parts with service clock recovering based on the adaptive method and SDH STM-1 physical layer for the ATM interface)
- An ATM Network Emulator [3] (ANE, equipment developed in Telefónica I+D able to emulate the behavior or a real ATM network) connected to the ATM interface
- A 2.048 Mb/s G.703 equipment used to provide the 2.048 Mb/s CBR
- A jitter and wander time interval analyzer

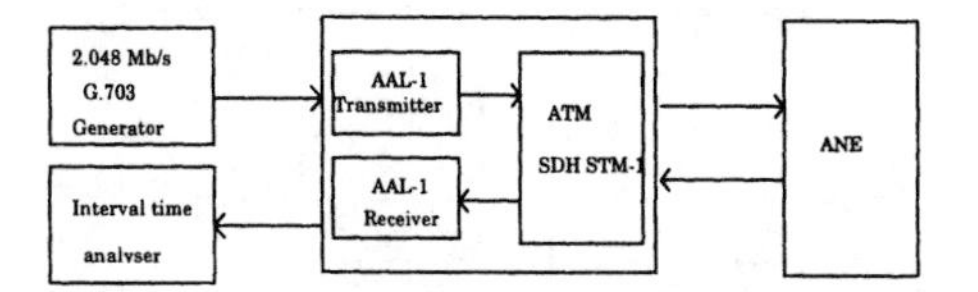

Figure 2. Configuration for circuit emulation testing

The recovered 2.048 MHz clock power spectrum was measured at the output of the AAL-1 Receiver both under ideal (no variable delay, no losses) and stressed network conditions: variable delay following a normal distribution -approximated by a binomial- N(185,10) cell interval units (2.7 ms at 155.52 Mb/s) and burst cell losses (exponential 10^{-10} distribution, 90% of single losses, 5% of two consecutive losses, 3% of three consecutive losses, 1% of four consecutive losses and 1% of five consecutive losses) programmed in the ANE.

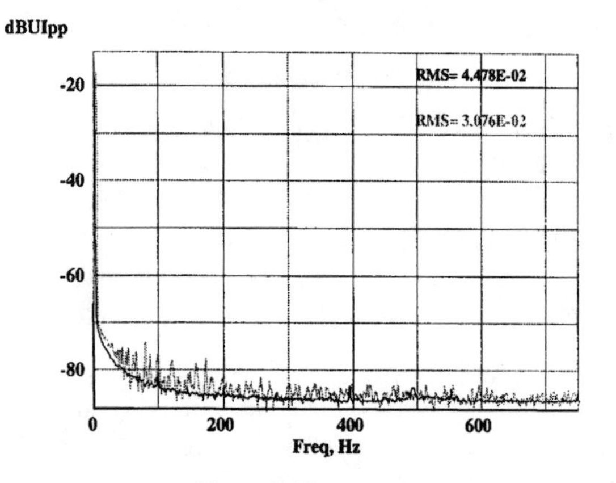

Figure 3. Jitter spectrum

Figure 3 shows the recovered clock jitter under ideal (UIrms=$3.076*10^{-2}$) and stressed network conditions (Uirms=$4.478*10^{-2}$). In both cases the jitter is significantly smaller than the maximum value accepted by the G.823 recommendation in this range of frequencies (3.5 dBUIpp).

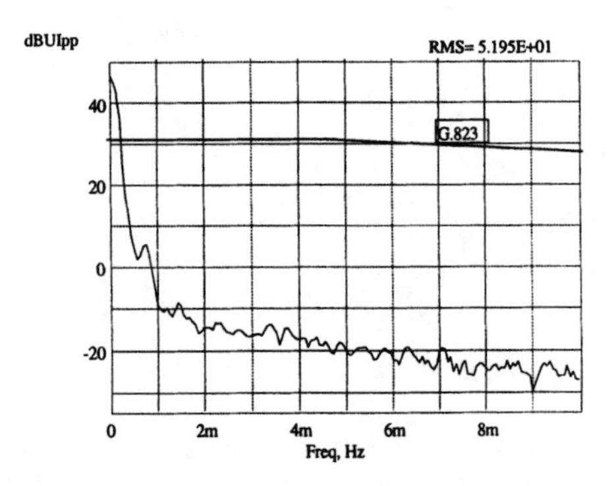

Figure 4. Wander spectrum sampled at 125 mHz under ideal network conditions

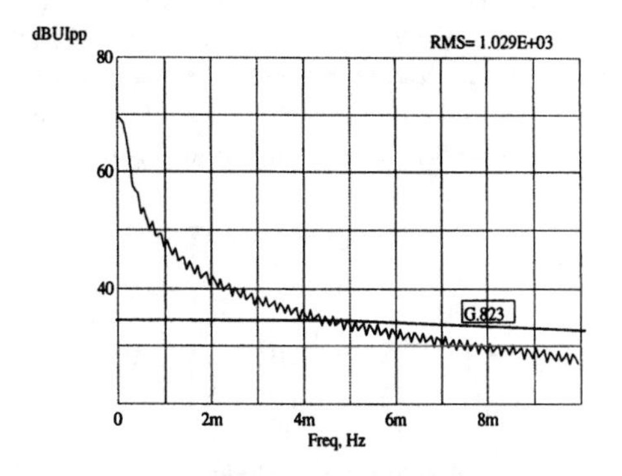

Figure 5. Wander spectrum sampled at 125 mHz under stressed network conditions

The wander of the recovered clock, measured under both ideal and stressed network conditions, is shown in figures 4 and 5. These wander values are compared with the G.823 recommendation. It can be seen that, under the related test conditions, the adaptive method fulfills the ITU-T G.823 recommendation for frequencies above 0,25 mHz and 5 mHz respectively.

Jitter and wander measurements were taken using an interval time analyzer HP7552 complemented with a pos-processing software package. This equipment measures time intervals related to an internal reference clock, being 0.2 ns its maximum achievable precision. In the Fourier domain, the maximum resolution is 30 mHz that corresponds to a measure time of 9 hours 6 minutes 8 seconds. Therefore no further wander measurements are achievable with this system.

On the other hand, the ITU-T Recommendation G.823 points out the lower wander frequency to 0 Hz. The measurements of the lower range are impossible to obtain, due to the need of an infinite measurement interval.

5.-Conclusions

In this paper the two clock recovering mechanism SRTS and Adaptive are compared. An implementation of the Adaptive method is detailed and evaluated in both ideal and stressed network conditions. The evaluation results show that the Adaptive method provides an adequate recovered clock that fulfills the ITU-T G.823 Recommendation for frequencies above 5 mHz, clock that, in our opinion, satisfies the the majority of CBR application requirements on clock stability. The use of the Adaptive avoids all the SRTS inconvenients: need of a high quality common reference clock and SRTS patent protection.

In Telefónica I+D we are working in order to improve the performances of the current implementation in two different ways. One is the use of a low pass digital filter in order to allow more flexibility in the filter parameters selection. The second possibility, apparently very promissing, that we are examining, is the use of algorithms that decrease the wander value increasing the jitter, in line with the method proposed in [4].

6. -References

[1] **ITU-T SG13/2 Recommendation I.363.1**. *B-ISDN ATM Adaptation Layer Specification, types 1 and 2*. Geneve 1995.

[2] **C.L. Lau and P.E. Fleischer**. *Synchronous Techniques for Timing Recovery in BISDN*. IEEE Globecom'92

[3] **L. Merayo, A. Alonso and L. Mola**. *ANE (ATM Network Emulator): Characteristics and Application Examples*. ISS'95

[4] **M. J. Klein and R.S. Urbansky**. *SONET/SDH Pointer Processor Implementations*. IEEE Globecom'94

52

Clock Recovery for Circuit Emulation Services over ATM

Luís Castel-Branco [(1)], *Mário Serafim Nunes* [(2)]

(1) PORTUGAL TELECOM - Business and Main Customers Dept.
R. Entrecampos 28, 1700 Lisboa, Portugal
Telephone: 351-1- 500 34 13 Fax: 351-1- 500 35 61
E-mail: castel-x-branco@telecom.pt

(2) INESC - Switching and Terminal Equipment Group
R. Alves Redol 9, 1017 Lisboa, Portugal
Telephone: 351-1- 3100 256 Fax: 351-1- 314 58 43
E-mail: Mario.Nunes@inesc.pt

Abstract

Different Source Clock Frequency Recovery techniques for support of Circuit Emulation Services in ATM networks are presented. The architectures of the Adaptive method are analysed, in particular its application to the 2 Mbit/s CBR services emulated over ATM. One technique based on the level of the receiver buffer is modelled and designed. It depends on the fill level of the receiver buffer (FIFO), which is maintained between two limits in order to prevent buffer overflow and underflow. The system consists essentially of a PLL with an input linear dependent on the FIFO fill level and an output which is used to generate the local service clock for reading the receiver buffer. The development of the algorithm is explained and the comparison with other Source Clock Frequency Recovery methods is made.

Keywords

ATM, Circuit Emulation, Clock Recovery, Adaptive Method

1 INTRODUCTION

The ATM (Asynchronous Transfer Mode) based networks have been specified to integrate all the existing services, according with one of the main objectives of the B-ISDN.

Presently there are several ATM field trials and pilot projects all over the world, from which it is worth mentioning the "Pan-European ATM-Virtual Path (VP) Network" in Europe.

One of the services supported by all these pilots is the Constant Bit Rate (CBR) Circuit Emulation Service (CES). The CES is defined as the capability of emulating CBR services, namely Nx64 Kbit/s circuits up to 2.048 Mbit/s, over an ATM network.

The adaptation of CBR services to ATM is defined in the ITU-T I.363 (ITU-T Rec. I.363, 1993) recommendation, in the AAL type 1 section. One of the services that must be provided by the AAL1 is the capability of recovering the source clock frequency at the receiver, with a specified quality. The different methods of asynchronous clock recovery for the support of CBR services defined in this recommendation are analysed and compared in the following sections.

2 CIRCUIT EMULATION OVER ATM

The Circuit Emulation over ATM has been an important subject studied in the main standard bodies.

Since the objective of the near-future Broadband ISDN is to integrate all existing and emerging services, a main effort has been made to find the best methods to reach a synchronous emulation connection over ATM.

In Europe the ETSI NA5 (Network Aspects Group 5) is a standardisation group very active in this area (Lecuit, 1993).

EURESCOM (European Institute for Research and Strategic Studies in Telecommunications) is another organisation where the main members are Public Telecommunication Operators (PNO) with an implementation-oriented view. In the EURESCOM Project P105 - Pan-European ATM-VP Network Studies the second Deliverable presents a detailed specification of the ATM/AAL Layer, including the study of the Adaptive Clock Recovery Method, (EU-P105 Deliverable 2, 1993).

The ATM Forum is another main standard body strongly involved in ATM specifications, namely in ATM Forum (July 1994), where the CES is specified, namely the DS1/E1 Nx64 Kbit/s and unstructured CBR services.

The ITU-T (former CCITT) has defined in I.363 - "B-ISDN ATM Adaptation Layer (AAL) Specification", (ITU-T Rec. I.363, 1993) the basic concepts of two clock recovery techniques, the Synchronous Residual Time Stamp (SRTS) and the Adaptive Clock method.

The SRTS method uses the Residual Time Stamp (RTS) to measure and convey information about the frequency difference between a common reference clock derived from the network and a service clock. The same derived network clock is assumed to be available at both the transmitter and the receiver. The SRTS method is capable of meeting the jitter specifications of the 2.048 Mbit/s hierarchy as defined in ITU-T Rec. G.823 (ITU-T Rec. G.823, 1988).

If the common network reference clock is not available (e.g. when two interconnected networks are not synchronised) then the source clock recovery method will be in a mode of operation associated with "plesiochronous network operation".

The Adaptive Clock method is based on the fact that the amount of transmitted cells is an indication of the source frequency and this information can be used at the receiver to recover the source clock frequency. By averaging the amount of received cells over a large period of time, the cell delay variation (CDV) effects are minimised. The receiver writes the received information into a buffer and then reads it with a local clock.

In the FIFO based method the fill level of the buffer is used to control the frequency of the local clock. The control is performed by continuously measuring the fill level around its medium position, and using this measure to drive the Phased-Locked Loop (PLL) which provides the local clock with a VCXO. The fill level of the buffer should be maintained between two limits in order to prevent buffer overflow and underflow.

Other types of adaptive clock recovery implementations are based on the determination of the average value of the inter-arrival time of the received cells.

3 ADAPTIVE TECHNIQUES

The main advantages of the Adaptive Clock Recovery techniques are that they do not need explicit timing information between source and destination, do not need a common network clock and are simple to implement, since the clock recovery is based on the frequency estimation based on the arrival rate information.

Comparing the adaptive method with others, as the SRTS method, it is possible to say that the Adaptive methods are more subject to deviations between the sender and receiver clocks, although they are simpler and do not require timing network services.

The adaptive methods can be divided in several types, namely:

- sliding window mean method, based on the last N cell arrival periods;
- weighted mean method, based on the last cell inter-arrival period;
- FIFO level method, with linear variation;
- FIFO level method, with non-linear variation.

3.1 Sliding window method

The sliding window method is conceptually simple and only needs basic components in its hardware implementation. This method is based on the information obtained directly by the determination of the time gap between two consecutive cells (inter-arrival time), which is a variable value dependent on the random delays that each cell suffers in the ATM network, the CDV.

The period of the recovered clock is calculated as the arithmetical mean of the last N cell inter-arrival times:

$$T_{read} = \frac{\sum_{i=1}^{N} T_{arrival_i}}{N} \quad (1)$$

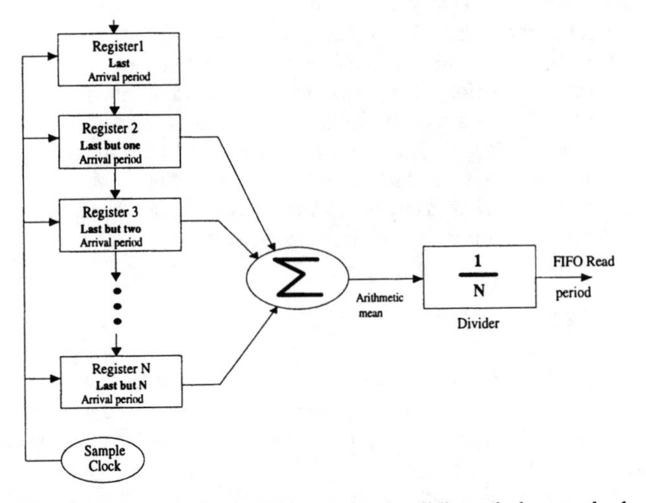

Figure 1 Simplified block diagram for the sliding window method.

The simplified block diagram for this technique is shown in Figure 1, where its main components can be seen: N registers, a sample clock, a sum block and a divider by N.

3.2 Weighted mean method

The sliding window method just described, based on the calculation of the last N cell inter-arrival periods, has the disadvantage of requiring a large number of registers (N) to comply with the G.823 Recommendation.

In the Weighted Mean method the clock recovered period is based on the calculation of a weighted mean with the last cell arrival period, then it only requires two registers and a simple arithmetic block, equivalent to a first order filter, as shown in the following expression:

$$Tread_i = \frac{Tread_{i-1} \times (N-1) + Tarrival}{N} \qquad (2)$$

The value N is obtained from the following expression:

$$N = \frac{1}{T_{cell} \times \omega_{pole}} \qquad (3)$$

The value of N is chosen in order to accomplish with the requirements of the jitter and wander specified in G.823.

3.3 FIFO level method

In the FIFO level based method two variants can be considered. In the first one the clock frequency output varies linearly with the fill level of the FIFO around its medium position (linear method). In the second method the clock frequency output varies linearly with the FIFO level near the half-full level and non-linearly when the FIFO is almost empty or almost full, to prevent underflow or overflow (non-linear method).

The first method is analysed and modelled in detail in the next sections.

4 FIFO LEVEL LINEAR METHOD FREQUENCY ANALYSIS

The FIFO level linear method is based on the variation of the FIFO level, given by the expression:

$$Length_{fifo} = \int \left(f_{arrival} - f_{read}\right) \cdot dt \tag{4}$$

This leads to the block diagram of the FIFO based method shown in Figure 2.

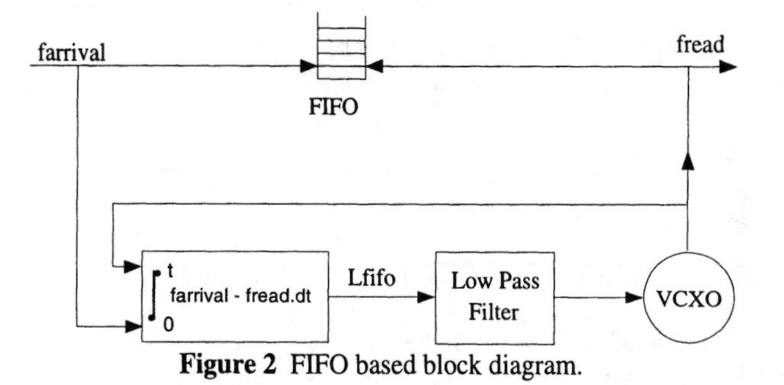

Figure 2 FIFO based block diagram.

From this model and using mathematical developments the timing block diagram can be obtained, as presented in Figure 3.

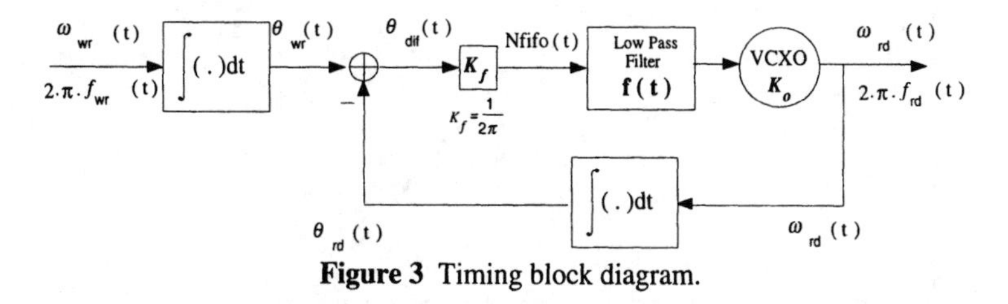

Figure 3 Timing block diagram.

From this, a more useful diagram can be derived, the frequency block diagram, which is shown in Figure 4.

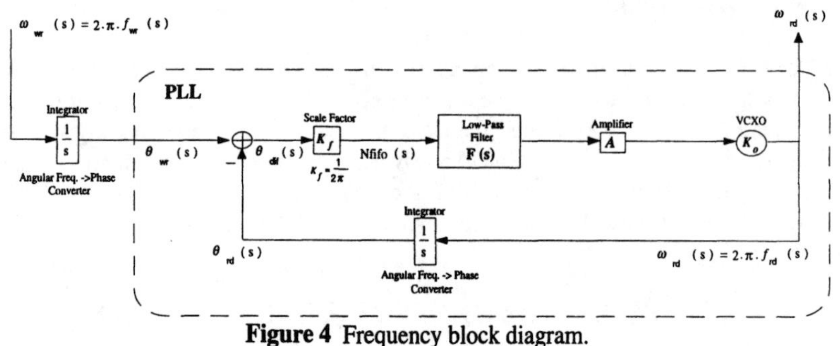

Figure 4 Frequency block diagram.

The length of the receiver FIFO is dependent on the maximum value of the Cell Delay Variation and is calculated by the following expression:

$$Length_{fifo} = 2 \times \left[Integer \geq \left(f_{cell} \times CDV \right) \right] \tag{5}$$

The maximum value of the CVD is not yet well defined. The value proposed in [4] is used, $CDV = 3$ ms.

Considering that for the 2.048 Mbit/s bit-rate the cell frequency is $f_{cell} = 5{,}447\text{kHz}$ the minimum FIFO length necessary to accommodate a CDV of 3 ms is:

$$Length_{fifo} \approx 40 \cdot cells \tag{6}$$

5 SYSTEM TRANSFER FUNCTIONS

In Figure 5 a comparison of the G.823 with the FIFO based adaptive clock recovery method is shown, using a Lag and a Lag-Lead low-pass filters.

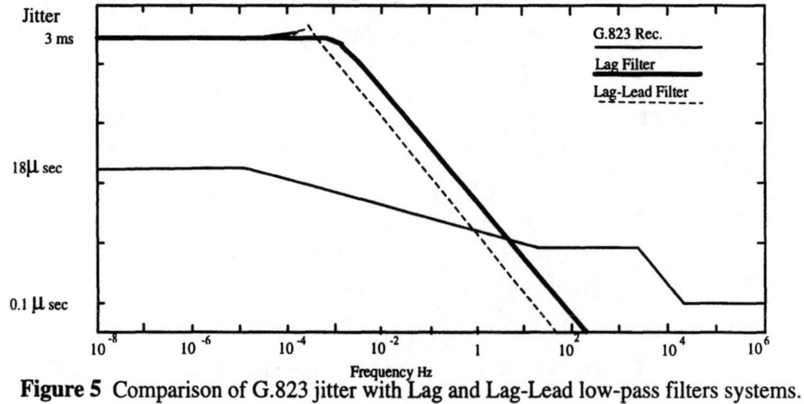

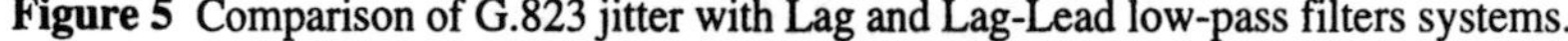

Figure 5 Comparison of G.823 jitter with Lag and Lag-Lead low-pass filters systems.

This figure shows that the output satisfies the jitter requirements specified in G.823 Recommendation for Constant Bit Rate Services at 2.048 Mbit/s for the jitter frequency range defined in the G.823, from 20 Hz up to 10^5 Hz. The wander, defined for frequencies below 20 Hz with a maximum of 18 μs is not satisfied for a CDV of 3ms in the ATM network.

6 DESIGN OF A FIFO LEVEL BASED CLOCK RECOVERY SYSTEM

The different modules of the FIFO level based clock recovery system are calculated in this section.

Phase Detector

It is implemented with an up-down counter with a constant or scale factor of:

$$K_f = 1/(2 \cdot \pi) = 0.159 Volt / rad \qquad (7)$$

VCXO

A VCXO with a free-run frequency of 16.384 MHz was chosen (8 x 2.048 Mbit/s), which implies a frequency divider with the value:

$$K_{0_{cell}} = 0.684 rad / (s \cdot Volt) \qquad (8)$$

Amplifier

This component guarantees that the input is always less then the VCXO voltage limit:

$$A = 0.153 \cdot [non - \dim ensional] \qquad (9)$$

Loop Gain

The Loop Gain is obtained by the product of the three last constants:

$$K = K_f \cdot K_0 \cdot A = 0.017 \sec^{-1} \qquad (10)$$

Transfer Function

The Lag Filter was chosen in order that the transfer function of the system comply with G.823 recommendation.

The system transfer function is given by the following expression:

$$T(s) = \frac{1}{(\frac{s}{\omega_n})^2 + 2 \cdot \xi \cdot \frac{s}{\omega_n} + 1} \qquad (11)$$

with:

$$\xi = 1/\sqrt{2} = 0.707 \quad (12)$$

$$f_n = 2 \cdot 10^{-1} Hz \quad (13)$$

7 SIMULATION

The test of the system was made using the VHDL language. The simulation with VHDL permits a very detailed and precise time simulation, with a time scale of nanoseconds, and a detailed analysis of the different system variables and of the FIFO level evolution with time.

Two other systems were simulated, the sliding window and the weighted mean methods. Concerning the sliding window method with the last N cell inter-arrival periods, it needs a very high number of registers to guarantee the compliment with the G.823, which is a strong drawback for a practical implementation. The weighted mean method with the last cell arrival period requires special hardware to establish precision time intervals between two consecutive arrival cells.

The results of the simulation of the FIFO based method agree with the analytical results, showing a better performance and lower jitter of the FIFO based method than the other adaptive methods.

8 CONCLUSIONS

Several clock recovery adaptive methods have been analysed and simulated.

The FIFO level based method was described in detail, including its time and frequency models analysis. The calculation of the FIFO length and the type of Low Pass Filter used in the PLL were two important factors carefully analysed and studied in order to obtain the best performance characteristics.

A FIFO level system was designed and simulated. Both the analytical and simulation approaches proved that this method gives the best performance and minimum jitter of the recovered clock when compared with the other adaptive techniques.

The main characteristics of the FIFO level based method are: the FIFO level signal is obtained using an up-down cell counter; the system is based on a well known component, a PLL (Phase Lock Loop); the VCXO, Low-Pass Filter and Amplifier characteristics are well determined.

The clock recovered output satisfies the jitter requirements specified in Recommendation G.823 for Constant Bit Rate Services at 2.048 Mbit/s in all the frequency range defined in this recommendation, from 20 Hz up to 10^5 Hz.

The G.823 wander requirements are not satisfied in this design example, however it is easy to satisfy the wander requirements for frequencies below 20 Hz simply by decreasing the low-pass filter natural frequency. For very low frequencies however the realisation of a low-pass filter is limited by the physical realisation of this component concerning stability with temperature and time. For these very low frequencies a digital low-pass filter is advisable.

The initialisation of the system requires further analysis, in order to optimise the synchronisation of the recovered clock and minimise the capture time of the system.

9 REFERENCES

ATM Forum (July 1994), "ATM Circuit Emulation Services, Version 4.0".

EU-P105 Deliverable 2 (January 1993): "ATM/ AAL Specification - Adaptive Clock Method".

ITU-T, Rec. I.363 (January 1993) - "B-ISDN ATM Adaptation Layer (AAL) Specification".

ITU-T, Rec. G.823 (December 1988)- "The Control of Jitter and Wander within Digital Networks Based on the 2048 Kbit/s Hierarchy.

Lecuit, J. Alonso (September 1993): "Timing Issues of AAL1 for SRTS Plesiochronous Operation", ETSI-NA5.

INDEX OF CONTRIBUTORS

KEYWORD INDEX